Topics in Mathematics

CALCULUS AND ORDINARY DIFFERENTIAL EQUATIONS

B.A./B.Sc. Ist Year

Topics in Mathematics

CALCULUS AND ORDINARY DIFFERENTIAL EQUATIONS

(B.A./B.Sc. Ist Year)

PAPER II

(As per Various University Syllabus)

By

OM P. CHUG
Formerly, Head Deptt. of Mathematics
Hindu College, Sonepat
Haryana

PARMANAND GUPTA
M.Sc. M.Phil (Maths)
M.A. (Economics)
Head, Deptt. of Mathematics
Indira Gandhi National College, Ladwa
Kurukshetra University
Haryana

R.S. DAHIYA, *M.A.*
Formerly, Head Deptt. of Mathematics
A.I. Jat College, Rohtak
Haryana

UNIVERSITY SCIENCE PRESS
(An Imprint of Laxmi Publications Pvt. Ltd.)
An ISO 9001:2008 Company
BENGALURU • CHENNAI • COCHIN • GUWAHATI • HYDERABAD
JALANDHAR • KOLKATA • LUCKNOW • MUMBAI • RANCHI • NEW DELHI
BOSTON (USA) • ACCRA (GHANA) • NAIROBI (KENYA)

TOPICS IN MATHEMATICS: CALCULUS AND ORDINARY DIFFERENTIAL EQUATIONS

Printed and bound in India

Typeset at : Goswami Associates, Delhi

Fourth Edition : 2008, *Reprint* : 2009, 2013, 2015

ISBN : 978-81-7008-659-8

Published in India by

UNIVERSITY SCIENCE PRESS

(An Imprint of Laxmi Publications Pvt. Ltd.)

An ISO 9001:2008 Company

113, GOLDEN HOUSE, DARYAGANJ, NEW DELHI-110002, INDIA

Telephone : 91-11-4353 2500, 4353 2501

Fax : 91-11-2325 2572, 4353 2528

www.laxmipublications.com info@laxmipublications.com

Branches

Bengaluru	080-26 75 69 30
Chennai	044-24 34 47 26
Cochin	0484-237 70 04, 405 13 03
Guwahati	0361-254 36 69, 251 38 81
Hyderabad	040-27 55 53 83, 27 55 53 93
Jalandhar	0181-222 12 72
Kolkata	033-22 27 43 84
Lucknow	0522-220 99 16
Mumbai	022-24 91 54 15, 24 92 78 69
Ranchi	0651-220 44 64

C—9901/015/03

Printed at : Ajit Printing Press, Delhi.

PREFACE

This book on "Calculus and Ordinary Differential Equations" has been written according to new guidelines and syllabus issued by UGC for the B.A./B.Sc. I class of various universities of India and strictly covering the syllabus of K.U. Kurukshetra and M.D.U. Rohtak.

In writing this book, the authors have placed great emphasis on the fundamental concepts and methods of proofs of articles. We have provided sufficient number of illustrative examples on each concept. To reinforce the concepts developed in the definitions and articles, several exercises have been given at proper places.

Questions from latest annual examination papers of B.A./B.Sc. I of K.U. and M.D.U. have also been included.

Salient features of the present book are :

1. Well explained concepts.
2. Sufficient number of solved examples.
3. Well graded exercises.
4. Hints for selected questions.

Suggestions for the improvement of the book from dear students and teacher colleagues are welcome.

—AUTHORS

SYLLABUS

K.U. Kurukshetra

B.A./B.Sc. I (Paper II)

CALCULUS AND ORDINARY DIFFERENTIAL EQUATIONS

Section I (3 Questions)

ε-δ definition of the limit of a function. Basic properties of limits. Continuous functions and classification of discontinuities. Differentiability. Successive differentiation. Leibnitz theorem. Maclaurin and Taylor series expansions. Asymptotes. Curvature. Tests for concavity and convexity. Points of inflexion. Multiple points. Tracing of curves in Cartesian and polar coordinates.

Section II (3 Questions)

Reduction formulae. Quadrature. Rectification. Volumes and surfaces of solids of revolution.

Section III (2 Questions)

Exact differential equations. First order higher degree equations solvable for *x, y, p.* Clairaut's form and singular solutions. Geometrical meaning of a differential equation. Orthogonal trajectories. Linear differential equations with constant coefficients. Homogeneous linear ordinary differential equations.

Section IV (2 Questions)

Linear differential equations of second order. Transformation of the equation by changing the dependent variable/the independent variable. Method of variation of parameters. Ordinary simultaneous differential equations.

Note. The examiner is requested to set ten questions in all selecting questions sectionwise as indicated in the syllabus. The candidate is required to attempt five questions selecting at least one question from each section.

SYLLABUS

M.D.U. Rohtak

B.A./B.Sc. I (Paper II)

CALCULUS AND ORDINARY DIFFERENTIAL EQUATIONS

Section I (2 Questions)

Successive differentiation. Leibnitz theorem. Maclaurin and Taylor series expansions. Asymptotes. Curvature. Tests for concavity and convexity. Points of inflexion. Multiple points. Tracing of curves in Cartesian and polar coordinates.

Section II (3 Questions)

Reduction formulae. Quadrature. Rectification. Volumes and surfaces of solids of revolution.

Section III (3 Questions)

Exact differential equations. First order higher degree equations solvable for x, y, p. Clairaut's form and singular solutions. Geometrical meaning of a differential equation. Orthogonal trajectories. Linear differential equations with constant coefficients. Homogeneous linear ordinary differential equations. Total differential equations.

Section IV (2 Questions)

Linear differential equations of second order. Transformation of the equation by changing the dependent variable/the independent variable, reduction of order. Method of variation of parameters and unknown coefficients. Ordinary simultaneous differential equations.

Note. The examiner is requested to set ten questions in all selecting questions sectionwise as indicated in the syllabus. The candidate is required to attempt five questions selecting at least one question from each section.

CONTENTS

Chapters *Pages*

PART A : CALCULUS

SECTION I

1*. Limits and Continuity ... 3
2*. Derivability ... 35
3. Successive Differentiation ... 44
4. General Theorems ... 75
5. Expansions ... 82
6. Asymptotes ... 104
7. Curvature ... 131
8. Multiple Points, Tests for Concavity and Convexity, Points of Inflexion ... 167
9. Curve Tracing ... 193

SECTION II

10. Reduction Formulae ... 244
11. Rectification ... 265
12. Quadrature (Areas of Curves) ... 287
13. Volumes and Surfaces of Solids of Revolution ... 309

PART B : ORDINARY DIFFERENTIAL EQUATIONS

SECTION III

14. Geometrical Meaning of a Differential Equation ... 333
15. Exact Differential Equations ... 336
16. Differential Equations of the First Order and Higher Degree ... 354
17. Singular Solutions ... 372
18. Orthogonal Trajectories ... 387
19. Linear Differential Equations with Constant Coefficients ... 394
20. Homogeneous Linear Ordinary Differential Equations ... 429
21*. Total Differential Equations ... 438

SECTION IV

22. Linear Differential Equations of the Second Order ... 466
23. Ordinary Simultaneous Differential Equations ... 501
Examination Papers ... 525

*__Chapter 1__ and __Chapter 2__ are in the syllabus of **K.U. Kurukshetra** and not in the syllabus of **M.D.U. Rohtak.**

__Chapter 21__ is in the syllabus of **M.D.U. Rohtak and not in the syllabus of **K.U. Kurukshetra.**

PART A : CALCULUS

SECTION I

1*

Limits and Continuity

1. Neighbourhood of a Point on the Real Line

Let c be a real number.

Let a and b be two other real numbers such that $a < c < b$. Then the open interval (a, b) is called a **neighbourhood** of the point c.

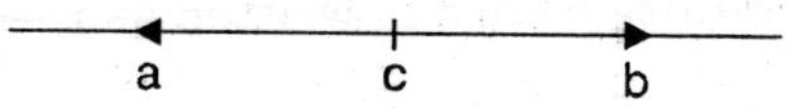

In fact, any open interval containing the given point c is **a neigh-bourhood** of c.

In general, a set A is called a neighbourhood of a point c if there exists an open interval (a, b) such that $c \in (a, b) \subseteq A$.

Remark. Each real number has an infinite number of neighbourhoods and each neighbourhood is an infinite set.

Examples 1. Open interval (1, 4) is a neighbourhood of the point 3.

2. Open interval (1, 3) is a neighbourhood of the point 2.

3. The closed interval $[a, b]$ is the neighbourhood of all of its points other than the end points a and b.

Symmetric Neighbourhood. The open interval (a, b) containing the point c is said to be a **symmetric neighbourhood** of c if $c - a = b - c$. Thus, (1, 3) is a **symmetric neighbourhood** of 2 because $2 - 1 = 3 - 2$.

Similarly, $(c - \delta, c + \delta)$ is a symmetric neighbourhood of c.

This symmetric neighbourhood $(c - \delta, c + \delta)$ of c is also expressed as : $\{\mathbf{x} : |\ \mathbf{x} - \mathbf{c}\ | < \delta\}$.

Left neighbourhood. The interval $(a, c]$ or $(c - \delta, c]$ is called a left *neighbourhood* of the point c.

Right neighbourhood. The interval $[c, b)$ or $[c, c + \delta)$ is called a *right neighbourhood* of the point c.

In the light of the above idea of one-sided neighbourhoods, we can say that the closed interval $[a, b]$ is the neighbourhood of each point of the open interval (a, b) and is the left neighbourhood of the point b and the right neighbourhood of the point a.

Deleted Neighbourhood

The open interval (a, b) with $a < c < b$ and excluding the point c is called *a* **deleted neighbourhood** of the point c.

or open interval $(c - \delta, c + \delta)$ excluding the point c is a deleted neighbourhood of the point c and is written as $\{x : |\ x - c\ | < \delta \text{ and } x \neq c\}$ or $\{x : 0 < |\ x - c\ | < \delta\}$.

2. (*a*) Meaning of *x* tends to *c*

When a variable number x is sufficiently close to c and is continuously approaching c but is never equal to c, we say that x tends to c.

---

*This chapter is in the syllabus of **K.U. Kurukshetra** and not in the syllabus of **M.D.U. Rohtak.**

We write it as $x \to c$ or Limit $x = c$.

Remark. We can also say that $x \to c$ means a deleted neighbourhood of c, *i.e.*, $0 < |x - c| < \delta$.

For example, $x \to 2$ means an open interval say (1.99, 2.01) but $x \neq 2$.

(*b*) x tends to c from left. When x approaches c through values less than c, we say that x tends to c from left and is denoted as $x \to c - 0$. In fact, $x \to c - 0$ means $c - \delta < x < c$.

(*c*) x tends to c from Right. When x approaches c through values greater than c, we say that x tends to c from right and is denoted as $x \to c + 0$. In fact, $x \to c + 0$ means $c < x < c + \delta$.

(*d*) The meaning of 'x approaches Infinity'. Let a variable x take up a series of positive values such that each value is greater than the preceding one and let this process continue. Evidently, the values of x will become greater and can be made greater than a given positive number, however large. In fact x is finally greater than the greatest number, we can imagine. This continuous increase in the value of x is expressed by saying that x approaches infinity, which is written as $x \to \infty$.

3. Division by Zero

Here we remind the reader of the following results.

We know that $\dfrac{0}{\text{Non - zero}} = 0$ and $\dfrac{\text{Non - zero}}{0} = \infty$.

Again, $0 \times$ any number $= 0$.

$\therefore \quad \dfrac{0}{0}$ = Any number and hence is called an **Indeterminate Form**

Similarly, $\dfrac{\infty}{\infty}$ is another indeterminate form.

SECTION A

4. Limit of a Function

The concept of limit is one of the key notions in calculus. Once the understanding of this concept is achieved, it becomes easy to grasp all other concepts in calculus, because the very underlying idea behind all processes in calculus, is the concept of limit.

We begin with an intuitive discussion of limit and then give a formal definition of it.

Consider the function f defined by the rule $f(x) = \begin{cases} 2x+1 & \text{if } x \neq 1 \\ 1 & \text{if } x = 1 \end{cases}$.

The graph of f is as shown in the figure. The fact that $f(1) = 1$ is shown by the isolated heavy point at (1, 1).

Let us investigate the functional values $f(x)$, when x is *close* to 1, but not equal to 1. First let x take values 0, 0.25, 0.50, 0.75, 0.9, 0.99, 0.999, 0.9999,......

We are taking values of x closer and closer to 1 but less than 1, in other words x is approaching 1 through values that are less than 1.

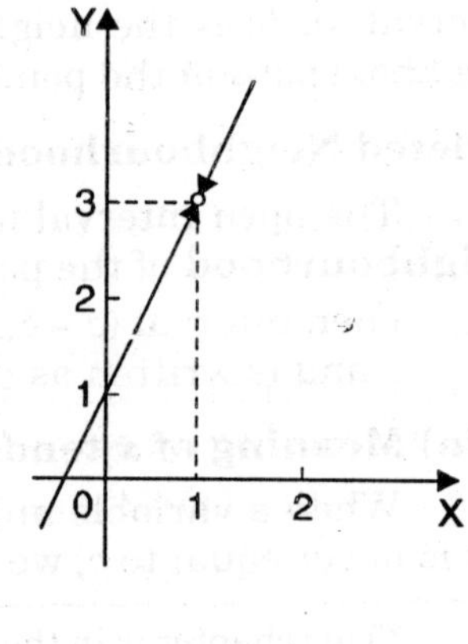

We illustrate this in the following table :

x	0	0.25	0.5	0.75	0.9	0.99	0.999	0.9999	0.99999
$f(x) = 2x + 1$ $(x \neq 1)$	1	1.5	2	2.5	2.8	2.98	2.998	2.9998	2.99998

Now let x approach 1 through values which are greater than 1 ; that is, let x take values 1.1, 1.01, 1.001, 1.0001, ... This is illustrated by the following table.

x	1.1	1.01	1.001	1.0001	1.00001
$f(x) = 2x+1$ $(x \neq 1)$	3.2	3.02	3.002	3.0002	3.00002

From the above table, it is clear that as x gets closer and closer to 1, $f(x)$ gets closer and closer to 3 ; and the closer x is to 1, the closer $f(x)$ is to 3. For instance, from the first table, when $x = 0.9$, $f(x) = 2.8$, *i.e.*, when x is 0.1 less than 1, $f(x)$ is 0.2 less than 3. When $x = 0.999$, $f(x) = 2.998$ *i.e.*, when x is 0.001 less than 1, $f(x)$ is 0.002 less than 3. Furthermore when $x = 0.99999$, $f(x) = 2.99998$ *i.e.*, when x is 0.00001 less than 1, $f(x)$ is 0.00002 less than 3.

From the second table, we observe that when $x = 1.1$, $f(x) = 3.2$, *i.e.*, when x is 0.1 greater than 1, $f(x)$ is 0.2 greater than 3. When $x = 1.001$, $f(x) = 3.002$, *i.e.*, when x is 0.001 greater than 1, $f(x)$ is 0.002 greater than 3. When $x = 1.00001$, $f(x) = 3.00002$ *i.e.*, when x is 0.00001 greater than 1, $f(x)$ is 0.00002 greater than 3.

Thus, from the two tables, we see that when x differs from 1 by $\pm$ 0.001, (that is $x = 0.999$ or $x = 1.001$), $f(x)$ differs from 3 by $\pm$ 0.002 (that is $f(x) = 2.998$ or $f(x) = 3.002$) and when x differs from 1 by $\pm$ 0.00001, $f(x)$ differs from 3 by $\pm$ 0.00002.

Now looking at the situation another way, we consider the values of $f(x)$ first. We see that we can make the value of $f(x)$ as *close to 3 as we please* by taking x *closer enough to* 1. Another way of saying this is that we can make the absolute value of the difference $f(x)$ and 3 *as small as we please*, by making the absolute difference of x and 1 small enough. This means that $|f(x) - 3|$ can be made as small as we please by making $|x - 1|$ small enough.

A more precise way of noting this is by using two symbols ε (epsilon and δ delta) for these small differences. Thus, we state that $|f(x) - 3|$ will be less than ε whenever $|x - 1|$ is less than δ and $|x - 1| \neq 0$ (since x is close to 1 and not equal to 1). Another way of saying this is : Given any positive number ε, we can make. $|f(x) - 3| < \varepsilon$ by taking $|x - 1|$ small enough, that is there is some sufficiently small positive number δ such that

$$|f(x) - 3| < \varepsilon$$

whenever $$0 < |x - 1| < \delta.$$

From the above two tables, we observe that $|f(x) - 3| = 0.002$ when $|x - 1| = 0.001$, so given $\varepsilon = 0.002$, we take $\delta = 0.001$ and say that

$$|f(x) - 3| < 0.002$$

whenever $$0 < |x - 1| < 0.001. \qquad (\because\ x \neq 1 \ \therefore\ 0 < |x - 1| < 0.001)$$

Thus, given any $\varepsilon > 0$, we can find a $\delta > 0$ such that $|f(x) - 3| < \varepsilon$ whenever $0 < |x - 1| < \delta$. This fact is expressed by saying that limit of $f(x)$ as x approaches 1 is equal to 3 and in symbols we write $\underset{x \to 1}{\text{Lt}}\ f(x) = 3$

5. Definition of Limit of a Function

Def. *Let f be a function which is defined at every point in some open interval containing the point c, except possibly at the number c itself and let l be a real number.* **Then *f* is said to tend to the limit *l* as *x* approaches** *c, if the difference between the value of f (x) and l can be made as small as we please by taking values of x sufficiently close to c and $x \neq c$.* In symbols, we write it as $\underset{x \to c}{\text{Lt}}\ f(x) = l$

or $$f(x) \to l \quad \text{as} \quad x \to c$$

In the light of the above discussion, we can also define the limit of a function as follows : ($\varepsilon - \delta$ **definition**)

A function f is said to tend to a limit l as x tends to c. If given $\varepsilon > 0$, however small. We can find a number $\delta > 0$, depending on ε such that

$$|\mathbf{f(x) - l}| < \varepsilon$$

whenever $0 < |x - c| < \delta$ *i.e.*, $f(x)$ lies in the open interval $(l - \varepsilon, l + \varepsilon)$ whenever x lies in the open interval $(c - \delta, c + \delta)$ and $x \neq c$, as shown in the adjoining figure

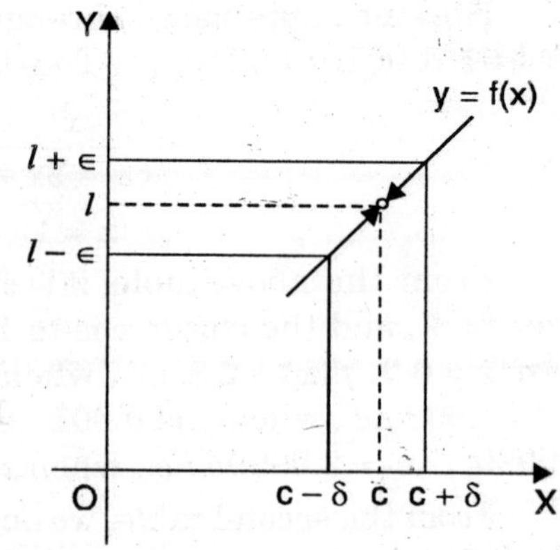

Note. $|x - c| < \delta \Rightarrow -\delta < x - c < \delta$

$(\because |x - a| < b \Rightarrow -b < x - a < b)$

Adding c to all sides $c - \delta < x < c + \delta$.

Remarks 1. If there is a number l for which any one of the above statements is true, we say that *limit of f as $x \to c$ exists.* If there is no number l for which any one of the statements is true, we say that the limit of f as $x \to c$ *does not exist.*

2. The limit of a function f as x tends to c, **if it exists is unique.**

3. For a function f to have limit as $x \to c$, *it is not necessary that the function be defined at the point $x = c$.* But the function must be defined for values of x close to c.

Definition. (*i*) Right-handed Limit. A function $f(x)$ is said to tend to a limit l as $x \to c$ from the right (from above) if, to any positive number, ε, however small, there corresponds a positive number, δ, such that $|f(x) - l| < \varepsilon$, for $c < x < c + \delta$.

In symbols, we write $\underset{x \to c+}{\text{Lt}} f(x) = l$.

Here l is the right-handed limit of $f(x)$.

(*ii*) Left-handed Limit. A function $f(x)$ is said to tend to a limit l as $x \to c$ from the left (from below) if, to any positive number ε, however small, there corresponds a positive number δ, such that $|f(x) - l| < \varepsilon$, for $c - \delta < x < c$

Symbolically, we write it as $\underset{x \to c-}{\text{Lt}} f(x) = l$.

Here l is the left-handed limit of $f(x)$.

The statement $\underset{x \to c}{\text{Lt}} f(x) = l$ implies that both the right-handed and left-handed limits exist and each is equal to l.

6. If $\underset{x \to c}{\text{Lt}} f(x)$ exists, then it is Unique

Proof. If possible, let $\underset{x \to c}{\text{Lt}} f(x) = l_1$ and $\underset{x \to c}{\text{Lt}} f(x) = l_2$.

$\because$ $\underset{x \to c}{\text{Lt}} f(x) = l_1$, therefore, (By Def. Art. 5) for $\varepsilon > 0$ (however small), there exists a positive number δ_1 s.t. $|f(x) - l_1| < \varepsilon$ whenever $0 < |x - c| < \delta_1$ ...(1)

Again, because $\underset{x \to c}{\text{Lt}} f(x) = l_2$, therefore, for $\varepsilon > 0$, there exists a positive number δ_2 s.t.

$$|f(x) - l_2| < \varepsilon \text{ whenever } 0 < |x - c| < \delta_2 \quad ...(2)$$

Let δ = Minimum* of δ_1 and δ_2

$\therefore$ Results (1) and (2) are both true even when we replace δ_1 and δ_2 by δ.

***δ = Minimum of δ_1 and δ_2**

$\Rightarrow$ Either $\delta = \delta_1$ or $\delta = \delta_2$.

$\therefore \qquad |f(x) - l_1| < \varepsilon$ and $|f(x) - l_2| < \varepsilon$

whenever $\qquad 0 < |x - c| < \delta \qquad$ ...(3)

Now $\qquad |l_1 - l_2| = |l_1 - f(x) + f(x) - l_2| \le |l_1 - f(x)| + |f(x) - l_2|$

$[\because \ |a + b| \le |a| + |b|]$

i.e., $\qquad |l_1 - l_2| \le |f(x) - l_1| + |f(x) - l_2| \qquad (\because |-t| = |t|)$

$< \varepsilon + \varepsilon$ whenever $0 < |x - c| < \delta$ *i.e.*, $|l_1 - l_2| < 2\varepsilon$.

Since ε is *arbitrarily* small *i.e.*, ε is smaller than the smallest positive number we can Imagine.

$\therefore \qquad |l_1 - l_2| = 0$ *i.e.*, $l_1 - l_2 = 0$ *i.e.*, $l_1 = l_2$.

$\therefore \qquad \underset{x \to c}{\text{Lt}} f(x)$ if it exists is unique.

7. Sandwich Theorem on Limits

Or **Squeeze Principle on Limits**

Statement. If $f(x) \le g(x) \le h(x)$ in a deleted neighbourhood of c and

$$\underset{x \to c}{\text{Lt}} f(x) = \underset{x \to c}{\text{Lt}} h(x) = l;$$

then $\qquad \underset{x \to c}{\text{Lt}} g(x) = l.$

Proof. Because $\underset{x \to c}{\text{Lt}} f(x) = \underset{x \to c}{\text{Lt}} h(x) = l,$

$\therefore$ (By Def., Art. 5) for $\varepsilon > 0$, however small, there exist $\delta_1 > 0$, $\delta_2 > 0$ such that

$|f(x) - l| < \varepsilon \quad$ for $\quad 0 < |x - c| < \delta_1$

and $\qquad |h(x) - l| < \varepsilon \quad$ for $\quad 0 < |x - c| < \delta_2$

$\because \qquad f(x) \le g(x) \le h(x)$

in a deleted neighbourhood of c.

Let this deleted neighbourhood of c be $0 < |x - c| < \delta_3$.

Let $\qquad \delta = \text{Minimum } \{\delta_1, \delta_2, \delta_3\}$.

$\therefore$ The above results are true even when $\delta_1, \delta_2, \delta_3$ are replaced by δ.

Thus $\delta > 0$ and for $0 < |x - c| < \delta$, we have

$|f(x) - l| < \varepsilon$, $|h(x) - l| < \varepsilon$ and $f(x) \le g(x) \le h(x)$

$\therefore \qquad l - \varepsilon < f(x) < l + \varepsilon$, $l - \varepsilon < h(x) < l + \varepsilon$ and $f(x) \le g(x) \le h(x)$

[$\because$ We know that $|\mathbf{x} - \mathbf{a}| < \mathbf{b} \Rightarrow \mathbf{a} - \mathbf{b} < \mathbf{x} < \mathbf{a} + \mathbf{b}$]

$\therefore \qquad l - \varepsilon < g(x) < l + \varepsilon$ for all x such that $0 < |x - c| < \delta$

or $\qquad |g(x) - l| < \varepsilon$ for $0 < |x - c| < \delta$

$\therefore$ By Def. of Limit Art. 5,

$$\underset{x \to c}{\text{Lt}} g(x) = l.$$

Example 1. *Show by definition of limit that the function $f(x) = 3x + 4$ has the limit 10 as $x \to 2$.*

Sol. We know (By Art. 5) that $\underset{x \to c}{\text{Lt}} f(x) = l$ iff given $\varepsilon > 0$, however small, there exists a positive real number δ (depending on ε) such that $|f(x) - l| < \varepsilon$ for $0 < |x - c| < \delta$

Here $\qquad f(x) = 3x + 4$ and $l = 10 \qquad$ (given)

$\therefore \qquad |f(x) - l| = |3x + 4 - 10| = |3x - 6| = |3(x - 2)|$

$= |3|\,|x - 2| = 3\,|x - 2|$

Let us take $\delta = \dfrac{\varepsilon}{3}$*

$\therefore$ For $0 < |x - 2| < \delta \left(= \dfrac{\varepsilon}{3}\right)$, we have

$$|f(x) - l| = 3\,|x - 2| < 3\delta \left(= 3 \cdot \frac{\varepsilon}{3} = \varepsilon\right)$$

$\therefore$ $|f(x) - 10| < \varepsilon$ for $0 < |x - 2| < \delta$

$\therefore$ $\underset{x \to 2}{\text{Lt}}\, f(x)$ where $f(x) = 3x + 4$ is 10.

Example 2. *Using the definition of limit, show that*

(i) $\underset{x \to c}{Lt} \dfrac{x^2 - c^2}{x - c} = 2c$ (ii) $\underset{x \to 0}{Lt}\, x^2 \sin \dfrac{1}{x} = 0.$

Sol. (i) We know (By Art. 5) that $\underset{x \to c}{\text{Lt}}\, f(x) = l$ iff given $\varepsilon > 0$, however small, there exists a positive real number δ (depending on ε) such that $|f(x) - l| < \varepsilon$ for $0 < |x - c| < \delta$

Here $f(x) = \dfrac{x^2 - c^2}{x - c}$, $x \neq c$ and $l = 2c$

$$\therefore \quad |f(x) - l| = \left|\frac{x^2 - c^2}{x - c} - 2c\right| = \left|\frac{(x - c)(x + c)}{x - c} - 2c\right|$$

$$= |x + c - 2c| = |x - c|$$

Let us take $\delta = \varepsilon$*

$\therefore$ For $0 < |x - c| < \delta\,(= \varepsilon)$, we have $|f(x) - l| = |x - c| < \delta = \varepsilon$

i.e., $|f(x) - l| < \varepsilon$ for $0 < |x - c| < \delta$

$\therefore$ $\underset{x \to c}{\text{Lt}}\, f(x) = l$ *i.e.*, $\underset{x \to c}{\text{Lt}} \dfrac{x^2 - c^2}{x - c} = 2c.$

(ii) *To prove* $\underset{x \to 0}{Lt}\, x^2 \sin \dfrac{1}{x} = 0$

Here $f(x) = x^2 \sin \dfrac{1}{x}$, $l = 0$

Let $\varepsilon > 0$, however small.

$$|f(x) - l| = \left|x^2 \sin \frac{1}{x} - 0\right| = \left|x^2 \sin \frac{1}{x}\right|$$

$$= |x^2| \left|\sin \frac{1}{x}\right| = |x|^2 \left|\sin \frac{1}{x}\right| \qquad (\because\ |x^n| = |x|^n)$$

---

*Because we are to free to take any value of δ depending on ε

(i) For example 1 on looking at $3\,|x - 2|$, we have taken $\delta = \dfrac{\varepsilon}{3}$.

(ii) For example 2 (i) on looking at $|x - c|$ above, we have taken δ [$= \varepsilon$].

$$\leq |x|^2 \qquad \left(\because \left|\sin\frac{1}{x}\right| \leq 1 \text{ as } |\sin\theta| \leq 1\right)$$

Let us take $\delta = {}^*\sqrt{\varepsilon}$

$\therefore$ For $0 < |x - c| < \delta$ *i.e.,* $0 < |x| < \delta$ $\quad (\because \; x \to 0 \text{ means } c = 0)$

$$|f(x) - l| \leq |x|^2 < \delta^2 < \varepsilon \qquad (\because \; \delta = \sqrt{\varepsilon})$$

$\therefore$ $\underset{x \to 0}{\text{Lt}} \; f(x) = l$ *i.e.,* $\underset{x \to 0}{\text{Lt}} \; x^2 \sin\frac{1}{x} = 0.$

Example 3. *Using* $\varepsilon - \delta$ *definition of limit, prove that* $\underset{x \to c}{Lt} \sin x = \sin c$

Sol. Let $\varepsilon > 0$, However small.

Let us choose a positive number $\delta (= \varepsilon)$ such that

$$0 < |x - c| < \delta \qquad ...(1) \qquad (x \neq c \Rightarrow 0 < |x - c|)$$

Let $f(x) = \sin x \quad \therefore \quad f(c) = \sin c$

$$|f(x) - f(c)| = |\sin x - \sin c|$$

$$= \left|2\cos\frac{x+c}{2}\sin\frac{x-c}{2}\right| \qquad \text{(C – D formula)}$$

$$= 2\left|\cos\frac{x+c}{2}\right|\left|\sin\frac{x-c}{2}\right| \qquad (\because \; |\alpha\beta| = |\alpha||\beta|)$$

$$\leq 2(1)\left|\frac{x-c}{2}\right| \qquad |\because \; \cos\theta \leq 1 \text{ and } |\sin\theta| \leq |\theta|$$

or $|f(x) - f(c)| \leq |x - c| < \delta \, (= \varepsilon)$ (By (1)) whenever $0 < |x - c| < \delta$

$\therefore$ By Def. $\underset{x \to c}{\text{Lt}} \; f(x) = f(c)$

i.e., $\underset{x \to c}{\text{Lt}} \; \sin x = \sin c.$

Example 4. *Using* ε-δ *definition, prove that* $\underset{x \to 1}{Lt} \dfrac{x^3 - 1}{x^2 - 1} = \dfrac{3}{2}$.

Sol. Here $f(x) = \dfrac{x^3 - 1}{x^2 - 1}$ when $x \to 1$ *i.e.* $x \neq 1$ and $l = \dfrac{3}{2}$ (given)

Let $\varepsilon > 0$, However small.

$$|f(x) - l| = \left|\frac{x^3 - 1}{x^2 - 1} - \frac{3}{2}\right| = \left|\frac{(x-1)(x^2 + x + 1)}{(x-1)(x+1)} - \frac{3}{2}\right|$$

$$(\because \; a^3 - b^3 = (a - b)(a^2 + b^2 + ab)$$

$$= \left|\frac{x^2 + x + 1}{x + 1} - \frac{3}{2}\right| = \left|\frac{2x^2 + 2x + 2 - 3x - 3}{2(x+1)}\right| = \left|\frac{2x^2 - x - 1}{2(x+1)}\right|$$

$$= \left|\frac{(x-1)(2x+1)}{2x+2}\right|$$

$$|\because \; 2x^2 - x - 1 = 2x^2 - 2x + x - 1 = 2x(x - 1) + (x - 1) = (x - 1)(2x + 1)$$

*Because we are free to take any value of δ depending on ε.

Here on looking at $|x|^2$, we have taken $\delta^2 = \varepsilon$ *i.e.,* $\delta = \sqrt{\varepsilon}$.

$$= |x-1|\left|\frac{2x+1}{2x+2}\right|$$

$$< |x-1| \qquad \left[\because |2x+1|<|2x+2| \text{ for all real } x \ \therefore \left|\frac{2x+1}{2x+2}\right|<1\right.$$

Let us take $\delta = \varepsilon$ (See footnotes examples 1 and 2 on page 8)

$\therefore$ For $\quad 0 < |x-1| < \delta\ (= \varepsilon)$,

we have $\quad |f(x) - l| \left(= \left|f(x) - \frac{3}{2}\right|\right) < |x-1| < \delta\ (= \varepsilon)$

$\therefore$ By Def. of limit, $\underset{x\to 1}{\text{Lt}}\ f(x) = \frac{3}{2}$.

8. Next we consider the behaviour of a function $f(x)$ which increases continuously as x tends to c. If the value of $f(x)$ becomes and remains greater than any positive number (however large) when $x \to c$, then $f(x)$ increases beyond limit and we say that it approaches to the limit $+\infty$.

Def. 1. *A function $f(x)$ is said to tend to plus infinity $(+\infty)$, if corresponding to any positive number G (however large), we can find a positive number δ, such that*

$$f(x) < G \text{ for } 0 < |x-c| < \delta. \quad \text{i.e.,} \quad \underset{x\to c}{Lt}\ f(x) = +\infty.$$

Def. 2. *A function $f(x)$ is said to tend to minus infinity $(-\infty)$, if corresponding to any positive number G (however large), we can find a positive number δ, such that*

$$f(x) < -G \text{ for } 0 < |x-c| < \delta. \quad \text{i.e.,} \quad \underset{x\to c}{Lt}\ f(x) = -\infty.$$

Def. 3. *If $f(x)$ does not tend to a unique limit (finite or infinite) as $x \to c$, $f(x)$ is said to* **oscillate.**

If for all values of x in the neighbourhood of c, $f(x)$ remains numerically less than a finite constant k, we say that $f(x)$ oscillates finitely, otherwise it is said to oscillate infinitely.

Def. 4. *A function $f(x)$ is said to tend to the limit L as $x \to \infty$, if corresponding to any given positive number ε [however small], we can find a positive number N, such that,*

$$|f(x) - L| < \varepsilon \quad \text{for} \quad x > N. \quad \text{i.e.,} \quad \underset{x\to +\infty}{Lt}\ f(x) = L.$$

Def. 5. *A function $f(x)$ is said to tend to plus infinity as $x \to \infty$, if corresponding to any given positive number G [however large], we can find a positive N, such that*

$$f(x) > G \quad \text{for} \quad x > N \quad \text{i.e.,} \quad \underset{x\to +\infty}{Lt}\ f(x) = +\infty.$$

9. *To prove that $\underset{x\to 0}{Lt}\ x^m = 0$, where m is a positive integer.*

Let $\varepsilon > 0$, However small. Let us choose a positive number $\delta\ (= \varepsilon)$ such that $|x - 0| = |x| < \delta$.

Sol. Case I. Let $x \to 0$ through positive values.

$\because\ x \to 0$, we may take $x < 1$

$\therefore$
$$|x^m - 0| = x^m$$
$$\le x \ (\because\ x < 1) = |x| \ (\because\ x > 0)$$
$$< \delta = \varepsilon$$

Hence given $\varepsilon > 0$, we can find a positive number δ [equal to ε in this case], such that

$$|x^m| < \varepsilon \quad \text{for} \quad |x| < \varepsilon \quad \therefore \quad \underset{x \to 0+}{\text{Lt}} \ x^m = 0.$$

Case II. Let x be negative

Put $x = -y$, $\therefore$ y is positive

$$|x^m - 0| = |(-1)^m y^m| = y^m \le y \qquad \text{(By Case I)}$$

$$< \varepsilon \text{ for } y < \varepsilon$$

$\therefore$ $|x^m| < \varepsilon$ for $-\delta < x < 0$ or $|x| < \delta$ (and $x < 0$) $\therefore$ $\underset{x \to 0-}{\text{Lt}} \ x^m = 0.$

Hence, right and left hand limits being equal, $x^m \to$ to the unique limit zero as $x \to 0$.

EXERCISE 1

Using the $\varepsilon - \delta$ definition of limit, prove the following :

1. $\underset{x \to 2}{\text{Lt}} \ (4x - 5) = 3$
2. $\underset{x \to 1}{\text{Lt}} \ \dfrac{x^2 - 1}{x - 1} = 2$
3. $\underset{x \to 2}{\text{Lt}} \ \dfrac{x^2 - 4}{x - 2} = 4$
4. $\underset{x \to 3}{\text{Lt}} \ (1 - 3x) = -8$
5. $\underset{x \to -1}{\text{Lt}} \ (2 - 3x) = 5$
6. $\underset{x \to 2}{\text{Lt}} \ \dfrac{x^3 - 8}{x^2 - 4} = 3$ *(M.D.U. 2005 A)*
7. $\underset{x \to 0}{\text{Lt}} \ x \sin \dfrac{1}{x} = 0$ *(K.U. 2003)*
8. $\underset{x \to 0}{\text{Lt}} \ x^2 \cos \dfrac{1}{x} = 0$
9. $\underset{x \to c}{\text{Lt}} \ \cos x = \cos c.$

10. Some Standard Results on Limits

Here we remind the reader of some of the standard results, he or she has learnt in class XII and these results are

(*i*) $\underset{x \to 0}{\text{Lt}} \ \dfrac{\sin x}{x} = 1$ and $\underset{x \to 0}{\text{Lt}} \ \dfrac{\tan x}{x} = 1$ where x is the angle measured in Radians.

(*ii*) $\underset{x \to 0}{\text{Lt}} \ \dfrac{a^x - 1}{x} = \log a$ (*iii*) $\underset{x \to 0}{\text{Lt}} \ (1 + x)^{\frac{1}{x}} = e$

(*iv*) $\underset{x \to \infty}{\text{Lt}} \left(1 + \dfrac{1}{x}\right)^x = e$ (*v*) $\underset{x \to 0}{\text{Lt}} \ \dfrac{\sin^{-1} x}{x} = 1$ and $\underset{x \to 0}{\text{Lt}} \ \dfrac{\tan^{-1} x}{x} = 1.$

Here we also remind the reader about the left hand limit and right hand limit as learnt in class XII.

Left hand limit = $\underset{x \to c-}{\text{Lt}} \ f(x).$

$x \to c- \ \Rightarrow \ x$ is slightly less than c.

Put $x = c - h$ where $h \to 0^+$.

Right hand limit = $\underset{x \to c+}{\text{Lt}} \ f(x).$

$x \to c+ \ \Rightarrow \ x$ is slightly $> c$.

Put $x = c + h$ where $h \to 0^+$.

If Left Hand Limit $\underset{x \to c-}{\text{Lt}} \ f(x)$ = Right Hand Limit $\underset{x \to c+}{\text{Lt}} \ f(x)$, then we say that $\underset{x \to c}{\text{Lt}} \ f(x)$ exists.

Note. In problems on modulus function, greatest integer function, functions with partitioned domain *i.e.* $f(x)$ is defined differently for $x < c$ (say) and for $x > c$ and $f(x)$ involving $e^{\frac{1}{x}}$, we are to find left hand limit and right hand limit seperately.

Here, we explain the results of this Art. 10 by the help of some examples.

11. Algebra of Limits

If $f(x)$ and $\phi(x)$ are two functions of x, such that

$$\underset{x \to c}{\text{Lt}}\ f(x) = l \text{ and } \underset{x \to c}{\text{Lt}}\ \phi(x) = m.$$

Then

(i) $\underset{x \to c}{\text{Lt}}\ [f(x) \pm \phi(x)] = \underset{x \to c}{\text{Lt}}\ f(x) \pm \underset{x \to c}{\text{Lt}}\ \phi(x) = l \pm m$

(ii) $\underset{x \to c}{\text{Lt}}\ af(x) = a\ \underset{x \to c}{\text{Lt}}\ f(x) = a.l.$

(iii) $\underset{x \to c}{\text{Lt}}\ [f(x)\,.\,\phi(x)] = \underset{x \to c}{\text{Lt}}\ f(x)\,.\,\underset{x \to c}{\text{Lt}}\ \phi(x) = l.m$

(iv) $\underset{x \to c}{\text{Lt}} \left[\dfrac{f(x)}{\phi(x)}\right] = \underset{x \to c}{\text{Lt}}\ f(x) \cdot \underset{x \to c}{\text{Lt}}\ \dfrac{1}{\phi(x)} = \dfrac{l}{m}.$

12. Absolute Function and Greatest Integer Function

(*a*) **Absolute function** $|x| = x$ if $x \geq 0$ and $= -x$ if $x < 0$

i.e., $|x|$ is the Numerical value of x.

Note. $|x|$ is also defined as $\sqrt{x^2}$.

(*b*) **Greatest integer function**

If x is any real number,

then the greatest integer $\leq x$ is denoted by $[x]$.

e.g., if $x = 3.9$; $[x] = 3$

if $x = 3.1$, $[x] = 3$

if $x = 5$, $[x] = 5$

if $x = -4.3$, $[x] = -5$.

Let $x = c$ be any real number.

Case I. c is an integer i.e., c ∈ Z or I

As $h \to 0^+$, $[c - h] = c - 1$ and $[c + h] = c$.

Case II. c ∈ R – Z i.e., c is a real number not an integer.

As $h \to 0^+$, then $[c - h] = [c]$ and $[c + h] = [c]$.

Note. $x - 1 < [x] \leq x$.

Example 1. *Evaluate the following limits :*

(i) $\underset{x \to 2}{Lt}\ \dfrac{x^3 - 8}{x^2 - 4}$ (ii) $\underset{x \to 2}{Lt}\ \dfrac{x^2 - 4}{\sqrt{3x - 2} - \sqrt{x + 2}}$ (iii) $\underset{x \to 0}{Lt}\ \dfrac{(1 + x)^n - 1}{x}$.

Sol. (*i*) $\displaystyle\lim_{x \to 2} \frac{x^3 - 8}{x^2 - 4} = \lim_{x \to 2} \frac{x^3 - 2^3}{x^2 - 2^2}$ | Form $\dfrac{0}{0}$

$$= \lim_{x \to 2} \frac{(x - 2)(x^2 + 2x + 4)}{(x - 2)(x + 2)} \quad | \because\ a^3 - b^3 = (a - b)(a^2 + ab + b^2)$$

$$= \lim_{x \to 2} \frac{x^2 + 2x + 4}{x + 2} \quad [\because\ x \to 2 \Rightarrow x \neq 2]$$

$$= \frac{2^2 + 2 \times 2 + 4}{2 + 2} = \frac{12}{4} = 3$$

(ii) $\lim_{x \to 2} \dfrac{x^2 - 4}{\sqrt{3x-2} - \sqrt{x+2}}$ | Form $\frac{0}{0}$

Rationalising, $= \lim_{x \to 2} \dfrac{x^2 - 4}{\sqrt{3x-2} - \sqrt{x+2}} \times \dfrac{\sqrt{3x-2} + \sqrt{x+2}}{\sqrt{3x-2} + \sqrt{x+2}}$

$$= \lim_{x \to 2} \frac{(x^2-4)(\sqrt{3x-2} + \sqrt{x+2})}{(3x-2)-(x+2)} = \lim_{x \to 2} \frac{(x^2-4)(\sqrt{3x-2} + \sqrt{x+2})}{2x-4}$$

$$= \lim_{x \to 2} \frac{(x+2)(x-2)(\sqrt{3x-2} + \sqrt{x+2})}{2(x-2)}$$

Cancelling $(x-2)$,

$$= \lim_{x \to 2} \frac{(x+2)(\sqrt{3x-2} + \sqrt{x+2})}{2} = \frac{(2+2)(\sqrt{3 \times 2 - 2} + \sqrt{2+2})}{2}$$

$$= \frac{(4)(2+2)}{2} = 8.$$

(iii) $\mathop{\text{Lt}}_{x \to 0} \dfrac{(1+x)^n - 1}{x}$ | Form $\frac{0}{0}$

Expanding by Binomial Theorem

$$= \mathop{\text{Lt}}_{x \to 0} \frac{1 + nx + \frac{n(n-1)}{2!} x^2 + \frac{n(n-1)(n-2)}{3!} x^3 + \ldots - 1}{x}$$

$$= \mathop{\text{Lt}}_{x \to 0} \frac{nx + \frac{n(n-1)}{2!} x^2 + \frac{n(n-1)(n-2)}{3!} x^3 + \ldots}{x}$$

$$= \mathop{\text{Lt}}_{x \to 0} \frac{nx\left(1 + \frac{(n-1)}{2!} x + \frac{(n-1)(n-2)}{3!} x^2 + \ldots\right)}{x}$$

$$= \mathop{\text{Lt}}_{x \to 0} n\left(1 + \frac{(n-1)}{2!} x + \frac{(n-1)(n-2)}{3!} x^2 + \ldots\right)$$

$$= n\,(1 + 0) = n.$$

Example 2. *Evaluate the following limits*

(i) $\mathop{Lt}_{\theta \to 0} \dfrac{1 - \cos\theta}{\theta^2}$ (ii) $\mathop{Lt}_{x \to 0} \dfrac{\sin x^\circ}{x}$ (iii) $\mathop{Lt}_{\theta \to 0} \dfrac{1 - \cos m\theta}{1 - \cos n\theta}$

where m, n are fixed non-zero real numbers.

Sol. (i) $\mathop{\text{Lt}}_{\theta \to 0} \dfrac{1 - \cos\theta}{\theta^2} = \mathop{\text{Lt}}_{\theta \to 0} \dfrac{2\sin^2 \frac{\theta}{2}}{\theta^2} = 2 \mathop{\text{Lt}}_{\theta \to 0} \left(\dfrac{\sin \frac{\theta}{2}}{2 . \frac{\theta}{2}}\right)^2$

$$= \frac{2}{4} \mathop{\text{Lt}}_{\theta \to 0} \left(\frac{\sin \frac{\theta}{2}}{\frac{\theta}{2}}\right)^2 = \frac{1}{2}\left[\mathop{\text{Lt}}_{\frac{\theta}{2} \to 0} \frac{\sin \frac{\theta}{2}}{\frac{\theta}{2}}\right]^2 = \frac{1}{2}(1)^2 = \frac{1}{2}.$$

(ii) $\underset{x \to 0}{\text{Lt}} \dfrac{\sin x^0}{x} = \underset{x \to 0}{\text{Lt}} \dfrac{\sin \dfrac{\pi x}{180}}{x}$ $\left[\begin{array}{l} \because \ \ 180 \text{ degrees} = \pi \text{ radians} \\ \therefore \ \ x \text{ degrees} = \dfrac{\pi x}{180} \text{ radians} \end{array}\right]$

$$= \underset{x \to 0}{\text{Lt}} \frac{\sin \dfrac{\pi x}{180}}{\dfrac{\pi x}{180}} \times \frac{\pi}{180} = \frac{\pi}{180}\left[\underset{\pi x/180 \to 0}{\text{Lt}} \frac{\sin \dfrac{\pi x}{180}}{\dfrac{\pi x}{180}}\right] = \frac{\pi}{180}(1) = \frac{\pi}{180}.$$

(iii) $\underset{\theta \to 0}{\text{Lt}} \dfrac{1-\cos m\theta}{1-\cos n\theta} = \underset{\theta \to 0}{\text{Lt}} \dfrac{2\sin^2 \dfrac{m\theta}{2}}{2\sin^2 \dfrac{n\theta}{2}} = \underset{\theta \to 0}{\text{Lt}} \left[\dfrac{\sin \dfrac{m\theta}{2}}{\dfrac{m\theta}{2}} \cdot \dfrac{\dfrac{m\theta}{2}}{\dfrac{n\theta}{2}} \cdot \dfrac{\dfrac{n\theta}{2}}{\sin \dfrac{n\theta}{2}}\right]^2$

$$= \left[\underset{\frac{m\theta}{2} \to 0}{\text{Lt}} \frac{\sin \dfrac{m\theta}{2}}{\dfrac{m\theta}{2}}\right]^2 \left[\frac{m}{n}\right]^2 \left[\underset{\frac{n\theta}{2} \to 0}{\text{Lt}} \frac{\dfrac{n\theta}{2}}{\sin \dfrac{n\theta}{2}}\right]^2 = (1)^2 \left(\frac{m^2}{n^2}\right)(1)^2 = \frac{m^2}{n^2}.$$

Example 3. *Evaluate the following limits :*

(i) $\underset{x \to 0}{Lt} \dfrac{\sqrt{1+x}-\sqrt{1-x}}{\sin^{-1} x}$ (ii) $\underset{x \to 1}{Lt} \dfrac{1-\sqrt{x}}{(\cos^{-1} x)^2}$.

Sol. (i) $\underset{x \to 0}{\text{Lt}} \dfrac{\sqrt{1+x}-\sqrt{1-x}}{\sin^{-1} x}$

Rationalising, $= \underset{x \to 0}{\text{Lt}} \dfrac{(\sqrt{1+x}-\sqrt{1-x})(\sqrt{1+x}+\sqrt{1-x})}{\sin^{-1} x(\sqrt{1+x}+\sqrt{1-x})}$

$= \underset{x \to 0}{\text{Lt}}\ 2\dfrac{x}{\sin^{-1} x} \cdot \dfrac{1}{\sqrt{1+x}+\sqrt{1-x}}$ $[\because\ 1+x-(1-x)=2x]$

$= 2(1)\dfrac{1}{1+1} = 1.$ $\left|\ \because\ \underset{x \to 0}{\text{Lt}} \dfrac{\sin^{-1} x}{x} = 1\right.$

(ii) Put $\cos^{-1} x = \theta$, then $x = \cos\theta$ so that as $x \to 1$, $\theta \to 0$

$\therefore\ \underset{x \to 1}{\text{Lt}} \dfrac{1-\sqrt{x}}{(\cos^{-1} x)^2} = \underset{\theta \to 0}{\text{Lt}} \dfrac{1-\sqrt{\cos\theta}}{\theta^2}$

Rationalising, $= \underset{\theta \to 0}{\text{Lt}} \dfrac{(1-\sqrt{\cos\theta})(1+\sqrt{\cos\theta})}{\theta^2(1+\sqrt{\cos\theta})} = \underset{\theta \to 0}{\text{Lt}} \dfrac{1-\cos\theta}{\theta^2(1+\sqrt{\cos\theta})}$

$$= \underset{\theta \to 0}{\text{Lt}} \frac{2\sin^2 \dfrac{\theta}{2}}{\theta^2(1+\sqrt{\cos\theta})} = \underset{\theta \to 0}{\text{Lt}} \frac{2\left(\dfrac{\sin \dfrac{\theta}{2}}{2 \cdot \dfrac{\theta}{2}}\right)^2}{1+\sqrt{\cos\theta}} = \frac{2 \cdot \left(\dfrac{1}{2}\right)^2}{1+1} = \frac{1}{4}$$

Remark. In problems on limits of Inverse Trigonometric functions, you can always put inverse function equal to θ.

Example 4. *Evaluate the following limits :*

(i) $\underset{x \to \infty}{Lt} \dfrac{x}{\sqrt{4x^2+1}-1}$

(ii) $\underset{n \to \infty}{Lt} \dfrac{1^2 + 2^2 + 3^2 + \ldots\ldots + n^2}{n^3}$ or $\underset{n \to \infty}{Lt} \dfrac{\Sigma n^2}{n^3}$.

Sol. (i) $\underset{x \to \infty}{\text{Lt}} \dfrac{x}{\sqrt{4x^2+1}-1}$ Dividing the num. and denom. by x,

$$= \underset{x \to \infty}{\text{Lt}} \frac{1}{\sqrt{4+\frac{1}{x^2}} - \frac{1}{x}} = \frac{1}{\sqrt{4+0}-0} = \frac{1}{2}$$

(ii)
$$\underset{n \to \infty}{\text{Lt}} \frac{1^2+2^2+3^2+\ldots\ldots+n^2}{n^3} = \underset{n \to \infty}{\text{Lt}} \frac{\frac{n(n+1)(2n+1)}{6}}{n^3}$$

$$= \underset{n \to \infty}{\text{Lt}} \frac{(n+1)(2n+1)}{6n^2}$$ Dividing the num. and denom. by n^2,

$$= \underset{n \to \infty}{\text{Lt}} \frac{\left(1+\frac{1}{n}\right)\left(2+\frac{1}{n}\right)}{6} = \frac{(1+0)(2+0)}{6} = \frac{1}{3}.$$

Example 5. *Evaluate the following limits :*

(i) $\underset{x \to 0}{Lt} \dfrac{3^x - 2^x}{x}$ (ii) $\underset{x \to 0}{Lt} \dfrac{(ab)^x - a^x - b^x + 1}{x^2}$

(iii) $\underset{x \to 1}{Lt} \dfrac{\log x}{x-1}$ (iv) $\underset{x \to \infty}{Lt} \left(\dfrac{x+6}{x+1}\right)^{x+4}$.

Sol. (i) $\underset{x \to 0}{\text{Lt}} \dfrac{3^x - 2^x}{x}$ Subtract and add 1,

$$= \underset{x \to 0}{\text{Lt}} \frac{3^x - 1 - 2^x + 1}{x} = \underset{x \to 0}{\text{Lt}} \frac{(3^x-1)-(2^x-1)}{x} = \underset{x \to 0}{\text{Lt}} \left[\frac{3^x-1}{x} - \frac{(2^x-1)}{x}\right]$$

$$= \underset{x \to 0}{\text{Lt}} \left(\frac{3^x-1}{x}\right) - \underset{x \to 0}{\text{Lt}} \left(\frac{2^x-1}{x}\right) = \log 3 - \log 2 = \log \frac{3}{2}.$$

(ii) $(ab)^x - a^x - b^x + 1 = a^x b^x - a^x - b^x + 1 = a^x (b^x - 1) - (b^x - 1) = (a^x - 1)(b^x - 1)$

$\therefore$ Required Limit $= \underset{x \to 0}{\text{Lt}} \dfrac{(ab)^x - a^x - b^x + 1}{x^2} = \underset{x \to 0}{\text{Lt}} \dfrac{(a^x-1)(b^x-1)}{x \cdot x}$

$$= \underset{x \to 0}{\text{Lt}} \frac{a^x - 1}{x} \cdot \underset{x \to 0}{\text{Lt}} \frac{b^x-1}{x} = (\log a)(\log b).$$

(iii) Put $x = 1 + h$ so that as $x \to 1, h \to 0$

$$\therefore \underset{x \to 1}{\text{Lt}} \frac{\log x}{x-1} = \underset{h \to 0}{\text{Lt}} \frac{\log(1+h)}{1+h-1} = \underset{h \to 0}{\text{Lt}} \frac{\log(1+h)}{h} = \underset{h \to 0}{\text{Lt}} \frac{1}{h} \log(1+h)$$

$$= \underset{h \to 0}{\text{Lt}} \log(1+h)^{1/h} = \log e = 1.$$ $\left[\because \underset{x \to 0}{\text{Lt}} (1+x)^{1/x} = e\right]$

(iv)
$$\underset{x \to \infty}{\text{Lt}} \left(\frac{x+6}{x+1}\right)^{x+4} = \underset{x \to \infty}{\text{Lt}} \left(\frac{x+1+5}{x+1}\right)^{x+4}$$

$$= \underset{x \to \infty}{\text{Lt}} \left(1 + \frac{5}{x+1}\right)^{x+4} = \underset{x \to \infty}{\text{Lt}} \left[\left(1 + \frac{5}{x+1}\right)^{\frac{x+1}{5}}\right]^{(x+4) \cdot \frac{5}{x+1}}$$

$$= e^5. \quad \left[\because \underset{x \to \infty}{\text{Lt}} \left(1 + \frac{1}{x}\right)^x = e\right.$$

and
$$\underset{x \to \infty}{\text{Lt}} \frac{5(x+4)}{x+1} = \underset{x \to \infty}{\text{Lt}} \frac{5x\left(1 + \frac{4}{x}\right)}{x\left(1 + \frac{1}{x}\right)} = \underset{x \to \infty}{\text{Lt}} \frac{5\left(1 + \frac{4}{x}\right)}{1 + \frac{1}{x}} = 5\left.\right]$$

Example 6. *Show that* $\underset{x \to 2}{Lt} \frac{|x-2|}{x-2}$ *does not exist.*

Sol. $\text{L.H.L.} = \underset{x \to 2-}{\text{Lt}} \frac{|x-2|}{x-2} = \underset{h \to 0+}{\text{Lt}} \frac{|2-h-2|}{2-h-2}$ [by putting $x = 2 - h, h > 0$]

$$= \underset{h \to 0+}{\text{Lt}} \frac{|-h|}{-h} = \underset{h \to 0+}{\text{Lt}} \frac{h}{-h} = \underset{h \to 0+}{\text{Lt}} (-1) = -1$$

$\text{R.H.L.} = \underset{x \to 2+}{\text{Lt}} \frac{|x-2|}{x-2} = \underset{h \to 0+}{\text{Lt}} \frac{|2+h-2|}{2+h-2}$ [by putting $x = 2 + h, h > 0$]

$$= \underset{h \to 0+}{\text{Lt}} \frac{|h|}{h} = \underset{h \to 0+}{\text{Lt}} \frac{h}{h} = \underset{h \to 0+}{\text{Lt}} 1 = 1.$$

Since L.H.L. ≠ R.H.L.

$\therefore \quad \underset{x \to 2}{\text{Lt}} \frac{|x-2|}{x-2}$ does not exist.

Example 7. *Find* $\underset{x \to 0}{Lt} \frac{x}{|x| + x^2}$.

Sol. $\text{L.H.L.} = \underset{x \to 0-}{\text{Lt}} \frac{x}{|x| + x^2} = \underset{x \to 0-}{\text{Lt}} \frac{x}{-x + x^2}$ [$\because$ $x \to 0- \Rightarrow x < 0 \Rightarrow |x| = -x$]

$$= \underset{x \to 0-}{\text{Lt}} \frac{x}{x(-1+x)} = \underset{x \to 0-}{\text{Lt}} \frac{1}{-1+x} = \frac{1}{-1+0} = -1$$

$$\text{R.H.L.} = \underset{x \to 0+}{\text{Lt}} \frac{x}{|x| + x^2} = \underset{x \to 0+}{\text{Lt}} \frac{x}{x + x^2}$$

$$= \underset{x \to 0+}{\text{Lt}} \frac{x}{x(1+x)} = \underset{x \to 0+}{\text{Lt}} \frac{1}{1+x} = \frac{1}{1+0} = 1$$

Since L.H.L. ≠ R.H.L.

$\therefore \quad \underset{x \to 0}{\text{Lt}} \frac{x}{|x| + x^2}$ does not exist.

Example 8. *For what value of p does* $\underset{x \to 1}{Lt} f(x)$ *exist where f is defined by*

$$f(x) = \begin{cases} 2px + 3 & \text{if } x < 1 \\ 1 - px^2 & \text{if } x > 1. \end{cases}$$

Sol. $\underset{x \to 1-}{\text{Lt}} f(x) = \underset{x \to 1-}{\text{Lt}} (2px + 3)$ [$\because$ when $x < 1, f(x) = 2px + 3$]

$$= 2p(1) + 3 = 2p + 3$$

$$\underset{x \to 1+}{\text{Lt}} f(x) = \underset{x \to 1+}{\text{Lt}} (1 - px^2) \qquad [\because \text{ when } x > 1, f(x) = 1 - px^2]$$
$$= 1 - p(1)^2 = 1 - p.$$

Now, for the existence of $\underset{x \to 1}{\text{Lt}} f(x)$, we must have

$$\underset{x \to 1-}{\text{Lt}} f(x) = \underset{x \to 1+}{\text{Lt}} f(x)$$

i.e., $$2p + 3 = 1 - p \quad \text{or} \quad 3p = -2$$

Hence, $$p = -\frac{2}{3}.$$

Example 9. *Show that* $\underset{x \to c}{Lt}$ $[x]$ *where* $[x]$ *denotes the greatest integer* $\leq x$ *does not exist if* c *is an integer.*

Sol.
$$\underset{x \to c-}{\text{Lt}} [x] = \underset{h \to 0+}{\text{Lt}} [c - h] \qquad [\text{Putting } x = c - h, h > 0]$$
$$= \underset{h \to 0+}{\text{Lt}} (c - 1) \qquad [\because \ c - 1 < c - h < c]$$
$$= c - 1 \qquad (\text{Also see Art. 12}(b))$$
$$\underset{x \to c+}{\text{Lt}} [x] = \underset{h \to 0+}{\text{Lt}} [c + h] \qquad [\text{Putting } x = c + h, h > 0]$$
$$= \underset{h \to 0+}{\text{Lt}} c \qquad [\because c < c + h < c + 1]$$
$$= c \qquad (\text{Also see Art. 12}(b))$$

Since $$\underset{x \to c-}{\text{Lt}} [x] \neq \underset{x \to c+}{\text{Lt}} [x]$$

$\therefore$ $\underset{x \to c}{\text{Lt}} [x]$ does not exist.

Example 10. *Prove that* $\underset{x \to c}{Lt}$ $[x]$ *where* $[x]$ *denotes the greatest integer* $\leq x = [c]$*, if* c *is not an integer.*

Sol. Since c is not an integer, let n be the greatest integer less than c. *i.e.,* let $[c] = n$
Then $[c] = n$ where $n < c < n + 1$

$$\underset{x \to c-}{\text{Lt}} [x] = \underset{h \to 0+}{\text{Lt}} [c - h] \qquad [\text{Putting } x = c - h, h > 0]$$
$$= \underset{h \to 0+}{\text{Lt}} [c] = [c] = n \qquad [\text{By Art. 12}(b)]$$
$$\lim_{x \to c+} [x] = \lim_{h \to 0+} [c + h] \qquad [\text{Putting } x = c + h, h > 0]$$
$$= \underset{h \to 0+}{\text{Lt}} [c] = [c] = n \qquad [\text{By Art. 12}(b)]$$

Since $$\underset{x \to c-}{\text{Lt}} [x] = \underset{x \to c+}{\text{Lt}} [x] = n \quad \therefore \quad \underset{x \to c}{\text{Lt}} [x] = n = [c].$$

EXERCISE 2

Evaluate the following limits : (1 – 9)

1. (*i*) $\underset{x \to 4}{\text{Lt}} \dfrac{x^2 - 16}{x^3 - 64}$ (*ii*) $\underset{x \to a}{\text{Lt}} \dfrac{\sqrt{x} - \sqrt{a}}{x - a}$

(*iii*) $\underset{x \to 1}{\text{Lt}} \dfrac{\sqrt{x^2 - 1} + \sqrt{x - 1}}{\sqrt{x^2 - 1}}$ (*iv*) $\underset{x \to 3}{\text{Lt}} \dfrac{x^2 - 9}{\sqrt{3x - 2} - \sqrt{x + 4}}$.

2. (i) $\underset{h \to 0}{\text{Lt}} \frac{1}{h}\left(\frac{1}{x+h} - \frac{1}{x}\right)$ (ii) $\underset{h \to 0}{\text{Lt}} \frac{1}{h}\left(\frac{1}{\sqrt{x+h}} - \frac{1}{\sqrt{x}}\right)$

3. (i) $\underset{\theta \to 0}{\text{Lt}} \frac{1-\cos\theta}{2\theta^2}$ (ii) $\underset{t \to 0}{\text{Lt}} \frac{1-\cos 6t}{t}$

(iii) $\underset{x \to 0}{\text{Lt}} \frac{\tan x - \sin x}{\sin^3 x}$ (iv) $\underset{x \to 0}{\text{Lt}} \frac{\tan x - \sin x}{x^3}$

4. (i) $\underset{x \to \frac{\pi}{2}}{\text{Lt}} \frac{1+\cos 2x}{(\pi - 2x)^2}$ (ii) $\underset{x \to \frac{\pi}{4}}{\text{Lt}} \frac{1-\tan x}{x - \frac{\pi}{4}}$ (iii) $\underset{x \to \frac{\pi}{2}}{\text{Lt}} (\sec x - \tan x)$

5. (i) $\underset{x \to 1}{\text{Lt}} \frac{1-x}{(\cos^{-1} x)^2}$ (ii) $\underset{x \to 1}{\text{Lt}} \frac{1-x}{\pi - 2\sin^{-1} x}$

6. (i) $\underset{n \to \infty}{\text{Lt}} \frac{1+2+3+\ldots+n}{1-n^2}$ (ii) $\underset{n \to \infty}{\text{Lt}} \frac{\Sigma n^3}{n^4}$

7. (i) $\underset{x \to 0}{\text{Lt}} \frac{a^x - 1}{b^x - 1}, a > 0, b > 0$ (ii) $\underset{x \to 0}{\text{Lt}} \frac{2^x - 1}{\sqrt{1+x} - 1}$ (*M.D.U. 2006*)

8. (i) $\underset{x \to 0}{\text{Lt}} \frac{\log(3+x) - \log(3-x)}{x}$ (ii) $\underset{x \to 0^+}{\text{Lt}} \frac{\log(1+x^3)}{\sin^3 x}$

9. (i) $\underset{x \to 0}{\text{Lt}} (1+3x)^{\frac{1}{x}}$ (ii) $\underset{x \to \infty}{\text{Lt}} \left(\frac{x+2}{x+3}\right)^x$.

10. Show that $\underset{x \to c}{\text{Lt}} \frac{|x-c|}{x-c}$ does not exist.

11. Show that the following limits do not exist :

(i) $\underset{x \to 3}{\text{Lt}} [x]$ (ii) $\underset{x \to 2}{\text{Lt}} (x - [x])$

where $[x]$ denotes the greatest integer $\le x$.

12. Find $\underset{x \to 3}{\text{Lt}} f(x)$ where $f(x) = \begin{cases} 4 & \text{if } x \ge 3 \\ x+1 & \text{if } x \le 3 \end{cases}$.

13. Show that $\underset{x \to 0}{\text{Lt}} e^{-1/x}$ does not exist.

ANSWERS

1. (i) $\frac{1}{6}$ (ii) $\frac{1}{2\sqrt{a}}$ (iii) $1 + \frac{1}{\sqrt{2}}$ (iv) $6\sqrt{7}$

2. (i) $-\frac{1}{x^2}$ (ii) $-\frac{1}{2x^{3/2}}$

3. (i) $\frac{1}{4}$ (ii) 0 (iii) $\frac{1}{2}$ (iv) $\frac{1}{2}$

4. (i) $\frac{1}{2}$ (ii) -2 (iii) 0

5. (i) $\frac{1}{2}$ (ii) 0

6. (i) $\frac{-1}{2}$ (ii) $\frac{1}{4}$ 7. (i) $\log_b a$ (ii) $2 \log 2$

8. (i) $\frac{2}{3}$ (ii) 1

9. (i) e^3 (ii) $\frac{1}{e}$ 12. 4.

SECTION B
CONTINUITY

13. (*i*) Definition of Continuity at a Point

A function f (x) is said to be continuous at a point x = c if for $\varepsilon > 0$, however small, there exists a positive number δ depending on ε such that $|f(x) - f(c)| < \varepsilon$ for all x s.t. $|x - c| < \delta$.

Note. Comparing with the Definition of limit in Art. 5, we have $l = f(c)$ and instead of a deleted neighbourhood of c $(0 < |x - c| < \delta)$ we have neighbourhood of c (*i.e.*, $|x - c| < \delta$).

From the above definition of continuity, we can say that

A function f (x) is said to be continuous at a point x = c, if and only if

$$\mathop{\text{Lt}}_{x \to c} f(x) = f(c)$$

i.e., the limit of the function = value of the function at that point.

This definition is based on the assumption that the function $f(x)$ has a well-defined value at the point, and the limit of $f(x)$ as $x \to c$ in any way is equal to its value.

Further $\mathop{\text{Lt}}_{x \to c} f(x)$ exists, if and only if both the limits of $f(x)$ as $x \to c$ from the right and $x \to c$ from the left exist and are equal.

Hence a function is continuous, if

$$\underset{\text{L.H. Lt}}{\mathop{\text{Lt}}_{x \to c-}} f(x) = f(c) = \underset{\text{R.H. Lt}}{\mathop{\text{Lt}}_{x \to c+}} f(x)$$

Cor. If $\mathop{\text{Lt}}_{x \to c-} f(x) = f(c)$, while either $\mathop{\text{Lt}}_{x \to c+} f(x)$ does not exist or is not equal to $f(c)$, then $f(x)$ is said to be **continuous on the left of c.**

If $f(c) = \mathop{\text{Lt}}_{x \to c+} f(x)$, while $\mathop{\text{Lt}}_{x \to c-} f(x)$ either does not exist or is not equal to $f(c)$, then $f(x)$ is said to be **continuous on the right of c.**

(*ii*) Definition of Continuity in an Interval. *A function $f(x)$ is said to be continuous in the closed interval $[a, b]$, if it is continuous for every value of x lying between a and b, continuous to the right of a, (i.e., $\mathop{Lt}_{x \to a+} f(x) = f(a)$) and continuous to the left of b (i.e., $\mathop{Lt}_{x \to b-} f(x) = f(b)$).*

14. Algebra of Continuous Functions

Theorem 1. The algebraic sum and product of a finite number of functions which are continuous at x = c, are continuous at x = c.

Proof. Let $f(x)$, $\phi(x)$, $\psi(x)$, ..., be finite number of functions which are continuous at $x = c$

$$\therefore \qquad \mathop{\text{Lt}}_{x \to c} f(x) = f(c), \ \mathop{\text{Lt}}_{x \to c} \phi(x) = \phi(c),$$

$$\mathop{\text{Lt}}_{x \to c} \psi(x) = \psi(c), \text{ and so on.}$$

From limits, we know that

$$\mathop{\text{Lt}}_{x \to c} [f(x) + \phi(x) - \psi(x) + \ldots\ldots]$$

$$= \underset{x \to c}{\text{Lt}} f(x) + \underset{x \to c}{\text{Lt}} \phi(x) - \underset{x \to c}{\text{Lt}} \psi(x) + \ldots\ldots$$

$$= f(c) + \phi(c) - \psi(c) + \ldots\ldots$$

$= \text{value of the function } [f(x) + \phi(x) - \psi(x) + \ldots\ldots]$ at $x = c$.

$\therefore$ $[f(x) + \phi(x) - \psi(x) + \ldots]$ is also continuous at $x = c$.

Hence the theorem.

Similarly, the theorem can be proved for the product.

Theorem 2. The quotient of two functions which are continuous at x = c is continuous at x = c, provided the denominator is not zero.

Proof. Let $f(x)$ and $\phi(x)$ be any two functions which are continuous at $x = c$.

$$\therefore \quad \underset{x \to c}{\text{Lt}} f(x) = f(c) \text{ and } \underset{x \to c}{\text{Lt}} \phi(x) = \phi(c)$$

From limits, we know that

$$\underset{x \to c}{\text{Lt}} \frac{f(x)}{\phi(x)} = \frac{\underset{x \to c}{\text{Lt}} f(x)}{\underset{x \to c}{\text{Lt}} \phi(x)} \text{ provided } \underset{x \to c}{\text{Lt}} \phi(x) \neq 0 = \frac{f(c)}{\phi(c)}$$

$$= \text{value of the function } \frac{f(x)}{\phi(x)} \text{ at } x = c.$$

$$\therefore \quad \frac{f(x)}{\phi(x)} \text{ is also continuous at } x = c.$$

Hence the theorem.

Theorem 3. If f is continuous, | f | is also continuous,

Proof. Let f *i.e.*, $f(x)$ be continuous at $x = c$.

$\therefore$ By Def. Art. 13(*i*), for $\varepsilon > 0$, (however small), there exists a positive real number δ (depending on ε) s.t.

$$|f(x) - f(c)| < \varepsilon \text{ whenever } |x - c| < \delta \qquad \ldots(1)$$

We know that $||f(x)| - |f(c)|| \leq |f(x) - f(c)|$ $\quad (\because \ ||\alpha| - |\beta|| \leq |\alpha - \beta|)$

$< \varepsilon$ whenever $|x - c| < \delta$ *by (1)*

$$\therefore \quad \underset{x \to c}{\text{Lt}} |f(x)| = |f(c)|.$$

$\therefore$ $|f|$ is continuous at $x = c$.

Remark. But the converse is not true.

Note. (*i*) $a + bx$ is a continuous function of x, where a and b are real numbers.

(*ii*) Polynomials in x are continuous functions of x.

(*iii*) A rational function in x is a continuous function of x at all those values of x which do not make the denominator zero.

(*iv*) $\frac{1}{x}$ is a continuous function of x, at all $x \neq 0$.

(*v*) e^x, e^{-x}, $\sin x$ and $\cos x$ are continuous functions of x for all real values of x.

(*vi*) $\tan x$ is a continuous function except at $x = \pm\frac{\pi}{2}, \pm\frac{3\pi}{2}, \ldots\ldots$

Example 1. *Using the $\varepsilon - \delta$ definition of continuity, show that sin x is continuous for every value of x.* (K.U. 2007)

Sol. Let $f(x) = \sin x$

Let c be any real number.

$\therefore$ $$f(c) = \sin c.$$

Let $\varepsilon > 0$ be any positive number, however small. For $f(x)$ to be continuous at $x = c$,

$$| f(x) - f(c) | < \varepsilon \text{ whenever } | x - c | < \delta. \qquad \text{(By Def. Art. 13}(i))$$

We have $$| f(x) - f(c) | = | \sin x - \sin c |$$

$$= \left| 2 \cos \frac{x+c}{2} \sin \frac{x-c}{2} \right|$$

$$= 2 \left| \cos \frac{x+c}{2} \right| \left| \sin \frac{x-c}{2} \right| \qquad [\because \ | ab | = | a | \ | b |]$$

$$\leq 2 \left| \sin \frac{x-c}{2} \right| \qquad \left(\because \left| \cos \frac{x+c}{2} \right| \leq 1 \text{ as } |\cos \theta| \leq 1 \right)$$

$$\leq 2 \left| \frac{x-c}{2} \right|. \qquad [\because \ | \sin \theta | \leq | \theta |]$$

$\therefore$ $$| f(x) - f(c) | \leq | x - c | \qquad ...(1)$$

Let us take $\delta = \varepsilon$.

$\therefore$ For $| x - c | < \delta \, (= \varepsilon)$, we have from (1) that $| f(x) - f(c) | \leq | x - c | < \delta \, (= \varepsilon)$

i.e., $$| f(x) - f(c) | < \varepsilon \text{ whenever } | x - c | < \delta.$$

$\therefore$ $f(x)$ is continuous at $x = c$.

But c is any real value of x.

$\therefore$ $f(x)$ is continuous for all values of x.

Example 2. *Using the $\varepsilon - \delta$ definition of continuity, show that $\cos^2 x$ is continuous for all real values of x.* (M.D.U. 2003 S ; K.U. 2005, 2006)

Sol. Let $$f(x) = \cos^2 x$$

Let c be any real number.

$\therefore$ $$f(c) = \cos^2 c$$

Let $\varepsilon > 0$ be any positive number, however small.

$$| f(x) - f(c) | = | \cos^2 x - \cos^2 c |$$
$$= | (1 - \sin^2 x) - (1 - \sin^2 c) | = | - \sin^2 x + \sin^2 c |$$
$$= | - (\sin^2 x - \sin^2 c) | = | \sin^2 x - \sin^2 c |$$
$$= | \sin (x + c) \sin (x - c) | \qquad (\because \ \sin^2 A - \sin^2 B = \sin (A + B) \sin (A - B))$$
$$= | \sin (x + c) | \ | \sin (x - c) |$$
$$\leq 1 . | x - c | \qquad | \because \ | \sin (x + c) | \leq 1 \text{ and } | \sin \theta | \leq | \theta |$$

or $$| f(x) - f(c) | \leq | x - c | \qquad ...(1)$$

Let us take $\delta = \varepsilon$.

$\therefore$ For $| x - c | < \delta \, (= \varepsilon)$, we have from (1) that $| f(x) - f(c) | \leq | x - c | < \delta \, (= \varepsilon)$

i.e., $$| f(x) - f(c) | < \varepsilon \quad \text{whenever} \quad | x - c | < \delta.$$

$\therefore$ $f(x)$ is continuous at $x = c$.

But c is any real value of x.

$\therefore$ $f(x)$ is continuous for all real values of x.

Example 3. $f(x) = x \sin \dfrac{1}{x}$; $x \neq 0$.

$= 0$ when $x = 0$

Using the ε – δ definition of continuity, show that f(x) is continuous at x = 0.

Sol. Let $\varepsilon > 0$ be any positive number, however small.

$$| f(x) - f(0) | = \left| x \sin \frac{1}{x} - 0 \right| = \left| x \sin \frac{1}{x} \right| = | x | \left| \sin \frac{1}{x} \right|$$

$$\leq | x | (1) \qquad | \because \ | \sin \theta | \leq 1 \text{ for all real } \theta$$

or $$| f(x) - f(0) | \leq | x | \qquad ...(1)$$

Let us choose a positive number $\delta (= \varepsilon)$

s.t. $| x - 0 | < \delta$ *i.e.*, $| x | < \delta$

$\therefore$ From (1) $| f(x) - f(0) | \leq | x | < \delta (= \varepsilon)$

whenever $| x - 0 | < \delta$.

$\therefore$ $f(x)$ is continuous at $x = 0$.

15. Discontinuous Function (*M.D.U. 2003*)

A function which is not continuous is said to be **discontinuous at the point.**

A function is discontinuous :

(*i*) If a function is not defined at the point.

(*ii*) If one or both the limits [L.H. Lt. and R.H. Lt.] do not exist as x approaches the point.

(*iii*) If both these limits exist and are either unequal or if equal, not equal to the value of the function at the point.

(*iv*) If either or both the limits are infinite.

16. Kinds of Discontinuity (*M.D.U. 2003*)

(*i*) **Discontinuity of the First Kind. Def.** If $\underset{x \to a+}{\text{Lt}} f(x)$ and $\underset{x \to a-}{\text{Lt}} f(x)$ exist but are not equal to one another, then $f(x)$ is said to have a discontinuity of the first kind at $x = a$.

(*ii*) **Discontinuity of the Second Kind. Def.** If neither $\underset{x \to a+}{\text{Lt}} f(x)$ nor $\underset{x \to a-}{\text{Lt}} f(x)$ exists, then the function $f(x)$ is said to have discontinuity of the second kind at $x = a$.

(*iii*) **Mixed Discontinuity. Def.** If only one of the limits $\underset{x \to a^+}{\text{Lt}} f(x)$ or $\underset{x \to a-}{\text{Lt}} f(x)$ exists, then the function $f(x)$ is said to have a mixed discontinuity at $x = a$.

(*iv*) **Removable Discontinuity. Def.** If $\underset{x \to a+}{\text{Lt}} f(x) = \underset{x \to c-}{\text{Lt}} f(x)$ *i.e.*, $\underset{x \to a}{\text{Lt}} f(x)$ exists and $\neq f(a)$, then $f(x)$ is said to have a removable discontinuity at $x = a$.

(*v*) **Infinite Discontinuity.** If either or both limits $\underset{x \to a+}{\text{Lt}} f(x)$ and $\underset{x \to c-}{\text{Lt}} f(x)$ are infinite, the function $f(x)$ is said to have an infinite discontinuity at $x = a$.

Examples on each of the five types of discontinuities. *(M.D.U. 2003)*

(*i*) Let $$f(x) = \frac{|x|}{x} \quad \text{when } x \neq 0$$
$$= 0 \quad \text{when } x = 0$$

$$\text{Left Hand Limit} = \underset{x \to c-}{\text{Lt}} f(x) = \underset{x \to 0+}{\text{Lt}} \frac{|x|}{x}$$

put $$x = 0 - h = -h \text{ as } h \to 0+ = \underset{h \to 0+}{\text{Lt}} \frac{|-h|}{-h} = \underset{h \to 0+}{\text{Lt}} \frac{h}{-h} = \underset{h \to 0+}{\text{Lt}} -1 = -1 \quad \text{...(1)}$$

$$\text{Right Hand Limit} = \underset{x \to 0+}{\text{Lt}} f(x) = \underset{x \to 0+}{\text{Lt}} \frac{|x|}{x}$$

put $$x = 0 + h = h \text{ as } h \to 0+ = \underset{h \to 0+}{\text{Lt}} \frac{|h|}{h} = \underset{h \to 0+}{\text{Lt}} \frac{h}{h} = \underset{h \to 0+}{\text{Lt}} 1 = 1 \quad \text{...(2)}$$

From (1) and (2) we can say that $\underset{x \to c-}{\text{Lt}} f(x)$ and $\underset{x \to a+}{\text{Lt}} f(x)$ both exist but are not equal.

$\therefore$ $f(x)$ is discontinuous at $x = 0$ and *discontinuity is of first kind* (By Art. 16)

(*ii*) Let $$f(x) = \sin \frac{1}{x} \text{ when } x \neq 0 \text{ and } = 0 \quad \text{when} \quad x = 0.$$

Here neither $\underset{x \to c-}{\text{Lt}} f(x)$ nor $\underset{x \to a+}{\text{Lt}} f(x)$ exists (For proof see example 9, page 29)

$\therefore$ $f(x)$ is discontinuous at $x = 0$ and *discontinuity is of second kind.*

(*iii*) Let $$f(x) = 1 \quad \text{when} \quad x \leq 0 = \sin \frac{1}{x} \quad \text{when} \quad x > 0$$

$$\underset{x \to 0-}{\text{Lt}} f(x) = \underset{x \to 0-}{\text{Lt}} 1 = 1 \text{ exists}$$

$$\underset{x \to c-}{\text{Lt}} f(x) = \underset{x \to a+}{\text{Lt}} \sin \frac{1}{x} \text{ does not exist}$$

(See example 9, page 29)

$\therefore$ $f(x)$ is discontinuous at $x = 0$ and discontinuity is a **mixed discontinuity** (By Art. 16)

(*iv*) Let $$f(x) = \frac{x + \sin x}{x} \quad \text{...(1)}$$

$$\underset{x \to 0}{\text{Lt}} f(x) = \underset{x \to 0}{\text{Lt}} \frac{x + \sin x}{x} = \underset{x \to 0}{\text{Lt}} \left(1 + \frac{\sin x}{x}\right) = 1 + 1 = 2 \quad \text{...(2)}$$

Putting $$x = 0 \text{ in (1)}, \quad f(0) = \frac{0}{0} \text{ which is undefined.} \quad \text{...(3)}$$

From (2) and (3), we can say that $\underset{x \to 0}{\text{Lt}} f(x) \neq f(0)$.

$\therefore$ $f(x)$ is discontinuous at $x = 0$ and $x = 0$ is a point of **removable discontinuity** (By Art. 16)

To remove this discontinuity, let us define $f(x)$ as below :

$$f(x) = \frac{x + \sin x}{x} \quad \text{if} \quad x \neq 0 \text{ and } = 2 \quad \text{if} \quad x = 0.$$

$\therefore$ $f(x)$ is continuous at $x = 0$ $\quad [\because$ By (2) above $\underset{x \to 0}{\text{Lt}} f(x) = 2 = f(0)]$

(*v*) Let $$f(x) = \frac{1}{x}$$

$$\underset{x \to 0+}{\text{Lt}} f(x) = \underset{x \to 0+}{\text{Lt}} \frac{1}{x} = \infty \text{ and hence does not exist.}$$

$$\underset{x \to 0-}{\text{Lt}} f(x) = \underset{x \to 0-}{\text{Lt}} \frac{1}{x} = -\infty \text{ and hence does not exist.}$$

$\therefore$ $x = 0$ is a point of *infinite discontinuity* of $f(x) = \dfrac{1}{x}$

17. Graphical Representation of Continuity

Let $y = f(x)$ be continuous at $x = a$, and let P$[a, f(a)]$, and Q$[a + h, f(a + h)]$ be two points on the graph.

Since $f(x)$ is continuous at $x = a$; $\therefore$ $\underset{x \to a}{\text{Lt}} f(x) = f(a)$

or $$f(a + h) \to f(a) \text{ as } h \to 0.$$

i.e., as $h \to 0$ the abscissa and ordinate of the points Q tend to the abscissa and ordinate respectively of the point P.

i.e., as $h \to 0$, point Q $\to$ P

Thus the graph of the function $y = f(x)$ is **without break** at P.

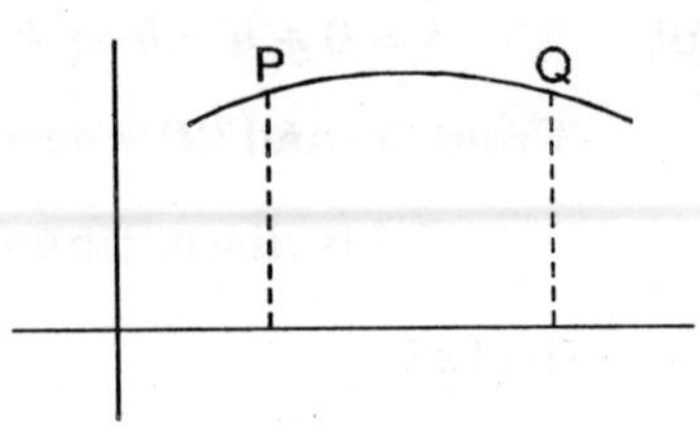

Hence if a function is continuous throughout in an interval (a, b), the graph of the function in this interval is without break, *e.g.*, graphs of $\sin x$, $\cos x$.

If a function is discontinuous at a point $x = a$, then there is necessarily a break in its graph at the point corresponding to $x = a$.

If the function is not defined at $x = a$, then there is a break in the graph as there is no point on the graph corresponding to $x = a$, *e.g.*, graphs of $\tan x$, $\cot x$, $\sec x$, $\text{cosec } x$.

Note. If the value of the function at the point is not given in question the function may be said as **not defined** and evidently, the function will be discontinuous function.

Example 1. *Discuss the continuity of* $f(x) = x \sin \dfrac{1}{x}$, *when* $x \neq 0$ *and* $f(0) = 0$ *at* $x = 0$.

Sol. Here, we have $f(x) = x \sin \dfrac{1}{x}$, for $x \neq 0$ and $f(0) = 0$.

Continuity at x = 0.

$$\underset{x \to 0}{\text{Lt}} f(x) = \underset{x \to 0}{\text{Lt}} x \sin \frac{1}{x}$$

Put $x = 0 + h = h$ as $h \to 0 = \underset{h \to 0}{\text{Lt}} h \sin \dfrac{1}{h}$

$= 0 \times$ [A finite quantity between -1 and $+1$]

$= 0$ $\therefore$ $\underset{x \to 0}{\text{Lt}} f(x)$ exists and $= 0$.

Also $f(0) = 0$ (given)

$\therefore$ $\underset{x \to 0}{\text{Lt}} f(x)$ exists and $= f(0)$.

Hence the function is continuous at $x = 0$.

Remark. $\underset{x \to 0}{\text{Lt}} x \sin \dfrac{1}{x} = \underset{x \to 0}{\text{Lt}} \dfrac{\sin \frac{1}{x}}{\frac{1}{x}} \neq 1.$

Here $\dfrac{\sin \frac{1}{x}}{\frac{1}{x}}$ is of the form $\dfrac{\sin \theta}{\theta}$ but $\theta = \dfrac{1}{x} \to \infty$ as $x \to 0$. This $\theta = \dfrac{1}{x}$ does not tend to 0.

Example 2. *Test for continuity at $x = a$, the function*

$$f(x) = \frac{1}{x-a} \operatorname{cosec} \frac{1}{x-a} \text{ for } x \neq a \text{ and } f(a) = 0.$$

Sol. Here, $f(x) = \dfrac{1}{x-a} \operatorname{cosec} \dfrac{1}{x-a}$, for $x \neq a$ and $f(a) = 0$.

To test continuity at x = a.

$$\underset{x \to a}{\text{Lt}}\ f(x) = \underset{x \to a}{\text{Lt}}\ \frac{1}{x-a} \operatorname{cosec} \frac{1}{x-a}$$

Put $x = a + h,\ h \to 0$

$$= \underset{h \to 0}{\text{Lt}}\ \frac{1}{(a+h-a)} \operatorname{cosec} \frac{1}{a+h-a} = \underset{h \to 0}{\text{Lt}}\ \frac{1}{h} \operatorname{cosec} \frac{1}{h}$$

which oscillates between $-\infty$ and $+\infty$

$\therefore\ \underset{x \to a}{\text{Lt}}\ f(x)$ does not exist.

Hence the function is discontinuous at $x = a$.

($\because$ In fact, $\underset{x \to a-}{\text{Lt}}\ f(x)$ and $\underset{x \to a+}{\text{Lt}}\ f(x)$ both do not exist)

Remark. This discontinuity is of *second* kind.

Example 3. *Show that the function, $f(x) = \dfrac{e^{1/x} - 1}{e^{1/x} + 1}$ for $x \neq 0$ and $f(0) = 0$ is discontinuous at $x = 0$.* (K.U. 2003)

Sol. L.H. Limit $= \underset{x \to 0-}{\text{Lt}}\ f(x) = \underset{x \to 0-}{\text{Lt}}\ \dfrac{e^{1/x} - 1}{e^{1/x} + 1}$

Put $x = 0 - h = -h,\ h \to 0+$

$$\text{L.H. Limit} = \underset{h \to 0+}{\text{Lt}}\ \frac{e^{\frac{1}{-h}} - 1}{e^{\frac{1}{-h}} + 1} = \underset{h \to 0+}{\text{Lt}}\ \frac{e^{\frac{-1}{h}} - 1}{e^{\frac{-1}{h}} + 1} \qquad \left[\because \frac{1}{-h} = \frac{-1}{h}\right]$$

(Put $h = 0$) $$= \frac{e^{-\infty} - 1}{e^{-\infty} + 1} = \frac{0-1}{0+1} \qquad \left[\because e^{-\infty} = \frac{1}{e^{\infty}} = \frac{1}{\infty} = 0\right]$$

$\therefore$ $$\text{L.H. Limit} = \frac{-1}{1} = -1$$

$$\text{R.H. Limit} = \underset{x \to 0+}{\text{Lt}}\ f(x) = \underset{x \to 0+}{\text{Lt}}\ \frac{e^{1/x} - 1}{e^{1/x} + 1}$$

Put $x = 0 + h = h,\ h \to 0+$

$$= \underset{h \to 0+}{\text{Lt}}\ \frac{e^{1/h} - 1}{e^{1/h} + 1} \qquad \left| \frac{\infty}{\infty} \text{ Form} \right.$$

Dividing every term by $e^{1/h}$, *i.e.*, multiplying every term by $e^{-1/h}$

$\therefore$ L.H. Limit $\neq$ R.H. Limit

$\therefore$ $\underset{x \to 0}{\text{Lt}} f(x)$ does not exist.

Hence $f(x)$ is discontinuous at $x = 0$.

Remark. This discontinuity is of *first* kind.

$$\left(\because \underset{x \to 0-}{\text{Lt}} f(x) \text{ and } \underset{x \to 0+}{\text{Lt}} f(x) \text{ both exist but are not equal}\right)$$

Example 4. *A function f(x) is defined as under :*

$$f(x) = \frac{x^2 - x - 6}{x^2 - 2x - 3}, \textit{ when } x \neq 3,\ f(3) = \frac{5}{3}.$$

Prove that f(x) is discontinuous at x = 3. Can the definition of f(x) for x = 3 be modified so as to make it continuous there ?

Sol. $$f(x) = \frac{x^2 - x - 6}{x^2 - 2x - 3} \text{ when } x \neq 3.$$

$$\underset{x \to 3}{\text{Lt}} f(x) = \underset{x \to 3}{\text{Lt}} \frac{x^2 - x - 6}{x^2 - 2x - 3}.$$

$$\underset{x \to 3}{\text{Lt}} \frac{(x-3)(x+2)}{(x-3)(x+1)} = \underset{x \to 3}{\text{Lt}} \frac{x+2}{x+1} \qquad (\because x - 3 \neq 0.)$$

Put $x = 3 + h,\ h \to 0 = \underset{h \to 0}{\text{Lt}} \dfrac{3+h+2}{3+h+1}$

(Put $h = 0$) $$= \frac{3+0+2}{3+0+1} = \frac{5}{4}$$

$\therefore$ $\underset{x \to 3}{\text{Lt}} f(x)$ exists and $= \dfrac{5}{4}$.

$\therefore$ $$\underset{x \to 3}{\text{Lt}} f(x) = \frac{5}{4} \neq f(3) \left(= \frac{5}{3}\right)$$

$\therefore$ $f(x)$ is discontinuous at $x = 3$ and the discontinuity is **removable.**

In order to make the given function continuous at $x = 3$, the limit of the function as $x \to 3$ must be equal to its value at $x = 3$.

i.e., $$f(3) = \underset{x \to 3}{\text{Lt}} f(x) = \frac{5}{4}$$

Hence to remove the discontinuity, the function $f(x)$ is defined as under

$$f(x) = \frac{x^2 - x - 6}{x^2 - 2x - 3}, \text{ when } x \neq 3,\ f(3) = \frac{5}{4}.$$

Example 5. *A function f(x) is defined as follows :*

$$f(x) = x, \textit{ when } 0 \le x < 1,\ f(x) = 2, \textit{ when } x = 1,$$

and $$f(x) = x + 1, \textit{ when } 1 < x \le 2.$$

Show that it is discontinuous at x = 1.

Sol. Here, we have
$f(x) = x$ for $0 \le x < 1$
$f(x) = 2$ for $x = 1$
$f(x) = x + 1$ for $1 < x \le 2$

at $x = 1$.

$$\text{L.H. Limit} = \underset{x \to 1-0}{\text{Lt}} f(x) = \underset{x \to 1-0}{\text{Lt}} x \qquad [\because f(x) = x \text{ for } 0 \le x < 1]$$

Put $x = 1 - h,\ h \to 0+ = \underset{h \to 0+}{\text{Lt}} (1 - h) = 1 - 0 = 1$ ...(*i*)

$$\text{R.H. Limit} = \underset{x \to 1+0}{\text{Lt}} f(x) = \underset{x \to 1+0}{\text{Lt}} (x + 1)$$

$$[\because f(x) = x + 1 \text{ for } 1 < x \le 2]$$

Put $x = 1 + h, h \to 0+ = \underset{h \to 0+}{\text{Lt}} (1 + h + 1) = 1 + 0 + 1 = 2$ ...(*ii*)

From (*i*) and (*ii*), L.H. Limit ≠ R.H. Limit

$\therefore \underset{x \to 1}{\text{Lt}} f(x)$ does not exist.

Hence $f(x)$ is discontinuous at $x = 1$.

Remark. This discontinuity is of *first* kind.

Example 6. $f(x) = \dfrac{x - |x|}{x}$ $x \ne 0$

$f(0) = 0.$

Discuss continuity at x = 0.

Sol. Left Limit $= \underset{x \to 0-}{\text{Lt}} f(x) = \underset{x \to 0-}{\text{Lt}} \dfrac{x - |x|}{x}$

Put $x = 0 - h = -h, h \to 0+$

$$= \underset{h \to 0+}{\text{Lt}} \frac{-h - |-h|}{-h} = \underset{h \to 0+}{\text{Lt}} \frac{-h-h}{-h}$$

$$= \underset{h \to 0+}{\text{Lt}} \frac{-2h}{-h} = \underset{h \to 0+}{\text{Lt}} 2 = 2$$

Right Limit $= \underset{x \to 0+}{\text{Lt}} f(x) = \underset{x \to 0+}{\text{Lt}} \dfrac{x - |x|}{x}$

Put $x = 0 + h = h$, where $h \to 0+$

$$= \underset{h \to 0+}{\text{Lt}} \frac{h - |h|}{h} = \underset{h \to 0+}{\text{Lt}} \frac{h-h}{h} = \underset{h \to 0+}{\text{Lt}} \frac{0}{h} = \underset{h \to 0+}{\text{Lt}} 0 = 0$$

$$\underset{x \to 0-}{\text{Lt}} f(x) \ne \underset{x \to 0+}{\text{Lt}} f(x)$$

$\therefore \underset{x \to 0}{\text{Lt}} f(x)$ does not exist.

Hence $f(x)$ is discontinuous at $x = 0$.

Note. This discontinuity is of *First* kind.

Example 7. *Determine the constants a and b so that the function defined below is continuous everywhere* :

$$f(x) = \begin{cases} 2x + 1 & \text{for} \quad x \le 1 \\ ax^2 + b & \text{for} \quad 1 < x < 3 \\ 5x + 2a & \text{for} \quad x \ge 3. \end{cases}$$

Sol. $\because$ $f(x)$ is continuous everywhere (given)

$\therefore$ $f(x)$ is continuous at $x = 1$ and also at x* $= 3$ also.

Continuity at x = 1.

---

*Because $x = 1$ and $x = 3$ are the points where the domain of the function is partitioned (divided). $f(x)$ being a polynomial for all other values of x is clearly continuous (even, if it was not mentioned).

Left Limit = $\underset{x \to 1-}{\text{Lt}} f(x) = \underset{x \to 1-}{\text{Lt}} (2x + 1)$ $\quad | \because f(x) = 2x + 1$ for $x \le 1$

Put $\quad x = 1 - h, h \to 0+ = \underset{h \to 0+}{\text{Lt}} [2(1-h) + 1] = 2(1-0) + 1 = 3$

Right Limit = $\underset{x \to 1+}{\text{Lt}} f(x) = \underset{x \to 1+}{\text{Lt}} [ax^2 + b]$

$| \because f(x) = ax^2 + b$ for $1 < x < 3$

Put $\quad x = 1 + h, h \to 0+ = \underset{h \to 0+}{\text{Lt}} [a(1+h)^2 + b] = a + b$

Puting. $\quad x = 1$ in $f(x) = 2x + 1 \quad x \le 1$;

we have $\quad f(1) = 2 + 1 = 3$

But $f(x)$ is continuous at $x = 1$

$\therefore \quad \underset{x \to 1-}{\text{Lt}} f(x) = \underset{x \to 1+}{\text{Lt}} f(x) = f(1) \quad \therefore \quad a + b = 3$ ...(1)

Continuity at $x = 3$.

Left Limit = $\underset{x \to 3-0}{\text{Lt}} f(x) = \underset{x \to 3-}{\text{Lt}} [ax^2 + b]$

Put $\quad x = 3 - h, h \to 0+ = \underset{h \to 0+}{\text{Lt}} [a(3-h)^2 + h] = 9a + b.$

Right Limit = $\underset{x \to 3+}{\text{Lt}} f(x) = \underset{x \to 3+}{\text{Lt}} [5x + 2a]$

Put $\quad x = 3 + h, h \to 0+ = \underset{h \to 0+}{\text{Lt}} [5(3+h) + 2a] = 15 + 2a$

Put $\quad x = 3$ in $f(x) = 5x + 2a$ for $x \ge 3$, $f(3) = 15 + 2a$

But $f(x)$ is continuous at $x = 3$

$\therefore \quad \underset{x \to 3-}{\text{Lt}} f(x) = \underset{x \to 3+}{\text{Lt}} f(x) = f(3)$

$\therefore \quad 9a + b = 15 + 2a$ or $7a + b = 15$ ...(2)

Let us solve (1) and (2) for a and b.

(2) – (1) gives $6a = 12$ *i.e.,* $a = 2$

Put $\quad a = 2$ in (1), $2 + b = 3$ or $b = 1$.

$\therefore \quad a = 2, b = 1.$

Example 8. *Let a function f be defined on R by*

$$f(x) = \begin{cases} 1 \text{ when } x \text{ is rational} \\ -1 \text{ when } x \text{ is irrational.} \end{cases}$$

Show that f is discontinuous everywhere.

Sol. Let c be any real number.

We know that each neighbourhood of c namely $| x - c | < \delta$ *i.e.,* $(c - \delta, c + \delta)$ contains an infinity of rational numbers and another infinity of irrational numbers.

Now it is given that $f(x) = 1$ at each rational x and $f(x) = -1$ at each irrational x.

$\therefore \quad \underset{x \to c}{\text{Lt}} f(x)$ does not exist.

$\therefore \quad f(x)$ is discontinuous at $x = c$.

But c is any real number.

$\therefore \quad f(x)$ is discontinuous everywhere.

Example 9. *Test the continuity of the function $f(x) = \sin \frac{1}{x}$ when $x \neq 0$ and $f(0) = 0$ at $x = 0$.* (M.D.U. 2006 A)

Sol. Let us take two values of x namely

$$x = \frac{1}{2n\pi - \frac{\pi}{2}} \quad \text{and} \quad x = \frac{1}{2n\pi + \frac{\pi}{2}},$$

where n is any integer.

Let us take n very large, *i.e.*, $n \to \infty$. Therefore, the above two values of x tend to 0.

For $x = \frac{1}{2n\pi - \frac{\pi}{2}}$, $\sin \frac{1}{x} = \sin\left(2n\pi - \frac{\pi}{2}\right) = -\sin \frac{\pi}{2} = -1$

For $x = \frac{1}{2n\pi + \frac{\pi}{2}}$, $\sin \frac{1}{x} = \sin\left(2n\pi + \frac{\pi}{2}\right) = \sin \frac{\pi}{2} = 1$

$\therefore$ $\underset{x \to 0}{\text{Lt}} \sin \frac{1}{x}$ does not exist.

Hence $f(x)$ is discontinuous at $x = 0$.

Remark. For this example 9, neither $\underset{x \to 0-}{\text{Lt}} \sin \frac{1}{x}$ nor $\underset{x \to 0+}{\text{Lt}} \sin \frac{1}{x}$ exists.

18. Definition. Composite of two Functions

Let $f : A \to B$ and $g : B \to C$ be two functions such that $f(x) = y$ and $g(y) = z$ where x, y, z belong to A, B, C respectively.

Then the function $h : A \to C$ such that $h(x) = z = g(y) = g(f(x))$ for all x in A is called the **composite** of functions f and g and is denoted by gof.

i.e., gof : A → C is defined by

$$\mathbf{(gof)x = g[f(x)] \text{ for all } x \text{ in } A.}$$

Similarly, fog is defined as

$$\mathbf{(fog)x = f(g(x)).}$$

Remark. In general $gof \neq fog$ and both are very different from fg *i.e.*, product of two functions.

For example (*i*) Let $f : R \to R$ be defined by $f(x) = x^2$

and $g : R \to R$ be defined by $g(x) = x + 3$

Then $(gof)x = g(f(x)) = g(x^2) = x^2 + 3$

and $(fog)\,x = f(g(x)) = f(x + 3) = (x + 3)^2$

Also $(fg)x = f(x)g(x) = x^2(x + 3)$.

(*ii*) Let $f : R \to R$ be defined by $f(x) = x^2$

and $g : R \to R$ be defined by $g(x) = \sin x$

Then $(gof)x = g(f(x)) = g(x^2) = \sin x^2$

and $(fog)x = f(g(x)) = f(\sin x) = (\sin x)^2 = \sin^2 x$

Also $(fg)x = f(x)g(x) = x^2 \sin x$.

19. Theorem. *Composite of two continuous functions is a continuous function i.e., if a function $f(x)$ is continuous at $x = a$ and g is a continuous function at $f(a)$, then gof is a continuous function at $x = a$.*

Proof. Let $\varepsilon > 0$ be a positive number, however small.

$\because$ g is continuous at $f(a)$ (given)

$\therefore$ There exists $\delta > 0$ such that

$$| g(f(x)) - g(f(a)) | < \varepsilon \text{ whenever } | f(x) - f(a) | < \delta$$

i.e., $$| (gof)\, x - (gof)\, a | < \varepsilon \text{ whenever } | f(x) - f(a) | < \delta \quad \ldots(1)$$

Also f is continuous at $x = a$ (given)

$\therefore$ For $\delta > 0$, there exists $\delta_1 > 0$ such that

$$| f(x) - f(a) | < \delta \text{ whenever } | x - a | < \delta_1 \quad \ldots(2)$$

$\therefore$ From (1) and (2), we have

$$| (gof)x - (gof)a | < \varepsilon \text{ whenever } | x - a | < \delta_1$$

$\therefore$ gof is continuous at $x = a$.

Example. *Prove that the following functions are continuous for all real x.*

(*i*) *sin* $|x|$ (*ii*) $|1 - x + |x||$.

Sol. (*i*) Let $f(x) = \sin x$ and $g(x) = |x|$

We know that $\sin x$ and $|x|$ are continuous functions

$\therefore$ f and g are continuous.

$$(fog)\,(x) = f(g(x)) = \sin\,(g(x)) = \sin |x|$$

Since composite of two continuous functions is continuous (By Art. 19)

$\therefore$ fog is continuous.

Hence, $\sin |x|$ is continuous.

(*ii*) Let $f(x) = |x|$ and $g(x) = 1 - x + |x|$

We know that $|x|$ and $1 - x$ are continuous functions

$\therefore$ f and g are continuous.

$$(fog)\,(x) = f(g(x)) = |g(x)| = |1 - x + |x||$$

Since composite of two continuous functions is continuous

$\therefore$ fog is continuous.

Hence, $|1 - x + |x||$ is continuous.

EXERCISE 3

1. Define the continuity of a function $f(x)$,
   (*i*) at a point $x = a$
   (*ii*) in the interval (a, b)
2. Use the $\varepsilon - \delta$ definition of continuity to prove that the following functions are continuous for all real values of x :
   (*i*) $\cos x$ (*ii*) $\sin^2 x$
3. Using $\varepsilon - \delta$ definition, prove that
   (*i*) $f(x) = (x - a) \cos \dfrac{1}{x - a}$ if $x \neq a$ and $= 0$ if $x = a$ is continuous at $x = a$.
   (*ii*) $f(x) = x \cos \dfrac{1}{x}$ if $x \neq 0$ and $= 0$ if $x = 0$ is continuous at $x = 0$.

4. Using $\varepsilon - \delta$ definition, prove that

(*i*) $f(x) = x^2 \sin \frac{1}{x}$ if $x \neq 0$ and $= 0$ if $x = 0$ is coutinuous at $x = 0$

(*ii*) $f(x) = x^2 \cos \frac{1}{x}$ if $x \neq 0$ and $= 0$ if $x = 0$ is continuous at $x = 0$.

5. (*i*) Examine the following function for continuity at $x = 2$:

$$f(x) = \frac{x^2 - 4}{x - 2} \quad \text{when } x \neq 2 \quad \text{and} \quad = 0 \quad \text{when } x = 2.$$

(*ii*) Examine the continuity at $x = 3$ of the following function :

$$f(x) = \frac{x^2 - 9}{x - 3} \quad \text{for } x \neq 3 \quad \text{and} \quad = 6 \quad \text{for } x = 3.$$

6. (*i*) Examine the continuity of the function :

$$f(x) = \frac{\sin 2x}{x} \quad \text{when } x \neq 0$$

$$f(x) = 1 \quad \text{when } x = 0 \text{ at } x = 0.$$

(*ii*) $$f(x) = \frac{\sin 3x}{x} \quad \text{when } x \neq 0 \quad \text{and} \quad = k \quad \text{when } x = 0$$

Find the value of k so that $f(x)$ is continuous at $x = 0$.

(*iii*) A function $f(x)$ is defined by

$$f(x) = \begin{cases} \dfrac{\tan 3x}{x}, & x \neq 0 \\ A, & x = 0 \end{cases}$$

Find the value of A for which $f(x)$ is continuous at $x = 0$.

7. Discuss the continuity of

$$f(x) = \frac{1 - x^n}{1 - x}, \ x \neq 1 \quad \text{and} \quad = n - 1, \quad \text{if} \quad x = 1 \quad \text{at} \quad x = 1.$$

8. Examine the following function for continuity :

$$f(x) = (1 + x)^{\frac{1}{x}} \quad \text{when } x \neq 0$$

and $$f(x) = 1 \quad \text{when } x = 0 \quad \text{at } x = 0.$$

9. Discuss the continuity of the function

$$f(x) = \frac{x}{\sqrt{x^2}} \quad \text{when } x \neq 0 \text{ and } f(0) = 0 \text{ at the origin.}$$

[Hint. By Def. $\sqrt{x^2} = |x|$]

10. (*i*) Discuss the continuity of the function :

$$f(x) = \frac{\sin ax}{x}, \quad \text{when } x \neq 0$$

$$f(0) = 1.$$

(*ii*) Examine for continuity at $x = 0$, the function,

$$f(x) = \frac{\sin^2 ax}{x^2}, \quad \text{when } x \neq 0 \text{ and } f(0) = 1.$$

11. Is the function $f(x) = \dfrac{3x + 4 \tan x}{x}$ continuous at $x = 0$. If not how may the function be defined to make it continuous at this point.

12. Discuss the continuity at $x = 0$, for the function f defined by

$$f(x) = \begin{cases} \dfrac{1 - \cos x}{x^2}, & x \neq 0 \\ 1 & \text{when } x = 0 \end{cases}$$

13. Test for continuity the function :

(i) $f(x) = (x - a) \sin \dfrac{1}{x - a}$, when $x \neq a$

and $f(a) = 0$ at $x = a$

(ii) $f(x) = (x - a) \cos \dfrac{1}{x - a}$ when $x \neq a$

and $f(a) = 0$ at $x = a$

(iii) $f(x) = x \cos \dfrac{1}{x}$, when $x \neq 0$

and $f(0) = 0$ at $x = 0$

(iv) $f(x) = x \sin \dfrac{1}{x} - 1$ when $x \neq 0$

$f(0) = 0$ at $x = 0$.

14. (i) Show that the function $f(x) = |x|$ is continuous at $x = 0$.

(ii) Examine for continuity at $x = a$, the function

$$f(x) = \frac{1}{x - a}.$$

15. (i) Discuss the continuity of $f(x) = \sqrt{|x| - x}$ at $x = 0$.

(ii) Let $f(x) = \begin{cases} \dfrac{2|x| + x^2}{x} & x \neq 0 \\ 0 \text{ at } & x = 0 \end{cases}$

Examine the continuity of $f(x)$ at $x = 0$.

16. Examine the continuity at $x = 1$ of the function defined as :

$f(x) = 5x - 4$ when $0 < x \leq 1$

$= 4x^2 - 3x$ when $1 < x < 2$.

17. Discuss the continuity of the following function at $x = 1$:

$$f(x) = \begin{bmatrix} 2x & \text{when} & 0 \leq x < 1 \\ 3 & \text{when} & x = 1 \\ 4x & \text{when} & 1 < x \leq 2 \end{bmatrix}.$$

18. Examine the following functions for continuity :

(i) $f(x) = \dfrac{1}{1 + e^{1/x}}$ at the origin

(ii) $f(x) = e^{\frac{-1}{(x-2)^2}}$ for $x \neq 2$ and $= 0$ for $x = 2$ at $x = 2$.

19. Examine whether the following function is continuous at the origin.

$$f(x) = \frac{x \cdot e^{1/x}}{1 + e^{1/x}} \text{ when } x \neq 0$$

$f(0) = 0$.

20. A function $f(x)$ is defined as :

$$f(x) = \begin{cases} x^2 + A & x \geq 0 \\ -x^2 - A & x < 0 \end{cases}$$

What should be A so that $f(x)$ is continuous at $x = 0$?

21. For what choices of a and b is $f(x)$ continuous for all x where

(i) $$f(x) = \begin{cases} ax^2 + b & x < 2 \\ 2 & x = 2 \\ 2ax - b & x > 2 \end{cases}$$

(ii) $$f(x) = \begin{cases} 1 & x \leq 3 \\ ax + b & 3 < x < 5 \\ 7 & x \geq 5 \end{cases}$$

Determine the constants a and b so that f is continuous for all x.

22. Find the points of discontinuity of the following functions :

(i) $\dfrac{x^2}{x^2 - 9}$ (ii) $\dfrac{x^2 + x}{x^3 - x}$ (iii) $\dfrac{\log x}{x - 1}$

(iv) $\tan x$ (v) $\dfrac{\sin x}{1 - 2\cos x}$.

23. Let $f(x + y) = f(x) f(y)$ for all x, y and $f(0) \neq 0$, show that $f(0) = 1$.
[**Hint** Put $x = 0$ and $y = 0$ in the given relation]

24. Discuss the continuity at the *origin* of the function defined as :

$$f(x) = \cos \frac{1}{x} \text{ when } x \neq 0 \text{ and } f(0) = 0.$$

[**Hint.** Take $x = \dfrac{1}{2n\pi}$ and $x = \dfrac{1}{2n\pi + \pi}$ and proceed as in Example 9 page 29]

25. The function $f(x)$ is given by $f(x) = \begin{bmatrix} 0 & \text{when} & x \leq 0 \\ 1 & \text{when} & 0 < x \leq 1 \\ 2 & \text{when} & x > 1 \end{bmatrix}$.

Show that $f(x)$ has two points of discontinuity.

26. A function f is defined by the rule $f(x) = [1 - x] + [x - 1]$ where $[t]$ denotes the greatest integer $\leq t$. Discuss the continuity of $f(x)$ at $x = 1$.

27. A function $f(x)$ is defined as below :

$$f(x) = \frac{x\,(e^{1/x} - 1)}{e^{1/x} + 1} \quad x \neq 0.$$
$$f(0) = 0.$$

Prove that $f(x)$ is continuous at $x = 0$. *(K.U. 2004 S)*

28. Prove that the following functions are continuous for all real x :

(i) $\cos |x + 2|$ (ii) $\cos (x^2 - x + 1)$
(iii) $e^{\sin x}$ (iv) $\sin (e^x)$.

29. If $f(x) = \begin{bmatrix} 1 & \text{if} & x > 0 \\ 0 & \text{if} & x = 0 \\ -1 & \text{if} & x < 0 \end{bmatrix}$ and $g(x) = 1 + x^2$.

Investigate the continuity of the composite functions *fog* and *gof*.

ANSWERS

5. (*i*) Discontinuous (*ii*) Continuous.
6. (*i*) Discontinuous (*ii*) $K = 3$ (*iii*) $A = 3$.
7. Discontinuous.
8. Discontinuous.
9. Discontinuous.
10. (*i*) Discontinuous unless $a = 1$. (*ii*) Discontinuous unless $a = \pm 1$.
11. Discontinuous ; $f(x) = \dfrac{3x + 4\tan x}{x}$ when $x \neq 0$

    $f(x) = 7$ when $x = 0$.
12. Discontinuous.
13. (*i*)Continuous (*ii*) Continuous (*iii*) Continuous (*iv*) Discontinuous.
14. (*ii*) Discontinuous.
15. (*i*) Continuous. (*ii*) Discontinuous.
16. Continuous.
17. Discontinuous.
18. (*i*) Discontinuous (*ii*) Continuous.
19. Continuous.
20. $A = 0$.
21. (*i*) $a = \frac{1}{2}, b = 0$ (*ii*) $a = 3, \quad b = -8$.
22. (*i*) $x = \pm 3$ (*ii*) $x = -1, 0, 1$ (*iii*) $x = 0, 1$

    (*iv*) $x = (2n + 1)\dfrac{\pi}{2}$ where n is any integer

    (*v*) $x = 2n\pi \pm \dfrac{\pi}{3}$ where n is any integer.
24. Discontinuous.
26. Discontinuous.
29. *fog* is continuous everywhere and *gof* is discontinuous at $x = 0$ only.

2*

Derivability

1. Derivability at an Interior Point

A function $f(x)$ defined on closed interval $[a, b]$ is said to be derivable at a point $c \in (a, b)$ if

$$\underset{x \to c}{\text{Lt}} \frac{f(x) - f(c)}{x - c} \text{ exists} \qquad (\textit{i.e.}, \text{ is unique and finite})$$

$\Big($Putting $x = c + h$ where $h \to 0$, we have

$$\text{this limit} = \underset{h \to 0}{\text{Lt}} \frac{f(c+h) - f(c)}{c + h - c} = \underset{h \to 0}{\text{Lt}} \frac{f(c+h) - f(c)}{h}\Bigg)$$

In case, this limit exists, the same is called the derivative of $f(x)$ at $x = c$ or differential coefficient of $f(x)$ at $x = c$ and is denoted by $f'(c)$ or $\frac{d}{dx} f(x)$ at $x = c$.

The process of getting $f'(c)$ is called the **differentiation.**

2. (i) Left Hand Derivative

If $$\underset{x \to c-}{\text{Lt}} \frac{f(x) - f(c)}{x - c}$$

$\left(\text{Putting } x = c - h \text{ where } h \to 0+, \text{ this limit} = \underset{h \to 0+}{\text{Lt}} \frac{f(c-h) - f(c)}{-h}\right)$

exists, then it is called **Left Hand derivative** of $f(x)$ at $x = c$ and is denoted by $\text{L}f'(c)$.

(ii) Right Hand Derivative

If $$\underset{x \to c+}{\text{Lt}} \frac{f(x) - f(c)}{x - c}$$

$\left(\text{Putting } x = c + h \text{ where } h \to 0+, \text{ this limit} = \underset{h \to 0+}{\text{Lt}} \frac{f(c+h) - f(c)}{h}\right)$

exists, then it is called **Right Hand derivative** of $f(x)$ at $x = c$ and is denoted by $\text{R}f'(c)$

If $\text{L}f'(c) = \text{R}f'(c)$, then we say that $f(x)$ is **derivable** at the point $x = c$.

The common value of $\text{L}f'(c)$ and $\text{R}f'(c)$ is denoted by $f'(c)$ and is called the **derivative** of $f(x)$ at $x = c$.

If $\text{L}f'(c) \neq \text{R}f'(c)$, then $f(x)$ is not differentiable at $x = c$.

---

*This chapter is in the syllabus of **K.U. Kurukshetra** and not in the syllabus of **M.D.U. Rohtak.**

3. If a function is derivable at a point, prove that it is continuous at that point

Proof. Let $f(x)$ be derivable at a point $x = c$.

$$\therefore \quad \underset{x \to c}{\text{Lt}} \frac{f(x) - f(c)}{x - c} \text{ exists } i.e., \text{ is unique and finite and } = f'(c) \qquad ...(1)$$

Now $\underset{x \to c}{\text{Lt}} f(x)$

Subtract and add $f(c)$

$$= \underset{x \to c}{\text{Lt}} (f(x) - f(c) + f(c)) = \underset{x \to c}{\text{Lt}} \left((x - c) \left(\frac{f(x) - f(c)}{x - c} \right) + f(c) \right)$$

(Putting $x = c$) $= (c - c) f'(c) + f(c)$ [By (1)]

$$= 0.f'(c) + f(c) = f(c)$$

or $$\underset{x \to c}{\text{Lt}} f(x) = f(c)$$

$\therefore$ $f(x)$ is continuous at $x = c$

Hence every differentiable function is continuous.

Note. The converse of the above theorem is not true, *i.e.*, if $f(x)$ is continuous at $x = c$, then it is not necessarily derivable at $x = c$.

For example, consider the function $f(x) = | x |$ at $x = 0$.

$$\therefore \quad f(0) = | 0 | = 0$$

Continuity of f at $x = 0$

$$\lim_{x \to 0-} f(x) = \lim_{x \to 0-} |x| = \lim_{x \to 0-} (-x) = 0 \qquad |\because \; x \to 0- \Rightarrow x < 0 \Rightarrow |x| = -x$$

$$\lim_{x \to 0+} f(x) = \lim_{x \to 0+} |x| = \lim_{x \to 0+} x = 0 \qquad |\because \; x \to 0+ \Rightarrow x > 0 \Rightarrow |x| = x$$

Since $$\lim_{x \to 0-} f(x) = \lim_{x \to 0+} f(x) = 0$$

$\therefore$ $\lim_{x \to 0} f(x) = 0 = f(0)$. Hence, f is continuous at $x = 0$.

Derivability of f at $x = 0$

By Art. 2 ; $$Lf'(0) = \lim_{x \to 0-} \frac{f(x) - f(0)}{x - 0} = \lim_{h \to 0+} \frac{f(-h) - f(0)}{-h} \qquad (\because \; c = 0)$$

$$= \lim_{h \to 0+} \frac{|-h| - 0}{-h} = \lim_{h \to 0+} = \frac{h}{-h} = \lim_{h \to 0+} -1 = -1$$

By Art. 2 ; $$Rf'(0) = \lim_{x \to 0+} \frac{f(x) - f(0)}{x - 0} = \lim_{h \to 0+} \frac{f(h) - f(0)}{h}$$

$$= \lim_{h \to 0+} \frac{|h| - 0}{h} = \lim_{h \to 0+} = \frac{h}{h} = \lim_{h \to 0+} 1 = 1$$

Thus $$Lf'(0) \neq Rf'(0) \Rightarrow f'(0) \text{ does not exist.}$$

$\therefore$ $f(x)$ is not derivable at $x = 0$.

$\therefore$ $f(x) = | x |$ is continuous at $x = 0$, but it is not differentiable at $x = 0$.

Remark. It follows from Art. 3 that if $f(x)$ is discontinuous at $x = a$, then it is not derivable at $x = a$ because if $f(x)$ is derivable at $x = a$, then it must be continuous by Art 3.

Note. We learn from Art. 3 that

(*i*) **derivability implies continuity**

(*ii*) **continuity need not imply derivability**

(*iii*) **discontinuity implies non-derivability.** (See Remark page 36)

Example 1. *Prove that* $f(x) = \begin{cases} x^2 - 1 & \text{if } x \geq 1 \\ 1 - x & \text{if } x < 1 \end{cases}$ *is not derivable at* $x = 1$.

Sol. $f(x) = x^2 - 1$ if $x \geq 1$ ...(1) | (given)

$= 1 - x$ if $x < 1$ ...(2)

Putting $x = 1$ in (1), $f(1) = 1^2 - 1 = 1 - 1 = 0$

(By Art. 2) $$Lf'(1) = \underset{x \to 1-}{\text{Lt}} \frac{f(x) - f(1)}{x - 1}$$

$$= \underset{x \to 1-}{\text{Lt}} \frac{(1 - x) - 0}{x - 1} \quad [\because \text{ By (2) } f(x) = 1 - x \text{ for } x < 1]$$

$$= \underset{x \to 1-}{\text{Lt}} \frac{-(x - 1)}{x - 1} = \underset{x \to 1-}{\text{Lt}} -1 = -1 \quad ...(3)$$

(By Art. 2) $$Rf'(1) = \underset{x \to 1+}{\text{Lt}} \frac{f(x) - f(1)}{x - 1}$$

$$= \underset{x \to 1+}{\text{Lt}} \frac{x^2 - 1 - 0}{x - 1} \quad [\because \text{ By (1) } f(x) = x^2 - 1 \text{ if } x \geq 1]$$

$$= \underset{x \to 1+}{\text{Lt}} \frac{(x - 1)(x + 1)}{x - 1} = \underset{x \to 1+}{\text{Lt}} (x + 1)$$

Put $x = 1 + h$ where $h \to 0+$

$$= \underset{h \to 0+}{\text{Lt}} (1 + h + 1)$$

Put h = 0 $= 1 + 0 + 1 = 2$ ...(4)

From (3) and (4), we have $Lf'(1) \neq Rf'(1)$.

$\therefore$ $f(x)$ is not derivable at $x = 1$.

Example 2. *Prove that the function* $f(x) = \dfrac{x\left(e^{\frac{1}{x}} - e^{\frac{-1}{x}}\right)}{e^{\frac{1}{x}} + e^{\frac{-1}{x}}}$ if $x \neq 0$

$= 0$ *if* $x = 0$

is not derivable at $x = 0$ *but is continuous at* $x = 0$.

Sol. Derivability at x = 0

$$Lf'(0) = \underset{x \to 0-}{\text{Lt}} \frac{f(x) - f(0)}{x - 0} = \underset{x \to 0-}{\text{Lt}} \frac{\dfrac{x\left(e^{\frac{1}{x}} - e^{\frac{-1}{x}}\right)}{e^{\frac{1}{x}} + e^{\frac{-1}{x}}}}{x} \quad (\because f(0) = 0 \text{ (given)})$$

$$= \underset{x \to 0-}{\text{Lt}} \frac{e^{\frac{1}{x}} - e^{\frac{-1}{x}}}{e^{\frac{1}{x}} + e^{\frac{-1}{x}}}$$

Put $x = 0 - h = -h$ as $h \to 0+$

$$= \underset{h \to 0+}{\text{Lt}} \frac{e^{\frac{-1}{h}} - e^{\frac{1}{h}}}{e^{\frac{-1}{h}} + e^{\frac{1}{h}}}$$

Multiplying every term by $e^{\frac{-1}{h}}$

$$= \underset{h \to 0+}{\text{Lt}} \frac{e^{\frac{-2}{h}} - 1}{e^{\frac{-2}{h}} + 1}$$

(Put h = 0) $= \dfrac{e^{-\infty} - 1}{e^{-\infty} + 1} = \dfrac{0 - 1}{0 + 1} = -1$ ...(1)

$$\text{R} f'(0) = \underset{x \to 0+}{\text{Lt}} \frac{f(x) - f(0)}{x - 0}$$

$$= \underset{x \to 0+}{\text{Lt}} \frac{\dfrac{x\left(e^{\frac{1}{x}} - e^{\frac{-1}{x}}\right)}{e^{\frac{1}{x}} + e^{\frac{-1}{x}}}}{x} \qquad | \because \ f(0) = 0 \text{ (given)}$$

$$= \underset{x \to 0+}{\text{Lt}} \frac{e^{\frac{1}{x}} - e^{\frac{-1}{x}}}{e^{\frac{1}{x}} + e^{\frac{-1}{x}}}$$

Put $x = 0 + h = h$ where $h \to 0+$

$$= \underset{h \to 0+}{\text{Lt}} \frac{e^{\frac{1}{h}} - e^{\frac{-1}{h}}}{e^{\frac{1}{h}} + e^{\frac{-1}{h}}}$$

Multiplying every term by $e^{\frac{-1}{h}}$

$$= \underset{h \to 0+}{\text{Lt}} \frac{1 - e^{\frac{-2}{h}}}{1 + e^{\frac{-2}{h}}}$$

(Put h = 0) $= \dfrac{1 - e^{-\infty}}{1 + e^{-\infty}} = \dfrac{1 - 0}{1 + 0} = 1$ ...(2)

From (1) and (2) $\text{L} f'(0) \neq \text{R} f'(0)$.

$\therefore$ $f(x)$ is not derivable at $-x = 0$.

Continuity at x = 0

Left Hand Limit $= \underset{x \to 0-}{\text{Lt}} f(x) = \underset{x \to 0-}{\text{Lt}} \dfrac{x\left(e^{\frac{1}{x}} - e^{\frac{-1}{x}}\right)}{e^{\frac{1}{x}} + e^{\frac{-1}{x}}}$

Put $x = 0 - h = -h$ where $h \to 0^+$

$$= \underset{h \to 0^+}{\text{Lt}} \frac{-h\left(e^{\frac{1}{h}} - e^{\frac{-1}{h}}\right)}{e^{\frac{-1}{h}} + e^{\frac{1}{h}}}$$

Multiplying every term by $e^{\frac{-1}{h}}$, we have

$$\underset{x \to 0-}{\text{Lt}} f(x) = \underset{h \to 0+}{\text{Lt}} \frac{-h\left(e^{\frac{-2}{h}} - 1\right)}{e^{\frac{-2}{h}} + 1}$$

(Put $h = 0$)

$$= \frac{0\,(e^{-\infty} - 1)}{e^{-\infty} + 1} = \frac{0\,(0-1)}{0+1} = 0 \qquad ...(3)$$

Proceeding on the same lines, we have

$$\underset{x \to 0+}{\text{Lt}} f(x) = 0 \qquad ...(4)$$

From (3) and (4), we have $\underset{x \to 0}{\text{Lt}} f(x)$ exists and = 0.

Also $\quad f(0) = 0 \quad$ (given)

$\therefore \quad \underset{x \to 0}{\text{Lt}} f(x) = f(0)$

$\therefore \quad f(x)$ is continuous at $x = 0$.

Example 3. *Prove that the function f defined by*

$$f(x) = x^2 \sin \frac{1}{x} \quad \text{if} \quad x \neq 0$$
$$= 0 \quad \text{if} \quad x = 0$$

is continuous and derivable at $x = 0$ but its derivative is not continuous at $x = 0$.

(M.D.U. 2003)

Sol. $\quad f(x) = x^2 \sin \frac{1}{x} \quad$ if $\quad x \neq 0 \qquad ...(1)$

$= 0 \quad$ if $\quad x = 0 \qquad ...(2)$ (given)

Continuity of f(x) at x = 0

$$\underset{x \to 0}{\text{Lt}} f(x) = \underset{x \to 0}{\text{Lt}} x^2 \sin \frac{1}{x} \qquad \text{[By (1)]}$$

Putting $x = 0$

= 0 × (A finite quantity between – 1 and + 1)

= 0

$= f(0)$ [By (2)]

$\therefore \quad f(x)$ is continuous at $x = 0$.

Derivability of f(x) at x = 0

(By Art. 1) $\quad \underset{x \to 0}{\text{Lt}} \frac{f(x) - f(0)}{x - 0} = \underset{x \to 0}{\text{Lt}} \frac{x^2 \sin \frac{1}{x} - 0}{x - 0} \qquad [\because \text{ By (2) } f(0) = 0]$

$$= \underset{x \to 0}{\text{Lt}}\ x \sin \frac{1}{x}$$

Put $x = 0$

$$= 0 \times \text{(A finite quantity between} - 1 \text{ and } + 1)$$
$$= 0 \text{ which exists (because it is unique and finite)}$$

$\therefore$ By Art. 1, $f(x)$ is derivable

at $x = 0$ and $f'(0) = 0$ ...(3)

Continuity of f ′(x) at x = 0

Diff. both sides of Eqn. (1) w.r.t. x

$$f'(x) = x^2 \cos \frac{1}{x} \frac{d}{dx}\left(\frac{1}{x}\right) + \sin \frac{1}{x} \frac{d}{dx} x^2 \quad \text{if} \quad x \neq 0$$

$$= x^2 \cos \frac{1}{x}\left(\frac{-1}{x^2}\right) + \left(\sin \frac{1}{x}\right) 2x$$

or
$$f'(x) = -\cos \frac{1}{x} + 2x \sin \frac{1}{x} \quad \text{if} \quad x \neq 0$$

$$= 2x \sin \frac{1}{x} - \cos \frac{1}{x} \quad \text{if} \quad x \neq 0 \qquad \text{...(4)}$$

$$\underset{x \to 0}{\text{Lt}}\ f'(x) = \underset{x \to 0}{\text{Lt}} \left(2x \sin \frac{1}{x} - \cos \frac{1}{x}\right)$$

$$= \underset{x \to 0}{\text{Lt}}\ 2x \sin \frac{1}{x} - \underset{x \to 0}{\text{Lt}}\ \cos \frac{1}{x} = 2(0)$$

(A finite quantity between – 1 and 1)

$$- \underset{x \to 0}{\text{Lt}}\ \cos \frac{1}{x} = 0 - \underset{x \to 0}{\text{Lt}}\ \cos \frac{1}{x}$$

which does not exist as $\cos \frac{1}{x}$ oscillates between – 1 and 1 as $x \to 0$.

(See Hint for Q. N. 24, Exercise 3, Page 33 Chapter 1)

$\therefore$ $\underset{x \to 0}{\text{Lt}}\ f'(x)$ does not exist.

$\therefore$ $f'(x)$ is discontinuous at $x = 0$.

Example 4. *A function f is defined as follows :*

$$f(x) = \begin{cases} x^p \cos \dfrac{1}{x} & \text{when } x \neq 0 \\ 0 & \text{when } x = 0. \end{cases}$$

What condition should be imposed on p so that

(i) f may be continuous at x = 0

(ii) f may be derivable at x = 0. (M.D.U. 2004 S)

Sol. (*i*)
$$\underset{x \to 0}{\text{Lt}}\ f(x) = \underset{x \to 0}{\text{Lt}}\ x^p \cos \frac{1}{x}$$

Put $x = 0$

$$= (0)^p \text{ [A finite quantity between} - 1 \text{ and } 1]$$
$$= 0 \quad \text{if } p > 0 \qquad [\because \ (0)^{\text{positive}} = 0]$$
$$= f(0) \qquad [\because \ f(0) = 0 \text{ (given)}]$$

$\therefore$ $f(x)$ is continuous at $x = 0$ if $p > 0$.

(*ii*) Because $f(x)$ is derivable at $x = 0$ (given), therefore

$$\underset{x \to 0}{\text{Lt}} \frac{f(x) - f(0)}{x - 0} \text{ exists}$$

i.e., $$\underset{x \to 0}{\text{Lt}} \frac{x^p \cos \frac{1}{x} - 0}{x} = \underset{x \to 0}{\text{Lt}}\, x^{p-1} \cos \frac{1}{x} \text{ exists}$$

Put $x = 0$

$= (0)^{p-1}$ [A finite quantity between -1 and 1]

$= 0$ *if* $p - 1 > 0$ *i.e.*, if $p > 1$

$\therefore$ $f(x)$ is derivable at $x = 0$ if $p > 1$.

Example 5. *Discuss the continuity and differentiability of the function* $f(x) = |x - 1| + |x - 2|$ *in [0, 3].* (M.D.U. 2005 A)

Sol. $|x - 1| = 0$ gives $x = 1$ and $|x - 2| = 0$ gives $x = 2$.

So let us break the given interval [0, 3] into subintervals [0, 1], [1, 2], and [2, 3].

On the interval [0, 1] ; $(x - 1) \le 0$ and $x - 2 \le 0$

$\therefore$ $|x - 1| = -(x - 1)$ and $|x - 2| = -(x - 2)$

$\therefore$ $f(x) = |x - 1| + |x - 2| = -(x - 1) - (x - 2)$

$= -x + 1 - x + 2 = 3 - 2x.$

On the interval [1, 2] ; $x - 1 \ge 0$ and $x - 2 \le 0$

$\therefore$ $|x - 1| = x - 1$ and $|x - 2| = -(x - 2)$

$\therefore$ $f(x) = |x - 1| + |x - 2| = x - 1 - (x - 2) = 1$

On the interval [2, 3] ; $x - 1 \ge 0$ and $x - 2 \ge 0$

$\therefore$ $|x - 1| = x - 1$ and $|x - 2| = x - 2$

$\therefore$ $f(x) = |x - 1| + |x - 2| = x - 1 + x - 2 = 2x - 3$

$\therefore$ $f(x) = 3 - 2x, \quad 0 \le x \le 1$ ...(1)

$= 1 \quad , 1 \le x \le 2$ ...(2)

$= 2x - 3, 2 \le x \le 3$ ...(3)

The three values of $f(x)$ given by (1), (2) and (3) are either linear polynomial functions or constant function and hence are continuous and derivable for all real x in the closed interval [0, 3] **except possibly** at breaking points $x = 1$ and $x = 2$.

i.e., $x = 1$ and $x = 2$ are the only doubtful points for continuity and derivability in [0, 3].

To check continuity at x = 1

Left hand limit $= \underset{x \to 1^-}{\text{Lt}} f(x) = \underset{x \to 1^-}{\text{Lt}} (3 - 2x)$ [By (1)]

Put $x = 1 - h, h \to 0^+$

$= \underset{h \to 0^+}{\text{Lt}} (3 - 2(1 - h)) = 3 - 2 = 1$

Right hand limit $= \underset{x \to 1^+}{\text{Lt}} f(x) = \underset{x \to 1^+}{\text{Lt}} = 1$ [By (2)]

$\therefore$ $\underset{x \to 1}{\text{Lt}} f(x)$ exists and $= 1$ [$\because$ Left Hand Limit = Right Hand Limit]

Putting $x = 1$ in (1) or (2) ; $f(1) = 3 - 2 = 1$ ...(4)

$\therefore$ $\underset{x \to 1}{\text{Lt}} f(x) = f(1)$

$\therefore$ $f(x)$ is continuous at $x = 1$

To check derivability at x = 1

$$Lf'(1) = \underset{x \to 1^-}{\text{Lt}} \frac{f(x) - f(1)}{x - 1}$$

$$= \underset{x \to 1^-}{\text{Lt}} \frac{3 - 2x - 1}{x - 1} \quad \text{[By (1) and (4)]}$$

$$= \underset{x \to 1^-}{\text{Lt}} \frac{-2x + 2}{x - 1} = \underset{x \to 1^-}{\text{Lt}} \frac{-2(x - 1)}{x - 1} = \underset{x \to 1^-}{\text{Lt}} (-2) = -2$$

$$Rf'(1) = \underset{x \to 1^+}{\text{Lt}} \frac{f(x) - f(1)}{x - 1} = \underset{x \to 1^+}{\text{Lt}} \frac{1 - 1}{x - 1} \quad \text{[By (2) and (4)]}$$

$$= \underset{x \to 1^+}{\text{Lt}} \frac{0}{x - 1}$$

Put $x = 1 + h, h \to 0^+$

$$= \underset{h \to 0^+}{\text{Lt}} \frac{0}{1 + h - 1} = \underset{h \to 0^+}{\text{Lt}} \frac{0}{h} = \underset{h \to 0^+}{\text{Lt}} 0 = 0$$

$$Lf'(1) \neq Rf'(1)$$

$\therefore$ $f(x)$ is not derivable at $x = 1$.

To check continuity at x = 2

Left hand limit $= \underset{x \to 2^-}{\text{Lt}} f(x) = \underset{x \to 2^-}{\text{Lt}} 1 = 1$ [By (2)]

Right hand limit $= \underset{x \to 2^+}{\text{Lt}} f(x) = \underset{x \to 2^+}{\text{Lt}} (2x - 3)$ [By (3)]

Put $x = 2 + h, h \to 0^+$

$$= \underset{h \to 0^+}{\text{Lt}} (2(2 + h) - 3) = 4 - 3 = 1$$

$$\therefore \quad \underset{x \to 2^-}{\text{Lt}} f(x) = \underset{x \to 2^+}{\text{Lt}} f(x) = 1$$

$\therefore$ $\underset{x \to 2}{\text{Lt}} f(x)$ exists and $= 1$.

Putting $x = 2$ in (2) or (3), $f(2) = 1$ ...(5)

$$\therefore \quad \underset{x \to 2}{\text{Lt}} f(x) = f(2) = 1$$

$\therefore$ $f(x)$ is continuous at $x = 2$

$\therefore$ **f(x) is continuous in the closed interval [0, 3].**

To check derivability at x = 2

$$Lf'(2) = \underset{x \to 2^-}{\text{Lt}} \frac{f(x) - f(2)}{x - 2} = \underset{x \to 2^-}{\text{Lt}} \frac{1 - 1}{x - 2} \quad \text{[By (2) and (5)]}$$

$$= \underset{x \to 2^-}{\text{Lt}} \frac{0}{x - 2} = \underset{h \to 0^+}{\text{Lt}} \frac{0}{2 - h - 2} = \underset{h \to 0^+}{\text{Lt}} \frac{0}{-h}$$

$$= \underset{h \to 0^+}{\text{Lt}} 0 = 0$$

$$Rf'(2) = \underset{x \to 2^+}{\text{Lt}} \frac{f(x) - f(2)}{x - 2} = \underset{x \to 2^+}{\text{Lt}} \frac{2x - 3 - 1}{x - 2} \quad \text{[By (3) and (5)]}$$

$$= \underset{x \to 2^+}{\text{Lt}} \frac{2x - 4}{x - 2} = \underset{x \to 2^+}{\text{Lt}} \frac{2(x - 2)}{x - 2} = \underset{x \to 2^+}{\text{Lt}} 2 = 2$$

$$Lf'(2) \neq Rf'(2)$$

$\therefore$ $f(x)$ is not derivable at the point $x = 2$.

$\therefore$ $f(x)$ is derivable in the closed interval [0, 3] except at the points $x = 1$ and $x = 2$.

EXERCISE 1

1. (*a*) Show that $f(x) = | x - 3 |$ is not derivable at $x = 3$.
(*b*) Show that $f(x) = | x - c |$ is not derivable at $x = c$.

Prove that the following functions (Q. No. 2 to 4) are continuous but not derivable :

2. $f(x) = \begin{cases} 2 + x & \text{if } x \geq 0 \\ 2 - x & \text{if } x < 0 \end{cases}$ at $x = 0$

3. $f(x) = \begin{cases} x \tan^{-1} \dfrac{1}{x} & \text{if } x \neq 0 \\ 0 & \text{if } x = 0 \end{cases}$ at $x = 0$

[**Hint.** We know that $\tan\left(-\dfrac{\pi}{2}\right) = -\tan \dfrac{\pi}{2} = -\infty$

$\therefore$ $\tan^{-1}(-\infty) = -\dfrac{\pi}{2}$. Also $\tan^{-1} \infty = \dfrac{\pi}{2}$]

4. $f(x) = \begin{cases} x \sin \dfrac{1}{x} & \text{if } x \neq 0 \\ 0 & \text{if } x = 0 \end{cases}$ at $x = 0$ (*K.U. 2006 S*)

5. Prove that the function

$$f(x) = \begin{cases} \dfrac{x(e^{1/x} - 1)}{e^{1/x} + 1} & \text{if } x \neq 0 \\ 0 & \text{if } x = 0 \end{cases}$$

is continuous at $x = 0$ but is not derivable at $x = 0$.

6. Discuss the continuity and derivability of the function

$$f(x) = \begin{cases} (x - a) \sin \dfrac{1}{x - a} & \text{when } x \neq a \\ 0 & \text{when } x = a \end{cases} \text{ at } x = a.$$

7. If $f(x)$ is differentiable at $x = a$, find $\underset{x \to a}{\text{Lt}} \dfrac{x f(a) - a f(x)}{x - a}$. (*M.D.U. 2004*)

[**Hint.** Subtract and add $a f(a)$ in the numerator.

$$\therefore \quad \underset{x \to a}{\text{Lt}} \frac{xf(a) - af(a) - af(x) + af(a)}{x - a}$$

$$= \underset{x \to a}{\text{Lt}} \frac{(x - a) f(a) - a(f(x) - f(a))}{x - a}$$

$$= \underset{x \to a}{\text{Lt}} \left(f(a) - a \frac{(f(x) - f(a))}{x - a} \right) = f(a) - af'(a) \Big]$$

ANSWERS

6. Continuous at $x = a$ but not derivable at $x = a$.

7. $f(a) - a f'(a)$.

3

Successive Differentiation

1. If y be a function of x, its derivative $\frac{dy}{dx}$ is itself a function of x. In general, we assume that it also possesses a derivative if it is differentiated further. The derivative $\frac{dy}{dx}$ is called the first differential co-efficient or first derivative of y w.r.t. x. The differential co-efficient of $\frac{dy}{dx}$, *i.e.*, $\frac{d}{dx}\left(\frac{dy}{dx}\right)$ is called second differential co-efficient or **second derivative** of y w.r.t. x, which we denote as $\frac{d^2y}{dx^2}$ [read as *dee two y* over *dee x squared*]. In like manner the third differential co-efficient or third derivative of y w.r.t. x means the differential co-efficient of $\frac{d^2y}{dx^2}$, *i.e.*, $\frac{d}{dx}\left(\frac{d^2y}{dx^2}\right)$ and is represented by $\frac{d^3y}{dx^3}$ and so on. In general, the nth differential co-efficient of y w.r.t. x is denoted by $\frac{d^ny}{dx^n}$. This process of finding the differential co-efficient of a function is called **Successive Differentiation.**

Thus, if $y = f(x)$, the successive differential co-efficients of $f(x)$ are

$$\frac{dy}{dx}, \frac{d^2y}{dx^2}, \frac{d^3y}{dx^3}, \ldots\ldots\ldots, \frac{d^ny}{dx^n}.$$

These are also denoted by :

(*i*) $y_1, y_2, y_3, \ldots\ldots\ldots, y_n$.

(*ii*) $y', y'', y''', \ldots\ldots\ldots, y^n$.

(*iii*) $Dy, D^2y, D^3y, \ldots\ldots\ldots D^ny$.

(*iv*) $f(x), f'(x), f''(x), \ldots\ldots\ldots, f^n(x)$.

Note 1. The following formulae on trigonometry will be helpful in learning this chapter.

1. $2 \sin A \cos B = \sin (A + B) + \sin (A - B)$

2. $2 \cos A \sin B = \sin (A + B) - \sin (A - B)$

3. $2 \cos A \cos B = \cos (A + B) + \cos (A - B)$

4. $2 \sin A \sin B = \cos (A - B) - \cos (A + B)$

5. $\sin^2 \theta = \frac{1 - \cos 2\theta}{2}$

6. $\cos^2 \theta = \frac{1 + \cos 2\theta}{2}$

7. $\sin 3\theta = 3 \sin \theta - 4 \sin^3 \theta$

8. $\cos 3\theta = 4 \cos^3 \theta - 3 \cos \theta$.

Note. 2. The following formulae on differentiation learn in Class XII will be helpful in learning this chapter and subsequent chapters.

1. $\dfrac{d}{dx}(c)$ where c is a constant.

2. (a) $\dfrac{d}{dx}(x^n) = nx^{n-1}$

   (b) $\dfrac{d}{dx}(u^n)$ where u is a function of $x = nu^{n-1}\dfrac{du}{dx}$

3. $\dfrac{d}{dx}(cu) = c\dfrac{du}{dx}$ where c is a constant and u is a function x.

4. $\dfrac{d}{dx}(uv) = u\dfrac{dv}{dx} + v\dfrac{du}{dx}$ where u and v are functions of x.

5. $\dfrac{d}{dx} = \left(\dfrac{u}{v}\right) = \dfrac{v\dfrac{du}{dx} - u\dfrac{dv}{dx}}{v^2}$

6. (a) $\dfrac{d}{dx}(\sin x) = \cos x$ (b) $\dfrac{d}{dx}\sin u = \cos u\dfrac{du}{dx}$

7. (a) $\dfrac{d}{dx}(\cos x) = -\sin x$ (b) $\dfrac{d}{dx}(\cos u) = -\sin u\dfrac{du}{dx}$

8. (a) $\dfrac{d}{dx}(\tan x) = \sec^2 x$ (b) $\dfrac{d}{dx}\tan u = \sec^2 u\dfrac{du}{dx}$

9. (a) $\dfrac{d}{dx}(\cot x) = -\operatorname{cosec}^2 x$ (b) $\dfrac{d}{dx}(\cot u) = -\operatorname{cosec}^2 u\dfrac{du}{dx}$

10. (a) $\dfrac{d}{dx}(\sec x) = \sec x\tan x$ (b) $\dfrac{d}{dx}(\sec u) = \sec u\tan u\dfrac{du}{dx}$

11. (a) $\dfrac{d}{dx}(\operatorname{cosec} x) = -\operatorname{cosec} x\cot x$ (b) $\dfrac{d}{dx}(\operatorname{cosec} u) = -\operatorname{cosec} u\cot u\dfrac{du}{dx}$

12. (a) $\dfrac{d}{dx}(\sin^{-1} x) = \dfrac{1}{\sqrt{1-x^2}}$ (b) $\dfrac{d}{dx}(\sin^{-1} u) = \dfrac{1}{\sqrt{1-u^2}}\dfrac{du}{dx}$

13. (a) $\dfrac{d}{dx}(\cos^{-1} x) = \dfrac{-1}{\sqrt{1-x^2}}$ (b) $\dfrac{d}{dx}\cos^{-1} u = \dfrac{-1}{\sqrt{1-u^2}}\dfrac{du}{dx}$

14. (a) $\dfrac{d}{dx}(\tan^{-1} x) = \dfrac{1}{1+x^2}$ (b) $\dfrac{d}{dx}\tan^{-1} u = \dfrac{1}{1+u^2}\dfrac{du}{dx}$

15. (a) $\dfrac{d}{dx}\cot^{-1} x = \dfrac{-1}{1+x^2}$ (b) $\dfrac{d}{dx}\cot^{-1} u = \dfrac{-1}{1+u^2}\dfrac{du}{dx}$

16. (a) $\dfrac{d}{dx}\sec^{-1} x = \dfrac{1}{x\sqrt{x^2-1}}$ (b) $\dfrac{d}{dx}\sec^{-1} u = \dfrac{1}{u\sqrt{u^2-1}}\dfrac{du}{dx}$

17. (a) $\dfrac{d}{dx}\operatorname{cosec}^{-1} x = \dfrac{-1}{x\sqrt{x^2-1}}$ (b) $\dfrac{d}{dx}\operatorname{cosec}^{-1} u = \dfrac{-1}{u\sqrt{u^2-1}}\dfrac{du}{dx}$

18. (a) $\dfrac{d}{dx}e^x = e^x$ (b) $\dfrac{d}{dx}e^u = e^u\dfrac{du}{dx}$

19. (a) $\dfrac{d}{dx}a^x = a^x\log a$ (b) $\dfrac{d}{dx}a^u = a^u\log a\dfrac{du}{dx}$

20. (a) $\frac{d}{dx}\log x = \frac{1}{x}$ (b) $\frac{d}{dx}\log u = \frac{1}{u}\frac{du}{dx}$

21. (a) $\frac{d}{dx}\log_a x = \frac{1}{x}\log_a e$ (b) $\frac{d}{dx}\log_a u = \frac{1}{u}\log_a e$

22. (a) $\sinh x = \frac{e^x - e^{-x}}{2}$ (b) $\cosh x = \frac{e^x + e^{-x}}{2}$

(c) $\cosh^2 x - \sinh^2 x = 1$

23. $\frac{d}{dx}\sinh x = \cosh x$

24. $\frac{d}{dx}\cosh x = \sinh x.$

Example 1. *Find* $\frac{d^3y}{dx^3}$, *when* $y = 4x^3 + 4x + 2$.

Sol. Here $y = 4x^3 + 4x + 2$

$\therefore$ $\frac{dy}{dx} = 12x^2 + 4$ $\qquad \frac{d^2y}{dx^2} = \frac{d}{dx}(12x^2 + 4) = 24x$

$$\frac{d^3y}{dx^3} = \frac{d}{dx}(24x) = 24.$$

Example 2. *If* $y = \frac{x}{\sqrt{1-x^2}}$, *find* $\frac{d^3y}{dx^3}$.

Sol.
$$y = \frac{x}{\sqrt{1-x^2}}.$$

$$\therefore \quad \frac{dy}{dx} = \frac{\sqrt{1-x^2}\,.\,1 - x\,.\,\frac{1}{2}(1-x^2)^{-1/2}\times(-2x)}{(1-x^2)}$$

$$= \frac{\sqrt{1-x^2} + \frac{x^2}{\sqrt{1-x^2}}}{1-x^2} = \frac{1-x^2+x^2}{(1-x^2)^{3/2}} = (1-x^2)^{-3/2}$$

$$\therefore \quad \frac{d^2y}{dx^2} = -\frac{3}{2}(1-x^2)^{-5/2}\times(-2x) = 3x(1-x^2)^{-5/2}$$

and $\therefore$
$$\frac{d^3y}{dx^3} = 3\left[x\left(-\frac{5}{2}\right)(1-x^2)^{-7/2}\times(-2x) + (1-x^2)^{-5/2}\right]$$

$$= 15x^2(1-x^2)^{-7/2} + 3(1-x^2)^{-5/2}$$

$$= \frac{15x^2}{(1-x^2)^{7/2}} + \frac{3}{(1-x^2)^{5/2}} = \frac{15x^2+3(1-x^2)}{(1-x^2)^{7/2}}$$

$$= \frac{3+12x^2}{(1-x^2)^{7/2}} = \frac{3(1+4x^2)}{(1-x^2)^{7/2}}.$$

Example 3. *If* $y = \sin(a\sin^{-1}x)$, *prove that*

$$(1-x^2)\frac{d^2y}{dx^2} = x\frac{dy}{dx} - a^2y.$$ (K.U. 1996 A)

Sol. $$y = \sin (a \sin^{-1} x) \quad ...(1)$$

$$\therefore \quad \frac{dy}{dx} = \cos (a \sin^{-1} x) \frac{a}{\sqrt{1-x^2}} = \frac{a . \cos (a \sin^{-1} x)}{\sqrt{1-x^2}}$$

Cross-multiplying

$$\sqrt{1-x^2} \frac{dy}{dx} = a \cos (a \sin^{-1} x)$$

Squaring both sides,

$$(1-x^2) \left(\frac{dy}{dx}\right)^2 = a^2 \cos^2 (a \sin^{-1} x)$$

or $$(1-x^2) \left(\frac{dy}{dx}\right)^2 = a^2 [1 - \sin^2 (a \sin^{-1} x)] \quad (\because \cos^2 \theta = 1 - \sin^2 \theta)$$

or $$(1-x^2) \left(\frac{dy}{dx}\right)^2 = a^2 \ (1-y^2) \quad \text{[By (1)]}$$

Again differentiating w.r.t. x, we have

$$(1-x^2) \frac{d}{dx} \left(\frac{dy}{dx}\right)^2 + \left(\frac{dy}{dx}\right)^2 \frac{d}{dx} (1-x^2) = a^2 \left[0 - \frac{d}{dx} y^2\right]$$

or $$(1-x^2)\, 2 \frac{dy}{dx} \frac{d^2 y}{dx^2} - 2x \left(\frac{dy}{dx}\right)^2 = -\, 2a^2 y \frac{dy}{dx}$$

$$\left[\because \ \frac{d}{dx}\left(\frac{dy}{dx}\right)^2 = 2 \frac{dy}{dx} \frac{d}{dx}\left(\frac{dy}{dx}\right) = 2 \frac{dy}{dx} \frac{d^2 y}{dx^2} \quad \text{and} \quad \frac{d}{dx} y^2 = 2y \frac{d}{dx} (y)\right]$$

Dividing every term by $2\dfrac{dy}{dx}$,

$$(1-x^2) \frac{d^2 y}{dx^2} - x \frac{dy}{dx} = -\, a^2 y \quad \text{or} \quad (1-x^2) \frac{d^2 y}{dx^2} = x \frac{dy}{dx} - a^2 y.$$

Example 4. *If $y = a \cos (\log x) + b \sin (\log x)$, prove that $x^2 y_2 + x y_1 + y = 0$.*

Sol. $$y = a \cos (\log x) + b \sin (\log x) \quad ...(1)$$

Differentiating w.r.t. x,

$$y_1 = -\, a \sin (\log x) . \frac{1}{x} + b \cos (\log x) . \frac{1}{x}$$

Multiplying every term by L.C.M. $= x$,

$$x y_1 = -\, a \sin (\log x) + b \cos (\log x)$$

Again differentiating both sides w.r.t. x

$$\frac{d}{dx} (x y_1) = -\, a \cos (\log x) . \frac{1}{x} - b \sin (\log x) . \frac{1}{x}$$

or $$x . y_2 + y_1 . 1 = -\frac{[a \cos (\log x) + b \sin (\log x)]}{x}$$

Cross-multiplying $x^2 y_2 + x y_1 = -\, y$ [By (1)]

or $$x^2 y_2 + x y_1 + y = 0.$$

Example 5. *If* $p^2 = a^2 \cos^2 \theta + b^2 \sin^2 \theta$, *prove that*

$$p + \frac{d^2p}{d\theta^2} = \frac{a^2b^2}{p^3}. \qquad \text{(K.U. 1997 S)}$$

Sol. $$p^2 = a^2 \cos^2 \theta + b^2 \sin^2 \theta$$

Differentiating w.r.t. θ, we have

$$2p \cdot \frac{dp}{d\theta} = a^2 \cdot 2 \cos \theta (-\sin \theta) + b^2 \cdot 2 \sin \theta \cos \theta$$

$$= (b^2 - a^2) \sin 2\theta \qquad ...(1)$$

Differentiating again, we have

$$2p \cdot \frac{d^2p}{d\theta^2} + 2 \frac{dp}{d\theta} \cdot \frac{dp}{d\theta} = (b^2 - a^2) \times 2 \cos 2\theta$$

$$\therefore \quad p \cdot \frac{d^2p}{d\theta^2} = (b^2 - a^2) \cos 2\theta - \left(\frac{dp}{d\theta}\right)^2$$

$$= (b^2 - a^2) \cos 2\theta - \frac{(b^2 - a^2)^2 \sin^2 2\theta}{4p^2} \qquad \text{[From (1)]}$$

Adding p^2 to both sides, we have

$$p \frac{d^2p}{d\theta^2} + p^2 = (b^2 - a^2) \cos 2\theta - \frac{(b^2 - a^2)^2 \sin^2 2\theta}{4p^2} + p^2$$

or $$p\left(\frac{d^2p}{d\theta^2} + p\right) = \frac{4(b^2 - a^2) \cos 2\theta.\, p^2 - (b^2 - a^2)^2 \sin^2 2\theta + 4p^4}{4p^2}$$

Putting the value of p^2 in R.H.S.

$$= \frac{4(b^2 - a^2)(\cos^2 \theta - \sin^2 \theta)(a^2 \cos^2 \theta + b^2 \sin^2 \theta) - 4(b^2 - a^2)^2 \sin^2 \theta \cos^2 \theta + 4(a^2 \cos^2 \theta + b^2 \sin^2 \theta)^2}{4p^2}$$

or $$p^3\left(\frac{d^2p}{d\theta^2} + p\right) = a^2b^2 \cos^4 \theta + 2a^2b^2 \sin^2 \theta \cos^2 \theta + a^2b^2 \sin^4 \theta$$

$$= a^2b^2 (\cos^2 \theta + \sin^2 \theta)^2 = a^2b^2.$$

$$\therefore \quad \frac{d^2p}{d\theta^2} + p = \frac{a^2b^2}{p^3}.$$

Example 6. *If* $x = a(\cos \theta + \theta \sin \theta)$, $y = a(\sin \theta - \theta \cos \theta)$, *prove that*

$$a\theta \frac{d^2y}{dx^2} = \sec^3 \theta.$$

Sol. We have, $$\frac{dx}{d\theta} = a(-\sin \theta + \theta \cos \theta + \sin \theta) = a\theta \cos \theta \qquad ...(1)$$

and $$\frac{dy}{d\theta} = a(\cos \theta + \theta \sin \theta - \cos \theta) = a\theta \sin \theta$$

$$\therefore \qquad \frac{dy}{dx} = \frac{dy}{d\theta}\cdot\frac{d\theta}{dx} = \frac{a\theta \sin\theta}{a\theta\cos\theta} = \tan\theta$$

Again, differentiating both sides w.r.t. x,

$$\frac{d^2y}{dx^2} = \sec^2\theta \ \frac{d\theta}{dx} = \sec^2\theta \cdot \frac{1}{a\theta\cos\theta} \qquad \text{(By (1))}$$

$$= \frac{\sec^3\theta}{a\theta}$$

Hence $\qquad a\theta\dfrac{d^2y}{dx^2} = \sec^3\theta.$

EXERCISE 1

1. (*i*) If $y = x^3 \log x$, prove that $y_4 = \dfrac{6}{x}$.

(*ii*) If $f(x) = \tan x$, find $f^{iv}(x)$, when $x = \dfrac{\pi}{4}$.

(*iii*) If $y = \dfrac{\log x}{x}$, prove that $\dfrac{d^2y}{dx^2} = \dfrac{2\log x - 3}{x^3}$.

(*iv*) If $f(x) = x^3 \sin x$, find $f^{iv}(x)$.

2. $y = A\sin nx + B\cos nx$, prove that $\dfrac{d^2y}{dx^2} + n^2y = 0$.

3. (*i*) If $y = (\sin^{-1}x)^2$, show that $(1-x^2)\dfrac{d^2y}{dx^2} - x\dfrac{dy}{dx} = 2$.

[**Hint.** Differentiate w.r.t. x ; cross-multiply and square, again differentiate.]

(*ii*) If $y = \tan^{-1}x$, show that $(1+x^2)\dfrac{d^2y}{dx^2} + 2x\dfrac{dy}{dx} = 0$.

(*iii*) If $y = (\tan^{-1}x)^2$, prove that $(x^2+1)^2y_2 + 2x(x^2+1)y_1 = 2$.

4. (*i*) $y = a\,e^{mx} + b\,e^{-mx}$, prove that $\dfrac{d^2y}{dx^2} - m^2y = 0$

(*ii*) If $y = e^{ax}\sin bx$, prove that $\dfrac{d^2y}{dx^2} - 2a\dfrac{dy}{dx} + (a^2+b^2)y = 0$.

5. (*i*) If $y = \cos^{-1}x$, show that $y_2(1-x^2) - xy_1 = 0$.

(*ii*) If $x = (a+bt)e^{-nt}$, prove that $\dfrac{d^2x}{dt^2} + 2n\dfrac{dx}{dt} + n^2x = 0$.

(*iii*) If $y = a\cosh\dfrac{x}{a}$, prove that $\dfrac{d^2y}{dx^2} = \dfrac{1}{a}\sqrt{1+\left(\dfrac{dy}{dx}\right)^2}$

6. If $y = e^{m\sin^{-1}x}$, show that $(1-x^2)y_2 - xy_1 = m^2y$.

7. (*a*) If $y = A[x + \sqrt{1+x^2}]^n$, prove that $(x^2+1)y_2 + xy_1 - n^2y = 0$.

(*b*) If $y = A(x + \sqrt{x^2-1})^n$; prove that $(x^2-1)y_2 + xy_1 - n^2y = 0$.

8. If $y = [\log (x + \sqrt{1+x^2})]^2$, show that $(1 + x^2) \dfrac{d^2y}{dx^2} + x\dfrac{dy}{dx} = 2$.

9. If $y = \dfrac{\sin^{-1} x}{\sqrt{1-x^2}}$, show that

(*i*) $(1 - x^2)\, y_1 - xy = 1$ (*ii*) $(1 - x^2)\, y_2 - 3xy_1 - y = 0$.

10. (*a*) If $y = \sin (\sin x)$, prove that $\dfrac{d^2y}{dx^2} + \tan x \dfrac{dy}{dx} + y \cos^2 x = 0$.

(*b*) If $y = x + \cot x$, show that $\sin^2 x \dfrac{d^2y}{dx^2} - 2y + 2x = 0$.

(*c*) If $y = \tan^{-1}\left(\dfrac{\sqrt{1+x^2}-1}{x}\right) + \tan^{-1}\left(\dfrac{2x}{1-x^2}\right)$, show that $y_2 = -\dfrac{5x}{(1+x^2)^2}$.

(*d*) If $y = \tan x + \sec x$, prove that $\dfrac{d^2y}{dx^2} = \dfrac{\cos x}{(1 - \sin x)^2}$.

11. (*i*) If $y = \dfrac{ax+b}{cx+d}$, show that $2y_1y_3 = 3y_2^{\,2}$.

(*ii*) If $y = \log (1 + \cos x)$, prove that $y_1y_2 + y_3 = 0$.

[Hint for (*i*) **and** (*ii*). Differentiate three times and put the values in L.H.S. and R.H.S.]

(*iii*) If $y = a\, e^x + b\, e^{2x} + c\, e^{3x}$, prove that $y_3 - 6y_2 + 11y_1 - 6y = 0$.

12. If $x = at^2, y = 2at$, find $\dfrac{d^2y}{dx^2}$.

13. (*i*) If $x = a \cos t, y = b \sin t$, find $\dfrac{d^2y}{dx^2}$.

(*ii*) If $x = a\,(\theta + \sin \theta), y = a(1 + \cos \theta)$, find $\dfrac{d^2y}{dx^2}$ at $\theta = \dfrac{\pi}{2}$.

14. If $x = 2 \cos t - \cos 2t$ and $y = 2 \sin t - \sin 2t$, find the value of $\dfrac{d^2y}{dx^2}$, when $t = \dfrac{\pi}{2}$.

15. (*a*) If $x = \sin t, y = \sin pt$, prove that $(1 - x^2) \dfrac{d^2y}{dx^2} - x \dfrac{dy}{dx} + p^2y = 0$. (*M.D.U. 2005*)

(*b*) If $= \log t$ and $y = \dfrac{1}{t}$ $(t > 0)$, prove that $\dfrac{d^2y}{dx^2} + \dfrac{dy}{dx} = 0$.

16. If $\dfrac{x^2}{a^2} + \dfrac{y^2}{b^2} = 1$, prove that $\dfrac{d^2y}{dx^2} = -\dfrac{b^4}{a^2y^3}$.

17. If $x^{1/2} + y^{1/2} = a^{1/2}$, find the value of $\dfrac{d^2y}{dx^2}$ for $x = a$.

18. If $ax^2 + 2hxy + by^2 = 1$, show that $\dfrac{d^2y}{dx^2} = \dfrac{h^2 - ab}{(hx + by)^3}$.

19. If $y^3 + x^3 - 3\, axy = 0$, show that $\dfrac{d^2y}{dx^2} = -\dfrac{2a^3xy}{(y^2 - ax)^3}$.

20. If $y = e^{-x} \cos x$, show that $\dfrac{d^4y}{dx^4} + 4y = 0$.

ANSWERS

1. *(ii)* 80   *(iv)* $x^3 \sin x - 12x^2 \cos x - 36x \sin x + 24 \cos x$

12. $\dfrac{-1}{2at^3}$.   **13.** *(i)* $\dfrac{-b}{a^2} \operatorname{cosec}^3 t$ *(ii)* $\dfrac{-1}{a}$

14. $\dfrac{-3}{2}$ **17.** $\dfrac{1}{2a}$.

HINTS AND SOLUTIONS

4. *(ii)* $$y = e^{ax} \sin bx \quad \text{...(1)}$$

$\therefore$ $$\frac{dy}{dx} = b\, e^{ax} \cos bx + a\, e^{ax} \sin bx$$

or $$\frac{dy}{dx} = b\, e^{ax} \cos bx + ay \quad \text{[by (1)] ...(2)}$$

$\therefore$ $$\frac{d^2y}{dx^2} = -\, b^2 e^{ax} \sin bx + a\, b\, e^{ax} \cos bx + a\frac{dy}{dx}$$

$$= -\, b^2 y + a\left[\frac{dy}{dx} - ay\right] + a\frac{dy}{dx}. \quad \text{by (1) and by (2)}$$

9. $$y = \frac{\sin^{-1} x}{\sqrt{1-x^2}}$$

$\therefore$ $$y_1 = \frac{\sqrt{1-x^2}\dfrac{d}{dx}\sin^{-1} x - \sin^{-1} x \dfrac{d}{dx}\sqrt{1-x^2}}{(1-x^2)}$$

or $$y_1 = \frac{\sqrt{1-x^2}\dfrac{1}{\sqrt{1-x^2}} - \sin^{-1} x \,.\, \dfrac{1}{2}(1-x^2)^{-1/2}(-2x)}{1-x^2}$$

Cross-multiplying, $(1-x^2)\, y_1 = 1 + x\dfrac{\sin^{-1} x}{\sqrt{1-x^2}} = 1 + xy$

10. *(a)* $$y = \sin(\sin x) \quad \text{...(1)}$$

$\therefore$ $$\frac{dy}{dx} = \cos(\sin x) \,.\, \cos x \quad \text{...(2)}$$

$\therefore$ $$\frac{d^2y}{dx^2} = -\cos(\sin x) \sin x - \cos x \,.\, \sin(\sin x) \,.\, \cos x$$

$$= -\frac{\dfrac{dy}{dx}}{\cos x} \,.\, \sin x - \cos^2 x \,.\, y \quad \text{[by (2) and by (1)}$$

(c) Put $x = \tan\theta$

(d) $$\frac{dy}{dx} = \sec^2 x + \sec x \tan x = \frac{1}{\cos^2 x} + \frac{\sin x}{\cos^2 x} = \frac{1+\sin x}{1-\sin^2 x} = \frac{1}{1-\sin x}.$$

14. $$\frac{dy}{dx} = \frac{\frac{dy}{dt}}{\frac{dx}{dt}} = \frac{2 \cos t - 2 \cos 2t}{-2 \sin t + 2 \sin 2t}$$

$$= \frac{\cos t - \cos 2t}{\sin 2t - \sin t} = \frac{2 \sin \frac{t+2t}{2} \sin \frac{2t-t}{2}}{2 \cos \frac{2t+t}{2} \sin \frac{2t-t}{2}}$$ | C – D Formulae

$$= \frac{2 \sin \frac{3t}{2} \sin \frac{t}{2}}{2 \cos \frac{3t}{2} \sin \frac{t}{2}} = \tan \frac{3t}{2}$$

$$\therefore \quad \frac{d^2y}{dx^2} = \frac{d}{dx}\left(\tan \frac{3t}{2}\right) = \sec^2 \frac{3t}{2} \frac{d}{dx}\left(\frac{3t}{2}\right)$$

$$= \sec^2 \frac{3t}{2} \cdot \frac{3}{2}\frac{dt}{dx} = \frac{3}{2} \sec^2 \frac{3t}{2} \frac{1}{-2 \sin t + 2 \sin 2t}.$$

15. $$\frac{dy}{dx} = \frac{\frac{dy}{dt}}{\frac{dx}{dt}} = \frac{p \cos pt}{\cos t} \quad \text{or} \quad \frac{dy}{dx} = \frac{p\sqrt{1-\sin^2 pt}}{\sqrt{1-\sin^2 t}} = \frac{p\sqrt{1-y^2}}{\sqrt{1-x^2}}$$

Squaring both sides and cross-multiplying,

$$(1-x^2)\left(\frac{dy}{dx}\right)^2 = p^2(1-y^2).$$

Again differentiate w.r.t. x.

16. $$\frac{x^2}{a^2} + \frac{y^2}{b^2} = 1 \qquad ...(1)$$

$$\therefore \quad \frac{d}{dx}\left(\frac{x^2}{a^2}\right) + \frac{d}{dx}\left(\frac{y^2}{b^2}\right) = \frac{d}{dx}(1)$$

or $$\frac{1}{a^2}\frac{d}{dx}(x^2) + \frac{1}{b^2}\frac{d}{dx}(y^2) = 0 \quad \text{or} \quad \frac{2x}{a^2} + \frac{2y}{b^2}\frac{dy}{dx} = 0$$

$$\therefore \quad \frac{2y}{b^2}\frac{dy}{dx} = -\frac{2x}{a^2} \quad \text{or} \quad \frac{dy}{dx} = -\frac{2x}{a^2} \times \frac{b^2}{2y} = -\frac{b^2x}{a^2y} \qquad ...(2)$$

Again differentiating both sides w.r.t. x,

$$\frac{d^2y}{dx^2} = \frac{-b^2}{a^2}\frac{d}{dx}\left(\frac{x}{y}\right) = -\frac{b^2}{a^2}\frac{\left[y \cdot 1 - x\frac{dy}{dx}\right]}{y^2}$$

$$= \frac{-\frac{b^2}{a^2}\left[y + \frac{b^2x^2}{a^2y}\right]}{y^2} \qquad \text{[by (2)]}$$

$$= -\frac{b^2(a^2y^2 + b^2x^2)}{a^4y^3} = \frac{-b^2 \cdot a^2b^2\left(\frac{x^2}{a^2} + \frac{y^2}{b^2}\right)}{a^4y^3}$$

$$= -\frac{b^4}{a^2y^3} \times 1. \qquad \text{[by (1)]}$$

2. We shall now find the derivatives of the nth order of some **Standard Functions** of x.

I. To find the nth differential of x^m.

Let $y = x^m$

then $y_1 = mx^{m-1}$

$y_2 = m(m-1)x^{m-2}$

$y_3 = m(m-1)(m-2)x^{m-3}$

....

....

$\therefore \quad y_n = [m(m-1)(m-2) \ldots\ldots \text{ upto } n \text{ factors}] \times x^{m-n}$

$= m(m-1)(m-2) \ldots\ldots (m-n+1) \cdot x^{m-n}$ where $n < m$.

[**Cor.** If m be a positive integer, and if $n = m$

then $y_m = m(m-1)(m-2) \ldots\ldots (m-m+1)\, x^{m-m}$

$= m(m-1)(m-2) \ldots\ldots 3.2.1. = m\,!$

i.e., $$\frac{d^m}{dx^m}(x^m) = m\,!$$

and $(m+1)$th, $(m+2)$th derivative ....etc., each will be $= 0$.

$\therefore\ y_m = m\,!$ which is constant and $y_{(m+1)}$ which is the derivative of y_m will be zero and so on.]

II. The nth differential co-efficient of $(ax+b)^m$, where $n < m$. *(G.N.D.U. 1976 S)*

Let $y = (ax+b)^m$

then $y_1 = m(ax+b)^{m-1} \cdot a$

$y_2 = m(m-1)(ax+b)^{m-2} \cdot a^2$

$y_3 = m(m-1)(m-2)(ax+b)^{m-3} \cdot a^3$

....

....

$\therefore \quad y_n = m(m-1)(m-2)(m-3) \ldots\ldots (m-n+1) \times (ax+b)^{m-n} \cdot a^n$

[**Cor.** If $n = m$, then

$y_m = m(m-1)(m-2) \ldots\ldots 3.2.1.\,(ax+b)^0 a^m = m\,!\, a^m$]

III. The nth differential co-efficient of $\dfrac{1}{ax+b}\left[x \neq -\dfrac{b}{a}\right]$

Let $$y = \frac{1}{ax+b} = (ax+b)^{-1}$$

then $y_1 = (-1)(ax+b)^{-2} \cdot a$

$y_2 = (-1)(-2)(ax+b)^{-3} \cdot a^2 = (-1)^2 \cdot 2\,!\,(ax+b)^{-3} \cdot a^2$

$y_3 = (-1)(-2)(-3)(ax+b)^{-4} \cdot a^3$

$= (-1)^3 \cdot 3\,!\,(ax+b)^{-4} \cdot a^3$

....

....

$\therefore \quad y_n = (-1)^n \cdot n\,!\,(ax+b)^{-(n+1)} \cdot a^n$

or $$\mathbf{y_n = \frac{(-1)^n n!\, a^n}{(ax+b)^{n+1}}}.$$

IV. The nth differential co-efficient of log (ax + b)

Let $$y = \log(ax+b)$$

then $$y_1 = \frac{1}{ax+b}\, a = (ax+b)^{-1} \,.\, a$$

$$y_2 = (-1)(ax+b)^{-2} \,.\, a^2$$

$$y_3 = (-1)(-2)(ax+b)^{-3} \,.\, a^3 = (-1)^2\, 2!\,(ax+b)^{-3} \,.\, a^3$$

....

....

$\therefore$ $$y_n = (-1)^{n-1}(n-1)!\,(ax+b)^{-n} \,.\, a^n$$

or $$\mathbf{y_n = \frac{(-1)^{n-1}(n-1)!\, a^n}{(ax+b)^n}}$$

$$\left[\text{Cor. If } a = 1; b = 0, y = \log x, \text{ then } y_n = \frac{(-1)^{n-1}(n-1)!}{x^n}.\right]$$

V. The nth differential co-efficient of a^{mx}

Let $$y = a^{mx}$$

$\therefore$ $$y_1 = m \,.\, a^{mx} (\log a)$$

$$y_2 = m^2 \,.\, a^{mx} (\log a)^2$$

$$y_3 = m^3 \,.\, a^{mx} (\log a)^3$$

....

....

$\therefore$ $$\mathbf{y_n = m^n \,.\, a^{mx} (\log a)^n}$$

[**Cor.** Put $a = e$, then $y = \mathbf{e^{mx}}$ and $\mathbf{y_n = m^n e^{mx}}$.]

VI. The nth differential co-efficient of sin (ax + b)

Let $$y = \sin(ax+b)$$

then $$y_1 = a\cos(ax+b)$$

$$= a\sin\left(ax+b+\frac{\pi}{2}\right) \qquad \because \sin\left(\frac{\pi}{2}+\theta\right) = \cos\theta \quad \text{[Note this step]}$$

$$y_2 = a^2\cos\left(ax+b+\frac{\pi}{2}\right) = a^2\sin\left(ax+b+\frac{\pi}{2}+\frac{\pi}{2}\right)$$

$$= a^2\sin\left(ax+b+2\,.\,\frac{\pi}{2}\right)$$

$$y_3 = a^3\cos\left(ax+b+2\,.\,\frac{\pi}{2}\right)$$

$$= a^3\sin\left(ax+b+2\,.\,\frac{\pi}{2}+\frac{\pi}{2}\right) = a^3\sin\left(ax+b+3\,.\,\frac{\pi}{2}\right)$$

....

....

$$\mathbf{y_n = a^n \sin\left(ax + b + n\,.\frac{\pi}{2}\right).}$$

$\left[\text{Cor. If } b = 0,\ y = \sin ax,\ y_n = a^n \sin\left(ax + n\frac{\pi}{2}\right)\right]$

VII. The nth differential co-efficient of cos (ax + b)

Let $\qquad y = \cos(ax + b)$

then $\qquad y_1 = -\sin(ax + b)\,.\,a = a\cos\left(ax + b + \frac{\pi}{2}\right)$ [Note this step]

$$\left[\because \quad \cos\left(\frac{\pi}{2} + \theta\right) = -\sin\theta\right]$$

$$y_2 = -a^2 \sin\left(ax + b + \frac{\pi}{2}\right) = a^2 \cos\left(ax + b + \frac{\pi}{2} + \frac{\pi}{2}\right)$$

$$= a^2 \cos\left(ax + b + 2\,.\frac{\pi}{2}\right)$$

$$y_3 = -a^3 \sin\left(ax + b + 2\,.\frac{\pi}{2}\right) = a^3 \cos\left(ax + b + 2\,.\frac{\pi}{2} + \frac{\pi}{2}\right)$$

$$= a^3 \cos\left(ax + b + 3\,.\frac{\pi}{2}\right)$$

....

....

$$\mathbf{y_n = a^n \cos\left(ax + b + n\frac{\pi}{2}\right).}$$

[**Cor.** If $\quad y = \cos ax$

$$y_n = a^n \cos\left(ax + n\,.\frac{\pi}{2}\right)\Bigg]$$

VIII. The nth differential co-efficient of e^{ax} sin (bx + c)

Let $\qquad y = e^{ax}\sin(bx + c)$

then $\qquad y_1 = e^{ax}\dfrac{d}{dx}[\sin(bx + c)] + \sin(bx + c)\dfrac{d}{dx}[e^{ax}]$

$$= e^{ax}\,.\cos(bx + c)\,.\,b + \sin(bx + c)\,.\,a\,.\,e^{ax}$$

$$= e^{ax}[a\sin(bx + c) + b\cos(bx + c)].$$

We determine two constants r and θ, to change the expression into a single sine which will enable us to make the required generalisation by putting, $\mathbf{a = r\cos\theta,\ b = r\sin\theta}$

$\therefore \qquad r = \sqrt{a^2 + b^2},\ \theta = \tan^{-1}\dfrac{b}{a}.$

Hence $\qquad y_1 = e^{ax}[r\cos\theta\,.\sin(bx + c) + r\sin\theta\,.\cos(bx + c)]$

$$= r\,e^{ax}\sin(bx + c + \theta).$$

Thus y_1 is obtained from y on multiplying it by the constant r and increasing the angle by the constant θ. Repeating the same rule to y_1, we have

$$y_2 = r^2 e^{ax} \sin (bx + c + 2\theta).$$

Similarly, $$y_3 = r^3 e^{ax} \sin (bx + c + 3\theta).$$

Hence, in general, $$y_n = \frac{d^n}{dx^n} [e^{ax} \sin (bx + c)] = r^n . e^{ax} \sin (bx + c + n\theta)$$

Putting values of r and θ,

$$\mathbf{y_n = (a^2 + b^2)^{n/2} e^{ax} . \sin \left(bx + c + n . \tan^{-1} \frac{b}{a} \right).}$$

IX. The nth differential co-efficient of $e^{ax} \cos (bx + c)$

Let $$y = e^{ax} \cos (bx + c)$$

then $$y_1 = a\, e^{ax} \cos (bx + c) - b\, e^{ax} \sin (bx + c)$$
$$= e^{ax} [\, a \cos (bx + c) - b \sin (bx + c)]$$

As in Art. VIII, put $a = r \cos \theta$, and $b = r \sin \theta$.

$$\therefore \quad r = \sqrt{a^2 + b^2}\,, \tan \theta = \frac{b}{a}, \quad i.e., \quad \theta = \tan^{-1} \frac{b}{a}$$

$$\therefore \quad y_1 = e^{ax}\, r\, [\cos \theta . \cos (bx + c) - \sin \theta \sin (bx + c)]$$
$$= r\, e^{ax} [\cos (bx + c + \theta)].$$

Thus y_1 is obtained from y on multiplying it by the constant r and increasing the angle by the constant θ.

Repeating the same rule to y_1, we have

$$y_2 = r^2 e^{ax} \cos (bx + c + 2\theta).$$

Similarly, $$y_3 = r^3 e^{ax} \cos (bx + c + 3\theta).$$

Hence, in general, $$y_n = \frac{d^n}{dx^n} [\, e^{ax} \cos (bx + c)] = r^n e^{ax} . \cos (bx + c + n\theta)$$

Putting values of r and θ,

$$\mathbf{y_n = (a^2 + b^2)^{n/2} e^{ax} . \cos \left(bx + c + n . \tan^{-1} \frac{b}{a} \right).}$$

Example 1. *Find the nth differential co-efficient of log $(ax + x^2)$.*

Sol. Let $$y = \log (ax + x^2) = \log x(a + x)$$
$$= \log x + \log (x + a)$$

Differentiating n times,

$$y_n = \frac{d^n}{dx^n} \log x + \frac{d^n}{dx^n} \log (x + a)$$

$$= \frac{(-1)^{n-1} (n-1)!\,.\,1^n}{x^n} + \frac{(-1)^{n-1} (n-1)!\,.\,1^n}{(x + a)^n} \qquad \text{[By Art. IV]}$$

$$= (-1)^{n-1} (n-1)! \left[\frac{1}{x^n} + \frac{1}{(x + a)^n} \right].$$

Example 2. *If $y = \cos^3 x$, find y_n.*

Sol. We know that $\cos 3x = 4 \cos^3 x - 3 \cos x$

$$\therefore \qquad \cos^3 x = \frac{1}{4} [\cos 3x + 3 \cos x]$$

i.e.,
$$y = \frac{1}{4} [\cos 3x + 3 \cos x]$$

Differentiating n times,

$$y_n = \frac{1}{4}\left[\frac{d^n}{dx^n} \cos 3x + 3 \frac{d^n}{dx^n} \cos x\right]$$

$$= \frac{1}{4}\left[3^n \cos\left(3x + \frac{n\pi}{2}\right) + 3.1^n \cos\left(x + \frac{n\pi}{2}\right)\right] \qquad \text{[By Art. VII]}$$

$$= \frac{1}{4}\left[3^n \cos\left(3x + \frac{n\pi}{2}\right) + 3 \cos\left(x + \frac{n\pi}{2}\right)\right].$$

Example 3. *If $y = \sin x \sin 3x$, find y_n.*

Sol.
$$y = \sin x \sin 3x$$

$$= \frac{1}{2} [2 \sin 3x \sin x] = \frac{1}{2} [\cos 2x - \cos 4x]$$

$$[\because \quad 2 \sin A \sin B = \cos (A - B) - \cos (A + B)]$$

$$\therefore \qquad y_n = \frac{1}{2}\left[\frac{d^n}{dx^n} \cos 2x - \frac{d^n}{dx^n} \cos 4x\right]$$

$$= \frac{1}{2}\left[2^n \cos\left(2x + \frac{n\pi}{2}\right) - 4^n \cos\left(4x + \frac{n\pi}{2}\right)\right].$$

Example 4. *Find nth derivative of $\sin^2 x \cos^3 x$.*

Sol. Let
$$y = \sin^2 x \cos^3 x = \sin^2 x \cos^2 x \,.\, \cos x$$

$$= (\sin x \cos x)^2 \cos x = \frac{1}{4} (2 \sin x \cos x)^2 \cos x$$

$$= \frac{1}{4} (\sin 2x)^2 \cos x = \frac{1}{4} \sin^2 2x \cos x$$

$$= \frac{1}{4}\left(\frac{1 - \cos 4x}{2}\right) \cos x \qquad \left|\because \sin^2 \theta = \frac{1 - \cos 2\theta}{2}\right.$$

$$= \frac{1}{8} (\cos x - \cos 4x \cos x) = \frac{1}{16} (2 \cos x - 2 \cos 4x \cos x)$$

or
$$y = \frac{1}{16} [2 \cos x - (\cos 5x + \cos 3x)]$$

$$[\because \quad 2 \cos A \cos B = \cos (A + B) + \cos (A - B)]$$

or
$$y = \frac{1}{16} [2 \cos x - \cos 5x - \cos 3x]$$

$$\therefore \qquad y_n = \frac{1}{16}\left[2.1^n \cos\left(x + \frac{n\pi}{2}\right) - 5^n \cos\left(5x + \frac{n\pi}{2}\right) - 3^n \cos\left(3x + \frac{n\pi}{2}\right)\right]$$

$$= \frac{1}{16}\left[2\cos\left(x+\frac{n\pi}{2}\right) - 5^n\cos\left(5x+\frac{n\pi}{2}\right) - 3^n\cos\left(3x+\frac{n\pi}{2}\right)\right]$$

Example 5. *Find the nth differential co-efficient of $e^x \sin^3 x$.*

Sol. We know that, $\sin 3x = 3\sin x - 4\sin^3 x$

$\therefore \quad \sin^3 x = \frac{1}{4}[3\sin x - \sin 3x]$ ...(1)

Let $\quad y = e^x \sin^3 x = e^x . \frac{1}{4}[3\sin x - \sin 3x]$ | From (1)

$= \frac{3}{4}e^x \sin x - \frac{1}{4} . e^x \sin 3x$

$\therefore \quad y_n = \frac{3}{4}(1^2 + 1^2)^{n/2} . e^x \sin(x + n\tan^{-1}\frac{1}{1})$

$- \frac{1}{4} . (1^2 + 3^2)^{n/2} e^x \sin[3x + n\tan^{-1}\frac{3}{1}]$

| Comparing to result in Art. (VIII)

$$= \frac{3}{4} . 2^{n/2} e^x \sin\left(x + n\frac{\pi}{4}\right) - \frac{1}{4} . 10^{n/2} . e^x \sin(3x + n\tan^{-1} 3).$$

Example 6. *If $y = \sqrt{x+a}$, find y_n.*

Sol. Here, $\quad y = (x+a)^{1/2}$

$\therefore \quad y_1 = \frac{1}{2}(x+a)^{-1/2}$

$y_2 = (\frac{1}{2})(-\frac{1}{2})(x+a)^{-3/2}$

$y_3 = \frac{1}{2}(-\frac{1}{2})(-\frac{3}{2})(x+a)^{-5/2} = (-1)^2 \frac{1.3}{2^3}(x+a)^{-5/2}$

....

....

$$\therefore \quad y_n = (-1)^{n-1}\frac{1.3.5 \ldots \text{upto } (n-1) \text{ terms}}{2^n}(x+a)^{-\left(\frac{2n-1}{2}\right)}$$

$$= (-1)^{n-1}\frac{1.3 \ldots (2n-3)}{2^n}(x+a)^{-\left(\frac{2n-1}{2}\right)}.$$

Remark. If none of the formulae of the IX Articles is applicable to find y_n in a problem, then we proceed as in the articles, *i.e.*, find y_1, y_2, y_3 and then generalise as done in the above example 6.

EXERCISE 2

Find the nth derivative of the following :

1. (*i*) 4^x (*ii*) 5^x (*P.U.*) (*iii*) $e^{2x} + e^{-2x}$ (*iv*) $\log 3x$.

2. (*i*) $\frac{1}{(3-2x)^3}$ (*ii*) $\frac{1}{(bx+a)^2}$.

3. (*i*) $\log\sqrt{\frac{2x+1}{x-2}}$ (*ii*) $\log\sqrt{\frac{3-2x}{5+4x}}$ (*iii*) $\log(x^2 - a^2)$ (*iv*) $\frac{3x+7}{x+2}$.

4. (*i*) $\sin^2 x$ (*P.U.*) (*ii*) $\cos^2 x$ (*iii*) $\sin^3 x$ (*iv*) $\cos^4 x$.

[**Hint.** (*iv*) $\cos^4 x = (\cos^2 x)^2 = \left(\frac{1+\cos 2x}{2}\right)^2 = \frac{1}{4} + \frac{\cos 2x}{2} + \frac{\cos^2 2x}{4}$

$$= \frac{1}{4} + \frac{1}{2}\cos 2x + \frac{1}{8}(1+\cos 4x) = \frac{3}{8} + \frac{1}{2}\cos 2x + \frac{1}{8}\cos 4x.]$$

(*v*) $\sin^4 x$.

5. (*i*) $\sin x \cos 2x$ (*ii*) $\cos x \cos 2x$.

6. (*i*) $\cos ax \sin bx$ (*ii*) $\sin x \sin 2x$ (*iii*) $\sin^3 x \cos^2 x$.

7. (*i*) $e^x \cos x$ (*ii*) $e^x \sin x$

(*iii*) $e^x \sin x \cos x$ (*iv*) $e^{3x} \sin^2 2x$.

8. (*i*) $e^{ax} \cos^2 x \sin x$ (*ii*) $e^x \cos^3 x$ (*iii*) $e^x \sin^2 x \sin 2x$

(*iv*) $a^x \cos x$. [**Hint.** $a^x = e^{\log a^x} = e^{x \log a}$]

9. If $\mu = \sin nx + \cos nx$, prove that $\mu_r = n^r [1 + (-1)^r \sin 2nx]^{1/2}$ where μ_r denotes the rth differential co-efficient of μ w.r.t. x.

[**Hint.**
$$\mu_r = n^r \sin\left(nx + \frac{r\pi}{2}\right) + n^r \cos\left(nx + \frac{r\pi}{2}\right)$$
$$= n^r\left[\sin\left(nx + \frac{r\pi}{2}\right) + \cos\left(nx + \frac{r\pi}{2}\right)\right]$$
$$= n^r\sqrt{\left[\sin\left(nx + \frac{r\pi}{2}\right) + \cos\left(nx + \frac{r\pi}{2}\right)\right]^2}$$
$$= n^r\sqrt{\sin^2\left(nx + \frac{r\pi}{2}\right) + \cos^2\left(nx + \frac{r\pi}{2}\right) + 2\sin\left(nx + \frac{r\pi}{2}\right)\cos\left(nx + \frac{r\pi}{2}\right)}$$
$$= n^r\sqrt{1 + \sin 2\left(nx + \frac{r\pi}{2}\right)} \qquad \because\ 2\sin\theta\cos\theta = \sin 2\theta$$
$$= n^r\sqrt{1 + \sin(r\pi + 2nx)} = n^r\sqrt{1 + (-1)^r \sin 2nx}\,\Big].$$

ANSWERS

1. (*i*) $4^x [\log 4]^n$ (*ii*) $[\log 5]^n . 5^x$

(*iii*) $2^n[e^{2x} + (-1)^n e^{-2x}]$ (*iv*) $\frac{[-1]^{n-1}(n-1)!}{x^n}$.

2. (*i*) $\frac{(n+2)!\,.\,2^{n-1}}{[3-2x]^{n+3}}$ (*ii*) $\frac{[-1]^n (n+1)!\,.\,b^n}{[bx+a]^{n+2}}$.

3. (*i*) $\frac{[-1]^{n-1}(n-1)!}{2}\left[\frac{2^n}{[2x+1]^n} - \frac{1}{(x-2)^n}\right]$ (*ii*) $\frac{1}{2}(-1)^{n-1}(n-1)!\left[\frac{[-2]^n}{[3-2x]^n} - \frac{4^n}{[5+4x]^n}\right]$

(*iii*) $(-1)^{n-1}(n-1)!\left[\frac{1}{(x+a)^n} + \frac{1}{(x-a)^n}\right]$ (*iv*) $\frac{(-1)^n n!}{(x+2)^{n+1}}$.

4. (*i*) $-2^{n-1}\cos[2x + n\,.\,\pi/2]$ (*ii*) $2^{n-1}\cos[2x + n\,.\,\pi/2]$

(*iii*) $\frac{3}{4}\sin[x + n\,.\,\pi/2] - \frac{1}{4}\,.\,3^n \sin[3x + n\,.\,\pi/2]$

(*iv*) $2^{n-1}\cos[2x + n\,.\,\pi/2] + 2^{2n-3}\cos[4x + n\,.\,\pi/2]$

(*v*) $-2^{n-1}\cos[2x + n\,.\,\pi/2] + 2^{2n-3}\cos[4x + n\,.\,\pi/2]$.

5. (*i*) $\frac{1}{2}[3^n\sin(3x + n\,.\,\pi/2) - \sin(x + n\,.\,\pi/2)]$ (*ii*) $\frac{1}{2}\left[3^n\cos\left(3x + n\frac{\pi}{2}\right) + \cos\left(x + n\frac{\pi}{2}\right)\right]$.

6. (*i*) $\frac{1}{2}\left[(a+b)^n\sin\left\{(a+b)x + \frac{n\pi}{2}\right\} + (b-a)^n\sin\left\{(b-a)x + \frac{n\pi}{2}\right\}\right]$

(*ii*) $\frac{1}{2}[\cos(x + n\,.\,\pi/2) - 3^n\cos(3x + n\,.\,\pi/2)]$

(*iii*) $\frac{1}{16}\left[2\sin\left(x + \frac{n\pi}{2}\right) - 5^n\sin\left(5x + \frac{n\pi}{2}\right) + 3^n\sin\left(3x + \frac{n\pi}{2}\right)\right]$

7. (*i*) $2^{n/2}\,.\,e^x\cos[x + n\,.\,\pi/4]$ (*ii*) $2^{n/2}\,.\,e^x\sin[x + n\,.\,\pi/4]$

(*iii*) $\frac{1}{2}\,.\,5^{n/2}e^x\sin(2x + n\tan^{-1}2)$ (*iv*) $\frac{1}{2}\left[3^ne^{3x} - 5^ne^{3x}\cos\left(4x + n\tan^{-1}\frac{4}{3}\right)\right]$.

8. (*i*) $\frac{1}{4}[(a^2+1)^{n/2}e^{ax}\sin(x + n\tan^{-1}1/a) + (a^2+9)^{n/2}e^{ax}\sin(3x + n\tan^{-1}3/a)]$

(*ii*) $\frac{3}{4}\,.\,2^{n/2}\,.\,e^x\cos[x + n\,.\,\pi/4] + \frac{1}{4}\,.\,10^{n/2}e^x\cos[3x + n\tan^{-1}3]$

(*iii*) $\frac{1}{2}\,.\,5^{n/2}\,.\,e^x\sin[2x + n\tan^{-1}2] - \frac{1}{4}[17]^{n/2}\,.\,e^x\sin[4x + n\tan^{-1}4]$

(*iv*) $a^x[(1 + \log a)^2]^{n/2}\,.\,\cos[x + n\tan^{-1}1/\log a]$.

---

3. Use of Partial Fractions

In order to determine the *n*th derivative, of any rational function, we have to split it into partial fractions.

Forming partial fractions is converse process of taking L.C.M.

To resolve a fraction into partial fractions, the degree of numerator must be less than the degree of denominator.

Partial fractions for :

(*i*) $\frac{f(x)}{(x-a)(x-b)(x-c)}$ are $\frac{A}{x-a} + \frac{B}{x-b} + \frac{C}{x-c}$

(*ii*) $\frac{f(x)}{(x-a)^2(x-b)}$ are $\frac{A}{x-a} + \frac{B}{(x-a)^2} + \frac{C}{x-b}$

(*iii*) $\frac{f(x)}{(x-a)^3(x-b)}$ are $\frac{A}{x-a} + \frac{B}{(x-a)^2} + \frac{C}{(x-a)^3} + \frac{D}{x-b}$

(*iv*) $\frac{f(x)}{(x-a)(x-b)(px^2+qx+r)} = \frac{A}{x-a} + \frac{B}{x-b} + \frac{Cx+D}{px^2+qx+r}$.

To find A, B, C, D, etc., we put each linear factor of L.C.M. equal to zero. The remaining constants are obtained by comparing coefficients of like powers of x on both sides.

Example 1. *Find nth derivative of* $\frac{1}{x^2 - a^2}$.

Sol. Let $y = \dfrac{1}{x^2 - a^2} = \dfrac{1}{(x-a)(x+a)} = \dfrac{A}{x-a} + \dfrac{B}{x+a}$ ...(1)

Multiplying both sides by L.C.M. = $(x-a)(x+a)$,

$$1 = A(x+a) + B(x-a).$$

Put $x - a = 0$, *i.e.*, $x = a$,

$\therefore$ $1 = A(a+a) + B(a-a)$ or $1 = 2aA$ $\therefore$ $A = \dfrac{1}{2a}$.

Put $x + a = 0$, *i.e.*, $x = -a$.

$\therefore$ $1 = A(-a+a) + B(-a-a)$, or $1 = -2aB$ $\therefore$ $B = \dfrac{-1}{2a}$.

Putting values of A and B in (1),

$$y = \frac{\frac{1}{2a}}{x-a} - \frac{\frac{1}{2a}}{x+a} = \frac{1}{2a}\left[\frac{1}{x-a} - \frac{1}{x+a}\right]$$

$$\therefore \quad y_n = \frac{1}{2a}\left[\frac{d^n}{dx^n}\frac{1}{x-a} - \frac{d^n}{dx^n}\frac{1}{x+a}\right]$$

$$= \frac{1}{2a}\left[\frac{(-1)^n n!\,1^n}{(x-a)^{n+1}} - \frac{(-1)^n n!\,.\,1^n}{(x+a)^{n+1}}\right]$$

$$\left(\because \text{ By Article III, } \frac{d^n}{dx^n}\frac{1}{ax+b} = \frac{(-1)^n n!\,a^n}{(ax+b)^{n+1}}\right)$$

$$= \frac{1}{2a}(-1)^n n!\left[\frac{1}{(x-a)^{n+1}} - \frac{1}{(x+a)^{n+1}}\right].$$

Example 2. *Find nth derivative of* $\dfrac{x^4}{(x-1)(x-2)}$.

Sol. Let $y = \dfrac{x^4}{(x-1)(x-2)}$

Dividing numerator namely x^4 by Denom.

$$(x-1)(x-2) = x^2 - 3x + 2.$$

$$
\begin{array}{l}
x^2 - 3x + 2)\ x^4 \quad (x^2 + 3x + 7 \\
\qquad\qquad x^4 - 3x^3 + 2x^2 \\
\qquad\qquad - \quad + \quad - \\
\hline
\qquad\qquad\quad 3x^3 - 2x^2 \\
\qquad\qquad\quad 3x^3 - 9x^2 + 6x \\
\qquad\qquad\quad - \quad + \quad - \\
\hline
\qquad\qquad\qquad 7x^2 - 6x \\
\qquad\qquad\qquad 7x^2 - 21x + 14 \\
\qquad\qquad\qquad - \quad + \quad - \\
\hline
\qquad\qquad\qquad\quad 15x - 14
\end{array}
$$

$$\therefore \quad y = \frac{x^4}{(x-1)(x-2)} = x^2 + 3x + 7 + \frac{A}{x-1} + \frac{B}{x-2} \quad ...(1)$$

Multiplying by L.C.M. $= (x-1)(x-2)$

$$x^4 = (x^2 + 3x + 7)(x-1)(x-2) + A(x-2) + B(x-1).$$

Put x = 1 $\quad 1 = A(1-2) = -A \quad$ or $\quad A = -1$

Put x = 2 $\quad 16 = B(2-1) = B.$

Putting values of A and B in (1), we get

$$y = \frac{x^4}{(x-1)(x-2)} = x^2 + 3x + 7 - \frac{1}{x-1} + \frac{16}{x-2}.$$

$$\therefore \quad y_n = \frac{d^n}{dx^n}(x^2 + 3x + 7) - \frac{d^n}{dx^n}\left(\frac{1}{x-1}\right) + 16\frac{d^n}{dx^n}\left(\frac{1}{x-2}\right)$$

or $$y_n\ (n \geq 3) = 0 - \frac{(-1)^n n!\, 1^n}{(x-1)^{n+1}} + 16\frac{(-1)^n n!\, 1^n}{(x-2)^{n+1}}$$

[$\because$ Third and higher order derivatives of $x^2 + 3x + 7$ are zero.]

$$= (-1)^n n! \left[\frac{-1}{(x-1)^{n+1}} + \frac{16}{(x-2)^{n+1}}\right].$$

Example 3. *Prove that the value of the nth differential co-efficient of* $\frac{x^3}{x^2-1}$ *for x = 0 is zero, if n is even ; and is* – (*n* !) *if n is odd and greater than 1.*

Sol. Let $$y = \frac{x^3}{x^2-1} = \frac{x^3}{(x-1)(x+1)}$$

$$x^2 - 1 \overline{)\ x^3\ } (x$$
$$x^3 - x$$
$$-\ \ +$$
$$x$$

or $$y = \frac{x^3}{(x-1)(x+1)} = x + \frac{A}{x-1} + \frac{B}{x+1} \quad ...(1)$$

Multiplying by L.C.M. $= (x-1)(x+1)$

$$x^3 = x(x-1)(x+1) + A(x+1) + B(x-1)$$

Put $x = 1$, $\quad 1 = A(2) \quad$ or $\quad A = \frac{1}{2}$

Put $x = -1$, $\quad -1 = B(-2) \quad$ or $\quad B = \frac{1}{2}.$

Putting values of A and B in (1),

$$y = x + \frac{1}{2}\left[\frac{1}{x+1} + \frac{1}{x-1}\right].$$

If $n > 1$, the second and higher derivatives of x are 0.

$\therefore$ For $n > 1$, $$y_n = \frac{1}{2}\left[\frac{(-1)^n n!}{(x+1)^{n+1}} + \frac{(-1)^n n!}{(x-1)^{n+1}}\right]$$

$$= \frac{1}{2}(-1)^n n!\left[\frac{1}{(x+1)^{n+1}} + \frac{1}{(x-1)^{n+1}}\right] \qquad ...(1)$$

Case I. When n is even, putting $x = 0$ in (1),

$$(y_n)_0 = \frac{(+1)n!}{2}\left[\frac{1}{1} + \frac{1}{(-1)}\right] = \frac{n!}{2}[1-1] = 0.$$

Case II. When n is odd,

$$(y_n)_0 = \frac{(-1)n!}{2}\left[\frac{1}{1} + \frac{1}{(+1)}\right] = -\frac{(n!)}{2}[1+1] = -n!.$$

EXERCISE 3

Find the nth derivative of the following :

1. (i) $\frac{1}{(x+2)(x-1)}$ (ii) $\frac{x}{1+3x+2x^2}$. *(M.D.U. 1999)*

2. (i) $\frac{x^2}{(x+2)(2x+3)}$ *(M.D.U. 2000)* (ii) $\frac{5x+12}{x^2+5x+6}$ (iii) $\frac{x+1}{6x^2-7x-3}$.

3. If $y = \frac{x+1}{x^2-4}$, $x \neq \pm 2$, prove that $y_n = \frac{(-1)^n n!}{4}\left[\frac{3}{(x-2)^{n+1}} + \frac{1}{(x+2)^{n+1}}\right]$.

4. (i) $\frac{1}{(x^2-a^2)(x^2-b^2)}$ (ii) $\frac{x^2}{(x-1)^3(x+1)}$.

ANSWERS

1. (i) $\frac{[-1]^n . n!}{3}\left[\frac{1}{(x-1)^{n+1}} - \frac{1}{(x+2)^{n+1}}\right]$ (ii) $(-1)^n n!\left[\frac{1}{(x+1)^{n+1}} - \frac{2^n}{(2x+1)^{n+1}}\right]$

2. (i) $\frac{[-1]^n n!}{2}\left[\frac{9.2^n}{(2x+3)^{n+1}} - \frac{8}{(x+2)^{n+1}}\right]$ (ii) $[-1]^n n!\left[\frac{2}{(x+2)^{n+1}} + \frac{3}{(x+3)^{n+1}}\right]$

(iii) $\frac{[-1]^n n!}{11}\left[\frac{2^n . 5}{(2x-3)^{n+1}} - \frac{3^n . 2}{(3x+1)^{n+1}}\right]$.

4. (i) $\frac{[-1]^n n!}{a^2-b^2}\left[\frac{1}{2a}\left\{\frac{1}{(x-a)^{n+1}} - \frac{1}{(x+a)^{n+1}}\right\} - \frac{1}{2b}\left\{\frac{1}{(x-b)^{n+1}} - \frac{1}{(x+b)^{n+1}}\right\}\right]$

(ii) $[-1]^n n!\left[\frac{(n+2)(n+1)}{4(x-1)^{n+3}} + \frac{3(n+1)}{4(x-1)^{n+2}} + \frac{1}{8(x-1)^{n+1}} - \frac{1}{8(x+1)^{n+1}}\right]$.

---

4. Leibnitz's Theorem

This theorem helps us to find the nth differential co-efficient of the product of two functions in terms of the successive derivatives of the functions.

Statement. If u, v be two functions of x, having derivatives of the nth order, then

$$\frac{d^n}{dx^n}(uv) = u_n v + {}^nC_1 u_{n-1} v_1 + {}^nC_2 u_{n-2} v_2 + \ldots\ldots + {}^nC_r u_{n-r} v_r + \ldots + {}^nC_n uv_n,$$

where suffixes of u and v denote differentiations w.r.t. x. (*M.D.U. 2002, 1997*)

Proof. We shall prove the theorem by Mathematical Induction.

Step I. Let $y = uv$

By actual differentiation, we have

$$y_1 = u_1 v + uv_1$$

and

$$y_2 = u_2 v + u_1 v_1 + u_1 v_1 + uv_2$$
$$= u_2 v + 2u_1 v_1 + uv_2 = u_2 v + {}^2C_1 u_1 v_1 + {}^2C_2 uv_2.$$

Thus, the theorem is true for $n = 1, 2$.

Step II. Let us assume that the theorem is true for a particular value of n, say m, so that, we have

$$y_m = u_m v + {}^mC_1 u_{m-1} v_1 + {}^mC_2 u_{m-2} v_2 + \ldots\ldots\ldots$$
$$+ {}^mC_{r-1} u_{m-r+1} v_{r-1} + {}^mC_r u_{m-r} v_r + \ldots\ldots\ldots + {}^mC_m uv_m.$$

Step III. Differentiating both sides, we have

$$y_{m+1} = u_{m+1} v + u_m v_1 + {}^mC_1 u_m v_1 + {}^mC_1 u_{m-1} v_2$$
$$+ {}^mC_2 u_{m-1} v_2 + {}^mC_2 u_{m-2} v_3 + \ldots\ldots$$
$$+ {}^mC_{r-1} u_{m-r+2} v_{r-1} + {}^mC_{r-1} u_{m-r+1} v_r$$
$$+ {}^mC_r u_{m-r+1} v_r + {}^mC_r u_{m-r} v_{r+1} + \ldots\ldots\ldots$$
$$+ {}^mC_m u_1 v_m + {}^mC_m uv_{m+1}$$
$$= u_{m+1} v + ({}^mC_1 + 1) v_m v_1 + ({}^mC_2 + {}^mC_1) u_{m-1} v_2 + \ldots\ldots$$
$$+ ({}^mC_r + {}^mC_{r-1}) u_{m-r+1} v_r + \ldots\ldots + {}^mC_m u\, v_{m+1}.$$

But, we know that

$${}^mC_{r-1} + {}^mC_r = {}^{m+1}C_r$$

Putting $r = 1, 2, 3, \ldots\ldots\ldots$

$${}^mC_0 + {}^mC_1 = {}^{m+1}C_1 \quad \text{or} \quad 1 + {}^mC_1 = {}^{m+1}C_1$$
$${}^mC_1 + {}^mC_2 = {}^{m+1}C_2$$

....

and

$${}^mC_m = 1 = {}^{m+1}C_{m+1}.$$

$\therefore$

$$y_{m+1} = u_{m+1} v + {}^{m+1}C_1 u_m v_1 + {}^{m+1}C_2 u_{m-1} v_2 + \ldots\ldots$$
$$+ {}^{m+1}C_r u_{m-r+1} v_r + \ldots\ldots\ldots + {}^{m+1}C_{m+1} uv_{m+1}$$

which is of exactly the same form as the theorem to be proved with $n = m + 1$.

$\therefore$ If the theorem is true for $n = m$, then it is also true for the next higher value $n = m + 1$.

But in Step I, we have proved that the theorem is true for $n = 1$ and $n = 2$.

$\therefore$ it must be true for next higher value $n = 2 + 1 = 3$.

Again $\because$ the theorem is true for $n = 3$.

$\therefore$ it must be true for next higher value $n = 3 + 1$, *i.e.*, 4 and so on.

Hence the theorem is true for any positive integer n.

Example 1. *Find the nth derivative of $x^2 \sin x$.*

Sol. Let $u = \sin x$ and $v = x^2$

$\therefore$

$$u_n = \sin\left(x + n\frac{\pi}{2}\right) \qquad v_1 = 2x$$

$$u_{n-1} = \sin\left[x + (n-1)\frac{\pi}{2}\right] \qquad v_2 = 2$$

$$u_{n-2} = \sin\left[x + (n-2)\frac{\pi}{2}\right] \qquad v_3 = 0$$

Now by Leibnitz's Theorem, we have

$$\frac{d^n}{dx^n}(uv) = u_n\, v + {}^nC_1\, u_{n-1}\, v_1 + {}^nC_2\, u_{n-2} v_2$$

or

$$\frac{d^n}{dx^n}(x^2 \sin x) = \sin\left(x + n\frac{\pi}{2}\right) . x^2 + {}^nC_1 \sin\left[x + (n-1)\frac{\pi}{2}\right] 2x + {}^nC_2 \sin\left[x + (n-2)\frac{\pi}{2}\right] 2$$

$$= x^2 \sin\left(x + \frac{n\pi}{2}\right) + 2nx \sin\left[x + (n-1)\frac{\pi}{2}\right] + n(n-1) \sin\left[x + (n-2)\frac{\pi}{2}\right].$$

Note. Generally, we take x^n as v.

Example 2. *Find the nth derivative of $e^x \log x$.*

Sol. Let $u = e^x$ and $v = \log x$

$\therefore$

$$u_n = e^x \qquad v_1 = \frac{1}{x}$$

$$u_{n-1} = u_{n-2} = \ldots\ldots. \qquad v_2 = \frac{-1}{x^2}$$

$$= u_2 = u_1 = e^x \qquad v_3 = \frac{2}{x^3}$$

....

$$v_n = \frac{(-1)^{n-1}(n-1)!}{x^n}$$

Now by Leibnitz's Theorem, we have

$$\frac{d^n}{dx^n}(e^x \log x) = e^x \log x + {}^nC_1\, e^x \left(\frac{1}{x}\right) + {}^nC_2\, e^x \left(\frac{-1}{x^2}\right) + \ldots.$$

$$\ldots + {}^nC_n \,.\, e^x \,.\, \frac{(-1)^{n-1}(n-1)!}{x^n}$$

$$= e^x \left[\log x + \frac{n}{x} - \frac{n(n-1)}{2} . \frac{1}{x^2} + \ldots + \frac{(-1)^{n-1}(n-1)!}{x^n}\right].$$

Example 3. *If* $y = x^2 e^x$, *show that*

$$\frac{d^n y}{dx^n} = \frac{1}{2} n(n-1) \frac{d^2 y}{dx^2} - n(n-2) \frac{dy}{dx} + \frac{1}{2} (n-1)(n-2) y.$$

Sol. $y = x^2 e^x = e^x \cdot x^2$

Differentiating n times by Leibnitz's Theorem, we get

$$y_n = u_n v + {}^nC_1 u_{n-1} v_1 + {}^nC_2 u_{n-2} v_2$$

or $$y_n = e^x \cdot x^2 + {}^nC_1 e^x \cdot 2x + {}^nC_2 e^x \cdot 2$$

$$= e^x[x^2 + 2nx + n(n-1)] \quad ...(1)$$

Put $n = 1, 2$ in (1), successively

$$y_1 = e^x[x^2 + 2x] \quad ...(2)$$

$$y_2 = e^x[x^2 + 4x + 2] \quad ...(3)$$

Now R.H.S. $= \frac{1}{2} n(n-1)\frac{d^2y}{dx^2} - n(n-2)\frac{dy}{dx} + \frac{1}{2}(n-1)(n-2)y$

$$= \frac{1}{2} n(n-1) \cdot e^x [x^2 + 4x + 2] - n(n-2) \cdot e^x (x^2 + 2x) + \frac{1}{2}(n-1)(n-2)x^2e^x$$

$$= e^x \left[x^2 \left\{ \frac{n(n-1)}{2} - n(n-2) + \frac{(n-1)(n-2)}{2} \right\} + x \{2n(n-1) - 2n(n-2)\} + n(n-1) \right]$$

$$= e^x [x^2 + 2nx + n(n-1)]$$

$= y_n$ from (1).

Example 4. *If* $y = \sin (m \sin^{-1} x)$, *then prove that*

$$(1 - x^2)y_2 - xy_1 + m^2y = 0$$

and $$(1 - x^2)y_{n+2} = (2n + 1) xy_{n+1} + (n^2 - m^2) y_n.$$

(M.D.U. 1999 ; K.U. 1999, 1996 A)

Sol. $y = \sin (m \sin^{-1}x) \quad ...(1)$

$\therefore$ $$y_1 = \cos (m \sin^{-1}x) \cdot \frac{m}{\sqrt{1-x^2}}$$

Cross-multiplying $\sqrt{1-x^2}\, y_1 = m \cos (m \sin^{-1} x)$.

Squaring both sides,

$$(1 - x^2)y_1^2 = m^2 \cos^2 (m \sin^{-1} x)$$

$$= m^2[1 - \sin^2 (m \sin^{-1} x)] \quad \left| \because \cos^2\theta = 1 - \sin^2\theta \right.$$

$$= m^2 (1 - y^2) \quad \text{[By (I)]}$$

Again differentiating both sides w.r.t. x,

$$(1 - x^2) \frac{d}{dx} y_1^2 + y_1^2 \frac{d}{dx} (1 - x^2) = m^2 \left[0 - \frac{d}{dx} y^2 \right]$$

or $$(1 - x^2)\, 2y_1y_2 - 2xy_1^2 = -2m^2yy_1$$

Dividing every term by $2y_1$,

$$(1 - x^2)y_2 - xy_1 = -m^2y$$

or $$(1 - x^2)y_2 - xy_1 + m^2y = 0.$$

Now differentiating every term n times by Leibnitz's Theorem, we have

$$(y_2)_n(1-x^2) + {}^nC_1(y_2)_{n-1}(-2x) + {}^nC_2(y_2)_{n-2}(-2) - [(y_1)_n x + {}^nC_1(y_1)_{n-1} \cdot 1] + m^2y_n = 0$$

or $$y_{n+2}(1-x^2) + {}^nC_1 y_{n+1}(-2x) + {}^nC_2 y_n(-2) - y_{n+1} \cdot x - {}^nC_1 y_n + m^2y_n = 0$$

or $$(1-x^2)y_{n+2} - 2nxy_{n+1} - n(n-1)y_n - xy_{n+1} - ny_n + m^2y_n = 0$$

or $$(1-x^2)y_{n+2} - (2n+1)xy_{n+1} - (n^2-m^2)y_n = 0$$

Hence $$(1-x^2)y_{n+2} = (2n+1)xy_{n+1} + (n^2-m^2)y_n.$$

Example 5. *If $y^{1/m} + y^{-1/m} = 2x$, prove that*

$$(x^2-1)y_{n+2} + (2n+1)xy_{n+1} + (n^2-m^2)y_n = 0.$$

(M.D.U. 2003, 2002 ; K.U. 1995 A)

Sol. $$y^{1/m} + y^{-1/m} = 2x$$

Put $$y^{1/m} = z$$

$$\therefore \quad z + \frac{1}{z} = 2x \quad \text{or} \quad z^2 - 2xz + 1 = 0 \qquad \therefore \quad z = \frac{2x \pm \sqrt{4x^2-4}}{2}$$

$$\therefore \quad z = \frac{2x \pm 2\sqrt{x^2-1}}{2} = x \pm \sqrt{x^2-1}$$

i.e., $$y^{1/m} = x \pm \sqrt{x^2-1} \quad \text{or} \quad y = (x \pm \sqrt{x^2-1})^m$$

If $$y = (x + \sqrt{x^2-1})^m, \text{ then}$$

$$y_1 = m(x+\sqrt{x^2-1})^{m-1}\left(1 + \frac{1}{2} \cdot \frac{1}{\sqrt{x^2-1}} \cdot 2x\right)$$

$$= m(x+\sqrt{x^2-1})^{m-1}\left(\frac{x+\sqrt{x^2-1}}{\sqrt{x^2-1}}\right) = \frac{m(x+\sqrt{x^2-1})^m}{\sqrt{x^2-1}} = \frac{my}{\sqrt{x^2-1}}$$

And, similarly, if $y = (x - \sqrt{x^2-1})^m$,

Then $$y_1 = \frac{-my}{\sqrt{x^2-1}}.$$

In either case, squaring, we get

$$y_1^2 = \frac{m^2y^2}{x^2-1} \quad \text{or} \quad (x^2-1)y_1^2 = m^2y^2.$$

Differentiating again,

$$(x^2-1) \cdot 2y_1y_2 + y_1^2 \cdot 2x = m^2 \cdot 2yy_1$$

Dividing both sides by $2y_1$, we have

$$(x^2-1)y_2 + xy_1 - m^2y = 0$$

Differentiating n times by Leibnitz's Theorem,

$$(y_2)_n(x^2-1) + {}^nC_1(y_2)_{n-1}(2x) + {}^nC_2(y_2)_{n-2} \cdot 2 + (y_1)_n x + {}^nC_1(y_1)_{n-1} \cdot 1 - m^2y_n = 0$$

or $$y_{n+2}(x^2-1) + n \cdot y_{n+1} \cdot 2x + \frac{n(n-1)}{2!} \cdot y_n \cdot 2 + [y_{n+1} \cdot x + ny_n \cdot 1] - m^2y_n = 0$$

or $$(x^2 - 1)\,y_{n+2} + (2n+1)\,xy_{n+1} + (n^2 - n + n - m^2)y_n = 0$$

or $$(x^2 - 1)y_{n+2} + (2n+1)\,xy_{n+1} + (n^2 - m^2)y_n = 0.$$

EXERCISE 4

Apply Leibnitz's Theorem to find y_n in the following cases :

1. (*i*) $x^3 e^{ax}$ (*ii*) $x^2 e^x$ (*iii*) $x^3 e^{2x}$

2. (*i*) $x^3 \cos x$ (*ii*) $x^3 \sin ax$

3. (*i*) $x^3 \log x$ (*ii*) $x^2 e^x \cos x$.

4. State Leibnitz's Theorem and hence or otherwise show that if $y = x^2 e^x$, then

$$\frac{d^8y}{dx^8} = 28\frac{d^2y}{dx^2} - 48\frac{dy}{dx} + 21y.$$

5. If $y = x^n \log x$, prove that $y_{n+1} = \dfrac{n!}{x}$. (*M.D.U. 1996*)

[**Hint.** We have $y_1 = x^n \cdot \dfrac{1}{x} + nx^{n-1} \log x$ Multiplying by x, we get

$$y_1 x = x^n + nx^n \log x \quad \text{or} \quad y_1 x = x^n + ny.$$

Now, diff. both sides n times and use $\dfrac{d^n}{dx^n}(x^n) = n!$]

6. If $y = x^2 \sin x$, prove that $\dfrac{d^ny}{dx^n} = (x^2 - n^2 + n)\sin\left(x + \dfrac{n\pi}{2}\right) - 2nx\cos\left(x + \dfrac{n\pi}{2}\right)$.

7. Differentiate n times the equation :

(*i*) $(1 - x^2)\dfrac{d^2y}{dx^2} - x\dfrac{dy}{dx} + a^2y = 0$ (*ii*) $x^2y_2 + xy_1 + y = 0$.

8. If $y = a\cos(\log x) + b\sin(\log x)$, show that :

$$x^2y_{n+2} + (2n+1)\,xy_{n+1} + (n^2+1)\,y_n = 0.$$ (*M.D.U. 2000 A, 2006 A*)

9. If $y = \sin^{-1} x$, prove that $(1 - x^2)y_{n+2} - (2n+1)\,xy_{n+1} - n^2y_n = 0$.

10. If $y = [\log(x + \sqrt{1 + x^2})]^2$, prove that

$$(1 + x^2)y_{n+2} + (2n+1)\,xy_{n+1} + n^2y_n = 0.$$ (*K.U. 2000 A ; M.D.U. 2007*)

11. (*a*) If $y = e^{m\sin^{-1}x}$, prove that :

(*i*) $(1 - x^2)y_2 - xy_1 = m^2y$ (*ii*) $(1 - x^2)y_{n+2} - (2n+1)\,xy_{n+1} - (n^2 + m^2)y_n = 0$.

(*K.U. 1996 S, 1997 S, 1999 S*)

(*b*) If $x = \sin\left(\dfrac{\log y}{a}\right)$, prove that $(1 - x^2)\,y_{n+2} - (2n+1)x\,y_{n+1} - (n^2 + a^2)\,y_n = 0$. (*M.D.U. 2004*)

(*c*) If $y = e^{m\cos^{-1}x}$, show that

$$(1 - x)^2\, y_{n+2} - (2n+1)xy_{n+1} - (n^2 + m^2)y_n = 0.$$ (*K.U. 2006*)

12. If $\cos^{-1}\left(\dfrac{y}{b}\right) = \log\left(\dfrac{x}{n}\right)^n$, prove that $x^2y_{n+2} + (2n+1)x\,y_{n+1} + 2n^2y_n = 0$. (*M.D.U. 2005*)

13. If $y = \dfrac{\sin^{-1}x}{\sqrt{1 - x^2}}$, prove that $(1 - x^2)y_{n+2} - (2n+3)x\,y_{n+1} - (n+1)^2\,y_n = 0$. (*K.U. 2003 ; M.D.U. 2007*)

14. If $y = (x^2 - 1)^n$, prove that $(x^2 - 1)\, y_{n+2} + 2xy_{n+1} - n(n+1)y_n = 0$. (*M.D.U. 2003 S ; K.U. 2004 S*)

15. If $y = e^{\tan^{-1} x}$ (or $x \tan(\log y)$), prove that $(1 + x^2)\, y_{n+2} + [(2n+2)x - 1]y_{n+1} + n(n+1)\, y_n = 0$.

(*M.D.U. 2001*)

ANSWERS

1. (*i*) $e^{ax} . a^{n-3} [a^3x^3 + 3n . a^2x^2 + 3n(n-1)ax + n(n-1)(n-2)]$

(*ii*) $e^x [x^2 + 2nx + n(n-1)]$ (*iii*) $2^{n-3} e^{2x} [8x^3 + 12nx^2 + 6n(n-1)x + n(n-1)(n-2)]$.

2. (*i*) $x^3 \cos\left(x + \frac{n}{2}\pi\right) + 3nx^2 \cos\left(x + \frac{n-1}{2}\pi\right) + 3n(n-1)\, x \cos\left(x + \frac{n-2}{2}\pi\right)$

$+ n(n-1)(n-2) \cos\left(x + \frac{n-3}{2}\pi\right)$.

(*ii*) $a^{n-3}\left[a^3 x^3 \sin\left(ax + \frac{n\pi}{2}\right) + 3na^2x^2 \sin\left\{ax + (n-1)\frac{\pi}{2}\right\} + 3n\,(n-1)\, ax \sin\left\{ax + (n-2).\frac{\pi}{2}\right\}\right.$

$\left. + n(n-1)(n-2) \sin\left\{ax + (n-3)\frac{\pi}{2}\right\}\right]$.

3. (*i*) $\frac{[-1]^{n-1} n!}{x^{n-3}}\left[\frac{1}{n} - \frac{3}{n-1} + \frac{3}{n-2} - \frac{1}{n-3}\right]$.

(*ii*) $e^x\left[2^{n/2} x^2 \cos\left(x + \frac{n\pi}{4}\right) + 2^{\frac{n+1}{2}} nx \cos\left(x + \overline{n-1}\frac{\pi}{4}\right) + 2^{\frac{n-2}{2}} n(n-1) \cos\left(x + \overline{n-2}\frac{\pi}{4}\right)\right]$.

7. (*i*) $[1 - x^2]y_{n+2} - [2n+1]xy_{n+1} - [n^2 - a^2]y_n = 0$

(*ii*) $x^2y_{n+2} + [2n+1]\, xy_{n+1} + [n^2 + 1]y_n = 0$.

HINTS AND SOLUTIONS

3. (*ii*) Let $u = e^x \cos x$ and $v = x^2$.

$$u_n = (a^2 + b^2)^{n/2} e^x \cos\left(x + n \tan^{-1}\frac{b}{a}\right)$$

Here $a = 1$ and $b = 1$

$\therefore$ $a^2 + b^2 = 1 + 1 = 2$ and $\tan^{-1}\frac{b}{a} = \tan^{-1} 1 = \frac{\pi}{4}$.

$\therefore$ $u_n = 2^{n/2} e^x \cos\left(x + \frac{n\pi}{4}\right)$.

5. $y = x^n \log x$ ...(1)

Diff. w.r.t. x, $y_1 = x^n . \frac{1}{x} + n\, x^{n-1} \log x$

Multiplying every term by x,

$y_1 x = x^n + n\, x^n \log x$ or $y_1 x = x^n + n\, y$ [by (1)]

Diff. both sides n times w.r.t. x,

${}^nC_0\, (y_1)_n . x + {}^nC_1\, (y_1)_{n-1} . 1 = n! + n\, y_n$ $\left(\because \frac{d^n}{dx^n}(x^n) = n!\right)$

or $\qquad y_{n+1}x + n\,y_n = n\,! + n\,y_n \quad$ or $\quad y_{n+1}x = n\,!$

$\therefore \qquad y_{n+1} = \dfrac{n!}{x}.$

6. $$y_n = x^2\,\frac{d^n}{dx^n}(\sin x) + {}^nC_1\,.\,2x\,\frac{d^{n-1}}{dx^{n-1}}(\sin x) + {}^nC_2\,.\,2.\,\frac{d^{n-2}}{dx^{n-2}}(\sin x) \qquad ...(1)$$

Now $\quad \dfrac{d^n}{dx^n}(\sin x) = 1^n \sin\left(\dfrac{n\pi}{2} + x\right) = \sin\left(\dfrac{n\pi}{2} + x\right) \qquad ...(2)$

Changing n to $(n-1)$, $\quad \dfrac{d^{n-1}}{dx^{n-1}}(\sin x) = \sin\left((n-1)\dfrac{\pi}{2} + x\right)$

$$= \sin\left(\frac{n\pi}{2} - \frac{\pi}{2} + x\right) = \sin -\left(\frac{\pi}{2} - \frac{n\pi}{2} - x\right) = -\sin\left[\frac{\pi}{2} - \left(\frac{n\pi}{2} + x\right)\right] = -\cos\left(\frac{n\pi}{2} + x\right)$$

Changing n to $(n-2)$ in (2)

$$\frac{d^{n-2}}{dx^{n-2}}(\sin x) = \sin\left[(n-2)\frac{\pi}{2} + x\right]$$

$$= \sin\left(\frac{n\pi}{2} - \pi + x\right) = \sin\left[-\left(\pi - \frac{n\pi}{2} + x\right)\right] = -\sin\left(\frac{n\pi}{2} + x\right).$$

11. (*b*) $\qquad x = \sin\left(\dfrac{\log y}{a}\right) \quad \Rightarrow \quad \sin^{-1}x = \dfrac{\log y}{a}$

$\Rightarrow \qquad \log y = a\sin^{-1}x \quad \Rightarrow \quad y = e^{a\sin^{-1}x}.$

12. $\qquad \cos^{-1}\left(\dfrac{y}{b}\right) = \log\left(\dfrac{x}{n}\right)^n$

$\therefore \qquad \dfrac{y}{b} = \cos\log\left(\dfrac{x}{n}\right)^n$

$\therefore \qquad y = b\cos\left(n\log\dfrac{x}{n}\right).$

14. $\qquad y = (x^2 - 1)^n \qquad ...(1)$

Diff. both sides of (1) w.r.t. x,

$$y_1 = n(x^2-1)^{n-1}\,.\,2x \text{ or } y_1 = 2nx\,\frac{(x^2-1)^n}{(x^2-1)}$$

Cross-multiplying $\quad (x^2 - 1)\,y_1 = 2n\,x\,y.$ $\qquad$ [By (1)]

---

5. Determination of the value of the nth derivative of a function for x = 0

Sometimes it is required to find the nth derivative of a function for $x = 0$. The **working rule** to find $(y_n)_{x=0}$ is being given below :

1. Put the given function equal to y.
2. Find $y_1 = \dfrac{dy}{dx}$.

Then (*i*) Take L.C.M. (if possible).

(*ii*) Square both sides if square Roots are there.

(*iii*) Try to get y in R.H.S. (if possible).

3. Again differentiate both sides w.r.t. x to get an equation in y_2, y_1, y.
4. Differentiate both sides n times w.r.t. x by Leibnitz Theorem.

Leibnitz Theorem is

$$(uv)_n = u_n \cdot v + {}^nC_1 \cdot u_{n-1} \cdot v_1 + {}^nC_2 \cdot u_{n-2} \cdot v_2 + \ldots + {}^nC_n\, uv_n.$$

5. Put $x = 0$ in equations of steps 1, 2, 3, 4.
6. Put $n = 1, 2, 3, 4$ in last equation of step 5.
7. Discuss the two cases when n is even and when n is odd.

Example 1. *If* $y = e^{m \cos^{-1} x}$, *show that*

$$(1 - x^2)\, y_{n+2} - (2n + 1)\, xy_{n+1} - (n^2 + m^2)\, y_n = 0$$

and calculate $y_n(0)$. (M.D.U. 1997)

Sol. Here $\qquad y = e^{m \cos^{-1} x}$ ...(1)

$\therefore \qquad y_1 = e^{m \cos^{-1} x} \cdot \dfrac{-m}{\sqrt{1-x^2}} = -\dfrac{ym}{\sqrt{1-x^2}}$ ... (2)

Squaring both sides, we have

$$(1 - x^2)\, y_1^2 = m^2 y^2$$

Differentiating w.r.t. x, we have

$$(1 - x^2)\, 2y_1 y_2 - 2xy_1^2 = 2m^2 yy_1$$

Dividing by $2y_1$, we get

$$(1 - x^2)\, y_2 - xy_1 = m^2 y \qquad \text{...(3)}$$

Differentiating n times by Leibnitz's Theorem, we get

$$(1 - x^2) y_{n+2} - 2nxy_{n+1} - n(n-1) y_n - xy_{n+1} - ny_n = m^2 y_n$$

i.e., $\qquad (1 - x^2) y_{n+2} - (2n + 1) xy_{n+1} - (n^2 + m^2) y_n = 0$ ...(4)

By putting $x = 0$, in (1), (2), (3) and (4), we get

$$y(0) = e^{m \cdot \pi/2}$$

$$y_1(0) = -m \cdot e^{m \cdot \pi/2}$$

$$y_2(0) = m^2 \cdot y(0) = m^2 \cdot e^{m \cdot \pi/2}$$

$$y_{n+2}(0) = (n^2 + m^2) y_n(0) \qquad \text{...(5)}$$

Putting $n = 1, 2, 3, 4$ ... in (5), we have

$$y_3(0) = (1^2 + m^2) y_1(0) = -m(1^2 + m^2) e^{m \cdot \pi/2}$$

$$y_4(0) = (2^2 + m^2) y_2(0) = m^2(2^2 + m^2) e^{m \cdot \pi/2}$$

$$y_5(0) = (3^2 + m^2) y_3(0) = -m(1^2 + m^2)(3^2 + m^2) e^{m \cdot \pi/2}$$

$$y_6(0) = (4^2 + m^2) y_4(0) = m^2(2^2 + m^2)(4^2 + m^2) e^{m \cdot \pi/2}$$

In general,

$$y_n(0) = \begin{cases} -m \cdot e^{m \cdot \pi/2} (1^2 + m^2)(3^2 + m^2) \ldots [(n-2)^2 + m^2] & \text{when } n \text{ is odd.} \\ m^2 \cdot e^{m \cdot \pi/2} (2^2 + m^2)(4^2 + m^2) \ldots [(n-2)^2 + m^2], & \text{when } n \text{ is even.} \end{cases}$$

Example 2. *If $y = (\sin^{-1} x)^2$, prove that :*

(i) $(1 - x^2)\dfrac{d^2y}{dx^2} - x\dfrac{dy}{dx} - 2 = 0$

(ii) $(1 - x^2)\, y_{n+2} - (2n + 1)\, xy_{n+1} - n^2 y_n = 0.$ (M.D.U. 1998, 2006 ; K.U. 2006 S)

Also find the value of nth derivative of y for x = 0.

Sol. $y = (\sin^{-1} x)^2$

Differentiating both sides w.r.t. x,

$$y_1 = 2(\sin^{-1} x) \cdot \frac{1}{\sqrt{1-x^2}} \quad ...(1)$$

Cross-multiplying and squaring both sides, $(1 - x^2)y_1^2 = 4\,(\sin^{-1} x)^2 = 4y.$

Differentiating again, $2(1 - x^2)\, y_1 y_2 - 2xy_1^2 = 4y_1$

Dividing both sides by $2y_1$, $(1 - x^2)\, y_2 - xy_1 - 2 = 0$ ...(2)

Differentiating n times by Leibnitz's Theorem,

$$(1 - x^2)y_{n+2} + {}^nC_1\, y_{n+1}(-2x) + {}^nC_2\, y_n(-2) - xy_{n+1} - {}^nC_1\, y_n = 0$$

or
$$(1 - x^2)y_{n+2} - (2n + 1)xy_{n+1} - n^2 y_n = 0 \quad ...(3)$$

Put $x = 0$ in (1), (2) and (3), then

$$y_1(0) = 0 \quad \text{and} \quad y_2(0) = 2,$$

and
$$y_{(n+2)}(0) = n^2 y_n(0) \quad ...(4)$$

Putting $n = 1, 2, 3, 4, ...$ in (4), we get

$$y_3(0) = 1^2 \,.\, y_1(0) = 0$$
$$y_4(0) = 2^2 \,.\, y_2(0) = 2 \,.\, 2^2$$
$$y_5(0) = 3^2 \,.\, y_3(0) = 0$$
$$y_6(0) = 4^2 \,.\, y_4(0) = 2 \,.\, 2^2 \,.\, 4^2$$
$$y_7(0) = 5^2 \,.\, y_5(0) = 0$$
$$y_8(0) = 6^2 \,.\, y_6(0) = 2 \,.\, 2^2 \,.\, 4^2 \,.\, 6^2.$$

....

....

In general,

when n is odd, $y_n(0) = 0$

when n is even, $y_n(0) = 2 \,.\, 2^2 \,.\, 4^2 \,.\, 6^2 \,.....\, (n - 2)^2.$

Example 3. *If $y = \tan^{-1} x$, prove that*

$$(1 + x^2)y_{n+1} + 2nxy_n + n(n - 1)\, y_{n-1} = 0.$$

Hence determine the values of all the derivatives of y with respect to x when x = 0.

Sol. Here $y = \tan^{-1} x$ ...(1)

$$\therefore \quad y_1 = \frac{1}{1+x^2} \quad ...(2)$$

or
$$y_1(1 + x^2) = 1$$

Differentiating n times by Leibnitz's Theorem, we have

$$y_{n+1}(1 + x^2) + ny_n \,.\, 2x + \frac{n(n-1)}{2}\, y_{n-1} \,.\, 2 = 0$$

or $$(1 + x^2)\, y_{n+1} + 2nxy_n + n(n-1)\, y_{n-1} = 0 \quad ...(3)$$

Putting $x = 0$ in (1), (2) and (3), we have

$$y(0) = 0,\ y_1(0) = 1$$

and $$y_{n+1}(0) = -n(n-1)y_{n-1}(0) \quad ...(4)$$

Putting $n = 1, 2, 3, 4$ .... in (4), we get

$$y_2(0) = -1\,.\,(0)\,.\,y(0) = 0$$
$$y_3(0) = -2\,.\,(1)\,.\,y_1(0) = -2 = (-1)^1\, 2\,!$$
$$y_4(0) = -3\,.\,(2)\,.\,y_2(0) = 0$$
$$y_5(0) = -4\,.\,(3)\,.\,y_3(0) = -4\,.\,(3)\,.\,(-2) = (-1)^2\, 4\,!$$

... ...

... ...

In general,

When n is even, $y_n(0) = 0$.

When n is odd, $y_n(0) = (-1)^{\frac{n-1}{2}}\,(n-1)\,!$.

EXERCISE 5

1. If $y = \sin^{-1} x$, prove that $(1 - x^2)\, y_{n+2} - (2n+1)xy_{n+1} - n^2 y_n = 0$.
   Also find the value of y_n when $x = 0$.

2. If $y = \dfrac{\sin^{-1} x}{\sqrt{1 - x^2}}$, prove that $y_n = (n-1)^2 y_{n-2}$ for $x = 0$.

3. If $y = [x + \sqrt{1 + x^2}\,]^m$, find $y_n(0)$. *(M.D.U. 1995)*

4. Find $y_n(0)$, when $y = \log(x + \sqrt{1 + x^2}\,)$. *(K.U. 2005 S)*

5. (*i*) If $y = [\log(x + \sqrt{1 + x^2}\,)]^2$, find all the derivatives of y w.r.t. x, when $x = 0$.
   (*ii*) If $y = (\sinh^{-1} x)^2$, prove that $(1 + x^2)\, y_{n+2} + (2n+1)\, xy_{n+1} + n^2 y_n = 0$.
   Hence find y_n when $x = 0$. *(K.U. 2005, 2007)*

   [**Hint.** $\sinh^{-1} x = \log[x + \sqrt{1 + x^2}\,]$].

6. If $y = e^{a \sin^{-1} x}$, prove that $(1 - x^2)\, y_{n+2} - (2n+1)\, xy_{n+1} - (n^2 + a^2)\, y_n = 0$.

   Deduce that $\displaystyle\operatorname*{Lt}_{x \to 0} \frac{y_{n+2}}{y_n} = n^2 + a^2$. Hence find $y_n(0)$.

7. If $y = \sin(m \sin^{-1} x)$, find $y_n(0)$. *(K.U. 1998 A)*

8. If $y = x \log\left(\dfrac{x-1}{1+x}\right)$, prove that $y_m = (-1)^{m-2}(m-2)\,!\left[\dfrac{x-m}{(x-1)^m} - \dfrac{(x+m)}{(x+1)^m}\right]$. *(M.D.U. 2004 S)*

ANSWERS

1. When n is even, $y_n(0) = 0$

When n is odd, $y_n(0) = 1^2 . 3^2 . 5^2 ..... (n-2)^2$

3. When n is even, $y_n(0) = m^2 (m^2 - 2^2)(m^2 - 4^2) .... [m^2 - (n-2)^2]$

When n is odd, $y_n(0) = m(m^2 - 1^2)(m^2 - 3^2) ...... [m^2 - (n-2)^2]$

4. When n is even, $y_n(0) = 0$

When n is odd, $y_n(0) = (-1)^{\frac{n-1}{2}} 1^2 . 3^2 . 5^2 ...... (n-2)^2$

5. (*i*), (*ii*) When n is even, $y_n(0) = (-1)^{\frac{n}{2}-1} . 2 . 2^2 . 4^2 . 6^2 ... (n-2)^2$

When n is odd, $y_n(0) = 0$

6. When n is even, $y_n(0) = a^2 (2^2 + a^2)(4^2 + a^2) .... [(n-2)^2 + a^2]$

When n is odd, $y_n(0) = a(1^2 + a^2)(3^2 + a^2) .... [(n-2)^2 + a^2]$

7. When n is even, $y_n(0) = 0$

When n is odd, $y_n(0) = m(1^2 - m^2)(3^2 - m^2) ..... [(n-2)^2 - m^2]$.

4

General Theorems

1. Introduction

In the present chapter, we shall consider a few general theorems which play a very important role in the subsequent development of the subject matter of Differential Calculus.

The student has already learnt two General Theorems namely Rolle's Theorem and Lagrange's Theorem in class XII. Here we remind the reader of **Rolle's Theorem** which we are to apply in this chapter.

Rolle's Theorem. If a function $f(x)$ is

(*i*) continuous in the closed interval $[a, b]$;

(*ii*) derivable in the open interval (a, b) ; and

(*iii*) $f(a) = f(b)$.

Then there exists at least one number c the open interval (a, b) such that $f'(c) = 0$ or there exists at least one real number θ $(0 < \theta < 1)$ such that $f'(a + \theta h) = 0$.

2. Taylor's Theorem with Lagrange's Form of Remainder after n terms

(M.D.U. 1999, 98, 96 ; K.U. 1999 A, 96 S, 96)

If a function $f(x)$ be such that

(*i*) $f(x), f'(x), f''(x) \ldots\ldots f^{n-1}(x)$ are continuous in the closed interval $[a, a + h]$ and

(*ii*) $f^n(x)$ exists in the open interval $(a, a + h)$, then there exists at least one number θ between 0 and 1 such that

$$f(a + h) = f(a) + hf'(a) + \frac{h^2}{2!} f''(a) + \ldots\ldots + \frac{h^{n-1}}{(n-1)!} f^{n-1}(a) + \frac{h^n}{n!} f^n(a + \theta h)$$

Proof. Consider the function

$$F(x) = f(x) + (a + h - x) f'(x) + \frac{(a + h - x)^2}{2!} f''(x) + \ldots\ldots$$

$$\ldots + \frac{(a + h - x)^{n-1}}{(n-1)!} f^{n-1}(x) + \frac{(\mathbf{a} + \mathbf{h} - \mathbf{x})^{\mathbf{n}}}{\mathbf{n}!} . \mathbf{A} \quad \ldots(1)$$

where A is a constant to be so chosen that

$$F(a + h) = F(a).$$

Put $x = a + h$ in (1),

$\therefore \quad F(a + h) = f(a + h)$ (All other terms becomes zero)

Put $x = a$ in (1),

$$F(a) = f(a) + hf'(a) + \frac{h^2}{2!} f''(a) + \ldots + \frac{h^{n-1}}{(n-1)!} f^{n-1}(a) + \frac{h^n}{n!} A$$

Putting these values of F(*a* + *h*) and F(*a*) in

F(*a* + *h*) = F(*a*), we have

$$f(a+h)=f(a)+hf'(a)+\frac{h^2}{2!}f''(a)+\ldots\ldots+\frac{h^{n-1}}{(n-1)!}f^{n-1}(a)+\frac{h^n}{n!}\text{A} \quad \ldots(2)$$

which gives the value of A.

Now let us apply Rolle's Theorem on F(x) in [a, a + h]

It is given that $f(x), f'(x), f''(x), \ldots\ldots f^{n-1}(x)$ are continuous in the closed interval $[a, a+h]$ and their derivatives $f'(x), f''(x), \ldots\ldots f^n(x)$ all exist in the open interval $(a, a+h)$.

Also $(a+h-x), \dfrac{(a+h-x)^2}{2!}, \ldots\ldots \dfrac{(a+h-x)^n}{n!}$ being polynomials and A are all continuous in the closed interval $[a, a+h]$ and derivable in open interval $(a, a+h)$. Therefore, F(*x*) is continuous in the closed interval $[a, a+h]$ and derivable in open interval $(a, a+h)$.

Also F(*a* + *h*) = F(*a*).

Thus, F(*x*) satisfies all the three conditions of Rolle's theorem, and therefore, there must exist at least one positive number θ less than 1 such that

$$F'(a+\theta h)=0 \quad [0<\theta<1] \quad \ldots(3)$$

Differentiating both sides of (1) w.r.t. *x* ;

$$F'(x)=f'(x)+[(a+h-x)f''(x)-f'(x)]+\frac{1}{2!}[(a+h-x)^2f'''(x)-2(a+h-x)f''(x)]$$

$$\ldots\ldots+\frac{1}{(n-1)!}[(a+h-x)^{n-1}f^n(x)-(n-1)(a+h-x)^{n-2}f^{n-1}(x)]$$

$$-\frac{n\text{A}}{n!}(a+h-x)^{n-1}$$

or

$$F'(x)=\frac{1}{(n-1)!}(a+h-x)^{n-1}f^n(x)-\frac{\text{A}}{(n-1)!}(a+h-x)^{n-1}$$

[∵ other terms cancel in pairs]

$$=\frac{(a+h-x)^{n-1}}{(n-1)!}[f^n(x)-\text{A}]$$

Putting $x = a + \theta h$,

$$F'(a+\theta h)=\frac{(h-\theta h)^{n-1}}{(n-1)!}[f^n(a+\theta h)-\text{A}]=0 \quad \text{[By (3)]}$$

i.e.,

$$\frac{h^{n-1}}{(n-1)!}(1-\theta)^{n-1}[f^n(a+\theta h)-\text{A}]=0$$

$$\therefore \quad f^n(a+\theta h)=\text{A} \quad [\because \text{ neither } h=0, \text{ nor } (1-\theta)=0 \text{ as } 0<\theta<1]$$

Putting this value of A in (2), we have

$$f(a+h)=f(a)+hf'(a)+\frac{h^2}{2!}f''(a)+\ldots\ldots+\frac{h^{n-1}}{(n-1)!}f^{n-1}(a)+\frac{h^n}{n!}f^n(a+\theta h)$$

$$[0<\theta<1]$$

Note. The (*n* + 1)th term $\dfrac{h^n}{n!}f^n(a+\theta h)$ in the above expansion is in general, denoted by R_n and is called *Lagrange's remainder after n terms.*

Cor. If we put $n = 1$ in the above theorem, it takes the form $f(a + h) = f(a) + hf'(a + \theta h)$ where $0 < \theta < 1$

which is Lagrange's mean value theorem and hence this shows *that mean-value theorem is a particular case of Taylor's theorem.*

Alternative form of Taylor's theorem with Lagrange's form of remainder after n terms

If we put $a + h = b$, then interval $[a, a + h]$ becomes $[a, b]$ and $a + \theta h = a + (b - a)\theta = c$, where c lies between a and b, and theorem becomes

$$f(b) = f(a) + (b - a)f'(a) + \frac{(b-a)^2}{2!} f''(a) + \ldots\ldots + \frac{(b-a)^{n-1}}{(n-1)!} f^{n-1}(a) + \frac{(b-a)^n}{n!} f^n(c)$$

$$[a < c < b]$$

Cor. Maclaurin's theorem with Lagrange's form of remainder after n terms.

(M.D.U. 2002 ; K.U. 1996 A)

Proof. Firstly state and prove Taylor's Theorem with Lagrange's form of Remainder as in Art. 2. Then put $a = 0$ and $h = x$.

Hence we get

$$\mathbf{f(x) = f(0) + x.f'(0) + \frac{x^2}{2!} f'(0) + \ldots\ldots + \frac{x^{n-1}}{(n-1)!} f^{n-1}(0) \ldots + \frac{x^n}{n!} f^n(\theta x)}$$

$$[0 < \theta < 1]$$

which is called **Maclaurin's theorem with Lagrange's form of remainder** and holds good when

(*i*) $f(x)$, $f'(x)$, $f''(x) \ldots\ldots f^{n-1}(x)$ are continuous in the closed interval $[0, x]$

and (*ii*) $f^n(x)$ exists in the open interval $(0, x)$.

3. Taylor's theorem with Cauchy's form of remainder.

(M.D.U. 2000, 2007 ; K.U. 1995, 2007)

If a function $f(x)$ be such that

(i) $f(x), f'(x), f''(x) \ldots\ldots f^{n-1}(x)$ are continuous in the closed interval $[a, a + h]$.

(ii) $f^n(x)$ exists in the open interval $(a, a + h)$; then there exists at least one number θ, between 0 and 1, such that

$$f(a + h) = f(a) + hf'(a) + \frac{h^2}{2!} f''(a) + \ldots\ldots + \frac{h^{n-1}}{(n-1)!} f^{n-1}(a)$$

$$+ \frac{h^n}{(n-1)!} (1 - \theta)^{n-1} f^n(a + \theta h) \quad [0 < \theta < 1]$$

Proof. Let $F(x) = f(x) + (a + h - x) f'(x) + \frac{(a+h-x)^2}{2!} f''(x)$

$$+ \ldots\ldots + \frac{(a+h-x)^{n-1}}{(n-1)!} f^{n-1}(x) + \mathbf{(a + h - x)\, A} \quad \ldots(1)$$

where A is a constant such that $F(a) = F(a + h)$.

Putting x = a in (1),

$$F(a) = f(a) + hf'(a) + \frac{h^2}{2!} f''(a) + \ldots\ldots + \frac{h^{n-1}}{(n-1)!} f^{n-1}(a) + hA \quad \ldots(2)$$

Putting x = a + h in (1),

$F(a + h) = f(a + h) + 0 + 0 + ... = f(a + h)$ ...(3)

But $F(a + h) = F(a)$ (assumption)

Putting values from (3) and (2),

$$f(a + h) = f(a) + hf'(a) + \frac{h^2}{2!} f''(a) + ... + \frac{h^{n-1}}{(n-1)!} f^{n-1}(a) + hA \quad ...(4)$$

Now Let us apply Rolle's Theorem on F(x) in [a, a + h].

(*i*) $F(x)$ is continuous in the closed interval $[a, a + h]$.

[$\because$ $f(x), f'(x), .f''(x) ..., f^{n-1}(x)$ are continuous in the closed interval $[a, a + h]$ (given) Also $(a + h - x), (a + h - x)^2 ...... (a + h - x)^{n-1}$ being polynomials in x are continuous in $[a, a + h]$]

(*ii*) Similarly $F(x)$ is derivable in the open interval $(a, a + h)$

(*iii*) Also $F(a) = F(a + h)$ [Assumption]

$\therefore$ By Rolle's Theorem, there exists at least one real number $\theta (0 < \theta < 1)$ such that $F'(a + \theta h) = 0$.

Differentiating both sides of (1) w.r.t. x,

$$F'(x) = f'(x) + [(a + h - x) f''(x) - f'(x)] + \frac{1}{2!} [(a + h - x)^2 f'''(x) - 2(a + h - x) f''(x)] ...$$

$$+ \frac{1}{(n-1)!} [(a + h - x)^{n-1} f^n(x) - (n-1)(a + h - x)^{n-2} f^{n-1}(x)] - A$$

$$\left[\because \frac{d}{dx}(uv) = u\frac{d}{dx}(v) + \frac{du}{dx}.v \text{ and } \frac{d}{dx}(a + h - x) = -1\right]$$

or $$F'(x) = \frac{1}{(n-1)!} (a + h - x)^{n-1} f^n(x) - A.$$

Put $$x = a + \theta h,$$

$$F'(a + \theta h) = \frac{1}{(n-1)!} (a + h - a - \theta h)^{n-1} f^n(a + \theta h) - A$$

$$= \frac{1}{(n-1)!} (h - \theta h)^{n-1} f^n(a + \theta h) - A$$

$$= \frac{h^{n-1}}{(n-1)!} (1 - \theta)^{n-1} f^n(a + \theta h) - A.$$

But $$F'(a + \theta h) = 0$$

$\therefore$ $$\frac{h^{n-1}}{(n-1)!} (1 - \theta)^{n-1} f^n(a + \theta h) - A = 0$$

or $$A = \frac{h^{n-1}}{(n-1)!} (1 - \theta)^{n-1} f^n(a + \theta h)$$

Putting this value of A in (4),

$$f(a + h) = f(a) + hf'(a) + \frac{h^2}{2!} f''(a) + ... + \frac{h^{n-1}}{(n-1)!} f^{n-1}(a)$$

$$+ h . \frac{h^{n-1}}{(n-1)!} (1 - \theta)^{n-1} f^n(a + \theta h)$$

or $$f(a+h) = f(a) + hf'(a) + \frac{h^2}{2!} f''(a) + \ldots\ldots + \frac{h^{n-1}}{(n-1)!} f^{n-1}(a) + \frac{h^n}{(n-1)!} (1-\theta)^{n-1} f^n(a+\theta h)$$

Note. F(x) of Art. 3 has been written from

F(x) of Art. 2 replacing last term $\frac{(a+h-x)^n}{n!}$ A by $(a + h - x)$ A.

Cor. Maclaurin's theorem with Cauchy's form of remainder after n terms.

Put $a = 0$ and $h = x$ in Art. 3. (Taylor's theorem with Cauchy form of remainder, then we get

$$f(x) = f(0) + xf'(0) + \frac{x^2}{2!} f''(0) + \ldots\ldots + \frac{x^{n-1}}{(n-1)!} f^{n-1}(0) + \frac{x^n (1-\theta)^{n-1}}{(n-1)!} f^n(\theta x) \quad [0 < \theta < 1]$$

which is called **Maclaurin's theorem with Cauchy's form of remainder** and holds good when

(i) $f(x), f'(x), f''(x) \ldots\ldots f^{n-1}(x)$ *are all continuous in the closed interval* $[0, x]$

and *(ii)* $f^n(x)$ *exists in the open interval* $(0, x)$.

Note. The $(n+1)$th term $\frac{x^n(1-\theta)^{n-1}}{(n-1)!} f^n(\theta x)$ $[0 < \theta < 1]$ is known as *Cauchy remainder after n terms* in Maclaurin's development of $f(x)$.

Example 1. *Expand a^x by Maclaurin's theorem with Lagrange's form of remainder after n terms.*

Sol. Here $f(x) = a^x$.

By successive differentiation, we get

$$f'(x) = a^x . \log a,\ f''(x) = a^x . (\log a)^2, \ldots\ldots, f^{n-1}(x) = a^x . (\log a)^{n-1}$$

and $$f^n(x) = a^x . (\log a)^n$$

Putting $x = 0$, we have

$$f(0) = 1, f'(0) = \log a, f''(0) = (\log a)^2, \ldots\ldots f^{n-1}(0) = (\log a)^{n-1} \quad [\because\ a^0 = 1]$$

and $$f^n(\theta x) = a^{\theta x} (\log a)^n$$

But by Maclaurin's theorem with Lagrange's form of remainder, we have

$$f(x) = f(0) + x . f'(0) + \frac{x^2}{2!} f''(0) + \ldots\ldots + \frac{x^{n-1}}{(n-1)!} f^{n-1}(0) + \frac{x^n}{n!} f^n(\theta x) \quad [0 < \theta < 1]$$

Substituting the values in the above result, we get

$$a^x = 1 + x . \log a + \frac{x^2}{2!} (\log a)^2 + \ldots\ldots + \frac{x^{n-1}}{(n-1)!} (\log a)^{n-1} + \frac{x^n}{n!} a^{\theta x} . (\log a)^n.$$

Here Lagrange's remainder after n terms $= \frac{x^n}{n!} a^{\theta x} . (\log a)^n$, where $0 < \theta < 1$.

Cor. Putting $a = e$, the expansion of e^x is

$$e^x = 1 + x + \frac{x^2}{2!} + \ldots\ldots + \frac{x^{n-1}}{(n-1)!} + \frac{x^n}{n!} e^{\theta x} \; [0 < \theta < 1].$$

Example 2. *Expand $e^{ax} \sin bx$ by Maclaurin's Theorem with Lagrange's form of remainder after n terms.*

Sol. Let $\quad f(x) = e^{ax} \sin bx$

$\therefore \quad f'(x) = e^{ax} \cos bx \,.\, b + \sin bx \,.\, e^{ax} \,.\, a = e^{ax} (b \cos bx + a \sin bx)$

$f''(x) = e^{ax} (-b^2 \sin bx + ab \cos bx) + (b \cos bx + a \sin bx) e^{ax} \,.\, a$

$= e^{ax} [-b^2 \sin bx + ab \cos bx + ab \cos bx + a^2 \sin bx]$

or $\quad f''(x) = e^{ax} [(a^2 - b^2) \sin bx + 2ab \cos bx]$

$\therefore \quad f'''(x) = e^{ax} [(a^2 - b^2) \cos bx \,.\, b - 2ab^2 \sin bx]$

$+ [(a^2 - b^2) \sin bx + 2ab \cos bx] e^{ax} \,.\, a$

....

and we know that $\quad f^n(x) = (a^2 + b^2)^{n/2} e^{ax} \sin \left(bx + n \tan^{-1} \frac{b}{a} \right)$

Putting $\quad x = 0$ in $f(x), f'(x), f''(x), f'''(x)$ and changing x to θx in $f^n(x)$

we have $\quad f(0) = 0, f'(0) = b, f''(0) = 2ab,$

$f'''(0) = (a^2 - b^2) b + (2ab)a = a^2 b - b^3 + 2a^2 b = 3a^2 b - b^3 = b(3a^2 - b^2)$

and $\quad f^n(\theta x) = (a^2 + b^2)^{n/2} e^{a\theta x} \sin \left(b\theta x + n \tan^{-1} \frac{b}{a} \right).$

Putting these values in Maclaurin's theorem,

$$f(x) = f(0) + xf'(0) + \frac{x^2}{2!} f''(0) + \frac{x^3}{3!} f'''(0) + \ldots\ldots + \frac{x^n}{n!} f^n(\theta x),$$

we have

$$e^{ax} \sin bx = bx + \frac{x^2}{2!} (2ab) + \frac{x^3}{3!} b(3a^2 - b^2) + \ldots\ldots$$

$$\ldots\ldots + \frac{x^n}{n!} (a^2 + b^2)^{n/2} e^{a\theta x} \sin \left(b\theta x + n \tan^{-1} \frac{b}{a} \right).$$

EXERCISE 1

1. (*a*) State and prove Taylor's theorem with Lagrange's form of remainder after n terms. (*K.U. 1999 A, 96 A ; M.D.U. 1999, 98, 96*)

   (*b*) State and prove Taylor's development of a function with Cauchy's form of remainder and hence deduce the Maclaurin's expansion. (*K.U. 1995 A ; M.D.U. 2000*)

   (*c*) State and prove Maclaurin's theorem (As special case of Taylor's theorem).

   (*d*) State and prove Maclaurin's theorem with Lagrange's form of remainder. (*M.D.U. 2002*)
   [**Hint.** Cor. Art. 2.]

2. Show that

   (*i*) $\log (1 + x) = x - \frac{x^2}{2} + \frac{x^3}{3} - \ldots\ldots + (-1)^{n-2} \frac{x^{n-1}}{n-1} + (-1)^{n-1} \frac{x^n}{n(1 + \theta x)^n}.$

$$\left[\textbf{Hint. } \frac{d^n}{dx^n} \log (ax+b) = \frac{(-1)^{n-1}(n-1)!\,a^n}{(ax+b)^n}.\right]$$

(ii) $\log(1-x) = -x - \dfrac{x^2}{2} - \dfrac{x^3}{3} - \dots - \dfrac{x^{n-1}}{n-1} - \dfrac{x^n}{n(1-\theta x)^n}$.

3. Show that $e^{ax}\cos bx = 1 + ax + (a^2-b^2)\dfrac{x^2}{2!} + \dfrac{x^3}{3!}a(a^2-3b^2) + \dots$

$$+ \frac{x^n}{n!}(a^2+b^2)^{n/2} e^{a\theta x} \cos\left(b\theta x + n\tan^{-1}\frac{b}{a}\right)$$

$$\left[\textbf{Hint. } \frac{d^n}{dx^n}(e^{ax}\cos bx) = (a^2+b^2)^{n/2} e^{ax}\cos\left(bx + n\tan^{-1}\frac{b}{a}\right)\right]$$

4. Show that for every value of x,

(i) $\sin x = x - \dfrac{x^3}{3!} + \dfrac{x^5}{5!} - \dots + (-1)^{n-1}\dfrac{x^{2n-1}}{(2n-1)!} + (-1)^n \dfrac{x^{2n}}{(2n)!}\sin(\theta x)$.

(K.U. 2000 A)

$$\left[\textbf{Hint. } \frac{d^n}{dx^n}\sin(ax+b) = a^n \sin\left(ax+b+\frac{n\pi}{2}\right) \text{ Also } \sin(n\pi+\theta) = (-1)^n \sin\theta.\right]$$

(ii) $\cos x = 1 - \dfrac{x^2}{2!} + \dfrac{x^4}{4!} - \dots + (-1)^n \dfrac{x^{2n}}{(2n)!} + (-1)^{n+1}\dfrac{x^{2n+1}}{(2n+1)!}\sin(\theta x)$

$$\left[\textbf{Hint. } \frac{d^n}{dx^n}\cos(ax+b) = a^n\cos\left(ax+b+\frac{n\pi}{2}\right) \text{ and } \cos(n\pi+\theta) = (-1)^n\cos\theta.\right]$$

5. Show that $\sin x + \cos x = 1 + x - \dfrac{1}{2!}x^2 - \dfrac{1}{3!}x^3 + \dfrac{1}{4!}x^4(\sin\theta x + \cos\theta x)$. *(K.U. 2004 S)*

6. Show that $\log(x+h) = \log x + \dfrac{h}{x} - \dfrac{h^2}{2x^2} + \dots + (-1)^{n-1}\dfrac{h^n}{n(x+\theta h)^n}$.

[**Hint.** Use Taylor's Theorem with Lagrange's form of Remainder.]

7. Expand $e^{ax}\sin px$ by Maclaurin's Theorem with Cauchy's form of remainder after n terms.

(K.U. 1999 S, 98 A)

$$\left[\textbf{Hint. } \frac{d^n}{dx^n} e^{ax}\sin bx = (a^2+b^2)^{\frac{n}{2}} e^{ax}\sin\left(bx + n\tan^{-1}\frac{b}{a}\right)\right]$$

8. Expand $b^x\cos ax$ by Maclaurin's Theorem with Lagrange's form of remainder after r terms.

(M.D.U. 2003 S)

[**Hint.** $b^x = e^{\log b^x}$ $(\because\ e^{\log f(x)} = f(x)) = e^{x\log b}$ $\therefore$ $b^x\cos ax = e^{x\log b}\cos ax$.]

ANSWERS

7. $px + \dfrac{x^2}{2!}(2ap) + \dfrac{x^3}{3!}p(3a^2-p^2) + \dots + x^n\dfrac{(1-\theta)^{n-1}}{(n-1)!}e^{a\theta x}(a^2+p^2)^{n/2}\sin\left(p\theta x + n\tan^{-1}\dfrac{p}{a}\right)$.

8. $1 + x\log b + \dfrac{x^2}{2!}[(\log b)^2 - a^2] + \dfrac{x^3}{3!}[(\log b)^2 - 3a^2]\log b$

$$+ \dots + \frac{x^r}{r!}[(\log b)^2 + a^2]^{r/2} b^{\theta x}\cos\left(a\theta x + r\tan^{-1}\frac{a}{\log b}\right).$$

5

Expansions

1. A set of terms formed and arranged according to some *definite law* and connected by the signs + ve or – ve is called a **series.** There are two kinds of series ; **finite** and **infinite.** A series is said to be *finite* when the number of terms it contains is limited and definite. For example, 2 + 4 + 6 + 8 + 10 is finite series of 5 terms whose sum is 30. The sum of a finite series is the sum of all the terms it is made up of.

Infinite Series. A series is said to be infinite when the number of terms it contains is unlimited *i.e.,* infinitely large so that it has no last term. For example, $1 + \frac{1}{3} + \frac{1}{9} + \frac{1}{27} + \ldots\ldots$ to infinity, it an infinite series each term of which is equal to one-third of the term preceding it. It is an infinite G.P.

2. Convergent and Divergent Series

If S_n the sum of first n terms of an infinite series tends to a definite limit S, as n approaches infinity, the infinite series is said to be convergent and this limit S is called the sum of the infinite series.

Illustration. The infinite series $1 + \frac{1}{3} + \frac{1}{9} + \frac{1}{27} + \ldots\ldots$ is convergent, for

If S_n denotes the sum to n terms

then $$S_n = 1 + \frac{1}{3} + \frac{1}{9} + \frac{1}{27} + \ldots\ldots \text{ to } n \text{ terms}$$

$$= \frac{1\left(1 - \frac{1}{3^n}\right)}{1 - \frac{1}{3}} = \frac{3}{2}\left(1 - \frac{1}{3^n}\right) = \frac{3}{2} - \frac{1}{2} \cdot \frac{1}{3^{n-1}}$$

Now if limit of S_n when $n \to \infty$, be denoted by S, then

$$S = \frac{3}{2} \qquad \left(\because \text{ as } n \to \infty, \frac{1}{3^{n-1}} \to 0\right)$$

Since the limit of S_n when $n \to \infty$ is definite and finite ; the series is convergent.

The second kind of the infinite series in which S_n, the sum of first n terms does not tend to a finite limit but increases beyond all bounds as n increases indefinitely, is called a **Divergent Series.** It is evident that the question of the sum of a non-convergents series does not arise.

Illustration 1. The series 1 + 2 + 3 + 4 + ... ∞ is a divergent series.

Here $$S_n = \frac{n(n+1)}{2} \quad \text{and} \quad \underset{n \to \infty}{\text{Lt}} S_n = \underset{n \to \infty}{\text{Lt}} \frac{n(n+1)}{2} = \infty$$

Since the limit of S_n when $n \to \infty$ is not definite and finite, the series is divergent.

Note. If we do not take into consideration convergency and divergency of series we may be led to erroneous conclusions as the next illustration will show.

Illustration 2. $\dfrac{1}{x-1}$ is a function of x and we can represent it by an infinite series :

By actual division, we have $\dfrac{1}{1-x} = 1 + x + x^2 + x^3 + \ldots\ldots + x^n + \ldots$ ...(1)

an equality which may be expected to hold good for all values of x.

Putting $x = 3$, we get $\dfrac{1}{1-3} = 1 + 3 + 3^2 + 3^3 + \ldots\ldots$

or $$-\tfrac{1}{2} = 1 + 3 + 3^2 + 3^3 + \ldots\ldots$$

This is an absurd result for a small negative quantity is equal to a very large positive quantity.

The absurdity is due to the fact that the infinite series on the right hand side of (1) which is a G.P. is convergent only if x lies between -1 and 1, and therefore we are not justified in putting $x = 3$ on both sides of (1).

3. Taylor's Infinite Series

If a function $f(x)$ *possesses derivatives of all orders* in the interval $(a, a + h)$, then for every integer n however large, there corresponds a Taylor's development with Lagrange's form of remainder namely

$$f(a + h) = f(a) + h f'(a) + \frac{h^2}{2!} f''(a) + \ldots\ldots + \frac{h^{n-1}}{(n-1)!} f^{n-1}(a) + R_n$$

where $$R_n = \frac{h^n}{n!} f^n(a + \theta h) \qquad [0 < \theta < 1]$$

or $$f(a + h) = S_n + R_n$$

where $$S_n = f(a) + hf'(a) + \frac{h^2}{2!} f''(a) + \ldots\ldots + \frac{h^{n-1}}{(n-1)!} f^{n-1}(a)$$

Now, suppose $R_n \to 0$, as $n \to \infty$, then

$$\operatorname*{Lt}_{n \to \infty} S_n = f(a + h)$$

or $$f(a) + hf'(a) + \frac{h^2}{2!} f''(a) + \ldots\ldots + \frac{h^{n-1}}{(n-1)!} f^{n-1}(a) + \ldots\ldots$$

converges and its sum is equal to $f(a + h)$.

This leads to

Theorem. *If $f(x)$ be a function*

(i) possessing derivatives of **all orders** *in the interval $(a, a + h)$ and*

(ii) Taylor's remainder $R_n = \dfrac{h^n}{n!} f^n(a + \theta h)$ tends to zero as $n \to \infty$, then

$$\mathbf{f(a + h) = f(a) + hf'(a) + \frac{h^2}{2!} f''(a) + \ldots\ldots + \frac{h^n}{n!} f^n(a) + \ldots + \text{ad inf.}} \qquad \textbf{...(A)}$$

This theorem expands $f(a + h)$ in an infinite series of ascending integral powers of h and the series is called **Taylor's infinite series.**

Other Forms of Taylor's Infinite Series :

(*i*) Writing x for a, in (A), we have

$$\mathbf{f(x + h) = f(x) + hf'(x) + \frac{h^2}{2!} f''(x) + \ldots\ldots + \frac{h^n}{n!} f^n(x) + \ldots\ldots}$$

(*ii*) Putting $a + h = b$ or $h = b - a$, in (A), we get

$$f(b) = f(a) + (b - a) f'(a) + \frac{(b - a)^2}{2!} f''(a) + \ldots\ldots + \frac{(b - a)^n}{n!} f^n(a) + \ldots\ldots$$

(*iii*) Changing $a + h$ to x *i.e.,* h to $x - a$, in (A), we have

$$f(x) = f(a) + (x - a) f'(a) + \frac{(x - a)^2}{2!} f''(a) + \ldots\ldots + \frac{(x - a)^n}{n!} f^n(a) + \ldots\ldots$$

which expands $f(x)$ in ascending integral powers of $(x - a)$.

Cor. Maclaurin's Infinite Series.

When $a = 0$ and $h = x$ in the above theorem, we notice that if

(*i*) *$f(x)$ possesses derivative of* **all orders** *in the interval (0, x) and*

(*ii*) *Maclaurin's remainder* $R_n = \frac{x^n}{n!} f^n(\theta x)$ *tends to zero as* $n \to \infty$, *then*

$$\mathbf{f(x) = f(0) + xf'(0) + \frac{x^2}{2!} f''(0) + \ldots\ldots + \frac{x^n}{n!} f^n(0) + \ldots\ldots \text{ ad. inf.}}$$

which is known as Maclaurin's infinite series.

Note. If the function $f(x)$ is denoted by y, then the expansion may be written in the form

$$y = y(0) + x \cdot y_1(0) + \frac{x^2}{2!} y_2(0) + \ldots\ldots + \frac{x^n}{n!} y_n(0) + \ldots\ldots$$

where $y(0), y_1(0), y_2(0), \ldots\ldots, y_n(0)$ etc. denote values of $y, y_1, y_2, \ldots\ldots, y_n$ respectively for $x = 0$.

4. Failure of Taylor's and Maclaurin's Theorem

(*a*) Taylor's theorem *fails* to expand $f(x + h)$ in an infinite series in the following situations :

(*i*) if any of functions $f(x), f'(x), f''(x)$ .... becomes infinite or does not exist for any value of x in the interval under consideration,

or (*ii*) if R_n does not tend to zero as $n \to \infty$.

(*b*) Maclaurin's theorem *fails* to expand $f(x)$ in an infinite series in the following situations :

(*i*) if any of $f(x), f'(x), f''(x)$ ...... becomes infinite or does not exist in the closed interval $[0, x]$,

or (*ii*) if R_n does not tend to zero as $n \to \infty$.

Note. We observe that before we can expand a given function as an infinite Taylor's or Maclaurin's series 2 it is essential to examine the behaviour of R_n as n tends to infinity. But this is not simple in many cases. We, therefore, generally obtain the expansion by *assuming* the possibility of expanding it in an infinite series by assuming that R_n tends to 0 as n tends to infinity.

5. Application of Maclaurin's Theorem

Working Rule

1st Step. Put the given function equal to $f(x)$.

2nd Step. Differentiate $f(x)$, a number of times and find $f'(x), f''(x), f'''(x)$ .... and so on.

3rd Step. Put $x = 0$ in the results obtained in 2nd Step and find $f(0), f'(0), f''(0), f'''(0)$ ...... and so on.

4th Step. Now substitute the values of $f(0), f'(0), f''(0), f'''(0)$, ... in,

$$f(x) = f(0) + xf'(0) + \frac{x^2}{2!} f''(0) + \frac{x^3}{3!} f'''(0) + ...$$

Example 1. *Expand a^x and e^x in powers of x, by Maclaurin's theorem.*

Sol. Let $f(x) = a^x$ $\quad\therefore\quad f(0) = a^0 = 1$

then $f'(x) = a^x \log a$ $\quad f'(0) = \log a$

$f''(x) = a^x (\log a)^2$ $\quad f''(0) = (\log a)^2$

$f'''(x) = a^x (\log a)^3$ $\quad f'''(0) = (\log a)^3$

and so on ....

Now Maclaurin's expansion is $f(x) = f(0) + xf'(0) + \frac{x^2}{2!} f''(0) + \frac{x^3}{3!} f'''(0) + ...$

Putting values of $f(0), f'(0), f''(0), f'''(0)$, we have

$$\mathbf{a^x = 1 + x \log a + \frac{x^2}{2!}(\log a)^2 + \frac{x^3}{3!}(\log a)^3 + ...} \qquad ...(1)$$

Putting $a = e$, we have $\mathbf{e^x = 1 + x + \frac{x^2}{2!} + \frac{x^3}{3!} + ...}$ $\qquad ...(2)$

Note. The series (1) and (2) are called exponential series and are convergent for all values of x.

Example 2. *Expand $\log(1 + x)$ in powers of x.*

Sol. Let $f(x) = \log(1 + x)$ $\quad\therefore\quad f(0) = \log 1 = 0$

then $f'(x) = \frac{1}{1+x} = (1 + x)^{-1}$ $\quad\therefore\quad f'(0) = 1$

$f''(x) = (-1)(1 + x)^{-2}$ $\quad\therefore\quad f''(0) = -1$

$f'''(x) = (-1)(-2)(1 + x)^{-3}$ $\quad\therefore\quad f'''(0) = 2$

$f^{iv}(x) = (-1)(-2)(-3)(1 + x)^{-4}$ $\quad\therefore\quad f^{iv}(0) = -2.3 = -6$

and so on.

Putting these values of $f(0), f'(0), f''(0)$ etc. in Maclaurin's expansion,

$$f(x) = f(0) + xf'(0) + \frac{x^2}{2!} f''(0) + \frac{x^3}{3!} f'''(0) + \frac{x^4}{4!} f^{iv}(0) + ...,$$

we have $\quad \log(1 + x) = 0 + x - \frac{x^2}{2} + \frac{x^3}{3} - \frac{x^4}{4} + ...$

or $\quad \mathbf{\log(1 + x) = x - \frac{x^2}{2} + \frac{x^3}{3} - \frac{x^4}{4} + ...}$

Changing x into $-x$, we have

$$\mathbf{\log(1 - x) = -x - \frac{x^2}{2} - \frac{x^3}{3} - \frac{x^4}{4} - ......}$$

Cor. (*i*) $\log(1 + x) - \log(1 - x)$

$$= \left[x - \frac{x^2}{2} + \frac{x^3}{3} - \frac{x^4}{4} + ...\right] - \left[-x - \frac{x^2}{2} - \frac{x^3}{3} - \frac{x^4}{4} - ......\right]$$

$$\therefore \quad \mathbf{\log\frac{1+x}{1-x} = 2\left[x + \frac{x^3}{3} + \frac{x^5}{5} + \frac{x^7}{7} + ...\right]}$$

Cor. (*ii*) $\log (1+x)^m = m \log (1+x) = m\left[x - \frac{x^2}{2} + \frac{x^3}{3} - \frac{x^4}{4} + \ldots\ldots\right]$

Note. The above series for log $(1 + x)$ is called logarithmic series and is convergent *i.e.*, valid for $|x|$ less than 1 and also for $x = 1$.

Example 3. *Expand sin x and cos x in powers of x and hence find cos 18°.*

Sol. Let $f(x) = \sin x \quad \therefore \quad f(0) = 0$

then
$$f'(x) = \cos x \quad \therefore \quad f'(0) = 1$$
$$f''(x) = -\sin x \quad \therefore \quad f''(0) = 0$$
$$f'''(x) = -\cos x \quad \therefore \quad f'''(0) = -1$$
$$f^{iv}(x) = \sin x \quad \therefore \quad f^{iv}(0) = 0$$
$$f^{v}(x) = \cos x \quad \therefore \quad f^{v}(0) = 1$$

and so on, ...

Putting these values of $f(0), f'(0), f''(0), f'''(0)$ in

$$f(x) = f(0) + xf'(0) + \frac{x^2}{2!}f''(0) + \frac{x^3}{3!}f'''(0) + \ldots,$$

we have
$$\sin x = 0 + x(1) + \frac{x^2}{2!}(0) + \frac{x^3}{3!}(-1) + \frac{x^4}{4!}(0) + \frac{x^5}{5!}(1) - \ldots\ldots$$

or
$$\mathbf{\sin x = x - \frac{x^3}{3!} + \frac{x^5}{5!} - \ldots\ldots} \qquad \ldots(1)$$

In a like manner, we get $\mathbf{\cos x = 1 - \frac{x^2}{2!} + \frac{x^4}{4!} - \ldots\ldots} \qquad \ldots(2)$

Note. Series (1) and (2) in Example 3 are convergent for all values of x.

To find cos 18°.

Now
$$x = 18° = \frac{\pi}{10} = \frac{3.14}{10} = 0.314$$

Putting $x = \frac{\pi}{10}$ in (2), we get

$$\cos 18° = \cos\frac{\pi}{10} = 1 - \left(\frac{\pi}{10}\right)^2 \cdot \frac{1}{2!} + \left(\frac{\pi}{10}\right)^4 \frac{1}{4!} - \ldots\ldots$$
$$= 1 - \frac{(0.314)^2}{2!} + \frac{(0.314)^4}{4!} - \ldots\ldots \qquad \left[\because \frac{\pi}{10} = .314 \text{ approx.}\right]$$
$$= 1 - 0.04929 + 0.00040 = 0.95111 = 0.9511 \text{ nearly.}$$

Example 4. *Expand* $\sin^{-1} x$ *upto four terms in powers of x.*

(M.D.U. 2000 ; K.U. 2006 S)

Sol. Let $f(x) = \sin^{-1} x \quad \therefore \quad f(0) = 0$

Also $f'(x) = \frac{1}{\sqrt{1-x^2}} = (1-x^2)^{-1/2} \quad \therefore \quad f'(0) = 1$

Expanding by Binomial theorem,

$$f'(x) = 1 + \left(-\frac{1}{2}\right)(-x^2) + \frac{\left(-\frac{1}{2}\right)\left(-\frac{3}{2}\right)(-x^2)^2}{2!} + \frac{\left(-\frac{1}{2}\right)\left(-\frac{3}{2}\right)\left(-\frac{5}{2}\right)}{3!}(-x^2)^3 + \ldots\ldots$$

$$\left[\because \text{ By Binomial theorem } (1+x)^n = 1 + nx + \frac{n(n-1)}{2!}x^2 + \frac{n(n-1)(n-2)}{3!}x^3 + \ldots\ldots\right]$$

or $$f'(x) = 1 + \frac{1}{2}x^2 + \frac{3}{8}x^4 + \frac{15}{48}x^6 + \ldots\ldots$$

$$f''(x) = x + \frac{3}{2}x^3 + \frac{15}{8}x^5 + \ldots\ldots \qquad \therefore \quad f''(0) = 0$$

$$f'''(x) = 1 + \frac{9}{2}x^2 + \frac{75}{8}x^4 + \ldots\ldots \qquad \therefore \quad f'''(0) = 1$$

$$f^{iv}(x) = 9x + \frac{75}{2}x^3 + \ldots\ldots \qquad f^{iv}(0) = 0$$

$$f^{v}(x) = 9 + \frac{225}{2}x^2 + \ldots\ldots \qquad f^{v}(0) = 9$$

$$f^{vi}(x) = 225x + \ldots\ldots \qquad f^{vi}(0) = 0$$

and proceeding further, $f^{vii}(0) = 225$, etc.

Putting these values in Maclaurin's expansion,

$$f(x) = f(0) + xf'(0) + \frac{x^2}{2!}f''(0) + \frac{x^3}{3!}f'''(0) + \ldots\ldots,$$

we have $$\sin^{-1}x = 0 + x + \frac{x^2}{2!}(0) + \frac{x^3}{3!}(1) + \frac{x^4}{4!}(0) + \frac{x^5}{5!}(9) + \frac{x^6}{6!}(0) + \frac{x^7}{7!}(225) + \ldots\ldots$$

$$= x + \frac{x^3}{3!} + \frac{9x^5}{5!} + \frac{225x^7}{7!} + \ldots\ldots$$

$$= x + \frac{1}{2}\cdot\frac{x^3}{3} + \frac{1.3}{2.4}\cdot\frac{x^5}{5} + \frac{1.3.5}{2.4.6}\frac{x^7}{7} + \ldots\ldots$$

Note. To expand an alone inverse function ; find its first derivative. *Expand by Binomial Theorem* and then find other derivatives.

Example 5. *Expand* $\tan^{-1}x$.

Sol. Let $$f(x) = \tan^{-1}x, \qquad \therefore \quad f(0) = 0$$

then $$f'(x) = \frac{1}{1+x^2} = (1+x^2)^{-1} \qquad \therefore \quad f'(0) = 1$$

$$= 1 - x^2 + x^4 - x^6 + \ldots\ldots \qquad \text{[By Binomial Theorem]}$$

$$f''(x) = -2x + 4x^3 - 6x^5 + \ldots.. \qquad f''(0) = 0$$

$$f'''(x) = -2 + 12x^2 - 30x^4 + \ldots\ldots \qquad f'''(0) = -2$$

$$f^{iv}(x) = 24x - 120x^3 + \ldots\ldots \qquad f^{iv}(0) = 0$$

$$f^{v}(x) = 24 - 360x^2 + \ldots\ldots \qquad f^{v}(0) = 24$$

and so on ....

Putting these values of $f(0), f'(0), f''(0)$ etc. in Maclaurin's expansion,

$$f(x) = f(0) + xf'(0) + \frac{x^2}{2!}f''(0) + \frac{x^3}{3!}f'''(0) + \ldots\ldots, \text{ we have}$$

$$\tan^{-1}x = 0 + x + \frac{x^2}{2!}(0) + \frac{x^3}{3!}(-2) + \frac{x^4}{4!}(0) + \frac{x^5}{5!}(24) + \ldots\ldots$$

or $$\tan^{-1}x = x - \frac{x^3}{3} + \frac{x^5}{5} - \frac{x^7}{7} + \ldots\ldots$$

Note 1. The expansion for $\tan^{-1} x$ is valid only if $-1 < x \le 1$.

2. This expansion for $\tan^{-1} x$ known as **Gregory's series** is very useful for finding the value of π.

Example 6. *Expand tan x by Maclaurin's Theorem as far as x^5 and hence find the value of tan 46° 30′ upto four decimal places.* (K.U. 1998, 2005 ; M.D.U. 1997)

Sol. Let $f(x) = \tan x$ $\therefore$ $f(0) = 0$

then $f'(x) = \sec^2 x = 1 + \tan^2 x$ **Note this step** $\therefore$ $f'(0) = 1$

$f''(x) = 2 \tan x \sec^2 x = 2 \tan x (1 + \tan^2 x)$

$= 2 \tan x + 2 \tan^3 x$ $\therefore$ $f''(0) = 0$

$f'''(x) = 2 \sec^2 x + 6 \tan^2 x \sec^2 x$

$= 2(1 + \tan^2 x) + 6 \tan^2 x (1 + \tan^2 x)$

$= 2 + 8 \tan^2 x + 6 \tan^4 x$ $\therefore$ $f'''(0) = 2$

$f^{iv}(x) = 16 \tan x \sec^2 x + 24 \tan^3 x \sec^2 x$

$= 8 \sec^2 x(2 \tan x + 3 \tan^3 x)$

$= 8(1 + \tan^2 x)(2 \tan x + 3 \tan^3 x)$

$= 16 \tan x + 40 \tan^3 x + 24 \tan^5 x$ $\therefore$ $f^{iv}(0) = 0$

$f^{v}(x) = 16 \sec^2 x + 120 \tan^2 x \sec^2 x + 120 \tan^4 x \sec^2 x$

$= 8 \sec^2 x (2 + 15 \tan^2 x + 15 \tan^4 x)$ $\therefore$ $f^{v}(0) = 16$

Putting these values in Maclaurin's expansion,

$$f(x) = f(0) + xf'(0) + \frac{x^2}{2!} f''(0) + \frac{x^3}{3!} f'''(0) + \frac{x^4}{4!} f^{iv}(0) + \frac{x^5}{5!} f^{v}(0) + \ldots\ldots$$

we have $$\tan x = 0 + x + \frac{x^3}{3!} . 2 + \frac{x^5}{5!} 16 \ldots\ldots$$

or $$\mathbf{\tan x = x + \frac{x^3}{3} + \frac{2}{15} x^5 + \ldots\ldots} \qquad \ldots(1)$$

Deduction. Here $x = 46° \, 30' = \left(46\frac{1}{2}\right)^\circ$ $\quad | \because \; 1° = 60' \;\; \therefore \;\; 30' = \left(\frac{1}{2}\right)^\circ$

$$= \left(\frac{93}{2}\right)^\circ = \frac{93}{2} \times \frac{\pi}{180} \text{ Radians}$$

$$= \frac{31}{120} \times \frac{22}{7} = \frac{31 \times 11}{60 \times 7} = \frac{314}{420} = 0.812$$

Putting $x = 46° \, 30' = 0.812$ in eqn. (1), we have

$$\tan 46° \, 30' = 0.812 + \frac{(0.812)^3}{3} + \frac{2}{15} (0.812)^5$$

$$= 0.812 + \frac{0.5353}{3} + \frac{2}{15} (0.3530)$$

$$= 0.812 + 0.1784 + 0.047 = \mathbf{1.0374.}$$

Note. The result of above eqn. (1) can be used as a formula also.

Example 7. *Expand by Maclaurin's Theorem* $\frac{e^x}{e^x + 1}$, *as far as x^3.*

Sol. Let $f(x) = \frac{e^x}{e^x + 1}$ $\therefore$ $f(0) = \frac{e^0}{e^0 + 1} = \frac{1}{2}$

then $$f'(x) = \frac{(e^x+1)e^x - e^x . e^x}{(e^x+1)^2} = \frac{e^x}{(e^x+1)^2} \quad \therefore \quad f'(0) = \frac{e^0}{(e^0+1)^2} = \frac{1}{2^2}$$

$$f''(x) = \frac{(e^x+1)^2 e^x - e^x\, 2(e^x+1)e^x}{(e^x+1)^4} = \frac{(e^x+1)e^x - 2e^{2x}}{(e^x+1)^3}$$

$$= \frac{e^{2x} + e^x - 2e^{2x}}{(e^x+1)^3} = \frac{e^x - e^{2x}}{(e^x+1)^3} \quad \therefore \quad f''(0) = \frac{e^0 - e^0}{(e^0+1)^3} = 0$$

$$f'''(x) = \frac{(e^x+1)^3 (e^x - 2e^{2x}) - (e^x - e^{2x}) . 3(e^x+1)^2 e^x}{(e^x+1)^6}$$

$$= \frac{(e^x+1)^2 [(e^x+1)(e^x - 2e^{2x}) - 3e^x(e^x - e^{2x})]}{(e^x+1)^6}$$

$$= \frac{e^x - 4e^{2x} + e^{3x}}{(e^x+1)^4} \quad \therefore \quad f'''(0) = -\frac{1}{2^3}$$

Putting these values of $f(0)$, $f'(0)$, $f''(0)$, $f'''(0)$ in Maclaurin's expansion

$$f(x) = f(0) + xf'(0) + \frac{x^2}{2!} f''(0) + \frac{x^3}{3!} f'''(0) + \ldots\ldots$$

We have $\dfrac{e^x}{e^x+1} = \dfrac{1}{2} + \dfrac{x}{4} - \dfrac{x^3}{48} + \ldots\ldots$

Example 8. *Prove by Maclaurin's Theorem that*

$$e^{\sin x} = 1 + x + \frac{x^2}{1.2} - \frac{3.x^4}{1.2.3.4} + \ldots\ldots \qquad \text{(K.U. 1999, 96 S)}$$

Sol. Let $f(x) = e^{\sin x}$ $\therefore$ $f(0) = e^0 = 1$

$f'(x) = e^{\sin x} \cos x$ $\therefore$ $f'(0) = e^0 \cos 0 = 1$

$f''(x) = e^{\sin x}(-\sin x) + \cos x\, e^{\sin x} \cos x$

$= e^{\sin x} [\cos^2 x - \sin x]$ $\therefore$ $f''(0) = e^0 [1 - 0] = 1$

$f'''(x) = e^{\sin x} [2 \cos x(-\sin x) - \cos x] + e^{\sin x} \cos x . [\cos^2 x - \sin x]$

$= e^{\sin x} \cos x . [-2 \sin x - 1 + \cos^2 x - \sin x]$

$= -e^{\sin x} \cos x[3 \sin x + \sin^2 x]$ $\therefore$ $f'''(0) = 0$

$f^{iv}(x) = -e^{\sin x} \cos x[3 \cos x + 2 \sin x \cos x] + e^{\sin x} . \sin x[3 \sin x + \sin^2 x]$
$- [3 \sin x + \sin^2 x] \cos x\, e^{\sin x} . \cos x$

$$\left[\because \frac{d}{dx}(uvw) = uv\frac{dw}{dx} + uw\frac{dv}{dx} + vw\frac{du}{dx}\right]$$

$\therefore$ $f^{iv}(0) = -3$

Putting these values in Maclaurin's Theorem, namely

$$f(x) = f(0) + xf'(0) + \frac{x^2}{2!} f''(0) + \frac{x^3}{3!} f'''(0) + \frac{x^4}{4!} f^{iv}(0) + \ldots\ldots,$$

we have $$e^{\sin x} = 1 + x.1 + \frac{x^2}{2!} . 1 + \frac{x^3}{3!} . 0 + \frac{x^4}{4!}(-3) + \ldots\ldots = 1 + x + \frac{x^2}{1.2} - \frac{3x^4}{1.2.3.4} + \ldots\ldots$$

Example 9. *Expand $e^{ax} \cos bx$, as an infinite series of ascending powers of x. Give also the (n + 1)th term of the above series.*

Sol. Let $f(x) = e^{ax} \cos bx,$

$f'(x) = a . e^{ax} \cos bx - e^{ax} \sin bx.b = e^{ax} (a \cos bx - b \sin bx)$

$$f''(x) = ae^{ax}(a\cos bx - b\sin bx) + e^{ax}(-ab\sin bx - b^2\cos bx)$$
$$= e^{ax}[(a^2 - b^2)\cos bx - 2ab\sin bx]$$
$$f'''(x) = ae^{ax}[(a^2 - b^2)\cos bx - 2ab\sin bx] + e^{ax}[(a^2 - b^2)(-b\sin bx) - 2ab^2\cos bx]$$
$$= e^{ax}[a(a^2 - b^2)\cos bx - 2a^2b\sin bx - b(a^2 - b^2)\sin bx - 2ab^2\cos bx]$$
$$= e^{ax}[a^3\cos bx + b^3\sin bx - 3ab^2\cos bx - 3a^2b\sin bx]$$
....

$$f^n(x) = (a^2 + b^2)^{n/2} e^{ax}\cos\left(bx + n\tan^{-1}\frac{b}{a}\right)$$

Putting $x = 0$ in the above

$$f(0) = 1, f'(0) = a, f''(0) = a^2 - b^2$$
$$f'''(0) = a(a^2 - 3b^2), \ldots\ldots\ldots f^n(0) = (a^2 + b^2)^{n/2}\cos\left(n\tan^{-1}\frac{b}{a}\right)$$

Putting these values of $f(0), f'(0)$ etc. in Maclaurin's expansion

$$f(x) = f(0) + xf'(0) + \frac{x^2}{2!}f''(0) + \frac{x^3}{3!}f'''(0) + \ldots\ldots + \frac{x^n}{n!}f^n(0) + \ldots\ldots,$$

we have $e^{ax}\cos bx = 1 + ax + \dfrac{a^2 - b^2}{2!}x^2 + \dfrac{a(a^2 - 3b^2)}{3!}x^3 + \ldots\ldots$

$$\ldots\ldots + \frac{(a^2 + b^2)^{n/2}}{n!}x^n\cos\left(n\tan^{-1}\frac{b}{a}\right) + \ldots\ldots$$

Example 10. *Expand sin $(e^x - 1)$ upto and including the term x^4.*

(M.D.U. 1996 ; K.U. 2006)

Sol. Let $f(x) = \sin(e^x - 1)$.

Put $x = 0, f(0) = \sin(e^0 - 1) = \sin 0 = 0$

$\therefore$ $f'(x) = \cos(e^x - 1) \,.\, e^x = e^x\cos(e^x - 1)$

Put $x = 0, f'(0) = e^0\cos(1 - 1) = 1\cos 0 = 1.$

Again diff. w.r.t. x, $f''(x) = e^x[-\sin(e^x - 1)]\,e^x + \cos(e^x - 1)\,.\,e^x$

or $f''(x) = -e^{2x}\sin(e^x - 1) + e^x\cos(e^x - 1)$

Put $x = 0, f''(0) = 0 + 1\cos 0 = 1.$

Again diff. w.r.t. x,

$$f'''(x) = -[e^{2x}\cos(e^x - 1)\,.\,e^x + \sin(e^x - 1)\,.\,e^{2x}\,.\,2]$$
$$+ e^x[-\sin(e^x - 1)\,.\,e^x] + \cos(e^x - 1)\,.\,e^x$$

or $f'''(x) = -e^{3x}\cos(e^x - 1) - 3e^{2x}\sin(e^x - 1) + e^x\cos(e^x - 1)$

Put x = 0, $f'''(0) = -1 - 0 + 1 = 0.$

Again diff. w.r.t. x,

$$f^{iv}(x) = -[-e^{3x}\sin(e^x - 1)\,.\,e^x + \cos(e^x - 1)\,.\,e^{3x}\,.\,3]$$
$$-3[e^{2x}\cos(e^x - 1)\,.\,e^x + \sin(e^x - 1)\,.\,e^{2x}\,.\,2]$$
$$-e^x\sin(e^x - 1)\,.\,e^x + \cos(e^x - 1)\,.\,e^x.$$

Put x = 0, $f^{iv}(0) = -3 - 3(1 + 0) - 0 + 1 = -5$

Putting these values of $f(0), f'(0), f''(0), f'''(0), f^{iv}(0)$ in Maclaurin's expansion

$$f(x) = f(0) + xf'(0) + \frac{x^2}{2!}f''(0) + \frac{x^3}{3!}f'''(0) + \frac{x^4}{4!}f^{iv}(0) + \ldots\ldots,$$

We have $\sin(e^x - 1) = 0 + x(1) + \frac{x^2}{2!}(1) + \frac{x^3}{3!}(0) + \frac{x^4}{4!}(-5) + ...$

or $\sin(e^x - 1) = x + \frac{x^2}{2} - \frac{5x^4}{24} + ...$

Example 11. *Show that* $\log \frac{\tan x}{x} = \frac{x^2}{3} + \frac{7}{90}x^4 + ...$ (M.D.U. 2002)

Sol. L.H.S. $= \log \frac{\tan x}{x}$

We know that $\tan x = x + \frac{x^3}{3} + \frac{2}{15}x^5 + ...$ (By Eqn. (1) Example 6, Page 88)

$$\therefore \quad \text{L.H.S.} = \log\left(\frac{x + \frac{x^3}{3} + \frac{2}{15}x^5 + ...}{x}\right) = \log\left[1 + \frac{x^2}{3} + \frac{2}{15}x^4 + ...\right]$$

$$= \log(1+z) \quad \text{where} \quad z = \frac{x^2}{3} + \frac{2}{15}x^4 + ... = z - \frac{z^2}{2} + \frac{z^3}{3} - ......$$

Putting $z = \frac{x^2}{3} + \frac{2}{15}x^4 + ...$

$$= \left(\frac{x^2}{3} + \frac{2}{15}x^4 + ...\right) - \frac{1}{2}\left(\frac{x^2}{3} + \frac{2}{15}x^4 + ...\right)^2 + \frac{1}{3}\left(\frac{x^2}{3} + \frac{2}{15}x^4 + ...\right)^3$$

$$= \frac{x^2}{3} + \frac{2}{15}x^4 + ... - \frac{1}{2}\left(\frac{x^4}{9} + ...\right) = \frac{x^2}{3} + \frac{2}{15}x^4 - \frac{x^4}{18} - ......$$

$$= \frac{x^2}{3} + \frac{7x^4}{90} + ... = \text{R.H.S.}$$

EXERCISE 1

Apply Maclaurin's Theorem to prove (1 – 4)

1. (*a*) $\log \sec x = \frac{x^2}{2} + \frac{x^4}{12} + \frac{x^6}{45} + ...$ (*b*) $\log \cos x = -\frac{x^2}{2!} - 2.\frac{x^4}{4!} - 16.\frac{x^6}{6!} - .....$

2. (*a*) $\sec x = 1 + \frac{x^2}{2!} + \frac{5x^4}{4!} + ...$ (*b*) $\cos^2 x = 1 - x^2 + \frac{x^4}{3} + ...$

3. $\tan^{-1}(1+x) = \frac{\pi}{4} + \frac{x}{2} - \frac{x^2}{4} + \frac{x^3}{12} + ...$

4. Prove that $e^{ax}\sin bx = bx + abx^2 + \frac{b(3a^2 - b^2)}{3!}x^3 + ...$

$$+ ...... \frac{(a^2 + b^2)^{n/2}}{n!} x^n \sin\left(n\tan^{-1}\frac{b}{a}\right) + ...$$

(*M.D.U. 1995*)

5. (a) Show that $\tan^{-1}\dfrac{2x}{1-x^2} = 2\left(x - \dfrac{x^3}{3} + \dfrac{x^5}{5} - \ldots\right)$.

[**Hint.** Put $x = \tan\theta$.]

(b) Expand $\sin^{-1}\left(\dfrac{2x^2}{1+x^4}\right)$ in series of ascending powers of x. *(M.D.U. 2006)*

[**Hint.** Put $x^2 = \tan\theta$.]

(c) Show that $\tan^{-1}\dfrac{\sqrt{1+x^2}-1}{x} = \dfrac{x}{2} - \dfrac{x^3}{6} + \dfrac{x^5}{10} - \dfrac{x^7}{14} + \ldots$

[**Hint.** Put $x = \tan\theta$.]

6. Use Maclaurin's Theorem to find the expansion in the ascending powers of x of $\log_e(1 + e^x)$ upto the term including x^4.

7. Use Maclaurin's theorem to prove that $\log(1 + \sin x) = x - \dfrac{x^2}{2} + \dfrac{x^3}{6} - \dfrac{x^4}{12} + \ldots$

Point out the case of failure, if any.

[**Hint.** See Note Example 2 Page 85.]

8. Prove that $e^x \log(1 + x) = x + \dfrac{x^2}{2!} + \dfrac{2x^3}{3!} + \ldots$

9. (a) Apply Maclaurin's theorem to show that $e^x \cos x = 1 + x - \dfrac{2x^3}{3!} - \dfrac{2^2x^4}{4!} - \dfrac{2^2x^5}{5!} + \ldots$

(b) Prove that $e^x \sec x = 1 + x + 2.\dfrac{x^2}{2!} + 4.\dfrac{x^3}{3!} + \ldots$

10. Expand by Maclaurin's theorem $\tan^2 x$ in ascending powers of x as far as x^4.

11. Expand $\log(1 + \sin^2 x)$ in powers of x as far as the term containing x^4.

12. (a) By Maclaurin's theorem expand $\log[1 - \log(1 - x)]$ in powers of x as far as the term in x^3.

(b) Expand by Maclaurin's series $\log[1 + \log(1 + x)]$ in powers of x as far as term containing x^3.

13. (a) By Maclaurin's theorem find the first four terms in the expansion of $\log(1 + \tan x)$.

(b) Prove that $\log(1 + \tan x) = x - \dfrac{x^2}{2} + \dfrac{2x^3}{3} - \dfrac{7x^4}{12} + \ldots$ *(K.U. 1998 S)*

14. Show that $\log(1 - x + x^2) = -x + \dfrac{x^2}{2} + \dfrac{2x^3}{3} + \dfrac{x^4}{4} - \dfrac{x^5}{5} - \dfrac{2x^6}{6} - \dfrac{x^7}{7} + \dfrac{x^8}{8} \ldots$

$\left[\textbf{Hint. } \log(1-x+x^2) = \log\left(\dfrac{1+x^3}{1+x}\right) = \log(1+x^3) - \log(1+x).\right]$

ANSWERS

4. (i) $1 + ax + \dfrac{a^2-b^2}{2!}x^2 + \dfrac{a(a^2-3b^2)}{3!}x^3 + \ldots$

5. (b) $2\left[x^2 - \dfrac{1}{3}x^6 + \dfrac{1}{5}x^{10} - \ldots\right]$

6. $\log 2 + \dfrac{x}{2} + \dfrac{x^2}{8} - \dfrac{x^4}{192} + \ldots$

7. Case of failure is $\sin x = -1$

10. $x^2 + \dfrac{2x^4}{3} + \ldots$

11. $x^2 - \frac{5}{6}x^4 + ...$ 12. (a) $x + \frac{x^3}{6} + ...$ (b) $x - x^2 + \frac{7}{6}x^3 + ...$

13. $x - \frac{x^2}{2} + \frac{2x^3}{3} - \frac{7x^4}{12} + ...$

6. Application of Taylor's Theorem

Working Rule *can be stated as :*

Step 1. Put the given function equal to $f(x + h)$.

Step 2. Put $h = 0$ and write $f(x)$.

Step 3. Differentiate $f(x)$ a number of times and obtain $f'(x), f''(x), f'''(x)$, etc. ....

Step 4. Now substitute the values of $f(x), f'(x), f''(x)$, ... in Taylor's expansion

$$f(x + h) = f(x) + hf'(x) + \frac{h^2}{2!}f''(x) + \frac{h^3}{3!}f'''(x) + ...$$

Example 1. *Apply Taylor's Theorem to find the expansion of log sin (x + h).*

Sol. (*i*) Let $f(x + h) = \log \sin (x + h)$

(*ii*) Putting $h = 0$, we have $f(x) = \log \sin x$

(*iii*) $\therefore$ $f'(x) = \frac{\cos x}{\sin x} = \cot x$

$f''(x) = -\text{cosec}^2 x$

$f'''(x) = -2 \text{ cosec } x(-\text{cosec } x \cot x) = 2 \text{ cosec}^2 x \cot x.$

....

Substituting these values of $f(x), f'(x), f''(x)$ and $f'''(x)$ in Taylor's expansion,

$$f(x + h) = f(x) + hf'(x) + \frac{h^2}{2!}f''(x) + \frac{h^3}{3!}f'''(x) + ..., \text{ we get}$$

$$\log \sin (x + h) = \log \sin x + h \cot x - \frac{h^2}{2!}(\text{cosec}^2 x) + \frac{h^3}{3!}(2 \text{ cosec}^2 x \cot x) + ...$$

$$= \log \sin x + h \cot x - \frac{h^2}{2}\text{cosec}^2 x + \frac{h^3}{3}\text{cosec}^2 x \cot x + ...$$

Example 2. *Expand $\sin^{-1}(x + h)$ in powers of h.*

Sol. (*i*) Let $f(x + h) = \sin^{-1}(x + h)$

(*ii*) Putting $h = 0$, we have $f(x) = \sin^{-1} x$.

(*iii*) $\therefore$ $f'(x) = \frac{1}{\sqrt{1 - x^2}}$

$$f''(x) = -\frac{1}{2}(1 - x^2)^{-3/2}(-2x) = \frac{x}{(1 - x^2)^{3/2}}$$

$$f'''(x) = \frac{(1 - x^2)^{3/2} - \frac{3}{2}(1 - x^2)^{1/2}(-2x) . x}{(1 - x^2)^3}$$

$$= \frac{(1 - x^2)^{1/2}(1 - x^2 + 3x^2)}{(1 - x^2)^3} = \frac{1 + 2x^2}{(1 - x^2)^{5/2}}$$

Putting these values of $f(x), f'(x), f''(x), f'''(x)$ in Taylor's expansion,

(iv) $$f(x+h) = f(x) + hf'(x) + \frac{h^2}{2!} f''(x) + \ldots,\text{ we have}$$

$$\sin^{-1}(x+h) = \sin^{-1} x + \frac{h}{\sqrt{1-x^2}} + \frac{x}{(1-x^2)^{3/2}} \cdot \frac{h^2}{2!} + \frac{1+2x^2}{(1-x^2)^{5/2}} \cdot \frac{h^3}{3!} + \ldots$$

Example 3. (a) *Expand sin (x + y) in powers of y and deduce that sin (x + y) = sin x cos y + sin y cos x.*

(b) *Obtain the value of sin 31° correct to four places of decimals.*

Sol. (a) (i) Put $f(x + y) = \sin(x + y)$

(ii) Putting $y = 0$, we have $f(x) = \sin x$

(iii) $\therefore$ $f'(x) = \cos x$; $f''(x) = -\sin x$,

$f'''(x) = -\cos x$, $f^{iv}(x) = \sin x$, ...

..............., and $f^n(x) = \sin(x + n\pi/2)$

By Taylor's expansion,

$$f(x+y) = f(x) + yf'(x) + \frac{y^2}{2!} f''(x) + \ldots\ldots + \frac{y^n}{n!} f^n(x) + \ldots\ldots$$

Putting values of $f(x), f'(x), \ldots\ldots f^n(x)$,

we have, $$\sin(x+y) = \sin x + y\cos x - \frac{y^2}{2!}\sin x - \frac{y^3}{3!}\cos x$$

$$+ \frac{y^4}{4!}\sin x + \ldots\ldots + \frac{y^n}{n!}\sin\left(x + \frac{n\pi}{2}\right) + \ldots\ldots \quad \ldots(1)$$

$$= \sin x\left[1 - \frac{y^2}{2!} + \frac{y^4}{4!} - \ldots\ldots\right] + \cos x\left[y - \frac{y^3}{3!} + \frac{y^5}{5!} - \ldots\ldots\right]$$

$$= \sin x \cos y + \cos x \sin y.$$

(b) Put $x = 30° = \frac{\pi}{6}$ and $y = 1° = \frac{\pi}{180} = .0175$ radians in (1),

Then $$\sin 31° = \sin\frac{\pi}{6} + \frac{\pi}{180}\cos\frac{\pi}{6} - \left(\frac{\pi}{180}\right)^2 \frac{1}{2!}\sin\frac{\pi}{6} \ldots\ldots$$

$$= .5 + (.0175)(.866) - \frac{(.0175)^2}{2!}(.5) \ldots\ldots$$

$$= .5 + .0151550 - .0000765 = .5150785 = .5151,$$

correct to four places of decimals.

Example 4. (i) *If $f(x) = x^3 + 8x^2 + 15x - 24$, calculate the value of $f(\frac{11}{10})$ by the application of Taylor's series.* (M.D.U. 1999 ; K.U. 1998 S)

(ii) *$f(x) = x^3 - 2x + 5$, find the value of f(2.001) with the help of Taylor's Theorem. Find the approximate change in the value of f(x) when x changes from 2 to 2.001.*

Sol. (i) By Taylor's Theorem, $$f(x+h) = f(x) + hf'(x) + \frac{h^2}{2!} f''(x) + \frac{h^3}{3!} f'''(x) + \ldots\ldots \quad \ldots(1)$$

To find $f\left(\frac{11}{10}\right)$ *i.e.,* $f\left(1 + \frac{1}{10}\right)$

Put $\quad x = 1$ and $h = \dfrac{1}{10}$ in the series (1),

$$f\left(\frac{11}{10}\right) = f\left(1+\frac{1}{10}\right) = f(1) + \frac{1}{10} f'(1) + \frac{1}{10^2} \cdot \frac{1}{2!} f''(1) + \frac{1}{3!} \cdot \frac{1}{(10)^3} f'''(1) + \ldots\ldots \qquad \ldots(i)$$

Now $\quad f(x) = x^3 + 8x^2 + 15x - 24 \quad \therefore \quad f(1) = 0$

$f'(x) = 3x^2 + 16x + 15 \quad \therefore \quad f'(1) = 34$

$f''(x) = 6x + 16 \quad \therefore \quad f''(1) = 22$

$f'''(x) = 6 \quad \therefore \quad f'''(1) = 6$

$f^{iv}(x) = 0 \quad \therefore \quad f^{iv}(1) = 0$

Substituting values of $f(1), f'(1), f''(1)$ ..... etc. in (i), we get

$$f\left(1+\frac{1}{10}\right) = 0 + \frac{1}{10} 34 + \frac{11}{100} + \frac{1}{1000} = 3.4 + .11 + .001 = 3.511.$$

(ii) Here put $x = 2$ and $h = .001$ in Taylor's series

$$f(x + h) = f(x) + hf'(x) + \frac{h^2}{2!} f''(x) + \frac{h^3}{3!} f'''(x) + \ldots..,$$

we have $$f(2.001) = f(2) + (.001) f'(2) + \frac{(.001)^2}{2!} f''(2) + \frac{(.001)^3}{3!} f'''(2) + \ldots.. \qquad \ldots(ii)$$

Now $\quad f(x) = x^3 - 2x + 5 \quad \therefore \quad f(2) = 9$

$f'(x) = 3x^2 - 2 \quad \therefore \quad f'(2) = 10$

$f''(x) = 6x \quad f''(2) = 12$

$f'''(x) = 6 \quad f'''(2) = 6$

$f^{iv}(x) = 0 \quad f^{iv}(2) = 0$

Putting these values in (ii),

$$f(2.001) = 9 + (.001)\,10 + \frac{1}{2!} (.001)^2 (12) + \frac{1}{3!} (.001)^3 (6) + \ldots\ldots$$

$$= 9 + .01 + .000006 + .000000001 = 9.010006001 = 9.01 \text{ approximately}$$

$\therefore$ Approximate change in the value of $f(x)$ as x changes from 2 to 2.001 $= f(2.001) - f(2)$

$= 9.01 - 9 = .01$ approximately.

EXERCISE 2

Apply Taylor's Theorem to prove the following expansions :

1. (a) $a^{x+h} = a^x \left[1 + h \log a + \dfrac{h^2}{2!} (\log a)^2 + \dfrac{h^3}{3!} (\log a)^3 + \ldots..\right]$

(b) $e^{x+h} = e^x + he^x + \dfrac{h^2}{2!} e^x + \ldots\ldots$ $\qquad$ (c) $\dfrac{1}{x+h} = \dfrac{1}{x} - \dfrac{h}{x^2} + \dfrac{h^2}{x^3} - \dfrac{h^3}{x^4} + \ldots\ldots$

(d) $\tan (x + h) = \tan x + h \sec^2 x + h^2 \sec^2 x \tan x + \dfrac{h^3}{3} \sec^2 x(1 + 3 \tan^2 x) + \ldots\ldots$

2. (a) $\cos\left(\dfrac{\pi}{4} + h\right) = \dfrac{1}{\sqrt{2}} \left(1 - h - \dfrac{h^2}{2!} + \dfrac{h^3}{3!} + \ldots\ldots\right)$

(*b*) $\sin\left(\frac{\pi}{4}+\theta\right)=\frac{1}{\sqrt{2}}\left(1+\theta-\frac{\theta^2}{2!}-\frac{\theta^3}{3!}+\ldots\ldots\right)$

3. (*a*) $\sec^{-1}(x+h)=\sec^{-1}x+\frac{h}{x\sqrt{x^2-1}}-\frac{2x^2-1}{x^2(x^2-1)^{3/2}}\cdot\frac{h^2}{2!}+\ldots\ldots$

(*b*) $\tan^{-1}(x+h)=\tan^{-1}x+\frac{h}{(1+x^2)}-\frac{xh^2}{(1+x^2)^2}\ldots\ldots$

4. Prove that $\frac{f(x+h)+f(x-h)}{2}=f(x)+\frac{h^2}{2!}f''(x)+\frac{h^4}{4!}f^{iv}(x)+\ldots\ldots$

5. (*a*) Expand a^x by Taylor's Theorem.

[**Hint.** First expand a^{x+h} in ascending powers of h as in Q. 1(*a*). Then put $x = 0$ and change h into x.]

(*b*) Expand $\sin x$ and $\cos x$ by Taylor's Theorem.

6. (*a*) Find the approximate change in the value of $5x^3 - 3x^2 + 7x - 8$ as x changes from 3 to 3.001. *(M.D.U. 2004 A)*

(*b*) If $f(x) = x^3 - 6x^2 + 7$; find the value of $f\left(\frac{21}{20}\right)$ by Taylor's Theorem. Also find the change in the value of $f(x)$ when x changes from 2 to 2.1. *(M.D.U. 2005 A)*

(*c*) If $f(x) = 3x^3 - 5x^2 + 7$, find the value of $f\left(\frac{21}{20}\right)$ by Taylor's theorem and find the change in the value of $f(x)$ when x changes from 2 to 2.1. *(M.D.U. 1998)*

(*d*) If $f(x) = x^3 + 2x^2 - 5x + 11$, find the value of $f\left(\frac{9}{10}\right)$ with the help of Taylor's series for $f(x + h)$. *(M.D.U. 1995)*

(*e*) If $f(x) = x^3 + 6x^2 + 9$, find the value of $f\left(\frac{11}{10}\right)$ by Taylor's theorem. *(M.D.U. 1999 S)*

7. Calculate the approximate value of $\sqrt{10}$ to four decimal places by taking the first four terms of an appropriate Taylor's expansion.

[**Hint.** Let $f(x)=\sqrt{x}$ $\therefore$ $f(10)=\sqrt{10}$

or $\sqrt{10}=f(10)=f(9+1)=f(a+h)$ where $a = 9$, $h = 1$

$$=f(a)+h\,f'(a)+\frac{h^2}{2!}f''(a)+\frac{h^3}{3!}f'''(a)+\ldots\ldots$$

$$=f(9)+f'(9)+\frac{1}{2!}f''(9)+\frac{1}{3!}f'''(9)+\ldots\ldots]$$

8. (*a*) Calculate the approximate value of $\sqrt{17}$ to four decimal places by taking first three terms of a Taylor's Expansion.

(*b*) Calculate the approximate value of $\sqrt{26}$ to three decimal places by Taylor's expansion.

9. Prove that $f\left(\frac{x^2}{1+x}\right)=f(x)-\frac{x}{1+x}f'(x)+\left(\frac{x}{1+x}\right)^2\frac{1}{2!}f''(x)-\left(\frac{x}{1+x}\right)^3\frac{1}{3!}f'''(x)+\ldots\ldots$

$\left[\textbf{Hint.}\text{ Write }\frac{x^2}{1+x}=\frac{x^2+x-x}{1+x}=\frac{x(x+1)-x}{1+x}=x-\frac{x}{1+x}=x+h,\text{ where }h=-\frac{x}{1+x}.\right]$

ANSWERS

5. (*a*) $a^x = 1 + x \log a + \dfrac{x^2}{2!}(\log a)^2 + \dfrac{x^3}{3!}(\log a)^3 + ...$

(*b*) $\sin x = x - \dfrac{x^3}{3!} + \dfrac{x^5}{5!} - ..., \cos x = 1 - \dfrac{x^2}{2!} + \dfrac{x^4}{4!} - ...$

6. (*a*) 0.124 (*b*) 1.5426 (*c*) 4.960375, 1.733

(*d*) 8.849 (*e*) 17.591

7. 3.1623 **8.** (*a*) 4.123 (*b*) 5.099.

7. Another Form of Taylor's Series

In this form $f(x)$ is expressed as a series in ascending integral powers of $x - a$.

We have $f(x) = f(a + x - a)$. Let $x - a = h$.

$$\therefore \quad f(x) = f(a+h) = f(a) + hf'(a) + \frac{h^2}{2!}f''(a) + \frac{h^3}{3!}f'''(a) + ...$$

Now replacing h by $x - a$, we get

$$f(x) = f(a) + (x-a)f'(a) + \frac{(x-a)^2}{2!}f''(a) + \frac{(x-a)^3}{3!}f'''(a) + ...$$

Example 1. *Expand sin x in ascending powers of* $\left(x - \dfrac{\pi}{2}\right)$. (K.U. 2005 S)

Sol. Here $f(x) = \sin x$

We know that $f(x) = f\left(\dfrac{\pi}{2} + x - \dfrac{\pi}{2}\right)$ $\left[\text{We have done this step to get} \left(x - \dfrac{\pi}{2}\right)\right]$

$$= f(a+h), \text{ where } a = \frac{\pi}{2} \text{ and } h = x - \frac{\pi}{2}$$

i.e.,
$$f(x) = f(a) + hf'(a) + \frac{h^2}{2!}f''(a) + \frac{h^3}{3!}f'''(a) + \frac{h^4}{4!}f^{iv}(a) + ...$$

(By Taylor's Expansion)

Putting values of a and h, we get

$$\sin x = f\left(\frac{\pi}{2}\right) + \left(x - \frac{\pi}{2}\right)f'\left(\frac{\pi}{2}\right) + \frac{\left(x - \frac{\pi}{2}\right)^2}{2!}f''\left(\frac{\pi}{2}\right)$$
$$+ \frac{\left(x - \frac{\pi}{2}\right)^3}{3!}f'''\left(\frac{\pi}{2}\right) + \frac{\left(x - \frac{\pi}{2}\right)^4}{4!}f^{iv}\left(\frac{\pi}{2}\right) + ... \quad ...(1)$$

We have $f(x) = \sin x$ $\quad \therefore \quad f\left(\dfrac{\pi}{2}\right) = 1$

$f'(x) = \cos x$ $\quad \therefore \quad f'\left(\dfrac{\pi}{2}\right) = 0$

$f''(x) = -\sin x$ $\quad \therefore \quad f''\left(\dfrac{\pi}{2}\right) = -1$

$$f'''(x) = -\cos x \qquad \therefore \qquad f'''\left(\frac{\pi}{2}\right) = 0$$

$$f^{iv}(x) = \sin x \qquad \therefore \qquad f^{iv}\left(\frac{\pi}{2}\right) = 1$$

Putting these values of $f\left(\frac{\pi}{2}\right), f'\left(\frac{\pi}{2}\right), f''\left(\frac{\pi}{2}\right), \ldots\ldots$ in (*i*), we get

$$\sin x = 1 - \frac{\left(x - \frac{\pi}{2}\right)^2}{2!} + \frac{\left(x - \frac{\pi}{2}\right)^4}{4!} - \ldots\ldots$$

Example 2. *Obtain the first four terms in the expansion of log sin x in powers of (x – 3).* **(K.U. 1997)**

Sol. Let $f(x) = \log \sin x$

We know that

$$f(x) = f(3 + x - 3) \qquad \textbf{(Note this step)}$$

$$= f(a + h) \text{ where } a = 3 \text{ and } h = x - 3$$

$$= f(a) + hf'(a) + \frac{h^2}{2!} f''(a) + \frac{h^3}{3!} f'''(a) + \ldots\ldots \qquad \text{(By Taylor's Expansion)}$$

Putting values of a and h, we have

$$f(x) = f(3) + (x - 3) f'(3) + \frac{(x-3)^2}{2!} f''(3) + \frac{(x-3)^3}{3!} f'''(3) + \ldots\ldots \qquad \ldots(1)$$

But $f(x) = \log \sin x$; Putting $x = 3$; $f(3) = \log \sin 3$

$$\therefore \quad f'(x) = \frac{1}{\sin x} (\cos x) = \cot x ; \qquad f'(3) = \cot 3$$

$$f''(x) = -\operatorname{cosec}^2 x = -[1 + \cot^2 x] ; f''(3) = -\operatorname{cosec}^2 3$$

$$f'''(x) = -[2 \cot x (-\operatorname{cosec}^2 x)] ;$$

$$= 2 \cot x \operatorname{cosec}^2 x \qquad f'''(3) = 2 \cot 3 \operatorname{cosec}^2 3$$

Putting these values in (1), we get

$$\log \sin x = \log \sin 3 + (x - 3) \cot 3 - \frac{(x-3)^2}{2} \operatorname{cosec}^2 3 + \frac{(x-3)^3}{3} \cot 3 \operatorname{cosec}^2 3 + \ldots\ldots$$

EXERCISE 3

1. (*i*) Expand e^x in power of $(x - 2)$. (*ii*) Expand $\log x$ in powers of $x - k$. (*M.D.U. 1997*)
   (*iii*) Expand a^x in powers of $x - a$. (*iv*) Expand $4x^2 + 7x + 5$ in powers of $x - 3$.
2. Expand $\log \sin x$ in powers of $(x - 2)$. (*K.U. 2000 A*)
3. Use Taylor's Theorem to express the polynomial $2x^3 + 7x^2 + x - 6$ in powers of $(x - 2)$.
4. Expand $\tan x$ in powers of $\left(x - \frac{\pi}{4}\right)$ upto first four terms.

(*K.U. 1996 A*)

5. Show that $f(ax) = f(x) + (a-1)\, x f'(x) + \dfrac{(a-1)^2 x^2}{2!} f''(x) + \ldots\ldots$

[**Hint.** $f(ax) = f(x + ax - x) = f[x + (a-1)x] = f(a+h)$ where $a = x,\ h = (a-1)x$.]

6. If $0 < x \le 2$, then prove that

$$\log x = (x-1) - \frac{(x-1)^2}{2} + \frac{(x-1)^3}{3} - \frac{(x-1)^4}{4} + \ldots\ldots$$

ANSWERS

1. (*i*) $e^2\left[1 + (x-2) + \dfrac{(x-2)^2}{2!} + \ldots\ldots\right]$

(*ii*) $\log k + \dfrac{1}{k}(x-k) - \dfrac{1}{2k^2}(x-k)^2 + \dfrac{1}{3k^3}(x-k)^3 - \ldots\ldots$

(*iii*) $a^a\left[1 + (x-a)\log a + \dfrac{(x-a)^2}{2!}(\log a)^2 + \dfrac{(x-a)^3}{3!}(\log a)^3 + \ldots\ldots\right]$

(*iv*) $62 + 31(x-3) + 4(x-3)^2$.

2. $\log \sin 2 + (x-2)\cot 2 - \dfrac{(x-2)^2}{2}\operatorname{cosec}^2 2 + \dfrac{(x-2)^3}{3}\cot 2 \operatorname{cosec}^2 2 + \ldots\ldots$

3. $40 + 53(x-2) + 19(x-2)^2 + 2(x-2)^3$.

4. $1 + 2\left(x - \dfrac{\pi}{4}\right) + 2\left(x - \dfrac{\pi}{4}\right)^2 + \dfrac{8}{3}\left(x - \dfrac{\pi}{4}\right)^3 + \ldots\ldots$

---

8. Method of Differential Equation

Here below we shall explain a method to expand functions like $(\sin^{-1}x)^2$, $e^{a\sin^{-1}x}$, $\sin(m\sin^{-1}x)$, $\cos(m\sin^{-1}x)$ etc.

These functions can also be expanded by the working rule given in Art. 5, but the finding of successive derivatives for these functions is very complicated and hence avoided.

The working rule for the above mentioned functions is being given below :

1. Put the given function equal to y.

2. Find $y_1 = \dfrac{dy}{dx}$

Then (*i*) Take L.C.M. (if possible)

(*ii*) Square both sides if square roots are there.

(*iii*) Try to get y in R.H.S. (if possible).

3. Again differentiate both sides w.r.t. x to get an equation in y_2, y_1, y.

4. Differentiate both sides n times w.r.t. x by Leibnitz Theorem.

[**Leibnitz Theorem** is

$(uv)_n = {}^nC_0\, u_n \cdot v + {}^nC_1\, u_{n-1} v_1 + {}^nC_2 \cdot u_{n-2} v_2 + \ldots\ldots + {}^nC_n\, uv_n$.]

5. Put $x = 0$ in equations of steps 1, 2, 3, 4.

6. Put $n = 1, 2, 3, 4$ in last equation of step 5.

7. Now to find $y_n(0)$, discuss the two cases when n is even and when n is odd.

Example 1. *Prove that*

$$\sin (m \sin^{-1} x) = mx + \frac{m(1^2 - m^2)}{3!} x^3 + \frac{m(1^2 - m^2)(3^2 - m^2)}{5!} x^5 + \ldots\ldots$$

$$+ \ldots\ldots + \frac{m(1^2 - m^2)(3^2 - m^2)\ldots\ldots[(2n-3)^2 - m^2]x^{2n-1}}{(2n-1)!} + \ldots.. \qquad \text{(K.U. 1996 S)}$$

Sol. Let $y = \sin (m \sin^{-1} x)$ ...(1)

Differentiating w.r.t. x, $y_1 = \cos (m \sin^{-1} x) \dfrac{m}{\sqrt{1-x^2}}$ ...(2)

$$\therefore \qquad \sqrt{1-x^2}\; y_1 = m \cos (m \sin^{-1} x)$$

Squaring both sides $(1 - x^2)y_1^{\,2} = m^2 \cos^2 (m \sin^{-1} x) = m^2 [1 - \sin^2 (m \sin^{-1} x)]$

or $\qquad (1 - x^2)y_1^{\,2} = m^2(1 - y^2)$ [By (1)]

Again differentiating w.r.t. x,

$$(1 - x^2)2y_1y_2 + y_1^{\,2}(-2x) = m^2(-2yy_1) \quad \text{or} \quad 2y_1y_2(1 - x^2) - 2xy_1^2 + 2m^2yy_1 = 0$$

Cancelling $2y_1$; $y_2(1 - x^2) - xy_1 + m^2y = 0$ ...(3)

Differentiating both sides n times ;

$$\frac{d^n}{dx^n}[y_2(1-x^2)] - \frac{d^n}{dx^n}[y_1 \cdot x] + \frac{d^n}{dx^n}(m^2y) = 0$$

or $\quad {}^nC_0 (y_2)_n (1 - x^2) + {}^nC_1(y_2)_{n-1}(-2x) + {}^nC_2(y_2)_{n-2}(-2)$

$$- [{}^nC_0(y_1)_n \cdot x + {}^nC_1 \cdot (y_1)_{n-1} \cdot 1] + m^2y_n = 0$$

or $\quad y_{n+2}(1 - x^2) - 2nxy_{n+1} - n(n-1)y_n - xy_{n+1} - ny_n + m^2y_n = 0$

$$\left(\because \ {}^nC_0 = 1, {}^nC_1 = n, {}^nC_2 = \frac{n(n-1)}{2!}\right)$$

or $\quad (1 - x^2)y_{n+2} - (2n + 1)xy_{n+1} - (n^2 - m^2)y_n = 0$ ...(4)

Putting $\quad x = 0$ in (1), (2), (3) and (4), we get

$y = \sin (0) = 0 = f(0)$

$y_1 = m \cos 0 = m = f'(0)$

$y_2 = 0 = f''(0)$

$y_{n+2} = (n^2 - m^2)y_n$

Putting $\quad n = 1, 2, 3, 4$ in $y_{n+2} = (n^2 - m^2)y_n$; we get

$y_3 = (1^2 - m^2)y_1 = m(1^2 - m^2) = f'''(0)$

$y_4 = (2^2 - m^2)y_2 = 0 = f^{iv}(0)$

$y_5 = (3^2 - m^2)y_3 = m(1^2 - m^2)(3^2 - m^2) = f^{v}(0)$

$y_6 = (4^2 - m^2)y_4 = 0 = f^{vi}(0)$

Generalising

When n is even $\quad y_n$ (at $x = 0$) $= 0 \quad (\because \quad y_2 = 0, y_4 = 0, y_6 = 0)$

and when n is odd

$$y_n \text{ (at } x = 0) = m(1^2 - m^2)(3^2 - m^2) \ldots\ldots [(n-2)^2 - m^2]$$

In particular y_{2n-1} (Replacing n by $(2n - 1)$ in the above equation

$$= m(1^2 - m^2)(3^2 - m^2) \ldots\ldots [(2n-3)^2 - m^2]$$

We know by Maclaurin's Theorem that

$$f(x) = f(0) + xf(0) + \frac{x^2}{2!} f''(0) + \frac{x^3}{3!} f'''(0) + \frac{x^4}{4!} f^{iv}(0) + \frac{x^5}{5!} f^{v}(0) + \ldots\ldots + \frac{x^{2n-1}}{(2n-1)!} f^{2n-1}(0) + \ldots\ldots$$

Putting values

$$\sin(m \sin^{-1} x) = mx + \frac{m(1^2 - m^2)}{3!} x^3 + \frac{m(1^2 - m^2)(3^2 - m^2)}{5!} x^5 + \ldots + \frac{m(1^2 - m^2)(3^2 - m^2) \ldots\ldots [(2n-3)^2 - m^2]}{(2n-1)!} x^{2n-1} + \ldots\ldots$$

Example 2. *If* $y = (\sin^{-1} x)^2$, *show that*

(*a*) $(1 - x^2)\, y_2 - xy_1 = 2$

(*b*) $\dfrac{(\sin^{-1} x)^2}{2} = \dfrac{x^2}{2!} + \dfrac{2^2}{4!} \cdot x^4 + \dfrac{2^2 \cdot 4^2}{6!} x^6 + \ldots\ldots$

Sol. Let $\quad y = (\sin^{-1} x)^2 \quad \ldots(1)$

$$\therefore \quad y_1 = 2 \sin^{-1} x \times \frac{1}{\sqrt{1-x^2}} \quad \ldots(2)$$

i.e., $\sqrt{1-x^2} \cdot y_1 = 2\sqrt{y}$. Squaring both sides $(1 - x^2) y_1^2 = 4y$.

Differentiating again,

$$(1 - x^2) \cdot 2y_1 y_2 + y_1^2 (-2x) = 4y_1$$

Dividing by $2y_1$, $\quad (1 - x^2) y_2 - xy_1 = 2 \quad \ldots(3)$

which proves part (*a*).

Differentiating (3), *n* times by Leibnitz's Theorem,

$$(1 - x^2) y_{n+2} + {}^nC_1 y_{n+1} (-2x) + {}^nC_2 y_n (-2) - [y_{n+1} \cdot x + {}^nC_1 y_n \cdot 1] = 0$$

or $\quad (1 - x^2) y_{n+2} - 2nxy_{n+1} - n(n-1) y_n - xy_{n+1} - ny_n = 0$

or $\quad (1 - x^2) y_{n+2} - (2n+1)\, xy_{n+1} = n^2 y_n \quad \ldots(4)$

Putting $\quad x = 0$, in (1), (2), (3), (4), we get

$$y = 0, \quad y_1 = 0, \quad y_2 = 2$$

$$(y_{n+2})_0 = n^2 (y_n)_0$$

Putting $\quad n = 1, 2, 3, 4$ in the last equation, we get

$$y_3 = 1^2 y_1 = 0$$
$$y_4 = 2^2 y_2 = 2^2 \cdot 2 = 2.2^2$$
$$y_5 = 3^2 y_3 = 0$$
$$y_6 = 4^2 y_4 = 2.2^2 \cdot 4^2$$

....

$\therefore$ By Maclaurin's Theorem,

$$f(x) = f(0) + xf'(0) + \frac{x^2}{2!} f''(0) + \frac{x^3}{3!} f'''(0) + \ldots\ldots$$

or $$y = (y)_0 + x(y_1)_0 + \frac{x^2}{2!}(y_2)_0 + \frac{x^3}{3!}(y_3)_0 + \frac{x^4}{4!}(y_4)_0 + \frac{x^5}{5!}(y_5)_0 + \frac{x^6}{6!}(y_6)_0 + \ldots$$

$$\therefore \quad (\sin^{-1}x)^2 = 0 + x(0) + \frac{x^2}{2!}.\,2 + \frac{x^3}{3!}.\,(0) + \frac{x^4}{4!}(2.2^2) + \frac{x^5}{5!}(0) + \frac{x^6}{6!}(2.2^2.\,4^2) + \ldots\ldots$$

or $$\frac{(\sin^{-1}x)^2}{2} = \frac{x^2}{2!} + \frac{2^2}{4!}x^4 + \frac{2^2.\,4^2}{6!}x^6 + \ldots\ldots \text{ which proves part } (b).$$

EXERCISE 4

1. Show that $e^{a\sin^{-1}x} = 1 + ax + \frac{a^2x^2}{2!} + \frac{a(a^2+1)}{3!}x^3 + \frac{a^2(a^2+2^2)}{4!}x^4 + \frac{a(a^2+1)(a^2+3^2)}{5!}x^5 + \ldots\ldots$

   Hence deduce that $e^{\theta} = 1 + \sin\theta + \frac{\sin^2\theta}{2!} + \frac{2}{3!}\sin^3\theta + \ldots\ldots$ *(M.D.U. 2005)*

   [**Hint.** Put $a = 1$ and $\sin^{-1}x = \theta$ *i.e.,* $x = \sin\theta$ in the above expansion]

2. If $y = \cos(m\sin^{-1}x)$, prove that $(y_{n+2})_0 = (n^2 - m^2)(y_n)_0$.

   Hence evaluate $(y_n)_0$ and expand $y = \cos m(\sin^{-1}x)$.

3. Prove that $(x + \sqrt{1+x^2})^n = 1 + nx + \frac{n^2x^2}{2!} + n\frac{(n^2-1^2)}{3!}x^3 + \frac{n^2(n^2-2^2)}{4!}x^4 + \ldots\ldots$

   (M.D.U. 2004 S ; K.U. 1997 S)

ANSWERS

2. If n is odd ; $y_n = 0$

   If n is even, $y_n = -m^2(2^2-m^2)(4^2-m^2)\ldots\ldots[(n-2)^2-m^2]$

   $$\cos(m\sin^{-1}x) = 1 - \frac{m^2}{2!}x^2 - \frac{m^2(2^2-m^2)}{4!}x^4 - \frac{m^2(2^2-m^2)(4^2-m^2)}{6!}x^6 - \ldots\ldots$$

9. Method of Differentiation and Integration

This method of treatment is best indicated by solved examples.

All the problems which we are going to do under this section can also be done by the methods explained earlier in this chapter.

Example 1. *Expand $\sin^{-1}x$ in ascending powers of x.*

Sol. (*i*) $\frac{d}{dx}(\sin^{-1}x) = \frac{1}{\sqrt{1-x^2}} = (1-x^2)^{-1/2}$

Expanding by Binomial Theorem

(*ii*) $= 1 + (-\frac{1}{2})(-x^2) + \frac{-\frac{1}{2}(-\frac{3}{2})}{2!}(-x^2)^2 + \ldots\ldots = 1 + \frac{1}{2}x^2 + \frac{1.3}{2.4}x^4 + \ldots\ldots$

(*iii*) Integrating *w.r.t. x within the limits 0 and x*

$$\sin^{-1}x = x + \frac{1}{2}.\frac{x^3}{3} + \frac{1.3}{2.4}.\frac{x^5}{5} + \ldots\ldots$$

Example 2. *Prove that* $\log\left(x+\sqrt{1+x^2}\right) = x - \frac{1.x^3}{2.3} + \frac{1.3}{2.4}\cdot\frac{x^5}{5}\ldots\ldots$

Sol. We know that

$$\frac{d}{dx}[\log(x+\sqrt{1+x^2})] = \frac{1}{x+\sqrt{1+x^2}}\left(1+\frac{x}{\sqrt{1+x^2}}\right)$$

$$= \frac{1}{\sqrt{1+x^2}} = (1+x^2)^{-1/2} = 1 + (-\tfrac{1}{2})\,x^2 + \frac{(-\frac{1}{2})(-\frac{1}{2}-1)}{2!}(x^2)^2\ldots\ldots$$

(Expanding by Binomial)

$$= 1 - \frac{1}{2}x^2 + \frac{1.3}{2.4}x^4 - \ldots\ldots$$

Integrating w.r.t. x within the limits 0 to x, we have

$$\log(x+\sqrt{1+x^2}) = x - \frac{1}{2}\cdot\frac{x^3}{3} + \frac{1.3}{2.4}\cdot\frac{x^5}{5} - \ldots\ldots$$

Note. Alternative Method

After finding $f'(x) = 1 - \frac{1}{2}x^2 + \frac{1.3}{2.4}x^4 - \ldots\ldots$

we may find $f''(x)$, $f'''(x)$, $f^{iv}(x)$ and $f^{v}(x)$.

Put $x = 0$ and then apply Maclaurin's Expansion.

EXERCISE 5

1. Obtain the expansion in powers of x of

(*a*) $\sin^{-1}\frac{2x}{1+x^2}$ **[Hint.** Put $x = \tan\theta$] (*b*) $\tan^{-1}\frac{x}{\sqrt{1-x^2}}$ **[Hint.** Put $x = \sin\theta$]

(*c*) $\tan^{-1}\frac{\sqrt{1+x^2}-\sqrt{1-x^2}}{\sqrt{1+x^2}+\sqrt{1-x^2}}$.

[Hint. Put $x^2 = \cos 2\theta$.]

ANSWERS

1. (*a*) $2\left(x - \frac{x^3}{3} + \frac{x^5}{5} - \ldots\ldots\right)$ (*b*) $x + \frac{1^2 . x^3}{3!} + \frac{1^2 . 3^2}{5!}x^5 + \ldots\ldots$ (*c*) $\frac{x^2}{2} + \frac{x^6}{12} + \ldots\ldots$

6

Asymptotes

1. Branches of a Curve

Let us consider the equation of the ellipse $\frac{x^2}{a^2} + \frac{y^2}{b^2} = 1$.

Solving for y, we get

$$y = b\sqrt{1 - \frac{x^2}{a^2}} \quad \ldots (i) \quad \text{or} \quad y = -b\sqrt{1 - \frac{x^2}{a^2}} \quad \ldots (ii)$$

These two equations (*i*) and (*ii*) represent explicity the two branches of the ellipse which in ordinary language we call the upper and lower half of the ellipse. It is to be noticed that the ellipse lies inside the rectangle whose sides are $x = \pm a$ and $y = \pm b$. Thus we see that both the branches of the ellipse lie wholly within a *finite part* of the *x-y* plane and we say, therefore, that both branches of the ellipse are **finite.**

Now let us consider the rectangular hyperbola $x^2 - y^2 = a^2$.

Solving for y, we get $\quad y = \sqrt{x^2 - a^2} \quad$ or $\quad y = -\sqrt{x^2 - a^2}$.

If $x \to \pm\infty$, y also tends of $\pm\infty$. In this case both the branches *extend to infinity* and are said to be the **infinite branches** of the rectangular hyperbola.

Again, let the curve $x^2y^2 = x^2 - y^2$, be considered. Solving for y, we have

$$y = \frac{x}{\sqrt{x^2+1}} \quad \text{or} \quad y = -\frac{x}{\sqrt{x^2+1}}.$$

Now as $x \to \infty$, $y \to 1$ and as $x \to -\infty$, $y \to -1$, along the first branch. Also in the case of second branch, as $x \to \infty$, $y \to -1$ and as $x \to -\infty$, $y \to 1$. Here both the branches are infinite, and x is capable of taking arbitrarily large values whereas y remains finite.

We are already familiar with the symbols $x \to \pm\infty$ and $y \to \pm\infty$. But what does $P \to \infty$ stand for, P being a point on an infinite branch of curve. We give the following definition :

Def. 1. *A point P(x, y) on an infinite branch of a curve is said to tend to infinity along the curve if either x or y or both tend to* $+\infty$ *or* $-\infty$ *as P travels along the branch of the curve.*

2. Asymptote

Def. 1. *A straight line, at a finite distance from the origin is said to be a* (**rectilinear**) **asymptote** *to an infinite branch of a curve, if the perpendicular distance of a point P on that branch from the straight line tends to zero, as P tends to infinity along the branch.*

Def. 2. *Another definition* of an asymptote is as follows :

A (**rectilinear**) **asymptote** *to an infinite branch of a curve is the limiting position of the tangent whose point of contact tends to infinity along the branch, but which itself remains at a finite distance from the origin.*

Def. 3. Another definition of asymptote.

If a st. line cuts a curve in two points at an infinite distance from the origin and yet is not itself wholly at infinity is called an asymptote to the curve.

Note 1. In the discussion which follows, we drop the word *rectilinear* and use simply *asymptote* to mean rectilinear asymptote, unless otherwise stated.

2. The asymptotes may be parallel to either x-axis or the y-axis and accordingly they are called *horizontal* and *vertical* asymptotes. If an asymptote is not parallel to y-axis ; it is called and *oblique asymptote.*

3. Asymptotes parallel to axes of Coordinates

(*i*) *To find the asymptote parallel to y-axis, of the curve $y = f(x)$.*

Let the line $x = k$ ...(1)

parallel to y-axis be an asymptote to the curve $y = f(x)$.

Then it is required to determine the value of k.

Let PM be the distance of a point P(x, y) on the curve from the line (1), then PM = $(x - k)$*

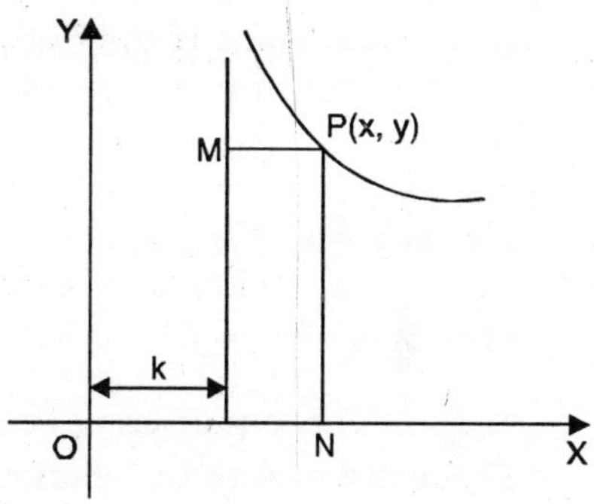

Now by definition of asymptote, if line (1) is an asymptote to the curve, then PM $\to 0$, as P $\to \infty$.

$\therefore$ As P $\to \infty$, PM $= (x - k) \to 0$ or $x \to k$.

If P(x, y) tends to infinity as $x \to k$, only y coordinate tends to infinity (*i.e.*, $+\infty$ or $-\infty$), and this gives

$$\underset{y \to \infty}{\text{Lt}}\ x = k \qquad i.e., \qquad x \to k \text{ as } y \to \infty \qquad \text{...(2)}$$

Hence to find the asymptotes parallel to y-axis, we find from the given equation, the definite values $k_1, k_2, k_3, \ldots\ldots$ to which x tends, as $y \to +\infty$ or $-\infty$. Then $x = k_1$, $x = k_2$, $x = k_3$, ... are the asymptotes parallel to y-axis.

(*ii*) *Asymptotes parallel to x-axis*. Proceeding as above, we arrive at the following method of finding asymptotes parallel to x-axis. *From the given equation, find the definite values $d_1, d_2, d_3, \ldots\ldots$ to which y tends as $x \to +\infty$ or $-\infty$; then $y = d_1$, $y = d_2$, $y = d_3$, ... are asymptotes parallel to the x-axis.*

Asymptotes parallel to coordinate axes for algebraic curve f(x, y) = 0.

Case (*i*) *Asymptotes parallel to y-axis* :

Now let the equation of the algebraic curve $f(x, y) = 0$, after arranging in descending powers of y, be

$$y^n\phi_0(x) + y^{n-1}\phi_1(x) + y^{n-2}\phi_2(x) + \ldots\ldots + \phi_n(x) = 0 \qquad \text{...(3)}$$

where $\phi_0(x), \phi_1(x), \phi_2(x), \ldots\ldots \phi_n(x)$, are polynomials in x.

Dividing equation (3) throughout by y^n ($\because\ y \to \infty$), we get

$$\phi_0(x) + \frac{1}{y}\phi_1(x) + \frac{1}{y^2}\phi_2(x) + \ldots\ldots\ldots + \frac{1}{y^n}\phi_n(x) = 0$$

Taking limits as $y \to \infty$ and $x \to k$ [From (2)]

we get
$$\phi_0(k) = 0 \qquad \text{...(4)}$$

---

*Perpendicular distance of the point (x_1, y_1) from the straight line

$ax + by + c = 0$ is $\dfrac{(ax_1 + by_1 + c)}{\sqrt{a^2 + b^2}}$.

Eliminating k from (1) and (4), combined equation of asymptotes parallel to y-axis is $\phi_0(x) = 0$.

Thus we arrive at the following *working* rule for finding asymptotes || to y-axis.

Rule. *To find asymptotes parallel to y-axis, equate to zero the coefficient of the highest power of y, present in the given equation of the curve. Resolve it now into real linear factors.*

If the coefficient of the highest power of y is either a *constant* or *not resolvable into real linear factors* ; then there are no asymptotes parallel to y-axis.

Case. *(ii) Asymptotes parallel to x-axis for the curve f(x, y) = 0.*

By interchanging y and x in Case (i), we arrive at the following rule.

Rule. *To find asymptotes parallel to x-axis, equate to zero the coefficient of the highest power of x, present in given equation of the curve.* Resolve L.H.S. into real linear factors.

If the coefficient of the highest power of x is either a *constant* or *not resolvable into real linear factors,* there are *no* asymptotes parallel to x-axis.

Note. In the chapter on asymptotes, care should be taken to write the equation of the curve s.t. R.H.S. is zero.

Example 1. *Find the asymptotes, parallel to the axes for the curve :*

(i) $x^2y^2 = a^2(x^2 + y^2)$ (ii) $\dfrac{a^2}{x^2} - \dfrac{b^2}{y^2} = 1.$

Sol. (i) The equation of the curve is (Making R.H.S. zero) $x^2y^2 - a^2(x^2 + y^2) = 0$.

Equating to zero the coefficient of x^2, the highest power of x, we get $y^2 - a^2 = 0$ or $y = \pm a$, which are the asymptotes parallel to x-axis.

Again equating to zero the coefficient of y^2, the highest power of y, we get $x^2 - a^2 = 0$ or $x = \pm a$, which are the asymptotes parallel to y-axis.

(ii) The equation of the curve is $\dfrac{a^2}{x^2} - \dfrac{b^2}{y^2} = 1$

Multiplying by L.C.M. $= x^2y^2$.

or $$a^2y^2 - b^2x^2 = x^2y^2 \quad \text{or} \quad x^2y^2 + b^2x^2 - a^2y^2 = 0$$

Equating to zero the coefficient of x^2, the highest power of x, we get

$$y^2 + b^2 = 0 \quad \text{or} \quad y^2 = -b^2 \quad \text{or} \quad y = \pm ib,$$

which gives imaginary values of y, and therefore there is no asymptote parallel to x-axis.

Again equating to zero the coefficient of y^2, the highest power of y, we get

$$x^2 - a^2 = 0 \quad \text{or} \quad x = \pm a,$$

which are the asymptotes parallel to y-axis.

EXERCISE 1

Find the asymptotes, parallel to axes, of the following curves :

1. $x^2y^2 - a^2(x^2 + y^2) - a^3(x + y) + a^4 = 0.$
2. $\dfrac{a^2}{x^2} + \dfrac{b^2}{y^2} = 1.$
3. $xy^3 - x^3 = a(x^2 + y^2).$
4. $x^2y^2 - y^2 - 2 = 0.$
5. $y^3 - xy^2 = x^2 + 1.$
6. $y = x(x - 2)(x - 3).$

ANSWERS

1. $y = \pm a, x = \pm a$	**2.** $y = \pm b\ ;\ x = \pm a$	**3.** $x = 0$
4. $y = 0, x = \pm 1$	**5.** No.	**6.** No.

4. Oblique Asymptotes

If y = mx + c is an oblique asymptote to any curve f(x, y) = 0 ; then $m = \underset{x \to \infty}{\text{Lt}} \dfrac{y}{x}$ **and** $c = \underset{x \to \infty}{\text{Lt}} (y - mx)$**.**

Let the straight line, $y = mx + c$ *i.e.,* $y - mx - c = 0$ ...(*i*)

be an oblique asymptote of the curve $f(x, y) = 0$, not parallel to y-axis so that m and c are both finite.

Let $P(x, y)$ be any point on the infinite branch of the curve, which corresponds to asymptote (*i*). If $p = PM$ be perpendicular distance of point P from st. line (*i*), then

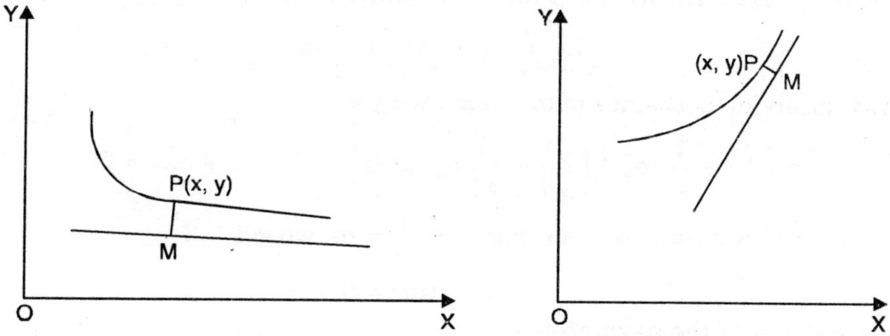

$$p = \pm \frac{y - mx - c}{\sqrt{1 + m^2}} \quad \text{or} \quad y - mx = c \pm p\sqrt{1 + m^2} \qquad ...(ii)$$

dividing both side by x, $\dfrac{y}{x} = m + \dfrac{c}{x} \pm \dfrac{p}{x}\sqrt{1 + m^2}$

Since st. line (*i*) is an asymptote, it follows, from definition, that $p \to 0$, as $x \to \infty$. Taking limits

$$\therefore \quad \underset{x \to \infty}{\text{Lt}} \frac{y}{x} = \underset{x \to \infty}{\text{Lt}} \left[m + \frac{c}{x} \pm \frac{p}{x}\sqrt{1 + m^2} \right] = m$$

Hence $\underset{x \to \infty}{\text{Lt}} \dfrac{y}{x} = m$ ...(*iii*)

Again from (*ii*) $\underset{x \to \infty}{\text{Lt}} (y - mx) = \underset{x \to \infty}{\text{Lt}} \left[c \pm p\sqrt{1 + m^2} \right] = c \pm \sqrt{1 + m^2} \underset{x \to \infty}{\text{Lt}} p = c + 0 = c$

Hence $\underset{x \to \infty}{\text{Lt}} (y - mx) = c$ ...(*iv*)

Note 1. Thus from (*iii*) and (*iv*), m and c are determined, and putting these values in (*i*), the oblique asymptotes can be found out.

2. The above article requires m and c to be finite. In particular m may be zero and therefore the asymptotes parallel to x-axis can also be found by using this article. Thus by the method of this article all the asymptotes of a given curve can be determined, excepting those which are parallel to y-axis.

5. Oblique asymptotes of the general rational algebraic equation

Let the general algebraic equation of nth degree be

$$(a_0y^n + a_1y^{n-1}x + a_2y^{n-2}x^2 + ... + a_{n-1}yx^{n-1} + a_nx^n) + (b_1y^{n-1} + b_2y^{n-2}.x + ... + b_{n-1}yx^{n-2} + b_nx^{n-1}) + ... + (l_{n-1}y + l_nx) + k_n = 0 \quad ...(i)$$

which can be written in the form

$$x^n\phi_n\left(\frac{y}{x}\right) + x^{n-1}\phi_{n-1}\left(\frac{y}{x}\right) + ... + x\phi_1\left(\frac{y}{x}\right) + \phi_0\left(\frac{y}{x}\right) = 0 \quad ...(ii)$$

where $\phi_r\left(\frac{y}{x}\right)$ is a polynomial in $\frac{y}{x}$ of degree r.

[We know that a homogeneous expression of degree n in x and y can be written as $x^n\ \phi_n\left(\frac{y}{x}\right)$]

Let the st. line $\quad y = mx + c \quad ...(iii)$

be an asymptote of curve (ii), where m and c are finite ;

then $\quad m = \underset{x\to\infty}{\text{Lt}} \frac{y}{x}\ ;\ c = \underset{x\to\infty}{\text{Lt}} (y - mx) \quad$ [By Art. 4]

To find m, divide bothsides of (ii) by x^n, we get

$$\phi_n\left(\frac{y}{x}\right) + \frac{1}{x}.\phi_{n-1}\left(\frac{y}{x}\right) + \frac{1}{x^2}\phi_{n-2}\left(\frac{y}{x}\right) + ... + ... = 0$$

Proceeding to limit as $x \to \infty$, so that $\underset{x\to\infty}{\text{Lt}} \frac{y}{x} = m$, we get

$$\boldsymbol{\phi_n(m) = 0} \quad ...(iv)$$

which gives the slopes of the asymptotes.

To find c, corresponding to a value of m.

Let $y - mx = p$, so that as $x \to \infty, p \to c \quad \left[\because\ c = \underset{x\to\infty}{\text{Lt}} (y - mx)\right]$

Then $\frac{y}{x} = m + \frac{p}{x}$. Putting this value for $\frac{y}{x}$ in (ii), we get

$$x^n\phi_n\left(m + \frac{p}{x}\right) + x^{n-1}\phi_{n-1}\left(m + \frac{p}{x}\right) + ... + x\phi_1\left(m + \frac{p}{x}\right) + \phi_0\left(m + \frac{p}{x}\right) = 0$$

Now expanding each term by Taylor's theorem, we have

$$x^n\left[\phi_n(m) + \frac{p}{x}\phi_n{}'(m) + \frac{p^2}{2!\,x^2}\phi_n''(m) + ...\right] + x^{n-1}\left[\phi_{n-1}(m) + \frac{p}{x}\phi_{n-1}'(m) + ...\right] + x^{n-2}\left[\phi_{n-2}(m) + \frac{p}{x}.\phi_{n-2}'(m) + ...\right] + ... = 0 \quad ...(v)$$

But $\phi_n(m) = 0$, from (iv). Putting $\phi_n(m) = 0$ and arranging the remaining terms in (v) is descending powers of x, we get

$$x^{n-1}\,[p\phi_n{}'(m) + \phi_{n-1}(m)] + x^{n-2}\left[\frac{p^2}{2!}.\phi_n''(m) + p.\phi_{n-1}'(m) + \phi_{n-2}(m)\right] + ... = 0 \quad ...(vi)$$

Dividing throughout by x^{n-1}, we have

$$p\phi_n{}'(m) + \phi_{n-1}(m) + \frac{1}{x}\left[\frac{p^2}{2!}.\phi_n''(m) + p.\phi_{n-1}'(m) + \phi_{n-2}(m)\right] + ... = 0$$

Now proceeding to limits as $x \to \infty$, so that $p \to c$, we have

$$c\phi'_n(m) + \phi_{n-1}(m) = 0 \qquad ..(vii)$$

$\therefore$ If $\phi_n'(m) \neq 0$, which will be case, if equation $\phi_n(m) = 0$, has no repeated roots, then we have $c = -\dfrac{\phi_{n-1}(m)}{\phi'_n(m)}$.

$\therefore$ If $m_1, m_2, m_3, ...$ be the *non-repeated* roots of $\phi_n(m) = 0$, and $c_1, c_2, c_3, ...$ are the corresponding values of c determined from (vii), then $y = m_1x + c_1, y = m_2x + c_2, y = m_3x + c_3$ .... are the asymptotes.

Exceptional Case. When $\phi_n'(m) = 0$, but $\phi_{n-1}(m) \neq 0$, the finite value of c cannot be determined from (vii), and there is *no asymptote* in this case.

Case of parallel asymptotes. If $\phi_n'(m) = 0$ and $\phi_{n-1}(m) = 0$, which is generally there when two roots of $\phi_n(m) = 0$ *i.e.*, two values of m given by $\phi_n(m) = 0$ are equal, then equation (vii) becomes an identify and in this case, the equation (vi) reduces to

$$x^{n-2}\left[\frac{p^2}{2!}\phi''_n(m) + p\phi'_{n-1}(m) + \phi_{n-2}(m)\right] + x^{n-3}[+ ...] + ... = 0$$

Dividing throughout by x^{n-2}, we get

$$\frac{p^2}{2!}\phi_n''(m) + p\phi'_{n-1}(m) + \phi_{n-2}(m) + \frac{1}{x}[......] + ... = 0$$

Now proceeding to limits as $x \to \infty$, so that $p \to c$, we have

$$\frac{c^2}{2!}\phi_n''(m) + c\phi'_{n-1}(m) + \phi_{n-2}(m) = 0$$

which gives two values say c', c'' *for* c, provided $\phi_n''(m) \neq 0$.

Thus we have two asymptotes

$y = mx + c'$; $y = mx + c''$, corresponding to a given slope m. These are evidently *parallel*.

The above discussion leads us to the following **working rule for finding oblique asymptotes of an algebraic curve of nth degree.**

(*i*) *Find the polynomial* $\phi_n(m)$, *which can be obtained by putting* $x = 1, y = m$ *in the highest degree terms of the given equation of the curve.*

(*ii*) *Put* $\phi_n(m)$ *equal to zero, solve for m, and let the roots be* $m_1, m_2, m_3, ...$

(*iii*) *Find the polynomial* $\phi_{n-1}(m)$ *by putting* $x = 1, y = m$ *in next lower degree terms of the given equation. Similarly, polynomial* $\phi_{n-2}(m)$ *can be found out by putting* $x = 1, y = m$ *in the next lower degree terms of the equation, and so on.*

(*iv*) *Find the values* c_1, c_2, c_3 ...... *corresponding to values* m_1, m_2, m_3, ... *by using the relation* $c = -\dfrac{\phi_{n-1}(m)}{\phi'_n(m)}$. [*provided* $\phi_n'(m) \neq 0$]

(*v*) *The required asymptotes are* $y = m_1x + c_1, y = m_2x + c_2, y = m_3x + c_3$ ....

(*vi*) *If* $\phi_n'(m) = 0$ *for some value of m, but* $\phi_{n-1}(m) \neq 0$, *then corresponding to that value of m there is no asymptote.*

(*vii*) *If* $\phi_n'(m) = 0 = \phi_{n-1}(m)$, *for some value of m (i.e., two roots of* $\phi_n(m) = 0$ *are equal), then the values of c are found from the equation*

$$\frac{c^2}{2!}\phi_n''(m) + c.\phi'_{n-1}(m) + \phi_{n-2}(m) = 0.$$

This gives two values of c, and therefore there are two parallel asymptotes corresponding to this value of m.

(viii) *If the three roots of* $\phi_n(m) = 0$ *are equal ; [i.e.,* $\phi_n''(m) = 0$, $\phi'_{n-1}(m) = 0$, $\phi_{n-2}(m) = 0$], *then values of c corresponding to that value of m are obtained from the equation*

$$\frac{c^3}{3!}\phi_n'''(m) + \frac{c^2}{2!}\phi''_{n-1}(m) + c\phi'_{n-2}(m) + \phi_{n-3}(m) = 0.$$

Note 1. Asymptote corresponding to $m = 0$, as a root of $\phi_n(m) = 0$, is parallel to x-axis, and are directly obtained by method of Article 3 case (*ii*).

2. When it is required to find the asymptotes of a curve, it is advisable to first find the asymptotes parallel to axes, if there be any, and *then* search for the oblique asymptotes.

Example 1. *Find all the asymptotes of the following curves :*

(*a*) $x^3 + 2x^2y - xy^2 - 2y^3 + 3xy + 3y^2 + x + 1 = 0.$

(*b*) $x^3 + 4x^2y + 4xy^2 + 5x^2 + 15xy + 10y^2 - 2y + 1 = 0.$

Sol. (*a*) The equation of the curve is $x^3 + 2x^2y - xy^2 - 2y^3 + 3xy + 3y^2 + x + 1 = 0$ ...(*i*)

Since the coefficients of x^3 and y^3, the highest degree terms in x and y, are constant, $\therefore$ there are no asymptotes parallel to x-axis or y-axis. *Now to find oblique asymptotes* ; putting $x = 1, y = m$ in the third and second degree terms in (*i*), we get

$$\phi_3(m) = 1 + 2m - m^2 - 2m^3 \text{ and } \phi_2(m) = 3m + 3m^2$$

The slopes of asymptotes are the roots of $\phi_n(m) = 0$ *i.e.*, $\phi_3(m) = 0$, *i.e.*, of $1 + 2m - m^2 - 2m^3 = 0$ or $2m^3 + m^2 - 2m - 1 = 0$.

$m = 1$ is a root of this equation by inspection. Using synthetic division

1	2	1	−2	−1
		2	3	1
	2	3	1	0

$\therefore$ The reduced equation is $2m^2 + 3m + 1 = 0$

Solving for m, we get $$m = -1, \frac{-1}{2}$$

$\therefore$ $$m = -1, 1, -\frac{1}{2}$$

Also $$\phi_3'(m) = 2 - 2m - 6m^2 = -(6m^2 + 2m - 2)$$

Now c is given by $$c\phi_3'(m) + \phi_2(m) = 0 \qquad \mid \mathbf{c\phi_n'(m) + \phi_{n-1}(m) = 0}$$

$\therefore$ $$c = -\frac{\phi_2(m)}{\phi_3'(m)} = -\frac{3m + 3m^2}{-(6m^2 + 2m - 2)} = \frac{3m + 3m^2}{6m^2 + 2m - 2}$$

When $m = -1$, $$c = \frac{-3+3}{6-2-2} = \frac{0}{2} = 0$$

When $m = 1$, $$c = \frac{3+3}{6+2-2} = \frac{6}{6} = 1$$

and when $m = -\frac{1}{2}$, $$c = \frac{-\frac{3}{2}+\frac{3}{4}}{\frac{3}{2}-1-2} = \frac{1}{2}.$$

Putting these values of m and c in $y = mx + c$, the corresponding asymptotes are

$$y = -x + 0, \qquad y = x + 1 \qquad \text{and} \qquad y = -\frac{1}{2}x + \frac{1}{2}$$

or $$x + y = 0, \qquad x - y + 1 = 0 \qquad \text{and} \qquad x + 2y - 1 = 0.$$

(*b*) The equation of the curve is $x^3 + 4x^2y + 4xy^2 + 5x^2 + 15xy + 10y^2 - 2y + 1 = 0$ ...(*i*)

Since the coefficient of x^3, the highest degree term in x, is constant, $\therefore$ there is no asymptote parallel to x-axis. Again equating to zero the coefficient of y^2, the highest degree term in y, the asymptote parallel to y-axis is given by the equation

$$4x + 10 = 0 \qquad \text{or} \qquad 2x + 5 = 0$$

Now to find oblique asymptotes, putting $x = 1$, $y = m$ in the third and second degree terms in the equation of the curve, we get

$$\phi_3(m) = 1 + 4m + 4m^2,$$

and $$\phi_2(m) = 5 + 15m + 10m^2$$

Slopes of asymptotes are given by the roots of equation $\phi_3(m) = 0$ *i.e.*, by roots of

$$1 + 4m + 4m^2 = 0 \quad \text{or} \quad (1 + 2m)^2 = 0$$

$\therefore$ $$m = -\frac{1}{2}, -\frac{1}{2} \qquad \text{(Equal)}$$

$\therefore$ c is now given by the equation

$$\mathbf{\frac{c^2}{2!}\phi_n''(m) + c\phi'_{n-1}(m) + \phi_{n-2}(m) = 0}$$

Here $$\frac{1}{2!}c^2\,\phi_3''(m) + c\phi'_2(m) + \phi_1(m) = 0 \qquad ...(ii)$$

Now $\phi_3'(m) = 4 + 8m$, $\phi_3''(m) = 8$, $\phi_2'(m) = 15 + 20m$ and $\phi_1(m) = -2m$

$\therefore$ Equation (*ii*) becomes $\frac{1}{2!}c^2 \,.\, 8 + c(15 + 20m) - 2m = 0$

For $m = -\frac{1}{2}$, it becomes $4c^2 + c(15 - 10) + 1 = 0$

or $$4c^2 + 5c + 1 = 0 \quad \text{or} \quad (4c + 1)(c + 1) = 0$$

$\therefore$ $$c = -1, \quad c = -\frac{1}{4}$$

Hence the corresponding asymptotes are

$$y = -\frac{1}{2}x - 1 \qquad \text{and} \qquad y = -\frac{1}{2}x - \frac{1}{4}$$

or $$x + 2y + 2 = 0 \qquad \text{and} \qquad 2x + 4y + 1 = 0.$$

Example 2. *Show that the parabola $y^2 - 4ax = 0$ has no asymptotes.*

Sol. Since the coefficient of y^2, the highest degree term in y, in the equation of the parabola is a constant, $\therefore$ there are no asymptotes parallel to y-axis. Again, since the coefficient of x, the highest degree term in x is also a constant, $\therefore$ there is no asymptote parallel to x-axis.

For *finding oblique asymptotes* putting $x = 1$, $y = m$ in the second degree and first degree terms in the equation of the parabola, we get

$$\phi_2(m) = m^2 \quad \text{and} \quad \phi_1(m) = -4a.$$

Slopes of asymptotes are given by the roots of equation $\phi_2(m) = 0$ *i.e.*, by the equation $m^2 = 0$ or $m = 0, 0$

But asymptotes (if any) corresponding to $m = 0$ are parallel to x-axis. But it has been proved above that there are no asymptotes parallel to x-axis.

Hence the curve has no oblique asymptote.

Since the parabola has neither oblique asymptotes nor asymptotes parallel to the axes,

$\therefore$ it has no asymptotes.

Example 3. *Find asymptotes of the curve* $(x - y)^2(x + 2y - 1) = 3x + y - 7$.

Sol. The equation of the curve is $(x - y)^2 (x + 2y - 1) = 3x + y - 7$

or $$(x - y)^2 (x + 2y) - (x - y)^2 - 3x - y + 7 = 0 \textbf{ (Note)} \qquad ...(i)$$

Since the coefficients of x^3 and y^3, the highest degree terms in x and y are constants, therefore there are no asymptotes parallel to x-axis or y-axis.

Now to find oblique asymptotes ; putting $x = 1, y = m$ in the third, second and first degree terms in (i), we get

$$\phi_3(m) = (1 - m)^2 (1 + 2m),$$
$$= (1 + m^2 - 2m)(1 + 2m)$$
$$= 2m^3 - 3m^2 + 1$$
$$\phi_2(m) = -(1 - m)^2$$
$$\phi_1(m) = -3 - m$$

$$\phi_3'(m) = 6m^2 - 6m,$$
$$\phi_3''(m) = 12m - 6$$
$$\phi_2'(m) = 2(1 - m)$$

Slopes of the asymptotes are given by $\phi_n(m) = 0$ *i.e.,* $\phi_3(m) = 0$

i.e., by $(1 - m)^2 (1 + 2m) = 0$

$$\therefore \qquad m = 1, 1, -\frac{1}{2}$$

c is given by the equation $c\phi_3'(m) + \phi_2(m) = 0$ or $c(6m^2 - 6m) - (1 - m)^2 = 0$ ...(ii)

For m $= -\frac{1}{2}$, we have

$$c\left(\frac{6}{4} + \frac{6}{2}\right) - \left(1 + \frac{1}{2}\right)^2 = 0$$

or $$\frac{9c}{2} - \frac{9}{4} = 0 \quad \therefore \quad c = \frac{1}{2}$$

$\therefore$ Equation of the asymptote is $y = mx + c$

or $$y = -\frac{1}{2}x + \frac{1}{2} \quad \text{or} \quad 2y + x - 1 = 0$$

For **equal** values $m = 1$, values of c are given by

$$\frac{c^2}{2!}\phi_3''(m) + c\phi_2'(m) + \phi_1(m) = 0$$

or $$\frac{c^2}{2} . (12m - 6) + 2c(1 - m) + (-3 - m) = 0$$

For m = 1, we have $3c^2 - 4 = 0$

or $$c^2 = \frac{4}{3} \quad \therefore \quad c = \pm\frac{2}{\sqrt{3}}$$

$\therefore$ Equation of asymptotes is $y = mx + c$ or $y = x \pm \frac{2}{\sqrt{3}}$. $(\because \ m = 1)$

Example 4. *Find the asymptotes of* $(x-y)^2(x^2+y^2)-10(x-y)x^2+12y^2+2x+y=0$.

Sol. The equation of the curve is $(x-y)^2(x^2+y^2)-10(x-y)x^2+12y^2+2x+y=0$ ...(*i*)

Since coefficients of x^4 and y^4; the highest degree terms in x and y are constants, therefore there are no asymptotes parallel to x-axis or y-axis.

Now *to find oblique asymptotes* putting $x=1$, $y=m$ in the fourth, third and second degree terms in (*i*), we get

$$\phi_4(m)=(1-m)^2(1+m^2)$$
$$=(1+m^2-2m)(1+m^2)$$
$$=m^4-2m^3+2m^2-2m+1$$
$$\phi_3(m)=-10(1-m)$$
$$\phi_2(m)=12m^2$$

$$\phi_4'(m)=4m^3-6m^2+4m-2$$
$$\phi_4''(m)=12m^2-12m+4$$
$$\phi_3'(m)=10$$

Slopes of the asymptotes are given by $\phi_4(m)=0$

i.e., by $(1-m)^2(1+m^2)=0$.

$\therefore$ $m=1, 1$ $\quad$ ($\because$ $m^2+1=0$ gives only imaginary values of m)

For equal values of $m=1$,

values of c are given by $\frac{c^2}{2!}\phi_4''(m)+c\phi_3'(m)+\phi_2(m)=0$

or by $c^2(6m^2-6m+2)+10c+12m^2=0$

For $m=1$, we have $2c^2+10c+12=0$

or $c^2+5c+6=0$ *i.e.*, $(c+2)(c+3)=0$

$\therefore$ $c=-2$, $c=-3$

$\therefore$ Asymptotes are $y=mx+c$

i.e., $y=x-2$ and $y=x-3$.

Example 5. *Find all the asymptotes of the curve* $(x+y)^2(x+y+2)=x+9y-2$.

(K.U. 1998 S, 2004 S, 2006)

Sol. The equation of the curve is

$$(x+y)^2(x+y+2)=x+9y-2$$

or
$$(x+y)^3+2(x+y)^2-x-9y+2=0 \quad ...(i)$$

Since the coefficients of x^3 and y^3, the highest degree terms in x and y are constants, therefore there are no asymptotes parallel to x-axis or y-axis.

Now *to find oblique asymptotes* ; putting $x=1$, $y=m$ in the third, second, first degree terms and constant terms in (*i*), we have

$$\phi_3(m)=(1+m)^3$$
$$\phi_2(m)=2(1+m)^2$$
$$\phi_1(m)=-1-9m$$
$$\phi_0(m)=2$$

$$\phi_3'(m)=3(1+m)^2$$
$$\phi_3''(m)=6(1+m)$$
$$\phi_3'''(m)=6$$
$$\phi_2'(m)=4(1+m)$$
$$\phi_2''(m)=4$$
$$\phi_1'(m)=-9$$

Slopes of the oblique asymptotes are given by $\phi_3(m)=0$

i.e., by $(1+m)^3=0$ $\therefore$ $m=-1,-1,-1$

For **these three equal values** of $m=-1$, values of c are given by

$$\mathbf{\frac{c^3}{3!}\phi_3'''(m)+\frac{c^2}{2!}\phi_2''(m)+c\phi_1'(m)+\phi_0(m)=0}$$

or
$$\frac{c^3}{6}\cdot 6+\frac{c^2}{2}(4)+c(-9)+2=0 \quad \text{or} \quad c^3+2c^2-9c+2=0$$

$c = 2$ is a root of this equation by inspection.

Using synthetic division,

$$\begin{array}{c|cccc} 2 & 1 & 2 & -9 & 2 \\ & & 2 & 8 & -2 \\ \hline & 1 & 4 & -1 & 0 \end{array}$$

$\therefore$ The reduced equation is $c^2 + 4c - 1 = 0$

Solving for c, $\quad c = \dfrac{-4 \pm \sqrt{16+4}}{2} = \dfrac{-4 \pm 2\sqrt{5}}{2}$

or $\quad c = -2 \pm \sqrt{5}$

$\therefore$ For $m = -1$, the three different values of c are

$$2, -2 + \sqrt{5} \text{ and } -2 - \sqrt{5}$$

$\therefore$ Equations of the three asymptotes are

$$y = mx + c$$

i.e., $\quad y = -x + 2, y = -x - 2 + \sqrt{5}$

and $\quad y = -x - 2 - \sqrt{5}$.

EXERCISE 2

Find all the asymptotes of the following curves (Q. No. 1—16) :

1. $x^2y^2 = a^2(x^2 + y^2)$ 2. $xy^2 = 4a^2(2a - x)$ 3. $x^3 + y^3 = 3axy$
4. (*a*) $x^3 + 3xy^2 - x^2y - 3y^3 + x^2 - 2xy + 3y^2 + 4x + 7 = 0$
   (*b*) $x^3 - 6x^2y + 11xy^2 - 6y^3 + x + y + 1 = 0$ (*M.D.U. 1995*)
5. $y^3 - x^2y - 2xy^2 + 2x^3 - 7xy + 3y^2 + 2x^2 + 2x + 2y + 1 = 0$ (*K.U. 1999 ; M.D.U. 2006*)
6. $x^3 + 4x^2y + 5xy^2 + 2y^3 + 2x^2 + 4xy + 2y^2 - x - 9y + 1 = 0$ (*K.U. 2000*)
7. $y^3 - 5xy^2 + 8x^2y - 4x^3 - 3y^2 + 9xy - 6x^2 + 2y - 2x + 1 = 0$
8. (*a*) $4x^3 - 3xy^2 - y^3 + 2x^2 - xy - y^2 - 1 = 0$ (*K.U. 1996 S*)
   (*b*) $x^3 + 2x^2y + xy^2 - x^2 - xy + 2 = 0$
   (*c*) $x^3 - 2x^2y + xy^2 + x^2 - xy + 2 = 0$ (*M.D.U. 1999*)
9. (*a*) $x(y - x)^2 = x(y - x) + 2$ (*M.D.U. 2007*)
   (*b*) $4x^2(y - x) + y(y - 2)(x - y) = 4x + 4y - 7$
10. $x^2(x - y)^2 + a^2(x^2 - y^2) = a^2xy$
11. $(x - 1)(x - 2)(x + y) + x^2 + x + 1 = 0$
12. $x^3 + 3x^2y - 4y^3 - x + y + 3 = 0$ (*K.U. 1995*)
13. $x^3 - x^2y - xy^2 + y^3 + 2x^2 - 4y^2 + 2xy + x + y + 1 = 0$ (*M.D.U. 2001*)
14. $x^2y^2(x^2 - y^2)^2 = (x^2 + y^2)^3$
15. $(x + y)^2(x^2 + xy + y^2) = a^2x^2 + a^3(y - x)$
16. $(x - y)^2(x - 2y)(x - 3y) - 2a(x^3 - y^3) - 2a^2(x - 2y)(x + y) = 0$.
17. Prove that the asymptotes of the curve $x^2y^2 = a^2(x^2 + y^2)$ are the sides of a square.
18. Show that the asymptotes of $x^2y^2 - a^2(x^2 + y^2) - a^3(x + y) + a^4 = 0$ form a square, through two of whose vertices the curve passes.

19. Find rectangular asymptotes of the curve $y = \dfrac{x}{x^2 - 1}$.

[**Hint.** Rectangular Asymptotes $\Rightarrow$ Asymptotes at right angles to each other. Cross-mutiplying $y(x^2 - 1) = x$.]

20. Find the asymptotes of the curve $(x + y + 1)(x^2 + y^2 - xy) - 3xy + x^2 + y^2 + 2x - 3y + 5 = 0$.

[**Hint.** Simplifying $(x + y)(x^2 + y^2 - xy) + (x^2 + y^2 - xy) - 3xy + x^2 + y^2 + 2x - 3y + 5 = 0$

or $\qquad (x + y)(x^2 + y^2 - xy) + 2x^2 + 2y^2 - 4xy + 2x - 3y + 5 = 0$.]

21. Find the asymptotes of the curve $(x + y + 1)^2(x^2 + y^2 - xy) + 3xy - 7x^2 - 2y^2 - 7x + 8 = 0$.

Remark. The above equation can be written as

$$[(x + y)^2 + 2(x + y) + 1]\,(x^2 + y^2 - xy) + 3xy - 7x^2 - 2y^2 - 7x + 8 = 0$$

or $\quad (x + y)^2 (x^2 + y^2 - xy) + 2(x + y)(x^2 + y^2 - xy) + x^2 + y^2 - xy + 3xy - 7x^2 - 2y^2 - 7x + 8 = 0$

or $\quad (x + y)^2 (x^2 + y^2 - xy) + 2(x + y)(x^2 + y^2 - xy) - 6x^2 - y^2 + 2xy - 7x + 8 = 0$.

ANSWERS

1. $x = \pm a, y = \pm a$. **2.** $x = 0$. **3.** $x + y + a = 0$.

4. (*a*) $2x - 2y + 1 = 0$. (*b*) $x - y = 0, x - 2y = 0, x - 3y = 0$.

5. $y = -x - 2, y = x - 1, y = 2x$. **6.** $x + 2y + 2 = 0, x + y \pm 2\sqrt{2} = 0$.

7. $x - y = 0, 2x - y + 1 = 0, \quad 2x - y + 2 = 0$.

8. (*a*) $2x + y = 0, 2x + y + 1 = 0, \quad x - y = 0$

(*b*) $x = 0, \quad x + y = 0, \quad x + y = 1$.

(*c*) $x = 0, \quad y - x = 0, \quad y - x = 1$.

9. (*a*) $x = 0, \quad y = x, y = x + 1$.

(*b*) $x - y = 0, 2x - y + 1 = 0$ and $2x + y - 1 = 0$.

10. $x = \pm a, y = x \pm a$. **11.** $x = 1, \quad x = 2, \quad x + y + 1 = 0$.

12. $y - x = 0, \quad 2y + x - 1 = 0$ and $2y + x + 1 = 0$.

13. $x + y - 1 = 0, 2x - 2y + (3 + \sqrt{5}) = 0, \quad 2x - 2y + (3 - \sqrt{5}) = 0$.

14. $x \pm y = \pm\sqrt{2}, \quad x = \pm 1, \quad y = \pm 1$. **15.** $x + y \pm a = 0$.

16. $x = y + a, \quad x = y + 2a, \quad 2y = x + 14a, \quad 3y = x - 13a$.

19. $x = \pm 1, \quad y = 0$. **20.** $3x + 3y + 8 = 0$.

21. $x + y - 1 = 0$ and $x + y + 3 = 0$.

---

6. Total number of asymptotes

To show that a curve of degree n can never have more than n asymptotes.

Consider the equation of algebraic curve of nth degree of the form

$$x^n \phi_n\left(\frac{y}{x}\right) + x^{n-1}\phi_{n-1}\left(\frac{y}{x}\right) + x^{n-2}\phi_{n-2}\left(\frac{y}{x}\right) + \ldots = 0,$$

where $\phi_r\left(\dfrac{y}{x}\right)$ is a polynomial in $\dfrac{y}{x}$ of degree r.

The slopes of oblique asymptotes are given by $\phi_n(m) = 0$, which is an equation of degree n and thus has *at the most* n real roots.

Since for each value of m, we have, in general, one value of c determined by the equation $c\phi_n'(m) + \phi_{n-1}(m) = 0$, $\quad\therefore\quad$ it follows that a *curve of degree n has, in general, at the most n asymptotes*. In case the curve has one or more asymptotes parallel to y-axis, then the degree of equation $\phi_n(m) = 0$, is smaller than n by at least the same number.

Further when equation determining c is an equation of second degree ; then $\phi_n'(m) = 0$ and $\therefore$ it follows that $\phi_n(m) = 0$, has two equal roots. Thus, there are two values of c corresponding to two equal roots and there would be at the most $(n - 2)$ other asymptotes corresponding to the remaining roots. Hence it follows that a curve of degree n can never have more than n asymptotes.

7. Asymptotes by Inspection

If equation of a curve of degree n, is of the form $F_n + F_{n-2} = 0$ where F_n is of degree n [i.e., contains terms of degree n and may also contain lower degree terms], and F_{n-2} is of degree (n – 2) at the most, then every linear factor of F_n equated to zero, will be an asymptote, provided no two linear factors of F_n are either coincident or differ by a constant.

Let $(ax + by + c)$ be a non-repeated factor of F_n and let

$F_n = (ax + by + c)\, F_{n-1}$ where F_{n-1} is of degree $(n - 1)$.

The asymptote parallel to $ax + by + c = 0$, is given by

$$ax + by + c + \underset{\substack{x \to \infty \\ \frac{y}{x} \to -\frac{a}{b}}}{\text{Lt}} \frac{F_{n-2}}{F_{n-1}} = 0$$

Now to find the limit $\dfrac{F_{n-2}}{F_{n-1}}$, the numerator as well as denominator is divided by x^{n-1} and it would be seen that $\dfrac{1}{x}$ appears as a factor

$$\therefore \qquad \text{Lt}\ \frac{F_{n-2}}{F_{n-1}} \to 0, \text{ as } x \to \infty.$$

Hence $ax + by + c = 0$, is an asymptote.

Cor. If the equation of the curve is of the form $U_1 . U_2 . U_3 \ldots\ldots U_n + F_{n-2} = 0$ where U_1, U_2, U_3,......, U_n are linear factor, then the curve has $U_1 = 0$, $U_2 = 0$, $U_3 = 0$, ..., $U_n = 0$ as asymptotes provided no two of the factors are coincident or differ by a constant.

Note. Conversely, a curve having straight lines $U_1 = 0$, $U_2 = 0$ ...... $U_n = 0$, as asymptotes, must have an equation of the form $U_1 . U_2 . U_3 \ldots\ldots\ldots\ldots U_n + F_{n-2} = 0$.

Example. (*a*) *Find the asymptotes of* $\dfrac{x^2}{a^2} - \dfrac{y^2}{b^2} = 1$.

(*b*) *Find asymptotes of the curve* $xy(x^2 - y^2) + 2x^2 + 2y^2 + 1 = 0$.

Sol. (*a*) The equation of the curve is $\left(\dfrac{x^2}{a^2} - \dfrac{y^2}{b^2}\right) - 1 = 0$

This is of the form $F_n + F_{n-2} = 0$ $[n = 2]$, where F_2 can be split up into non-repeated linear factors.

Thus the asymptotes are given by $F_2 = 0$

$$\text{or} \qquad \frac{x^2}{a^2} - \frac{y^2}{b^2} = 0 \quad \text{or} \quad \left(\frac{x}{a} + \frac{y}{b}\right)\left(\frac{x}{a} - \frac{y}{b}\right) = 0$$

$$\therefore \qquad \frac{x}{a} + \frac{y}{b} = 0 \quad \text{and} \quad \frac{x}{a} - \frac{y}{b} = 0,$$

are the required asymptotes.

(*b*) The equation of the curve is $xy(x^2 - y^2) + 2x^2 + 2y^2 + 1 = 0$

This is of the form $F_n + F_{n-2} = 0 \ (n = 4)$.

The linear factors of $F_4 = xy(x^2 - y^2)$ are $x, y, (x - y)$ and $(x + y)$.

Since none of them is repeated, the four asymptotes of the curve are

$$x = 0, \quad y = 0, \quad x - y = 0 \quad \text{and} \quad x + y = 0.$$

8. Intersections of a curve with its asymptotes

To prove that any asymptote of an algebraic curve of the nth degree cuts the curve in (n – 2) points.

Let
$$y = mx + c \qquad \text{...(i)}$$
be an asymptote of the algebraic curve

$$x^n \phi_n\left(\frac{y}{x}\right) + x^{n-1}\phi_{n-1}\left(\frac{y}{x}\right) + x^{n-2}\phi_{n-2}\left(\frac{y}{x}\right) + \ldots = 0 \qquad \text{...(ii)}$$

To find points of intersections of asymptote (*i*) and curve (*ii*) let us solve them for x and y.

Putting value of y from (*i*) in (*ii*) $\left(\text{i.e., Putting } \frac{y}{x} = m + \frac{c}{x}\right)$

the abscissae of the points of intersection are the roots of the equation

$$x^n \phi_n\left(m + \frac{c}{x}\right) + x^{n-1}\phi_{n-1}\left(m + \frac{c}{x}\right) + x^{n-2}\phi_{n-2}\left(m + \frac{c}{x}\right) + \ldots\ldots = 0$$

Expanding by Taylor's theorem, we have

$$x^n\left[\phi_n(m) + \frac{c}{x}\phi_n'(m) + \frac{1}{2!}\frac{c^2}{x^2}\cdot\phi_n''(m) + \ldots\right] + x^{n-1}\left[\phi_{n-1}(m) + \frac{c}{x}\phi_{n-1}'(m) + \ldots\right]$$
$$+ x^{n-2}[\phi_{n-2}(m) + \ldots] + \ldots 0$$

Or arranging the terms according to descending power of x, we get

$$x^n\phi_n(m) + x^{n-1}[c \cdot \phi_n'(m) + \phi_{n-1}(m)]$$
$$+ x^{n-2}\left[\frac{c^2}{2!}\phi_n''(m) + c\phi_{n-1}'(m) + \phi_{n-2}(m)\right] + \ldots = 0 \qquad \text{...(iii)}$$

Since (*i*) is an asymptote of (*ii*), $\therefore$ m and c are given by

$$\phi_n(m) = 0 \quad \text{and} \quad c\phi_n'(m) + \phi_{n-1}(m) = 0$$

Substituting these values in (*iii*), it reduces to

$$x^{n-2}\left[\frac{c^2}{2!}\phi_n''(m) + c\phi_{n-1}'(m) + \phi_{n-2}(m)\right] + x^{n-3}[\ldots] + \ldots = 0$$

which is an equation of $(n - 2)$ th degree and determines $(n - 2)$ values of x.

Hence (*i*) cuts (*ii*) in $(n - 2)$ points.

Cor. 1. *If a curve of nth degree has n asymptotes, then they cut the curve in n(n – 2) points.*

Cor. 2. *If the equation of a curve of nth degree can be put in the form $F_n + F_{n-2} = 0$, where F_{n-2} is of degree (n – 2) at the most and F_n consists of n non-repeated linear factors, then the n(n – 2) points of intersection of the curve and its asymptotes lie on the curve $F_{n-2} = 0$.*

Proof. The joint equation of n asymptote of curve is $F_n = 0$, and the $n(n - 2)$ point of intersection of the curve and its asymptote satisfy the two equations $F_n + F_{n-2} = 0$ and $F_n = 0$, simultaneously.

$\therefore$ They also satisfy the equation, $(F_n + F_{n-2}) - F_n = 0$ *i.e.*, $F_{n-2} = 0$

Hence the result.

For example :

(*i*) The asymptotes of a *cubic* curve, cut the curve in 3(3 – 2) = 3 points which lie on a curve of degree 3 – 2 = 1 *i.e.*, on a *straight line*.

(*ii*) The asymptotes of a *biquadratic or quartic curve*, cut the curve in 4(4 – 2) = 8 points which lie on a curve of degree 4 – 2 = 2, *i.e.*, on a *conic*.

9. Method to find the equation of a curve joining the points of intersections of the given curve and its asymptotes.

Step I. Find the asymptotes of the given curve by the methods explained in Art. 3 and Art. 5.

Step II. Make the R.H.S. of the equation of each of these asymptotes as zero and multiply the L.H.S. of all these equations of the asymptotes to get the joint equation of asymptotes.

Note. But if the equation of the curve is of the form $F_n + F_{n-2} = 0$ as given in Art. 7 ; then there is no need of doing steps I and II ; as the joint equation of the asymptotes as explained in Art. 7 is $F_n = 0$ provided F_n is the product of non-repeated linear factors.

Step III. The number of points of intersections of the curve and the asymptotes is $n(n-2)$ (Cor. 1., Art. 8) where n is the degree of the curve.

Step IV. Then equation of the required curve is L.H.S. of equation of the given curve after making R.H.S. zero + λ (joint equation of the asymptotes after making R.H.S. zero) where λ = 1 or – 1 or 2 or – 2 etc.

Example 1. *Show that the points of intersection of the curve* $2y^3 - 2x^2y - 4xy^2 + 4x^3 - 14xy + 6y^2 + 4x^2 + 6y + 1 = 0$, *and its asymptotes lie on the straight line* $8x + 2y + 1 = 0$.

(M.D.U. 2000 ; K.U. 2005)

Sol. The equation of the curve is $2y^3 - 2x^2y - 4xy^2 + 4x^3 - 14xy + 6y^2 + 4x^2 + 6y + 1 = 0$ ...(*i*)

Since the coefficients of x^3 and y^3, the highest degree terms in x and y are constants,

∴ the curve has no asymptote parallel either to x-axis or y-axis.

Now to find oblique asymptotes, putting $x = 1$, $y = m$ in the third and second degree terms in (*i*), we get $\phi_3(m) = 2m^3 - 2m - 4m^2 + 4$ and $\phi_2(m) = -14m + 6m^2 + 4$.

The slopes of asymptotes are the roots of $\phi_3(m) = 0$

i.e., of $2m^3 - 2m - 4m^2 + 4 = 0$ or $(2m-4)(m^2-1) = 0$

$$\therefore \quad m = 2, 1, -1.$$

Also $\phi_3'(m) = 6m^2 - 2 - 8m = 2(3m^2 - 4m - 1)$

Now c is given by, $c\phi_3'(m) + \phi_2(m) = 0$

$$\therefore \quad c = -\frac{\phi_2(m)}{\phi_3'(m)} = -\frac{-14m + 6m^2 + 4}{2(3m^2 - 4m - 1)} = \frac{7m - 3m^2 - 2}{3m^2 - 4m - 1}$$

When $m = 2$, $$c = \frac{14 - 12 - 2}{12 - 8 - 1} = 0$$

When $m = 1$, $$c = \frac{7 - 3 - 2}{3 - 4 - 1} = \frac{2}{-2} = -1$$

and when $m = -1$, $$c = \frac{-7 - 3 - 2}{3 + 4 - 1} = \frac{-12}{6} = -2.$$

Putting values of m and c in $y = mx + c$, the corresponding asymptotes are

$$y = 2x + 0, \quad y = x - 1 \quad \text{and} \quad y = -x - 2$$

Making R.H.S. of each equation as zero,

or $$2x - y = 0, x - y - 1 = 0 \quad \text{and} \quad x + y + 2 = 0.$$

These three asymptotes will cut the curve again in 3(3 – 2) = 3 points. The joint equation of the asymptotes is $(2x - y)(x - y - 1)(x + y + 2) = 0$

or $$2x^3 - 2xy^2 - x^2y + y^3 - 7xy + 3y^2 + 2x^2 + 2y - 4x = 0$$

Multiplying by 2 throughout,

$$2y^3 - 2x^2y - 4xy^2 + 4x^3 - 14xy + 6y^2 + 4x^2 + 4y - 8x = 0 \qquad ...(ii)$$

Now equation of the given curve is

$$2y^3 - 2x^2y - 4xy^2 + 4x^3 - 14xy + 6y^2 + 4x^2 + 6y + 1 = 0 \qquad ...(iii)$$

Hence the 3 points of intersection of asymptotes (*ii*) and curve (*iii*) also lie on the curve (*ii*) – (*iii*)

i.e., on $$8x + 2y + 1 = 0$$

which is the required straight line.

Example 2. *Find the equation of the cubic which has the same asymptotes as the curve $x^3 - 6x^2y + 11xy^2 - 6y^3 + 4x + 5y + 7 = 0$, and which passes through the points (0, 0), (0, 2) and (2, 0).*

Sol. The equation of given curve is

$$x^3 - 6x^2y + 11xy^2 - 6y^3 + 4x + 5y + 7 = 0$$

or $$(x - y)(x - 2y)(x - 3y) + (4x + 5y + 7) = 0$$

which is of the form $F_n + F_{n-2} = 0$; and F_n $(= F_3)$ consists of non-repeated linear factors $(x - y)$, $(x - 2y)$ and $(x - 3y) = 0$.

∴ By the method of inspection, the asymptotes are given by

$$F_n = 0 \quad \text{or} \quad (x - y)(x - 2y)(x - 3y) = 0$$

or asymptotes are $(x - y) = 0$, $x - 2y = 0$ and $x - 3y = 0$.

The joint equation of the asymptotes is $F_3 \equiv (x - y)(x - 2y)(x - 3y) = 0$.

Since the equation of any curve having these asymptotes is of the form $F_3 + F_1 = 0$, where F_1 is of first degree in x and y.

∴ The equation of required cubic may be written as

$$F_3 + (ax + by + c) = 0 \text{ (see Note Art. 7)}$$

or $$(x - y)(x - 2y)(x - 3y) + ax + by + c = 0 \qquad ...(i)$$

where $ax + by + c$, is the general linear expression.

Since curve (*i*) passes through (0, 0), ∴ $c = 0$

Again ∵ curve (*i*) " " (2, 0), ∴ $8 + 2a + c = 0$

and ∴ curve (*i*) " " (0, 2), ∴ $-48 + 2b + c = 0$

which give, $c = 0$, $a = -4$ and $b = 24$.

Putting these values in (*i*), the required cubic is

$$(x - y)(x - 2y)(x - 3y) - 4x + 24y = 0$$

or $$x^3 - 6x^2y + 11x^2y - 6y^3 - 4x + 24y = 0.$$

Example 3. *Find the equation of the quartic curve which has $x = 0$, $y = 0$, $y = x$, and $y = -x$, for asymptotes and which passes through (a, b) and which cuts the asymptotes in eight points that lie on the circle $x^2 + y^2 = a^2$.*

Sol. The equations of the asymptotes are $x = 0$, $y = 0$, $y - x = 0$ and $y + x = 0$

The joint equation of the asymptotes is

$$xy(y - x)(y + x) = 0 \quad \text{or} \quad xy(y^2 - x^2) = 0 \qquad ...(i)$$

The equation of given circle is $\quad x^2 + y^2 - a^2 = 0 \qquad ...(ii)$

The equation of any quartic curve whose asymptotes are given by (*i*) and whose points of intersection with its asymptotes lie on (*ii*) is given by

$$xy(y^2 - x^2) + \lambda(x^2 + y^2 - a^2) = 0 \qquad ...(iii)$$

where λ is a constant.

This passes through the point (a, b).

$$\therefore \quad ab(b^2 - a^2) + \lambda(a^2 + b^2 - a^2) = 0 \quad \text{or} \quad \lambda = \frac{a(a^2 - b^2)}{b}$$

Putting this value of λ in (*iii*), the equation of required curve is

$$xy(y^2 - x^2) + \frac{a(a^2 - b^2)}{b}(x^2 + y^2 - a^2) = 0$$

or

$$bxy(y^2 - x^2) + a(a^2 - b^2)(x^2 + y^2 - a^2) = 0.$$

EXERCISE 3

1. (*a*) Find the asymptotes of the curve $x^2y - xy^2 + xy + y^2 + x - y = 0$, and show that they cut the curve in three points on the straight line $x + y = 0$.

   (*b*) Find the equation of the straight line on which lie the three points of intersections of the cubic $x^3 + 2x^2y - xy^2 - 2y^3 + 4y^2 + 2xy + y - 1 = 0$.

2. Show that the asymptotes of the cubic, $x^3 - 2y^3 + xy(2x - y) + y(x - y) + 1 = 0$, cut the curve in three points which again lie on the straight line $x - y + 1 = 0$. (*M.D.U. 1998, 2006*)

3. Find the asymptotes of the curve $x(x^2 - y^2) + y(3y - x) = 0$ and prove that the three points where these asymptotes cut the curve lies on $7x - 3y + 6 = 0$.

4. Show that the points of intersection of the curve $4x^3 - 2x^2y - 4xy^2 + 2y^3 + 6y^2 - 14xy + 4x^2 + 6y + 1 = 0$ and its asymptotes lie on $8x + 2y + 1 = 0$. (*M.D.U. 2007*)

5. Show that the asymptotes of the curve $(x^2 - y^2)(y^2 - 4x^2) + 6x^3 - 5x^2y - 3xy^2 + 2y^3 - x^2 + 3xy - 1 = 0$ cut the curve again in eight points which lie on a circle of radius unity. (*M.D.U. 2003 S*)

6. Find the asymptotes of the curve $xy(x^2 - y^2) + x^2 + y^2 = a^2$, and show that the eight points of intersection of the curve with its asymptotes lie on a circle whose centre is at the origin.

7. Show that the eight points of intersection of curve $x^4 - 5x^2y^2 + 4y^4 + x^2 - y^2 + x + y + 1 = 0$ and its asymptotes lie on a rectangular hyperbola.

   [**Hint.** A second degree curve is a **rectangular hyperbola** if its asymptotes are at right angles.]

8. Show that four asymptotes of the curve $xy(x^2 - y^2) + 25y^2 + 9x^2 - 144 = 0$ cut it again in eight points on an ellipse whose eccentricity is $\frac{4}{5}$.

   [**Hint.** Eccentricity e of the ellipse is given by $b^2 = a^2(1 - e^2)$.]

9. Find the equation of the hyperbola having $x + y - 1 = 0$, and $x - y + 2 = 0$, as its asymptotes, and passing through the origin. (*M.D.U. 1996*)

10. (*a*) Find the equation of the cubic, which has the same asymptotes as the curve $x^3 - 6x^2y + 11xy^2 - 6y^3 + x + y + 1 = 0$ and which passes through the points (0, 0), (1, 0) and (0, 1). (*K.U. 1999 S, 97 S*)

    (*b*) Find the equation of the cubic which has the same asymptotes as the curve $x^3 - 6x^2y + 11xy^2 - 6y^3 + x + y + 4 = 0$ and which passes through (0, 0), (2, 0) and (0, 2). (*M.D.U. 2004*)

11. Find the cubic which has the same asymptotes as the curve $x^3 - 6x^2y + 11xy^2 - 6y^3 + x + y + 1 = 0$ and which touches the axis of y at the origin and goes through the point (3, 2).

[**Hint.** Tangent at the origin is obtained by equating the lowest degree terms to zero.]

ANSWERS

1. (*a*) $y = 0, x - 1 = 0$ and $x - y + 2 = 0$ (*b*) $x + 3y = 1$

3. $x = 3, y = x + 1, x + y + 2 = 0$ **6.** $x = 0, y = 0, x \pm y = 0$

9. $(x + y - 1)(x - y + 2) + 2 = 0$

10. (*a*) $x^3 - 6x^2y + 11xy^2 - 6y^3 - x + 6y = 0$ (*b*) $x^3 - 6x^2y + 11xy^2 - 6y^3 - 4x + 2y = 0$

11. $x^3 - 6x^2y + 11xy^2 - 6y^3 - x = 0.$

10. Asymptotes in Polar Co-ordinates

Theorem 1. *The polar equation of any line is $p = r\cos(\theta - \alpha)$ where p is the length of the perpendicular from the pole to the line and α, is the angle which this perpendicular makes with x-axis.*

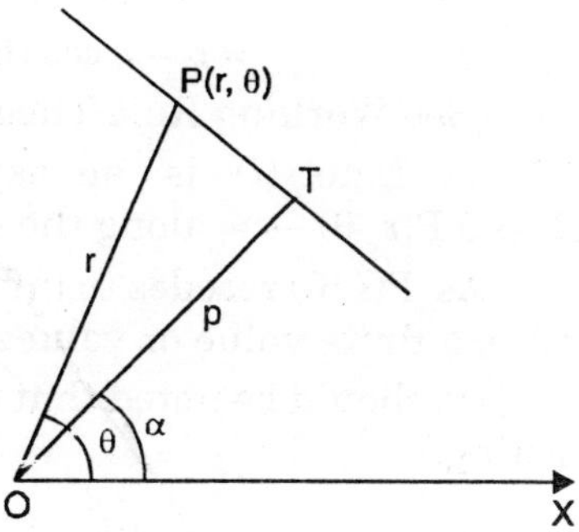

Sol. Let O be the pole and OX be the initial line.

Let P(r, θ) be any point on the given line.

$\therefore$ $OP = r$ and $\angle XOP = \theta$.

Let $OT = p$ be the length of perpendicular from pole on the given line and $\angle XOT = \alpha$.

$\therefore$ $\angle TOP = \angle XOP - \angle XOT = \theta - \alpha$.

Now in ΔOPT,

$$\cos(\theta - \alpha) = \frac{OT}{OP} = \frac{p}{r}$$

or $\quad$ $\mathbf{p = r\cos(\theta - \alpha)}$ **is the required equation of the line.**

Theorem 2. *The length of the perpendicular from a point $P(r_1, \theta_1)$ on the line $p = r\cos(\theta - \alpha)$ is $\pm (p - r_1 \cos(\theta_1 - \alpha))$.*

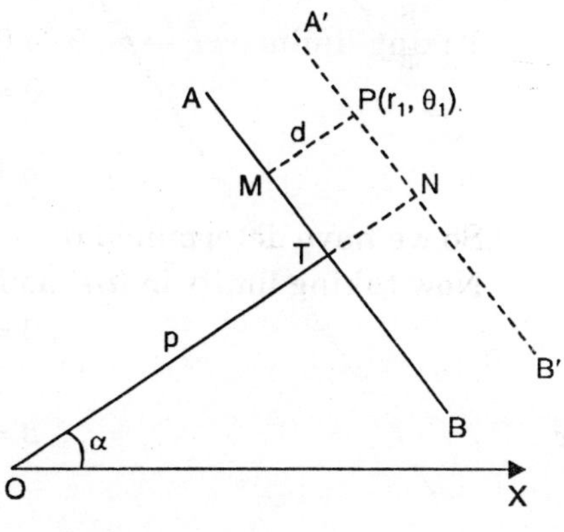

Sol. Let $p = r\cos(\theta - \alpha)$ be the equation of the line AB.

$\therefore$ $OT = p$ and $\angle XOT = \alpha$.

Through $P(r_1, \theta_1)$ draw a line A′ B′ parallel to AB and let PM = d be the length of perpendicular from 1 on the line AB.

$\therefore$ $ON = OT + TN = OT + PM = (p + d)$

and $\quad$ $\angle XON = \alpha$

$\therefore$ Equation of line A′ B′ is

$$p + d = r\cos(\theta - \alpha)$$

It passes through $P(r_1, \theta_1)$

$\therefore$ $p + d = r_1 \cos(\theta_1 - \alpha)$

$\therefore$ $d = -p + r_1\cos(\theta_1 - \alpha) = -(p - r_1\cos(\theta_1 - \alpha_1))$.

Similarly by taking P below AB, we could see that

$$d = (p - r_1\cos(\theta_1 - \alpha))$$

Combining the two ; d (the perpendicular distance of $P(r_1, \theta_1)$ from the line $p = r\cos(\theta - \alpha)$) is given by

$$\mathbf{d = \pm (p - r_1 \cos(\theta_1 - \alpha))}$$

Working rule for finding the perpendicular distance of a point from the line $p = r\cos(\theta - \alpha)$.

1. Make R.H.S. of equation of the line as zero.

2. Substitute the co-ordinates of the point in L.H.S. to get ⊥ distance.

11. *Polar equation of an asymptote to the curve $r = f(\theta)$ is*

$$\mathbf{p = r \sin(\theta_1 - \theta)} \textit{ where } \mathbf{p = \underset{\theta \to \theta_1}{Lt} \frac{-d\theta}{du}\left(u = \frac{1}{r}\right)}$$

and θ_1 is a root of the equation obtained by putting $u = 0$.)

Let $$p = r\cos(\theta - \alpha) \quad ...(i)$$

be an asymptote to the curve $r = f(\theta)$ ...(*ii*)

We are to determine p and α.

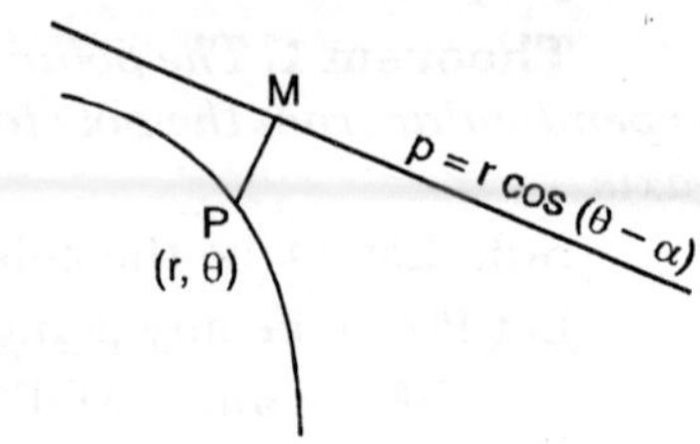

Let P(r, θ) be any point on the curve (*i*).

PM, the distance of P(r, θ) from the line (*i*)

$$= p - r\cos(\theta - \alpha) \quad ...(iii)$$

(See Working Rule Theorem 2, Art 10)

∵ Line (*i*) is an asymptote to curve (*ii*), therefore PM → 0 P$(r, \theta) \to \infty$ along the curve.

As P(r, θ) recedes to infinity along the curve ; $r \to \infty$ and θ tend to a finite value or values (say θ_1).

(It should be noted that $\theta \to \infty$ for circular asymptotes which are beyond the scope of our learning.)

Also $r \to \infty$ implies $u \to 0$ $\left(\because\ r = \frac{1}{u}\right)$

Dividing both sides of (*iii*) by r , we get

$$\frac{PM}{r} = \frac{p}{r} - \cos(\theta - \alpha)$$

Taking limits as $r \to \infty$, $\theta \to \theta_1$, PM → 0, so that

$$0 = 0 - \cos(\theta_1 - \alpha) \quad \text{or} \quad \cos(\theta_1 - \alpha) = 0$$

$$\therefore \quad \theta_1 - \alpha = \frac{\pi}{2} \quad \therefore \quad \alpha = \theta_1 - \frac{\pi}{2} \quad ...(iv)$$

So we have determined α.

Now taking limits in (*iii*) as PM → 0, $r \to \infty$, $\theta \to \theta_1$; we have

$$0 = p - \underset{\substack{r \to \infty \\ \theta \to \theta_1}}{\text{Lt}}\ r\cos(\theta - \alpha)$$

or $$p = \underset{\substack{r \to \infty \\ \theta \to \theta_1}}{\text{Lt}}\ r\cos(\theta - \alpha)$$

$$= \text{Lt}\ r\cos\left(\theta - \theta_1 + \frac{\pi}{2}\right) \qquad \left|\ \because \text{ From } (iv),\ \alpha = \theta_1 - \frac{\pi}{2}\right.$$

$$= \underset{\substack{r \to \infty \\ \theta \to \theta_1}}{\text{Lt}}\ -r\sin(\theta - \theta_1) \qquad |\ \text{It is of the form } \infty \cdot 0$$

or $$p = \underset{\substack{r \to \infty \\ \theta \to \theta_1}}{\text{Lt}}\ -\frac{\sin(\theta - \theta_1)}{\left(\frac{1}{r}\right)} \qquad \left|\ \text{Form } \frac{0}{0}\right.$$

or $$p = \underset{\substack{u \to 0 \\ \theta \to \theta_1}}{\text{Lt}} - \frac{\sin(\theta - \theta_1)}{u} = \underset{\substack{u \to 0 \\ \theta \to \theta_1}}{\text{Lt}} - \frac{\cos(\theta - \theta_1)}{\frac{du}{d\theta}} \quad \text{(By L' Hospital Rule)}$$

$$= \text{Lt} \frac{-1}{\frac{du}{d\theta}} \text{Lt} \cos(\theta - \theta_1) = \text{Lt}\left(-\frac{d\theta}{du}\right) \quad ...(v)$$

$$\left[\because \underset{\theta \to \theta_1}{\text{Lt}} \cos(\theta - \theta_1) = \cos 0 = 1\right]$$

So we have determined p.

Putting the value of α from (iv) in (i), equation of the asymptote is

$$p = r\cos\left(\theta - \theta_1 + \frac{\pi}{2}\right) \quad \text{or} \quad p = r\cos\left[\frac{\pi}{2} - (\theta_1 - \theta)\right]$$

or $$p = r\sin(\theta_1 - \theta) \text{ where } p \text{ as given by } (v) \text{ is } = \underset{\theta \to \theta_1}{\text{Lt}} - \frac{d\theta}{du}.$$

Working rule for finding polar asymptotes

Step I. *Put* $r = \frac{1}{u}$ *in the given equation. Also change all T-ratios if any into* $\sin\theta$ *and* $\cos\theta$.

Step II. *Find the limit of* θ *as* $u \to 0$. *Let* θ_1 *be this limit or one of the limits if more than one such limits exist.*

Step III. *Determine* $p = \underset{\substack{\theta \to \theta_1 \\ u \to 0}}{\text{Lt}} \left(-\frac{d\theta}{du}\right)$ *for value or values of* θ *obtained in Step II.*

Step IV. *Putting the values of* p *and* θ_1 *in the equation* $p = r\sin(\theta_1 - \theta)$ *we get the corresponding asymptote.*

Note. If $\underset{\theta \to \theta_1}{\text{Lt}} \left(-\frac{d\theta}{du}\right)$ does not tend to a finite limit ; then there is no asymptote corresponding to the value $\theta = \theta_1$.

Example 1. *Find the asymptotes of the curve* $r\theta = a$.

Sol. The equation of the curve is $r\theta = a$.

Put $$r = \frac{1}{u} \quad \therefore \quad \frac{\theta}{u} = a \quad \therefore \quad u = \frac{\theta}{a}.$$

As $$u \to 0 \quad \frac{\theta}{a} \to 0 \quad i.e., \quad \theta \to 0 \quad \therefore \quad \theta_1 = 0$$

Since $$u = \frac{\theta}{a}, \text{ therefore } \frac{du}{d\theta} = \frac{\theta}{a}$$

$$p = \underset{\theta \to \theta_1}{\text{Lt}} \left(-\frac{d\theta}{du}\right) = \underset{\theta \to 0}{\text{Lt}} - a = -a.$$

$\therefore$ Equation of the asymptote is

$$p = r\sin(\theta_1 - \theta) \quad \text{or} \quad -a = r\sin(-\theta) = -r\sin\theta$$

i.e., $$r\sin\theta = a.$$

Remark. The student should remember the following results from Trigonometry.

1. If $\sin\theta = 0$; then $\theta = n\pi$ where n is any integer.

2. If $\cos\theta = 0$, then $\theta = (2n + 1)\frac{\pi}{2}$.

3. $\sin(n\pi + \theta) = (-1)^n \sin\theta$.

4. $\cos(n\pi + \theta) = (-1)^n \cos\theta$.

5. If $\cos\theta = \cos\alpha$, then $\theta = 2n\pi \pm \alpha$.

6. If $\sin\theta = \sin\alpha$, then $\theta = n\pi + (-1)^n \alpha$.

7. If $\tan\theta = \tan\alpha$, then $\theta = n\pi + \alpha$.

8. $\dfrac{1}{(-1)^n} = (-1)^n$.

Example 2. *Find the asymptotes of the curve $r\cos\theta = a\sin^2\theta$.*

(M.D.U. 2005 A ; K.U. 2000 A, 2007 A)

Sol. The equation of the curve is $r\cos\theta = a\sin^2\theta$

Put $\quad r = \dfrac{1}{u} \quad \therefore \quad \dfrac{1}{u}\cos\theta = a\sin^2\theta$

$$\therefore \quad u = \frac{\cos\theta}{a\sin^2\theta}$$

As $\quad u \to 0, \quad \dfrac{\cos\theta}{a\sin^2\theta} \to 0, \quad$ *i.e.*, $\quad \cos\theta \to 0$

$$\therefore \quad \theta \to (2n+1)\frac{\pi}{2} \quad i.e., \quad \theta_1 = (2n+1)\frac{\pi}{2}$$

Since $\quad u = \dfrac{\cos\theta}{a\sin^2\theta}$

$$\therefore \quad \frac{du}{d\theta} = \frac{1}{a}\left[\frac{\sin^2\theta(-\sin\theta) - \cos\theta \,.\, 2\sin\theta\cos\theta}{\sin^4\theta}\right]$$

$$= \frac{(-\sin^3\theta - 2\sin\theta\cos^2\theta)}{a\sin^4\theta}$$

$$= \frac{-\sin\theta(\sin^2\theta + 2\cos^2\theta)}{a\sin^4\theta} = \frac{-(\sin^2\theta + 2\cos^2\theta)}{a\sin^3\theta}$$

$$p = \underset{\theta \to \theta_1}{\text{Lt}} \left(-\frac{d\theta}{du}\right) = \underset{\theta \to (2n+1)\pi/2}{\text{Lt}} \frac{a\sin^3\theta}{\sin^2\theta + 2\cos^2\theta}$$

or $\quad p = a\,\dfrac{((-1)^n)^3}{((-1)^n)^2 + 0}$

$$\left[\begin{aligned} \because \quad & \sin(2n+1)\frac{\pi}{2} = \sin\left(n\pi + \frac{\pi}{2}\right) = (-1)^n \sin\frac{\pi}{2} = (-1)^n \\ & \text{and } \cos(2n+1)\frac{\pi}{2} = \cos\left(n\pi + \frac{\pi}{2}\right) = (-1)^n \cos\frac{\pi}{2} = 0 \end{aligned}\right]$$

or $\quad p = a(-1)^n$

$\therefore$ Equation of the asymptote is

$$p = r\sin(\theta_1 - \theta)$$

or $\quad a(-1)^n = r\sin\left[(2n+1)\dfrac{\pi}{2} - \theta\right]$

or $\quad a(-1)^n = r\sin\left(n\pi + \dfrac{\pi}{2} - \theta\right)$

or $\quad a(-1)^n = (-1)^n r\sin\left(\dfrac{\pi}{2} - \theta\right) \quad$ or $\quad a = r\cos\theta$.

Example 3. *Find the asymptotes of the curve $r = a \sec\theta + b \tan\theta$.*

(K.U. 1998, 2006 S)

Sol. The equation of the curve is $r = a \sec\theta + b \tan\theta$.

Put $r = \dfrac{1}{u}$ and change all t-ratios in $\sin\theta$, $\cos\theta$.

$$\therefore \quad \frac{1}{u} = \frac{a}{\cos\theta} + \frac{b\sin\theta}{\cos\theta} = \frac{a + b\sin\theta}{\cos\theta}$$

$$\therefore \quad u = \frac{\cos\theta}{a + b\sin\theta}$$

As $$u \to 0, \quad \frac{\cos\theta}{a + b\sin\theta} \to 0 \quad i.e., \quad \cos\theta \to 0$$

$$\therefore \quad \theta \to (2n+1)\frac{\pi}{2} \qquad i.e., \quad \theta_1 = (2n+1)\frac{\pi}{2}$$

Since $$u = \frac{\cos\theta}{a + b\sin\theta}$$

$$\therefore \quad \frac{du}{d\theta} = \frac{(a + b\sin\theta)(-\sin\theta) - \cos\theta\,(b\cos\theta)}{(a + b\sin\theta)^2}$$

or $$\frac{du}{d\theta} = -\frac{(a\sin\theta + b)}{(a + b\sin\theta)^2}$$

$$p = \underset{\theta \to \theta_1}{\text{Lt}} \left(-\frac{d\theta}{du}\right) = \underset{\theta \to (2n+1)\pi/2}{\text{Lt}} \frac{(a + b\sin\theta)^2}{(a\sin\theta + b)}$$

$$= \frac{\left[a + b\sin(2n+1)\dfrac{\pi}{2}\right]^2}{\left[a\sin(2n+1)\dfrac{\pi}{2} + b\right]} = \frac{[a + b(-1)^n]^2}{[a(-1)^n + b]}$$

$$\because \quad \sin(2n+1)\frac{\pi}{2} = \sin\left(n\pi + \frac{\pi}{2}\right) = (-1)^n \sin\frac{\pi}{2} = (-1)^n.$$

$\therefore$ Equation of asymptotes is $p = r\sin(\theta_1 - \theta)$

or $$p = r\sin\left((2n+1)\frac{\pi}{2} - \theta\right) = r\sin\left(n\pi + \frac{\pi}{2} - \theta\right) = r(-1)^n \sin\left(\frac{\pi}{2} - \theta\right)$$

or $$p = r(-1)^n\cos\theta \quad \text{or} \quad \frac{[a + b(-1)^n]^2}{[a(-1)^n + b]} = (-1)^n r\cos\theta.$$

Note. $\because$ If n is *even* ; then $(-1)^n = 1$.

$\therefore$ Equation of asymptote is $\dfrac{(a+b)^2}{a+b} = r\cos\theta \quad i.e., \quad r\cos\theta = a + b$

If n is *odd* ; then $(-1)^n = -1$

$\therefore$ Equation of asymptote is $\dfrac{(a-b)^2}{(-a+b)} = -r\cos\theta$

or $$\frac{(a-b)^2}{a-b} = r\cos\theta \ i.e., \quad r\cos\theta = (a - b).$$

Example 4. *Prove that the curve* $r = \dfrac{a}{1 - \cos\theta}$ *has no asymptotes.*

Sol. The equation of the curve is $r = \dfrac{a}{1 - \cos\theta}$

Put $$r = \frac{1}{u} \quad \therefore \quad \frac{1}{u} = \frac{a}{1 - \cos\theta}$$

$$\therefore \qquad u = \frac{1 - \cos\theta}{a}$$

As $$u \to 0\,;\ \frac{1 - \cos\theta}{a} \to 0 \quad i.e., \quad \cos\theta \to 1$$

or $$\cos\theta \to \cos 0$$

$$\therefore \qquad \theta \to 2n\pi \pm 0 = 2n\pi$$

$$\therefore \qquad \theta_1 = 2n\pi, \text{ where } n \text{ is any integer}$$

Since $$u = \frac{1 - \cos\theta}{a} \quad \therefore \quad \frac{du}{d\theta} = \frac{\sin\theta}{a}$$

$$\therefore \qquad \underset{\theta \to \theta_1}{\text{Lt}} \left(-\frac{d\theta}{du}\right) = \underset{\theta \to 2n\pi}{\text{Lt}} -\frac{a}{\sin\theta} \to -\frac{a}{\sin 2n\pi} \to -\infty \qquad (\because \ \sin 2n\pi = 0)$$

$$\therefore \qquad \underset{\theta \to \theta_1}{\text{Lt}} -\frac{d\theta}{du} \text{ does not exist.}$$

$\therefore$ The curve $r = \dfrac{a}{1 - \cos\theta}$ has no asymptote.

Example 5. *Find asymptotes of the curve*

$$r^n f_n(\theta) + r^{n-1} f_{n-1}(\theta) + \ldots\ldots + f_0(\theta) = 0 \qquad \text{(M.D.U. 2005, 2003, 95)}$$

Sol. Equation of the given curve is

$$r^n f_n(\theta) + r^{n-1} f_{n-1}(\theta) + \ldots\ldots + f_0(\theta) = 0$$

Putting $r = \dfrac{1}{u}$, we have

$$\frac{f_n(\theta)}{u^n} + \frac{f_{n-1}(\theta)}{u^{n-1}} + \ldots\ldots + f_0(\theta) = 0$$

Multiplying by L.C.M. $= u^n$,

$$f_n(\theta) + u f_{n-1}(\theta) + \ldots\ldots + u^n f_0(\theta) = 0 \qquad \ldots(1)$$

Let $u \to 0$.

$\therefore$ Eqn. (1) becomes $f_n(\theta) = 0 \qquad \ldots(2)$

Let θ_1 be a root of this equation (2)

$$\therefore \qquad f_n(\theta_1) = 0 \qquad ...(3)$$

Diff. both sides of (1) w.r.t. θ,

$$f'_n(\theta) + u f'_{n-1}(\theta) + \frac{du}{d\theta} f_{n-1}(\theta) + ... + u^n f'_0(\theta) + nu^{n-1} \frac{du}{d\theta} f_0(\theta) = 0$$

Taking limits as $u \to 0$, and $\theta \to \theta_1$, we have

$$f'_n(\theta_1) + 0 + f_{n-1}(\theta_1) \underset{\substack{u \to 0 \\ \theta \to \theta_1}}{\text{Lt}} \frac{du}{d\theta} + 0 + 0 + ... + 0 = 0$$

$$\therefore \qquad f_{n-1}(\theta_1) \underset{\substack{u \to 0 \\ \theta \to \theta_1}}{\text{Lt}} \frac{du}{d\theta} = -f'_n(\theta_1)$$

$$\therefore \qquad \underset{\substack{u \to 0 \\ \theta \to \theta_1}}{\text{Lt}} \frac{du}{d\theta} = \frac{-f'_n(\theta_1)}{f_{n-1}(\theta_1)}$$

$$\therefore \qquad p = \underset{\substack{\theta \to \theta_1 \\ u \to 0}}{\text{Lt}} \left(\frac{-d\theta}{du} \right) = \frac{f_{n-1}(\theta_1)}{f'_n(\theta_1)} \qquad ...(4)$$

$\therefore$ Equation of the asymptote is

$$p = r \sin(\theta_1 - \theta)$$

where p and θ_1 are given by eqns. (4) and (3).

EXERCISE 4

Find the asymptotes of the following curves :

1. (*i*) $r = \dfrac{a\theta}{\theta - 1}$ (*ii*) $r^2\theta = a^2$

(*iii*) $r(\theta^2 - \pi^2) = 2a\theta$.

2. (*i*) $r = \dfrac{2a}{1 - 2\cos\theta}$.

(*ii*) $r = \dfrac{2a}{1 + 2\cos\theta}$. *(K.U. 1997 S)*

(*iii*) $r \cos\theta = a \sin\theta$ or $r = a \tan\theta$.

3. (*i*) $r \sin n\theta = a$ (*ii*) $r \sin 2\theta = a$.

4. (*i*) $r \cos 2\theta = a \sin 3\theta$ *(K.U. 1996 A ; M.D.U. 1996)*

(ii) $r \sin \theta = 2 \cos 2\theta$. (K.U. 1999)

(iii) $r \cos \theta = a \cos 2\theta$.

5. (i) $r^n \sin n\theta = a^n$ or $r^n = a^n \operatorname{cosec} n\theta$ where n is a positive integer > 1. (M.D.U. 2004 S)

(ii) $r = \dfrac{3a \sin \theta \cos \theta}{\sin^3 \theta + \cos^3 \theta}$.

[**Hint.** $u = 0$ gives $\sin^3 \theta = -\cos^3 \theta$ *i.e.*, $\tan \theta = -1$.]

6. (i) $r = 4(\sec \theta + \tan \theta)$

(ii) $r = a \operatorname{cosec} \theta + b$ (K.U. 1998 S)

(iii) $r^2 = a^2 (\sec^2 \theta + \operatorname{cosec}^2 \theta)$ (iv) $r = a + b \cot n\theta$

(v) $r = a \operatorname{cosec} \theta + b \cot \theta$. (K.U. 2005 S)

7. $r(1 - e^{\theta}) = a$. **8.** $r \log \theta = a$.

9. (i) $r(\pi + \theta) = ae^{\theta}$ (ii) $r\theta \cos \theta = a \cos 2\theta$.

ANSWERS

1. (i) $r \sin (\theta - 1) = a$ (ii) $\theta = 0$ (iii) $r \sin \theta = -a$.

2. (i) $r \sin \left(\dfrac{\pi}{3} \pm \theta\right) = \dfrac{-2a}{\sqrt{3}}$ (ii) $r \sin \left(\dfrac{2\pi}{3} \pm \theta\right) = \dfrac{2a}{\sqrt{3}}$ (iii) $r \cos \theta = (-1)^n a$.

3. (i) $r \sin \left(\theta - \dfrac{m\pi}{n}\right) = a \dfrac{(-1)^m}{n}$ where m is any integer

(ii) $-a \dfrac{(-1)^n}{2} = r \sin \left(\dfrac{n\pi}{2} - \theta\right)$

4. (i) $\dfrac{a}{2} \sin \left(\dfrac{n\pi}{2} + \dfrac{3\pi}{4}\right) = r \sin \left[(2n + 1) \dfrac{\pi}{4} - \theta\right]$

(ii) $r \sin \theta = 2$ (iii) $r \cos \theta + a = 0$.

5. (i) $\theta = \dfrac{m\pi}{n}$ where m is any integer (ii) $\sqrt{2}\, r \sin \left(\dfrac{\pi}{4} + \theta\right) + a = 0$.

6. (i) $4[1 + (-1)^n] = (-1)^n r \cos \theta$ (ii) $r \sin \theta$

(iii) $\pm a(-1)^n = r \sin \left(\dfrac{n\pi}{2} - \theta\right)$ (iv) r s[illegible] $- \dfrac{m\pi}{n}\Big) = \dfrac{b}{n}$ where m is an integer.

(v) $a(-1)^n + b = (-1)^n r \sin \theta$.

7. $r \sin \theta + a = 0$. **8.** $a = r \sin (\theta - 1)$.

9. (i) $r \sin \theta\, e^{\pi} + a = 0$

(ii) $r \sin \theta = a$, $(2n + 1)\pi r \cos \theta + 2a = 0$, where n is any *integer*.

---

12. Asymptotes by Expansion

To show that $y = mx + c$ *is an asymptote of the curve whose equation can be expressed in the form*

$$y = mx + c + \frac{A}{x} + \frac{B}{x^2} + \frac{C}{x^3} + \ldots.. \qquad \ldots.(i)$$

Dividing both sides of (i) by x, we get

$$\frac{y}{x} = m + \frac{c}{x} + \frac{A}{x^2} + \frac{B}{x^3} + \frac{C}{x^4} + \ldots\ldots.$$

so that when $x \to \infty$, we have $\text{Lt}\, \frac{y}{x} = m$ ...(ii)

Again from (i), we have $y - mx = c + \frac{A}{x} + \frac{B}{x^2} + \frac{C}{x^3} + \ldots..$

so that when $x \to \infty$, we have Lt $(y - mx) = c$ ...(iii)

Hence from (ii) and (iii), it follows that $y = mx + c$, is an asymptotes of curve (i).

REVISION EXERCISE

Find the asymptotes of the following curves :

1. $x^3 - 2y^3 + xy(2x - y) + y(x - 1) + 1 = 0.$
2. $(y - a)^2(x^2 - a^2) = x^4 + a^4.$
3. $(x^2 - y^2)^2 - 4x^2 + x = 0.$
4. $y^2(x - 2a) = x^3 - a^3.$
5. $(x - y)(2x + y)^2 + 2x^2 - xy - y^2 - 1 = 0.$
6. $(x^2 - y^2)(x + 2y + 1) + x + y + 1 = 0.$
7. $x^2y - xy^2 + xy + y^2 + y - 1 = 0.$
8. $y^3 + x^2y - 2xy^2 - y + 1 = 0.$
9. $2x^3 - x^2y - 2xy^2 + y^3 - 4x^2 + 8xy + 4x + 1 = 0.$ *(M.D.U. 1997)*
10. $(x + y)^2 (x + 2y)(x + 3y) - 2(x^3 + y^3) - 2(x + 2y)(x + y) = 0.$
11. $(x^2 - y^2)(2x + 3y)(3x + 2y) + 2x^2 + 3y^2 - 6 = 0.$
12. $(y - x)(y - 2x)^2 + (y + 3x)(y - 2x) - 2x + 4y - 3 = 0.$

ANSWERS

1. $6x - 6y + 1 = 0,\ 2x + 2y + 1 = 0,\ 3x + 6y - 2 = 0.$
2. $x = \pm a,\ y = x + a,\ y = -x + a.$
3. $y = x \pm 1,\ y = -x \pm 1.$
4. $y = x + a$ and $y = -x - a,\ x = 2a.$
5. $y = x,\ 2x + y = 0,\ 2x + y + 1 = 0.$
6. $x - y = 0,\ x + y = 0,\ x + 2y + 1 = 0.$
7. $y = 0,\ y = x + 2,\ x = 1.$

8. $y = 0, y = x \pm 1$.

9. $x - y + 2 = 0, x + y - 2 = 0, 2x - y - 4 = 0$.

10. $x + y = 0, x + y = 3, x + 2y + 14 = 0, x + 3y - 13 = 0$.

11. $x + y = 0, x - y = 0, 2x + 3y = 0, 3x + 2y = 0$.

12. $x - y + 4 = 0, 2x - y - 3 = 0, 2x - y - 2 = 0$.

7

Curvature

1. Introduction

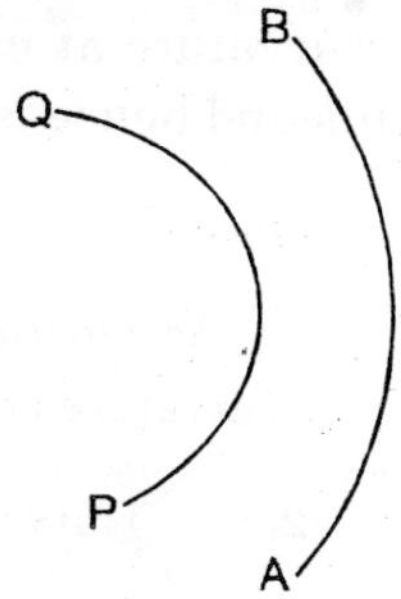

In the adjoining figure, curve PQ bends more sharply than the curve AB. The measure of the sharpness of the bending of a curve at a particular point is called *curvature* of the curve at the point. In this chapter, we shall find mathematical expressions for the curvature of a curve at a given point which will give a definite numerical measure of bending of the curve at that point.

2. Definitions

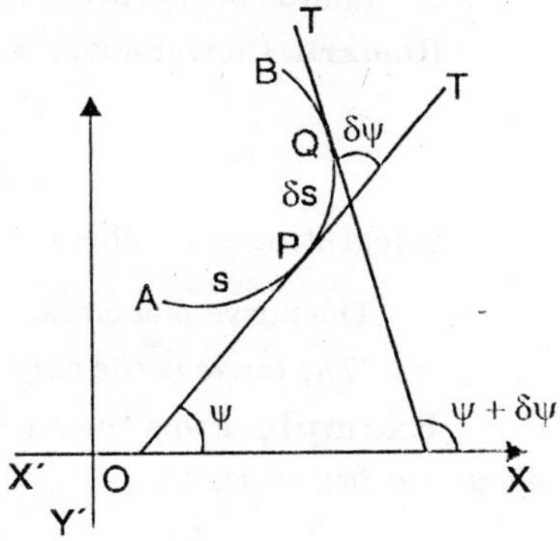

Let P, Q be two neighbouring points on a curve AB. Let arc AP = s, arc AQ = $s + \delta s$, so that arc PQ = δs, A being a fixed point on the curve, from which arcs are measured. Let the tangents to the curve at points P and Q make angles ψ and $\psi + \delta\psi$ respectively with a fixed line *say* x-axis. Then

(*i*) the *angle* $\delta\psi$ through which the tangent turns as its point of contact travels along the arc PQ is called the **total bending** or **total curvature** *of arc PQ* ;

(*ii*) *the ratio* $\frac{\delta\psi}{\delta s}$ *is called the* **mean** *or* **average curvature** of *arc* PQ ;

(*iii*) *the limiting value of the mean curvature when Q* $\to$ *P is called the* **curvature of the curve at the point P.** *Thus, the curvature* (k) *at point*

$$P = \underset{Q \to P}{Lt} \frac{\delta\psi}{\delta s} = \underset{\delta s \to 0}{Lt} \frac{\delta\psi}{\delta s} = \frac{d\psi}{ds}; \text{ and}$$

(*iv*) *the reciprocal of the curvature of the curve at P, provided this curvature is not zero, is called the* **radius of curvature of the curve at P.** This is usually denoted by ρ. Thus,

$$\rho = \frac{1}{k} = \frac{ds}{d\psi}.$$

3. Curvature of Circle

To show that the curvature of a circle is constant and equal to the reciprocal of its radius.

Consider any circle with centre C and radius r.

Let A, the lowest point of the circle be taken as the (fixed point) origin and tangent at A as x-axis.

Let P be any point on the circle s.t. arc AP = s and tangent at P make an angle ψ with x-axis.

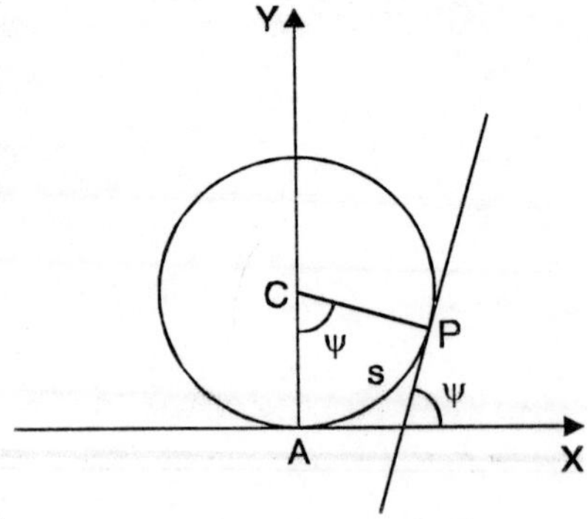

$\therefore \quad \angle ACP = \psi$ [$\because$ Angle between two lines = Angle between their perpendiculars]

$\therefore \quad s = r\psi$ [$\because$ From trigonometry $l = r\theta$ for a circle]

Differentiating w.r.t. ψ, $\dfrac{ds}{d\psi} = r$

$\therefore$ Curvature $= \dfrac{d\psi}{ds} = \dfrac{1}{r}$

i.e., curvature at *every* point of the circle is reciprocal of its radius and hence is constant.

Note 1. $\because$ Curvature at any point of the circle $= \dfrac{1}{r}$

$\therefore$ As r increases, curvature decreases

i.e., Curvature becomes smaller and smaller as the radius of circle becomes larger and larger.

2. $\because$ Curvature $= \dfrac{1}{r}$

$\therefore$ Radius of curvature $= r =$ Radius of circle.

Remark. Conversely, if curvature of a curve is constant ;

i.e., $$\frac{d\psi}{ds} = \frac{1}{r} \text{ (say)} \quad \therefore \quad d\psi = \frac{1}{r}\, ds$$

Integrating both sides, $\psi = \dfrac{1}{r} s$ or $s = r\psi$

$\therefore$ The curve is a circle.

$\therefore$ *"The circle is the only curve of constant curvature".*

Example. *Find the radius of curvature at the point (s, ψ) on the curve $s = a \log(\tan\psi + \sec\psi) + a \tan\psi \sec\psi$.*

Sol. $$\rho = \frac{ds}{d\psi} = a\,\frac{1}{\tan\psi + \sec\psi}(\sec^2\psi + \sec\psi\tan\psi) + a(\sec\psi\sec^2\psi + \tan\psi\sec\psi\tan\psi)$$

$$= a\sec\psi + a\sec\psi(\sec^2\psi + \tan^2\psi)$$

$$= a\sec\psi\,.\,[1 + \tan^2\psi + \sec^2\psi] = 2a\sec^3\psi.$$

4. The relation between the length of the *arc* s of a curve measured from a given fixed point on the curve and the *angle* ψ between the tangents at its extremities is called the *instrinsic equation of the curve.*

The expression $\dfrac{ds}{d\psi}$ for radius of curvature is suitable only for those curves whose intrinsic equations are given. We now proceed to find formulae for radius of curvature when equations of curves are given in other forms.

In this chapter, we shall need few results of chapter on "derivatives of arcs."

All these results can be obtained from the following figures :

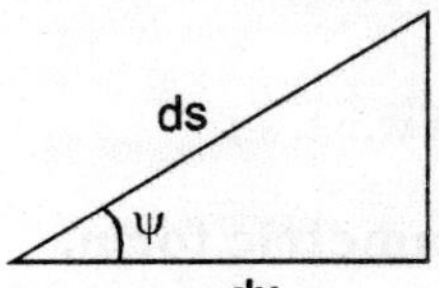

Fig. (*i*)

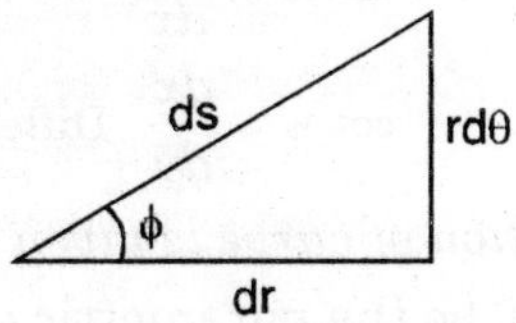

Fig. (*ii*)

$\therefore \qquad \cos\psi = \dfrac{dx}{ds}, \ \sin\psi = \dfrac{dy}{ds}, \ \tan\psi = \dfrac{dy}{dx}$

Again, $\qquad \cos\phi = \dfrac{dr}{ds}, \ \sin\phi = r\dfrac{d\theta}{ds}, \tan\phi = r\dfrac{d\theta}{dr}.$

5. Radius of Curvature for Cartesian Equation

When the equation of the curve is given in the explicit form $y = f(x)$.

We know that slope of the tangent at any point = $\tan\psi = \dfrac{dy}{dx}$

Differentiating both sides w.r.t. s,

$$\sec^2\psi \frac{d\psi}{ds} = \frac{d}{ds}\left(\frac{dy}{dx}\right)$$

or $$\sec^2\psi \cdot \frac{1}{\rho} = \frac{d}{dx}\left(\frac{dy}{dx}\right)\frac{dx}{ds}$$

or $$\sec^2\psi \cdot \frac{1}{\rho} = \frac{d^2y}{dx^2}\cos\psi \qquad \left[\because \text{ From Fig. } (i) \text{ of Art. 4, } \cos\psi = \frac{dx}{ds}\right]$$

$$\therefore \qquad \rho = \frac{\sec^2\psi}{\cos\psi\dfrac{d^2y}{dx^2}} = \frac{\sec^3\psi}{\dfrac{d^2y}{dx^2}} = \frac{(1+\tan^2\psi)^{3/2}}{\dfrac{d^2y}{dx^2}}$$

i.e., $$\rho = \frac{\left[1+\left(\dfrac{dy}{dx}\right)^2\right]^{3/2}}{\dfrac{d^2y}{dx^2}} = \frac{(1+y_1{}^2)^{3/2}}{y_2} \qquad \ldots(1)$$

Convention of Signs. The positive root is taken in numerator of (1), therefore radius of curvature, ρ, will be positive when y_2 is positive (*i.e.,* when curve is concave upwards) and *negative,* when y_2 is negative (*i.e.,* when curve is concave downwards). *In practice numerical value of ρ is taken.* Since at a point of inflexion y_2 is zero therefore curvature of a curve at a point of inflexion is zero.

Cor. When equation of curve is given in the form $x = f(y)$, then by interchanging x and y, we get

$$\rho = \frac{\left[1+\left(\dfrac{dx}{dy}\right)^2\right]^{3/2}}{\dfrac{d^2x}{dy^2}}$$

[**Hint for proof.** $\because$ $\tan\psi = \dfrac{dy}{dx}$

$\therefore$ $\cot\psi = \dfrac{dx}{dy}$. Differentiate w.r.t. s.]

(ii) When the equation of curve is given in **parametric form.**

Let $x = f(t), y = \phi(t)$, be the parametric equations of the curve, t being the parameter.

Then
$$\frac{dy}{dx} = \frac{dy}{dt}\bigg/\frac{dx}{dt} = \frac{y'}{x'}.$$
where dashes denote differentiations w.r.t. t.

Also
$$\frac{d^2y}{dx^2} = \frac{d}{dx}\left(\frac{y'}{x'}\right) = \frac{d}{dt}\left(\frac{y'}{x'}\right)\cdot\frac{dt}{dx}$$
$$= \frac{x'y'' - y'.x''}{(x')^2}\cdot\frac{1}{x'} = \frac{x'.y'' - y'x''}{(x')^3}$$

$\therefore$ Substituting these values of $\dfrac{dy}{dx}$ and $\dfrac{d^2y}{dx^2}$ in formula (1) and simplifying, we get
$$\rho = \frac{(\mathbf{x}'^2 + \mathbf{y}'^2)^{3/2}}{\mathbf{x}'\mathbf{y}'' - \mathbf{y}'\mathbf{x}''}.$$

(iv) When x and y are given as function of the length of arc s, measured from a fixed point on the curve.

Let the curve be given by the equations $x = f(s)$ and $y = \phi(s)$, where s denotes the length of arc measured from a *fixed point* on the curve.

We know [from Fig. (*i*) of Art. 4] that
$$\cos\psi = \frac{dx}{ds} \quad ...(i), \qquad \text{and} \qquad \sin\psi = \frac{dy}{ds} \qquad ...(ii)$$

Differentiating (*i*) w.r.t. s, we have $-\sin\psi\,.\,\dfrac{d\psi}{ds} = \dfrac{d^2x}{ds^2}$ ...(*iii*)

or
$$-\frac{dy}{ds}\cdot\frac{1}{\rho} = \frac{d^2x}{ds^2} \qquad [\because \text{ of } (ii)]$$

$\therefore$
$$\rho = -\frac{\mathbf{dy}}{\mathbf{ds}}\bigg/\frac{\mathbf{d^2x}}{\mathbf{ds^2}}.$$

Again differentiating (*ii*) w.r.t. s, we get $\cos\psi\dfrac{d\psi}{ds} = \dfrac{d^2y}{ds^2}$ ...(*iv*)

or
$$\frac{dx}{ds}\cdot\frac{1}{\rho} = \frac{d^2y}{ds^2} \qquad [\because \text{ of } (i)]$$

$\therefore$
$$\rho = \frac{\mathbf{dx}}{\mathbf{ds}}\bigg/\frac{\mathbf{d^2y}}{\mathbf{ds^2}}.$$

Squaring and adding (*iii*) and (*iv*), we get
$$\left(\frac{d\psi}{ds}\right)^2[\sin^2\psi + \cos^2\psi] = \left(\frac{d^2x}{ds^2}\right)^2 + \left(\frac{d^2y}{ds^2}\right)^2 \quad \text{or} \quad \frac{1}{\rho^2} = \left(\frac{\mathbf{d^2x}}{\mathbf{ds^2}}\right)^2 + \left(\frac{\mathbf{d^2y}}{\mathbf{ds^2}}\right)^2.$$

Example 1. *Prove that the radius of curvature for the catenary $y = c \cosh x/c$ is equal to the portion of the normal intercepted between the curve and the x-axis and that it varies as the square of the ordinate.*

Sol. Equation of the curve is $y = c \cosh \frac{x}{c}$ ...(1)

Diff. (1) w.r.t. x, $y_1 = c \sinh \frac{x}{c} \times \frac{1}{c}$ $\quad \left| \because \frac{d}{d\theta}(\cosh \theta) = \sinh \theta \right.$

or $$y_1 = \sinh \frac{x}{c}$$

Again diff. w.r.t. x, $$y_2 = \cosh \frac{x}{c} \times \frac{1}{c} = \frac{1}{c} \cosh \frac{x}{c}$$

$$\therefore \quad \rho = \frac{(1 + y_1^2)^{3/2}}{y_2} = \frac{\left(1 + \sinh^2 \frac{x}{c}\right)^{3/2}}{\frac{1}{c} \cosh \frac{x}{c}}$$

$$= \frac{\left(\cosh^2 \frac{x}{c}\right)^{3/2}}{\frac{1}{c} \cosh \frac{x}{c}} \qquad \left| \begin{array}{l} \because \cosh^2 \theta - \sinh^2 \theta = 1 \\ \therefore \cosh^2 \theta = 1 + \sinh^2 \theta \end{array} \right.$$

or $$\rho = \frac{c \cosh^3 x/c}{\cosh x/c} = c \cosh^2 \frac{x}{c} \qquad \text{...(2)}$$

Now portion of the normal intercepted between the curve and the x-axis.

$$= \text{Length of normal} = \mathbf{y} \sqrt{\mathbf{1} + \mathbf{y_1}^2}$$

$$= c \cosh \frac{x}{c} \sqrt{1 + \sinh^2 \frac{x}{c}} = c \cosh \frac{x}{c} \,.\, \cosh \frac{x}{c} = c \cosh^2 \frac{x}{c} \qquad \text{...(3)}$$

From (2) and (3), we have ρ = Length of Normal

Again $\frac{\rho}{(\text{Ordinate})^2} = \frac{\rho}{y^2} = \frac{c \cosh^2 x/c}{c^2 \cosh^2 x/c} = \frac{1}{c}$ = constant.

$\therefore$ ρ varies as y^2 *i.e.*, as square of the ordinate.

Remark. If the equation of the curve is given to be

$y = \frac{c}{2}(e^{x/c} + e^{-x/c})$, we should write it as

$$y = c\left(\frac{e^{x/c} + e^{-x/c}}{2}\right) = c \cosh \frac{x}{c} \qquad \left| \because \cosh \theta = \frac{e^{\theta} + e^{-\theta}}{2} \right.$$

Example 2. *In the cycloid $x = a(\theta + \sin \theta)$; $y = a(1 - \cos \theta)$, prove that $\rho = 4a \cos \frac{1}{2}\theta$.*

(K.U. 1997 S ; M.D.U. 1998)

Sol. Here $\quad x = a(\theta + \sin \theta), y = a(1 - \cos \theta)$

$\therefore \quad x' = a(1 + \cos \theta), y' = a \sin \theta$

and $\quad x'' = -a \sin \theta, y'' = a \cos \theta$

$$\therefore \quad \rho = \frac{(x'^2 + y'^2)^{3/2}}{x'y'' - y'x''} = \frac{a^3\left\{(1+\cos\theta)^2 + \sin^2\theta\right\}^{3/2}}{a^2(1+\cos\theta)\cos\theta + a^2\sin\theta \,.\, \sin\theta}$$

$$= \frac{a^3(1+\cos^2\theta + 2\cos\theta + \sin^2\theta)^{3/2}}{a^2(\cos\theta + \cos^2\theta + \sin^2\theta)} = \frac{a(1+1+2\cos\theta)^{3/2}}{(\cos\theta + 1)}$$

$$= \frac{a(2+2\cos\theta)^{3/2}}{(1+\cos\theta)} = a \,.\, 2^{3/2}\frac{(1+\cos\theta)^{3/2}}{(1+\cos\theta)} = a \,.\, 2^{3/2}(1+\cos\theta)^{1/2}$$

$$= a \,.\, 2^{3/2} \,.\, (2\cos^2 \tfrac{1}{2}\theta)^{1/2} = 4a\cos\tfrac{1}{2}\theta.$$

Remark. We can also use the formula $\rho = \dfrac{(1+y_1{}^2)^{3/2}}{y_2}$ for parametric equations.

Note on Parabola $y^2 = 4ax$

The shape of the parabola $y^2 = 4ax$ is as shown in the adjoining figure.

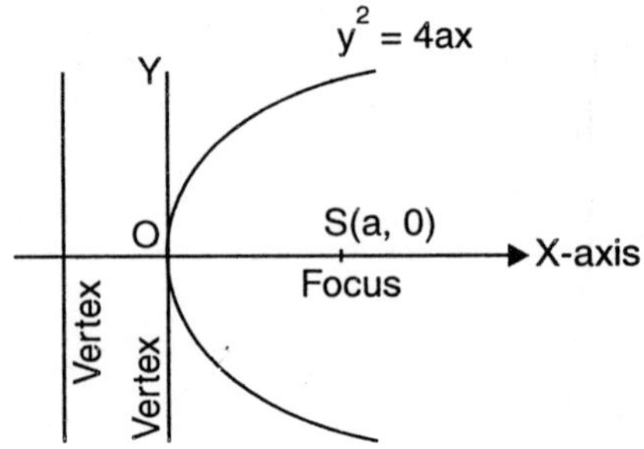

O(0, 0) is the VERTEX of this parabola.

x-axis is the AXIS of this parabola.

y-axis is the **tangent at the vertex** to this parabola.

For the parabola $y^2 = 4ax$; **4a,** the coefficient of x in this equation is called length of **Latus Rectum** of the parabola.

The point **(a, 0)** on the axis of parabola is called **Focus** of the parabola.

The line $x = -a$ is DIRECTRIX of the parabola.

Any chord of the parabola passing through the focus is called **a focal chord.**

Any point (x, y) on the parabola $y^2 = 4ax$ is $(\mathbf{at^2, 2at})$.

[$\because$ $(at^2, 2at)$ satisfies the equation of parabola]

This point $(at^2, 2at)$ is briefly written as point t on the parabola. t is also called "parameter".

Also $\mathbf{x = at^2, y = 2at}$ are called **Parametric Equations** of the parabola $y^2 = 4ax$.

Example 3. *Show that for the parabola $y^2 = 4ax$, ρ^2 varies as $(SP)^3$ where ρ is the radius of curvature at any point P of the parabola and S is the focus of the parabola.*

Sol. Equation of the parabola is $y^2 = 4ax$.

Any point $P(x, y)$ on the parabola is $(at^2, 2at)$

i.e., parametric equations of the parabola are

$$x = at^2, \quad y = 2at$$

$$\therefore \quad x' = 2at, \quad y' = 2a$$

$$x'' = 2a, \quad y'' = 0$$

$$\therefore \quad \rho = \frac{(x'^2 + y'^2)^{3/2}}{x'y'' - y'x''} = \frac{(4a^2t^2 + 4a^2)^{3/2}}{0 - 4a^2}$$

or $$\rho = -\frac{(4a^2)^{3/2}(t^2+1)^{3/2}}{4a^2} = -(4a^2)^{1/2}(1+t^2)^{3/2}$$

or $$\rho = 2a(1+t^2)^{3/2} \text{ (Numerically)}$$

Also SP (where S$(a, 0)$ is focus and P is $(at^2, 2at)$

$$= \sqrt{(at^2-a)^2+(2at-0)^2} \quad \Big| \; \sqrt{(x_2-x_1)^2+(y_2-y_1)^2}$$

$$= \sqrt{a^2t^4+a^2-2a^2t^2+4a^2t^2}$$

$$= \sqrt{a^2t^4+a^2+2a^2t^2} = \sqrt{(at^2+a)^2} = at^2+a = a(1+t^2)$$

$$\therefore \quad \frac{\rho^2}{(SP)^3} = \frac{[2a(1+t^2)^{3/2}]^2}{[a(1+t^2)]^3} = \frac{4a^2(1+t^2)^3}{a^3(1+t^2)^3} = \frac{4}{a} = \text{Constant.}$$

$\therefore$ ρ^2 varies as $(SP)^3$.

Note on Ellipse $\quad \frac{x^2}{a^2} + \frac{y^2}{b^2} = 1$

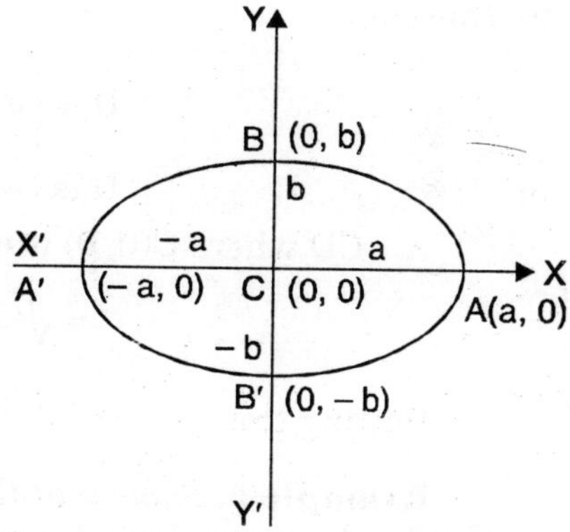

The shape of the ellipse $\frac{x^2}{a^2} + \frac{y^2}{b^2} = 1$ is as shown in the adjoining figure.

C(0, 0) is called CENTRE of the ellipse.

x-axis is called MAJOR AXIS of the ellipse and length of major axis is $2a$.

y-axis is called MINOR AXIS of the ellipse and length of minor-axis is $2b$.

Eccentricity e of the ellipse is given by

$$\mathbf{b^2 = a^2(1 - e^2).}$$

Any point (x, y) of the ellipse $\frac{x^2}{a^2} + \frac{y^2}{b^2} = 1$ is **$(a \cos \theta, b \sin \theta)$**

[$\because$ $(a \cos \theta, b \sin \theta)$ satisfies the equation of the ellipse]

The point $(a \cos \theta, b \sin \theta)$ is briefly written as point θ on the ellipse.

θ is called **Parameter** or **Eccentric Angle** of the point.

Conjugate Diameters of an Ellipse

Two diameters of an ellipse are said to be **conjugate** if each bisects chords parallel to the other.

If CP and CD are two semi-conjugate diameters of an ellipse and P is point θ *i.e.*, P is $(a \cos \theta, b \sin \theta)$; then the point D is $\frac{\pi}{2} + \theta$ *i.e.*, D is $\left[a \cos\left(\frac{\pi}{2}+\theta\right), b \sin\left(\frac{\pi}{2}+\theta\right)\right]$.

Example 4. *If CP, CD are a pair of semi-conjugate diameters of an ellipse of semi-axes of lengths a and b, prove that the radius of curvature at P =* $\frac{CD^3}{ab}$ *where C is the centre of the ellipse.*

Sol. Equation of the ellipse is $\frac{x^2}{a^2} + \frac{y^2}{b^2} = 1.$

Any point P(x, y) on the ellipse is $(a \cos \theta, b \sin \theta)$ *i.e.,* parametric equations of the ellipse are

$$x = a \cos \theta \qquad y = b \sin \theta$$

$$\therefore \quad x' = -a \sin \theta \qquad y' = b \cos \theta$$

$$x'' = -a \cos \theta \qquad y'' = -b \sin \theta$$

$$\therefore \quad \rho = \frac{(x'^2 + y'^2)^{3/2}}{x'y'' - y'x''} = \frac{(a^2 \sin^2 \theta + b^2 \cos^2 \theta)^{3/2}}{ab \sin^2 \theta + ab \cos^2 \theta}$$

or $$\rho = \frac{(a^2 \sin^2 \theta + b^2 \cos^2 \theta)^{3/2}}{ab} \qquad \ldots(1)$$

$\because$ CP and CD are semi-conjugate diameters of the ellipse and point P is $(a \cos \theta, b \sin \theta)$, therefore

$$\text{D is } \left[a \cos\left(\frac{\pi}{2} + \theta\right), b \sin\left(\frac{\pi}{2} + \theta\right)\right]$$

or $$\text{D is } (-a \sin \theta, b \cos \theta)$$

$\therefore$ CD where C(0, 0) is centre of the ellipse

$$= \sqrt{(-a \sin \theta - 0)^2 + (b \cos \theta - 0)^2} = \sqrt{a^2 \sin^2 \theta + b^2 \cos^2 \theta}$$

Putting this value of $\sqrt{a^2 \sin^2 \theta + b^2 \cos^2 \theta} = \text{CD}$ in (1), we have $\rho = \frac{\text{CD}^3}{ab}$.

Example 5. *Show that the radius of curvature at any point of the curve $x^{2/3} + y^{2/3} = a^{2/3}$ is equal to three times the length of the perpendicular from the origin on the tangent.*

Sol. Equation of the curve is $x^{2/3} + y^{2/3} = a^{2/3}$ ...(1)

Dividing by $a^{2/3}$; $$\left(\frac{x}{a}\right)^{2/3} + \left(\frac{y}{a}\right)^{2/3} = 1$$

or $$\left(\left(\frac{x}{a}\right)^{1/3}\right)^2 + \left(\left(\frac{y}{a}\right)^{1/3}\right)^2 = 1$$

Comparing with $\cos^2 \theta + \sin^2 \theta = 1$, we have

$$\left(\frac{x}{a}\right)^{1/3} = \cos \theta \quad \text{and} \quad \left(\frac{y}{a}\right)^{1/3} = \sin \theta$$

Cubing both equations $$\frac{x}{a} = \cos^3 \theta \quad \text{and} \quad \frac{y}{a} = \sin^3 \theta$$

or $$x = a \cos^3 \theta \quad \text{and} \quad y = a \sin^3 \theta \qquad \ldots(2)$$

Equations (2) are parametric equations of curve (1).

(These equations (2) of curve (1) should be preferred in this chapter)

Let us find ρ for curve (1) using Equations (2)

Differentiating eqns. (2) w.r.t. θ,

$$\frac{dx}{d\theta} = -3a\cos^2\theta\sin\theta \text{ and } \frac{dy}{d\theta} = 3a\sin^2\theta\cos\theta$$

$$\therefore \qquad y_1 = \frac{dy}{dx} = \frac{\frac{dy}{d\theta}}{\frac{dx}{d\theta}} = \frac{3a\sin^2\theta\cos\theta}{-3a\cos^2\theta\sin\theta} = -\frac{\sin\theta}{\cos\theta} = -\tan\theta$$

Again differentiating w.r.t. x,

$$y_2 = \frac{d^2y}{dx^2} = -\sec^2\theta\frac{d\theta}{dx} = \frac{-1}{\cos^2\theta}\cdot\frac{1}{-3a\cos^2\theta\sin\theta} = \frac{1}{3a\cos^4\theta\sin\theta}$$

We know that, $$\rho = \frac{(1+y_1^2)^{3/2}}{y_2}$$

Putting values of y_1 and y_2,

$$\rho = \frac{(1+\tan^2\theta)^{3/2}}{\left(\frac{1}{3a\cos^4\theta\sin\theta}\right)} = 3a\cos^4\theta\sin\theta\left(\sec^2\theta\right)^{3/2}$$

$$= 3a\cos^4\theta\sin\theta\sec^3\theta = 3a\cos^4\theta\sin\theta\frac{1}{\cos^3\theta}$$

or $$\rho = 3a\cos\theta\sin\theta \qquad ...(3)$$

Now let us find the equation of the tangent at any point (x, y) *i.e.*, $(a\cos^3\theta, a\sin^3\theta)$ (using equations (2) to curve (1).

We know that slope of the tangent at any point

$$(x, y) = \frac{dy}{dx} = -\tan\theta = -\frac{\sin\theta}{\cos\theta}$$

∴ Equation of the tangent at any point $(a\cos^3\theta, a\sin^3\theta)$

is $$y - a\sin^3\theta = -\frac{\sin\theta}{\cos\theta}(x - a\cos^3\theta) \qquad | \; y - y_1 = m(x - x_1)$$

Cross-multiplying $y\cos\theta - a\sin^3\theta\cos\theta = -x\sin\theta + a\cos^3\theta\sin\theta$

or $$x\sin\theta + y\cos\theta - a\sin^3\theta\cos\theta - a\cos^3\theta\sin\theta = 0$$

or $$x\sin\theta + y\cos\theta - a\sin\theta\cos\theta(\sin^2\theta + \cos^2\theta) = 0$$

or $$x\sin\theta + y\cos\theta - a\sin\theta\cos\theta = 0 \qquad ...(4)$$

p = length of ⊥ from origin (0, 0) on tangent (4)

$$= \frac{|0 + 0 - a\sin\theta\cos\theta|}{\sqrt{\sin^2\theta + \cos^2\theta}} \qquad \left|\; \frac{|ax_1 + by_1 + c|}{\sqrt{a^2+b^2}}\right.$$

$$= a\sin\theta\cos\theta \qquad ...(5)$$

From (3) and (5), we have $\rho = 3p$

i.e., radius of curvature at any point of the curve $x^{2/3} + y^{2/3} = a^{2/3}$ is equal to three times the length of the perpendicular from the origin on the tangent.

Example 6. *The tangents at two points P and Q on the cycloid $x = a(\theta - \sin\theta)$, $y = a(1 - \cos\theta)$ are at right angles. Show that if ρ_1, ρ_2 be the radii of curvature at these points, then $\rho_1^2 + \rho_2^2 = 16a^2$.*

Sol. Equations of the curve are $x = a(\theta - \sin\theta)$ and $y = a(1 - \cos\theta)$

Differentiating w.r.t. θ,

$$\frac{dx}{d\theta} = a(1-\cos\theta) \text{ and } \frac{dx}{d\theta} = a\sin\theta$$

$$\therefore \qquad y_1 = \frac{dy}{dx} = \frac{dy/d\theta}{dx/d\theta} = \frac{a\sin\theta}{a(1-\cos\theta)}$$

or

$$y_1 = \frac{2\sin\frac{\theta}{2}\cos\frac{\theta}{2}}{2\sin^2\frac{\theta}{2}} = \cot\frac{\theta}{2}$$

Again diff. w.r.t. x,

$$y_2 = -\operatorname{cosec}^2\frac{\theta}{2}\cdot\frac{1}{2}\frac{d\theta}{dx} = \frac{-1}{2\sin^2\frac{\theta}{2}}\,\frac{1}{a(1-\cos\theta)}$$

or

$$y_2 = \frac{-1}{2\sin^2\frac{\theta}{2}}\,\frac{1}{a\,.\,2\sin^2\frac{\theta}{2}} = \frac{-1}{4a\sin^4\frac{\theta}{2}}$$

We know that

$$\rho = \frac{(1+y_1^2)^{3/2}}{y_2}$$

Putting values of y_1 and y_2,

$$\rho = \frac{\left(1+\cot^2\frac{\theta}{2}\right)^{3/2}}{\left(\frac{-1}{4a\sin^4\frac{\theta}{2}}\right)} = -4a\sin^4\frac{\theta}{2}\left(\operatorname{cosec}^2\frac{\theta}{2}\right)^3$$

$$= -4a\sin^4\frac{\theta}{2}\operatorname{cosec}^3\frac{\theta}{2} = -4a\sin^4\frac{\theta}{2}\,\frac{1}{\sin^3\frac{\theta}{2}} = -4a\sin\frac{\theta}{2}$$

or

$$\rho = 4a\sin\frac{\theta}{2} \text{ (numerically)}$$

| (∵ Radius of curvature being length is never negative)

∴ Radius of curvature at any point θ of the given curve is $\rho = 4a\sin\frac{\theta}{2}$ ...(1)

Let $\theta = \theta_1$ and $\theta = \theta_2$ for points P and Q of the given curve where Radii of curvature are ρ_1 and ρ_2 (given)

∴ From (1), $\rho_1 = 4a\sin\frac{\theta_1}{2}$ and $\rho_2 = 4a\sin\frac{\theta_2}{2}$ ...(2)

We know that slope of the tangent at any point $\theta = \frac{dy}{dx} = \cot\frac{\theta}{2}$

∴ Slopes of the tangents at points P $(\theta = \theta_1)$ and Q $(\theta = \theta_2)$ are $\cot\frac{\theta_1}{2}$ and $\cot\frac{\theta_2}{2}$.

Because the tangents at points P and Q are at right angles (given), therefore $m_1m_2 = -1$.

$\therefore \qquad \cot \frac{\theta_1}{2} \cot \frac{\theta_2}{2} = -1$ or $\cot \frac{\theta_1}{2} = \frac{-1}{\cot \frac{\theta_2}{2}}$

or $\qquad \cot \frac{\theta_1}{2} = -\tan \frac{\theta_2}{2} = \cot \left(\frac{\pi}{2} + \frac{\theta_2}{2} \right)$

$$\therefore \qquad \frac{\theta_1}{2} = \frac{\pi}{2} + \frac{\theta_2}{2} \qquad \text{...(3)}$$

Putting values of ρ_1 and ρ_2 from Eqn. (2) in L.H.S. $= \rho_1^2 + \rho_2^2$

$$= 16a^2 \sin^2 \frac{\theta_1}{2} + 16\, a^2 \sin^2 \frac{\theta_2}{2}$$

Putting the value of $\frac{\theta_1}{2}$ from (3),

$$= 16a^2 \left[\sin^2 \left(\frac{\pi}{2} + \frac{\theta_2}{2} \right) + \sin^2 \frac{\theta_2}{2} \right]$$

$$= 16a^2 \left[\cos^2 \frac{\theta_2}{2} + \sin^2 \frac{\theta_2}{2} \right] = 16a^2 \times 1 = 16a^2 = \text{ R.H.S.}$$

Example 7. *Show that the radius of curvature of the curve* $\sqrt{x} + \sqrt{y} = \sqrt{a}$ *is minimum at the point* $\left(\frac{1}{4} a, \frac{1}{4} a \right)$. (K.U. 2006 S)

Sol. Equation of the curve is $\sqrt{x} + \sqrt{y} = \sqrt{a}$ ...(1)

Diff. both sides of Eqn. (1) w.r.t. x,

$$\frac{1}{2} x^{-1/2} + \frac{1}{2} y^{-1/2} \frac{dy}{dx} = 0 \quad \text{or} \quad \frac{1}{2} \frac{1}{\sqrt{y}} \frac{dy}{dx} = \frac{-1}{2} \frac{1}{\sqrt{x}}$$

$$\therefore \qquad \frac{dy}{dx} = -\frac{2\sqrt{y}}{2\sqrt{x}} = -\frac{\sqrt{y}}{\sqrt{x}} \qquad \text{...(2)}$$

Again diff. (2) w.r.t. x,

$$\frac{d^2y}{dx^2} = -\frac{\left[\sqrt{x} \, \frac{1}{2} y^{-1/2} \frac{dy}{dx} - \sqrt{y} \,.\, \frac{1}{2} x^{-1/2} \right]}{x} = -\frac{1}{2} \frac{\left[\frac{\sqrt{x}}{\sqrt{y}} \left(-\frac{\sqrt{y}}{\sqrt{x}} \right) - \frac{\sqrt{y}}{\sqrt{x}} \right]}{x} \qquad \text{[By (2)]}$$

$$= -\frac{1}{2} \frac{\left(-1 - \frac{\sqrt{y}}{\sqrt{x}} \right)}{x} = \frac{1}{2} \left(\frac{\sqrt{x} + \sqrt{y}}{x\sqrt{x}} \right) = \frac{1}{2} \cdot \frac{\sqrt{a}}{x\sqrt{x}} \qquad \text{[By (1)] ...(3)}$$

Putting these values of $\frac{dy}{dx}$ and $\frac{d^2y}{dx^2}$ in $\rho = \frac{\left(1 + \left(\frac{dy}{dx} \right)^2 \right)^{3/2}}{\frac{d^2y}{dx^2}}$, we have

$$\rho = \frac{\left(1 + \frac{y}{x} \right)^{3/2}}{\left(\frac{\sqrt{a}}{2x\sqrt{x}} \right)} = \frac{(x+y)^{3/2}}{x^{3/2}} \frac{2x\sqrt{x}}{\sqrt{a}} \quad \text{or} \quad \rho = \frac{2}{\sqrt{a}} (x+y)^{3/2} \qquad \text{...(4)}$$

To find minimum value of ρ

Diff. (4) w.r.t. x, $$\frac{d\rho}{dx} = \frac{2}{\sqrt{a}}\frac{3}{2}(x+y)^{1/2}\left(1+\frac{dy}{dx}\right)$$

or $$\frac{d\rho}{dx} = \frac{3}{\sqrt{a}}(x+y)^{1/2}\left(1+\frac{dy}{dx}\right) \quad ...(5)$$

Again diff. (5) w.r.t. x,

$$\frac{d^2\rho}{dx^2} = \frac{3}{\sqrt{a}}\left[(x+y)^{3/2}\frac{d^2y}{dx^2} + \frac{1}{2}(x+y)^{-1/2}\left(1+\frac{dy}{dx}\right)\left(1+\frac{dy}{dx}\right)\right] \quad ...(6)$$

Put $\mathbf{\frac{d\rho}{dx} = 0}$ $\therefore$ $\frac{3}{\sqrt{a}}(x+y)^{1/2}\left(1+\frac{dy}{dx}\right) = 0$

But $\frac{3}{\sqrt{a}} \neq 0$ also $x + y \neq 0$ [By (1)]

$\therefore$ $1+\frac{dy}{dx} = 0$ or $\frac{dy}{dx} = -1$

$\therefore$ From (2) $-\frac{\sqrt{y}}{\sqrt{x}} = -1$ or $\sqrt{y} = \sqrt{x}$ $\therefore$ $y = x$ ...(7)

Let us solve Eqns. (1) and (7) for x and y.

Putting $y = x$ from (7) in (1), $\sqrt{x} + \sqrt{x} = \sqrt{a}$ or $2\sqrt{x} = \sqrt{a}$

$\therefore$ $\sqrt{x} = \frac{\sqrt{a}}{2}$ $\therefore$ $x = \frac{a}{4}$

$\therefore$ From (7) $y = x = \frac{a}{4}$

$\therefore$ Turning point for ρ to be minimum is $\left(\frac{a}{4}, \frac{a}{4}\right)$

At $x = y = \frac{a}{4}$, from (3), $\frac{d^2y}{dx^2} = \frac{1}{2}\frac{\sqrt{a}}{\frac{a}{4}\cdot\sqrt{\frac{a}{4}}} = \frac{1}{2}\cdot\frac{4}{a}\cdot\sqrt{\frac{4}{a}}\sqrt{a} = \frac{4}{a}$

Putting $x = \frac{a}{4}, y = \frac{a}{4}, \frac{dy}{dx} = -1, \frac{d^2y}{dx^2} = \frac{4}{a}$ in (6), $\frac{d^2\rho}{dx^2} = \frac{3}{\sqrt{a}}\left[\left(\frac{a}{4}+\frac{a}{4}\right)^{3/2}\frac{4}{a}\right] > 0$

$\therefore$ ρ is minimum at the point $\left(\frac{a}{4}, \frac{a}{4}\right)$

EXERCISE 1

1. Find the radius of curvature at the point (s, ψ) on the following curves :

   (*i*) $s = a \log \tan\left(\frac{\pi}{4} + \frac{\psi}{2}\right)$ (*ii*) $s = 4a \sin \psi$ [cycloid].

2. Find the radius of curvature at the given point of the following curves :

   (*i*) Rectangular hyperbola $xy = c^2$ at the point (x, y).

[**Hint.** From the equation of the curve $y = \frac{c^2}{x}$.]

(*ii*) $y = 4 \sin x - \sin 2x$ at $x = \frac{\pi}{2}$. (*iii*) $\sqrt{x} + \sqrt{y} = 1$ at the point $(\frac{1}{4}, \frac{1}{4})$.

3. Find the radius of curvature at any point of the curve $x = a(\cos\theta + \theta\sin\theta)$, $y = a(\sin\theta - \theta\cos\theta)$.

4. Find the radius of curvature of the curve $y = e^x$, at the point where it crosses the y-axis.

5. Find the radius of curvature at the origin of the two branches of the curve given by

$$x = 1 - t^2,\ y = t - t^3.$$

[**Hint.** At the origin $x = 0, y = 0$

$\therefore \quad 1 - t^2 = 0$ and $t - t^3 = 0$.

Common values of t on solving are ± 1.]

6. Prove that the radius of curvature of the curve $y = \frac{a}{2}\left(e^{x/a} + e^{-x/a}\right)$ is $\frac{y^2}{a}$.

[**Hint.** We know that $\cosh\theta = \frac{e^{\theta} + e^{-\theta}}{2}$,

$\therefore$ Equation of the curve becomes $y = a \cosh \frac{x}{a}$. Now see example 1.]

7. For the curve $y = a\, e^{x/a}$, prove that $\rho = a \sec^2\theta \operatorname{cosec}\theta$, where $\theta = \tan^{-1}\frac{y}{a}$.

8. For the curve $y = \frac{ax}{a+x}$, prove that $\left(\frac{2\rho}{a}\right)^{2/3} = \left(\frac{y}{x}\right)^2 + \left(\frac{x}{y}\right)^2$

9. Find the radius of curvature for the curve $\sqrt{\frac{x}{a}} - \sqrt{\frac{y}{b}} = 1$ at the points where it touches the co-ordinate axes.

[**Hint.** The point where the curve touches x-axis ; $\frac{dy}{dx} = 0$. The point where the curve touches y-axis, $\frac{dx}{dy} = 0$.]

10. Show that for the curve $x = a\cos\theta(1 + \sin\theta)$ and $y = a\sin\theta(1 + \cos\theta)$, the radius of curvature is a, at the point for which the value of the parameter θ is $-\frac{\pi}{4}$.

11. Prove that for the curve $x^3 + y^3 = 3axy$, the radius of curvature at the point $\left(\frac{3a}{2}, \frac{3a}{2}\right)$ is numerically equal to $\frac{3a\sqrt{2}}{16}$.

12. Show that the radius of curvature at the point $(-2a, 2a)$ on the curve $x^2y = a(x^2 + y^2)$ is $2a$.

13. Find the points on the parabola $y^2 = 8x$ at which the radius of curvature is $7\frac{13}{16}$.

14. If ρ_1 and ρ_2 be the radii of curvature at the extremities of a focal chord of a parabola $y^2 = 4ax$, prove that $(\rho_1)^{-2/3} + (\rho_2)^{-2/3} = (2a)^{-2/3}$

(M.D.U. 2005, 2003 A ; K.U. 1999 A)

[**Hint.** If t_1 and t_2 are the parameters of the two extremities of a focal chord of a parabola, then $t_1t_2 = -1$.]

15. In an ellipse $\frac{x^2}{a^2} + \frac{y^2}{b^2} = 1$, show that the radius of curvature at the end of the major axis is equal to the semi-latus rectum of the ellipse.

$\left[\textbf{Hint.}\ \text{Ends of the major axis are } (\pm a, 0), \text{ and semi-latus rectum } = \frac{b^2}{a}.\right]$

16. If ρ_1, ρ_2 be the radii of curvature at the extremities of two conjugate diameters of an ellipse, prove that $(\rho_1^{2/3} + \rho_2^{2/3})(ab)^{2/3} = a^2 + b^2$. *(K.U. 2006)*

17. Prove that for the ellipse $\frac{x^2}{a^2} + \frac{y^2}{b^2} = 1$ (or ellipse $x = a \cos \theta, y = b \sin \theta$), $\rho = \frac{a^2b^2}{p^3}$ where p is the perpendicular from the centre upon the tangent at (x, y). *(M.D.U. 2004 S, 2003 S)*

[**Hint.** Length of the perpendicular from the point (x_1, y_1) on the st. line

$ax + by + c = 0$ is $\frac{\pm (ax_1 + by_1 + c)}{\sqrt{a^2 + b^2}}$.]

18. (*i*) Show that the radius of curvature at the point $(a \cos^3 \theta, a \sin^3 \theta)$ on the curve $x^{2/3} + y^{2/3} = a^{2/3}$ is $3a \sin \theta \cos \theta$. *(M.D.U. 1999 ; K.U. 2000, 2007)*

(*ii*) Find the radius of curvature at the point (x, y) of the curve $x^{2/3} + y^{2/3} = a^{2/3}$.
[**Hint.** Take parametric equations as $x = a \cos^3 \theta$, $y = a \sin^3 \theta$.]

(*iii*) Show that the radius of curvature ρ at any point (x, y) of the curve $x^{2/3} + y^{2/3} = a^{2/3}$ satisfies $\rho^3 = 27\, axy$.

19. Show that the radius of curvature at any point of the astroide $x = a \cos^3 \theta$, $y = a \sin^3 \theta$ is equal to three times the length of the perpendicular from the origin to the tangent.
[**Hint.** It is example 5]

20. Show that the radius of curvature at a point of the curve $x = a\, e^{\theta} (\sin \theta - \cos \theta)$, $y = a\, e^{\theta} (\sin \theta + \cos \theta)$ is twice the distance of the tangent at the point from the origin.

21. Show that the radius of curvature at each point of the curve $x = a\left(\cos t + \log \tan \frac{t}{2}\right)$, $y = a \sin t$ is inversely proportional to the length of the normal intercepted between the point on the curve and the x-axis. *(M.D.U. 2002)*

[**Hint.** formula for length of the normal intercepted between point of curve and x-axis

$= y \sqrt{1 + \left(\frac{dy}{dx}\right)^2}$]

22. Find the least value of $|\ \rho\ |$ for the curve $y = \log x$ $(x > 0)$.

ANSWERS

1. (*i*) $a \sec \psi$ (*ii*) $4a \cos \psi$

2. (*i*) $\frac{(x^4 + c^4)^{3/2}}{2c^2x^3}$ (*ii*) $\frac{5\sqrt{5}}{4}$ (*iii*) $\frac{1}{\sqrt{2}}$

3. $a\theta$ **4.** $\sqrt{8}$ **5.** $2\sqrt{2}$ **9.** $\frac{2a^2}{b}, \frac{2b^2}{a}$

13. $(\frac{9}{8}, 3)$; $(\frac{9}{8}, -3)$ **18.** (*ii*) $3a \sin \theta \cos \theta$ **22.** $\frac{3\sqrt{3}}{2}$.

6. Radius of Curvature for Polar Equations

To find the radius of curvature for the curve $r = f(\theta)$ or $f(r, \theta) = 0$ (K.U. 2000 A, 1995 A)

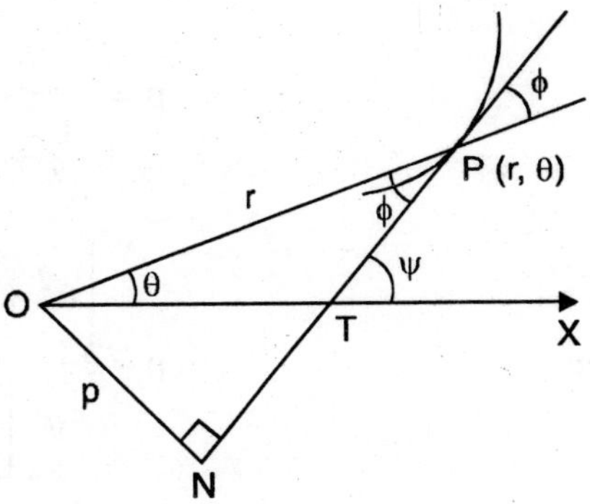

Let the tangent at any point $P(r, \theta)$ to curve make an angle ψ with the initial line OX.

Then from the figure, we have

$$\psi = \theta + \phi$$

$$\therefore \quad \frac{d\psi}{ds} = \frac{d\theta}{ds} + \frac{d\phi}{ds} = \frac{d\theta}{ds} + \frac{d\phi}{d\theta} \cdot \frac{d\theta}{ds}$$

or
$$\frac{1}{\rho} = \frac{d\theta}{ds}\left[1 + \frac{d\phi}{d\theta}\right] = \frac{\sin\phi}{r}\left[1 + \frac{d\phi}{d\theta}\right] \qquad ...(1)$$

$$\left[\because \text{ From Fig. } (ii) \text{ of Art. 4, } \sin\phi = r\frac{d\theta}{ds} \quad \therefore \quad \frac{d\theta}{ds} = \frac{\sin\phi}{r}\right]$$

Again from Fig. (*ii*) Art. 4, $\tan\phi = r \cdot \dfrac{d\theta}{dr} = \dfrac{r}{\dfrac{dr}{d\theta}} = \dfrac{r}{r_1}$ ...(2)

To find $\dfrac{d\phi}{d\theta}$

Differentiating (2) w.r.t. θ, $\quad \sec^2\theta \dfrac{d\phi}{d\theta} = \dfrac{r_1 \cdot r_1 - rr_2}{r_1^2} = \dfrac{r_1^2 - rr_2}{r_1^2}$

$$\therefore \quad \frac{d\phi}{d\theta} = \frac{r_1^2 - rr_2}{r_1^2 \sec^2\phi} = \frac{r_1^2 - rr_2}{r_1^2\left[1 + \dfrac{r^2}{r_1^2}\right]} = \frac{r_1^2 - rr_2}{r_1^2 + r^2} \qquad \left|\because \tan\phi = \frac{r}{r_1}\right.$$

Also $\quad \tan\phi = \dfrac{r}{r_1}$ gives $\sin\phi = \dfrac{r}{\sqrt{r^2 + r_1^2}}$

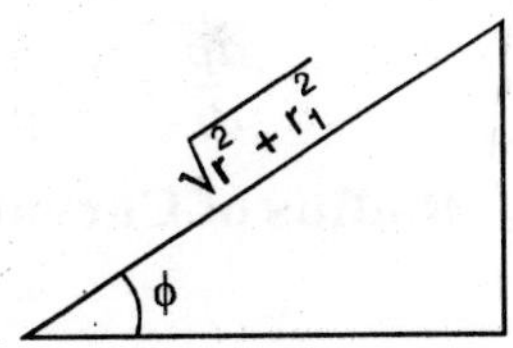

Substituting these values of $\dfrac{d\phi}{d\theta}$ and $\sin\phi$ in (1), we get

$$\frac{1}{\rho} = \frac{1}{\sqrt{r^2 + r_1^2}}\left[1 + \frac{r_1^2 - rr_2}{r_1^2 + r^2}\right]$$

or
$$\frac{1}{\rho} = \frac{r^2 + 2r_1^2 - rr_2}{(r^2 + r_1^2)^{3/2}} \quad \therefore \quad \rho = \frac{\mathbf{(r^2 + r_1^2)^{3/2}}}{\mathbf{r^2 + 2r_1^2 - rr_2}} \qquad ...(3)$$

Cor. *If equation of a curve is given in the form $u = f(\theta)$, where*

$$u = \frac{1}{r} \qquad \textit{i.e.,} \quad r = \frac{1}{u}, \text{ then we have } r_1 = -\frac{1}{u^2} \cdot \frac{du}{d\theta}$$

and
$$r_2 = -\frac{1}{u^2} \cdot \frac{d^2u}{d\theta^2} + \frac{2}{u^3} \cdot \left(\frac{du}{d\theta}\right)^2$$

Substituting these values of r_1 and r_2 in (3), we get

$$\rho = \frac{\left[\frac{1}{u^2} + \frac{1}{u^4}\left(\frac{du}{d\theta}\right)^2\right]^{3/2}}{\frac{1}{u^2} + 2\frac{1}{u^4}\left(\frac{du}{d\theta}\right)^2 - \frac{1}{u}\left[-\frac{1}{u^2}\cdot\frac{d^2u}{d\theta^2} + \frac{2}{u^2}\left(\frac{du}{d\theta}\right)^2\right]}$$

or $$\rho = \frac{\left[u^2 + \left(\frac{du}{d\theta}\right)^2\right]^{3/2}}{u^3\left[u + \frac{d^2u}{d\theta^2}\right]} = \frac{(\mathbf{u^2 + u_1{}^2})^{3/2}}{\mathbf{u^3(u + u_2)}}$$

where $$u_1 = \frac{du}{d\theta} \quad \text{and} \quad u_2 = \frac{d^2u}{d\theta^2}.$$

7. Radius of Curvature for Pedal Equations $\left(\textbf{To prove } \rho = \mathbf{r}\dfrac{\mathbf{dr}}{\mathbf{dp}}\right)$

Let the pedal equation of the curve be $p = f(r)$.

From the figure (*refer to figure of article 9.6*), we have

$$\psi = \theta + \phi \quad \therefore \quad \frac{d\psi}{ds} = \frac{d\theta}{ds} + \frac{d\phi}{ds} \quad \textit{i.e.,} \quad \frac{1}{\rho} = \frac{d\theta}{ds} + \frac{d\phi}{ds} \qquad \ldots(i)$$

But we know that $p = r \sin \phi$

Differentiating this w.r.t. **r**, we have

$$\frac{dp}{dr} = \sin\phi \,.\, 1 + r\cos\phi \,.\, \frac{d\phi}{dr} = r\,.\,\frac{d\theta}{ds} + r\,.\,\frac{dr}{ds}\frac{d\phi}{dr} \qquad \left[\because \ \sin\phi = r\,.\,\frac{d\theta}{ds}, \cos\phi = \frac{dr}{ds}\right]$$

$$= r\left[\frac{d\theta}{ds} + \frac{d\phi}{ds}\right] = r\,.\,\frac{1}{\rho} \qquad [\because \text{ of relation } (i)]$$

or $$\frac{dp}{dr} = r\,.\,\frac{1}{\rho} \qquad \therefore \qquad \rho = \frac{r}{dp/dr} = \mathbf{r}\,.\,\frac{\mathbf{dr}}{\mathbf{dp}}.$$

8. Radius of Curvature for Tangential Polar Equations p = f(ψ).

$$\left(\textbf{To prove } \rho = \mathbf{p} + \frac{\mathbf{d^2p}}{\mathbf{d\psi^2}}\right)$$

A *relation between* perpendicular p from the origin on any tangent to a curve and *angle* ψ which this tangent makes with x-axis, is called the **tangential polar equation** of the curve.

Let p be length of the perpendicular OL drawn from the origin on the tangent to curve at the point P(x, y), then OL makes an angle $\psi - \frac{\pi}{2}$ with the positive direction of x-axis.

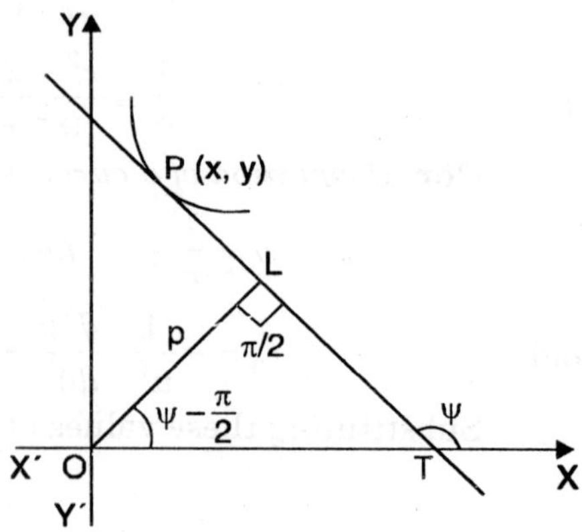

$\therefore$ Equation of the tangent PT is $p = X\cos\left(\psi - \frac{\pi}{2}\right) + Y\sin\left(\psi - \frac{\pi}{2}\right)$

[Normal form $x\cos\alpha + y\sin\alpha = p$]

or $$p = X\sin\psi - Y\cos\psi$$

where X, Y are Cartesian co-ordinates of any point on this tangent.

As point P(x, y) lies on this tangent

$\therefore$ $$p = x\sin\psi - y\cos\psi \quad ...(i)$$

Differentiating both sides of (i) w.r.t. ψ, we get

$$\frac{dp}{d\psi} = x\cos\psi + \sin\psi \cdot \frac{dx}{d\psi} + y\sin\psi - \cos\psi\,\frac{dy}{d\psi}$$

$$= x\cos\psi + y\sin\psi + \sin\psi \cdot \frac{dx}{ds}\cdot\frac{ds}{d\psi} - \cos\psi \cdot \frac{dy}{ds}\cdot\frac{ds}{d\psi}$$

$$= x\cos\psi + y\sin\psi + \sin\psi\,\rho \cdot \cos\psi - \cos\psi \cdot \rho\sin\psi$$

$$\left[\because \frac{dx}{ds} = \cos\psi, \frac{dy}{ds} = \sin\psi\right]$$

$$= x\cos\psi + y\sin\psi$$

Differentiating again w.r.t. ψ, we get

$$\frac{d^2p}{d\psi^2} = -x\sin\psi + \cos\psi \cdot \frac{dx}{d\psi} + y\cos\psi + \sin\psi \cdot \frac{dy}{d\psi}$$

$$= -\sin\psi + y\cos\psi + \cos\psi \cdot \frac{dx}{ds}\cdot\frac{ds}{d\psi} + \sin\psi \cdot \frac{dy}{ds}\cdot\frac{ds}{d\psi}$$

$$= (-x\sin\psi + y\cos\psi) + \cos\psi \cdot \cos\psi \cdot \rho + \sin\psi \cdot \sin\psi \cdot \rho$$

$$= -p + \rho[\cos^2\psi + \sin^2\psi] \quad [\because \text{ of relation } (i)]$$

Hence $$\rho = \mathbf{p} + \frac{\mathbf{d^2p}}{\mathbf{d\psi^2}}.$$

Example 1. *Find the radius of curvature for curve $r^n = a^n\cos n\theta$.* (M.D.U. 2007)

Sol. First Method. Taking logarithms of both sides, we have

$$n\log r = n\log a + \log\cos n\theta$$

Now differentiating, $$\frac{n}{r}\cdot\frac{dr}{d\theta} = 0 + \frac{1}{\cos n\theta}(-n\sin n\theta)$$

$$= -n\tan n\theta \quad ...(i)$$

$\therefore$ $$r_1 = -r\tan n\theta$$

Differentiating again, $$r_2 = -rn\cdot\sec^2 n\theta - r_1\cdot\tan n\theta$$

$$= -rn\sec^2 n\theta + r\tan^2 n\theta \quad ...(ii)$$

$\therefore$ $$\rho = \frac{(r^2 + r_1{}^2)^{3/2}}{r^2 + 2r_1{}^2 - rr_2}$$

$$= \frac{(r^2 + r^2\tan^2 n\theta)^{3/2}}{r^2 + 2r^2\tan^2 n\theta + r^2 n\sec^2 n\theta - r^2\tan^2 n\theta}$$

[$\because$ of (i) and (ii)]

$$= \frac{r^3 \sec^3 n\theta}{(n+1)r^2 \sec^2 n\theta} = \frac{r \cdot \sec n\theta}{n+1} = \frac{r}{n+1} \cdot \frac{1}{\cos n\theta}$$

$$= \frac{r}{(n+1)\dfrac{r^n}{a^n}} = \frac{a^n}{(n+1)r^{n-1}} \qquad \left[\because \frac{r^n}{a^n} = \cos n\theta, \text{from eq. of curve}\right]$$

Alternative Method. We first change of polar equation of the curve into the pedal one and find the value of $r \cdot \dfrac{dr}{dp}$.

As found in (i) above, we have $\dfrac{dr}{d\theta} = -r \tan n\theta$

$$\therefore \qquad \tan \phi = r\frac{d\theta}{dr} = \frac{r}{\dfrac{dr}{d\theta}} = -\cot n\theta = \tan\left(\frac{\pi}{2} + n\theta\right)$$

i.e.,

$$\phi = \frac{1}{2}\pi + n\theta \qquad \ldots(iii)$$

Now

$$p = r \sin \phi = r \sin\left(\frac{1}{2}\pi + n\theta\right) \qquad [\because \text{ of } (iii)]$$

or

$$p = r \cos n\theta = r \cdot \frac{r^n}{a^n} \qquad \text{[From the equation of the curve]}$$

Hence pedal equation is $\qquad p = \dfrac{r^{n+1}}{a^n}$

Differentiating, $\qquad \dfrac{dp}{dr} = \dfrac{(n+1)r^n}{a^n}$

$$\therefore \qquad \rho = r \cdot \frac{dr}{dp} = r \cdot \frac{a^n}{(n+1)\,r^n} = \frac{a^n}{(n+1)\,r^{n-1}}.$$

Note. The above question may also be put in the following form : *Show that for the curve $a^n p = r^{n+1}$; ρ varies inversely as the $(n-1)$th power of radius vector.*

Remark. To transform polar equation to pedal equation

1. Find ϕ from the formula $\tan \phi = \dfrac{r}{\left(\dfrac{dr}{d\theta}\right)}$.
2. Put this value of ϕ in $p = r \sin \phi$.
3. Eliminate θ.

Example 2. *If ρ_1, ρ_2 be the radii of curvature at the extremities of any chord through the pole of the cardioide $r = a(1 + \cos \theta)$; show that $\rho_1^2 + \rho_2^2 = \dfrac{16a^2}{9}$.*

Sol. Let ρ_1 and ρ_2 be the radii of curvature at the extremities P_1 and P_2 of the chord P_1OP_2 of the curve

$$r = a(1 + \cos \theta) \qquad \ldots(1)$$

Diff. (1) w.r.t. θ, $\qquad r_1 = a(0 - \sin \theta) = -a \sin \theta$

Again diff. w.r.t. θ, $r_2 = -a\cos\theta$

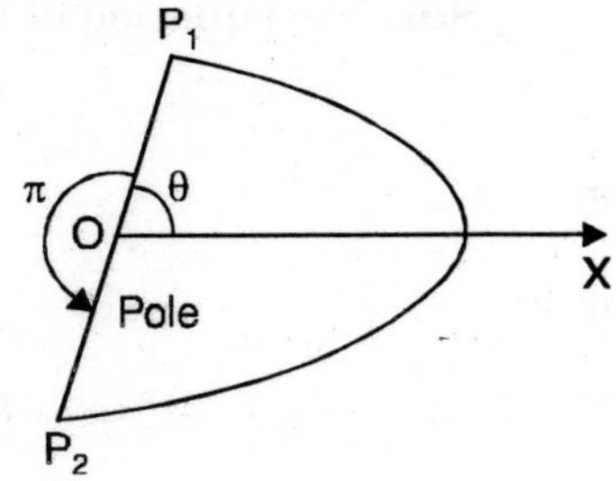

We know that $\rho = \dfrac{(r^2 + r_1^2)^{3/2}}{r^2 + 2r_1^2 - rr_2}$

$$= \frac{[a^2(1+\cos\theta)^2 + a^2\sin^2\theta]^{3/2}}{a^2(1+\cos\theta)^2 + 2a^2\sin^2\theta + a^2\cos\theta(1+\cos\theta)}$$

$$= \frac{a^3[1+\cos^2\theta + 2\cos\theta + \sin^2\theta]^{3/2}}{a^2[1+\cos^2\theta + 2\cos\theta + 2\sin^2\theta + \cos\theta + \cos^2\theta]}$$

$$= \frac{a[2+2\cos\theta]^{3/2}}{[3+3\cos\theta]} = \frac{a.2^{3/2}(1+\cos\theta)^{3/2}}{3(1+\cos\theta)}$$

$$= \frac{a.2^{3/2}(1+\cos\theta)^{1/2}}{3} = \frac{a.2^{3/2}\left(2\cos^2\dfrac{\theta}{2}\right)^{1/2}}{3}$$

$$= \frac{a.2^{3/2}\,.\,2^{1/2}\cos\dfrac{\theta}{2}}{2} = \frac{4a}{3}\cos\frac{\theta}{2}.$$

Let $\angle XOP_1 = \theta$, $\therefore$ $\angle XOP_2 = \pi + \theta$

$\therefore$ ρ_1 at $P_1 = \dfrac{4a}{3}\cos\dfrac{\theta}{2}$ ...(1)

Changing θ to π + θ in (1), we have

$$\rho_2 \text{ at } P_2 = \frac{4a}{3}\cos\left(\frac{\pi+\theta}{2}\right) = \frac{4a}{3}\cos\left(\frac{\pi}{2}+\frac{\theta}{2}\right)$$

$$= -\frac{4a}{3}\sin\frac{\theta}{2} \quad \text{...(2)}$$

Squaring and adding (1) and (2), we have

$$\rho_1^2 + \rho_2^2 = \frac{16a^2}{9}\left(\cos^2\frac{\theta}{2} + \sin^2\frac{\theta}{2}\right) = \frac{16a^2}{9}.$$

Example 3. *Find the radius of curvature at the point (p, r) of the ellipse*

$$\frac{1}{p^2} = \frac{1}{a^2} + \frac{1}{b^2} - \frac{r^2}{a^2b^2}.$$

Sol. Equation of curve is $\dfrac{1}{p^2} = \dfrac{1}{a^2} + \dfrac{1}{b^2} - \dfrac{r^2}{a^2b^2}$

or $\dfrac{a^2b^2}{p^2} = b^2 + a^2 - r^2$ *i.e.,* $r^2 = a^2 + b^2 - \dfrac{a^2b^2}{p^2}$

Differentiating w.r.t. p, $2r\dfrac{dr}{dp} = \dfrac{2a^2b^2}{p^3}$ $\therefore$ $\rho = r\dfrac{dr}{dp} = \dfrac{a^2b^2}{p^3}$.

Example 4. *Find the radius of curvature for the hyperbola*

$$p^2 = a^2\cos^2\psi - b^2\sin^2\psi.$$

Sol. The equation of the curve is

$$p^2 = a^2 \cos^2 \psi - b^2 \sin^2 \psi$$

or $$p^2 = \frac{a^2(1+\cos 2\psi)}{2} - \frac{b^2(1-\cos 2\psi)}{2}$$

or $$p^2 = \tfrac{1}{2}[a^2 + a^2 \cos 2\psi - b^2 + b^2 \cos 2\psi]$$

or $$p^2 = \tfrac{1}{2}[(a^2 - b^2) + (a^2 + b^2) \cos 2\psi] \quad ...(1)$$

Diff. w.r.t. ψ, $$2p\frac{dp}{d\psi} = \tfrac{1}{2}[-2(a^2 + b^2) \sin 2\psi] = -(a^2 + b^2) \sin 2\psi$$

$\therefore$ $$p\frac{dp}{d\psi} = -\tfrac{1}{2}(a^2 + b^2) \sin 2\psi \quad ...(2)$$

Now from (1), $$p^2 - \tfrac{1}{2}(a^2 - b^2) = \tfrac{1}{2}(a^2 + b^2) \cos 2\psi \quad ...(3)$$

Squaring and adding (2) and (3), we have

$$p^2\left(\frac{dp}{d\psi}\right)^2 + p^4 + \tfrac{1}{4}(a^2 - b^2)^2 - p^2(a^2 - o^2) = \tfrac{1}{4}(a^2 + b^2)^2$$

or $$p^2\left(\frac{dp}{d\psi}\right)^2 + p^4 - p^2(a^2 - b^2) = \tfrac{1}{4}(a^2 + b^2)^2 - \tfrac{1}{4}(a^2 - b^2)^2 = a^2b^2$$

$$[\because (a^2 + b^2)^2 - (a^2 - b^2)^2 = 4a^2b^2]$$

Dividing both sides by p^2, $$\left(\frac{dp}{d\psi}\right)^2 + p^2 - (a^2 - b^2) = \frac{a^2b^2}{p^2}$$

Diff. both sides w.r.t. ψ, $$2\frac{dp}{d\psi}\frac{d^2p}{d\psi^2} + 2p\frac{dp}{d\psi} = -\frac{2a^2b^2}{p^3}\frac{dp}{d\psi}$$

Cancelling $2\dfrac{dp}{d\psi}$, $$\frac{d^2p}{d\psi^2} + p = -\frac{a^2b^2}{p^3}$$

$\therefore$ $$\rho = \mathbf{p} + \frac{\mathbf{d^2p}}{\mathbf{d\psi^2}} = -\frac{a^2b^2}{p^3} = \frac{a^2b^2}{p^3} \text{ numerically.}$$

Example 5. *If ϕ be angle which the radius vector of the curve $r = f(\theta)$, makes with the tangent, prove that $\dfrac{r}{\rho} = \sin\phi\left(1 + \dfrac{d\phi}{d\theta}\right)$, where ρ is the radius of curvature.*

Apply this result to show that $\rho = a/2$ for circle $r = a\cos\theta$.

Sol. We know that $\psi = \theta + \phi$ [*Refer to Fig. of Art. 6 Page 189*]

$\therefore$ Differentiating w.r.t. s, we have

$$\frac{d\psi}{ds} = \frac{d\theta}{ds} + \frac{d\phi}{ds} = \frac{d\theta}{ds} + \frac{d\phi}{d\theta}\frac{d\theta}{ds} = \frac{d\theta}{ds}\left[1 + \frac{d\phi}{d\theta}\right]$$

or $$\frac{1}{\rho} = \frac{\sin\phi}{r}\left[1 + \frac{d\phi}{d\theta}\right] \qquad \left[\because \sin\phi = r\frac{d\theta}{ds}\right]$$

$$\therefore \qquad \frac{r}{\rho} = \sin\phi\left[1+\frac{d\phi}{d\theta}\right] \qquad ...(i)$$

Also equation of circle is $r = a\cos\theta$

Differentiating, $\quad r_1 = \dfrac{dr}{d\theta} = -a\sin\theta$

$$\therefore \qquad \tan\phi = r\frac{d\theta}{dr} = \frac{r}{r_1} = \frac{a\cos\theta}{-a\sin\theta} = -\cot\theta = \tan\left(\frac{\pi}{2}+\theta\right)$$

$$\therefore \qquad \phi = \frac{\pi}{2}+\theta \quad \text{and} \quad \frac{d\phi}{d\theta} = 1$$

But from (i), $\quad \rho = \dfrac{r\,\text{cosec}\,\phi}{(1+d\phi/d\theta)} = \dfrac{a\cos\theta\,\text{cosec}\,(\frac{1}{2}\pi+\theta)}{1+1}$

$$= \frac{a}{2}\cos\theta\,.\sec\theta = \frac{a}{2}.$$

EXERCISE 2

1. Show that for the curve $r = a(1+\cos\theta)$, the radius of curvature $\rho = \dfrac{4a}{3}\cos\dfrac{\theta}{2}$ and $\dfrac{\rho^2}{r} = \dfrac{8a}{9}$. *(M.D.U. 2006 ; K.U. 1999 S)*
2. Show that the radius of the curvature at any point of the cardioid $r = a(1-\cos\theta)$ is $\dfrac{2}{3}\sqrt{2ar}$ and prove that $\dfrac{\rho^2}{r}$ is constant.
3. If ρ_1, ρ_2 be the radii of curvature at the extremities of any chord through the pole of the cardioid $r = a(1-\cos\theta)$, show that $\rho_1^2+\rho_2^2 = \dfrac{16a^2}{9}$. *(K.U. 1997)*
4. Find the radius of curvature at any point of the curve $r = a\cos n\theta$ and show that at the point where $r = a$, its value is $\dfrac{a}{1+n^2}$.
5. Find the radius of curvature of the curve $r = a\sin n\theta$ at the pole.
6. Find the radius of curvature at any point (r, θ) of the following curves :

   (i) $r^m = a^m\sin m\theta$ $\qquad$ (ii) $r^2\cos 2\theta = a^2$

   (iii) $\theta = \dfrac{\sqrt{r^2-a^2}}{a} - \cos^{-1}\dfrac{a}{r}$ $\qquad$ (iv) $r = ae^{\theta\cot\alpha}$

   (v) $\dfrac{2a}{r} = 1+\cos\theta.$ *(M.D.U. 2000, 2007)*
7. Find the radius of curvature at any point (p, r) on the following curves :

   (i) $p^2 = ar$ $\qquad$ (ii) $r^2 = a^2 - b^2 + \dfrac{a^2b^2}{p^2}$

   (iii) $2ap^2 = r^3$ $\qquad$ (iv) $pa^2 = r^3.$
8. Find the radius of curvature for the curves :

   (i) $p = a(1+\sin\psi)$ $\qquad$ (ii) $p^2 = a^2\cos^2\psi + b^2\sin^2\psi.$

9. Find the radius of curvature to the curve $r = a(1 + \cos\theta)$ at the points where tangent is parallel to the initial line.

[**Hint.** Firstly do Q. 1 ; and then use $\theta + \phi = 180°$.]

10. Show that the radius of curvature of the lemniscate $r^2 = a^2 \cos 2\theta$ at the point where the tangent is parallel to x-axis is $\dfrac{\sqrt{2}\,.\,a}{3}$.

11. Prove that in the curve $r^2 = a^2 \sin 2\theta$

(*i*) The curvature varies as the radius vector.

(*ii*) The tangent turns three times as fast as the radius vector.

[**Hint.** Prove that $\psi = 3\theta$]

(*iii*) Also for the same curve find the points at which radii vectors are perpendicular to the tangents and find the radii of curvature at these points.

ANSWERS

5. $\dfrac{na}{2}$ **6.** (*i*) $\dfrac{a^m}{(m+1)\,r^{m-1}}$ (*ii*) $\dfrac{r^3}{a^2}$ (*iii*) $\sqrt{r^2 - a^2}$

(*iv*) $r\,.\,\text{cosec}\,\alpha$ (*v*) $2r\sqrt{\dfrac{r}{a}}$

7. (*i*) $\dfrac{2r^{3/2}}{\sqrt{a}}$ (*ii*) $\dfrac{a^2b^2}{p^3}$ (*iii*) $\frac{2}{3}\sqrt{2ar}$ (*iv*) $\dfrac{a^2}{3r}$

8. (*i*) a (*ii*) $\dfrac{a^2b^2}{p^3}$ **9.** $\dfrac{2a}{\sqrt{3}}$ **11.** (*iii*) $\left(\pm a, \dfrac{\pi}{4}\right); \dfrac{a}{3}$.

---

9. Radius of Curvature at the Origin

When the curve passes through the origin, the following methods may be used for finding the radius of curvature at the origin.

(*i*) **Method of direct substitution.** Calculate the values of y_1 and y_2 at origin, and then substitute these values direct in the formula,

$$\rho = \frac{(1 + y_1^{\,2})^{3/2}}{y_2}$$

(*ii*) **Method of expansion.** But the above method very often fails or becomes very labourious. So let us obtain the values of $y_1(0)$ and $y_2(0)$ by the following method :

Let $y = f(x)$ be the equation of the curve. Since it passes through origin, therefore $f(0) = 0$.

$\therefore$ By Maclaurin's Expansion,

$$y = f(0) + xf'(0) + \frac{x^2}{2!}f''(0) + \frac{x^3}{3!}f'''(0) + \ldots\ldots$$

i.e.,
$$y = xf'(0) + \frac{x^2}{2!}f''(0) + \frac{x^3}{3!}f'''(0) + \ldots\ldots$$

i.e.,
$$y = px + \frac{1}{2!}qx^2 + \frac{1}{3!}rx^3 + \ldots\ldots \qquad \ldots(1)$$

where $$p = f'(0) = y_1(0)\ ;\ q = f''(0) = y_2(0) \text{ etc.}$$

Diff. (1) w.r.t. x, $$y_1 = p + \frac{2qx}{2!} + \frac{3rx^2}{3!} + \ldots\ldots$$

Again diff. w.r.t. x, $$y_2 = \frac{2q}{2!} + \frac{6rx}{3!} + \ldots$$

$\therefore$ At the origin $y_1 = p$ (Putting $x = 0$ in y_1)

and $$y_2 = \frac{2q}{2!} = q$$

Putting these values of y_1 and y_2 in

$$\rho = \frac{(1+y_1^2)^{3/2}}{y_2}, \text{ we have}$$

$$\rho = \frac{(1+\mathbf{p}^2)^{3/2}}{\mathbf{q}}.$$

Remark. To find values of p and q.

Substituting the value of $y = px + \frac{qx^2}{2!} + \frac{rx^3}{3!} + \ldots\ldots$ in the equation of the curve and then equate the coefficients of the like powers of x in the identity thus obtained. This way we will get the values of p and q.

(*iii*) **Newton's method.** (*a*) **If a curve passes through the origin, and axis of x is the tangent at the origin, then**

$$\rho \text{ at the origin} = \underset{\substack{x \to 0 \\ y \to 0}}{\text{Lt}} \frac{x^2}{2y}.$$

(*b*) **If a curve passes through the origin and axis of y is the tangent there, then radius of curvature at the origin.**

$$= \underset{\substack{x \to 0 \\ y \to 0}}{\text{Lt}} \frac{y^2}{2x}.$$

(*a*) Since axis of x is the tangent at the origin, therefore its slope

$$y_1(0) = \left(\frac{dy}{dx}\right)_{(0,\,0)} = 0$$

Now $\frac{x^2}{2y}$ is of the indeterminate form $\frac{0}{0}$, as $x \to 0, y \to 0$.

$\therefore$ By Hospital's rule $$\underset{\substack{x \to 0 \\ y \to 0}}{\text{Lt}} \frac{x^2}{2y} = \underset{\substack{x \to 0 \\ y \to 0}}{\text{Lt}} \frac{2x}{2y_1} = \underset{\substack{x \to 0 \\ y \to 0}}{\text{Lt}} \frac{x}{y_1} \qquad \left[\text{Again of form } \frac{0}{0}, \because\ y_1(0) = 0\right]$$

$$= \underset{\substack{x \to 0 \\ y \to 0}}{\text{Lt}} \frac{1}{y_2} = \frac{1}{y_2(0)} \qquad \ldots(i)$$

But ρ at origin $$= \frac{[1+y_1^2(0)]^{3/2}}{y_2(0)} = \frac{(1+0)^{3/2}}{y_2(0)} = \frac{1}{y_2(0)} \qquad \ldots(ii)$$

$\therefore$ From (*i*) and (*ii*), ρ (*at origin*) $= \displaystyle\mathop{\text{Lt}}_{\substack{x \to 0 \\ y \to 0}} \frac{x^2}{2y}$.

(*b*) Proof is similar to part (*a*) and left as an exercise for the students. (Interchange the letters x and y in the above proof).

Cor. Curvature at the origin when polar equation of the curve is given. If the initial line is the tangent at the pole (origin), then

$$\rho \text{ (at the origin or pole)} = \mathop{\text{Lt}}_{x \to 0} \frac{x^2}{2y} = \mathop{\text{Lt}}_{\theta \to 0} \frac{r^2 \cos^2 \theta}{2r \sin \theta} = \mathop{\text{Lt}}_{\substack{\theta \to 0 \\ r \to 0}} \left(\frac{r}{2\theta} \cdot \frac{\theta}{\sin \theta} \cdot \cos^2 \theta \right)$$

$$= \mathop{\text{Lt}}_{\substack{\theta \to 0 \\ r \to 0}} \left(\frac{r}{2\theta} \right) \qquad \left[\because \frac{\theta}{\sin \theta} \to 1, \cos \theta \to 1, \text{ when } \theta \to 0 \right]$$

$$= \mathop{\text{Lt}}_{\substack{\theta \to 0 \\ r \to 0}} \frac{\frac{dr}{d\theta}}{2} \qquad \text{| L's Hospital Rule}$$

$$= \text{Lt} \left(\frac{1}{2} \frac{dr}{d\theta} \right).$$

Remark. The tangents to an algebraic curve at the origin are obtained by **equating the lowest degree terms to zero.** (See chapter on Singular Points Page 212)

Example 1. *Show that the radii of curvature of the curve* $y^2 = x^2 \dfrac{(a + x)}{(a - x)}$ *at the origin are* $a\sqrt{2}$. (K.U. 1998 S)

Sol. Equation of the curve is $y^2(a - x) = x^2 (a + x)$ ...(*i*)

i.e., $y^2(a - x) - x^2(a + x) = 0$

Equating to zero the lowest degree terms, we get

$$a(y^2 - x^2) = 0 \quad \therefore \quad y = \pm x \text{ are the tangents at origin.}$$

$\therefore$ Newton's Method is not applicable here.

Let $$y = px + \frac{1}{2!} qx^2 + \frac{1}{3!} rx^3 + ...$$

Substituting this in the equation of the curve, we get

$$(a - x)(px + \tfrac{1}{2} qx^2 + \tfrac{1}{6} rx^3 + ...)^2 = x^2(a + x)$$

Equating coefficients of x^2 and x^3 on both sides of the above identity, we have

$$ap^2 = a \quad ...(i) \qquad \text{and} \qquad apq - p^2 = 1 \qquad ...(ii)$$

From (*i*), $p^2 = 1$ or $p = \pm 1$

When $p = 1$, then (*ii*) gives $aq - 1 = 1$ or $q = \dfrac{2}{a}$

$$\therefore \quad \rho \text{ at the origin} = \frac{(1 + p^2)^{3/2}}{q} = \frac{(1 + 1)^{3/2}}{2/a} = a\sqrt{2}$$

And when $p = -1$, then (*ii*) gives $-aq - 1 = 1$ or $q = -2/a$

$$\therefore \quad \rho \text{ at the origin} = \frac{(1 + 1)^{3/2}}{-2/a} = -a\sqrt{2}.$$

Alternative Method

Equation of the curve is $y^2 = x^2 \dfrac{(a+x)}{a-x}$

$$\therefore \quad y = \pm \frac{x(a+x)^{1/2}}{(a-x)^{1/2}} = \pm x \frac{a^{1/2}\left(1+\frac{x}{a}\right)^{1/2}}{a^{1/2}\left(1-\frac{x}{a}\right)^{1/2}}$$

or
$$y = \pm x \frac{\left(1+\frac{x}{a}\right)^{1/2}}{\left(1-\frac{x}{a}\right)^{1/2}} \quad \text{or} \quad y = \pm x\left(1+\frac{x}{a}\right)^{1/2}\left(1-\frac{x}{a}\right)^{-\frac{1}{2}}$$

or
$$y = \pm x\left(1+\frac{x}{2a}+\ldots\right)\left(1+\frac{x}{2a}+\ldots\right)$$

[$\because$ By Binomial Theorem $(1+x)^n = 1 + nx + \ldots$]

or
$$y = \pm x\left(1+\frac{x}{2a}+\frac{x}{2a}+\frac{x^2}{4a^2}+\ldots\right) \quad \text{or} \quad y = \pm\left(x+\frac{x^2}{a}+\frac{x^3}{4a^2}+\ldots\right)$$

$$\therefore \quad y_1 = \pm\left(1+\frac{2x}{a}+\frac{3x^2}{4a^2}+\ldots\right) \quad \text{and} \quad y_2 = \pm\left(\frac{2}{a}+\frac{6x}{4a^2}+\ldots\right)$$

$$\therefore \quad \text{At } (0, 0)\, y_1 = \pm 1 \text{ and } y_2 = \pm\frac{2}{a}$$

$$\therefore \quad \rho = \frac{(1+y_1^{\,2})^{3/2}}{y_2} = \frac{(1+1)^{3/2}}{\pm\frac{2}{a}} = \pm 2\sqrt{2}\,\frac{a}{2}.$$

or
$$\rho = \pm\sqrt{2}a = \sqrt{2}a \text{ (Numerically)}.$$

Example 2. *Find the radius of curvature at the origin for the curve* $x^3 - 2x^2y + 3xy^2 - 4y^3 + 5x^2 - 6xy + 7y^2 - 8y = 0$.

Sol. The curve passes through the origin. Equating to zero the lowest degree terms, we find $y = 0$, *i.e.*, x-axis is the tangent at the origin.

$\therefore$ By Newton's formula ρ at $(0, 0) = \underset{\substack{x \to 0 \\ y \to 0}}{\text{Lt}} \dfrac{x^2}{2y}$

Dividing the equation of the curve by $2y$, we get

$$x \cdot \frac{x^2}{2y} - x^2 + \frac{3}{2}xy - 2y^2 + 5\frac{x^2}{2y} - 3x + \frac{7}{2}y - 4 = 0$$

Taking limits as x and y both tend to zero, we have

$$5 \underset{\substack{x \to 0 \\ y \to 0}}{\text{Lt}} \frac{x^2}{2y} - 4 = 0, \text{ other terms become zero.}$$

i.e.,
$$5\rho - 4 = 0 \quad \text{or} \quad \rho = \frac{4}{5}.$$

Example 3. *Apply Newton's formula to find the radius of curvature at the origin for the cycloid $x = a(\theta + \sin\theta)$, $y = a(1 - \cos\theta)$.*

Sol. When $\theta = 0$, $x = 0$ and $y = 0$, therefore origin corresponds to $\theta = 0$.

Also $$\frac{dy}{dx} = \frac{dy}{d\theta}\Big/\frac{dx}{d\theta} = \frac{a\sin\theta}{a(1+\cos\theta)} = \frac{\sin\theta}{1+\cos\theta}$$

At $\theta = 0$, $$\frac{dy}{dx} = \frac{\sin 0}{1+\cos 0} = \frac{0}{2} = 0$$

$\therefore$ Tangent at the origin is x-axis.

$$\therefore \quad \rho \text{ at origin} = \underset{\substack{x \to 0 \\ y \to 0}}{\text{Lt}} \frac{x^2}{2y} = \underset{\theta \to 0}{\text{Lt}} \frac{a^2(\theta + \sin\theta)^2}{2a(1-\cos\theta)} \qquad \left[\frac{0}{0}\text{ form}\right]$$

$$= \underset{\theta \to 0}{\text{Lt}} \frac{2a(\theta+\sin\theta)(1+\cos\theta)}{2\sin\theta} \qquad \left[\frac{0}{0}\text{ form again}\right]$$

$$= \underset{\theta \to 0}{\text{Lt}} \frac{a(\theta+\sin\theta)(-\sin\theta) + a(1+\cos\theta)^2}{\cos\theta} = \frac{0 + a.4}{1} = 4a.$$

EXERCISE 3

1. Find the radius of curvature at the origin for the curve $x^3 + y^2 - 2x^2 + 6y = 0$.
2. Find the radius of curvature at the origin for the curves :
   (*i*) $2x^4 + 3y^4 + 4x^2y + xy - y^2 + 2x = 0$
   (*ii*) $2x^3 + 4x^2y + xy^2 + 5y^3 - x^2 - 2xy + y^2 + 4x = 0$
   (*iii*) $2x^4 + 4x^3 + xy^2 + 6y^3 - 3x^2 - 2xy + y^2 - 4x = 0$.
3. Show that the radius of curvature at the origin for the curve $x^3 + y^3 = 3axy$ is equal to $\frac{3a}{2}$.
   [**Hint.** Divide both sides by $2xy$.]
4. Find the radius of curvature at the origin of the following curves :
   (*i*) $y = 6x + 5x^2 + x^3$
   (*ii*) $y - x = x^2 + 2xy + y^2$
   (*iii*) $a(y^2 - x^2) = x^3$.

ANSWERS

1. $\frac{3}{2}$ 2. (*i*) 1 (*ii*) 2 numerically (*iii*) 2.

4. (*i*) $\frac{37\sqrt{37}}{10}$ (*ii*) $\frac{1}{2\sqrt{2}}$ (*iii*) $2\sqrt{2}\,a$ (numerically).

10. Centre of Curvature, Circle of Curvature and Evolute

(*i*) **Centre of Curvature** *for any point P of a curve is the point on the positive direction of the normal at P (i.e., the direction on the concave side of curve) at a distance ρ from it.*

Let PD be the normal to the curve at P, and C be a point on it such that PC = ρ, then C is the centre of curvature of the curve at P.

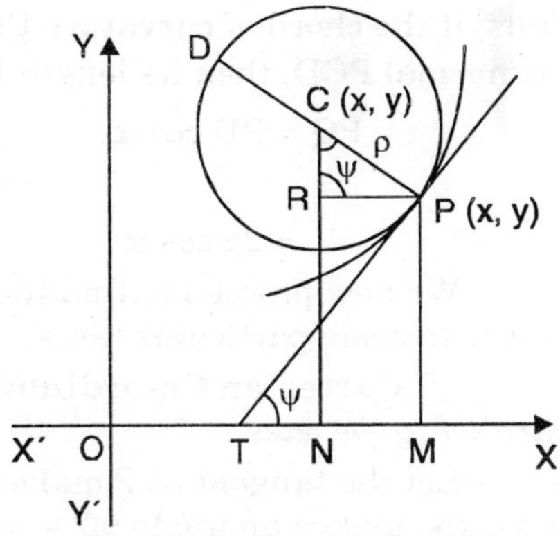

(*ii*) **Evolute of a Curve.** The locus of the centres of curvature of the given curve is called the **Evolute** of the curve.

(*iii*) *The circle with its centre at the centre of curvature C and radius equal to ρ is called the* **circle of curvature** *of the curve at the point P.*

Remark. Evidently the circle of curvature touches the curve at P and both the curve and the circle of curvature have the same curvature at this point.

(*iv*) *To find the co-ordinates of* **centre of curvature** *for any point P(x, y) of curve y = f(x).*

Let C(X, Y) be centre of curvature corresponding to any point P(x, y) on the curve, then PC = ρ. (above Fig.). Let tangent TP make an angle ψ with positive direction of x-axis. Draw PM and CN perpendicular on x-axis, and draw PR perpendicular to CN. Then

$$\angle PCN = 90° - \angle CPR = 90° - (90° - \angle RPT) = \angle RPT = \angle PTX = \psi$$

$$\therefore \quad X = ON = OM - NM = OM - RP = x - CP \sin\psi = x - \rho \sin\psi \qquad \ldots(1)$$

and $$Y = NC = NR + RC = MP + RC = y + CP \cos\psi = y + \rho \cos\psi \qquad \ldots(2)$$

But we know that $\tan\psi = y_1$

$$\therefore \quad \sin\psi = \frac{y_1}{\sqrt{1+y_1^2}} \quad \text{and} \quad \cos\psi = \frac{1}{\sqrt{1+y_1^2}}$$

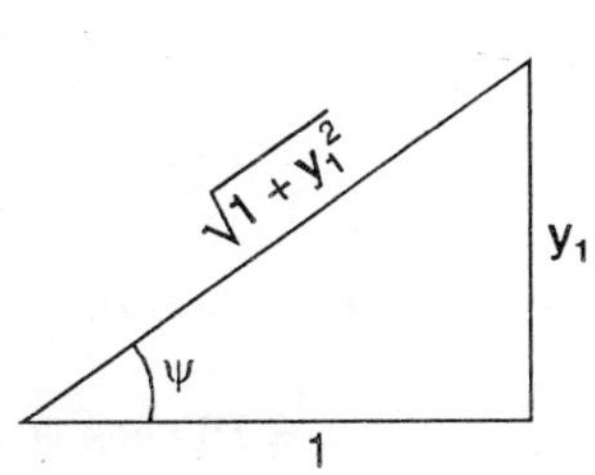

Also $$\rho = \frac{(1+y_1^2)^{3/2}}{y_2}$$

Hence substituting these values in (1) and (2), we get

$$\mathbf{X = x - \frac{y_1(1+y_1^2)}{y_2}, \; Y = y + \frac{1+y_1^2}{y_2}}.$$

Cor. 1. We have already proved that

$$\mathbf{X = x - \rho \sin\psi, \text{ and } Y = y + \rho \cos\psi.}$$

Since x, y, ρ, ψ depend upon s ; therefore the above equations may be treated as **parametric equations of the evolute.**

Cor. 2. $$\rho = \frac{ds}{d\psi}, \; \sin\psi = \frac{dy}{ds} \text{ and } \cos\psi = \frac{dx}{ds}$$

∴ Substituting these values, we get

$$X = x - \frac{ds}{d\psi} \cdot \frac{dy}{ds} = x - \frac{dy}{d\psi} \quad \text{and} \quad Y = y + \frac{ds}{d\psi} \cdot \frac{dx}{ds} = y + \frac{dx}{d\psi},$$

co-ordinates of centre of curvature in another form.

Cor. 3. To find the **equation of the circle of curvature** at a given point of the curve.

Let (α, β) be the centre of curvature and ρ be radius of curvature at the given point. Then equation of circle of curvature at the given point is $(x - \alpha)^2 + (y - \beta)^2 = \rho^2$.

11. Chord of Curvature

The length intercepted by the circle of curvature of the curve at P, on a straight line drawn through P in any given direction is called **chord of curvature** *through P in that direction.*

Thus, if the chord of curvature PQ, makes angle α, with the normal PCD, then its length PQ is given by

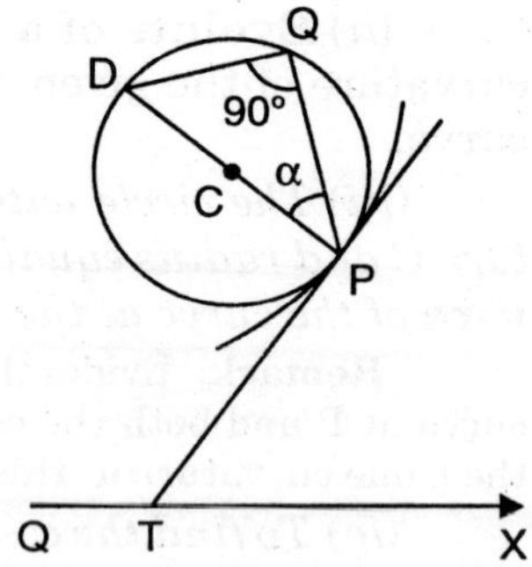

$$PQ = PD \cos \alpha \qquad [\because \angle DQP, \text{ being a semi-circle is art. angle}]$$

$$= 2\rho \cos \alpha$$

We now proceed to find the *length of chord of curvature* in some particular cases.

(*i*) **Cartesian Co-ordinates.** *Chords of curvature parallel to the axes.*

Let the tangent at P make angle ψ with *x*-axis, then the chord of curvature PA, parallel to *x*-axis, makes an angle 90° – ψ with the normal PCD, and chord of curvature PB, parallel to *y*-axis makes angle ψ with the normal PCD.

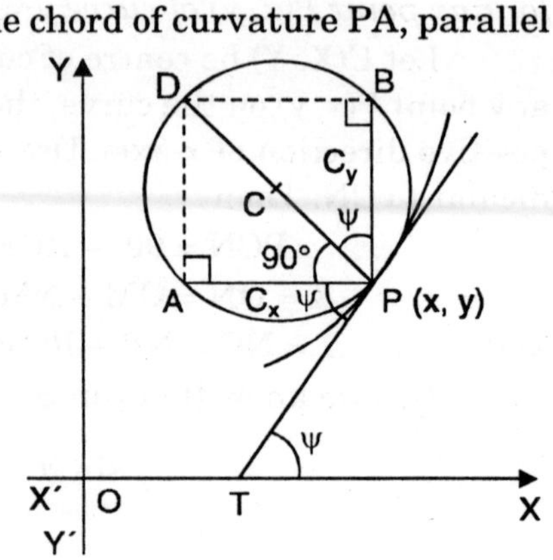

∴ C_x = *length of the chord of curvature PA, parallel to x-axis.*

$$= PD \cos (90° - \psi)$$

$$= \mathbf{2\rho \sin \psi}$$

$$= \frac{2\,.\,(1+y_1^2)^{3/2}}{y_2} \cdot \frac{y_1}{\sqrt{1+y_1^2}}$$

$$= \frac{\mathbf{2\, y_1\, (1+y_1^2)}}{\mathbf{y_2}}$$

and C_y = *length of the chord of curvature PB, parallel to y-axis*

$$= PD \cos \psi = \mathbf{2\rho \cos \psi} = \frac{2(1+y_1^2)^{3/2}}{y_2} \cdot \frac{1}{\sqrt{1+y_1^2}}$$

$$= \frac{\mathbf{2(1+y_1^2)}}{\mathbf{y_2}}.$$

(*ii*) **Polar Co-ordinates.** *Chord of curvature through the pole and perpendicular to the radius vector.*

PL, the chord of curvature through pole O, makes angle of 90° – φ, with PCD, the normal to the curve at P, and PM the chord of curvature ⊥ to the radius vector OP, makes angle φ with the normal PCD.

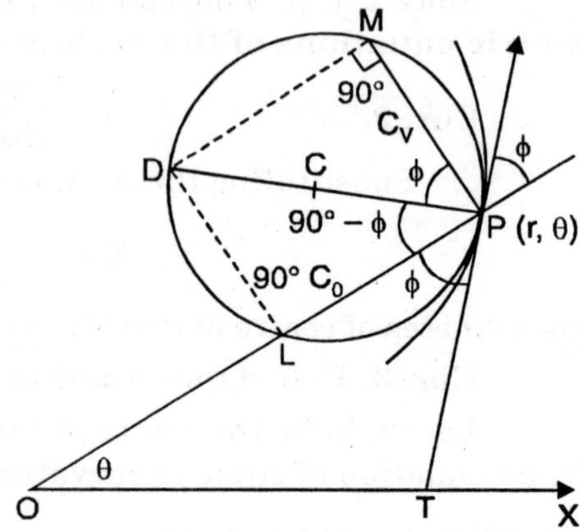

∴ $\mathbf{C_0}$ = *length of chord of curvature PL through the pole.*

$$= PD \cos (90° - \phi)$$

$$= \mathbf{2\rho \sin \phi}$$

$$= \frac{2(r^2+r_1^2)^{3/2}}{r^2+2r_1^2-rr_2} \cdot \frac{r}{\sqrt{r^2+r_1^2}}$$

$$= \frac{2r(r^2+r_1^2)}{r^2+2r_1^2-rr_2}$$

and $\mathbf{C_p}$ = *length of chord of curvature PM ⊥ to radius vector*

$$= \text{PD} \cos \phi = 2\rho \cos \phi = \frac{2 \cdot (r^2 + r_1^2)^{3/2}}{r^2 + 2r_1^2 - rr_2} \cdot \frac{r_1}{\sqrt{r^2 + r_1^2}}$$

$$= \frac{2r_1(r^2 + r_1^2)}{r^2 + 2r_1^2 - rr_2}.$$

(*iii*) **Pedal Equations.** When the pedal equation $p = f(r)$ of the curve is given, then C_0 = *length of chord of curvature through pole along radius vector.*

$$= \text{PD} \cos (90° - \phi) = 2\rho \sin \phi \qquad ...(1)$$

But $\rho = r \cdot \dfrac{dr}{dp}$ and $\sin \phi = \dfrac{p}{r}$

$\therefore$ Substituting in (1), we get

$$C_0 = 2r \frac{dr}{dp} \cdot \frac{p}{r} = 2p \cdot \frac{dr}{dp}.$$

Also from equation $p = f(r)$, we have

$$\frac{dp}{dr} = f'(r), \quad \text{and} \quad \sin \phi = \frac{p}{r} = \frac{f(r)}{r}$$

$\therefore$ From (1), $C_0 = 2\rho \sin \phi = 2 \cdot r \dfrac{dr}{dp} \cdot \sin \phi = 2r \cdot \dfrac{1}{f'(r)} \cdot \dfrac{f(r)}{r} = \dfrac{2f(r)}{f'(r)}.$

Also C_p = *length of chord* $\perp$ *to the radius vector*

$$= \text{DP} \cos \phi = 2\rho \cos \phi = 2 \cdot r \frac{dr}{dp} \cdot \frac{\sqrt{r^2 - p^2}}{r}$$

$$\left[\because \sin \phi = \frac{p}{r}, \quad \therefore \cos \phi = \frac{\sqrt{r^2 - p^2}}{r}\right]$$

$$= 2 \cdot \sqrt{r^2 - p^2} \frac{dr}{dp}.$$

Example 1. *Find the coordinates of the centre of curvature at any point (x, y) of the parabola $y^2 = 4ax$. Also find the equation of the evolute of the parabola.*

(K.U. 2005 ; M.D.U. 2006 A)

Sol. Equation of the parabola is $y^2 = 4ax$ the parametric equations of this parabola are

$$x = at^2, \qquad y = 2at.$$

$$\therefore \quad \frac{dx}{dt} = 2at \qquad \frac{dy}{dt} = 2a$$

$$y_1 = \frac{dy}{dx} = \frac{\left(\frac{dy}{dt}\right)}{\left(\frac{dx}{dt}\right)} = \frac{2a}{2at} = \frac{1}{t}$$

Again differentiating w.r.t. x,

$$y_2 = -\frac{1}{t^2} \frac{dt}{dx} = -\frac{1}{t^2} \cdot \frac{1}{2at} = -\frac{1}{2at^3}$$

$\therefore$ Co-ordinates of centre of curvature at any point (x, y) of the parabola are

$$X = x - y_1\frac{(1+y_1^2)}{y_2} = at^2 - \frac{1}{t}\frac{\left(1+\frac{1}{t^2}\right)}{-\frac{1}{2at^3}} = at^2 + 2at^2\left(1+\frac{1}{t^2}\right)$$

or $$X = 3at^2 + 2a \qquad ...(1)$$

and $$Y = y + \frac{1}{y_2}(1+y_1^2) = 2at - 2at^3\left(1+\frac{1}{t^2}\right)$$

$$= 2at - 2at^3 - 2at = -2at^3 = -2at^3 \qquad ...(2)$$

From (1) and (2),

The coordinates of centre of curvature at any point $(x, y) = (at^2, 2at)$ of the parabola are $(3at^2 + 2a, -2at^3)$.

Note, we know that evolute of a curve is the locus of centres of curvature.

In fact, equations (1) and (2) are parametric equations of the evolute of parabola.

To get cartesian equation of evolute of parabola, let us eliminate t from (1) and (2).

Form (1), $$t^2 = \frac{X - 2a}{3a} \qquad ...(3)$$

From (2), $$t^3 = -\frac{Y}{2a} \qquad ...(4)$$

Now cubing (3), squaring (4) and equating the two values of t^6, we get

$$\left(\frac{X-2a}{3a}\right)^3 = \frac{Y^2}{4a^2} \quad \text{or} \quad 27aY^2 = 4(X - 2a)^3$$

Changing X to x and Y to y,

we get $$27ay^2 = 4(x - 2a)^3,$$

which is the equation of the evolute of the parabola.

Example 2. *Find the coordinates of the centre of curvature of ellipse* $\frac{x^2}{a^2} + \frac{y^2}{b^2} = 1$ *or* $x = a \cos\theta$, $y = b \sin\theta$. *Hence show that the equation of its evolute is* $(ax)^{2/3} + (by)^{2/3} = (a^2 - b^2)^{2/3}$.

(K.U. 2005 S)

Sol. Let (X, Y) be the coordinates of the centre of curvature at point 'θ' of the given ellipse

$$x = a\cos\theta, \qquad y = b\sin\theta$$

$$\therefore \quad y_1 = \frac{dy}{d\theta}\Big/\frac{dx}{d\theta} = \frac{b\cos\theta}{-a\sin\theta} = -\frac{b}{a}\cot\theta$$

and $$y_2 = \frac{d}{dx}\left(-\frac{b}{a}\cot\theta\right) = \frac{d}{d\theta}\left(-\frac{b}{a}\cot\theta\right).\frac{d\theta}{dx} = \frac{b}{a}\operatorname{cosec}^2\theta.\frac{1}{-a\sin\theta} = -\frac{b}{a^2}\operatorname{cosec}^3\theta$$

$$\therefore \quad X = x - y_1\frac{(1+y_1^2)}{y_2} = a\cos\theta - \frac{-\frac{b}{a}\cot\theta\left(1+\frac{b^2}{a^2}\cot^2\theta\right)}{-\frac{b}{a^2}\operatorname{cosec}^3\theta}$$

$$= a\cos\theta - a\sin^3\theta\frac{\cos\theta}{\sin\theta}\left(1+\frac{b^2\cos^2\theta}{a^2\sin^2\theta}\right)$$

$$= a\cos\theta - a\cos\theta\,.\,\sin^2\theta - \frac{b^2}{a}\cos^3\theta$$

$$= a\cos\theta\,(1-\sin^2\theta) - \frac{b^2}{a}\cos^3\theta = a\cos^3\theta - \frac{b^2}{a}\cos^3\theta$$

$$= \frac{a^2-b^2}{a}\cos^3\theta \qquad \ldots(i)$$

and $$\mathrm{Y} = y + \frac{(1+y_1^{\,2})}{y_2} = b\sin\theta + \frac{\left(1+\dfrac{b^2}{a^2}\cot^2\theta\right)}{\dfrac{-b}{a^2}\operatorname{cosec}^3\theta}$$

$$= b\sin\theta - \frac{a^2\sin^3\theta}{b}\left(1+\frac{b^2\cos^2\theta}{a^2\sin^2\theta}\right)$$

$$= b\sin\theta - \frac{a^2}{b}\sin^3\theta - b\sin\theta\cos^2\theta$$

$$= b\sin\theta\,(1-\cos^2\theta) - \frac{a^2}{b}\sin^3\theta = b\sin^3\theta - \frac{a^2}{b}\sin^3\theta$$

$$= -\frac{a^2-b^2}{b}\sin^3\theta \qquad \ldots(ii)$$

(*i*) and (*ii*) give the co-ordinates (X,Y) of the centre of curvature of the ellipse. To find the equation of its evolute, we eliminate θ between (*i*) and (*ii*).

From (*i*), $a\mathrm{X} = (a^2-b^2)\cos^3\theta$, $\therefore\ (a\mathrm{X})^{2/3} = (a^2-b^2)^{2/3}\cos^2\theta$

and from (*ii*), $b\mathrm{Y} = -(a^2-b^2)\sin^3\theta$ $\therefore\ (b\mathrm{Y})^{2/3} = (a^2-b^2)^{2/3}\sin^2\theta$

Adding, we get $(a\mathrm{X})^{2/3} + (b\mathrm{Y})^{2/3} = (a^2-b^2)^{2/3}$

Changing X to x and Y to y, we get

$(ax)^{2/3} + (by)^{2/3} = (a^2-b^2)^{2/3}$, which is the equation of the evolute of the ellipse.

Example 3. *If C_x and C_y be the chords of curvature parallel to the axes of x and y respectively, at any point of the curve $y = ae^{x/a}$, prove that $\dfrac{1}{C_x^{\,2}} + \dfrac{1}{C_y^{\,2}} = \dfrac{1}{2aC_x}$.*

Sol. Equation of the curve is $y = ae^{x/a}$

Differentiating, $$y_1 = a\,.\,e^{x/a}\,.\,\frac{1}{a} = e^{x/a} = \frac{y}{a} \qquad [\because\ y = ae^{x/a}]$$

and $$y_2 = e^{x/a}\,\frac{1}{a} = \frac{y}{a}\cdot\frac{1}{a} = \frac{y}{a^2}$$

Now $$\rho = \frac{(1+y_1^2)^{3/2}}{y_2} = \frac{\left(1+\dfrac{y^2}{a^2}\right)^{3/2}}{\dfrac{y}{a^2}} = \frac{(a^2+y^2)^{3/2}}{ay}$$

Also $$\tan\psi = \frac{dy}{dx} = \frac{y}{a},\ \therefore\ \sin\psi = \frac{y}{\sqrt{a^2+y^2}} \text{ and } \cos\psi = \frac{a}{\sqrt{a^2+y^2}}$$

Now $$C_x = 2\rho\sin\psi = 2\,.\,\frac{(a^2+y^2)^{3/2}}{ay}\cdot\frac{y}{\sqrt{a^2+y^2}} = \frac{2(a^2+y^2)}{a} \qquad \ldots(i)$$

and $$C_y = 2\rho \cos \psi = 2\,\frac{(a^2+y^2)^{3/2}}{ay} \cdot \frac{a}{\sqrt{a^2+y^2}} = \frac{2(a^2+y^2)}{y}$$

$$\therefore \quad \frac{1}{C_x^{\,2}} + \frac{1}{C_y^{\,2}} = \frac{a^2}{4(a^2+y^2)^2} + \frac{y^2}{4(a^2+y^2)^2} = \frac{1}{4(a^2+y^2)}$$

$$= \frac{1}{2a}\left[\frac{a}{2(a^2+y^2)}\right] = \frac{1}{2a\,.\,C_x}. \qquad [\because \text{ of } (i)]$$

Example 4. *Show that the chord of curvature of the cardioide $r = a(1 - \cos \theta)$, through the pole is $4r/3$*

Sol. Equation of the curve is $r = a(1 - \cos \theta)$. ...(1)

We know that C_0 (chord of curvature through the pole) $= 2\rho \sin \phi$ ...(2)

Diff. (1) w.r.t. θ, $\quad r_1 = a \sin \theta$

Again diff. w.r.t θ, $\quad r_2 = a \cos \theta$

$$\therefore \quad \rho = \frac{(r^2 + r_1^{\,2})^{3/2}}{r^2 + 2r_1^{\,2} - rr_2}$$

$$= \frac{[a^2(1-\cos\theta)^2 + a^2\sin^2\theta]^{3/2}}{a^2(1-\cos\theta)^2 + 2a^2\sin^2\theta - a^2\cos\theta\,(1-\cos\theta}\,,$$

or $$\rho = \frac{a^3[1+\cos^2\theta - 2\cos\theta + \sin^2\theta]^{3/2}}{a^2\,[1+\cos^2\theta - 2\cos\theta + 2\sin^2\theta - \cos\theta + \cos^2\theta]}$$

or $$\rho = \frac{a(2-2\cos\theta)^{3/2}}{3-3\cos\theta} = \frac{a.2^{3/2}\,(1-\cos\theta)^{3/2}}{3(1-\cos\theta)}$$

or $$\rho = \frac{a.2^{3/2}\,(1-\cos\theta)^{1/2}}{3} = \frac{a\,.\,2^{3/2}\left(2\sin^2\dfrac{\theta}{2}\right)^{1/2}}{3}$$

$$= \frac{a\,.\,2^{3/2}\,.\,2^{1/2}\sin\dfrac{\theta}{2}}{3} = \frac{4a}{3}\sin\frac{\theta}{2} \qquad ...(3)$$

Also $$\tan\phi = r\frac{d\theta}{dr} = \frac{r}{r_1} = \frac{a(1-\cos\theta)}{a\sin\theta} = \frac{2\sin^2\dfrac{\theta}{2}}{2\sin\dfrac{\theta}{2}\cos\dfrac{\theta}{2}}$$

or $$\tan\phi = \tan\frac{\theta}{2} \quad \therefore \quad \phi = \frac{\theta}{2} \qquad ...(4)$$

Putting the values of ρ and ϕ from (3) and (4) in (2), we have

$$C_0 = \frac{2\,.\,4a}{3}\sin\frac{\theta}{2}\sin\frac{\theta}{2} = \frac{8a}{3}\sin^2\frac{\theta}{2}$$

$$= \frac{8a}{3}\left(\frac{1-\cos\theta}{2}\right) = \frac{4a}{3}(1-\cos\theta) = \frac{4a}{3}\cdot\frac{r}{a} \qquad \left|\because \text{ From (1), } 1-\cos\theta = \frac{r}{a}\right.$$

$$= \frac{4r}{3}.$$

EXERCISE 4

1. Find the centre of curvature for the point $\left(\frac{3a}{2}, \frac{3a}{2}\right)$ of the folium $x^3 + y^3 = 3axy$.
2. Find the evolute of the parabola $y^2 = 4ax$.
   [**Hint.** It is example 1.]
3. Find the evolute of the ellipse $\frac{x^2}{a^2} + \frac{y^2}{b^2} = 1$.
   [**Hint.** It is example 2.]
4. Find the evolute of the hyperbola $\frac{x^2}{a^2} - \frac{y^2}{b^2} = 1$.
   [**Hint.** Take the parametric equations of the hyperbola as
   $$x = a \sec \theta, y = b \tan \theta.$$
   Prove that centre of curvature (X, Y) is
   $$\left(\frac{a^2 + b^2}{a} \sec^3 \theta, -\frac{(a^2 + b^2)}{b} \tan^3 \theta\right).$$
   Then use $\sec^2 \theta - \tan^2 \theta = 1$.]
5. Define evolute of a curve and show that the evolute of the hypocycloid
   $x^{2/3} + y^{2/3} = a^{2/3}$ is $(x + y)^{2/3} + (x - y)^{2/3} = 2 . a^{2/3}$. *(M.D.U. 2005 A)*
   [**Hint.** Parametric equations of this curve are
   $$x = a \cos^3 \theta, y = a \sin^3 \theta.]$$
6. Find the centre of curvature for any point (x, y) on the rectangular hyperbola $xy = c^2$ and find the equation of its evolute.
7. (*a*) Show that the equation of the circle of curvature at point $\left(\frac{a}{4}, \frac{a}{4}\right)$ on the curve $\sqrt{x} + \sqrt{y} = \sqrt{a}$ is
   $$\left[x - \frac{3}{4}a\right]^2 + \left[y - \frac{3}{4}a\right]^2 = \frac{1}{2}a^2.$$
   (*b*) Show that the circle of curvature at the origin of the parabola
   $$y = mx + \frac{x^2}{a} \text{ is } x^2 + y^2 = a(1 + m^2)(y - mx).$$
8. Find the centre of curvature C for any point P on the catenary $y = c \cosh\left(\frac{x}{c}\right)$ and show that PC = PG where G is the point of intersection of normal at P with x-axis.
   [**Hint.** PG = Length of the normal = $y\sqrt{1 + y_1^2}$.]
9. Show that the evolute of the tractrix $x = c \cos t + c \log \tan \frac{1}{2} t$, $y = c \sin t$ is the catenary $y = c \cosh \frac{x}{c}$.
10. Prove that the chord of curvature parallel to y-axis for the curve $y = a \log \sec \frac{x}{a}$ is of constant length.
11. If C_x, C_y be chords of curvatures parallel to the axis of x and y respectively at any point of the curve $y = c \cosh \frac{x}{c}$; prove that $4c^2 (C_x^2 + C_y^2) = C_y^4$.

12. (*a*) Find the chords of curvature through the pole for the curves :
(*i*) $r = ae^{m\theta}$ (*ii*) $r^n = a^n \cos n\theta$. (*K.U. 1998 A*)
(*b*) Show that for the curve $p = ae^{br}$, the chord of curvature through the pole is of constant length.

13. (*a*) Show that the length of the chord of curvature through the pole of the curve $r = a(1 + \cos\theta)$ is $\frac{4r}{3}$. (*K.U. 1995 A, 2004 S*)
(*b*) If C_o and C_p denote lengths of chords of curvature of the cardioide $r = a(1 + \cos\theta)$, along and perpendicular to radius vector through any point, show that $3(C_o^2 + C_p^2) = 8aC_0$.

14. Show that at any point on the equiangular spiral $r = ae^{\theta\cot\alpha}$, $\rho = r\operatorname{cosec}\alpha$, and show that radius of curvature subtends a right angle at the pole. (*M.D.U. 1997*)

15. Prove that the points on the curve $r = f(\theta)$, the circle of curvature at which passes through the pole (origin) are given by the equation $f(\theta) + f''(\theta) = 0$.

ANSWERS

1. $\left(\frac{21}{16}a, \frac{21}{16}a\right)$. **2.** $27ay^2 = 4(x - 2a)^3$.

3. $(ax)^{2/3} + (by)^{2/3} = (a^2 - b^2)^{2/3}$ **4.** $(ax)^{2/3} - (by)^{2/3} = (a^2 + b^2)^{2/3}$.

6. $\left(\frac{3x}{2} + \frac{c^4}{2x^3}, \frac{3c^2}{2x} + \frac{x^3}{2c^2}\right)$; $(x + y)^{2/3} - (x - y)^{2/3} = (4c)^{2/3}$.

8. $\left(x - c\cosh\frac{x}{c}\sinh\frac{x}{c}, y + c\cosh\frac{x}{c}\right)$.

12. (*a*) (*i*) $2r$ (*ii*) $\frac{2r}{n+1}$.

12. Two Important Properties of the Evolute

(*a*) *The normal at any point of a curve is the tangent to its evolute at the corresponding centre of curvature.*

(*b*) *The length of the arc of the evolute between any two points is equal to the difference between the radii of curvature at the corresponding points of original curve.*

(*c*) Let C(X, Y) be the centre of curvature corresponding to any point P(x, y) of the given curve. Then C is a point on the evolute, and we have

$$X = x - \rho\sin\psi,\ Y = y + \rho\cos\psi.$$

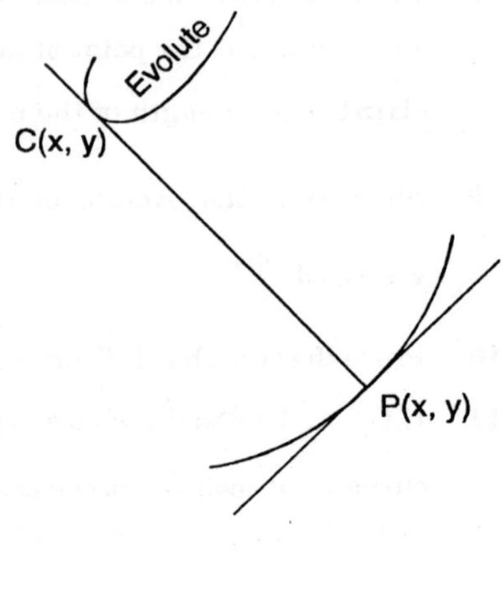

Since x, y, ρ, ψ depend upon s, therefore the above two equations may be considered as the *parametric equations of the evolute, s being the parameter.*

Differentiating, w.r.t. parameter s,

we have $$\frac{dX}{ds} = \frac{dx}{ds} - \rho\cos\psi\cdot\frac{d\psi}{ds} - \sin\psi\frac{d\rho}{ds}$$

$$= \cos\psi - \cos\psi - \sin\psi\frac{d\rho}{ds}$$

$$\left(\because \frac{dx}{ds} = \cos\psi \text{ and } \frac{d\psi}{ds} = \frac{1}{\rho}\right)$$

$$= -\sin\psi \cdot \frac{d\rho}{ds} \qquad \ldots(i)$$

and
$$\frac{dY}{ds} = \frac{dy}{ds} - \rho \sin\psi \cdot \frac{d\psi}{ds} + \cos\psi \frac{d\rho}{ds}$$

$$= \sin\psi - \sin\psi + \cos\psi \frac{d\rho}{ds} = \cos\psi \cdot \frac{d\rho}{ds} \qquad \ldots(ii)$$

$\therefore$ Dividing (*ii*) by (*i*).

$$\frac{dY}{ds} \Big/ \frac{dX}{ds} = -\cot\psi \text{ i.e., } \frac{dY}{dX} = -\cot\psi = \frac{-1}{\tan\psi}$$

i.e., slope of the tangent at C to evolute = slope of normal at P to the given curve.

Also the two lines namely tangent at C to evolute and normal at P to the given curve have point C(X, Y) in common. Therefore, the two lines are same.[By one-point Form]

Hence tangents to the evolute are normals to the original curve.

(*b*) Now to prove the second property, square and add (*i*) and (*ii*).

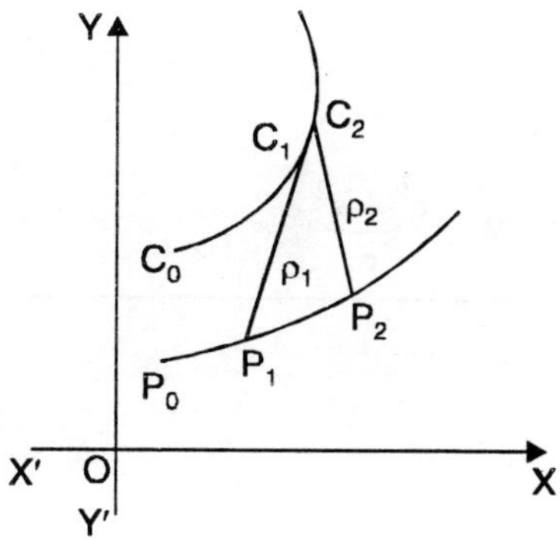

$$\therefore \qquad \left(\frac{dX}{ds}\right)^2 + \left(\frac{dY}{ds}\right)^2 = \left(\frac{d\rho}{ds}\right)^2$$

or
$$(dX)^2 + (dY)^2 = (d\rho)^2 \qquad \ldots(iv)$$

Let S be the length of the arc of the evolute, measured from some fixed point C_0 on it, upto point (X, Y).

$$\therefore \qquad (dX)^2 + (dY)^2 = (dS)^2 \qquad \ldots(v)$$

[See Fig. (*i*) Art. 4]

From (*iv*) and (*v*), $(dS)^2 = (d\rho)^2 \quad \therefore \quad dS = d\rho.$

Integrating, we get $\qquad S = \rho + A \qquad \ldots(vi)$

where A is the constant of integration.

Let P_1, P_2 be any two points on the original curve, and C_1, C_2 be the corresponding centres of curvature on the evolute. Let ρ_1 and ρ_2 be radii of curvature at P_1 and P_2, and S_1 and S_2 be lengths of the arcs C_0C_1 and C_0C_2.

Then by (*vi*), we have

$$S_1 = \rho_1 + A \quad \text{and} \quad S_2 = \rho_2 + A$$

$\therefore$ By subtracting, $S_2 - S_1 = \rho_2 - \rho_1$.

Hence length of arc C_1C_2 = difference between radii of curvature at points P_1 and P_2.

REVISION EXERCISE

1. Show that in the parabola $y^2 = 4ax$, the radius of curvature ρ at any point P is twice the part of the normal intercepted between the curve and the directrix.

2. Find the radius of curvature at any point of the catenary $x = c \log (s + \sqrt{c^2 + s^2})$, $y = \sqrt{c^2 + s^2}$.

3. Prove that if :

(*i*) $s = ae^{x/c}$, then $\rho = \frac{s}{c}\sqrt{s^2 - c^2}$ $\qquad$ (*ii*) $s = \sqrt{8\,ay}$, then $\rho = 4a\sqrt{1 - \frac{y}{2a}}$.

4. If the equation of a curve is given in polar co-ordinates and if $u = \frac{1}{r}$, prove that curvature is given by $\left(\frac{d^2u}{d\theta^2} + u\right) \sin^3 \phi$.

5. Prove that for any curve $\frac{d^2r}{ds^2} = \frac{\sin^2 \phi}{r} - \frac{\sin \phi}{\rho}$.

6. Show that the distance between the pole and centre of curvature corresponding to any point on the curve $r^m = a^m \cos m\theta$ is $\frac{[a^{2m} + (m^2 - 1)r^{2m}]^{1/2}}{(m + 1)r^{m-1}}$.

7. The curve $r = a\, e^{\theta \cot \alpha}$ cuts any radius vector in the consecutive points $A_1, A_2, A_3, \ldots\ldots A_n, \ldots\ldots\ldots\ldots$ If ρ_n denotes the radius of curvature at A_n, prove that

$$\frac{1}{m - n} \log \frac{\rho_m}{\rho_n} = 2\pi \cot \alpha.$$

ANSWERS

2. $\frac{c^2 + s^2}{c}$.

8

Multiple Points, Tests for Concavity and Convexity, Points of Inflexion

1. Singular Point

Definition. *A point on a curve at which the curve exhibits an extraordinary or unusual behaviour is called a* **singular point.**

The student is already familiar with one type of singular points namely points of maxima and minima.

In this chapter, two main types of the singular points (*i*) *multiple points* and (*ii*) *points of inflexion* shall be considered.

MULTIPLE POINTS

2. Definition

A point on a curve through which more than one branch of the curve passes, is called a multiple point of the curve. In particular a point on a curve is called a double point if two branches of the curve pass through it, a triple point if three branches pass through it and so on. In general, if r branches of the curve pass through a point, that point is called a *multiple point of rth order.* In general, a curve has r tangents (real, coincident or imaginary), at a multiple point of rth order, one for each of the r branches. Thus a curve has two tangents at a double point, one for each branch, and these tangents may be real and different, real and coincident or imaginary.

3. Types of Double Points

There are three kinds of double points.

(*i*) **Node.** A double point on a curve through which pass *two real branches of the curve* and the tangents at which are *real and distinct,* is called a **node.** Thus, point P in the adjoining figure is a node.

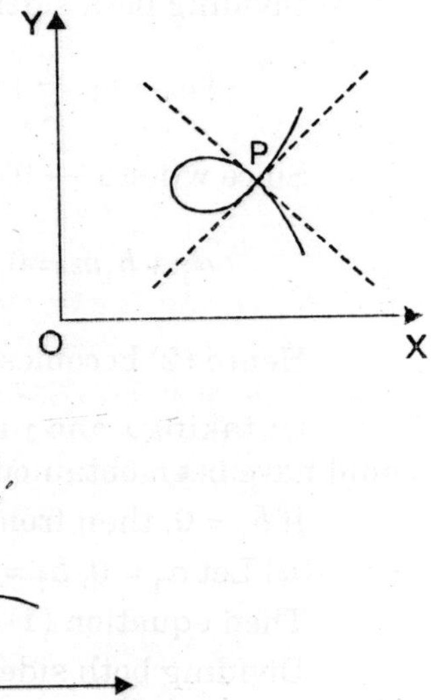

(*ii*) **Cusp.** A double point on a curve through which pass two real branches of the curve and the tangents at which are *real and coincident* is called a **cusp.** Thus, the point Q in the figures (*i*) and (*ii*) is a cusp.

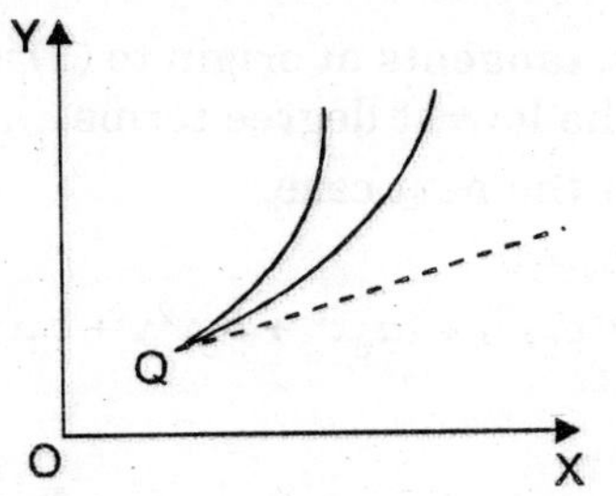

Fig. (*i*)

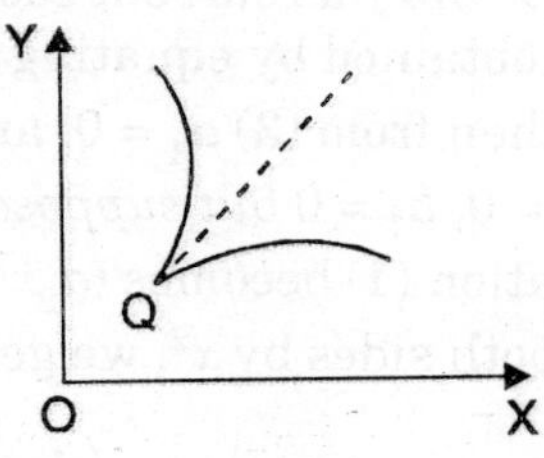

Fig. (*ii*)

(*iii*) **Conjugate point or Isolated point.** If there are no real points on the curve in the neighbourhood of a point R on the curve, then R is called a **Conjugate point** *or* **an isolated point.** For example, (2, 0) is a conjugate point on the curve $y^2 = (x-2)^2(x-5)$ because (2, 0) satisfies the equation of the curve, but for all value of x very close to 2, y^2 is negative and hence y is imaginary *i.e.,* there is no point on the curve is the neighbourhood of (2, 0).

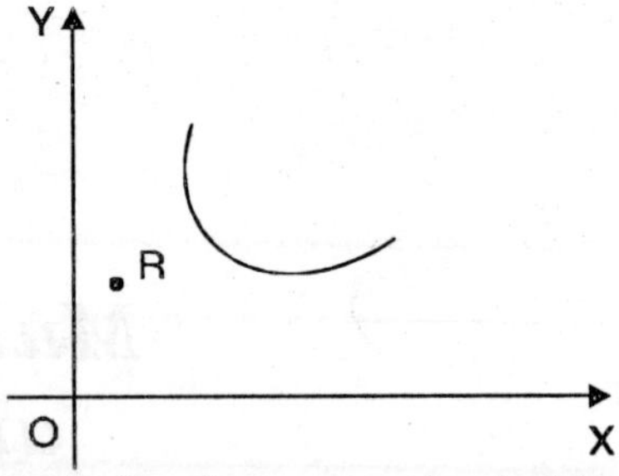

Remark. Since there is no other point in the neighbourhood of conjugate point R, the *tangents at R are generally imaginary,* (as in the above example also) but in some of the cases the *tangents at R may be real.* In such cases the nature of the curve in the neighbourhood of the point under consideration, has to be examined. If the branches of the curve are imaginary, then the point is a conjugate point.

4. Tangents at the Origin to an Algebraic Curve

When a curve passes through the origin, its equation can be written in the form

$$(a_1x + b_1y) + (a_2x^2 + b_2xy + c_2y^2) + (a_3x^3 + b_3x^2y + c_3xy^2 + d_3y^3) + \ldots = 0 \qquad \ldots(1)$$

Let P(x, y) be any point on the curve, near the origin O.

Then slope of the chord OP $= \dfrac{y-0}{x-0} = \dfrac{y}{x}$.

When P → origin O, the chord OP → the tangent at the origin O.

∴ Slope of the tangent at O $= \underset{\substack{x\to 0\\ y\to 0}}{\text{Lt}} \dfrac{y}{x} = m$ (say)

∴ Equation of the tangent at O is

$$Y - 0 = \underset{\substack{x\to 0\\ y\to 0}}{\text{Lt}} \left(\frac{y}{x}\right)[X - 0] \text{ or } Y = \underset{\substack{x\to 0\\ y\to 0}}{\text{Lt}} \frac{y}{x}.X = mX \qquad \ldots(2)$$

Case 1. *When m is finite i.e., y-axis is not the tangent at the origin.*

(*i*) Let us suppose $b_1 \neq 0$.

Dividing both sides of (1) by x, we get

$$\left(a_1 + b_1.\frac{y}{x}\right) + \left(a_2x + b_2y + c_2y.\frac{y}{x}\right) + (a_3x^2 + b_3xy + \ldots\ldots) + \ldots = 0$$

Since when $x \to 0, y \to 0, \dfrac{y}{x} \to m$, we get on taking limits,

$$a_1 + b_1m = 0, \text{ so that } m = -\frac{a_1}{b_1} \text{ [provided } b_1 \neq 0] \qquad \ldots(3)$$

Hence (2) becomes $Y = \dfrac{-a_1}{b_1}X$ or $a_1X + b_1Y = 0$

Or taking x and y as current coordinates, tangents at origin to (1) is $a_1x + b_1y = 0$ which could have been obtained by equating to zero the lowest degree terms.

If $b_1 = 0$, then from (3) $a_1 = 0$, and we get the next case.

(*ii*) Let $a_1 = 0$, $b_1 = 0$ *but suppose* $b_2 \neq 0$, $c_2 \neq 0$.

Then equation (1) becomes $(a_2x^2 + b_2xy + c_2y^2) + (a_3x^3 + b_3x^2y + c_3xy^2 + d_3y^3) + \ldots = 0$

Dividing both sides by x^2, we get

$$\left[a_2 + b_2.\frac{y}{x} + c_2\left(\frac{y}{x}\right)^2\right] + \left[a_3x + b_3y + c_3y.\frac{y}{x} + d_3y.\left(\frac{y}{x}\right)^2\right] + \ldots\ldots = 0$$

Taking limits as $x \to 0, y \to 0$, we get $a_2 + b_2 m + c_2 m^2 = 0$ ...(4)
which is a quadratic in m and its two roots are the slopes of the two tangents at the origin.

But by (2) equation of either tangent at origin is $Y = mX$ ...(5)
where m is a root of (4).

Eliminating m between (4) and (5), we get $a_2 X^2 + b_2 XY + c_2 Y^2 = 0$

On taking x, y as current coordinates, $a_2 x^2 + b_2 xy + c_2 y^2 = 0$ which is the joint equation of two tangents at the origin and could have been written down by equating to zero, the lowest degree terms in the equation of curve [*viz*. (1) with $a_1 = 0, b_1 = 0$]

If $b_2 = 0, c_2 = 0$, then by (4), $a_2 = 0$.

(*iii*) $a_1 = b_1 = a_2 = b_2 = c_2 = 0$, we can similarly show that the equation of tangents at the origin is obtained by equating to zero the lowest degree terms in the curve ; and so on.

Case 2. *When y-axis is tangent at origin.*

By supposing the axes of x and y interchanged and considering the slope of OP with respect to y-axis, it can be seen that tangent retains the same form.

Working rule for finding tangents at the origin to an algebraic curve.

Thus, in general the equation to the tangent or tangents at the origin is obtained by equating to zero the terms of lowest degree in the equation of the curve.

5. Working rule *for finding the nature of origin if it is a double point.*

(*a*) (*i*) Write the equation of tangents at the origin by equating to zero the lowest degree terms in the equation of the curve. If origin is a double point, then we shall get two tangents which may be real or imaginary.

(*ii*) *If the two tangents are imaginary, then origin, is a conjugate point i.e., an isolated point.*

For example tangents at the origin to the curve

$$a^2x^2 + b^2y^2 = (x^2 + y^2)^2 \quad \text{are}$$

$$a^2x^2 + b^2y^2 = 0 \qquad \text{(Equating lowest degree terms to zero)}$$

or $$a^2x^2 = -b^2y^2$$

$\therefore$ $$ax = \pm\sqrt{-b^2y^2} \text{ or } ax = \pm i\, by$$

which are two imaginary tangents.

Thus the origin is an isolated point of the given curve.

(*iii*) If the two tangents are *real and distinct,* then *origin is either a node* or *a conjugate point.* To be definite, nature of the curve in the neighbourhood of the origin is to be examined. If the curve has *two real branches passing through the origin, then it is a node otherwise* it is a *conjugate point.*

(*iv*) If the two tangents are *real and coincident,* then origin is either a cusp or a conjugate point.

Remark 1. The tangents may be real at a conjugate point.

For example, let the equation of the curve be $(y - x)^2 + x^6 = 0$ ...(1)

Origin (0, 0) satisfies this equation (1) of the curve

$\because$ O lies on curve (1)

From (1), $(y - x)^2 = -x^6 \quad \therefore \quad y - x = \pm\sqrt{-x^6}$.

or $$y - x = \pm ix^3 \qquad \text{where } i = \sqrt{-1}.$$

For small values of x (+ ve or – ve) y is imaginary.

Hence the two branches of the curve through (0, 0) are imaginary.

$\therefore$ (0, 0) is a conjugate point.

But the tangents at the origin are (By equating the lowest degree terms to zero)

$$(y - x)^2 = 0$$

i.e., $y - x = 0$ and $y - x = 0$ which are **real** (and coincident) tangents.

(*b*) *Method for testing the nature of the curve at the origin.*

If the tangents at the origin are $y^2 = 0$, solve the equation for y, neglecting the third and higher powers of y. If for small values of x, the values of y are real, the branches of curve through the origin are real otherwise they are imaginary.

In case the tangents at the origin are $x^2 = 0$, solve the equation of curve for x, neglecting terms involving powers of x above the second and proceed similarly.

In other cases, solve either for y or x, whichever is more convenient.

Remark 2. It is to be carefully *noted* that while neglecting higher powers of y or x, the reduced equation of curve, may not coincide with that of the tangents or two branches of the curve may not coincide.

Cor. Nature of Double points other than the origin.

If the point (h, k) is a double point of the curve known in advance, shift the origin to the point (h, k) (*i.e.,* put $x = X + h$ and $y = Y + k$). The double point is the origin now and its nature can be studied by proceeding as in the above article.

Example 1. *Examine the nature of the origin for the curve* $x^4 - ax^2y + axy^2 + a^2y^2 = 0$.

Sol. The equation of the curve is $x^4 - ax^2y + axy^2 + a^2y^2 = 0$ ...(1)

Equating to zero the lowest degree terms, the tangents at the origin are

$$a^2y^2 = 0 \quad \text{or} \quad y^2 = 0, \quad i.e., \quad y = 0, y = 0.$$

Because there are two tangents at the origin, therefore origin is a double point.

Since the two tangents are real and coincident, therefore the origin is either a cusp or a conjugate point.

Now (1) can be written as $ay^2(x + a) - ax^2y + x^4 = 0$

(we have arranged (1) as a quadratic in y)

Solving for y, $$y = \frac{ax^2 \pm \sqrt{a^2x^4 - 4ax^4(x + a)}}{2a(x + a)} = \frac{ax^2 \pm x^2\sqrt{-4ax - 3a^2}}{2a(x + a)}$$

Since, for small values of $x \neq 0$, $(-4ax - 3a^2)$ is – ve.

$\therefore$ y is imaginary in the neighbourhood of the origin.

Hence origin is a conjugate point.

Example 2. *Find the nature of the origin for the curve* $y^2(a^2 + x^2) = x^2(a^2 - x^2)$.

Sol. The equation of the curve is $x^4 + x^2y^2 + a^2y^2 - a^2x^2 = 0$.

Equating to zero the lowest degree terms, the tangents at the origin are

$$a^2y^2 - a^2x^2 = 0 \quad \text{or} \quad y^2 - x^2 = 0 \; i.e., \quad y = \pm x.$$

Because there are two tangents at the origin, therefore origin is a double point.

Since the two tangents are real and distinct, $\therefore$ the origin is either a node or a conjugate point.

Solving the equation of the curve for y, we get $y = \pm x\sqrt{\dfrac{a^2 - x^2}{a^2 + x^2}}$

Since for small values of $x \neq 0$, $(a^2 - x^2)$ is positive, $\therefore$ y is real in the neighbourhood of the origin. Hence origin is a node.

Example 3. *Prove that (a, 4a) is a node on the curve* $y(x-a)^2 = x(y-4a)^2$ *; and find the equations of the tangents at the point.*

Sol. Shift the origin to the point $(a, 4a)$, keeping the directions of the axes unaltered, by putting $x = X + a$ and $y = Y + 4a$.

The transformed equation is

$$(Y + 4a)X^2 = (X + a) Y^2 \quad \text{or} \quad X^2 Y - XY^2 + a(4X^2 - Y^2) = 0 \qquad ...(1)$$

Equating to zero the lowest degree terms, the tangents at the new origin are

$$4X^2 - Y^2 \quad \text{or} \quad Y = \pm 2X \qquad ...(2)$$

Since the two tangents are real and distinct, the new origin is either a node or a conjugate point.

Solving equation (1) for Y, we get

$$Y = \frac{X^2 \pm \sqrt{X^4 + 16aX^2 (X + a)}}{2(X + a)}$$

Since for small values of $X \neq 0$, $[X^4 + 16aX^2 (X + a)]$ is positive,

$\therefore$ Y is real and hence the two branches of the curve through new origin are real. Therefore, the new origin is a node.

Referred to the original axes, the equations of tangents (given by (2)) at the node $(a, 4a)$ are

$$y - 4a = \pm 2(x - a) \qquad [\text{Putting } X = x - a \text{ and } Y = y - 4a \text{ in (2)}]$$

i.e., $\quad 2x - y + 2a = 0 \quad \text{and} \quad 2x + y - 6a = 0.$

6. Conditions for the existence of a multiple point on a curve, and the classification of double point

Let $P(x, y)$ be any point on the curve $f(x, y) = 0$ ...(1)

Differentiating (1) w.r.t. x,

we have
$$\frac{df}{dx} = 0$$

i.e.,
$$\frac{\partial f}{\partial x} + \frac{\partial f}{\partial y}\frac{dy}{dx} = 0 \qquad ...(2)$$

which gives the slope $\frac{dy}{dx}$ of the tangent at P.

If $P(x, y)$ is a multiple point, then there must be at least two tangents to the curve at P. Thus $\frac{dy}{dx}$ must have at least two values at $P(x, y)$. But equation (2) is a first degree equation in $\frac{dy}{dx}$ and is satisfied by at least two values of $\frac{dy}{dx}$ which is possible only if it is an identity.

$$\therefore \qquad \frac{\partial f}{\partial x} = 0 \text{ and } \frac{\partial f}{\partial y} = 0.$$

Also $P(x, y)$ must be on the given curve $f(x, y) = 0$.

Thus, *the necessary and sufficient conditions for any point P(x, y) on the curve f(x, y) = 0 to be a multiple point are*

$$\frac{\partial f}{\partial x}=0 \text{ and } \frac{\partial f}{\partial y}=0 \qquad ...(3)$$

$\therefore$ *Those of the simultaneous solutions of equations (3) which satisfy equation (1) determine double points or multiple points on the curve.*

Classification of double points

Equation (2) can be written as

$$F+G\frac{dy}{dx}=0 \text{ where } F=\frac{\partial f}{\partial x} \text{ and } G=\frac{\partial f}{\partial y}.$$

Differentiating again w.r.t. x, we get $\frac{dF}{dx}+\frac{dG}{dx}\frac{dy}{dx}+G\frac{d^2y}{dx^2}=0$

or $$\left(\frac{\partial F}{\partial x}+\frac{\partial F}{\partial y}\frac{dy}{dx}\right)+\left(\frac{\partial G}{\partial x}+\frac{\partial G}{\partial y}\frac{dy}{dx}\right)\frac{dy}{dx}+G\frac{d^2y}{dx^2}=0 \qquad \left(\because \frac{df}{dx}=\frac{\partial f}{\partial x}+\frac{\partial f}{\partial y}\frac{dy}{dx}\right)$$

Putting $F=\frac{\partial f}{\partial x}$ and $G=\frac{\partial f}{\partial y}$, we have

$$\frac{\partial^2 f}{\partial x^2}+\frac{\partial^2 f}{\partial y\partial x}\cdot\frac{dy}{dx}+\left(\frac{\partial^2 f}{\partial x\partial y}+\frac{\partial^2 f}{\partial y^2}\cdot\frac{dy}{dx}\right)\cdot\frac{dy}{dx}+\frac{\partial f}{\partial y}\cdot\frac{d^2y}{dx^2}=0.$$

Since at a double point, $\frac{\partial f}{\partial x}=0$ and $\frac{\partial f}{\partial y}=0$, therefore the above equation reduces to

$$\frac{\partial^2 f}{\partial y^2}\cdot\left(\frac{dy}{dx}\right)^2+2\cdot\frac{\partial^2 f}{\partial x\partial y}\cdot\frac{dy}{dx}+\frac{\partial^2 f}{\partial x^2}=0 \qquad ...(4)$$

which is a quadratic in $\frac{dy}{dx}$, and gives the slopes of two tangents at the double point (x, y).

The tangents at point (x, y) will be real and distinct, real and coincident or imaginary according as the roots of equation (4) are different, equal or imaginary

i.e., according as discriminant of equation (4) $>=$ or <0

i.e., $$\text{as } 4\left(\frac{\partial^2 f}{\partial x\partial y}\right)^2-4\frac{\partial^2 f}{\partial x^2}\cdot\frac{\partial^2 f}{\partial y^2}>= \text{ or } <0$$

or $$\left(\frac{\partial^2 f}{\partial x\partial y}\right)^2-\frac{\partial^2 f}{\partial x^2}\cdot\frac{\partial^2 f}{\partial y^2}>, \text{ or } = \text{ or } <0$$

Hence, *in general ; a double point will be a node, cusp or a conjugate point, according as*

$$\left(\frac{\partial^2 f}{\partial x\partial y}\right)^2 >, = \text{ or } < \frac{\partial^2 f}{\partial x^2}\cdot\frac{\partial^2 f}{\partial y^2} \qquad ...(5)$$

Note 1. The condition (5) is *not a sure test* for the point (x, y) to be a node, a cusp or a conjugate point. This is, in fact, the condition for two tangents at the double point to be real and distinct, real and coincident or imaginary. This result may, *sometimes,* lead to an incorrect conclusion (see Example 4, Art. 7). To be absolutely correct, we have to examine the nature of curve in the immediate neighbourhood of the double point (x, y).

Note 2. If $\frac{\partial^2 f}{\partial y^2}=\frac{\partial^2 f}{\partial x\partial y}=\frac{\partial^2 f}{\partial x^2}=0$, at the point $P(x, y)$, then P will be a multiple point of order higher than two.

Note 3. Equation (4) namely $\frac{\partial^2 f}{\partial y^2}\left(\frac{dy}{dx}\right)^2 + 2\frac{\partial^2 f}{\partial x \partial y}\frac{dy}{dx} + \frac{\partial^2 f}{\partial x^2} = 0$ **gives us the slopes of the tangents at the double point (x, y).**

7. Working rule for finding the position and nature of double points on a curve

1. Write the equation of the curve in form $f(x, y) = 0$ *i.e.*, make the R.H.S. as zero.

2. Find five partial derivatives $\frac{\partial f}{\partial x}, \frac{\partial f}{\partial y}, \frac{\partial^2 f}{\partial x^2}, \frac{\partial^2 f}{\partial y^2}, \frac{\partial^2 f}{\partial x \partial y}$.

3. Solve the equations $\frac{\partial f}{\partial x} = 0$ and $\frac{\partial f}{\partial y} = 0$.

All the pairs of values of x, y obtained by solving the above two equations are the **possible** double points.

Out of these, only those which satisfy the equation of the curve are the double points.

4. At each double point, calculate

$$D = \left(\frac{\partial^2 f}{\partial x \partial y}\right)^2 - \frac{\partial^2 f}{\partial x^2}\frac{\partial^2 f}{\partial y^2}$$

(*a*) If D is + ve, double point is a node or conjugate point

(*b*) If D is 0, double point is a cusp or conjugate point

(*c*) If D is – ve, double point is a conjugate point.

5. For cases (*a*), (*b*) of step 4, when D is + ve or 0 ; test the nature of the curve in the immediate neighbourhood of the double point (x, y) by proceeding as in Cor. Art. 5.

Example 1. *Find the position and nature of the double points on the curve* $x^4 - 2ay^3 - 3a^2y^2 - 2a^2x^2 + a^4 = 0$.

Sol. The equation of the given curve is

$$f(x, y) \equiv x^4 - 2ay^3 - 3a^2y^2 - 2a^2x^2 + a^4 = 0 \qquad \ldots(1)$$

$$\therefore \quad \frac{\partial f}{\partial x} = 4x^3 - 4a^2x, \quad \frac{\partial f}{\partial y} = -6ay^2 - 6a^2y$$

For double points $\quad \frac{\partial f}{\partial x} = 0 \quad$ and $\quad \frac{\partial f}{\partial y} = 0$

$$\therefore \quad \frac{\partial f}{\partial x} = 4x^3 - 4a^2x = 0, \text{ gives } x = 0, a, -a$$

and

$$\frac{\partial f}{\partial y} = -6ay^2 - 6a^2y = 0, \text{ gives } y = 0, -a$$

Thus, the possible double points are (0, 0), (0, – *a*), (*a*, 0), (*a*, – *a*), (– *a*, 0) and (– *a*, – *a*). Out of these only (0, – *a*), (*a*, 0) and (– *a*, 0), satisfy the given equation (1) and hence these are the only double points.

Now,

$$\left.\begin{aligned} \frac{\partial^2 f}{\partial x^2} &= 12x^2 - 4a^2, \quad \frac{\partial^2 f}{\partial x \partial y} = 0, \\ \text{and} \quad \frac{\partial^2 f}{\partial y^2} &= -12ay - 6a^2 \end{aligned}\right] \qquad \ldots(2)$$

Now at $(0, -a)$, from (2), we have

$$\left.\begin{aligned} &\frac{\partial^2 f}{\partial x^2} = -4a^2, \frac{\partial^2 f}{\partial x \partial y} = 0 \text{ and } \frac{\partial^2 f}{\partial y^2} = 6a^2 \\ \therefore \quad &\left(\frac{\partial^2 f}{\partial x \partial y}\right)^2 - \frac{\partial^2 f}{\partial x^2} \cdot \frac{\partial^2 f}{\partial y^2} = 0 - 6a^2(-4a^2) = 24a^4 = +\text{ve} \end{aligned}\right\} \quad ...(A)$$

$\therefore$ Tangents at $(0, -a)$ are real and distinct.

$\therefore$ Double point $(0, -a)$ is a node or conjugate point.

Shift the origin to $(0, -a)$ [by changing x to $X + 0$, y to $Y - a$].

Equation (1) transform to

$$X^4 - 2a(Y - a)^3 - 3a^2(Y - a)^2 - 2a^2X^2 + a^4 = 0 \text{ or } X^4 - 2aY^3 + 3a^2Y^2 - 2a^2X^2 = 0 \quad ...(3)$$

Now solving (3) for Y [neglecting Y^3 and higher powers of Y], we get

$$3a^2Y^2 = 2a^2X^2 - X^4 \quad \text{or} \quad Y = \pm \frac{X}{a \cdot \sqrt{3}} \cdot \sqrt{2a^2 - X^2}$$

which gives real values of Y, for small enough positive or negative values of X.

$\therefore$ Two real branches of curve pass through the new origin, which therefore is a node. Hence $(0, -a)$ is a node.

At the point $(a, 0)$, from (2), we have

$$\left.\begin{aligned} &\frac{\partial^2 f}{\partial x^2} = 8a^2, \frac{\partial^2 f}{\partial x \partial y} = 0 \text{ and } \frac{\partial^2 f}{\partial y^2} = -6a^2 \\ \therefore \quad &\left(\frac{\partial^2 f}{\partial x \partial y}\right)^2 - \frac{\partial^2 f}{\partial x^2} \cdot \frac{\partial^2 f}{\partial y^2} = 0 - 8a^2(-6a^2) = 48a^4 = +\text{ve} \end{aligned}\right\} \quad ...(A)$$

$\therefore$ The tangents at the point $(a, 0)$ are real and distinct.

$\therefore$ Double point $(a, 0)$ is a node or conjugate point.

Shifting the origin to $(a, 0)$ [by substituting $x = X + a$, $y = Y + 0$], the equation (1) transforms to

$$(X + a)^4 - 2aY^3 - 3a^2Y^2 - 2a^2(X + a)^2 + a^4 = 0$$

or

$$X^4 + 4aX^3 - 2aY^3 + 4a^2X^2 - 3a^2Y^2 = 0 \quad ...(4)$$

Solving (4) for Y [neglecting Y^3 and higher powers of Y], we get

$$3a^2Y^2 = 4a^2X^2 + 4aX^3 + X^4 \quad \text{or} \quad Y = \pm \frac{1}{a \cdot \sqrt{3}} X \sqrt{4a^2 + 4aX + X^2}$$

which gives real values of Y, for small enough values of X, + ve or – ve.

Hence two real branches of the curve pass through the new origin. Thus, new origin or the point $(a, 0)$ is a node.

At the point $(-a, 0)$, from (2) we have

$$\left.\begin{aligned} &\frac{\partial^2 f}{\partial x^2} = 8a^2, \frac{\partial^2 f}{\partial x \partial y} = 0 \text{ and } \frac{\partial^2 f}{\partial y^2} = -6a^2 \\ \therefore \quad &\left(\frac{\partial^2 f}{\partial x \partial y}\right)^2 - \frac{\partial^2 f}{\partial x^2} \cdot \frac{\partial^2 f}{\partial y^2} = 0 - 8a^2(-6a^2) \\ &= 48a^4 = +\text{ve} \end{aligned}\right\} \quad ...(A)$$

∴ Tangents at $(-a, 0)$ are real and distinct.

∴ Double point $(-a, 0)$ is a node or conjugate point.

Shifting the origin to $(-a, 0)$ [by substituting $x = X - a$, $y = Y + 0$], equation (1) transforms to

$$(X - a)^4 - 2aY^3 - 3a^2Y^2 - 2a^2 (X - a)^2 + a^4 = 0$$

or
$$X^4 - 4aX^3 - 2aY^3 + 4a^2X^3 - 3a^2Y^2 = 0 \quad \ldots(5)$$

Solving (5) for Y [neglecting Y^3 and higher powers of Y], we get

$$3a^2Y^2 = 4a^2X^2 - 4aX^3 + X^4,$$

or
$$Y = \pm \frac{1}{a \cdot \sqrt{3}} X \sqrt{4a^2 - 4aX + X^2}$$

which gives real values of Y, for small enough values of X, + ve or – ve.

∴ Two real branches of the curve pass through the new origin. Thus, the new origin *i.e.*, the point $(-a, 0)$ is a node.

Hence the given curve has three nodes at the points $(0, -a)$, $(a, 0)$ and $(-a, 0)$.

Note. It is important to know the nature of curve in the neighbourhood of a double point. After determining the positions of the double points, the steps (A) could have been avoided, because from equations (3), (4) or (5) the tangents at the new origin are given by equations $3Y^2 - 2X^2 = 0$; $4X^2 - 3Y^2 = 0$; $4X^2 - 3Y^2 = 0$ respectively, and are evidently real and distinct.

Example 2. *Find the position and nature of the double points on the curve $(x - 2)^2 = y(y - 1)^2$. Also find tangents at the double point.* (K.U. 1999 A, 99 S)

Sol. The equation of the given curve is

$$f(x, y) \equiv y(y - 1)^2 - (x - 2)^2 = 0 \quad \ldots(1)$$

$$\therefore \quad \frac{\partial f}{\partial x} = -2(x - 2), \quad \frac{\partial f}{\partial y} = (y - 1)^2 + y\, 2(y - 1) = (y - 1)(3y - 1)$$

For double points $\frac{\partial f}{\partial x} = 0$ and $\frac{\partial f}{\partial y} = 0$

$$\therefore \quad -2(x - 2) = 0 \quad \text{and} \quad (y - 1)(3y - 1) = 0$$

$$\therefore \quad x = 2 \quad \text{and} \quad y = 1, \tfrac{1}{3}.$$

Thus, the possible double points are (2, 1) and $(2, \frac{1}{3})$. Out of these only (2, 1) satisfies the given equation (1) and hence is the only double point.

Now
$$\frac{\partial^2 f}{\partial x^2} = -2, \quad \frac{\partial^2 f}{\partial x \partial y} = 0 \quad \text{and} \quad \frac{\partial^2 f}{\partial y^2} = (y - 1) \cdot 3 + (3y - 1) = 6y - 4$$

Now at (2, 1)
$$\frac{\partial^2 f}{\partial x^2} = -2, \quad \frac{\partial^2 f}{\partial x \partial y} = 0 \quad \text{and} \quad \frac{\partial^2 f}{\partial y^2} = 2$$

$$\therefore \quad \left(\frac{\partial^2 f}{\partial x \partial y}\right)^2 - \frac{\partial^2 f}{\partial x^2} \cdot \frac{\partial^2 f}{\partial y^2} = 4 = +\text{ve}$$

∴ Tangents at (2, 1) are real and distinct.

∴ Point (2, 1) is a node or conjugate point.

Shifting the origin to (2, 1) [By changing x to $X + 2$, y to $Y + 1$] ; equation (1) transforms to $(Y + 1)Y^2 - X^2 = 0$ or $Y^3 + Y^2 - X^2 = 0$ ...(2)

$$\therefore \quad X = \pm Y\sqrt{Y + 1}$$ (see Remark 2, Art. 5)

For values of Y small enough + ve or – ve, X is real.

Hence two real branches of the curve pass through the new origin.

Hence the new origin or the point (2, 1) is a node.

Let us find tangents at the double point.

From (2), equating lowest degree terms to zero, the tangents at the double point (2, 1) *i.e.,* new origin are $Y^2 - X^2 = 0$ *i.e.,* $Y - X = 0$; $Y + X = 0$

Replacing X by $x - 2$ and Y by $y - 1$, equations of tangents at (2, 1) are

$$y - 1 - x + 2 = 0, y - 1 + x - 2 = 0$$

or

$$x - y - 1 = 0, \quad x + y - 3 = 0.$$

Note. In $f(x, y)$, there is no need to expand the brackets.

Remark. For an alternative method to find tangents at a double point use equation in Note 3, Art. 6.

Example 3. *Examine the nature of the multiple points on the curve*

$$(x + y)^3 - \sqrt{2}\,(y - x + 2)^2 = 0.$$

Sol. The equation of the given curve is $f(x, y) \equiv (x + y)^3 - \sqrt{2}\,(y - x + 2)^2 = 0$ ...(1)

$$\therefore \quad \frac{\partial f}{\partial x} = 3(x + y)^2 + 2\sqrt{2}\,(y - x + 2)$$

$$\frac{\partial f}{\partial y} = 3\,(x + y)^2 - 2\sqrt{2}\,(y - x + 2)$$

For double points $\quad \frac{\partial f}{\partial x} = 0$ and $\frac{\partial f}{\partial y} = 0$

$$\therefore \quad 3(x + y)^2 + 2\sqrt{2}\,(y - x + 2) = 0 \quad ...(2)$$

and

$$3(x + y)^2 - 2\sqrt{2}\,(y - x + 2) = 0 \quad ...(3)$$

Let us solve these two equations for x and y.

Adding (2) and (3), we get $6(x + y)^2 = 0$ *i.e.,* $x + y = 0$ ...(4)

Subtracting (3) from (2), we get $4\sqrt{2}\,(y - x + 2) = 0$ *i.e.,* $y - x + 2 = 0$

(4) + (5) gives $2y + 2 = 0$ *i.e.,* $y = -1$

Putting $y = -1$ in (4), we get $x - 1 = 0$ or $x = 1$.

Thus the possible double point is (1, – 1).

Also (1, – 1) satisfies equation (1).

$\therefore$ (1, – 1) is the only double point.

Now

$$\frac{\partial^2 f}{\partial x^2} = 6(x + y) - 2\sqrt{2}, \quad \frac{\partial^2 f}{\partial x \partial y} = 6(x + y) + 2\sqrt{2},$$

$$\frac{\partial^2 f}{\partial y^2} = 6(x + y) - 2\sqrt{2}.$$

Now at (1, – 1).

$$\frac{\partial^2 f}{\partial x^2} = -2\sqrt{2}, \quad \frac{\partial^2 f}{\partial x \partial y} = 2\sqrt{2}, \quad \frac{\partial^2 f}{\partial y^2} = -2\sqrt{2}$$

$$\therefore \quad \left(\frac{\partial^2 f}{\partial x \partial y}\right)^2 - \frac{\partial^2 f}{\partial x^2} \cdot \frac{\partial^2 f}{\partial y^2} = 8 - 8 = 0$$

$\therefore$ Double point (1, – 1) is a cusp or a conjugate point.

Shifting the origin to (1, – 1) [By changing x to X + 1 and y to Y – 1] equation (1) transforms to $(X + Y)^3 - \sqrt{2}\,(Y - X)^2 = 0$

or
$$X^3 + Y^3 + 3X^2Y + 3XY^2 - \sqrt{2}\,Y^2 - \sqrt{2}\,X^2 + 2\sqrt{2}\,XY = 0$$

Arranging it as a quadratic in Y (neglecting Y^3), we have

$$(3X - \sqrt{2}\,)Y^2 + Y(3X^2 + 2\sqrt{2}X) + (X^3 - \sqrt{2}\,X^2) = 0$$

Solving for Y, we get $Y = \dfrac{-(3X^2 + 2\sqrt{2}X) \pm \sqrt{28\sqrt{2}X^3 - 3X^4}}{2(3X - \sqrt{2})}$

When X is very small and + ve ; $(28\sqrt{2}\,X^3 - 3X^4)$* is + ve and hence Y is real.

Hence two real branches of the curve pass through the new origin.

Hence the new origin or the point (1, – 1) is a cusp.

Example 4. *Find the position and nature of double points on the curve* $(y - x)^2 + x^6 = 0$.

(K.U. 1996 S)

Sol. The equation of the given curve is $f(x, y) \equiv (y - x)^2 + x^6 = 0$ ...(1)

$$\therefore \quad \frac{\partial f}{\partial x} = -2(y - x) + 6x^5$$

$$\frac{\partial f}{\partial y} = 2(y - x)$$

For double points $\dfrac{\partial f}{\partial x} = 0$ and $\dfrac{\partial f}{\partial y} = 0$

$$\therefore \quad -2(y - x) + 6x^5 = 0 \qquad ...(2)$$

and
$$2(y - x) = 0 \qquad ...(3)$$

Let us solve these two equations for x, y.

Putting $y - x = 0$ from (3) in (2), we get $x = 0$.

Putting $x = 0$ in (3), we get $y = 0$.

Thus, the possible double point is (0, 0).

Also (0, 0) satisfies equation (1).

$\therefore$ (0, 0) is the only double point.

Now $\dfrac{\partial^2 f}{\partial x^2} = 2 + 30x^4,\ \dfrac{\partial^2 f}{\partial x \partial y} = -2,\ \dfrac{\partial^2 f}{\partial y^2} = 2$

At (0, 0) $\dfrac{\partial^2 f}{\partial x^2} = 2,\ \dfrac{\partial^2 f}{\partial x \partial y} = -2,\ \dfrac{\partial^2 f}{\partial y^2} = 2.$

$$\therefore \quad \left(\frac{\partial^2 f}{\partial x \partial y}\right)^2 - \frac{\partial^2 f}{\partial x^2} \cdot \frac{\partial^2 f}{\partial y^2} = 4 - 4 = 0$$

$\therefore$ Double point (0, 0) is a cusp or a conjugate point.

From (1), $\quad (y - x)^2 = -x^6$

$$\therefore \quad y - x = \pm\sqrt{-x^6} \quad \text{or} \quad y = x \pm x^3\sqrt{-1}$$

---

*The sign of this expression depends upon the smaller degree term namely $28\sqrt{2}X^3$.

For all small values of x (+ ve or – ve) y is imaginary.

Hence the two branches of the curve through (0, 0) are imaginary.

Hence the origin is a conjugate point.

EXERCISE 1

Find the position and nature of double points on the following curve :

1. $2(x^3 + y^3) - 3(3x^2 + y^2) + 12x - 4 = 0.$ (*K.U. 1997 A*)
2. $x^3 + x^2 + y^2 - x - 4y + 3 = 0.$ (*M.D.U. 2003 S*)
3. $a^2y^2 = a^2x^2 - 4x^3.$
4. (*a*) $x(x^2 + y^2) = ay^2$ (*b*) $y^3 = x^3 + ax^2.$
5. (*a*) $x^3 + y^3 = 3axy$

   (*b*) $x^4 + y^3 + 2x^2 + 3y^2 = 0.$
6. (*a*) $x^4 - 4ax^3 + 2ay^3 + 4a^2x^2 - 3a^2y^2 - a^4 = 0.$

   (*b*) $x^4 + 4ax^3 - 2ay^3 + 4a^2x^2 + 3a^2y^2 - a^4 = 0.$ (*K.U. 1997 S*)
7. $x^4 + 4ax^3 + 4a^2x^2 - b^2y^2 - 2b^3y - a^4 - b^4 = 0.$
8. $x^3 - y^2 - 7x^2 + 4y + 15x - 13 = 0$ (*K.U. 2000 A*)
9. $x^4 - 4y^3 - 12y^2 - 8x^2 + 16 = 0.$
10. $y(y - 6) = x^2 (x - 2)^3 - 9.$
11. $y^2 = (x - 1)(x - 2)^2.$ (*M.D.U. 1996*)
12. $x^4 - 2y^3 - 3y^2 - 2x^2 + 1 = 0$
13. $x^2y^2 = (a + y)^2(b^2 - y^2).$
14. $(y - x)^2 + x^7 = 0.$
15. $(2y + x + 1)^2 = 4(1 - x)^5.$
16. Examine the curve $x^3 + 2x^2 + 2xy - y^2 + 5x - 2y = 0$ for a singular point and show that it is a cusp.
17. (*a*) Show that at the point (1, – 1), there is a cusp on the curve

    $$x^3 + xy^2 + y^3 - 4x^2 + y^2 + 4x + y - 1 = 0.$$

    (*b*) Show that (0, 0) is a conjugate point on the curve $(x - y)^2 + x^4 = 0.$
18. Show that the origin is a node, a cusp or a conjugate point on the curve $y^2 = ax^2 + bx^3$ according as a is positive, zero or negative.
19. Prove that the curve $y^2 = (x - a)^2 (x - b)$, has at $x = a$, a node if $a > b$, a cusp if $a = b$ and a conjugate point if $a < b$. (*K.U. 1998 A*)
20. (*a*)Show that the curve $y^2 = bx \sin \frac{x}{a}$ has a node, or a conjugate point at the origin according as a and b have like or unlike signs. (*M.D.U. 2002 A, 2003 A*)

    (*b*) Determine the nature of double points of the curve $y^2 = 2x \sin x.$
21. Show that the curve $y^2 = bx \tan \frac{x}{a}$ has a node or a conjugate point at the origin according as a and b have like or unlike signs.
22. *Find the equations of the tangents at the multiple point of the following curve :*

    $$(y - 2)^2 = x(x - 1)^2.$$

    and hence show that the multiple point is a node. (*K.U. 1995 A*)
23. Find the equations of the tangents to the curve $(x - 2)^2 = y (y - 1)^2$ at (2, 1) and hence show that the point is a node.

ANSWERS

1. Node at (2, 0) and (1, 1) **2.** Node at (–1, 2)
3. Node at (0, 0) **4.** (*a*) Cusp at (0, 0), (*b*) Cusp at (0, 0)
5. (*a*) Node at (0, 0), (*b*) Conjugate point at (0, 0).
6. (*a*) Conjugate point at $(a, 0)$, (*b*) Nodes at points $(0, a)$, $(-2a, a)$ and $(-a, 0)$
7. Conjugate point at $(-a, -b)$.
8. Node at (3, 2) **9.** Nodes at (0, – 2) ; (2, 0) ; (– 2, 0).
10. Cusp at (2, 3), conjugate at (0, 3)
11. Node at (2, 0).
12. Nodes at (0, –1), (1, 0) and (–1, 0).
13. Node, Cusp or Conjugate point at $(0, -a)$ according as $b > a$, $b = a$ or $b < a$.
14. Cusp at (0, 0) **15.** Cusp at (1, –1).
20. (*b*) Node at (0, 0) **22.** $x - y + 1 = 0$, $x + y = 3$.

8. Kinds of Cusps

We know that at a cusp two branches of a curve have a common tangent and therefore a common normal. A cusp is called a **single cusp** or a **double cusp** according as the *two branches* of the curve lie entirely *on one side* or *on both sides of the common normal.*

Species of Cusps. A cusp, single or double, is said to be *of first species* or *of second species,* according as the *two branches* of the curve, *lie on opposite or same side of the common tangent.*

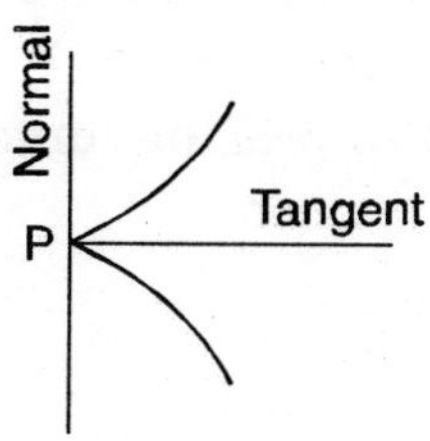

Fig. (*i*)
Single cusp of first species

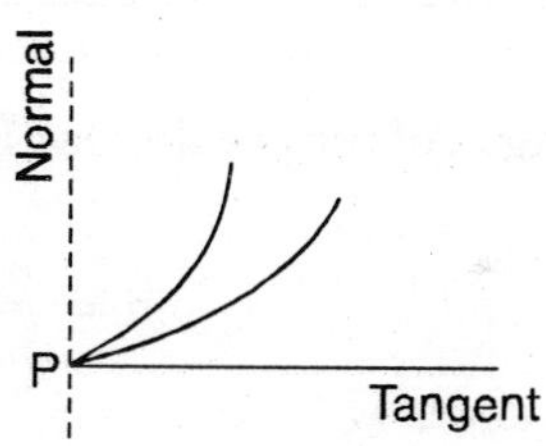

Fig. (*ii*)
Single cusp of second species

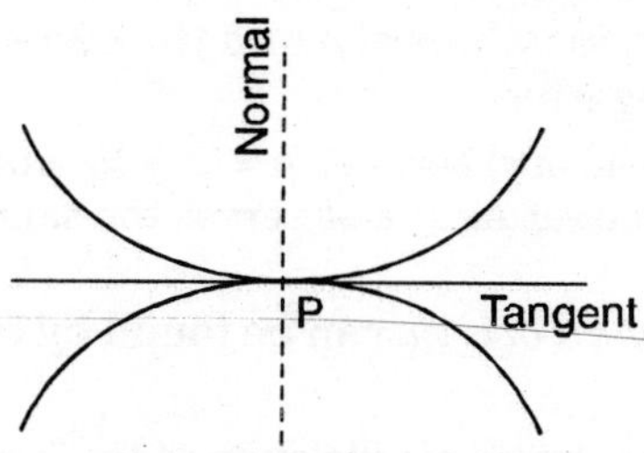

Fig. (*iii*)
Double cusp of first species

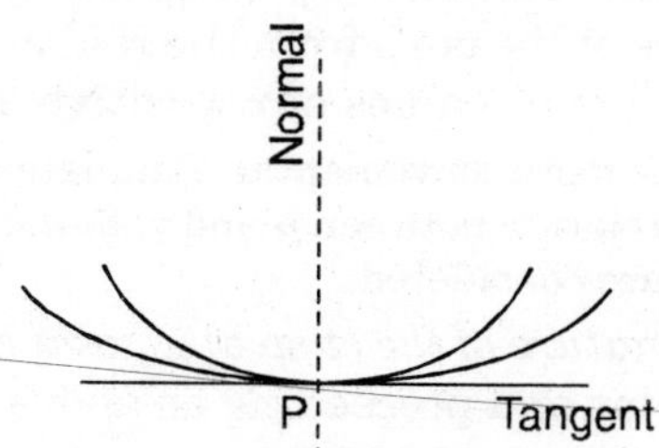

Fig. (*iv*)
Double cusp of second species

Point of oscul-inflexion. A double cusp *with the change of species on the two sides* of the *common normal* is called a *point of oscul-inflexion* [See Fig. (*v*)].

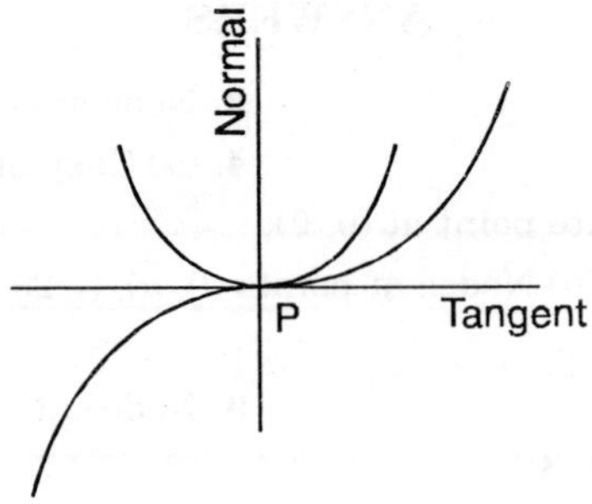

(Point of oscul-inflexion Fig. (v))

9. Working rule *for determining the nature of a cusp at the origin.*

Case 1. *When the cuspidal tangents are $y^2 = 0$, i.e., x-axis.*

Solve the equation of the curve for y, neglecting terms containing powers of y higher than two.

(i) If the *roots are real* for *only one sign of x,* the cusp is a *single cusp.*

(ii) If the *roots are real* for *both signs of x,* the cusp is a *double cusp.*

(iii) If the *roots are opposite in sign,* then the cusp is of the first species.

(iv) If the *roots are of the same sign,* then cusp is of the *second species.*

Case 2. *When the cuspidal tangents are $x^2 = 0$, i.e., y-axis.*

Solve the equation of curve for x, neglecting terms containing powers of x higher than two and distinguish between different cases that arise on similar lines, as in case 1.

Case 3. *When the cuspidal tangents are of the form $(ax + by)^2 = 0$.*

Let p be the length of perpendicular from any point (x, y) of the curve on the tangent line $ax + by = 0$.

Then $$p = \frac{ax + by}{\sqrt{a^2 + b^2}}$$

or p is proportional to $ax + by$.

Let us put $p = ax + by$ and then eliminate y between this equation and the equation of the curve, we shall get a relation between p and x.

Solving this relation for p (neglecting p^3 and higher powers of p), we shall be able to decide the nature of the cusp from the real or non-real values of p and the signs of p (like those of y in case I) for small values of x, positive and negative.

Note. If it is more convenient to eliminate x (instead of y) between $p = ax + by$, and the equation of the curve, then a relation between p and y, found by eliminating x, also serves the same purpose, when small values of y are considered.

Cor. The *nature of the cusp at a point other than origin,* can be found by transferring the origin to that point and proceeding as above.

Example 1. *Show that the curve $y^3 = x^3 + ax^2$, has a single cusp of the first species at the origin.*

Sol. The equation of the curve is $x^3 - y^3 + ax^2 = 0$ ...(i)

Equating to zero the lowest degree terms, the tangents at origin are

$$ax^2 = 0 \quad \text{or} \quad x^2 = 0$$

which are real and coincident. Hence origin is either a cusp or a conjugate point.

Solving (*i*) for x [after neglecting x^3], we get

$$x = \pm\sqrt{\frac{y^3}{a}} = \pm y\sqrt{\frac{y}{a}} \qquad \ldots(ii)$$

When y is small and positive, x is real, $\therefore$ real branches of curve pass through the origin. Hence origin is a cusp.

Also from (*ii*), the values of x are real only if y is positive *i.e.*, values of x are real only for one sign of y (namely + ve, because when $y < 0$, x is imaginary).

$\therefore$ the cusp is a single cusp.

Again, for any small positive value of y, the two values of x are of opposite signs, $\therefore$ the cusp is of first species.

Hence origin a single cusp of the first species.

Example 2. *Show that the curve $(x + y)^3 - \sqrt{2}\,(y - x + 2)^2 = 0$ has a single cusp of the first species at the point (1, –1).*

Sol. The equation of the curve is $(x + y)^3 - \sqrt{2}\,(y - x + 2)^2 = 0$ ...(1)

Transferring the origin to the point (1, –1), by putting $x = X + 1, y = Y - 1$ the equation (1) becomes, $(X + Y)^3 - \sqrt{2}\,(Y - X)^2 = 0$ ...(2)

Equating to zero the lowest degree terms in (2), the tangents at the new origin are $(Y - X)^2 = 0$, which are real and coincident.

$\therefore$ New origin is either a cusp or a conjugate point.

From (2), neglecting Y^3, we get $X^3 + 3X^2Y + 3XY^2 - \sqrt{2}\,(Y^2 - 2XY + X^2) = 0$

or
$$Y^2(3X - \sqrt{2}) + X(2\sqrt{2} + 3X)\,.\,Y - X^2(\sqrt{2} - X) = 0$$

Solving,

$$Y = \frac{-X(2\sqrt{2} + 3X) \pm \sqrt{X^2(2\sqrt{2} + 3X)^2 + 4X^2(\sqrt{2} - X)(3X - \sqrt{2})}}{2(3X - \sqrt{2})}$$

$$= \frac{-X(2\sqrt{2} + 3X) \pm X\sqrt{28\sqrt{2}X - 3X^2}}{2(3X - \sqrt{2})}$$

For very small positive values of X, $(28\sqrt{2}X - 3X^2)$ is positive and $\therefore$ Y is real.

Hence two real branches of the curve pass through the new origin.

Thus, the new origin is a cusp.

Put $\qquad p = Y - X \quad$ or $\quad Y = X + p$ ...(3)

Eliminating Y between (2) and (3), we get

$$(p + 2X)^3 - \sqrt{2}p^2 = 0 \quad \text{or} \quad p^3 + 6Xp^2 + 12X^2 p + 8X^3 - \sqrt{2}p^2 = 0$$

Neglecting p^3, we have $(6X - \sqrt{2})p^2 + 12X^2p + 8X^3 = 0$

Solving, $$p = \frac{-12X^2 \pm \sqrt{144X^4 - 32X^3(6X - \sqrt{2})}}{2(6X - \sqrt{2})}$$

$$= \frac{-6X^2 \pm \sqrt{8\sqrt{2}X^3 - 12X^4}}{6X - \sqrt{2}} = \frac{-6X^2 \pm \sqrt{8\sqrt{2}X^3}}{6X - \sqrt{2}}, \text{ neglecting } -12X^4$$

($\because$ Near the origin an expression has the same sign an its lowest degree term.)

Now p is real only when X is positive (*i.e.*, of one sign).

$\therefore$ New origin is a single cusp.

Also $\sqrt{8\sqrt{2}X^3}$ *i.e.*, $\sqrt{8\sqrt{2}X^{3/2}} > 6X^2$, for small values of X, because $X^{3/2} > X^2$, when X is small.

$\therefore$ Values of p have opposite sign, which shows that cusp is of first species.

Hence the new origin or point (1, –1) is a single cusp of first species.

Example 3. *Show that the curve $a^2y^2 - 2abx^2y + b^2(x^5 + x^6) = 0$ has a point of oscul-inflexion at the origin.*

Sol. Equation of the curve is $a^2y^2 - 2abx^2y + b^2(x^5 + x^6) = 0$ ...(1)

Equating to zero the lowest degree terms, tangents at origin are

$$a^2y^2 = 0 \quad \text{or} \quad y^2 = 0$$

which are real and coincident $\therefore$ Origin is a cusp or a conjugate point.

Neglecting x^6 and solving for y, we have

$$y = \frac{2abx^2 \pm \sqrt{4a^2b^2x^4 - 4a^2b^2x^5}}{2a^2}$$

$$= \frac{bx^2 \pm bx^2\sqrt{1-x}}{a} = \frac{bx^2}{a}[1 \pm \sqrt{1-x}]$$

For small values of x, positive or negative, y is real, $\therefore$ there is double cusp at the origin.

Now when x is small and positive $\sqrt{1-x} < 1$.

$\therefore$ $1 - \sqrt{1-x}$ and $1 + \sqrt{1-x}$ are both positive and hence both the values of y are positive, *i.e.*, on the right of the origin the cusp of the second species.

But when x is small and negative, $\sqrt{1-x} > 1$.

$\therefore$ $1 - \sqrt{1-x}$ and $1 + \sqrt{1-x}$ have opposite signs. Hence for small negative values of x, the values of y are both positive and negative. Hence the cusp is of first species on the left of the origin.

Thus, the origin is a double cusp with change of species and hence is a point of oscul-inflexion.

EXERCISE 2

Find the nature of the cusp at the origin on the following curves :

1. (*i*) $y^2 = x^3$ (*ii*) $y^2 - x^4 = 0$ (*iii*) $(y - 4x^2)^2 = x^7$.
2. (*a*) Show that the curve $y^2(2a - x) = x^3$, has a single cusp of the first species at the origin.
   (*b*) Show that the curve $a^4y^2 = x^5(2a - x)$ has a single cusp of first species at the origin.
3. (*a*) Show that the curve $x^3 + y^3 = ax^2$ has a single cusp of the first species at the origin.
   (*b*) Show that the curve $x^3 + y^3 - 2ay^2 = 0$ has a single cusp of first species at the origin.

4. Show that the following curves have single cusp of first species at the origin :
(a) $x^2(x - y) + y^2 = 0$ (b) $x^2(x + y) - y^2 = 0$ (c) $x^2(2a - y) = y^3$.

5. Show that the curve $x^4 - 2ax^2y - axy^2 + a^2y^2 = 0$, has a single cusp of the second species at the origin.

6. Show that curve $y^2 - 2x^2y + x^4 - x^5 = 0$, has a single cusp of the second species at the origin. (*M.D.U. 1995*)

7. Show that the curve $x^6 - ayx^4 - a^3x^2y + a^4y^2 = 0$ has a double cusp of second kind at the origin.

8. Show that the curve $y^3 = (x - a)^2(2x - a)$ has single cusp of first species at $(a, 0)$.

9. Show that the curve $(2x + y)^2 - 6xy(2x + y) - 7x^3 = 0$ has a single cusp of first species at the origin. (*K.U. 1998 S*)

10. (a) Show that the curve $x^5 + 16x^2y - 64y^2 = 0$ has a point of oscul-inflexion at the origin.
(b) Show that the curve $x^5 - ax^3y - a^2x^2y + a^3y^2 = 0$ has a point of oscul-inflexion at the origin.
(c) Show that the curve $x^7 + 2x^4 + 2x^3y + 2xy + y^2 + x^2 = 0$ has a double cusp with change of species *i.e.,* a point of oscul-inflexion at the origin.
(d) Show that the curve $y^5 - ay^3x - ay^2x + a^2x^2 = 0$ has a point of oscul-inflexion at the origin. (*M.D.U. 2006*)

ANSWERS

1. (*i*) Single cusp of first species (*ii*) Double cusp of first species
(*iii*) Single cusp of second species.

POINT OF INFLEXION

10. Concavity and Convexity

Definition. Let P be a point on curve $y = f(x)$, where that tangent PT is not parallel to y-axis. Then the curve is said to be

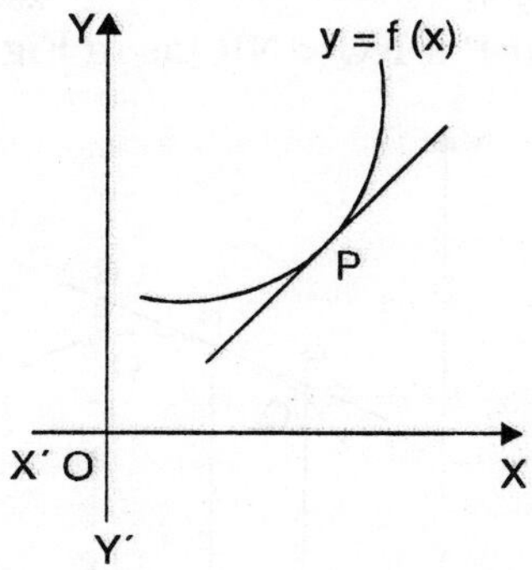

Fig. (*i*)

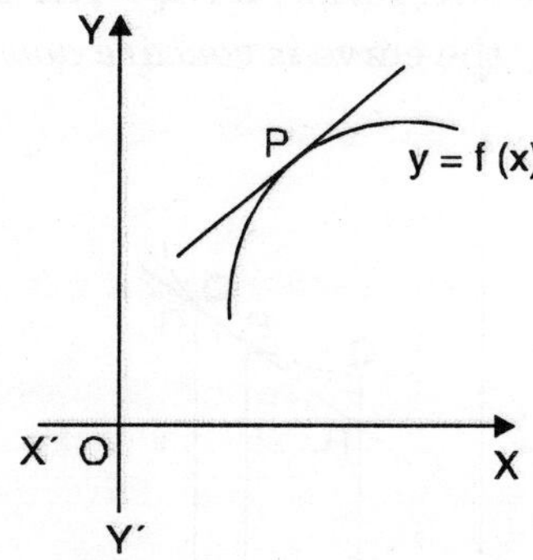

Fig. (*ii*)

(*i*) *Concave upwards or convex downwards* at P, if in the immediate neighbourhood of P, the curve lies above the tangent PT on both sides, [as in Fig. (*i*)] ; and

(*ii*) *Concave downwards or convex upwards* at P, if in the immediate neighbourhood of P, the curve lies below the tangent PT on both sides [as in Fig. (*ii*)].

Note. When the tangents at P are parallel to y-axis, we may define *concavity to the right or convexity to the left, and concavity to the left or convexity to the right* (as at P_1 and P_2 respectively in the above figure) on similar lines.

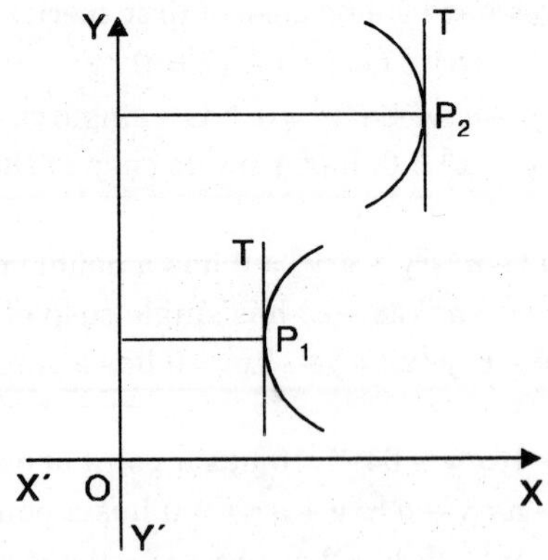

11. Criteria for Concavity and Convexity

To show that at a point P on the curve $y = f(x)$, the curve is

(i) Concave upwards (or in the positive direction of y-axis) if $f''(x)$ *or* $\frac{d^2y}{dx^2} > 0$, *and*

(ii) Concave downwards (or in the negative direction of y-axis) if $f''(x)$ *or* $\frac{d^2y}{dx^2}$ *is* < 0.

Let P(x, y) be the point on the curve $y = f(x)$. Take a neighbouring point Q$(x + h, y + k)$ on both sides of P. From Q draw QN ⊥ OX, meeting the tangent at P in R.

∴ Co-ordinates of R are $(x + h, \text{RN})$

(∵ Points Q, R have the same abscissa ON $= x + h$)

The curve will be *concave upwards* at P if the curve on either side of P lies above the tangent at P (Def. Art. 10) *i.e.*, if NQ > NR [as in Fig. (*i*)].

Similarly, the curve is *concave downwards at* P if NQ < NR [as in Fig. (*ii*)].

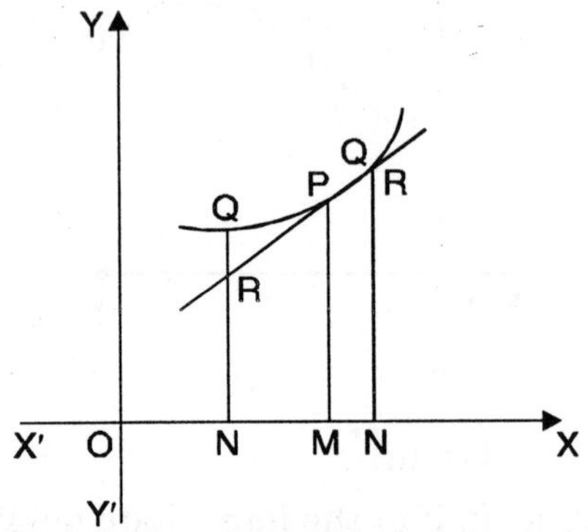

Fig. (*i*)

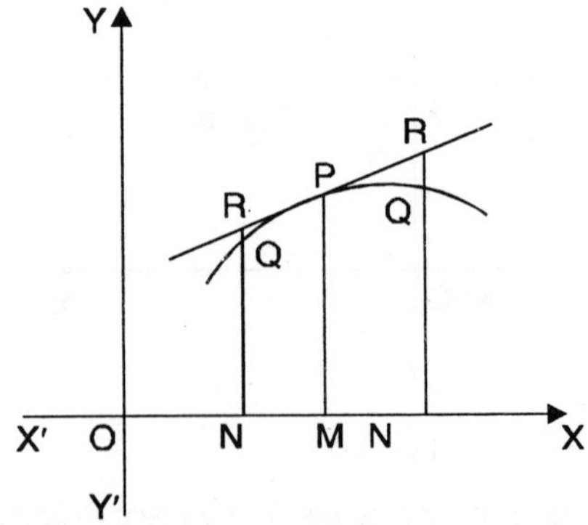

Fig. (*ii*)

Let us find NR and NQ.

The equation of tangent at P(x, y) is

$$Y - y = f'(x)\,(X - x)$$

or
$$Y = y + f'(x)\,.\,[X - x] = f(x) + f'(x)\,[X - x] \qquad (\because\ y = f(x))$$

The point R($x + h$, NR) lie on the tangent at P.

$$\therefore \qquad NR = f(x) + f'(x)\,[x + h - x] = f(x) + hf'(x) \qquad \ldots(1)$$

Also NQ = Ordinate of Q *i.e.*, ordinate corresponding to abscissa $x + h = f(x + h)$.

Subtracting (1) from (2),

$$NQ - NR = f(x + h) - f(x) - hf'(x)$$

$$= f(x) + hf'(x) + \frac{h^2}{2!} f''(x + \theta h) - f(x) - hf'(x), \text{ where } 0 < \theta < 1$$

[By expanding $f(x + h)$ by Taylor's theorem with remainder after two terms]

$$\text{or} \qquad NQ - NR = \frac{h^2}{2!} f''(x + \theta h) \qquad \ldots(i)$$

Assuming that $f''(x)$ is continuous and not zero, it is to be noted that $f''(x + \theta h)$ has the same sign as $f''(x)$ for arbitrarily small values of h, whatever be sign of h.

Also since h^2 is positive whether h be positive or negative, it follows from (*i*), that the sign of NQ – NR, depends upon that of $f''(x)$.

(*i*) The curve will be *concave upwards* at P [as in Fig. (*i*)], if NQ > NR (for both h positive and negative) *i.e.*, if NQ – NR is positive *i.e.*, if $f''(x)$ is positive or if $\frac{d^2y}{dx^2}$ is positive.

And (*ii*) the curve will be *concave downwards* (or *convex upwards*) at P [as in Fig. (*ii*)] if NQ < NR [for both h positive or negative] *i.e.*, if NQ – NR is negative *i.e.*, if $f''(x)$ is negative or if $\frac{d^2y}{dx^2}$ is negative.

Hence the curve is *concave upwards or downwards according as* $f''(x)$ or $\frac{d^2y}{dx^2}$ *is positive or negative.*

Cor. Generalization of criteria for concavity and convexity. The above result can be further generalized as given below :

If $y = f(x)$ be function of x, such that $f''(x) = f'''(x) = \ldots\ldots = f^{n-1}(x) = 0$, at the point P($x, y$) and $f^n(x) \neq 0$; then by Taylor's Theorem with remainder after n terms, we have

$$NQ = f(x + h) = f(x) + hf'(x) + \frac{h^2}{2!} f''(x) + \frac{h^2}{3!} f'''(x) + \ldots\ldots$$

$$\ldots\ldots + \frac{h^{n-1}}{(n-1)!} f^{n-1}(x) + \frac{h^n}{n!} f^n(x + \theta h), \text{ where } 0 < \theta < 1$$

$$= f(x) + h \,.\, f'(x) + \frac{h^n}{n!} f^n(x + \theta h)$$

[$\because$ all other derivatives are given to be zero]

$$\therefore \qquad NQ - NR = \frac{h^n}{n!} f^n(x + \theta h), \text{ where } 0 < \theta < 1 \qquad [\because \; NR = f(x) + hf'(x)]$$

Assuming that $f^n(x)$ is continuous function at point P and h is very small, then $f^n(x + \theta h)$ has the same sign as $f^n(x)$, whatever the sign of h.

As for concavity or convexity, (NQ – NR) is to have same sign for both signs of h and given $f^n(x)$, this is possible only if n is even.

$\therefore$ *if n be even* and thus $\dfrac{h^n}{n!}$ be positive, then curve is *concave upwards at* P, if NQ – NR > 0 or if $f^n(x) > 0$ (n even).

And the *curve is concave downwards at P,*

if $$\text{NQ} - \text{NR} < 0 \text{ or if } f^n(x) < 0 \ (n \text{ even})$$

Thus, the curve $y = f(x)$, is concave upwards or downwards point $P(x, y)$ according as $f^n(x)$ is positive or negative and n is even.

Example 1. *Show that the curve $y = e^x$ is everywhere concave upwards.*

Sol. We have $y = e^x \quad \therefore \quad \dfrac{dy}{dx} = e^x$ and $\dfrac{d^2y}{dx^2} = e^x$

Since $\dfrac{d^2y}{dx^2}$ is always positive, $\therefore$ the curve is concave upwards everywhere.

Example 2. *Find the intervals of x for which the graph of the function $f(x) = (x^2 + 4x + 5)e^{-x}$ is (i) concave upwards (ii) concave downwards.*

Sol.
$$f(x) = (x^2 + 4x + 5)\, e^{-x}$$

$\therefore$
$$f'(x) = (x^2 + 4x + 5)\,.\, e^{-x}(-1) + e^{-x}(2x + 4)$$
$$= e^{-x}(-x^2 - 4x - 5 + 2x + 4) = -(x^2 - 2x + 1)\, e^{-x}$$

and
$$f''(x) = -(x^2 + 2x + 1)\,.\, e^{-x}(-1) - e^{-x}(2x + 2)$$
$$= (x^2 + 2x + 1 - 2x - 2)\, e^{-x} = (x^2 - 1)\, e^{-x} = (x - 1)(x + 1)\, e^{-x}.$$

Since $e^{-x} > 0$ for each x, + ve or – ve, hence the sign of $f''(x)$ depends upon the sign of $(x - 1)(x + 1)$.

Now $f''(x) > 0$ implies $(x - 1)(x + 1) > 0$ which is true only if x does not lie between – 1 and 1 *i.e.*, if $x < -1$ or $x > 1$. $[\because \ (x - \alpha)(x - \beta) > 0 \ \Rightarrow \ x$ does not lie between α and $\beta]$

Further $f''(x) < 0$ implies $(x - 1)(x + 1) < 0$ which holds true if x lies between – 1 and 1 *i.e.*, if $-1 < x < 1$.

Thus, if $x < -1$ or $x > 1$ *i.e.*, if x is in the interval $(-\infty, -1)$ or $(1, \infty)$, $f''(x) > 0$ so that the graph of f is concave upwards.

Again, if $-1 < x < 1$ *i.e.*, if x is in the open interval $(-1, 1)$, then $f''(x) < 0$ so that the graph of f will be concave downwards.

Hence the graph of f is concave upwards in the intervals $(-\infty, -1)$, $(1, \infty)$ and is concave downwards in the interval $(-1, 1)$.

12. Point of Inflexion

Definition. *A point on a curve at which the curve changes from concavity to convexity or vice versa is called a* **point of inflexion.**

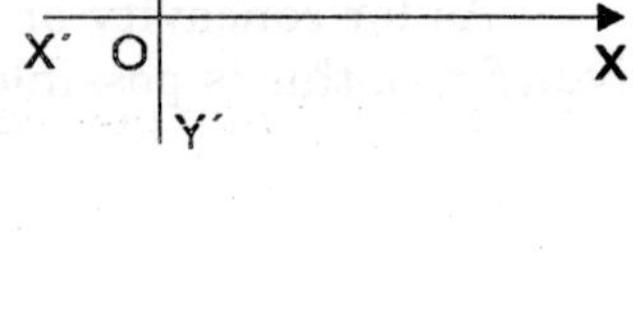

Since a change from concavity to convexity or *vice versa* at a point is possible only if the curve crosses the tangent at that point. Therefore, we may also define *the point of inflexion* as *a point on the curve at which the curve crosses the tangent.*

Remark. It is to be noted that the curve at a point of inflexion exhibits an extra-ordinary or unusual behaviour, for the curve does not ordinarily cross the tangent, as it does at a point of inflexion. Hence a point of inflexion on a curve is also a *singular point.*

13. Criteria for a Point of Inflexion

By definition at a point of inflexion, the curve changes from concavity to convexity or *vice versa*.

$\therefore \quad \dfrac{d^2y}{dx^2}$ is positive on one side of the point of inflexion P and negative on other side of P.

Also if $\dfrac{d^2y}{dx^2}$ is continuous at P, it can pass from positive values to negative values or *vice-versa,* only by passing through zero value of P.

Hence *a point P is a point of inflexion,* if

(*i*) $\dfrac{d^2y}{dx^2} = 0$ *at this point, and*

(*ii*) $\dfrac{d^2y}{dx^2}$ *changes sign in passing through this point*

i.e., $\quad \dfrac{d^3y}{dx^3} \neq 0$, *at this point.*

Analytical proof. Let P(x, y) be a point of inflexion on the curve $y = f(x)$. Take a neighbouring point Q($x + h, y + k$) on both sides of P. From Q draw QN ⊥ OX, meeting the tangent at P in R.

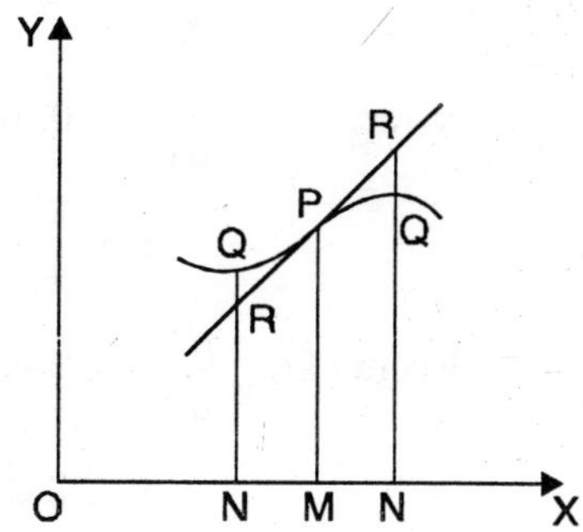

$\therefore$ Co-ordinates of R are ($x + h$, NR)

($\because$ Points Q and R have the same abscissa ON = $x + h$)

The equation of tangent at P(x, y) is

$$Y - y = f'(x)\,[X - x]$$

or
$$Y = y + f'(x)\,[X - x]$$
$$= f(x) + f'(x)\,[X - x] \qquad [\because \; y = f(x)]$$

The point R($x + h$, NR) lies on the tangent at P.

$$\therefore \qquad NR = f(x) + f'(x)\,[x + h - x] = f(x) + hf'(x) \qquad ...(1)$$

Also NQ = Ordinate of Q *i.e.,* ordinate corresponding to abscissa $x + h$

$$= f(x + h) = f(x) + hf'(x) + \frac{h^2}{2!}\,.\,f''(x) + \frac{h^3}{3!}\,.\,f'''(x + \theta h), \qquad ...(2)$$

where $0 < \theta < 1$.

[By Taylor's theorem with remainder after three terms]

Subtracting (1) from (2),

$$NQ - NR = \frac{h^2}{2!}\,.\,f''(x) + \frac{h^3}{3!}\,f'''(x + \theta h), \text{ where } 0 < \theta < 1$$

Let $\qquad f''(x) = 0$ at P, but $f'''(x) \neq 0$, (given)

$$\therefore \qquad NQ - NR = \frac{h^3}{3!}\,f'''(x + \theta h) \qquad ...(1)$$

Since $f'''(x)$ is continuous function at P and $f'''(x) \neq 0$, $\therefore$ $f'''(x + \theta h)$ has the sign as that of $f''''(x)$ for sufficiently small values of h, whatever be sign of h.

But $\dfrac{h^3}{3!}$ changes sign with h.

∴ It follows from (1) that NQ – NR changes sign with h, so that the curve is concave on one side of P and convex on the other side. Hence P is *a point of inflexion, if $f''(x) = 0$ and $f'''(x) \neq 0$, at P.*

Note. In the above discussion, it is assumed that $\frac{dy}{dx} = f'(x)$ is finite. If $\frac{dy}{dx}$ becomes infinite, then $\frac{d^2y}{dx^2}$ does not exist and the points of inflexion are found by considering $\frac{d^2x}{dy^2} = 0$ and its change of sign.

Example 1. *Discuss the concavity and convexity of the curve*

$$y = (\sin x + \cos x)\, e^x \text{ where } 0 \le x \le 2\pi.$$

Find also the points of inflexion.

Sol. The equation of the curve is $y = (\sin x + \cos x)\, e^x$

$$\therefore \quad \frac{dy}{dx} = (\sin x + \cos x)\, e^x + e^x (\cos x - \sin x) = 2e^x \cos x$$

and $$\frac{d^2y}{dx^2} = 2e^x (-\sin x) + 2 \cos x \,.\, e^x = 2e^x (\cos x - \sin x) \qquad \ldots(1)$$

Now $$\frac{d^2y}{dy^2} = 0, \text{ when } \cos x - \sin x = 0 \text{ or } \tan x = 1$$

or $$x = \frac{\pi}{4}, \frac{5\pi}{4} \text{ in the interval } 0 \le x \le 2\pi.$$

From (1), $$\frac{d^2y}{dx^2} = 2 (\cos x - \sin x)\, e^x$$

$$= 2\sqrt{2}\left[\frac{1}{\sqrt{2}} \cos x - \frac{1}{\sqrt{2}} \sin x\right] e^x$$

$$= 2\sqrt{2}\left[\cos\frac{\pi}{4} . \cos x - \sin\frac{\pi}{4} . \sin x\right] . e^x = 2\sqrt{2} \,.\, e^x \cos\left(x + \frac{\pi}{4}\right)$$

In the interval $0 \le x < \frac{\pi}{4}$, $\frac{d^2y}{dx^2} > 0$ ...(2)

∴ The curve is concave upwards.

In the interval $\frac{\pi}{4} < x < \frac{5\pi}{4}$, $\frac{d^2y}{dx^2} < 0$ ...(3)

∴ The curve is concave downwards.

In the interval $\frac{5\pi}{4} < x \le 2\pi$, $\frac{d^2y}{dx^2} > 0$ ...(4)

∴ The curve is concave upwards.

Also by (2) and (3), $\frac{d^2y}{dx^2}$ changes sign from positive to negative at $x = \frac{\pi}{4}$, which ∴ gives a *point of inflexion.*

When $x = \frac{\pi}{4}$, $y = \left(\sin\frac{\pi}{4} + \cos\frac{\pi}{4}\right) e^{\pi/4} = \sqrt{2}e^{\pi/4}$

∴ Point $\left(\frac{\pi}{4}, \sqrt{2}e^{\pi/4}\right)$ is a *point of inflexion.*

Similarly, from (3) and (4), $\frac{d^2y}{dx^2}$ changes sign from negative to positive at $x = \frac{5\pi}{4}$ which $\therefore$ also gives a *point of inflexion.*

When $x = \frac{5\pi}{4}$, $y = \left(\sin\frac{5\pi}{4} + \cos\frac{5\pi}{4}\right) e^{5\pi/4} = -\sqrt{2}e^{5\pi/4}$

$\therefore$ Point $\left(\frac{5\pi}{4}, -\sqrt{2}e^{5\pi/4}\right)$ is another *point of inflexion.*

Example 2. *Find the points of inflexion on the curve*

$$y(a^2 + x^2) = x^3.$$

Sol. The equation of the curve is $y = \frac{x^3}{a^2 + x^2}$ ...(*i*)

$$\therefore \quad \frac{dy}{dx} = \frac{(a^2 + x^2).3x^2 - x^3.2x}{(a^2 + x^2)^2} = \frac{3a^2x^2 + x^4}{(a^2 + x^2)^2}$$

and

$$\frac{d^2y}{dx^2} = \frac{(a^2 + x^2)^2.(6a^2x + 4x^3) - (3a^2x^2 + x^4)2(a^2 + x^2)2x}{(a^2 + x^2)^4}$$

$$= \frac{x[(a^2 + x^2)(6a^2 + 4x^2) - 4(3a^2x^2 + x^4)]}{(a^2 + x^2)^3}$$

$$= \frac{x[6a^4 + 10a^2x^2 + 4x^4 - 12a^2x^2 - 4x^4]}{(a^2 + x^2)^3}$$

$$= \frac{x[6a^4 - 2a^2x^2]}{(a^2 + x^2)^3} = \frac{2a^2x(3a^2 - x^2)}{(a^2 + x^2)^3}$$

$$= \frac{2a^2x(\sqrt{3}a - x)(\sqrt{3}a + x)}{(a^2 + x^2)^3} \quad ...(1)$$

Now for points of inflexion, put $\frac{d^2y}{dx^2} = 0$.

$\therefore \quad 2a^2x(\sqrt{3}a - x)(\sqrt{3}a + x) = 0$

But $\quad 2a^2 \neq 0$

$\therefore$ Either $x = 0$ or $\sqrt{3}\,a - x = 0$ or $\sqrt{3}\,a + x = 0$

i.e., $x = 0$ or $x = \sqrt{3}\,a$ or $x = -\sqrt{3}\,a$

Test x = 0. Just before $x = 0$, from (1),

$$\frac{d^2y}{dx^2} = \frac{(-)(+)(+)}{(+)} = (-)$$

Just after $x = 0$; from (1), $\frac{d^2y}{dx^2} = \frac{(+)(+)(+)}{+} = (+)$

$\therefore$ $\frac{d^2y}{dx^2}$ changes sign while passing through $x = 0$

$\therefore$ $x = 0$ gives a point of inflexion.

Similarly $\frac{d^2y}{dx^2}$ changes sign while passing through each of

$$x = \sqrt{3}\,a \quad \text{and} \quad x = -\sqrt{3}\,a$$

$$\therefore \qquad x = \sqrt{3}\,a \quad \text{and} \quad x = -\sqrt{3}\,a$$

also give points of inflexion.

From (*i*), when $x = 0, y = 0$; when $x = \sqrt{3}\,a, y = \frac{3\sqrt{3}a}{4}$, and when $x = -\sqrt{3}\,a, y = -\frac{3\sqrt{3}\,a}{4}$.

Hence (0, 0) $\left(\sqrt{3}\,a, \frac{3\sqrt{3}\,a}{4}\right)$ and $\left(-\sqrt{3}\,a, \frac{-3\sqrt{3}\,a}{4}\right)$ are the points of inflexion.

Remark. To find points of inflexion, we should factorise $\frac{d^2y}{dx^2}$.

If $(x - \alpha)^n$ where n is **odd** ; is a factor of $\frac{d^2y}{dx^2}$; then $x = \alpha$ gives a point of inflexion.

If $(x - \alpha)^n$ where n is **even** ; is a factor of $\frac{d^2y}{dx^2}$; then $x = \alpha$ does not give a point of inflexion.

Example 3. *Find the points of inflexion on the curve*

$$x = a\,(2\theta - \sin\theta),\; y = a\,(2 - \cos\theta). \qquad \text{(M.D.U. 2004 S)}$$

Sol. Here $\frac{dx}{d\theta} = a(2 - \cos\theta)$ and $\frac{dy}{d\theta} = a \sin\theta$

$$\therefore \quad \frac{dy}{dx} = \frac{dy}{d\theta}\Big/\frac{dx}{d\theta} = \frac{\sin\theta}{(2 - \cos\theta)}$$

$$\therefore \quad \frac{d^2y}{dx^2} = \frac{d}{d\theta}\left(\frac{\sin\theta}{2 - \cos\theta}\right)\cdot\frac{d\theta}{dx}$$

$$= \frac{(2 - \cos\theta)\,.\cos\theta - \sin\theta\,.\sin\theta}{(2 - \cos\theta)^2}\cdot\frac{1}{a(2 - \cos\theta)}$$

$$= \frac{2\cos\theta - (\cos^2\theta + \sin^2\theta)}{a(2 - \cos\theta)^3} = \frac{2\cos\theta - 1}{a(2 - \cos\theta)^3}$$

For points of inflexion $\frac{d^2y}{dx^2} = 0$ or $\frac{2\cos\theta - 1}{a(2 - \cos\theta)^3} = 0$

which gives $\cos\theta = \frac{1}{2} = \cos\frac{\pi}{3} \quad \therefore \quad \theta = 2n\pi \pm \frac{\pi}{3}$,

where n is 0 or any integer.

When θ passes through each of these values, $\frac{d^2y}{dx^2}$ changes sign, hence there are points of inflexion corresponding to every value of θ given above.

The coordinates of the points of inflexion are

$$\left[a\left(4n\pi \pm \frac{2\pi}{3} \mp \frac{\sqrt{3}}{2}\right), \frac{3a}{2}\right],$$

where n is any integer.

Example 4. *Find the points of inflexion, if any, of the curve* $x = (\log y)^3$. (K.U. 2003)

Sol. Equation of the curve is $x = (\log y)^3$.

Here y is the independent variable and x is the dependent variable.

(Also see note Art. 13).

$$\therefore \quad \frac{dx}{dy} = 3(\log y)^2 \cdot \frac{1}{y} = \frac{3(\log y)^2}{y}$$

$$\frac{d^2x}{dy^2} = \frac{3\left[y \cdot 2(\log y) \cdot \frac{1}{y} - (\log y)^2 \cdot 1\right]}{y^2} = 3y^{-2}\,[2\log y - (\log y)^2]$$

and
$$\frac{d^3x}{dy^3} = 3y^{-2}\left[2 \cdot \frac{1}{y} - 2(\log y) \cdot \frac{1}{y}\right] + [2\log y - (\log y)^2] \cdot (-6y^{-3})$$

$$= 6y^{-3}(1 - \log y) - 6y^{-3} \cdot \log y\,(2 - \log y)$$

$$= 6y^{-3}[(\log y)^2 - 3\log y + 1]$$

For points of inflexion $\frac{d^2x}{dy^2} = 0$ or $\frac{3}{y^2}[2\log y - (\log y)^2] = 0$

$\therefore \quad 2\log y - (\log y)^2 = 0$ or $\log y\,[2 - \log y] = 0$

$\therefore \quad \log y = 0$, which gives $y = e^0 = 1$

or $\quad \log y = 2$, which gives $y = e^2$

When $y = 1$, $\frac{d^3x}{dy^3} = \frac{6}{1^3}[(\log 1)^2 - 3\log 1 + 1] = 6 \neq 0$, and also

When $y = e^2$, $\frac{d^3x}{dy^3} = \frac{6}{e^6}[(\log e^2)^2 - 3\log e^2 + 1] = \frac{-6}{e^6} \neq 0$

$\therefore$ Points of inflexion correspond to $y = 1$ and $y = e^2$.

When $y = 1$, $x = (\log 1)^3 = 0$ and when $y = e^2$, $x = (\log e^2)^3 = (2 \times 1)^3 = 0$

$\therefore$ $(0, 1)$ and $(8, e^2)$ are the points of inflexion.

EXERCISE 3

1. Prove that the curve $y = \log x$ is everywhere concave downwards.
2. Find the range of the value of x for which the curve $y = x^4 - 6x^2 + 12x^2 + 5x + 7$ is concave upwards or downwards. Determine also its points of inflexion.
3. Find the range of values of x for which the curve $y = 3x^5 - 40x^3 + 3x - 20$, is concave upwards or downwards. Also find its points of inflexion. *(K.U. 1998 A)*
4. Find the points of inflexion on the following curves :

   (*i*) $y = 3x^4 - 4x^3 + 1$ *(M.D.U. 1999)* (*ii*) $y = 3x^5 - 40x^3 + 3x - 20$

   (*iii*) $54y = (x + 5)^2 (x^3 - 10)$ (*iv*) $y = (x - 2)^6 (x - 3)^5$ *(M.D.U. 2000, 2005, 2007)*

   (*v*) $y = \frac{1}{h\sqrt{\pi}} e^{-x^2/h^2}$ (*vi*) $y = (\log x)^3$

   (*vii*) $y = \frac{x^2}{a^2 + x^2}$ (*viii*) $y = \frac{x}{x^2 + 2x + 2}$.

5. Find the points of inflexion of the following curves :

(i) $x^2y = a^2(x - y)$ (K.U. 1996 A)

(ii) $x = (y - 1)(y - 2)(y - 3)$

(iii) $xy = a^2 \log \frac{y}{a}$.

6. Find the points of inflexion on the curve $x = a \tan t, y = \frac{a}{2} \sin 2t$.

7. Show that the abscissa of the point inflexion on the curve $x = a - b \cos \theta, y = a\theta - b \sin \theta$ is $(a^2 - b^2)/a$. (M.D.U. 2004 A)

8. Show that every point in which the sine curve $y = c \sin \frac{x}{a}$, meets the axis of x is a point of inflexion.

9. Show that the points inflexion of the curve $y^2 = (x - a)^2 (x - b)$, lie on the line $3x + a = 4b$. (M.D.U. 1998 ; K.U. 2004 S)

10. Show that the origin is a point of inflexion on the curve $a^{m-1} y = x^m$, if m is odd and greater than 2.

11. Show that the abscissae of the points of inflexion on the curve $y^2 = f(x)$, satisfy the equation.

$$[f'(x)]^2 = 2 f(x) f''(x).$$ (M.D.U. 2005 A)

12. For the curve $y = x^3 + bx^2 + c$, where $b < 0$, show that the point of inflexion is equidistant from the maximum and minimum points.

13. Show that the line joining the points of inflexion of the curve $y^2 (x - a) = x^2 (x + a)$ subtends an angle of $\pi/3$ at the origin.

ANSWERS

2. The curve is concave upwards in $(-\infty, 1)$ and $(2, \infty)$ and concave downwards in $(1, 2)$. The points of inflexion are $(1, 19)$ and $(2, 33)$.

3. The curve is concave downwards in $(-\infty, -2)$ and $(0, 2)$ and upwards in $(-2, 0)$ and $(2, \infty)$. The points of inflexion are $(0, -20)$; $(-2, 198)$ and $(2, -238)$.

4. (i) $(0, 1)$ and $(\frac{2}{3}, \frac{11}{27})$

(ii) $(-2, 198)$ $(0, -20)$, $(2, -238)$ give points of inflexion.

(iii) $x = -2, \frac{1}{2}(-4 \pm \sqrt{18})$ (iv) $x = 3, \frac{1}{11}(28 \pm \sqrt{3})$ (v) $x = \pm \frac{h}{\sqrt{3}}$.

(vi) $(1, 0), (e^2, 8)$ (vii) $x = 0, \pm a\sqrt{3}$ (viii) $x = -2, 1 \pm \sqrt{3}$.

5. (i) $(0, 0)\left(\sqrt{3}\,a, \frac{\sqrt{3}\,a}{4}\right), \left(-\sqrt{3}\,a, \frac{-\sqrt{3}\,a}{4}\right)$ (ii) $(0, 2)$

(iii) $(\frac{3}{2} a\, e^{-3/2}, a\, e^{3/2})$

6. $(0, 0), \left(\pm\sqrt{3a}, \pm\frac{\sqrt{3}a}{4}\right)$.

9

Curve Tracing

1. Introduction

Let us consider the problem of tracing curves *i.e.*, of finding approximate shape of curves from their cartesian, polar or parametric equations without having to plot a large number of, points on them. We shall consider mainly those curves whose equations can be solved for y x or r. A working knowledge of the topics on Maxima, Minima ; concavity, points of Inflexion and asymptotes is essential for learning this chapter. In the following article, we shall explain the various steps which are helpful in tracing a curve. But these steps and (particularly their order) is by no means rigid and some steps can be omitted and their order can be varied to suit the particular problem

2. Procedure for Tracing Curves having Cartesian Equations

I. Symmetry

(*i*) The curve is symmetrical about the **x-axis,** if the equation of the curve remains unchanged when **y is changed into – y** *i.e.,* if the equation of the curve contains only **even powers of y.**

e.g., $y^2 = x^2 + 4$ is symmetrical about x-axis where as $y^2 + yx = x^2$ is not symmetrical about x-axis.

(*ii*) The curve is symmetrical about the **y-axis** if the equation of the curve does not change when **x is changed into – x** *i.e.,* if the curve contains **only even powers of x.**

e.g., $x^2 = 4ay$ is symmetrical about y-axis where as $y = x^3 - 3ax^2$ is not symmetrical about y-axis.

(*iii*) The curve is symmetrical about the **Line y = x** if the equation of the curve remains unchanged when **x is changed to y** and **y is changed to x.** ($y = x$ is a st. line through origin and making an angle of 45° with x-axis).

e.g., $x^3 + y^3 = 3axy$ is symmetrical about the line $y = x$ where as $x^5 + y^5 = 5ax^2y$ is not symmetrical about the line $y = x$.

(*iv*) The curve is symmetrical about the **Line y = – x** if the equation of the curve remains unchanged when **x is changed to – y** and **y is changed to – x.** ($y = -x$ is a st. line $\perp$ to $y = x$).

e.g., $x^4 + y^4 = 4a^2xy$ is symmetrical about the line $y = -x$ where as $x^3 + y^3 = 3axy$ is not symmetrical about this line.

(*v*) The curve is symmetrical in **opposite quadrants** if the equation of the curve remains unchanged when **x is changed to – x** and **y is changed to – y.**

e.g., $x^5 + y^5 = 5ax^2y$ is symmetrical in opposite quadrants whereas $x^3 + y^3 = 3axy$ is not symmetrical in opposite quadrants.

II. Origin

(*a*) Find whether the curve passes through the origin. If the constant term is missing from the equation of the algebraic curve, then it passes through the origin.

If the algebraic curve passes through the origin, then write down the equation of the tangents at the origin **by equating the lowest degree terms to 0.**

If the origin is a *double point* (*i.e.*, there are two tangents at the origin) ; then find its nature whether a *node* (if the tangents are *real, distinct*) or a *cusp* (if the tangents are *real, coincident*) or a *conjugate* (or *isolated*) point if the tangents are *imaginary.* If a *cusp,* find its type.

A cusp is called a **single cusp** or a **double cusp** according as the two branches of the curve lie entirely *on one side* or *on both sides of the common normal.*

A cusp *single* or *double* is said to be of *first kind* or *second kind* according as the two branches of the curve , *lie on opposite or same side of the common tangent.*

(*b*) **Position of curve w.r.t. tangent at the origin**

Find $y_c - y_t$ (or $y_c^2 - y_t^2$ if square roots are there) where y_c and y_t are respectively the ordinates of a point on the curve and a point on the tangent at the origin for the same value of x.

Then discuss the two cases namely $x > 0$ and $x < 0$. (It should be noted that $x > 0$ in first and fourth quadrants and $x < 0$ in second and third quadrants).

If $(y_c - y_t)$ is + ve *i.e.*, $y_c > y_t$; then the curve lies above the tangent and if $(y_c - y_t)$ is – ve *i.e.*, $y_c < y_t$; then the curve lies below the tangent.

III. Asymptotes

(*a*) *Working Rule for finding asymptotes of an algebraic curve.*

(*i*) **Asymptotes || to x-axis.** Equate to zero the co-efficient of the highest power of x, present in the given equation of the curve. Resolve L.H.S. into real linear factors.

(*ii*) **Asymptotes || to y -axis.** (Replace x by y in (*i*)).

(*iii*) **Oblique asymptotes**

Putting $x = 1, y = m$ in the highest degree terms of the given equation, find $\phi_n(m)$.

Putting $x = 1, y = m$ in the next lower degree terms of the given equation, find $\phi_{n-1}(m)$. Similarly, find $\phi_{n-2}(m)$.

Now put $\phi_n(m)$ equal to zero and solve for m. These are the slopes of the asymptotes.

For *distinct* real values of m, find c from the equation

$$c = -\frac{\phi_{n-1}(m)}{\phi_n'(m)}.$$

For two *equal* real values of m, find c from the equation

$$\frac{c^2}{2!}\phi_n''(m) + c\phi'_{n-1}(m) + \phi_{n-2}(m) = 0.$$

Putting the values of m and corresponding values of c in $y = mx + c$, we get the required asymptotes.

(*b*) **Position of the curve w.r.t. asymptotes.**

(*i*) **Position of the curve w.r.t. asymptotes || to x-axis.**

Let the asymptote parallel to x-axis be $y = a$.

Find the value of $y - a$ from the equation of the curve.

(Generally, $(y - a)$ will occur as a factor in the equation of the curve.)

Now in the R.H.S. of this value of $y - a$, put $y = a$.

Then discuss the two cases when x is > 0 (Near ∞) and $x < 0$ (Near $-\infty$).

If $(y - a)$ is + ve, the curve lies above the asymptote $y = a$
and if $(y - a)$ is – ve, the curve lies below the asymptote $y = a$.

(*ii*) **Position of the curve w.r.t. asymptotes parallel to y-axis.**

Let the asymptote parallel to y-axis be $x = a$.

Find the value of $x - a$ from the equation of the curve.

(Generally $(x - a)$ will occur as a factor in the equation of the curve).

Now in the R.H.S. of this value of $x - a$, put $x = a$.

Then discuss the two cases when $y > 0$ and $y < 0$.

(It may be noted that $y > 0$ in First and Second Quadrants and $y < 0$ in Third and Fourth Quadrants.)

If $(x - a)$ is + ve, the curve lies to the right of the asymptote $x = a$ and if $(x - a)$ is – ve, the curve lies to the left of the asymptote $x = a$.

(*iii*) **Position of the curve w.r.t. oblique asymptotes.**

Write the equation of the curve or one of its branches is in the form $y = mx + c + \frac{A}{x} + \frac{B}{x^2} + \ldots\ldots.$, then

$y = mx + c$ is the asymptote to the curve.

$\therefore \quad y_c - y_a = \frac{A}{x} + \frac{B}{x^2} + \ldots\ldots$ where y_c and y_a stand for the ordinates of a point on the curve and a point on the asymptote (both having same abscissa).

Now discuss the two cases namely $x > 0$ and $x < 0$.

If $(y_c - y_a)$ is positive, curve lies above the asymptote.

If $(y_c - y_a)$ is negative, curve lies below the asymptote.

IV. Points of Intersections. (These can be obtained by solving the two equations).

Find intersections of the curve (*i*) with the x-axis $(y = 0)$; (*ii*) with the y-axis $(x = 0)$ (*iii*) with the line $y = x$ particularly if the curve is symmetrical about it.

(*iv*) With the line $y = -x$ particularly if the curve is symmetrical about it.

(*v*) With the asymptotes (if necessary).

(An asymptote of an nth degree curve cuts the curve in $(n - 2)$ points).

Also write down the equations of the tangents at these points of intersection by shifting the origin to the points.

V. Region. Find regions in the four quadrants to which the curve is limited. This is usually done by solving for y or for x (or for each of the two) separately and considering both positive and negative values of x (or y). Values of x (or y) which make y (or x) imaginary are to be rejected.

Again, such values* which make the left and right members of an equation opposite in sign are to be rejected.

(For example, see the solved examples).

---

* In Quadrant I, x is + ve and y is + ve.
In Quadrant II, x is – ve and y is + ve.
In Quadrant III, x is – ve and y is – ve.
In Quadrant IV, x is + ve and y is – ve.

Note. If $x^2 \le a^2$; then $-a \le x \le a$.
If $x^2 \ge a^2$; then either $x \ge a$ or $x \le -a$.
If $(x-\alpha)(x-\beta) \le 0$; then x lies between α and β.
If $(x-\alpha)(x-\beta) \ge 0$; then x does not lies between α and β.

VI. Special points (or singularities)

(*a*) **Find** $\frac{dy}{dx}$ $\left(\frac{dy}{dx}\right.$ represents the slope of the tangent to the curve at the point $(x, y)\Big)$

Find the points where

(*i*) $\frac{dy}{dx} = 0$ *i.e.,* tangents are || to x-axis

(*ii*) $\frac{dy}{dx} = \infty$ *i.e.,* tangent are || to y-axis

(*iii*) $\frac{dy}{dx}$ is positive *i.e.,* the function is increasing *i.e.,* the curve is rising.

(*iv*) $\frac{dy}{dx}$ is negative *i.e.,* the curve is falling.

Find $\frac{d^2y}{dx^2}$ (If not tedious)

Find the points where

(*i*) $\frac{d^2y}{dx^2}$ is + ve *i.e.,* the curve is concave upwards.

(*ii*) $\frac{d^2y}{dx^2}$ is – ve *i.e.,* the curve is concave downwards.

(*iii*) Also find **points of Inflexion.**

(At a point of Inflexion, curve crosses the tangent. At a point of inflexion $\frac{d^2y}{dx^2} = 0$ and $\frac{d^3y}{dx^3} \ne 0$ or $\frac{d^2y}{dx^2}$ changes sign while passing through it).

Remark. The steps of Art 2. can be remembered as :

R—SOAP

where S ≡ Symmetry, O ≡ origin, A ≡ Asymptotes ; P ≡ Points

(*i*) of intersection with $y = 0, x = 0, y = x, y = -x$

(*ii*) points where $\frac{dy}{dx} = 0$ or ∞ or > 0 or < 0,

and R ≡ Region.

Note. Sometimes inconvenient steps may be omitted without any disadvantage.

Example 1. *Trace the curve* $9ay^2 = (x-2a)(x-5a)^2$.

Sol. The equation of the curve is $9ay^2 = (x-2a)(x-5a)^2$ ...(1)

1. **Symmetry.** Since (1) contains only even powers of y.

$\therefore$ The curve is symmetrical about x-axis.

2. **Origin.** The curve does not pass through the origin because (0, 0) does not satisfy the given equation.

3. **Asymptotes.** The curve has no asymptotes.

4. **Points of Intersection**

(*i*) **Intersection with x-axis**

Putting $y = 0$ in (1), we get $(x - 2a)(x - 5a)^2 = 0$ or $x = 2a, 5a$.

Thus, the curve meets x-axis in the points A(2a, 0) and B(5a, 2).

Shifting the origin to (2a, 0) by putting $x = X + 2a, y = Y + 0$; the equation (1) transforms to $9aY^2 = X(X - 3a)^2$.

Equating to zero the lowest degree terms, the tangent at the new origin is $X = 0$ *i.e.*, new y-axis.

Thus, the tangent at A(2a, 0) is parallel to old y-axis.

Shifting the origin to B(5a, 0) by putting $x = X + 5a$ and $y = Y + 0$, the equation (1) transforms to $9aY^2 = (X + 3a)X^2$

$\therefore$ Tangents at new origin are $3aY^2 = 2aX^2$ *i.e.*, $Y^2 = \dfrac{X^2}{3}$ $\quad\therefore\quad Y = \pm \dfrac{X}{\sqrt{3}}$

$\therefore$ New origin is a Node.

(*ii*) **Intersections with y-axis**

Putting $x = 0$ in (1), we get $9ay^2 = -50a^3$, which gives imaginary values of y. Therefore, the curve does not meet y-axis.

5. **Region**

From (1), $$y = \frac{1}{\sqrt{9a}}(x - 5a)\sqrt{x - 2a} \qquad ...(2)$$

(Taking only positive sign with square root)

(This value of y, corresponds to the part of the curve in first two quadrants.)

When $x < 2a$, y is imaginary.

Thus, the curve does not lie to the left of the line $x = 2a$.

When $x > 2a$, y is real.

6. **Special points**

From (2), $$\frac{dy}{dx} = \frac{1}{\sqrt{9a}}\left[(x - 5a)\frac{1}{2\sqrt{x - 2a}} + \sqrt{x - 2a}\right]$$

or $$\frac{dy}{dx} = \frac{1}{\sqrt{9a}}\left[\frac{x - 5a + 2(x - 2a)}{2\sqrt{x - 2a}}\right] = \frac{3(x - 3a)}{2\sqrt{a}\,.\sqrt{x - 2a}}.$$

or $$\frac{dy}{dx} = \frac{x - 3a}{2\sqrt{a}\sqrt{x - 2a}} \qquad ...(3)$$

$$\frac{dy}{dx} = 0 \text{ gives } x = 3a.$$

But when $x = 3a$, then from (1) $9ay^2 = 4a^3$.

$\therefore$ $$y = \pm\frac{2a}{3}$$

$\therefore$ Tangents at the points $\left(3a, \pm\dfrac{2a}{3}\right)$ are parallel to x-axis.

When $x > 3a$, $\dfrac{dy}{dx}$ is positive | From (3)

$\therefore$ The curve rises for all values of $x > 3a$ (provided y is + ve) when $x < 3a$, $\frac{dy}{dx}$ is negative.

$\therefore$ The curve is falling for values of $x < 3a$.

Again $\frac{dy}{dx} \to \infty$ when $x = 2a$.

$\therefore$ The tangent at $(2a, 0)$ is $\parallel$ to y-axis. Thus, the shape of the curve is as shown in the figure.

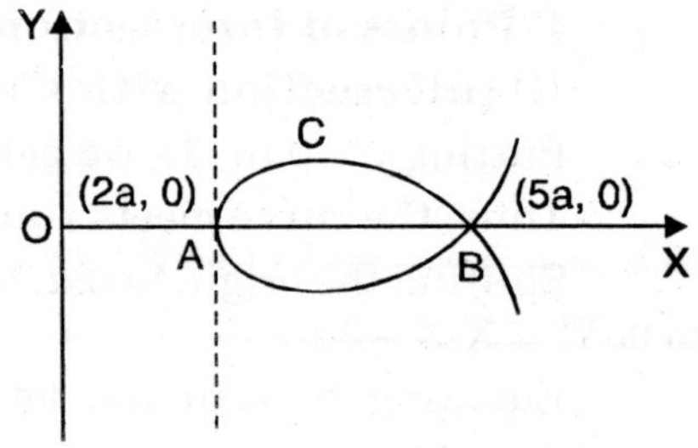

Note. Two important results from real analysis :

1. $\mathbf{x^2 < a^2 \Rightarrow -a < x < a.}$ **2.** $\mathbf{x^2 > a^2 \Rightarrow x > a \text{ or } x < -a.}$

Example 2. *Trace the curve* $x^2y^2 = x^2 - a^2$. (K.U. 1997 A)

Sol. The equation of the curve is $x^2y^2 = x^2 - a^2$ ...(1)

1. **Symmetry.** Since (1) contains only even powers of y and only even powers of x, therefore, the curve is symmetrical about both the axes. Also the curve is symmetrical in opposite Quadrants. ($\because$ On changing x to $-x$ and y to $-y$ in (1), equation (1) remains unchanged.)

2. **Origin.** The curve does not pass through the origin because (0, 0) does not satisfy the given equation.

3. **Asymptotes.**

Asymptotes $\parallel$ *to x-axis* are given by equating the co-efficient of highest power of x (here x^2) to zero *i.e.*, $y^2 - 1 = 0$ $\therefore$ $y = 1$ and $y = -1$ are the asymptotes.

Asymptotes $\parallel$ *to y-axis* is $x = 0$ (equating co-efficient of highest power of y to 0). But it will be seen below that $x = 0$ can't be an asymptote (see Equation (3) step 5 below).

Oblique Asymptotes. $\phi_4(m) = m^2$.

$\phi_4(m) = 0$ gives $m = 0$ and this value of m gives asymptotes parallel to x-axis (which we have already obtained).

4. **Points of intersection**

With x-axis. Putting $y = 0$ in (1), $x^2 = a^2$ $\therefore$ $x = \pm a$.

$\therefore$ The curve meets the x-axis in the points $(a, 0)$ and $(-a, 0)$.

Also $x = a$ and $x = -a$ are the tangents at the points $(a, 0)$ and $(-a, 0)$, respectively. [See Step (6)].

With y-axis. Putting $x = 0$ in (1), we get $0 = -a^2$ which is impossible.

$\therefore$ The curve does not meet the y-axis.

5. **Region**

From (1),
$$y = \frac{\sqrt{x^2 - a^2}}{x} \qquad \ldots(2)$$

(This value of y corresponds to the part of the curve in first two quadrants.)

For y to be real $x^2 - a^2$ must be positive *i.e.*, $x^2 > a^2$.

$$\therefore \qquad x > a \quad \text{or} \quad x < -a.$$

$\therefore$ No portion of the curve lies between the lines $x = a$ and $x = -a$ ...(3)

Again from (1), $x^2(1 - y^2) = a^2$

$$\therefore \qquad x = \frac{a}{\sqrt{1 - y^2}}.$$

For x to be real $1 - y^2 > 0$ *i.e.*, $y^2 < 1$. $\therefore$ $-1 < y < 1$.

$\therefore$ The curve lies entirely between the lines

$$y = -1 \quad \text{and} \quad y = 1.$$

6. **Special points.**

From (2), $\dfrac{dy}{dx} = \dfrac{a^2}{x^2\sqrt{x^2 - a^2}}$

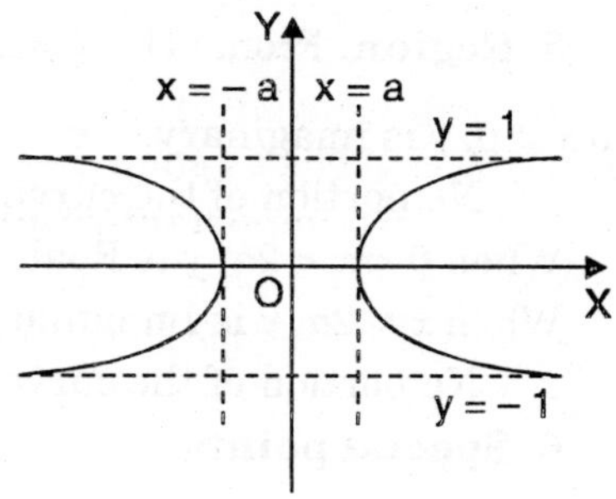

$\dfrac{dy}{dx}$ never becomes zero.

$\dfrac{dy}{dx}$ is always positive.

$\dfrac{dy}{dx} \to \infty$ when $x^2 - a^2 = 0$

i.e., $x = \pm a$

For $x = \pm a$, $y = 0$ from (1)

$\therefore$ Tangents at the points $(a, 0)$ and $(-a, 0)$ are parallel to y-axis.

Thus, the shape of the curve is as shown in the figure.

Note. We have not discussed the shape of the curve w.r.t. asymptotes parallel to x-axis namely $y = 1$ and $y = -1$ because in **Region** we could know that the curve lies between the lines $y = -1$ and $y = 1$ *i.e.*, curve does not lie above the line $y = 1$ and below the line $y = -1$.

But for the sake of discussion and clarity let us discuss the shape of the curve w.r.t. asymptotes $y = 1$ and $y = -1$.

Position of curve w.r.t. asymptote $y = 1$.

From (1), $\quad x^2(y^2 - 1) = -a^2 \quad$ or $\quad x^2(y - 1)(y + 1) = -a^2 \quad$ ...(4)

$$\therefore \quad y - 1 = \frac{-a^2}{x^2(y+1)} \quad \text{or} \quad y - 1 = \frac{-a^2}{2x^2} \quad \text{(Putting } y = 1 \text{ in R.H.S.)}$$

when x is positive or negative, R.H.S. is – ve.

$\therefore$ Curve lies below the asymptote $y = 1$.

Position of curve w.r.t. asymptote $y = -1$

Again from (4), $\quad y + 1 = \dfrac{-a^2}{x^2(y-1)}$

or $\quad y + 1 = \dfrac{-a^2}{-2x^2} \quad$ (Putting $y = -1$ in R.H.S.)

or $\quad y - (-1) = \dfrac{a^2}{2x^2}$

when x is positive or negative, R.H.S. is positive.

$\therefore$ Curve lies above the asymptote $y = -1$.

Example 3. *Trace the curve* $y^2(2a - x) = x^3$. (M.D.U. 1999 S ; K.U. 1998 S)

Sol. The equation of the curve is $y^2(2a - x) = x^3$ ...(1)

1. **Symmetry.** Since (1) contains only even powers of y, the curve is symmetrical about x-axis.

2. **Origin.** The tangents at the origin are given by $y^2 = 0$ *i.e.*, $y = 0, y = 0$. Since the two tangents are real and coincident, $\therefore$ origin is a cusp.

3. **Asymptotes.** Equating to zero, the co-eff. of y^2, the highest degree term in y, the asymptote parallel to y-axis is $x - 2a = 0$ *i.e.*, $x = 2a$. There is no other asymptote of the curve.

We can discuss the position of the curve w.r.t. asymptote $x = 2a$.

4. **Points of intersections.** The curve meets x-axis and y-axis at the origin only.

5. **Region.** From (1), $y = x\sqrt{\dfrac{x}{2a - x}}$

when $x < 0$, y is imaginary.

$\therefore$ No portion of the curve lies to the left of the line $x = 0$ *i.e.*, y-axis.

When $0 < x < 2a$, y is Real.

When $x > 2a$, y is Imaginary.

$\therefore$ No portion of the curve lies to the right of the line $x = 2a$.

6. **Special points**

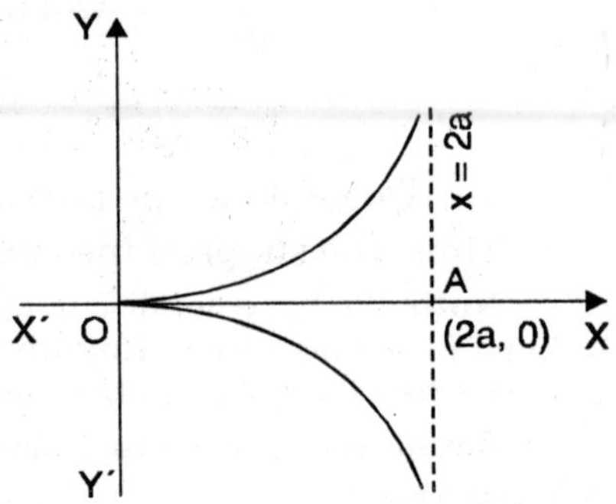

From (1), $$y = \frac{x^{3/2}}{\sqrt{2a - x}} \quad ...(2)$$

$$\therefore \quad \frac{dy}{dx} = \frac{\sqrt{x}\,(3a - x)}{(2a - x)^{3/2}}$$

$$\therefore \quad \frac{dy}{dx} = 0$$

when $$\sqrt{x}\,(3a - x) = 0$$

or $$x = 0, \quad x = 3a$$

Rejecting $x = 3a$, because when $x = 3a$, from (2) y is imaginary.

When $x = 0$, $y = 0$ $\therefore$ Tangent at (0, 0) *i.e.*, at origin is $\parallel$ to x-axis *i.e.*, the tangent at origin is x-axis (Also see step 2).

Again, $\dfrac{dy}{dx} \to \infty$ when $x \to 2a$. From (2), when $x \to 2a$, $y \to \infty$

Thus, $x = 2a$ is an asymptote (Also see step 3)

when $0 < x < 2a$, $\dfrac{dy}{dx}$ is positive.

$\therefore$ For positive values of y, y is an increasing function of x *i.e.*, the curve rises for values of x between 0 and $2a$.

Thus, the shape of curve is as shown in the figure.

Example 4. *Trace the curve* $y^2(x - a) = x^2\,(x + a)$.

Sol. The equation of the curve is $y^2(x - a) = x^2\,(x + a)$ ...(1)

1. **Symmetry.** The curve is symmetrical about the x-axis only.

2. **Origin.** The curve passes through the origin and the tangents at the origin are (equating to zero the lowest degree terms) given by $x^2 + y^2 = 0$ or $y = \pm ix$ which are imaginary.

$\therefore$ Origin is an isolated point.

3. **Asymptotes.** There is no asymptote parallel to x-axis ($\because$ co-eff. of x^3, highest power of x is a constant 1).

Equating to zero, the co-efficient of y^2, the asymptote parallel to y-axis is $x - a = 0$.

Let us find oblique asymptotes.

From (1),
$$y^2 = \frac{x^2(x+a)}{x-a} = \frac{x^2 . x\left(1+\frac{a}{x}\right)}{x\left(1-\frac{a}{x}\right)}$$

or
$$y^2 = x^2\left(1+\frac{a}{x}\right)\left(1-\frac{a}{x}\right)^{-1}$$

$\therefore$
$$y = \pm x\left(1+\frac{a}{x}\right)^{1/2}\left(1-\frac{a}{x}\right)^{-1/2}$$

or
$$y = \pm x\left[1+\frac{1}{2}\frac{a}{x}-\frac{1}{8}.\frac{a^2}{x^2}+......\right]\left[1+\frac{a}{2x}+\frac{3}{8}\frac{a^2}{x^2}+...\right]$$

or
$$y = \pm x\left[1+\frac{a}{x}+\frac{a^2}{2x^2}+...\right] = \pm\left[x+a+\frac{a^2}{2x^2}+...\right]$$

Taking + ve sign, equation of one branch is
$$y = x + a + \frac{a^2}{2x} + ...$$

$\therefore$ Equation of the asymptote is $y = x + a$

$\therefore$
$$y_c - y_a = \frac{a^2}{2x} + ...$$

which is > 0 for large $x > 0$ and < 0 for $x < 0$.

Thus, the curve lies above the asymptote $y = x + a$ for $x > 0$ and below the asymptote for $x < 0$.

Taking negative sign, equation of another branch is
$$y = -x - a - \frac{a^2}{2x} - ......$$

$\therefore$ Equation of the asymptote is $y = -x - a$.

$\therefore$
$$y_c - y_a = \frac{-a^2}{2x} ......$$

which is < 0 for large $x > 0$ and > 0 for $x < 0$.

$\therefore$ The curve lies below the asymptote $y = -x - a$ for $x > 0$ and above the asymptote for $x < 0$.

We can also discuss the position of the curve w.r.t. the asymptote $x = a$. (We have omitted this step here because step (5) will do the same job).

4. Points of intersection

The curve cuts the x-axis where putting $y = 0$, we get
$$x^2(x + a) = 0 \quad i.e., \quad x = 0 \text{ or } -a.$$

Thus, the curve passes through (0, 0) and $(-a, 0)$.

The curve meets y-axis, where putting $x = 0$, we get $-ay^2 = 0$

$\therefore$
$$y = 0 \quad i.e., \text{ at } (0, 0)$$

Shifting the origin to $(-a, 0)$, the equation (1) transforms to
$$Y^2(X - 2a) = (X - a)^2 X.$$

The tangent at the new origin is X = 0 *i.e.*, the new y-axis.

Thus, the tangent at $(-a, 0)$ is parallel to old y-axis.

5. **Region.** From (1), $y = x\sqrt{\dfrac{x+a}{x-a}}$

When $x < -a$, y is Real.

When $-a < x < a$, y is imaginary. Hence no portion of the curve lies between the lines $x = \pm a$.

When $x > a$, y is Real.

($-a$ and a are the numbers obtained on equating each of the factors in the square root to zero).

6. **Special points**

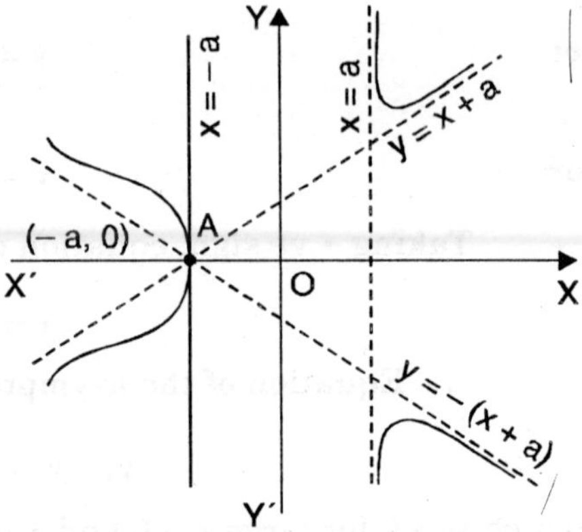

From (1), $$y = x\sqrt{\frac{x+a}{x-a}} \quad ...(2)$$

$$\therefore \quad \frac{dy}{dx} = \frac{x^2 - ax - a^2}{(x+a)^{1/2}\,(x-a)^{3/2}}$$

Now $\dfrac{dy}{dx} = 0$ when

$x^3 - ax - a^2 = 0$ or when

$$x = \tfrac{1}{2}(1 \pm \sqrt{5})\, a.$$

When $x = \left(\dfrac{1-\sqrt{5}}{2}\right) a$, from (2) y is imaginary.

So the tangent is || to x-axis when $x = \dfrac{1+\sqrt{5}}{2}\, a$.

Thus, the shape of the curve is as shown in the figure.

Example 5. *Trace the curve* $y = x^3 - 3ax^2$.

Sol. The curve is $y = x^3 - 3ax^2$ (1) or $y = x^2(x - 3a)$

1. **Symmetry.** No symmetry.

2. **Origin.** The curve passes through the origin.

The equation of tangent at the origin is $y = 0$ *i.e.*, x-axis.

3. **Asymptotes.** No asymptotes.

4. **Points of intersection**

Intersections with x-axis. Putting $y = 0$ (1), we get

$$x^3 - 3ax^2 = 0 \quad \text{or} \quad x^2(x - 3a) = 0$$

$$\therefore \quad x = 0 \quad \text{or} \quad x = 3a$$

$\therefore$ Intersections with x-axis are (0, 0) and ($3a$, 0)

Shifting the origin to ($3a$, 0), (1) becomes

$$Y = (X + 3a)^2 X$$

$\therefore$ Equation of the tangent at the new origin is $Y = 9a^2X$.

5. **Region**

From (1), $$y = x^2(x - 3a)$$

when $x < 0$, y is – ve.

$\therefore$ No portion of the curve lies in the second quadrant.

When $0 < x < 3a$ y is – ve.

$\therefore$ The curve lies in fourth quadrant for values of x s.t. $0 < x < 3a$.

When $x > 3a$, y is + ve.

$\therefore$ The curve lies in the first quadrant for values of $x > 3a$.

6. **Special points**

From (1), $$\frac{dy}{dx} = 3x^2 - 6ax = 3x(x - 2a) \qquad ...(2)$$

$$\frac{dy}{dx} = 0 \text{ gives } x = 0 \text{ and } x = 2a.$$

When $x = 0$, then from (1) $y = 0$

$\therefore$ Tangent at (0, 0) is x-axis.

when $x = 2a$, from (1) $y = 8a^3 - 12a^3 = -4a^3$

$\therefore$ Tangent at $(2a, -4a^3)$ is parallel to x-axis.

When $x < 0$, then from (2), $\frac{dy}{dx}$ is + ve.

$\therefore$ y increases as x increases *i.e.*, the curve is rising.

When $0 < x < 2a$, $\frac{dy}{dx}$ is – ve.

$\therefore$ The curve falls in this portion.

When $x > 2a$, $\frac{dy}{dx}$ is + ve.

$\therefore$ Again y increases as x increases. ($\forall\, x > 2a$)

From (2), $$\frac{d^2y}{dx^2} = 6x - 6a = 6(x - a)$$

$$\frac{d^2y}{dx^2} = 0 \text{ gives } x = a$$

$$\frac{d^3y}{dx^3} = 6.$$

At $x = a$, $$\frac{d^3y}{dx^3} = 6 \neq 0$$

$\therefore$ $x = a$ gives a point of inflexion.

When $x = a$, then from (1)

$$y = x^3 - 3ax^2 = a^3 - 3a^3 = -2a^3$$

$\therefore$ $(a, -2a^3)$ gives a point of inflexion.

When $x > a$, $\frac{d^2y}{dx^2}$ is positive.

$\therefore$ The curve is concave upwards for values of $x > a$.

When $x < a$, $\frac{d^2y}{dx^2}$ is negative.

$\therefore$ The curve is concave downwards for values of $x < a$.

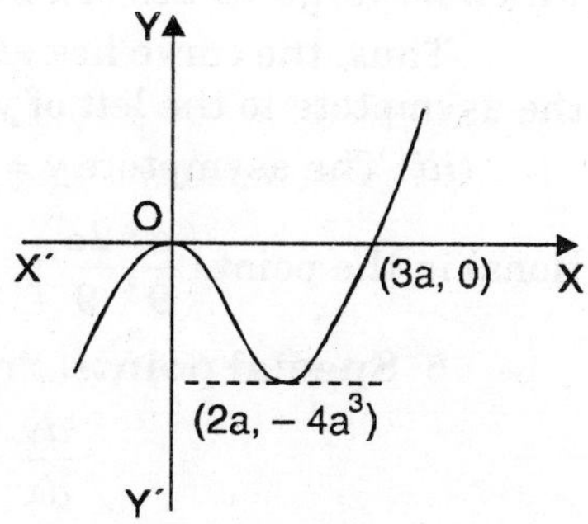

The shape of the curve is as shown in the figure.

Example 6. *Trace the curve $x^3 + y^3 = ax^2$ $(a > 0)$.*

Sol. The equation of the curve is $x^3 + y^3 = ax^2$ ...(1)

1. **Symmetry.** The curve is neither symmetrical about the coordinate axes nor about the line $y = x$.

2. **Origin.** It passes through the origin. The tangents at the origin are given by $x^2 = 0$ *i.e.,* $x = 0$ and $x = 0$.

$\therefore$ The origin is a cusp.

From (1), on neglecting x^3, we have $ax^2 = y^3$

or
$$x = \pm\sqrt{\frac{y^3}{a}}.$$

The roots are real only when y is positive. Thus, the cusp is a single cusp. Again, the two values of x are of opposite sign, so that it is a cusp of the first species.

Hence there is a single cusp of the first species at (0, 0).

3. **Axes Intersection.** The curve meets x-axis in the point $(a, 0)$. Transferring the origin to $(a, 0)$, the transformed equation is

$$Y^3 + (X + a)^3 = a(X + a)^2 \quad \text{or} \quad X^3 + Y^3 + 2aX^2 + a^2X = 0.$$

The tangent at the new origin is $X = 0$, *i.e.,* new y-axis.

$\therefore$ At $(a, 0)$ the tangent is parallel to old y-axis.

4. **Asymptotes.** (*i*) From (1), we have

$$y^3 = ax^2 - x^3 = -x^3(1 - a/x)$$

$$\therefore \quad y = -x(1 - a/x)^{1/3}$$

$$= -x\left[1 - \frac{a}{3x} - \frac{a^2}{9x^2} + \ldots\ldots\right] = -x + \frac{a}{3} + \frac{a^2}{9x} + \ldots\ldots$$

$\therefore$ The only real asymptote is $y = -x + a/3$.

($\because$ There are no asymptotes parallel to the co-ordinate axes.)

(*ii*) If y_c and y_a denote the ordinates of the curve and that of asymptote for the same value of x, then

$$y_c - y_a = \frac{a^2}{9x} - \ldots$$

which for large values of x is > 0 for $x > 0$ and < 0 for $x < 0$.

Thus, the curve lies above the asymptote to the right of y-axis (*i.e.,* for $x > 0$) and below the asymptote to the left of y-axis (*i.e.,* for $x < 0$).

(*iii*) The asymptote $y = -x + a/3$, meets the curve $x^3 + y^3 = ax^2$ (on solving the two equations) in the point $\left(\frac{a}{9}, \frac{2a}{9}\right)$.

5. **Special points.** From (1), $y = x^{2/3}(a - x)^{1/3}$.

$$\therefore \quad \frac{dy}{dx} = \frac{2}{3}x^{-1/3}(a - x)^{1/3} - \frac{1}{3}(a - x)^{-2/3} \cdot x^{2/3} = \frac{2a - 3x}{3x^{1/3}(a - x)^{2/3}}$$

$$\frac{dy}{dx} = 0 \text{ when } x = \frac{2a}{3}.$$

$\therefore$ The tangent is parallel to x-axis at the point $\left(\frac{2a}{3}, \frac{(4)^{1/3}\, a}{3}\right)$

Also $dy/dx = \infty$, when $x = a$ or $x = 0$.

$\therefore$ The tangent is parallel to y-axis at $(a, 0)$ as already shown.

6. **Region.** (*i*) x and y cannot be both negative, for this would make the L.H.S. of (1) negative and the R.H.S. positive. Hence no part of curve exists in the third quadrant.

(*ii*) The equation (1) of the curve can be written as $y^3 = x^2(a - x)$.

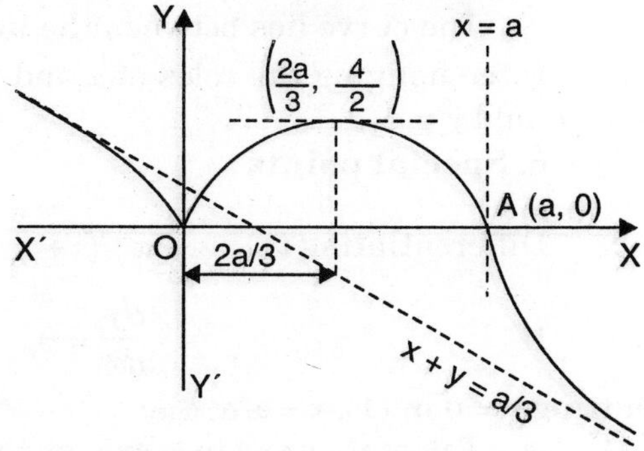

Now when x is negative, y is positive and as x decreases from 0 to $-\infty$, y increases from 0 to ∞.

When x is positive and lies between 0 and a, y is also positive.

When $x > a$, y is negative. This is possible only if the curve crosses the line $x = a$ which is tangent to curve, at the point $(a, 0)$.

$\therefore$ $(a, 0)$ is point of inflexion. As $x \to \infty$, y also $\to \infty$. The shape of the curve is as shown in the figure above.

Example 7. *Trace the curve* $x^{2/3} + y^{2/3} = a^{2/3}$ **(Astroid).** (M.D.U. 2002)

Sol. The curve is $x^{2/3} + y^{2/3} = a^{2/3}$ ...(1)

1. **Symmetry.**

(*i*) The curve is symmetrical about x-axis.

($\because$ on changing y to $-y$, the equation remains unchanged)

(*ii*) The curve is symmetrical about y-axis.

(*iii*) The curve is symmetrical about the line $y = x$.

(*iv*) The curve is symmetrical about the line $y = -x$.

(*v*) The curve is also symmetrical in opposite quadrants.

2. **Origin**

The curve does not pass through the origin.

3. **Asymptotes**

The curve has no asymptotes.

4. **Points of Intersection**

(*i*) *To find intersections with x-axis* (Putting $y = 0$ in (1), we have)

$$x^{2/3} = a^{2/3}$$

cubing,
$$x^2 = a^2 \quad \therefore \quad x = \pm a.$$

$\therefore$ Intersections with x-axis are $(a, 0)$ and $(-a, 0)$.

(*ii*) Similarly, intersections with y-axis are $(0, a)$ and $(0, -a)$.

(*iii*) *To find intersections with line y = x.*

Putting $y = x$ in (1), $x^{2/3} + x^{2/3} = a^{2/3}$ or $2x^{2/3} = a^{2/3}$

$\therefore$ Cubing, $8x^2 = a^2 \quad \therefore \quad x = \pm\dfrac{a}{2\sqrt{2}}$.

$\therefore$ Intersections are $\left(\pm\dfrac{a}{2\sqrt{2}}, \pm\dfrac{a}{2\sqrt{2}}\right)$.

(*iv*) Similarly, intersections with the line $y = -x$ are $\left(\pm\dfrac{a}{2\sqrt{2}}, \mp\dfrac{a}{2\sqrt{2}}\right)$.

5. **Region**

From (1), $\quad y^{2/3} = a^{2/3} - x^{2/3} \quad \therefore \quad y = ((a^{2/3} - x^{2/3})^3)^{1/2}$

For y to be real, $a^{2/3} - x^{2/3} > 0$ or $a^{2/3} > x^{2/3}$.

i.e., $x^{2/3} < a^{2/3}$ $\therefore$ $x^2 < a^2$ $\therefore$ $-a < x < a$.

$\therefore$ The curve lies between the lines $x = -a$ and $x = a$.

Interchanging the roles of x and y, we can prove that the curve lies between the lines $y = -a$ and $y = a$.

6. **Special points**

Differentiating (1), $\frac{2}{3}x^{-1/3} + \frac{2}{3}y^{-1/3}\frac{dy}{dx} = 0$ $\therefore$ $\frac{dy}{dx} = -\frac{x^{-1/3}}{y^{-1/3}} = -\frac{y^{1/3}}{x^{1/3}}$ ...(2)

If $\frac{dy}{dx} = 0$, then $y^{1/3} = 0$ or $y = 0$

putting $y = 0$ in (1), $x = \pm a$.

$\therefore$ Tangents are $\parallel$ to x-axis at the points $(\pm a, 0)$ *i.e.*, x-axis is a tangent at the points $(\pm a, 0)$.

Similarly, $\frac{dy}{dx} \to \infty$ when $x = 0$

putting $x = 0$ in (1), $y = \pm a$.

$\therefore$ Tangents are $\parallel$ to y-axis at the points $(0, \pm a)$.

Again from (2), $\frac{dy}{dx}$ is negative when x and y are both positive.

$\therefore$ y decreases as x increases. (In the first quadrant)

Diff. (2), $$\frac{d^2y}{dx^2} = -\left[\frac{x^{1/3}\cdot\frac{1}{3}y^{-2/3}\frac{dy}{dx} - y^{1/3}\cdot\frac{1}{3}x^{-2/3}}{x^{2/3}}\right]$$

or $$\frac{d^2y}{dx^2} = -\frac{1}{3}\left[\frac{\frac{x^{1/3}}{y^{2/3}}\left(-\frac{y^{1/3}}{x^{1/3}}\right) - \frac{y^{1/3}}{x^{2/3}}}{x^{2/3}}\right] = \frac{1}{3}\left[\frac{x^{2/3} + y^{2/3}}{x^{4/3}y^{1/3}}\right]$$

or $$\frac{d^2y}{dx^2} = \frac{a^{2/3}}{3x^{4/3}y^{1/3}}$$ [By (1)]

$\frac{d^2y}{dx^2}$ is positive when x and y are both positive.

$\therefore$ Curve is concave upwards in first quadrant.

The shape of the curve is as shown in the figure.

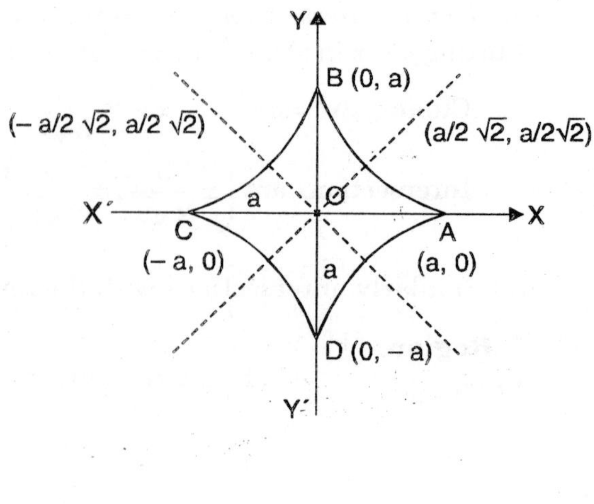

Note. The point where the line $y = x$ meets the astroid $x^{2/3} + y^{2/3} = a^{2/3}$ $\left(\text{i.e., the point } \left(\frac{a}{2\sqrt{2}}, \frac{a}{2\sqrt{2}}\right)\right)$ is called the **VERTEX** of the astroid.

Example 8. *Trace the curve $x^3 + y^3 = 3axy$.* (K.U. 1996, 2006 S ; M.D.U. 2006)

Sol. The equation of the curve is $x^3 + y^3 = 3axy$ ...(1)

1. **Symmetry**

The curve is symmetrical about the line $y = x$.

2. **Origin**

The curve passes through the origin.

Tangents at the origin (equating to zero the lowest degree terms), are given by

$$xy = 0 \quad \text{or} \quad x = 0, y = 0.$$

These tangents being real and distinct, origin is a *node*.

3. **Asymptotes**

The curve has no asymptotes parallel to the axes.

To find oblique asymptotes

Putting $x = 1, y = m$ in the highest degree terms :

$$\phi_3(m) = 1 + m^3$$

Similarly, $\phi_2(m) = -3am$

$$\phi_3(m) = 0 \text{ gives } m^3 + 1 = 0 \text{ or } m = -1$$

To find c, $c\phi_3'(m) + \phi_2(m) = 0$

or $c(3m^2) - 3am = 0$

when $m = -1, c = -a.$

$\therefore$ The asymptote is $y = mx + c$ or $y = -x - a$ or $x + y + a = 0$.

Position of the curve w.r.t. the asymptote $y = -x - a$.

Let the equation of the curve be $y = -x - a + \dfrac{A}{x} + \dfrac{B}{x^2} + \ldots$ ...(2)

Putting this value of y in (1),

$$x^3 + \left(-x - a + \frac{A}{x} + \frac{B}{x^2} + \ldots\ldots\right)^3 = 3ax\left(-x - a + \frac{A}{x} + \frac{B}{x^2} + \ldots\right)$$

Comparing co-effs. of x, $3A - 3a^2 = -3a^2 \quad \therefore \quad A = 0.$

Comparing constant terms $-a^3 + 3B = 0$ or $B = \dfrac{a^3}{3}$.

Putting these values of A and B in (2), the equation of the curve is

$$y = -x - a + \frac{a^3}{3} \cdot \frac{1}{x^2} + \ldots\ldots$$

$$\therefore \quad y_c - y_a = \frac{a^3}{3}\frac{1}{x^2},$$

which is positive for large $x > 0$ and $x < 0$.

$\therefore$ The curve lies above the asymptote.

4. **Points of Intersection**

The curve meets the axes only at the origin (0, 0). To find intersections with the line $y = x$ [Putting $y = x$ in (1)], we have

$$2x^3 = 3ax^2 \quad \text{or} \quad x^2(2x - 3a) = 0$$

$\therefore \qquad x = 0 \quad \text{and} \quad x = \frac{3a}{2}.$

$\therefore \qquad y = 0 \quad \text{and} \quad y = \frac{3a}{2}.$

$\therefore$ The two points are (0, 0) and $\left(\frac{3a}{2}, \frac{3a}{2}\right)$.

To find the slope of the tangent at $\left(\frac{3a}{2}, \frac{3a}{2}\right)$

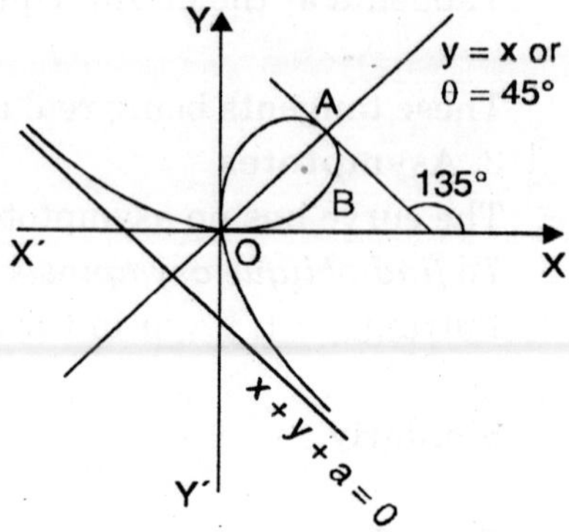

Diff. (1) w.r.t. x,

$$3x^2 + 3y^2 \frac{dy}{dx} = 3a\left(x\frac{dy}{dx} + y\right)$$

or
$$\frac{dy}{dx} = \frac{ay - x^2}{y^2 - ax}$$

$\left(\frac{dy}{dx} \text{ being the slope of the tangent at } (x, y)\right)$

At the point $\left(\frac{3a}{2}, \frac{3a}{2}\right)$, $\frac{dy}{dx} = -1$

i.e., tangent to the curve at $\left(\frac{3a}{2}, \frac{3a}{2}\right)$ will make an angle of 135° with the x-axis.

5. **Region**

x and y both can't be – ve because if x and y are both – ve, then L.H.S. of (1) is – ve and R.H.S. of (1) is + ve which is impossible.

$\therefore$ No portion of the curve lies in third quadrant.

EXERCISE 1

Trace the following curves :

1. $y^2 = x^3$.
2. $9ay^2 = x(x - 3a)^2$. *(M.D.U. 2000)*
3. $a^2y^2 = x^3(2a - x)$.
4. $a^4y^2 = a^2x^4 - x^6$.
5. $y^2(a^2 + x^2) = x^2(a^2 - x^2)$ or $x^2(x^2 + y^2) = a^2(x^2 - y^2)$. *(K.U. 1998)*
6. $a^2y^2 = x^2(a^2 - x^2)$. *(K.U. 1999)*
7. $y^2(x^2 + y^2) + a^2(x^2 - y^2) = 0$ or $x^2(y^2 + a^2) + y^2(y^2 - a^2) = 0$.
8. $y^2 = (x - 2)^2(x - 5)$. *(K.U. 2005 S)*
9. $a^2x^2 = y^3(2a - y)$
10. $y^2 = (x - 1)(x - 2)(x - 3)$.
11. $xy^2 = 4a^2(2a - x)$.
12. (*a*) $x^2y^2 = (a + y)^2(a^2 - y^2)$ (*b*) $x^2y^2 = (a^2 + y^2)(a^2 - y^2)$. *(K.U. 2000)*
13. $y(x^2 + 4a^2) = 8a^3$. *(M.D.U. 1998)*
14. $y^2x = a^2(x - a)$.
15. $y^2(x^2 - 1) = x$.
16. (*a*) $y = \frac{x^2 + 1}{x^2 - 1}$ (*b*) $x^2y^2 = x^2 + 1$.
17. $x^2y^2 = a^2(y^2 - x^2)$. *(M.D.U. 1999)*

18. $y^2(a + x) = x^2(a - x)$ or $x(x^2 + y^2) = a(x^2 - y^2)$.

19. $y^2(a - x) = x^2(a + x)$.

20. $y = x^3$.

21. $y = x^3 - x$.

22. $x^3y - x - 1 = 0$.

23. $(x^2 + a^2)y = a^2x$.

24. $x = (y - 1)(y - 2)(y - 3)$

25. $(x^2 - a^2)(y^2 - b^2) = a^2b^2$.

26. (a) $y^3 = x^3 + ax^2$ (b) $x^3 = y^3 + ay^2$.

27. (a) $x^3 + y^3 = a^2x$. (b) $x^3 + y^3 = a^2y$.

28. $y = (x - 2)(x + 1)^2$.

29. $\left(\frac{x}{a}\right)^{2/3} + \left(\frac{y}{b}\right)^{2/3} = 1$. *(M.D.U. 1995)*

30. $x^6 + y^6 = a^2x^2y^2$.

ANSWERS

1.

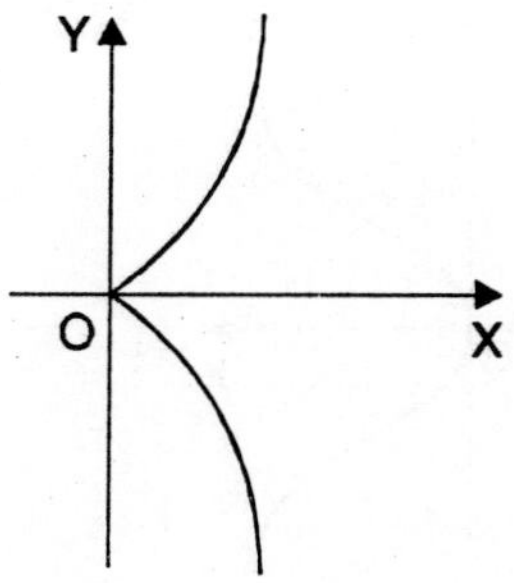

2.

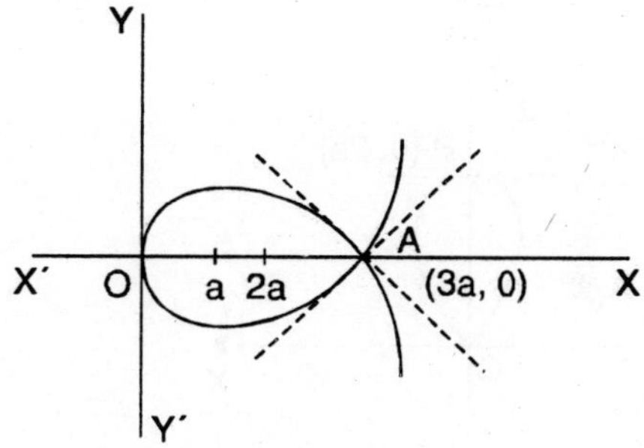

3.

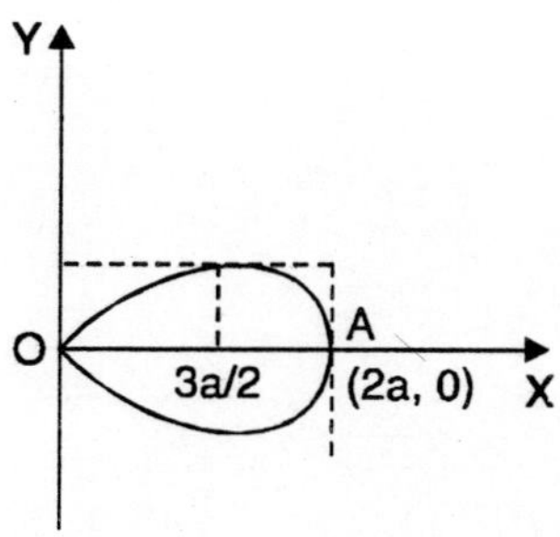

4.

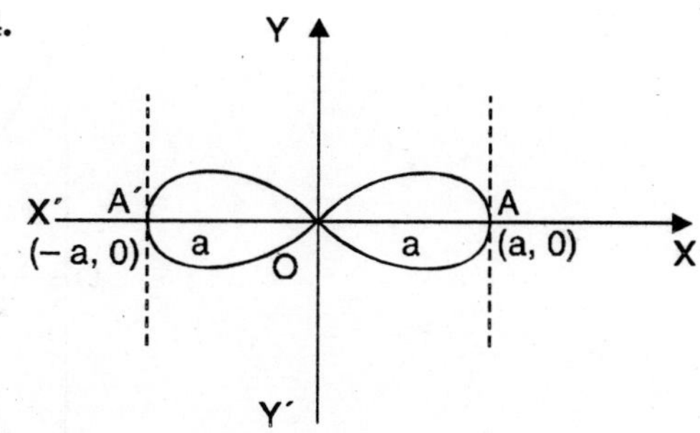

5.

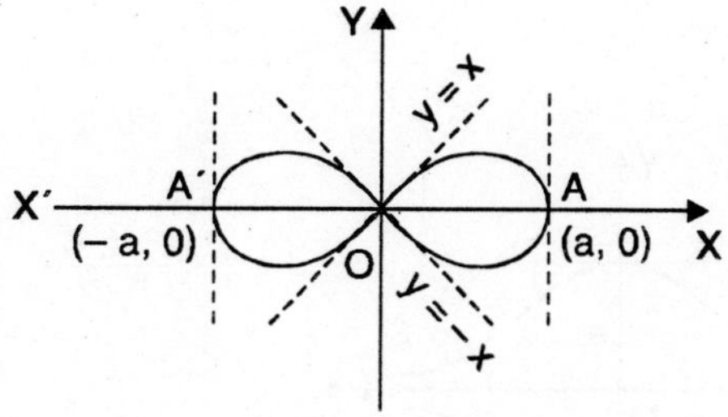

6.

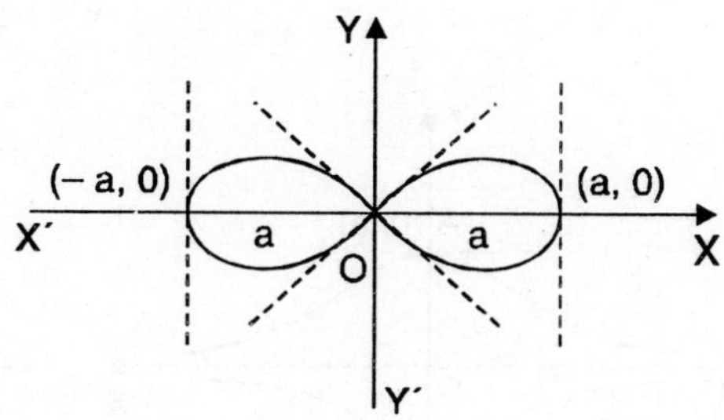

7.

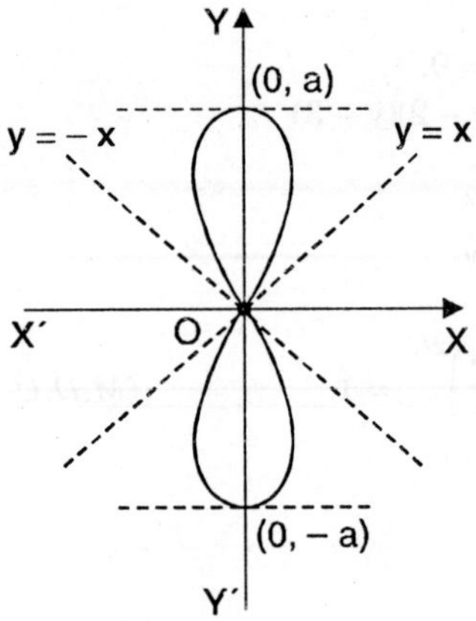
Y
(0, a)
y = – x
y = x
X′
O
X
(0, – a)
Y′

8.

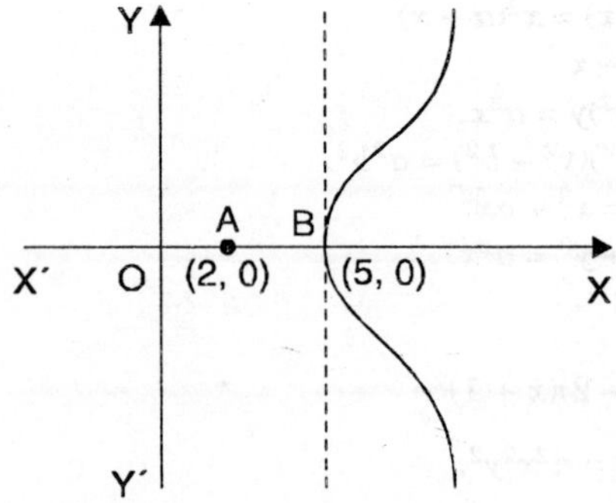
Y
A
B
X′
O
(2, 0)
(5, 0)
X
Y′

9.

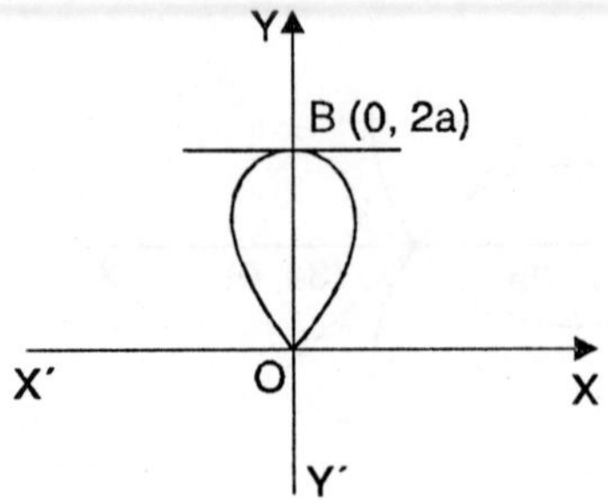
Y
B (0, 2a)
X′
O
X
Y′

10.

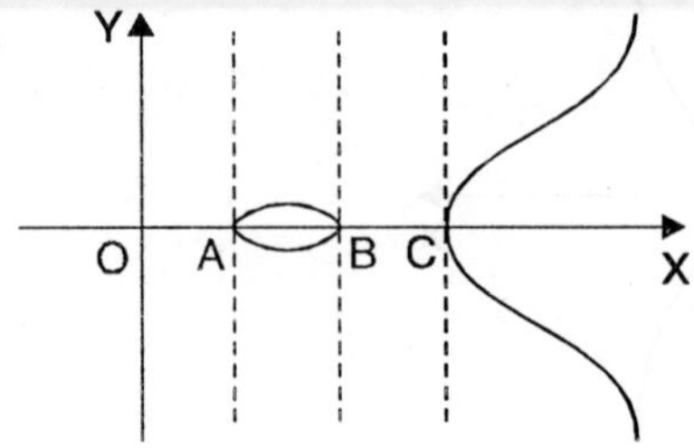
Y
O
A
B
C
X

11.

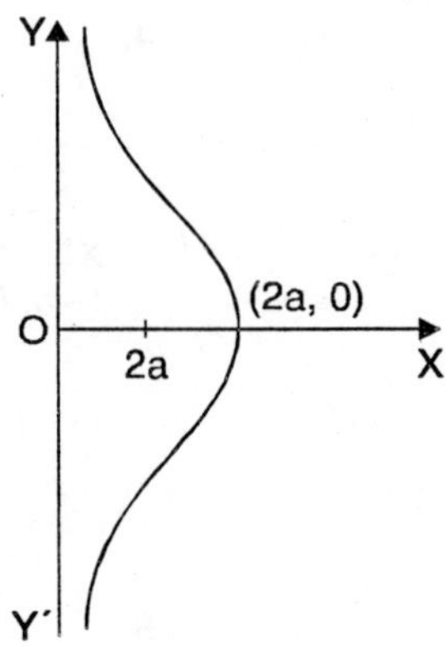
Y
(2a, 0)
O
2a
X
Y′

12.

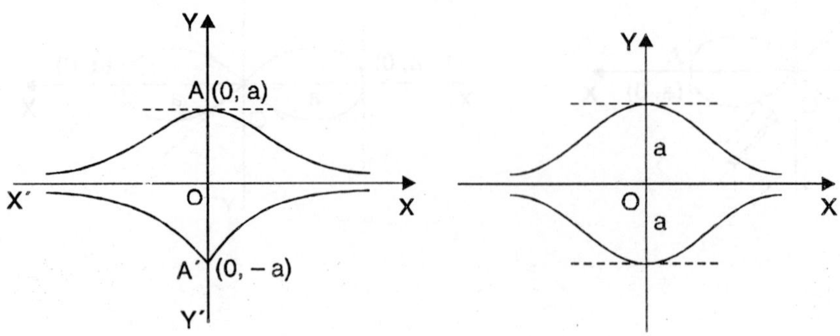
Y
A (0, a)
X′
O
X
A′ (0, – a)
Y′
Y
a
O
a
X

13.

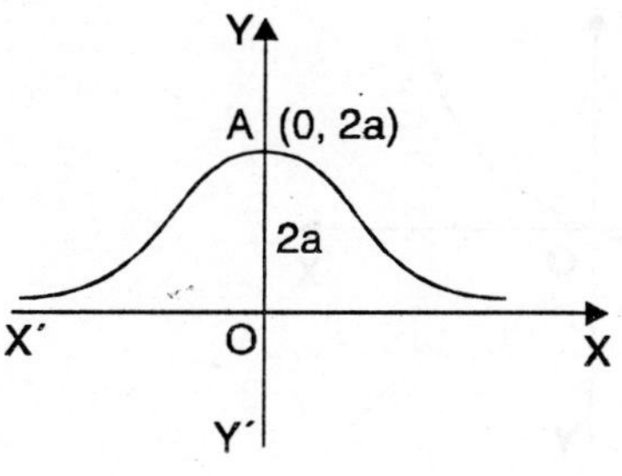

14.

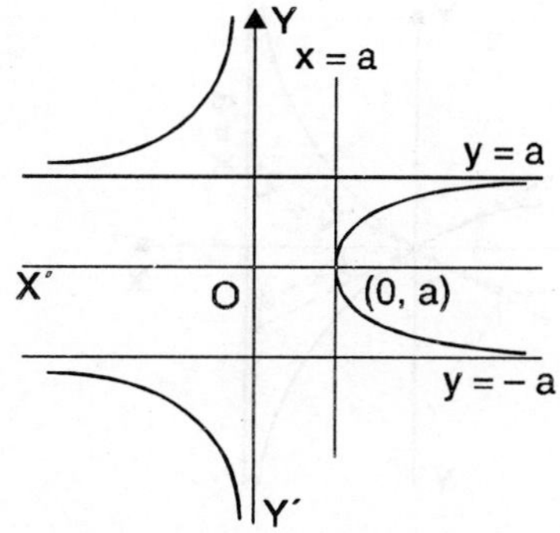

15.

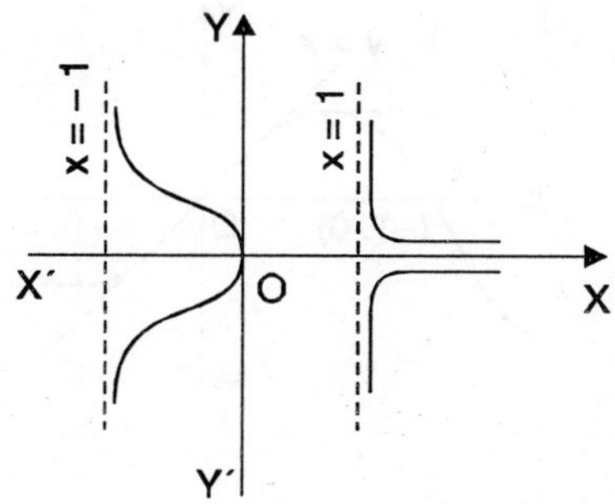

16.

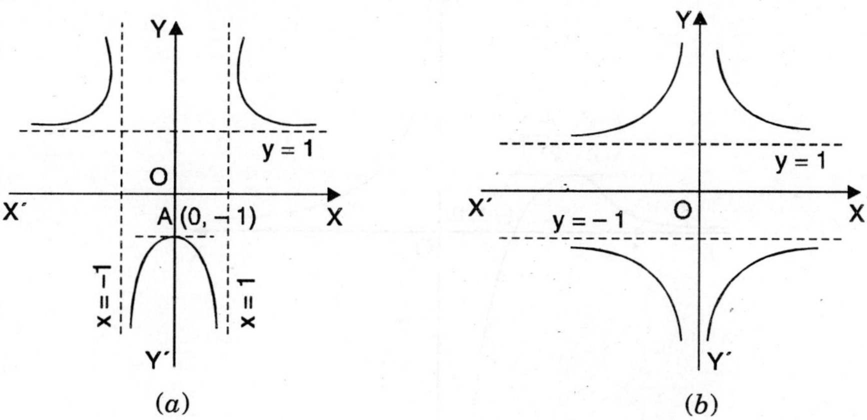

(*a*) (*b*)

17.

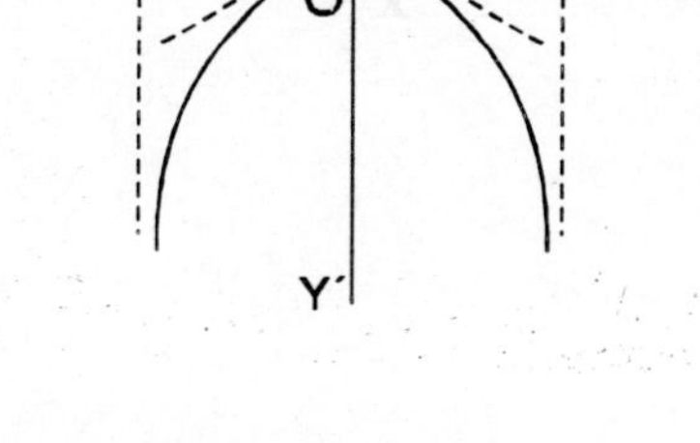

18.

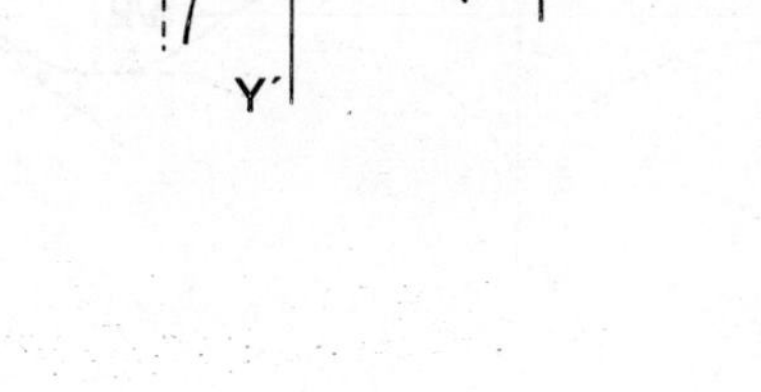

19.

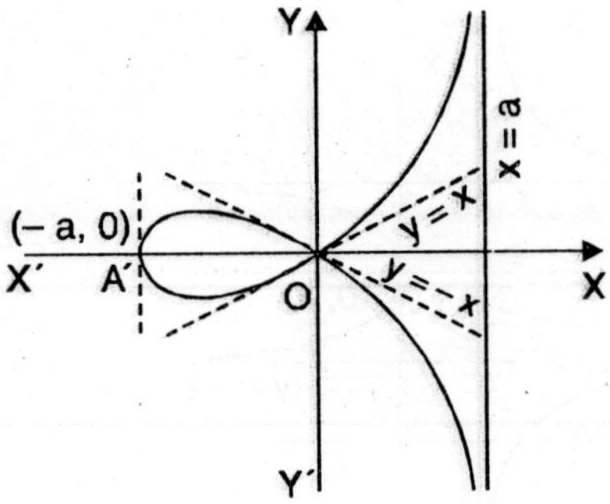
Y
X′
X
Y′
(– a, 0)
A′
O
y = x
y = – x
x = a

20.

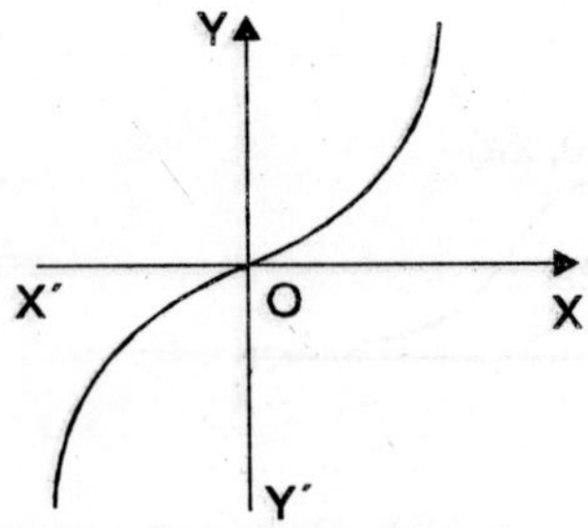
Y
X′
O
X
Y′

21.

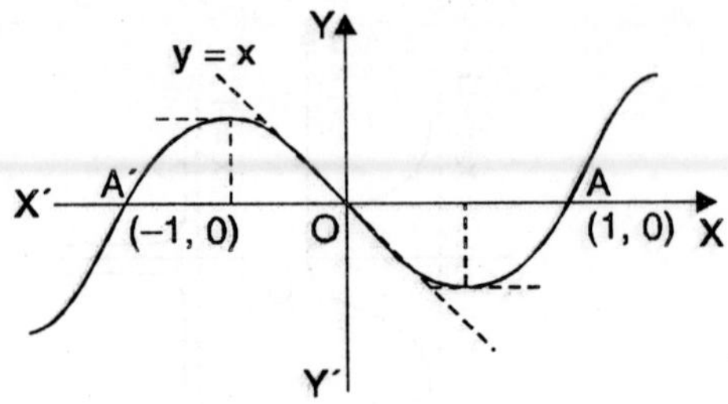
Y
y = x
X′
A
(–1, 0)
O
A
(1, 0)
X
Y′

22.

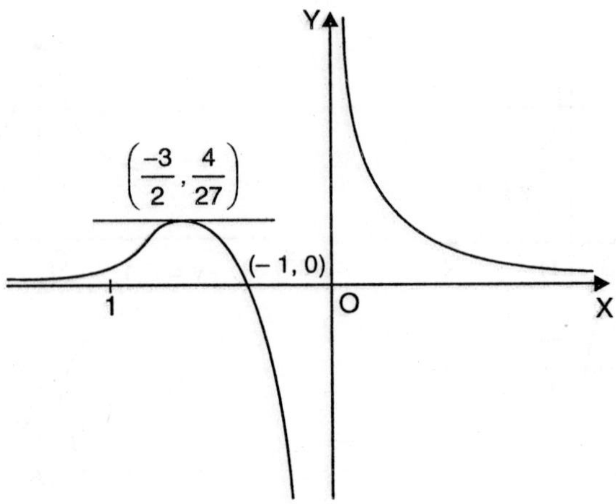
Y
$\left(\frac{-3}{2}, \frac{4}{27}\right)$
(– 1, 0)
1
O
X

23.

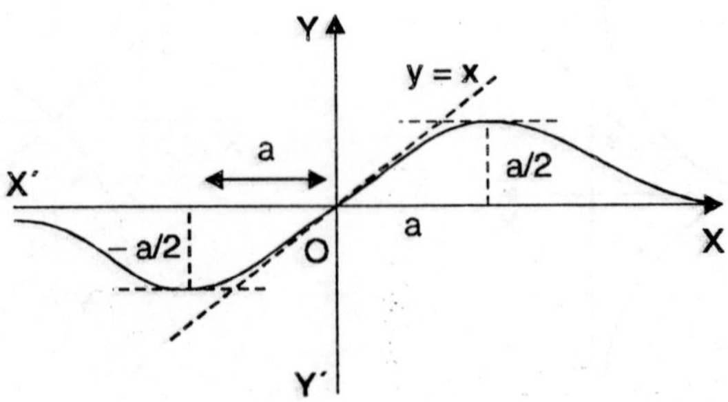
Y
y = x
a
a/2
X′
– a/2
O
a
X
Y′

24.

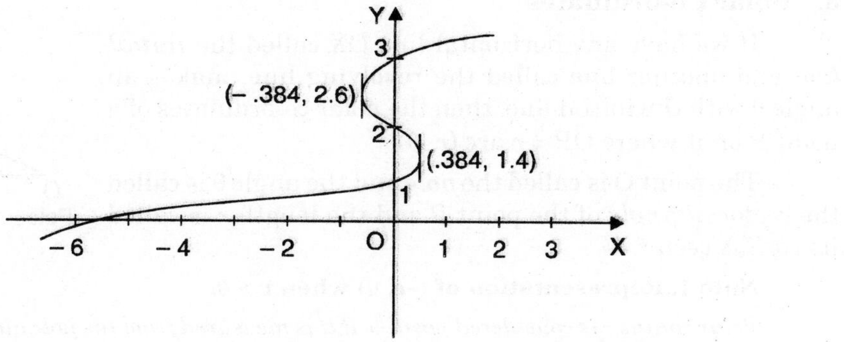

25.

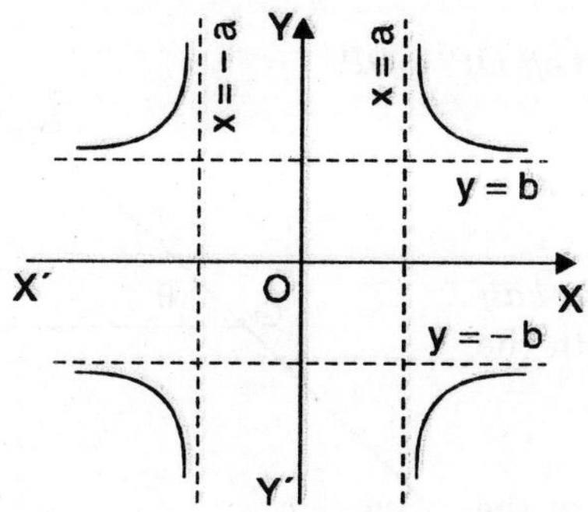

26. (*a*)

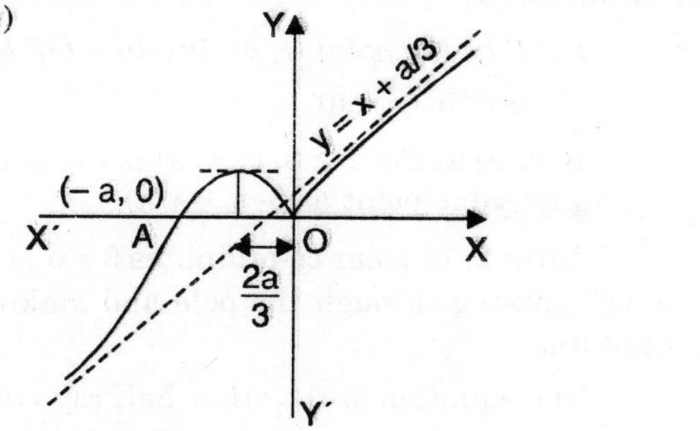

27. (*a*)

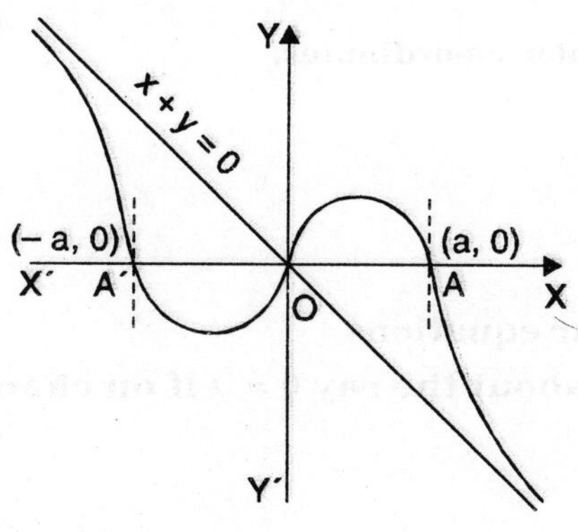

28.

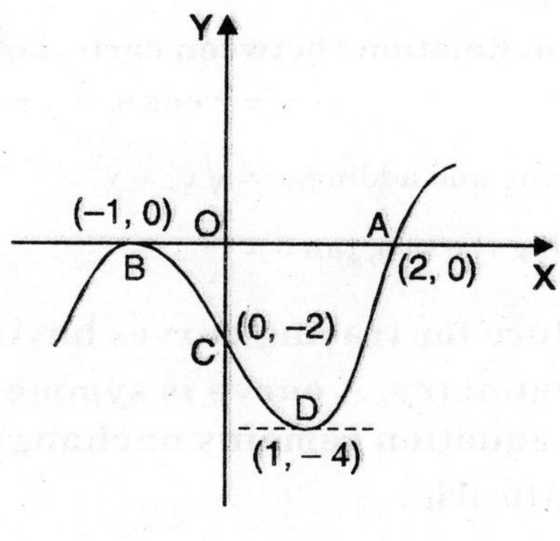

29.

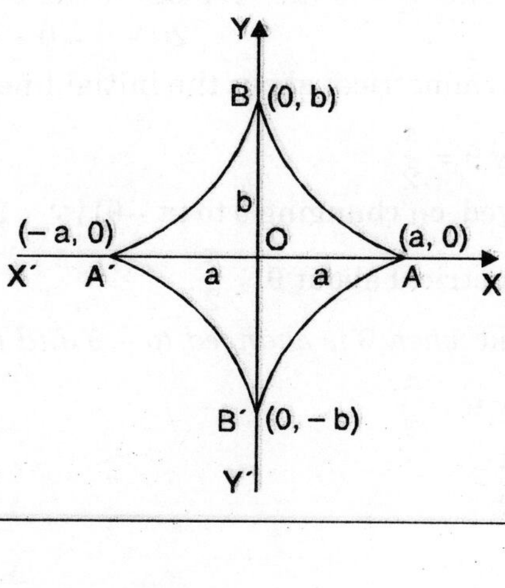

30.

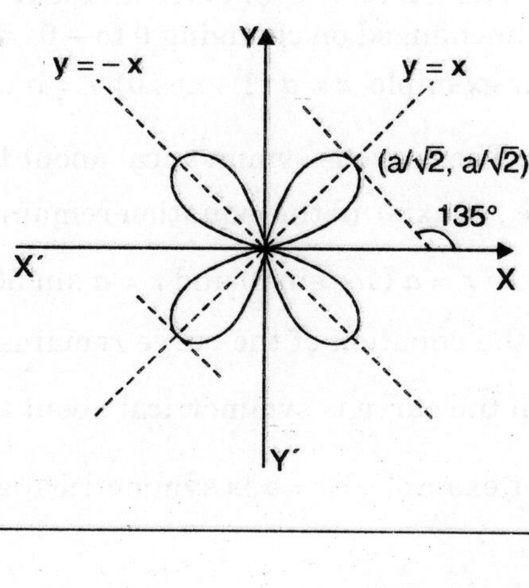

3. Polar Co-ordinates

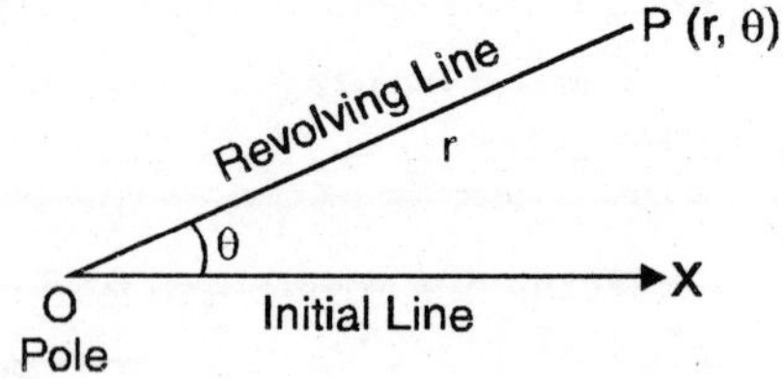

If we have any horizontal line OX called the *initial line* and another line called the revolving line, makes an angle θ with the initial line, then the polar co-ordinates of a point P on it where OP = r arc (r, θ).

The point O is called the *pole* and the angle θ is called the *vectorial angle* of the point P and the length r is called its *radius vector.*

Note 1. Representation of $(-r, \theta)$ when $r > 0$.

Polar radius r is considered positive if it is measured from the pole along the half ray bounding the vectorial angle.

Let P be the point (r, θ). Produce OP backwards and cut off $OP' = OP$.

Then P' is $(-r, \theta)$.

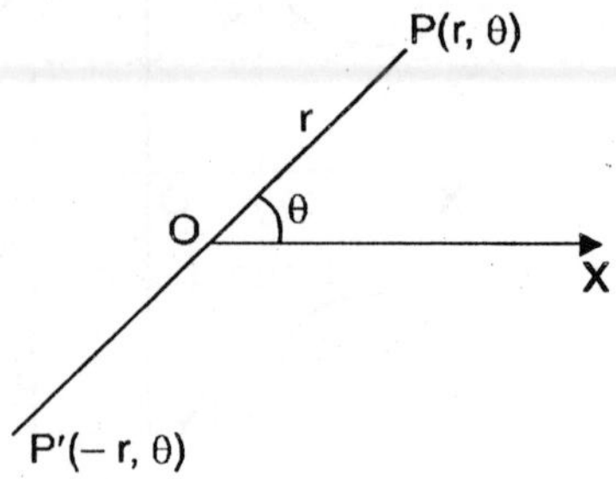

Remark. For $r \geq 0$, $(-r, \theta)$ is the same point as $(r, \theta + \pi)$ or (r, θ) is the same point as $(-r, \theta + \pi)$.

Note 2. In polar co-ordinates $\theta = \alpha$ is the equation of the *half ray* OP passing through the pole and making an angle α with the initial line.

The equation of the other half ray OP′ is $\theta = \pi + \alpha$.

Note 3. $r = a$ is the equation of a circle having centre at the pole and radius a.

Note 4. Relations between cartesian and polar co-ordinates,

$$\mathbf{x = r \cos \theta, \; y = r \sin \theta}$$

Squaring and adding, $r = \sqrt{x^2 + y^2}$.

Dividing the two, $\tan \theta = \dfrac{y}{x}$.

4. Procedure for tracing curves having polar equations

1. Symmetry. A curve is symmetrical about the ray $\theta = \alpha$ if on changing θ to $2\alpha - \theta$, the equation remains unchanged.

In particular :

(*i*) The curve is symmetrical about the initial line $\theta = 0$ (*i.e.,* x-axis) if the equation remains unchanged on changing θ to $-\theta$. $(\because \; 2\alpha - \theta = 0 - \theta = -\theta)$

For example, $r = a(1 + \cos \theta)$, $r = a \cos 3\theta$ are symmetrical about the initial line.

(*ii*) The curve is symmetrical about the half ray $\theta = \dfrac{\pi}{2}$.

(*i.e.,* y-axis) if the equation remains unchanged on changing θ to $(\pi - \theta)$ $[\because \; 2\alpha - \theta = \pi - \theta]$ *e.g.,* $r = a(1 + \sin \theta)$ and $r = a \sin 3\theta$ are symmetrical about $\theta = \dfrac{\pi}{2}$.

If the equation of the curve remains unchanged *when θ is changed to $-\theta$ and r to $-r$,* even then the curve is symmetrical about the half ray $\theta = \dfrac{\pi}{2}$.

For example, $r\theta = a$ is symmetrical about $\theta = \dfrac{\pi}{2}$.

(iii) The curve is symmetrical about the **pole** (*i.e.*, symmetrical in opposite quadrants) if the equation remains unchanged on changing r to $-r$ **or** θ to $\pi + \theta$.

2. **Pole or origin.** *(i) Find whether the curve passes through the pole or not.* This can be done by putting $r = 0$ in the equation and if on doing so, we get some real value (or values) of θ ; then the curve passes through the pole.

If it is not possible to find real value of θ for which $r = 0$, the curve does not pass through the pole.

(iii) Find the tangents at the pole. **Putting r = 0, the real values of θ give the tangents at the pole.**

For example, consider the curve $r = a\,(1 - \cos\theta)$. Putting $r = 0$, we get $\cos\theta = 1 \therefore \quad \theta = 0$.

Hence the curve passes through the pole and the line $\theta = 0$ *i.e.,* the initial line is the tangent at the pole.

Again, consider $r = a\,(3 - \sin\theta)$.

Putting $r = 0$, we get $\sin\theta = 3$ which does not give any real value of θ [$\because \quad |\sin\theta| \le 1$].

Hence the curve does not pass through the pole.

(iii) Find the points where the curve cuts the initial line and the line $\theta = \pi/2$.

3. **Asymptotes.** *(i)* If $\theta \to \theta_1$ (some finite value) when $r \to \infty$, then there is an asymptote.

Working Rule to find polar asymptotes :

Put $r = \dfrac{1}{u}$. and let $u \to 0$ so that $\theta \to \theta_1$ (θ_1 is a definite number)

Determine
$$p = \underset{\substack{\theta \to \theta_1 \\ u \to 0}}{\text{Lt}} \left(-\frac{d\theta}{du}\right).$$

Putting the values of p and θ_1 in the equation $p = r \sin(\theta_1 - \theta)$, we get the corresponding asymptote.

(ii) If as $\theta \to \infty$, $r \to a$, then there is a circular asymptote $r = a$.

Circular asymptotes of the curve $r = f(\theta)$ are given by $r = \underset{\theta \to \infty}{\text{Lt}} f(\theta)$.

4. **Points of Intersection.** Find some points on the curve for convenient values of θ. (especially for values of $\theta = \alpha$ for which the curve is symmetrical).

5. **Region.** Solve the given equation for r or θ (if possible). Find the regions in which the curve does not lie. This can be done in the following way :

(i) If for $\theta_1 < \theta < \theta_2$, r is imaginary, then there is no portion of the curve between the lines $\theta = \theta_1$ and $\theta = \theta_2$. Consider $r^2 = a^2 \cos 2\theta$.

For $\pi/4 < \theta < 3\pi/4$, $\cos 2\theta$ is negative. $\therefore \quad r^2$ is – ve and so r is imaginary. Hence the curve does not lie between the lines $\theta = \pi/4$ and $\theta = 3\pi/4$.

(ii) If the greatest and least numerical values of r be respectively a and b, the curve lies entirely within the circle $r = a$ and entirely outside the circle $r = b$.

(iii) Trace the variations of r when θ varies in the intervals $(0, \infty)$ and $(-\infty, 0)$ marking the values of θ for which $r = 0$ or attains maximum and minimum values. Plot the points so obtained.

[**Note.** When r is a periodic function of θ, the negative values of θ need not be considered. We may consider values from $\theta = 0$ to those values of θ where the values begin to repeat.]

6. **Value of ϕ.** Find $\tan \phi = r\dfrac{d\theta}{dr}$ and ϕ.

(ϕ is the angle between the tangent and radius vector.)

Find the points where ϕ is 0 or $\dfrac{\pi}{2}$.

Again, r increases as θ increases if $\dfrac{dr}{d\theta}$ as positive and r decreases as θ increases if $\dfrac{dr}{d\theta}$ is negative.

Note. Conversion into cartesians. Transform the equation into cartesians, when tracing of the curve becomes easy on transformation to cartesian system of coordinate.

Example 1. *Trace the curve $r = a(1 - \cos \theta)$.* (K.U. 1999 S, 2006)

Sol. The equation of the curve is $r = a(1 - \cos \theta)$ ...(1)

1. **Symmetry.** The curve is symmetrical about the initial line because the equation of curve remains unaltered when θ is changed to $-\theta$.

2. **Pole or origin.** (*i*) When $\theta = 0$, $r = 0$ hence the curve passes through the pole and the tangent at the pole is the line $\theta = 0$, *i.e.*, the initial line.

(*ii*) The curve meets the initial line $\theta = 0$ at (0, 0) and the lines $\theta = \pi/2$ and π in the points $(a, \pi/2)$ and $(2a, \pi)$ respectively.

3. **Asymptotes.** Since for any finite value of θ, r does not tend to infinity, $\therefore$ the curve has no asymptote.

4. **Point of Intersection.** The corresponding values of θ and r are given below :

$\theta = 0$	$\pi/3$	$\pi/2$	$2\pi/3$	π
$r = 0$	$a/2$	a	$3a/2$	$2a$.

[We need not trace the curve for values of θ from π to 2π as the curve is symmetrical about the initial line.]

5. **Region.** (*i*) $\because$ $|\cos \theta| \le 1$ $\therefore$ From (1), $r \le 2a$.

$\therefore$ The curve lies entirely within the circle $r = 2a$.

(*ii*) When θ increases from 0 to π, r remains positive and increases from 0 to $2a$.

When θ increases from π to 2π, r is positive and decreases from $2a$ to 0.

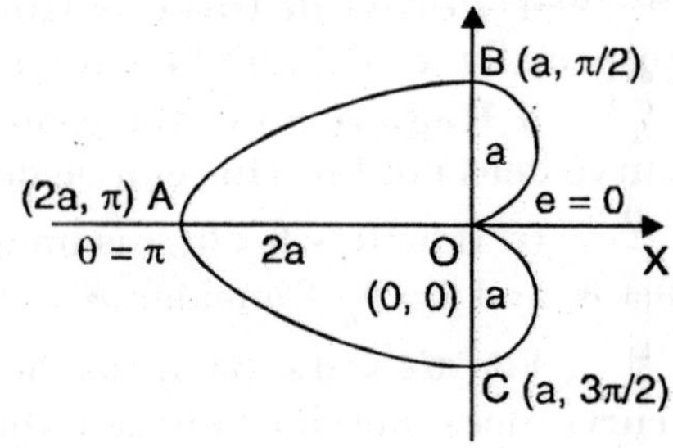

6. **Value of ϕ.**

From (1), $\dfrac{dr}{d\theta} = a \sin \theta$.

$$\therefore \quad \tan \phi = r\frac{d\theta}{dr} = \frac{r}{\left(\dfrac{dr}{d\theta}\right)} = \frac{a(1 - \cos \theta)}{a \sin \theta}$$

$$= \frac{a \cdot 2\sin^2 \frac{\theta}{2}}{a \cdot 2\sin \frac{\theta}{2}\cos \frac{\theta}{2}} = \tan \frac{\theta}{2}.$$

$\therefore$ $$\phi = \frac{\theta}{2}.$$

Now $\phi = 0$ when $\theta = 0$ and $\phi = \pi/2$ when $\theta = \pi$.

$\therefore$ At $(2a, \pi)$ the tangent is perpendicular to the line $\theta = \pi$.

The shape of the curve is as shown in the figure.

Note. We will not consider the values of θ after 2π here, because r is periodic with period 2π and therefore we do not get any new point on the curve.

Example 2. *Trace the curve $r = a \sin 3\theta$.*

Sol. The equation of the curve is $r = a \sin 3\theta$ ...(1)

1. **Symmetry**

From (1), $r = a \sin 3\theta = a \sin (\pi - 3\theta) = a \sin (3\pi - 3\theta)$ etc.

or $$r = a \sin 3\theta = a \sin 3\left(\frac{\pi}{3} - \theta\right) = a \sin 3(\pi - \theta) \text{ etc.}$$

$\therefore$ (1) remains unchanged when θ is changed to $\frac{\pi}{3} - \theta$, or $\pi - \theta$.

$\therefore$ The curve is symmetrical about the rays

$$\theta = \frac{\pi}{6}, \frac{\pi}{2}\left(=\frac{3\pi}{6}\right), \frac{5\pi}{6}, \frac{7\pi}{6}, \frac{9\pi}{6}, \frac{11\pi}{6}$$

($\because$ If on changing θ to $2\alpha - \theta$, the equation is unchanged, symmetry is about the ray $\theta = \alpha$).

2. **Pole or origin.** Putting $r = 0$, we get $\sin 3\theta = 0$

$\therefore$ $3\theta = n\pi$ where n is zero or any other integer.

$\therefore$ $$\theta = 0, \frac{\pi}{3}, \frac{2\pi}{3}, \frac{3\pi}{3}, \frac{4\pi}{3}, \frac{5\pi}{3}, \frac{6\pi}{3}$$

$\therefore$ The curve passes through the pole and the tangents at the pole are

$$\theta = 0, \theta = \frac{\pi}{3}, \theta = \frac{2\pi}{3}, \theta = \frac{3\pi}{3}, \theta = \frac{4\pi}{3}, \theta = \frac{5\pi}{3}, \theta = \frac{6\pi}{3}.$$

3. **Asymptotes.** Putting $r = \frac{1}{u}$; from (1), $u = \frac{1}{a \sin 3\theta}$

$u = 0$ does not give any finite value of θ.

$\therefore$ No asymptotes.

4. **Point of Intersection**

When $\theta = 0, \frac{\pi}{6}, \frac{3\pi}{6}, \frac{5\pi}{6}, \frac{7\pi}{6}, \frac{9\pi}{6}, \frac{11\pi}{6}$ (Rays of symmetry)

From (1), $r = 0, a, -a, a, -a, a, -a$.

5. **Region.** (*i*) From (1), $r \le a$ numerically [$\because$ $\sin 3\theta \le 1$ numerically]

$\therefore$ The curve lies entirely within the circle $r = a$.

(ii) Region w.r.t. tangents at the pole. Tangents at the pole are

$$\theta = 0, \theta = \frac{\pi}{3}, \theta = \frac{2\pi}{3}, \theta = \frac{3\pi}{3}, \theta = \frac{4\pi}{3}, \theta = \frac{5\pi}{3}$$

When $0 \le \theta \le \pi/3$; then $0 \le 3\theta \le \pi$ *i.e.,* 3θ lies in first or second quadrant.

$\therefore$ From (1), r is + ve [$\because$ $\sin \theta$ is + ve in I or II quadrant]

Similarly, when $\frac{\pi}{3} \le \theta \le \frac{2\pi}{3}$, r is – ve.

When $\frac{2\pi}{3} \le \theta \le \frac{3\pi}{3}$ r is + ve.

When $\frac{3\pi}{3} \le \theta \le \frac{4\pi}{3}$, r is – ve.

When $\frac{4\pi}{3} \le \theta \le \frac{5\pi}{3}$, r is + ve.

When $\frac{5\pi}{3} \le \theta \le \frac{6\pi}{3}$, r is – ve.

6. **Value of ϕ.** From (1), $r = a \sin 3\theta$

$$\therefore \quad \frac{dr}{d\theta} = 3a \cos 3\theta \quad \therefore \quad \tan \phi = \frac{r}{\left(\frac{dr}{d\theta}\right)} = \frac{1}{3} \tan 3\theta$$

$$\tan \phi = \infty \left(i.e., \quad \phi = \frac{\pi}{2} \right)$$

for $$\theta = \frac{\pi}{6}, \frac{3\pi}{6}, \frac{5\pi}{6}, \frac{7\pi}{6}, \frac{9\pi}{6}, \frac{11\pi}{6}$$

$\therefore$ Tangent is perpendicular to the radius vector at the points

$$\theta = \frac{\pi}{6}, \frac{3\pi}{6}, \frac{5\pi}{6}, \frac{7\pi}{6}, \frac{9\pi}{6}, \frac{11\pi}{6}.$$

Thus, the shape of the curve is as shown in the figure.

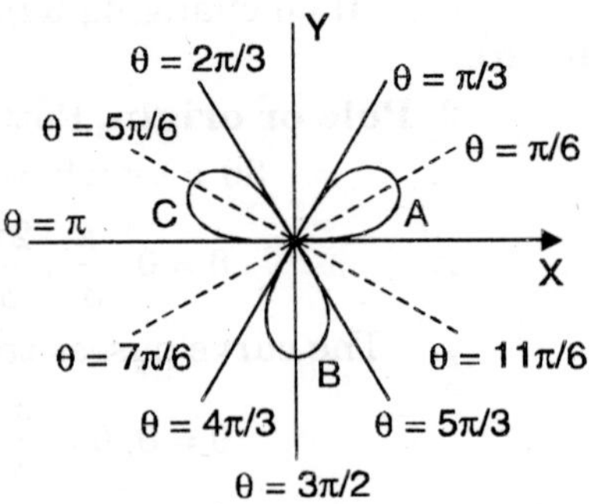

Remark. We have not considered values of θ outside $[0, 2\pi]$ because $r = a \sin 3\theta$ is periodic.

Note 1. $\cos x = \cos(-x) = \cos(2\pi - x) = \cos(4\pi - x) = \ldots\ldots$

Also $\cos \theta = 0$ gives $\theta = (2n + 1)\pi/2$.

Note 2. The curve r = a sin nθ or r = a cos nθ consists of n or 2n loops according as n is odd or even.

Example 3. *Trace the curve $r^2 \cos \theta = a^2 \sin 3\theta$.*

Sol. The equation of the curve is $r^2 \cos \theta = a^2 \sin 3\theta$ ...(1)

1. **Symmetry.** On changing r to $-r$, equation (1) remains unchanged.

$\therefore$ The curve is symmetrical about the pole, *i.e.,* the curve is symmetrical in opposite quadrants.

2. **Origin or Pole.**

Putting $r = 0$ in (1), we get $\sin 3\theta = 0$

$\therefore$ $3\theta = n\pi$ when n is zero or any other integer.

$\therefore$ $\theta = 0, \frac{\pi}{3}, \frac{2\pi}{3}, \frac{3\pi}{3}, \frac{4\pi}{3}, \frac{5\pi}{3}, \frac{6\pi}{3}$

$\therefore$ The curve passes through the pole and the tangents at the pole are

$$\theta = 0, \theta = \frac{\pi}{3}, \theta = \frac{2\pi}{3}, \theta = \frac{3\pi}{3}, \theta = \frac{4\pi}{3}, \theta = \frac{5\pi}{3}, \theta = \frac{6\pi}{3}.$$

3. **Asymptotes.** Putting $r = \frac{1}{u}$, from (1),

$$u^2 = \frac{\cos\theta}{a^2 \sin 3\theta} \qquad \ldots(2)$$

Let $u \to 0$. Therefore, $\cos\theta \to 0$, *i.e.*, $\theta \to \frac{\pi}{2}, \frac{3\pi}{2}$

Diff. (2) w.r.t. θ, $2u \frac{du}{d\theta} = \frac{1}{a^2}\left[\frac{-\sin 3\theta \sin\theta - \cos\theta (3\cos 3\theta)}{\sin^2 3\theta}\right]$

$\therefore$ $\frac{du}{d\theta} = -\frac{(\sin 3\theta \sin\theta + 3\cos\theta \cos 3\theta)}{2a^2 u \sin^2 3\theta}$

For $\theta_1 = \frac{\pi}{2}, p = \underset{\substack{\theta \to \theta_1 \\ u \to 0}}{\text{Lt}} \left(-\frac{d\theta}{du}\right) = \underset{\substack{\theta \to \pi/2 \\ u \to 0}}{\text{Lt}} \frac{2ua^2 \sin^2 3\theta}{\sin 3\theta \sin\theta + 3\cos\theta \cos 3\theta} = 0$

$\therefore$ The asymptote is $p = r\sin(\theta_1 - \theta)$

or $0 = r\sin\left(\frac{\pi}{2} - \theta\right)$ *i.e.*, $r\cos\theta = 0$ or $x = 0$ $[\because x = r\cos\theta]$

i.e., y-axis is an asymptote to the curve.

Again, for $\theta_1 = \frac{3\pi}{2}$, y-axis is an asymptote.

4. **Points of Intersection.** When $\theta = 0$; From (1), $r = 0$ (Also see step 2).

When $\theta = \frac{\pi}{6}, r^2 = \frac{2a}{\sqrt{3}}$ $\therefore$ $r = \pm\sqrt{\frac{2a}{\sqrt{3}}}$

$\therefore$ The two points are $\left(\pm\sqrt{\frac{2a}{\sqrt{3}}}, \frac{\pi}{6}\right)$

when $\theta = \frac{\pi}{3}, r = 0$ again.

5. **Region w.r.t. tangents at the pole**

From (1), $r^2 = \frac{a^2 \sin 3\theta}{\cos\theta}$

when $0 \le \theta \le \frac{\pi}{3}$, r^2 is + ve $\therefore$ r is real.

when $\frac{\pi}{3} \le \theta \le \frac{\pi}{2}$, r^2 is – ve $\therefore$ r is imaginary.

$\therefore$ No portion of the curve lies between the two half-rays $\theta = \frac{\pi}{3}$ and $\theta = \frac{\pi}{2}$.

when $\frac{\pi}{2} \le \theta \le \frac{2\pi}{3}$ r^2 is + ve $\therefore$ r is real

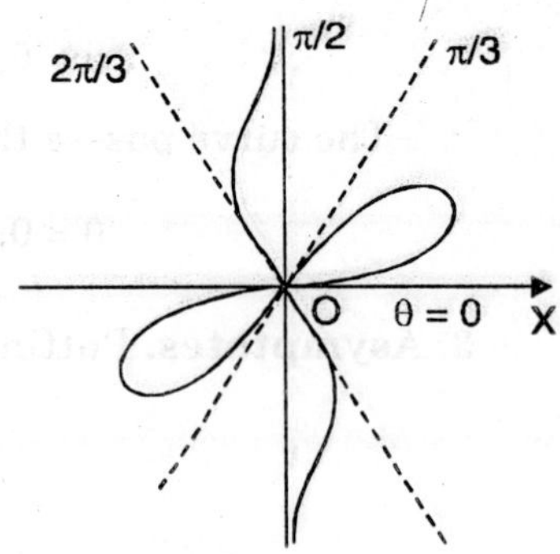

when $\frac{2\pi}{3} \le \theta \le \pi$ r^2 is – ve

$\therefore$ r is imaginary.

$\therefore$ No portion of the curve lies between the half rays

$$\theta = \frac{2\pi}{3} \text{ and } \theta = \pi.$$

Hence the shape of the curve is as shown.

Note. We have not discussed the values of θ between π and 2π because the curve is symmetrical in opposite quadrants.

Example 4. *Trace the curve* $r\theta = a$.

Sol. The equation of the curve is $r\theta = a$ ...(1)

Firstly, we consider positive values of θ, *i.e.,* values of θ in $(0, \infty)$.

1. **Symmetry.** Equation (1) remains unchanged where θ is changed to – θ and r to $-r$, $\therefore$ the curve is symmetrical about the line $\theta = \frac{\pi}{2}$ *i.e., y*-axis.

2. **Origin.** Since r does not become zero for any real finite value of θ, therefore the curve does not pass through the pole.

3. **Asymptotes.** Equation (1) is $r\theta = a$.

Putting $r = \frac{1}{u}$, we have $\frac{\theta}{u} = a$, $u = \frac{\theta}{a}$

$u \to 0$ gives $\theta \to 0$ $(\because \theta_1 = 0)$

$$\frac{du}{d\theta} = \frac{1}{a}$$

$$p = \underset{\substack{\theta \to \theta_1 \\ u \to 0}}{\text{Lt}} \left(-\frac{d\theta}{du}\right) = \underset{\theta \to 0}{\text{Lt}} \,(-a) = -a$$

Putting the values of p and θ_1 in $p = r \sin(\theta_1 - \theta)$, the equation of the asymptote is $-a = r \sin(-\theta)$ or $r \sin\theta = a$

or $$y = a \qquad (\because r\sin\theta = y)$$

4. **Points of intersection**

when $$\theta = 0, \frac{\pi}{6}, \frac{\pi}{3}, \frac{\pi}{2}, \frac{2\pi}{3}, \pi, \frac{3\pi}{2}, 2\pi, ...$$

$\therefore$ From (1);

$$r = \infty, \frac{6a}{\pi}, \frac{3a}{\pi}, \frac{2a}{\pi}, \frac{3a}{2\pi}, \frac{a}{\pi}, \frac{2a}{3\pi}, \frac{a}{2\pi}, ...$$

5. **Region.** (*i*) From (1), $r = \frac{a}{\theta}$

$$\therefore \quad r\sin\theta = \frac{a}{\theta}\sin\theta = a\,\frac{\sin\theta}{\theta} < a \qquad \left[\because \frac{\sin\theta}{\theta} < 1\right]$$

or $$y < a \qquad (\because y = r\sin\theta)$$

$\therefore$ The curve lies entirely below the asymptote

$$r\sin\theta = a \quad \text{or} \quad y = a.$$

(*ii*) For positive values of θ, r is a positive.

6. **Value of ϕ.** From (1), $r = \frac{a}{\theta}$ $\therefore$ $\frac{dr}{d\theta} = -\frac{a}{\theta^2}$

$\frac{dr}{d\theta}$ is negative for all values of θ.

$\therefore$ r decreases as θ increases.

As θ increases from 0 to ∞, r is + ve and decreases from ∞ to 0.

$$\tan \phi = \frac{r}{\left(\frac{dr}{d\theta}\right)} = \frac{a}{\theta}\left(-\frac{\theta^2}{a}\right) = -\theta.$$

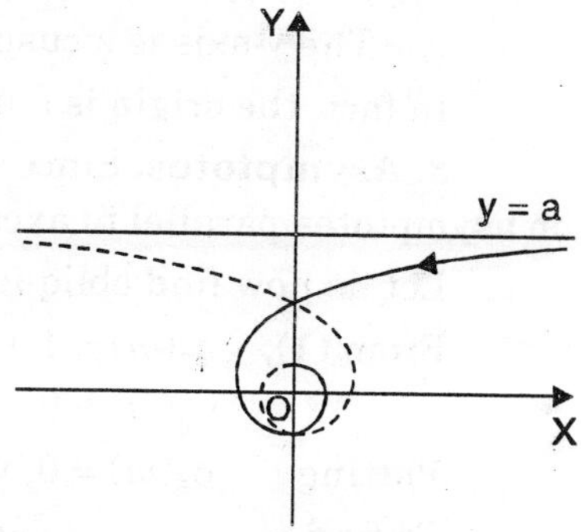

Hence for positive values of θ the shape of the curve as shown by thick lines in the figure.

The part of the curve for – ve values of θ (or here r also because from $r = \frac{a}{\theta}$, r is negative when θ is negative) is obtained by turning the curve for positive value of θ through two right angles. This branch is shown in the figure by the dotted line.

5. Sometimes it is helpful to change from the Cartesian to polar co-ordinates and vice versa.

Example 5. *Trace the following curves :*

(*a*) $r = a\,\frac{\sin^2\theta}{\cos\theta}$ (*b*) $r = a\,(\sec\theta + \cos\theta)$.

Sol. (*a*) The equation of the curve is $r\cos\theta = a\sin^2\theta$ ...(1)

Changing to cartesians by putting $\cos\theta = \frac{x}{r}$ and $\sin\theta = \frac{y}{r}$ (1) becomes,

$$r\frac{x}{r} = a\frac{y^2}{r^2} \quad \text{or} \quad r^2x = ay^2. \quad \text{or} \quad (x^2+y^2)x = ay^2 \qquad (\because r^2 = x^2+y^2)$$

or $y^2(a-x) = x^3$ which is the same equation as solved in example 3, Art. 2.

(*b*) The equation of the curve is $r = a\,(\sec\theta + \cos\theta)$ ...(2)

Changing to cartesians by putting

$\cos\theta = \frac{x}{r}$, (2) becomes

$$r = a\left(\frac{r}{x} + \frac{x}{r}\right)$$

or $$r = a\left(\frac{r^2+x^2}{rx}\right)$$

or $$r^2x = a(r^2+x^2)$$

or $$(x^2+y^2)x = a(2x^2+y^2) \qquad |\because r^2 = x^2+y^2$$

or $$y^2(x-a) = x^2(2a-x)$$

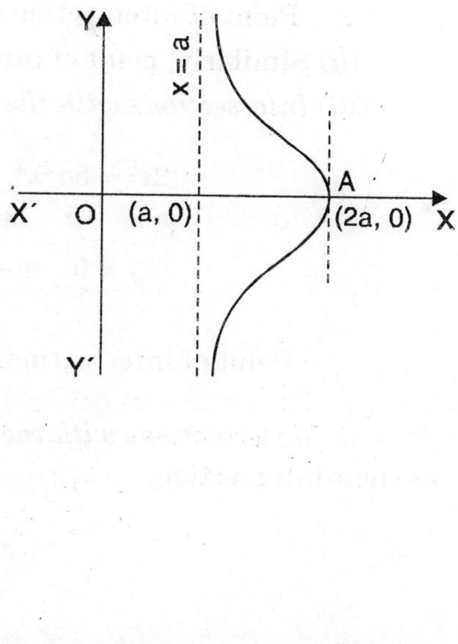

Now, the curve can be easily traced by the cartesian method and is being left as an exercise for the students.

The shape of the curve is as shown in the figure.

Example 6. *Trace the curve $x^5 + y^5 = 5a^2x^2y$.*

Sol. The equation of the curve is $x^5 + y^5 = 5a^2x^2y$ ...(1)

1. **Symmetry.** The curve is symmetrical in opposite quadrants because equation (1) remains unchanged when both x and y are changed to $-x$ and $-y$ respectively.

2. **Origin.** The curve passes through the origin. The tangents at the origin are $x^2y = 0$ *i.e.,* $x = 0, x = 0, y = 0$.

$\therefore$ The y-axis is a cuspidal tangent.

In fact, the origin is a node as well as a cusp.

3. **Asymptotes.** Since co-efficients of x^5 and y^5 are constants, therefore, the curve has no asymptotes parallel to axes.

Let us now find oblique asymptotes :

From (1), $\phi_5(m) = 1 + m^5$ [Putting $x = 1$ and $y = m$ in fifth degree terms]

$\phi_4(m) = 0$ [$\because$ There are no fourth degree terms in equation (1)]

Putting $\phi_5(m) = 0$, we have $m^5 + 1 = 0$ or $m^5 = -1$ $\therefore$ $m = -1$.

To find c, $c\phi_5'(m) + \phi_4(m) = 0$

or $a(5m^4) = 0$ or $c = \dfrac{0}{5m^4} = 0$.

$\therefore$ Equation of the asymptote is $y = mx + c$ or $y = -x$.

Position of the curve w.r.t. the asymptote. In the second quadrant, x is negative and y is positive.

$\therefore$ R.H.S. of (1) is + ve.

$\therefore$ L.H.S. of (1) namely $y^5 + x^5$ must also be + ve.

Hence y should be numerically $> x$.

Thus, the curve lies above the asymptote $y = -x$.

Because of symmetry in opposite quadrants, the curve approaches the other end (in the fourth quadrant) from below.

4. **Points of Intersections**

(*i*) *With x-axis.* Putting $y = 0$ in (1), $x^5 = 0$ or $x = 0$.

$\therefore$ Point of intersection is (0, 0)

(*ii*) Similarly, point of intersection with y-axis is also (0, 0).

(*iii*) *Intersections with the line $y = x$.* Putting $y = x$ in (1), we have

$$2x^5 = 5a^2x^3. \quad \therefore \quad x = 0 \quad \text{or} \quad x = \pm\sqrt{\frac{5}{2}}\cdot a$$

$$\therefore \quad y = 0 \quad \text{and} \quad x = \pm\sqrt{\frac{5}{2}}\cdot a$$

$\therefore$ Point of intersection are (0, 0) ; $\left(\pm\sqrt{\dfrac{5}{2}}\,a, \pm\sqrt{\dfrac{5}{2}}\,a\right)$.

(*iv*) *Intersections with the asymptote $y = -x$.* Solving (1) and $y = -x$, we again have (0, 0) as their intersection.

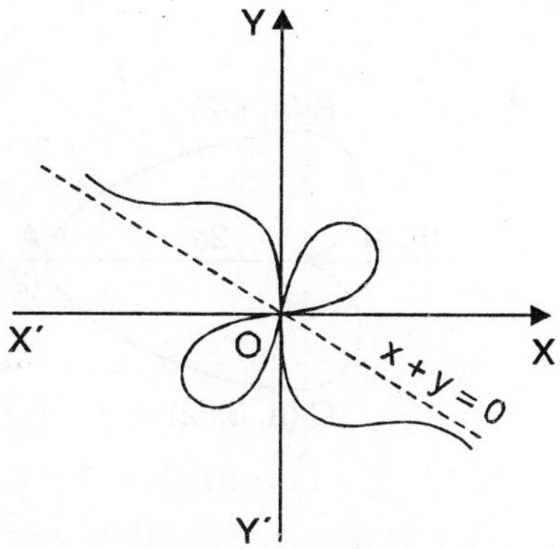

5. **Region.** Transforming to polars [by putting $x = r\cos\theta$ and $y = r\sin\theta$ in (1), we have]

$$r^2 = \frac{5a^2\cos^2\theta\sin\theta}{\cos^5\theta+\sin^5\theta}.$$

For values of θ between $\frac{3\pi}{4}$ and π, r^2 is negative and so r is Imaginary.

Thus, no portion of the curve lies between the lines $\theta = \frac{3\pi}{4}$ and $\theta = \pi$. Thus, the shape of the curve is as shown in the figure.

Note. We could do without this step 5 in the above solution [because the conclusion draw in step (5) was already obtained in step (3)].

EXERCISE 2

Trace the following curves :

1. (*i*) $r = a(1 + \cos\theta)$ (*ii*) $r = 3(1 + \cos\theta)$ (*iii*) $3r = 2(1 + \cos\theta)$.

2. $r = a(1 + \sin\theta)$. (*K.U. 2007*)

3. $r = a(1 - \sin\theta)$.

4. $r = 2a\cos\theta$.

5. $r = a + b\cos\theta$ when $a > b$.

6. $r^2 = a^2\cos 2\theta$.

7. $r = a\sin 2\theta$.

8. $r = a\cos 2\theta$.

9. $r = a\cos 3\theta$.

10. $r = a\cos 5\theta$.

11. $r = \frac{a\theta^2}{1+\theta^2}$.

[**Hint.** $r = a$ is a circular asymptote for this curve $\because \underset{\theta\to\infty}{\text{Lt}} \frac{a\theta^2}{1+\theta^2} = a.$]

12. (*a*) $r = ae^{m\theta}$, where a, m are + ve. (*b*) $r = ae^{\theta\cot\alpha}$.

[**Hint.** Curve in part (*b*) in the same as in (*a*) where $m = \cot\alpha$.]

13. $r = a\theta$.

14. $r^2\cos 2\theta = a^2$.

[**Hint.** Cartesian equation is $x^2 - y^2 = a^2$.]

15. $r = a(1 + \sec\theta)$.

16. $x^4 + y^4 = 4a^2xy$. (*K.U. 1996 S*)

[**Hint.** No portion of the curve lies in second and fourth quadrants.]

17. $y^4 - x^4 + xy = 0$.

18. $x^5 + y^5 = 5ax^2y^2$.

19. $x^4 + y^4 = 4axy^2$.

20. $\frac{2a}{r} = 1 + \cos\theta$ (Parabola).

ANSWERS

1.

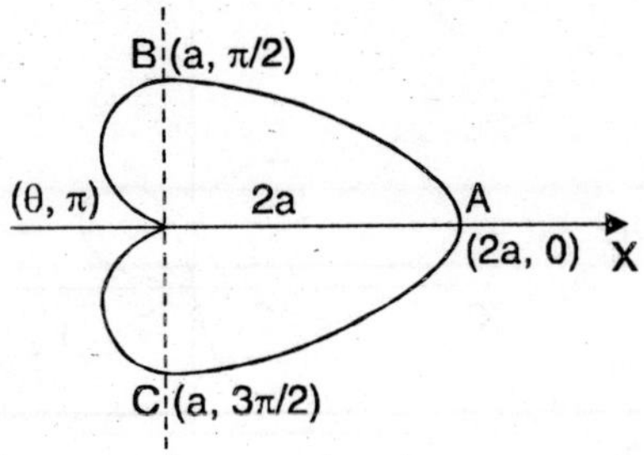

2.

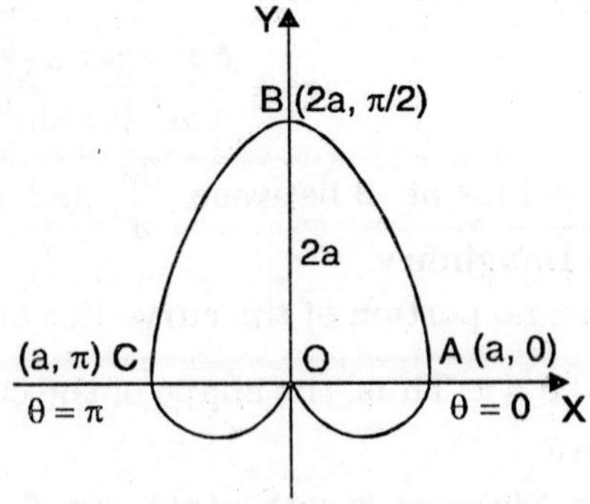

3.

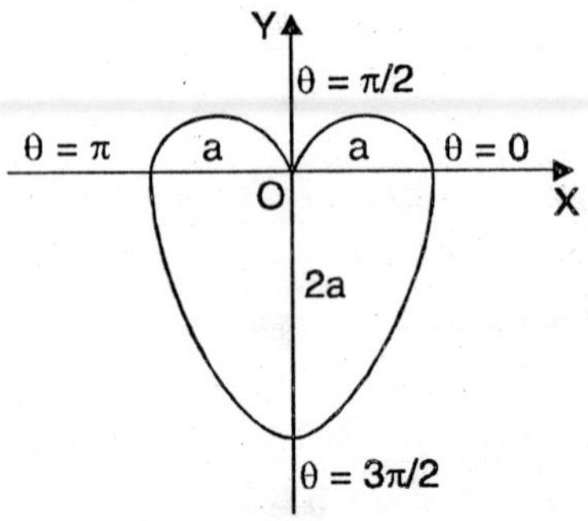

4.

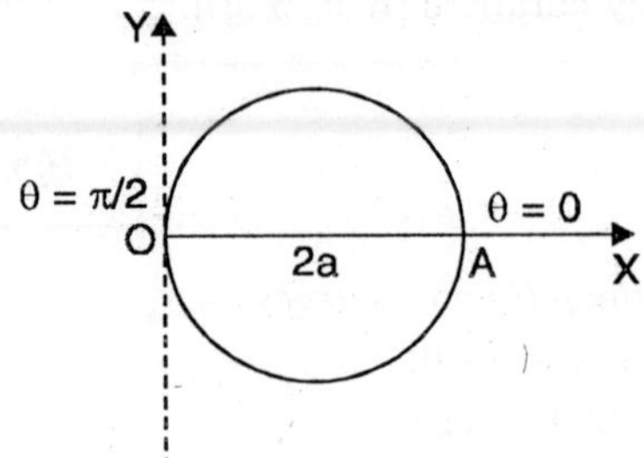

5.

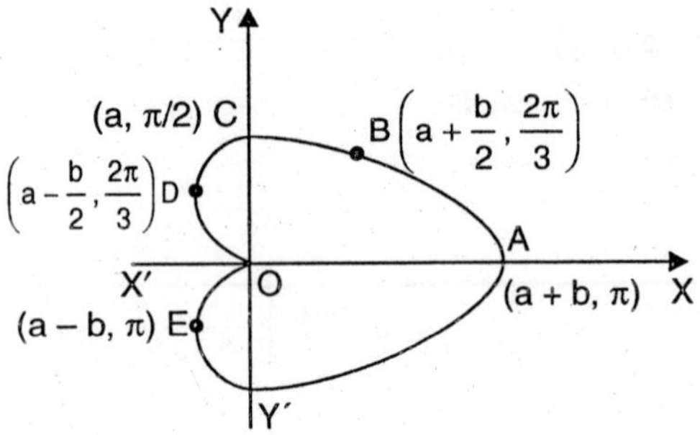

6.

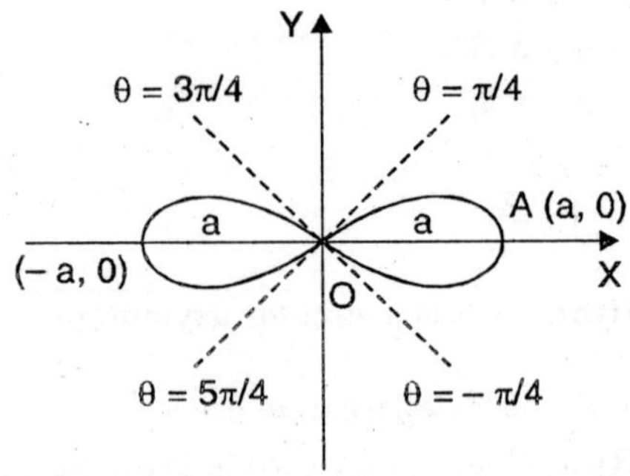

7.

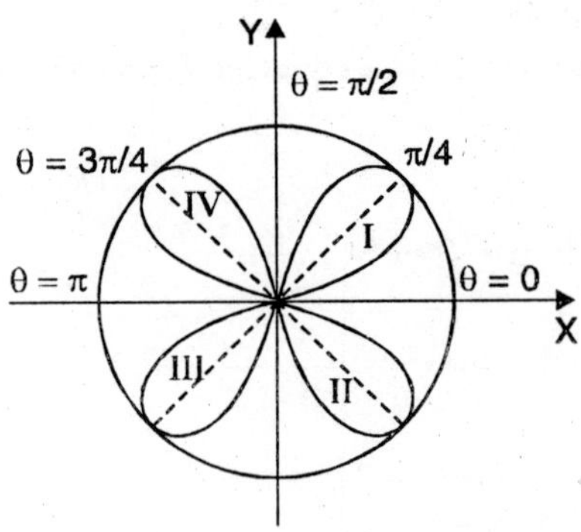

8.

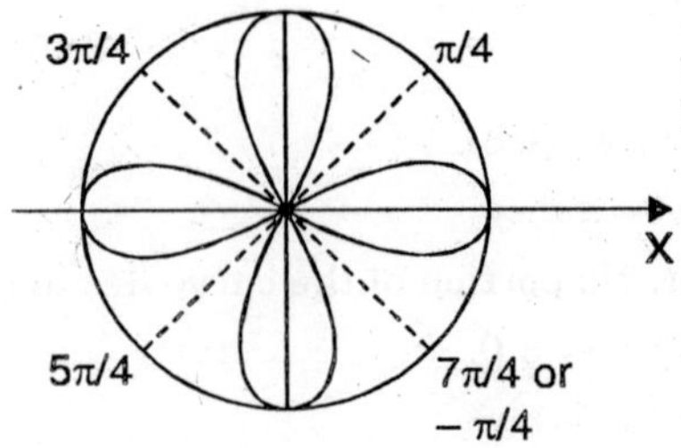

9.

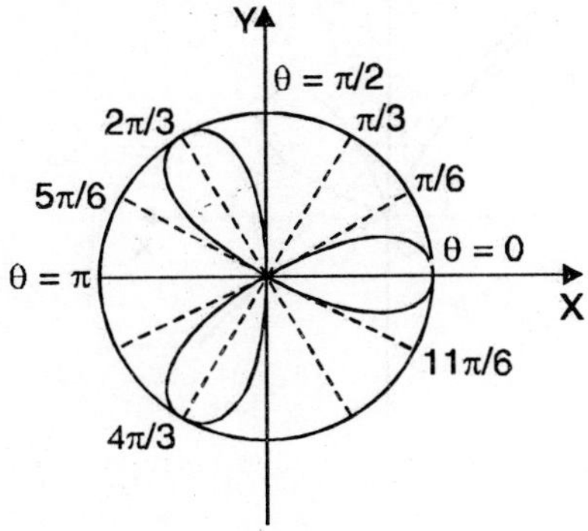

10.

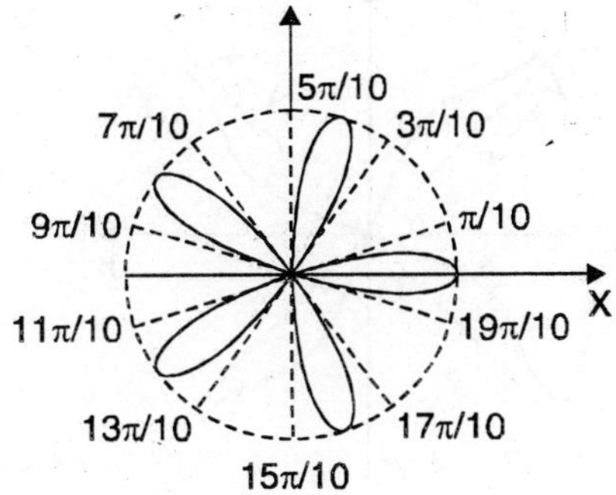

11.

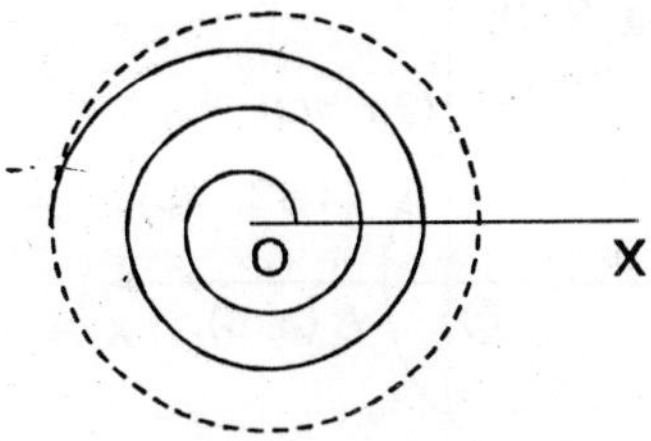

12.

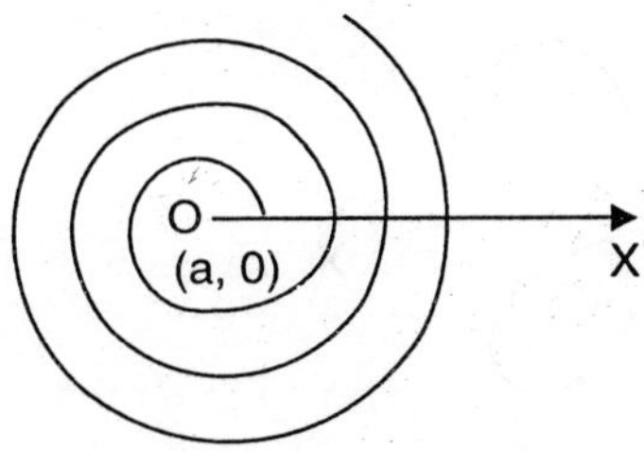

Circular Asymptote $r = a$ is shown as dotted. The part of the curve for negative values is its reflection in the initial line.

13.

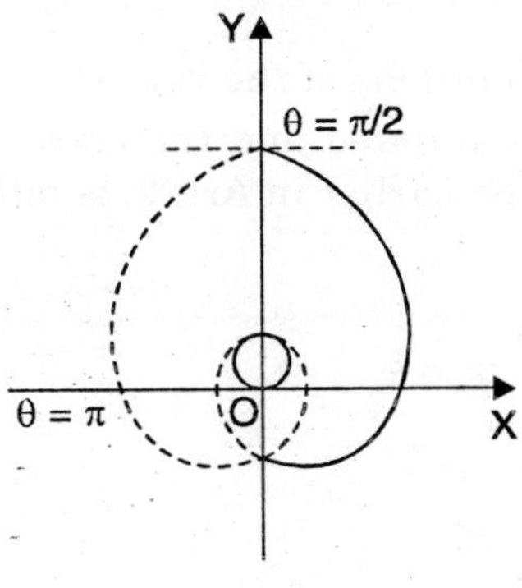

14.

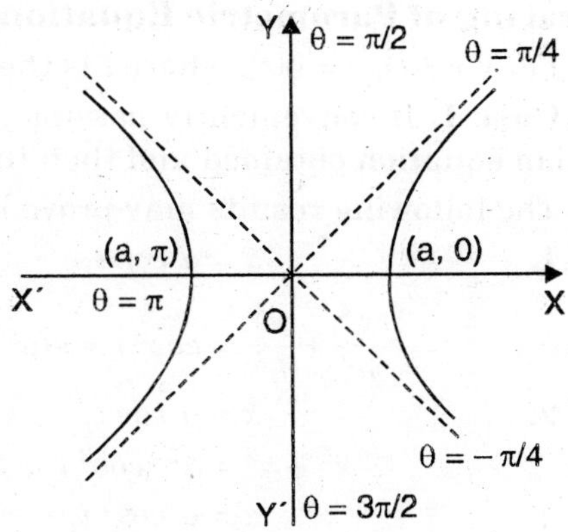

15.

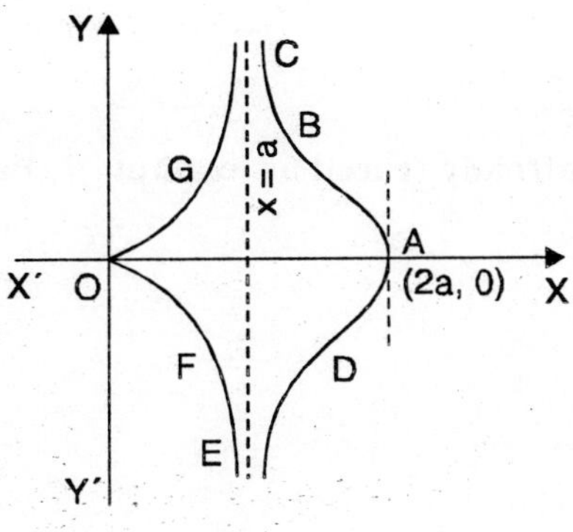

16.

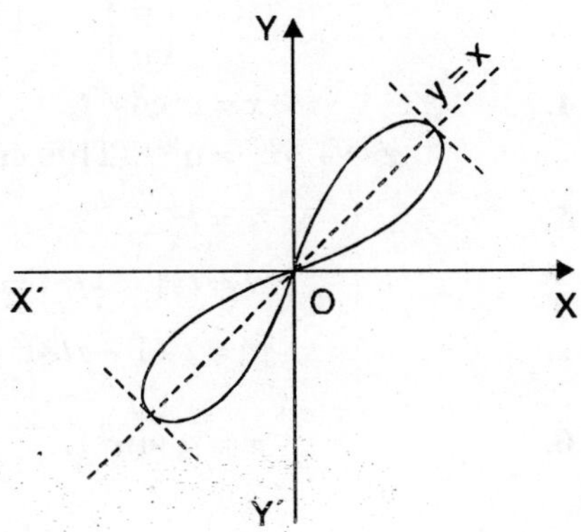

17.

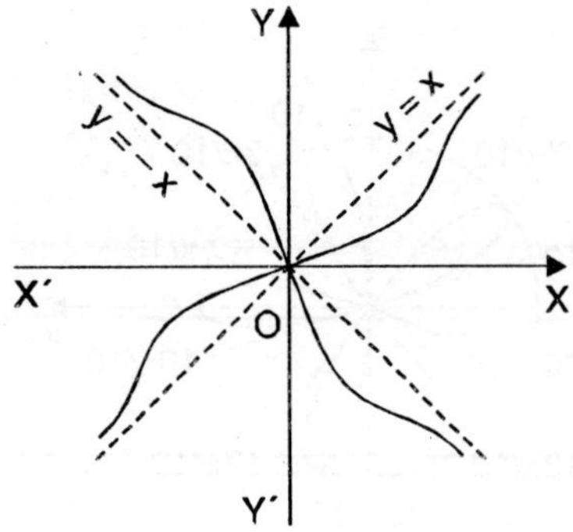

18.

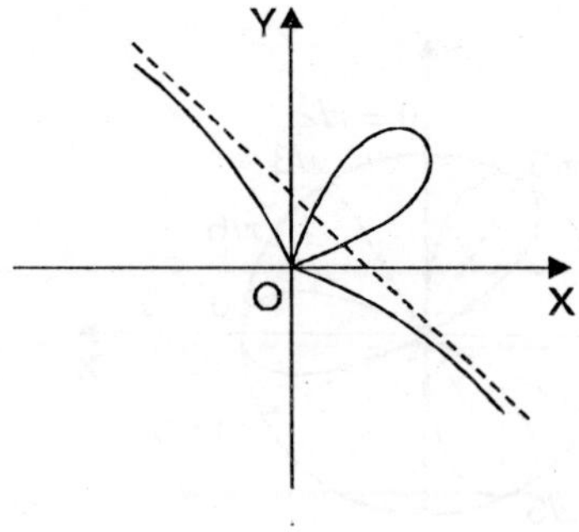

19.

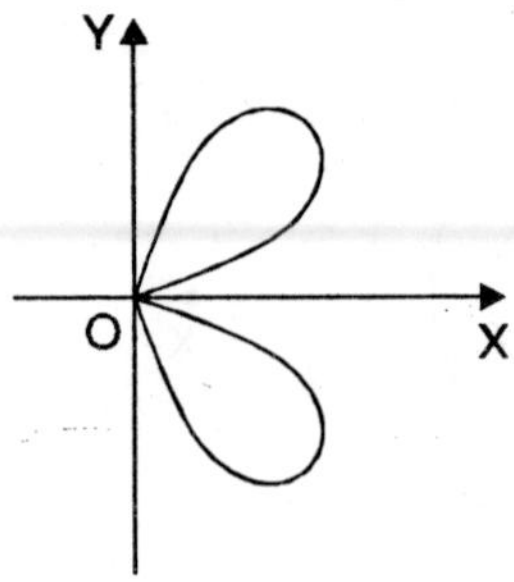

20.

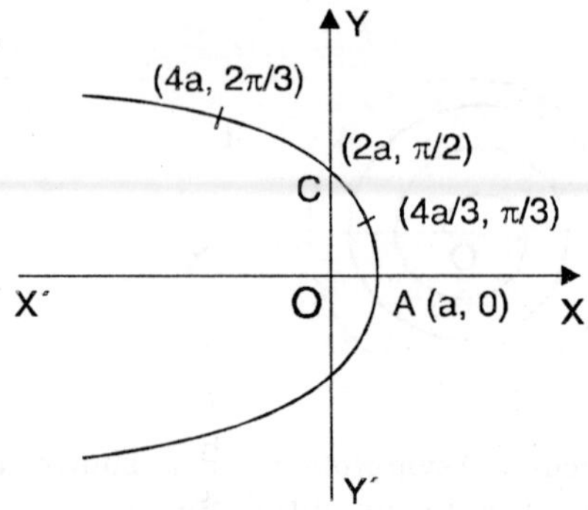

6. Tracing of Parametric Equations

Let $x = f(t)$, $y = \phi(t)$, where t is the parameter in the equations of the curve.

Case I. If conveniently possible, the parameter is eliminated and the corresponding cartesian equation obtained and then the procedure explained earlier in Art. 2. is followed.

The following results may prove helpful :

1. $x = a \cos t, \qquad y = b \sin t.$ (*Ellipse*)

$$\therefore \quad \frac{x^2}{a^2} + \frac{y^2}{b^2} = \cos^2 t + \sin^2 t = 1.$$

2. $x = a \cos t, \qquad y = a \sin t.$ (*Circle*)

$$\therefore \quad x^2 + y^2 = a^2 (\cos^2 t + \sin^2 t) = a^2$$

3. $x = a \cos^3 t, \qquad y = b \sin^3 t$ (*Hypo-cycloid*)

$$\therefore \quad \left(\frac{x}{a}\right)^{2/3} + \left(\frac{y}{b}\right)^{2/3} = 1.$$

4. $x = a \cos^3 t, \qquad y = a \sin^3 t$ (*Astroid*)

$x^{2/3} + y^{2/3} = a^{2/3}$.(This curve, we have already traced as example 7, Page 249.)

5. $x = t^2, \qquad y = t - \frac{1}{3} t^3$

$$\therefore \quad y = t(1 - \tfrac{1}{3}t^2) = t(1 - x/3)$$

$$\therefore \quad y^2 = t^2(1 - x/3)^2 = x(1 - x/3)^2.$$

6. $x = a \sin^2 t, \qquad y = a \dfrac{\sin^3 t}{\cos t}$ (*Cissoid*)

$$\therefore \quad y = \frac{\sin t}{\cos t} x \quad \therefore \quad y^2 = \frac{\sin^2 t}{1-\sin^2 t} . x^2 = \frac{x/a . x^2}{1-x/a}$$

or $$y^2 (a - x) = x^3.$$

7. $$x = \frac{1-t^2}{1+t^2}, y = \frac{2t}{1+t^2}$$ *(Circle)*

$$\therefore \quad x^2 + y^2 = \frac{(1-t^2)^2 + 4t^2}{(1+t^2)^2} = \frac{(1+t^2)^2}{(1+t^2)^2} = 1.$$

8. $$x = \frac{3at}{1+t^3}, y = \frac{3at^2}{1+t^3}.$$ *(Folium of Descarte's)*

$$x^3 + y^3 = 27a^3t^3 \frac{(1+t^3)}{(1+t^3)^3} = \frac{27a^3t^3}{(1+t^3)^2} = 3axy.$$

Case II. If it is not conveniently possible to eliminate the parameter, then the following procedure is followed :

1. **Symmetry.**

(*i*) If on changing t to $-t$ or (t to $\pi - t$) ; $x(=f(t))$ remains unchanged and y $(= \phi(t))$ changes to $-y$; then the curve is **symmetrical about x-axis.**

e.g., The parabola $x = at^2, y = 2at$ is symmetrical about x-axis.

(*ii*) Similarly, if on changing t to $-t$ (or t to $\pi - t$), x changes to $-x$ and y remains unchanged, then the curve is **symmetrical about y-axis.**

e.g., (1) the cycloid $x = a(t + \sin t), y = a(1 - \cos t)$ is symmetrical about y-axis.

(2) The ellipse $x = a \cos t, y = b \sin t$ is symmetrical about y-axis. (changing t to $\pi - t$)

(*iii*) If on changing t to $-t$; x changes to $-x$ an y to $-y$, the curve is symmetrical in *opposite quadrants.*

e.g., the rectangular hyperbola $x = ct, y = \frac{c}{t}$ is symmetrical in opposite quadrants.

2. **Origin**

If on putting $x = 0$ (or $y = 0$) a real value of t can be found out, which makes $y = 0$ (or $x = 0$), then the curve passes through the origin.

or Put both x and y equal to zero and find values of t. If there is any common value of t, then the curve passes through the origin.

3. **Asymptotes.** Find asymptotes if any.

GENERAL METHOD TO FIND ASYMPTOTES

Asymptotes || to x-axis. Find the definite values $d_1, d_2, \ldots\ldots$ to which y tends as $x \to +\infty$ or $-\infty$, then $y = d_1, y = d_2 \ldots\ldots$ are asymptotes || to x-axis.

Asymptotes || to y-axis. Find the definite values $k_1, k_2, k_3, \ldots\ldots$ to which x tends as $y \to +\infty$ or $-\infty$, then $x = k_1, x = k_2, \ldots\ldots$ are asymptotes || to y-axis.

Oblique asymptotes. $y = mx + c$ is an oblique asymptote to a curve where

$$m = \underset{\substack{x \to \infty \\ y \to \infty}}{\text{Lt}} \frac{y}{x} \quad \text{and} \quad c = \underset{\substack{x \to \infty \\ y \to \infty}}{\text{Lt}} (y - mx).$$

4. **Points of Intersection.** (*i*) The points of intersection of the curve with x-axis are given by the roots of the equation $y = \phi(t) = 0$.

(*ii*) The points of intersection of the curve with y-axis are given by the roots of the equation $x = f(t) = 0$.

5. **Region.** If easily possible, find the greatest and least values of x and y and therefore, the lines parallel to the axes between which the curve lies or does not lie.

For example. If the curve is $x = a \cos \theta, y = b \sin \theta$; then the greatest value of x is a and the least value is $-a$.

($\because$ greatest and least values of cos θ are 1 and – 1)

$\therefore$ The curve lies entirely between the lines $x = \pm a$.

Similarly, the curve lies entirely between the lines $y = \pm b$.

6. **Special points.** Find $\frac{dx}{dt}$ and $\frac{dy}{dt}$ and then consider the signs of $\frac{dx}{dt}$ and $\frac{dy}{dt}$ in suitable intervals of t.

Find $\frac{dy}{dx}\left(=\frac{dy/dt}{dx/dt}\right)$.

Give to the parameter t certain values and find the corresponding values of x, y and the slope of the tangent namely $\frac{dy}{dx}$. Plot these points (whose cartesian co-ordinates are known to us). Find those points on the curve for which $\frac{dy}{dx} = 0$ or ∞. Also find $\frac{d^2y}{dx^2}$ and discuss for concavity and points of inflexion (as explained in Step 6 of Art. 2).

Example 1. *Trace the cycloid* $x = a(\theta + \sin \theta)$. $y = a(1 + \cos \theta)$.

(M.D.U. 1996 ; K.U. 2005)

Sol. The parametric equations of the cycloid are

$$x = a(\theta + \sin \theta),\ y = a(1 + \cos \theta) \qquad \ldots(1)$$

Let us trace this curve firstly for values of θ in the interval $[-\pi, \pi]$.

1. **Symmetry.** On changing θ to – θ in (1), changes to $-x$ and y remains unchanged.

$\therefore$ The curve is symmetrical about y-axis.

2. **Origin.** Putting $y = 0,\ a(1 + \cos \theta) = 0$

$\therefore$ $\cos \theta = -1$ or $\theta = \pi$

For $\theta = \pi,\ x = a(\theta + \sin \theta) = a\pi \neq 0$

$\therefore$ The curve does not pass through the origin.

3. **Asymptotes.** The curve has no asymptotes.

4. **Points or Intersection**

(*i*) *Intersections with x-axis.* Putting $y = 0$ in (1), we get $\theta = \pi$ which gives $x = a\pi$.

$\therefore$ Intersection with x-axis is $(a\pi, 0)$.

(*ii*) *Intersection with y-axis.* Putting $x = 0$ in (1), we have $\theta + \sin \theta = 0$ which is satisfied by only $\theta = 0$ and for $\theta = 0$, $y = a(1 + \cos \theta) = 2a$.

$\therefore$ Intersection with y-axis is $(0, 2a)$.

5. **Region.** We know that $-1 \le \cos \theta \le 1$

$\therefore$ $1 - 1 \le 1 + \cos \theta \le 1 + 1$

or $0 \le (1 + \cos \theta) \le 2$

$\therefore$ $0 \le a(1 + \cos \theta) \le 2a$ or $0 \le y \le 2a$

$\therefore$ Curve lies entirely between the lines $y = 0$ and $y = 2a$.

6. Special points. From (1),

$$\frac{dx}{d\theta} = a(1+\cos\theta) \text{ and } \frac{dy}{d\theta} = -a\sin\theta$$

$$\because \quad \frac{dy}{dx} = \frac{dy/d\theta}{dx/d\theta} = -\frac{a\sin\theta}{a(1+\cos\theta)} = -\frac{2\sin\frac{\theta}{2}\cos\frac{\theta}{2}}{2\cos^2\frac{\theta}{2}} = -\tan\frac{\theta}{2} \qquad \ldots(2)$$

Corresponding values of x, y and dy/dx for different values of θ are given below

	$\theta = -\pi$	$-\pi/2$	0	$\pi/2$	π
From (1)	$x = -a\pi$	$-a\left(\frac{\pi}{2}+1\right)$	0	$a\left(\frac{\pi}{2}+1\right)$	$a\pi$
	$y = 0$	a	$2a$	a	0
From (2)	$\frac{dy}{dx} = \infty$	1	0	-1	$-\infty$

$\therefore$ **Points on the curve are** $(-a\pi, 0)$, $\left[-a\left(\frac{\pi}{2}+1\right), a\right]$, $[0, 2a]$, $\left[a\left(\frac{\pi}{2}+1\right), a\right]$, $[a\pi, 0]$.

Tangents at the points $[-a\pi, 0]$ and $[a\pi, 0)$ are parallel to y-axis.

($\because$ $dy/dx = \infty$ or $-\infty$)

Tangent at the point $(0, 2a)$ is || to x-axis ($\because$ $dy/dx = 0$)

Form (2), $\qquad \frac{dy}{dx} = -\tan\frac{\theta}{2}$

Again diff. w.r.t. x, $\qquad \frac{d^2y}{dx^2} = -\left(\sec^2\frac{\theta}{2}\right)\frac{1}{2}\frac{d\theta}{dx}$

$$\frac{d^2y}{dx^2} = \frac{-1}{2}\sec^2\frac{\theta}{2} \cdot \frac{1}{2a}\sec^2\frac{\theta}{2}$$

$$\because \quad \frac{d\theta}{dx} = \frac{1}{a(1+\cos\theta)} = \frac{1}{2a\cos^2\theta/2} = \frac{1}{2a}\sec^2\frac{\theta}{2}$$

or $$\frac{d^2y}{dx^2} = \frac{-1}{4a}\sec^4\frac{\theta}{2}$$

$\therefore$ $\frac{d^2y}{dx^2}$ **is negative for all values of θ.**

$\therefore$ **The curve is concave downwards.**

For values of $\theta > \pi$ or $< -\pi$, the same types of branches of the curve will be obtained.

The curve consists of congruent arches on both sides of y-axis which extend to infinity.

Hence the shape of the curve is as shown in the figure.

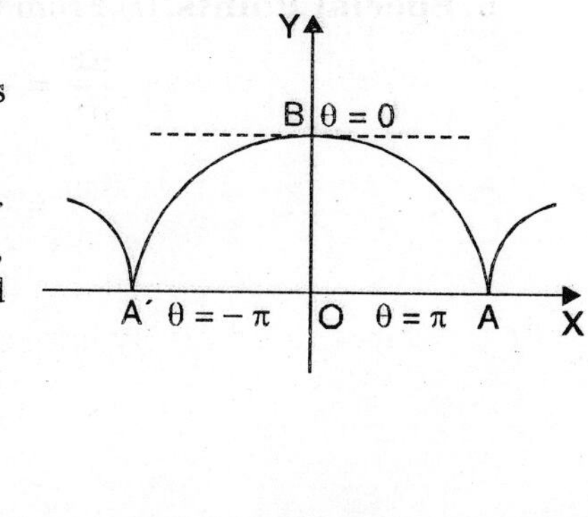

Note. B is called **vertex** of the cycloid and Line A′A *i.e.*, the line joining the two end points is called **base** of the cycloid and a is called the radius of the generating circle.

Remark. We have taken θ from $-\pi$ to π to get one **arch** of the cycloid or one complete cycloid. We will again take θ from $-\pi$ to π for the cycloid

$$x = a\,(\theta + \sin\theta),\ y = a\,(1 - \cos\theta)$$

i.e., **We take θ in $[-\pi, \pi]$ for the cycloids**

$$\mathbf{x = a\,(\theta + \sin\theta),\ y = a\,(1 \pm \cos\theta)}$$

We will take θ in $[0, 2\pi]$ for the cycloids

$$\mathbf{x = a\,(\theta - \sin\theta),\ y = a\,(1 \pm \cos\theta)}$$

($\because$ For these two cycloids, for values of $\theta < 0$ or $\theta > 2\pi$, the same types of branches of the curve will be obtained.)

Example 2. *Trace the curve*

$$x = a\left[\cos t + \frac{1}{2}\log\tan^2\frac{t}{2}\right],\ y = a\sin t. \qquad \textbf{(Tractrix)}$$

Sol. The equations of the curve are

$$x = a\cos t + \frac{a}{2}\log\tan^2\frac{t}{2},\ y = a\sin t \qquad \ldots(1)$$

1. **Symmetry.** (*i*) On changing t to $-t$ in (1), x remains unchanged and y changes to $-y$.

$\therefore$ The curve is symmetrical about x-axis.

(*ii*) On changing t to $\pi - t$ in (1), y is unchanged and x changes to $-x$.

$\therefore$ Symmetry about y-axis.

2. **Origin.** Putting $y = 0$, we get $\sin t = 0$ or $t = 0$ and then

$$x = a + \frac{a}{2}\log 0 = -\infty.$$

$\therefore$ The curve does not pass through the origin.

3. **Asymptotes.** When $t = 0$, $x \to -\infty$ and $y = 0$.

Thus $y = 0$, *i.e.,* x-axis is an asymptote to the curve. (*See step* (3) *Art.* 6)

4. **Points of Intersection.** The curve does not meet x-axis.

The curve meets y-axis (putting $x = 0$) where

$$x = a\cos t + \frac{1}{2}a\log\tan^2\frac{t}{2} = 0 \text{ which is satisfied by } t = \pm\frac{\pi}{2}.$$

and then $y = a\sin\left(\pm\frac{\pi}{2}\right) = \pm a$. Thus, the curve meets y-axis in the points $(0, \pm a)$.

5. **Region.** We know that $-1 \le \sin t \le 1$

$\therefore$ $$-a \le a\sin t \le a \quad \text{or} \quad -a \le y \le a$$

Thus, the curve lies entirely between the lines $y = \pm a$.

6. **Special Points.** (*i*) From (1),

$$\frac{dx}{dt} = -a\sin t + \frac{1}{2}a\,.\,\frac{1}{\tan^2 t/2}\,.\,2\tan\frac{t}{2}\,.\,\sec^2\frac{t}{2}\,.\,\frac{1}{2}$$

$$= a\left(-\sin t + \frac{1}{2\sin t/2\cos t/2}\right) = a\left(-\sin t + \frac{1}{\sin t}\right)$$

$$= a\left(\frac{1-\sin^2 t}{\sin t}\right) = a\,\frac{\cos^2 t}{\sin t} \text{ and } \frac{dy}{dt} = a\cos t$$

$$\therefore \qquad \frac{dy}{dx} = \frac{\frac{dy}{dt}}{\frac{dx}{dt}} = \frac{a\cos t}{\frac{a\cos^2 t}{\sin t}} = \tan t \qquad \text{...(2)}$$

$$\frac{dy}{dx} = \infty, \text{ when } t = \pm\frac{\pi}{2} \text{ and then } x = 0, y = \pm a.$$

Thus, at the points $(0, \pm a)$ the tangent is || to y-axis.

i.e., y-axis itself is a tangent at the points $(0, \pm a)$.

$\frac{dy}{dx} = 0$ when $t = 0$ and then $x \to \infty$, $y = 0$, showing that $y = 0$ or x-axis is an asymptote as proved earlier.

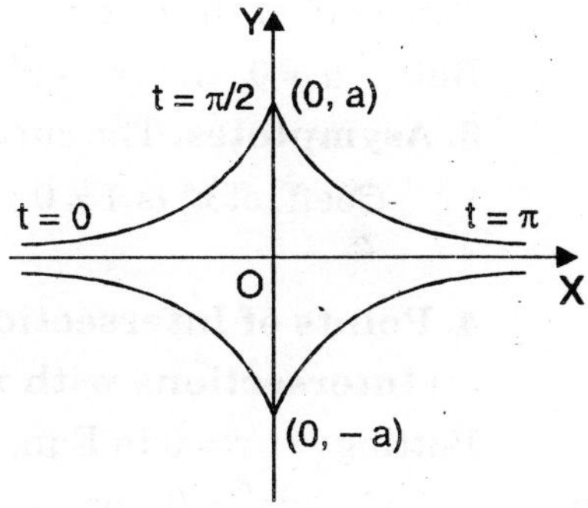

From (2), $\frac{dy}{dx} = \tan t$

Diff. w.r.t. x, $\frac{d^2y}{dx^2} = \sec^2 t\,\frac{dt}{dx}$

$$= \sec^2 t \,.\, \frac{\sin t}{a\cos^2 t} = \frac{\sin t}{a\cos^4 t}$$

$\therefore \quad \frac{d^2y}{dx^2}$ is + ve for values of t in $[0, \pi]$.

$\therefore$ The curve is concave upwards for values of t in $[0, \pi]$.

The shape of the curve is as shown in the figure.

Example 3. *Trace the curve* $x = \frac{a(t+t^3)}{1+t^4}$, $y = \frac{a(t-t^3)}{1+t^4}$. (M.D.U. 1997)

Sol. Equations of the curve are $x = \frac{a(t+t^3)}{1+t^4} = \frac{at\,(1+t^2)}{1+t^4}$ ...(1)

and $$y = \frac{a(t-t^3)}{1+t^4} = \frac{at\,(1-t^2)}{1+t^4} \qquad \text{...(2)}$$

Let us eliminate t from Equations (1) and (2) to form Cartesian Equation.

Dividing Eqn. (1) by Eqn. (2), we have

$$\frac{x}{y} = \frac{1+t^2}{1-t^2} \quad \text{or} \quad y + t^2y = x - t^2x \quad \text{or} \quad t^2(x+y) = x - y$$

$$\therefore \qquad t^2 = \frac{x-y}{x+y} \qquad \text{...(3)}$$

S.B.S. of Eqn. (1) and cross-multiplying $x^2(1+t^4)^2 = a^2t^2\,(1+t^2)^2$

Putting the value of t^2 from (3),

$$x^2\left[1+\frac{(x-y)^2}{(x+y)^2}\right]^2 = \frac{a^2\,(x-y)}{x+y}\left(1+\frac{x-y}{x+y}\right)^2$$

or $$x^2 \frac{[(x+y)^2+(x-y)^2]^2}{(x+y)^4} = \frac{a^2(x-y)}{(x+y)}\left[\frac{2x}{(x+y)}\right]^2$$

or $$\frac{4x^2(x^2+y^2)^2}{(x+y)^4} = \frac{4a^2x^2(x-y)}{(x+y)^3}$$

Dividing both sides by $\dfrac{4x^2}{(x+y)^3}$, $$\frac{(x^2+y^2)^2}{x+y} = a^2(x-y)$$

or cross-multiplying $$(x^2+y^2)^2 = a^2(x^2-y^2) \quad ...(4)$$

1. Symmetry. Curve (4) is symmetrical both about x-axis and y-axis [$\because$ It contains only even powers of y and only EVEN powers of x].

2. Origin. Curve (4) passes through the origin.

Tangents at the origin are $a^2(x^2-y^2) = 0$ (Equating lowest degree terms to zero)

But $a \neq 0 \quad \therefore \quad x^2 - y^2 = 0$ or $y^2 = x^2 \quad \therefore \quad y = \pm x$.

3. Asymptotes. The curve has no asymptotes.

[$\because$ Coeff. of x^4 is $1 \neq 0$; Coeff. of y^4 is $1 \neq 0$;

$\phi_4(m) = (1+m^2)^2 = 0$ gives no real values of m]

4. Points of Intersection

(*i*) **Intersections with x-axis**

Putting $y = 0$ in Eqn. (4), $x^4 = a^2x^2$

or $x^4 - a^2x^2 = 0$ or $x^2(x^2-a^2) = 0$

$\therefore$ Either $x = 0$ or $x^2 = a^2$

i.e., $x = \pm a$

$\therefore$ Intersections of the curve (4) with x-axis are $(-a, 0)$ $(0, 0)$ and $(a, 0)$.

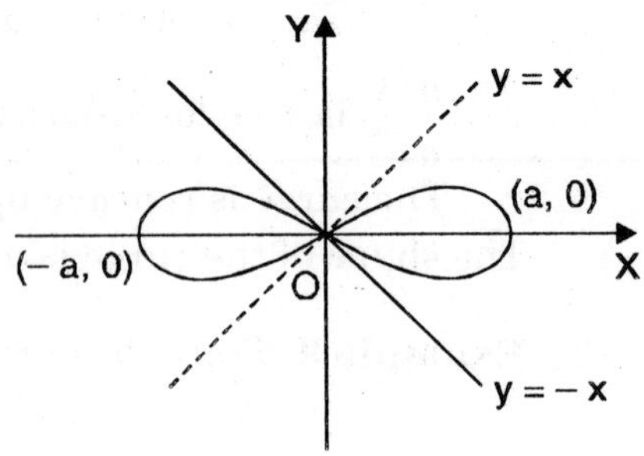

(*ii*) **Intersections with y-axis**

Putting $x = 0$ in Eqn. (4), $y^4 = -a^2y^2$ or $y^4 + a^2y^2 = 0$ or $y^2(y^2+a^2) = 0$

$\therefore \quad y = 0$ or $y = \pm ia$

$\therefore$ The only point of intersection of curve (4) with y-axis is (0, 0).

Thus the shape of the curve is as shown in the figure.

EXERCISE 3

Trace the following curves :

1. $x = a(\theta + \sin\theta), y = a(1 - \cos\theta)$.
[**Hint.** See Remark Example 1.]

2. $x = a(\theta - \sin\theta), y = a(1 + \cos\theta)$.
[**Hint.** See Remark Example 1.]

3. $x = a(\theta - \sin\theta), y = a(1 - \cos\theta)$

4. $x = t^2, y = t - \frac{1}{3}t^3$.

5. $x = \dfrac{a(1-t^2)}{1+t^2}, \quad y = \dfrac{at(1-t^2)}{1+t^2}$

[**Hint.** Eliminate t by dividing.]

6. $x = a\,[\cos\theta - \log(1+\cos\theta)],\ y = a\sin\theta.$

7. $x = a\sin 2\theta\,(1+\cos 2\theta),\ y = a\cos 2\theta\,(1-\cos 2\theta).$

ANSWERS

1.

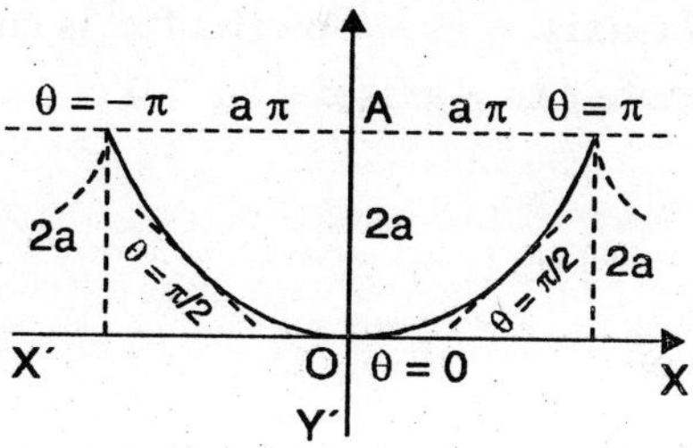

2.

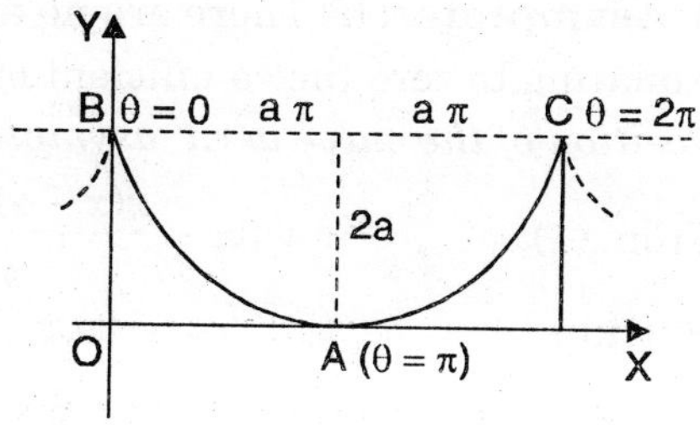

3.

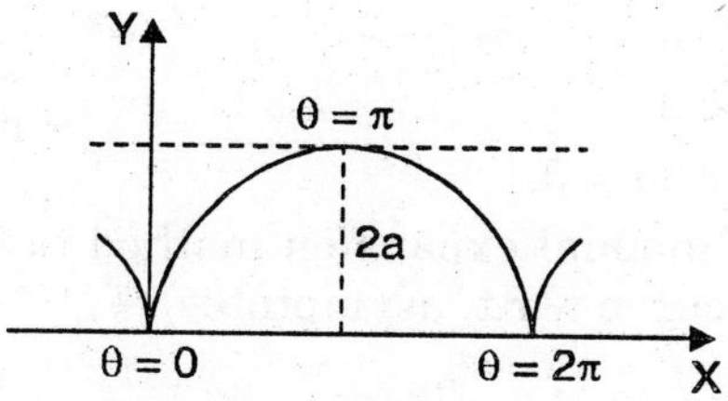

4.

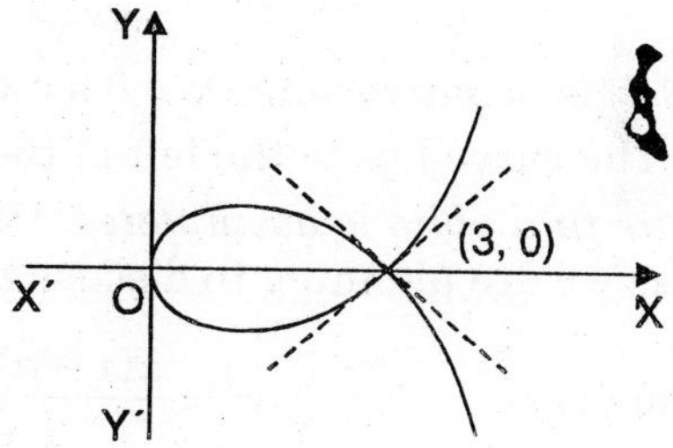

5.

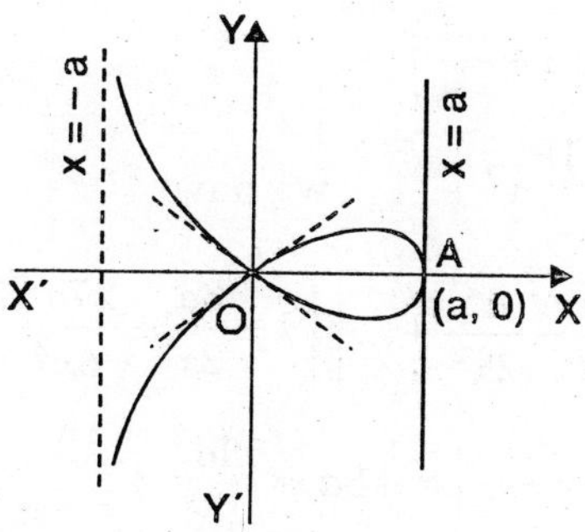

6.

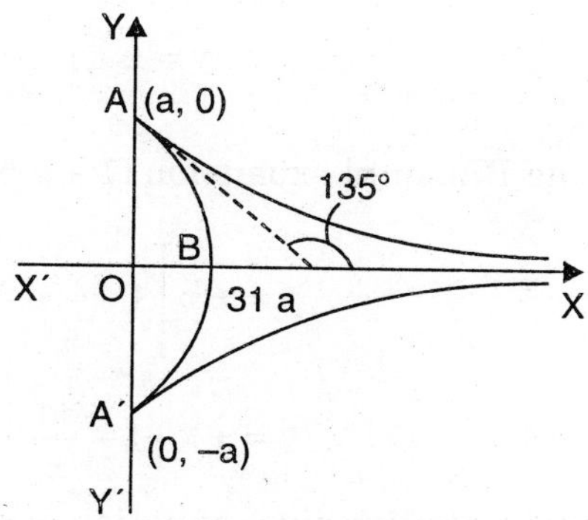

7.

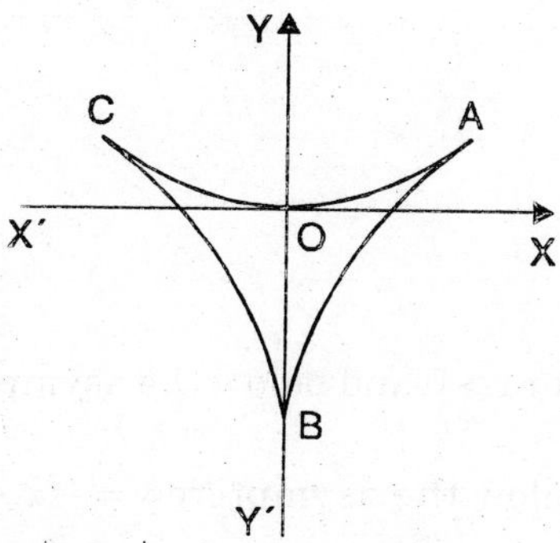

7. Miscellaneous Solved Examples on Curve Tracing

Example 1. *Trace the curve* $y^2(x + 3a) = x(x - a)(x - 2a)$.

Sol. Equation of the curve is $y^2(x + 3a) = x(x - a)(x - 2a)$ ...(1)

1. **Symmetry.** The curve is symmetrical about x-axis.

($\because$ There are even powers of y only in the equation)

2. **Origin.** The curve passes through the origin and the tangent at the origin is $x = 0$ *i.e.*, y-axis. (Equating lowest degree terms to 0)

3. **Asymptotes.** (*i*) There are no asymptotes $\parallel$ to x-axis. [$\because$ Co-eff. of x^3 is constant]

Equating to zero the co-efficient of y^2, the asymptote $\parallel$ to y-axis $x + 3a = 0$.

Position of the curve w.r.t. asymptote $x + 3a = 0$.

From (1), $$x + 3a = \frac{x(x-a)(x-2a)}{y^2}$$

Putting $x = -3a$ in R.H.S., we have

$$x + 3a = -\frac{-60a^3}{y^2}$$

R.H.S. is negative when $y > 0$ as well as when $y < 0$.

$\therefore$ The curve lies to the left of the asymptote $x + 3a = 0$.

Let us find oblique asymptotes. (We are using Binomial expansion method to find asymptotes as we are planning to discuss the position of curve w.r.t. asymptotes).

From (1), $$y^2 = \frac{x(x-a)(x-2a)}{x+3a} = x^2\left(1-\frac{a}{x}\right)\left(1-\frac{2a}{x}\right)\left(1+\frac{3a}{x}\right)^{-1}$$

$$\therefore \quad y = \pm x\left(1-\frac{a}{x}\right)^{1/2}\left(1-\frac{2a}{x}\right)^{1/2}\left(1+\frac{3a}{x}\right)^{-1/2}$$

Using Binomial expansion $(1 + x)^n = 1 + nx + \frac{n(n-1)}{2!}x^2 + \ldots\ldots$, we have

$$y = \pm x\left[1-\frac{a}{2x}-\frac{a^2}{8x^2}+\ldots\ldots\right]\left[1-\frac{a}{x}-\frac{a^2}{2x^2}+\ldots\right]\left[1-\frac{3a}{2x}+\frac{27a^2}{8x^2}+\ldots\right]$$

or $$y = \pm x\left(1-\frac{3a}{x}+\frac{11a^2}{2x^2}+\ldots\right) \quad \text{or} \quad y = \pm\left(x-3a+\frac{11a^2}{2x}+\ldots\right)$$

Taking positive sign, equation of one branch is

$$y = x - 3a + \frac{11a^2}{2x} + \ldots$$

$\therefore$ Equation of the asymptote is $y = x - 3a$.

Also $$y_c - y_a = \frac{11a^2}{2x}$$

which for large values of x is + ve if $x > 0$ and – ve if $x < 0$.

Thus, the curve lies above the asymptote $y = x - 3a$ for $x > 0$ and below the asymptote for $x < 0$.

Similarly, (by taking – ve sign) the curve lies below the asymptote $y = -x + 3a$ for $x > 0$ and above the asymptote for $x < 0$.

4. **Points of Intersection**

(*i*) *With x-axis.* Putting $y = 0$ in (1), we get $x(x - a)(x - 2a) = 0$, *i.e.,* $x = 0, a, 2a$.

Thus, the curve meet x-axis at O(0, 0), A(a, 0), B($2a$, 0).

Tangent at A(a, 0). Putting $x = a$ in (1), $y^2 = 0$, *i.e.,* $y = 0, y = 0$.

$\therefore$ The line $x = a$ meets the curve (1) in two coincident points.

$\therefore$ The line $x = a$ is a tangent to the curve (1) at the point A(a, 0).

(We could also find the tangent at (a, 0) by the general method of shifting the origin).

Similarly, tangent at B($2a$, 0), is $x = 2a$.

(*ii*) *Intersection with y-axis.* The curve meets the y-axis at the origin only.

(*iii*) *Intersection with the asymptotes* $y = x - 3a$ and $y = -x + 3a$. (It is being left as an exercise for the students).

5. **Region.** Solving (1) for y and taking + ve square root,

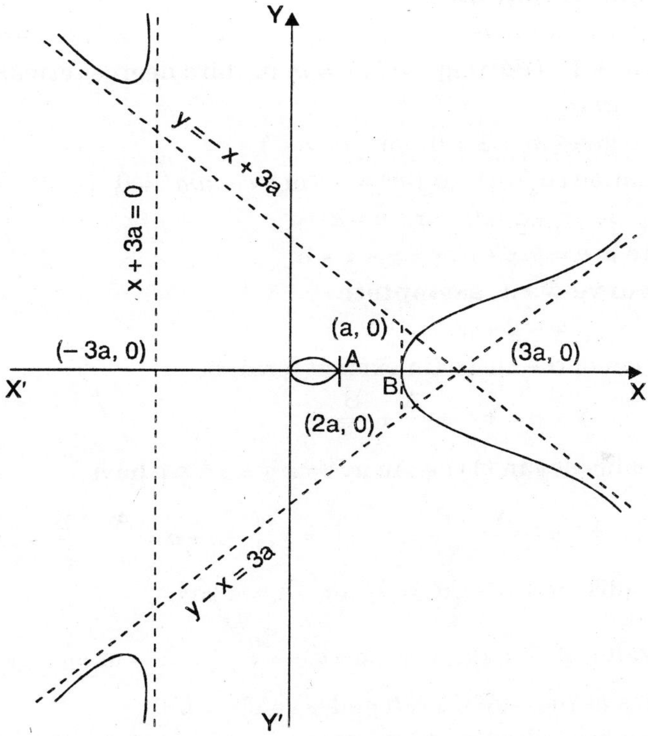

($\because$ The curve is symmetrical about x-axis), we have

$$y = \sqrt{\frac{x(x-a)(x-2a)}{x+3a}} \qquad \ldots(2)$$

when $x < -3a$ y is real,

when $-3a < x < 0$ y is imaginary.

$\therefore$ No portion of the curve lies between the lines $x = -3a$ and $x = 0$.

when $0 < x < a$ y is real,

when $a < x < 2a$ y is imaginary.

$\therefore$ No portion of the curve lies between the lines

$$x = a \quad \text{and} \quad x = 2a,$$

when $x > 2a$ $\quad y$ is real.

Hence the shape of the curve is as shown in the figure.

Example 2. *Trace the curve* $ay^2 = x^2y + x^3$.

Sol. The equation of the curve is $ay^2 = x^2y + x^3$ ...(1)

1. **Symmetry.** There is no symmetry.

2. **Origin.** The curve passes through the origin.

The tangents at the origin are given by $y^2 = 0$

i.e., $y = 0, \quad y = 0.$

$\therefore$ Origin is a cusp.

3. **Asymptotes.** There is no asymptote parallel to co-ordinate axes.

To find oblique asymptotes

Equation (1) of the curve is $x^2y + x^3 - ay^2 = 0$ (Making R.H.S. zero)

$\phi_3(m) = m + 1$ (Putting $x = 1, y = m$ in third degree terms)

$\phi_2(m) = -am^2$

$\phi_3(m) = 0$ gives $m + 1 = 0$ or $m = -1$.

c is given by the equation $c\phi_3'(m) + \phi_2(m) = 0$ or $c - am^2 = 0$

For $m = -1, \quad c - a = 0$ or $c = a$

$\therefore$ Asymptote is $y = mx + c$ or $y = -x + a$.

Position of curve w.r.t. asymptote

Asymptote is $y = -x + a$

Let the equation of the curve (in explicit form) be

$$y = -x + a + \frac{A}{x} + \frac{B}{x^2} + \ldots\ldots \qquad \text{..(2)}$$

Putting this value of y in (1) *i.e.,* in $ay^2 = x^2y + x^3$, we have

$$a\left(-x + a + \frac{A}{x} + \frac{B}{x^2} + \ldots\ldots\right)^2 = x^2\left(-x + a + \frac{A}{x} + \frac{B}{x^2} + \ldots\ldots\right) + x^3$$

Comparing co-effs. of x, $a(-2a) = A$ or $A = -2a^2$

Putting this value of A in (2), $y = -x + a - \dfrac{2a^2}{x} + \ldots\ldots$ or $y_c - y_a = -\dfrac{2a^2}{x}$

which for large values of x is $-$ ve if $x > 0$ and $+$ ve if $x < 0$.

Thus, the curve lies below the asymptote $y = -x + a$ for $x > 0$ and above the asymptote for $x < 0$.

4. **Points of intersection.** The curve (1) meets the co-ordinate axes only at the origin (0, 0).

Intersection with the asymptote

Putting $y = -x + a$ in (1) $ay^2 = x^2y + x^3$, we have

$$a(-x + a)^2 = x^2(-x + a) + x^3$$

or $$ax^2 - 2a^2x + a^3 = -x^3 + ax^2 + x^3$$

or $$2a^2x = a^3 \quad \therefore \quad x = \frac{a}{2}.$$

Putting $x = \frac{a}{2}$ in $y = -x + a$, we have $y = \frac{a}{2}$.

$\therefore$ Point of intersection of the curve with the asymptote is $\left(\frac{a}{2}, \frac{a}{2}\right)$.

5. **Region.** (*i*) (1) is $ay^2 - x^2y - x^3 = 0$

Solving for y, $\quad y = \dfrac{x^2 \pm \sqrt{x^4 + 4ax^3}}{2a} \quad$ or $\quad y = \dfrac{1}{2a}\left[x^2 \pm \sqrt{x^3(x + 4a)}\right]$

When $\quad x < -4a, \quad y$ is Real.

When $\quad -4a < x < 0 \quad y$ is Imaginary.

$\therefore$ The curve does not exist between the ordinates $x = -4a$ and $x = 0$.

When $\quad x > 0 \quad y$ is Real.

(*ii*) Again, in third quadrant x and y are both negative, which makes L.H.S. of (1) positive and R.H.S. of (1) negative, which is impossible. $\therefore$ No portion of the curve lies in third quadrant.

6. **Special points** Equation (1) is $ay^2 = x^2y + x^3$

Diff. w.r.t. x, $\quad 2ay \dfrac{dy}{dx} = x^2 \dfrac{dy}{dx} + y \,.\, 2x + 3x^2$

or $\quad (2ay - x^2) \dfrac{dy}{dx} = 2xy + 3x^2$

$$\therefore \quad \frac{dy}{dx} = \frac{2xy + 3x^2}{2ay - x^2} \qquad \ldots(3)$$

$\dfrac{dy}{dx} = 0$ at the points given by $2xy + 3x^2 = 0$ *i.e.*, $x(2y + 3x) = 0$

$\therefore \quad x = 0$ or $y = -\dfrac{3x}{2}$.

Putting $\quad x = 0, \quad y = 0$

Putting $\quad y = -\dfrac{3x}{2}, \; a\dfrac{9x^2}{4} = -\dfrac{3x^3}{2} + x^3$

or $\quad \dfrac{9ax^2}{4} + \dfrac{x^3}{2} = 0$ or $\dfrac{x^2}{2}\left(\dfrac{9a}{2} + x\right) = 0$

$\therefore \quad x = 0$ and $x = -\dfrac{9a}{2}$

For $\quad x = -\dfrac{9a}{2}, \quad y = \dfrac{27}{4}a$.

$\therefore$ Tangents at the points (0, 0) and $\left(-\dfrac{9a}{2}, \dfrac{27a}{4}\right)$ are parallel to x-axis.

From (3), $\dfrac{dy}{dx} \to \infty$ at the points where $2ay - x^2 = 0$ or $y = \dfrac{x^2}{2a}$.

Putting $y = \dfrac{x^2}{2a}$ in $ay^2 = x^2y + x^3$, we have

$$\frac{x^4}{4a} = \frac{x^4}{2a} + x^3$$

or $$x^4 = 2x^4 + 4ax^3$$

or $$x^4 + 4ax^3 = 0$$

or $$x^3(x + 4a) = 0$$

$\therefore$ $$x = 0, x = -4a$$

For $$x = 0, y = 0$$

For $$x = -4a,$$

$$y = \frac{16a^2}{2a} = 8a$$

$\therefore$ Tangent at $(-4a, 8a)$ is parallel to y-axis.

Thus, the shape of the curve is as shown in the figure.

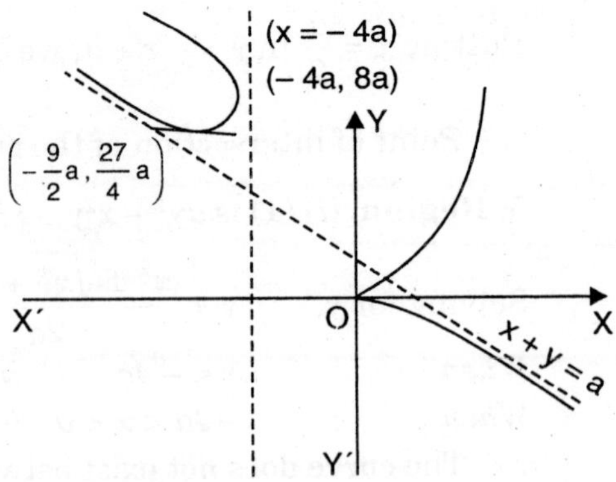

Example 3. *Trace the curve $r = a + b \cos \theta$ when $a < b$.*

Sol. The equation of the curve is $r = a + b \cos \theta$ ...(1) where $a < b$ *i.e.*, $(a - b)$ is negative.

1. **Symmetry.** The curve is symmetrical about the initial line.

($\because$ On changing θ to $-\theta$, equation (1) remains unchanged)

2. **Origin or Pole.** Where $r = 0$, $\cos \theta = \frac{-a}{b} < 1$ numerically ($\because$ $a < b$)

$\therefore$ The curve passes through the pole.

Now $$\cos \theta = \frac{-a}{b} = \cos \alpha \text{ (say)}.$$

$\therefore$ $\theta = \alpha$ is tangent to the curve at the pole where

$$\cos \alpha = \frac{-a}{b}.$$

3. **Asymptotes.** Putting $r = \frac{1}{u}$ in (1), $u = \frac{1}{a + b \cos \theta}$

$u = 0$ does not give any finite values of θ.

$\therefore$ No asymptotes.

4. **Points of Intersection.** Corresponding values of θ and r are given below

$\theta = 0$	$\frac{\pi}{4}$	$\frac{\pi}{3}$	$\frac{\pi}{2}$	$\frac{2\pi}{3}$	$\frac{3\pi}{4}$	π
From (1), $r = a + b$	$a + \frac{b}{\sqrt{2}}$	$a + \frac{b}{2}$	a	$a - \frac{b}{2}$	$a - \frac{b}{\sqrt{2}}$	$(a - b)$

It is given that $a < b$ $\therefore$ $a - b$ is – ve

$\therefore$ The point $(a - b, \pi)$ lies on the half ray $\theta = 0$.

Without **much** loss of generality, let $a > \frac{b}{2}$.

$\therefore$ r is positive for all values of θ from 0 to $\frac{2\pi}{3}$, but r is negative when $\theta = \frac{3\pi}{4}$ or π.

$\therefore$ r must vanish somewhere between $\theta = \frac{2\pi}{3}$ and $\theta = \frac{3\pi}{4}$

$\therefore$ *From Step 2,* α lies between $\frac{2\pi}{3}$ and $\frac{3\pi}{4}$.

5. **Region.** Since $|\cos\theta| \le 1 \quad \therefore \quad r \le (a+b)$

$\therefore$ Entire curve lies within the circle $r = (a+b)$.

6. **Value of ϕ.** From (1), $\dfrac{dr}{d\theta} = -b\sin\theta$

$\therefore$ $\dfrac{dr}{d\theta}$ is negative for all values θ between 0 and π.

$\therefore$ r decreases as θ increases from 0 to π,

$$\tan\phi = \frac{r}{\left(\dfrac{dr}{d\theta}\right)} = \frac{a + b\cos\theta}{-b\sin\theta}$$

$\tan\phi \to \infty$ when $\sin\theta = 0$ *i.e.,* when $\theta = 0$ or π.

$\therefore$ Tangents at the points $\theta = 0$ and $\theta = \pi$ *i.e.,* the points $(a+b, 0)$ and $(a-b, \pi)$ are $\perp$ to the radius vector *i.e.,* $\perp$ to the initial line.

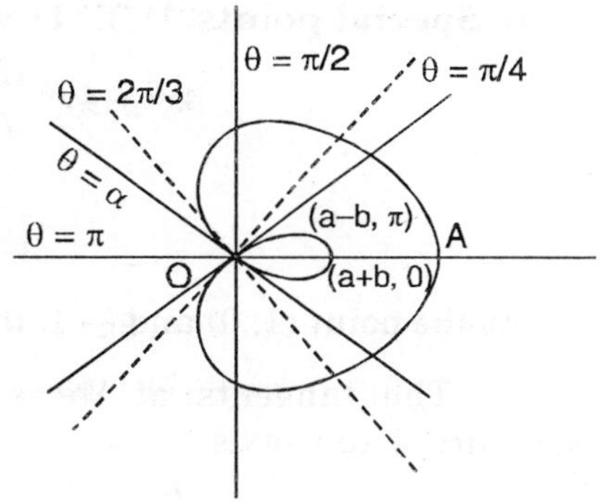

Exercise. *Trace the curve* $r = 2 + 3\cos\theta$

Here $a = 2$, $b = 3$ $(a < b)$.

Example 4. *Trace the curve* $x^4 + y^4 = x^2 + y^2$.

Sol. The equation of the curve is $x^4 + y^4 = x^2 + y^2$ ...(1)

1. **Symmetry.** (*i*) The curve is symmetrical about x-axis

($\because$ There are even powers of y only)

(*ii*) The curve is also symmetrical about y-axis.

(*iii*) The curve is also symmetrical about the lines $y = x$ and $y = -x$.

(*iv*) The curve is also symmetrical in opposite quadrants.

($\because$ On changing x to $-x$, and y to $-y$, equation (1) remains unchanged.)

2. **Origin.** The curve passes through the origin.

The tangents at the origin are given by (equating the lowest degree terms to zero) $x^2 + y^2 = 0$ or $y = \pm ix$ which are imaginary. Hence origin is an *isolated* point.

3. **Asymptotes.** The curve has no asymptotes.

4. **Points of Intersections.** (*i*) Putting $y = 0$ in (1), we get $x^4 = x^2$ or $x^2(x^2 - 1) = 0$.

$\therefore$ $x = 0, \pm 1$.

$\therefore$ Intersections of the curve with x-axis are (0, 0), (1, 0) and (– 1, 0).

(*ii*) Similarly, the curve meets y-axis in the points

(0, 0), (0, 1) and (0, – 1).

(*iii*) The line $y = x$ meets the curve (1) in the points (1, 1) and (– 1, –1).

(*iv*) The line $y = -x$ meets the curve (1) in the points

(1, – 1) and (– 1, 1).

5. **Region.** From (1), $y^4 - y^2 = x^2 - x^4$ or $y^2(y^2 - 1) = x^2(1 - x^2)$

i.e., $$y^2 - 1 = \frac{x^2}{y^2}(1 - x^2)$$

If $x^2 \le 1$ *i.e.,* $-1 < x \le 1$, then $y^2 - 1 \ge 0$ *i.e.,* $y^2 \ge 1$

$\therefore$ $y \ge 1$ or $y \le -1$.

i.e., when x lies between – 1 and 1 ; $y \ge 1\ y \le -1$.

If $x^2 \geq 1$ *i.e.,* $x \geq 1$ or $x \leq -1$, then $y^2 - 1$ is – ve

i.e., $\therefore$ $y^2 \leq 1$ or $-1 \leq y \leq 1$

i.e., when $x \geq 1$ or $x \leq -1$, y lies between -1 and 1.

6. **Special points.** Diff. (1) w.r.t. x, we get

$$4x^3 + 4y^3 \frac{dy}{dx} = 2x + 2y \frac{dy}{dx}$$

or

$$\frac{dy}{dx} = \frac{x - 2x^3}{2y^3 - y}$$

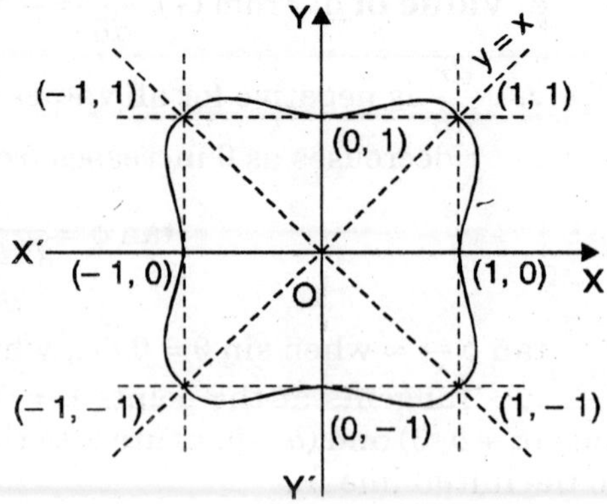

At the point (1, 0) and (– 1, 0) $\frac{dy}{dx} \to \infty$.

$\therefore$ The tangents at these points (1, 0) and (– 1, 0) are || to y-axis.

Again, at $(0, \pm 1)$, $\frac{dy}{dx} = 0$.

$\therefore$ The tangents at these points are || to x-axis.

At the points (1, 1), $\frac{dy}{dx} = -1$, showing that tangent at (1, 1) makes an angle of 135° with axis or is $\perp$ to the line $y = x$. Thus, the shape of the curve is as shown in the figure.

MISCELLANEOUS EXERCISE

Trace the follwing curves :

1. $y^2(x^2 + y^2) - 4x(x^2 + 2y^2) + 16x^2 = 0$.
   [**Hint.** The given equation is $y^4 + y^2(x^2 - 8x) + (16x^2 - 4x^3) = 0$.
   Solving as a quadratic in y^2, we have
   $y^2 = 4x$ and $x^2 + y^2 - 4x = 0$. Now trace the two curves separately.]
2. $xy^2 + (x + a)^2(x + 2a) = 0$
3. $2y^2 = x(4 - x^2)$.
4. $x(x - 2a)y^2 = a^2(x - a)(x - 3a)$.
5. $y = x + \frac{1}{x}$, $x \neq 0$
6. $x^2 \cdot = (1 + y)^2(4 - y)^2$.
7. $x^2(x - y) + y^2 = 0$.
8. $y = e^x$.
9. $y = \log x$.
10. $r = \ldots 2 \cos \theta + \cos 3\theta)$.
11. $x^2(y + 1) = y^2(x - 4)$.
12. $r^2\theta = a^2$.

ANSWERS

1.

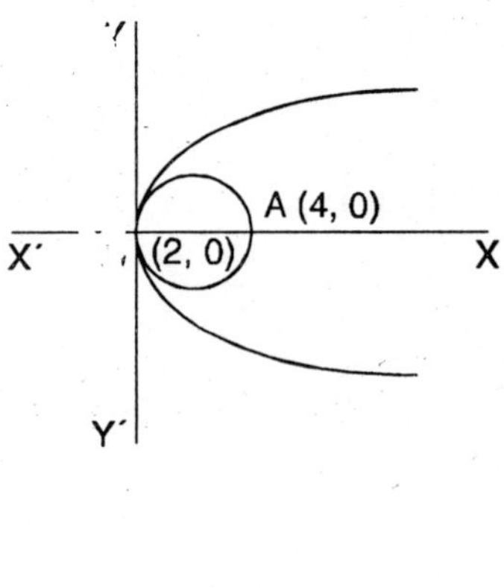

2.

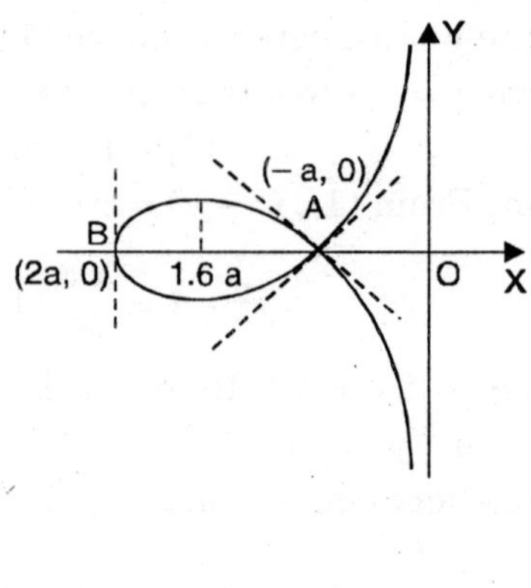

3.

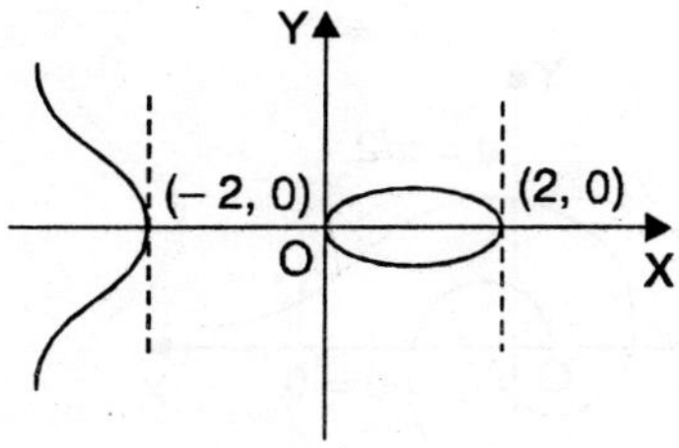
Y
(– 2, 0)
(2, 0)
O
X

4.

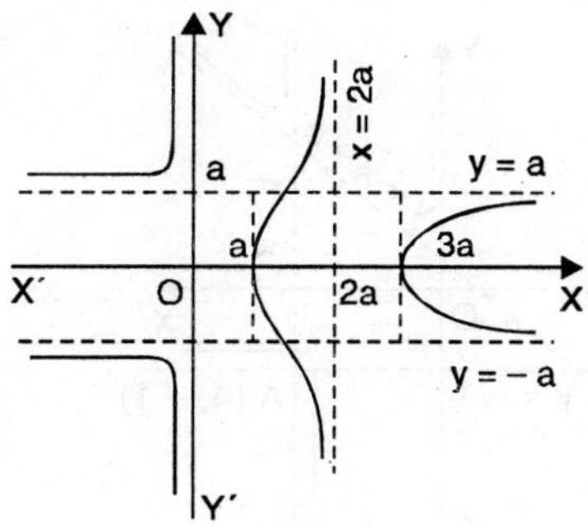
Y
x = 2a
a
y = a
a
3a
X′
O
2a
X
y = – a
Y′

5.

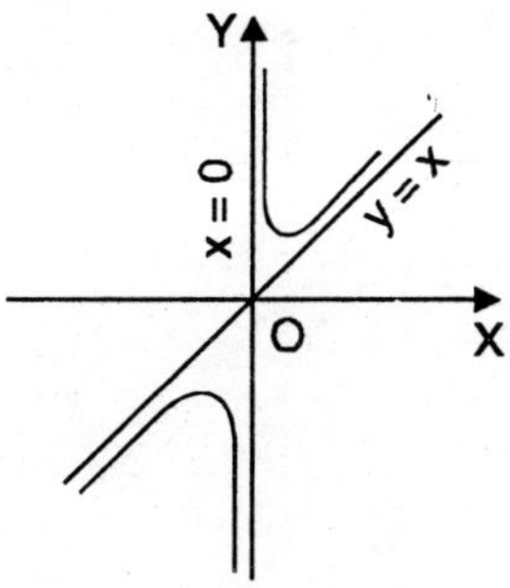
Y
x = 0
y = x
O
X

6.

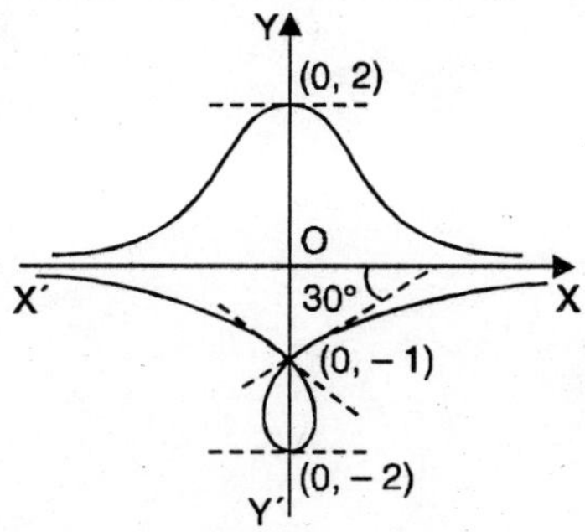
Y
(0, 2)
O
X′
30°
X
(0, – 1)
(0, – 2)
Y′

7.

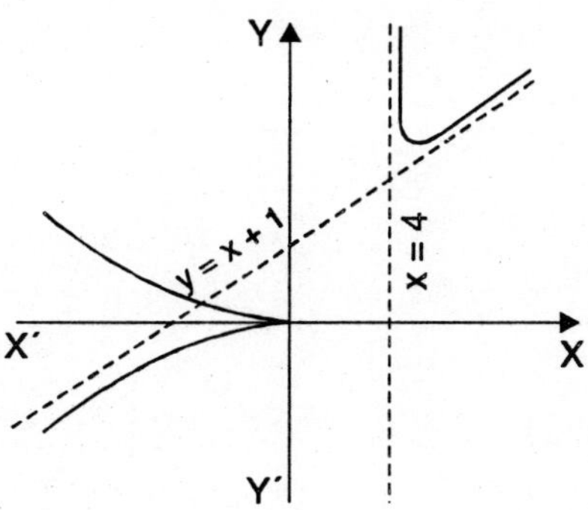
Y
y = x + 1
x = 4
X′
X
Y′

8.

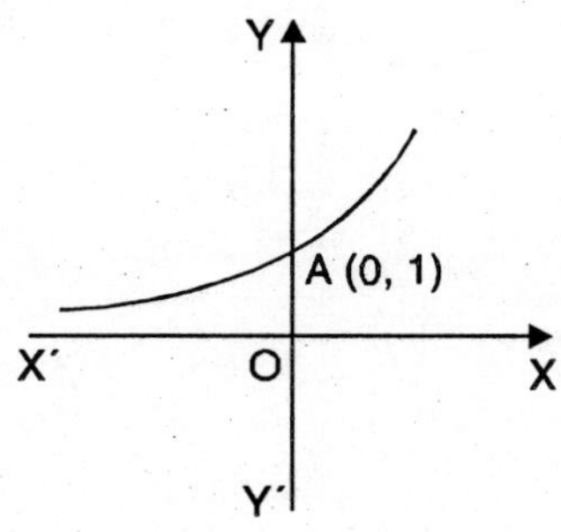
Y
A (0, 1)
X′
O
X
Y′

9.

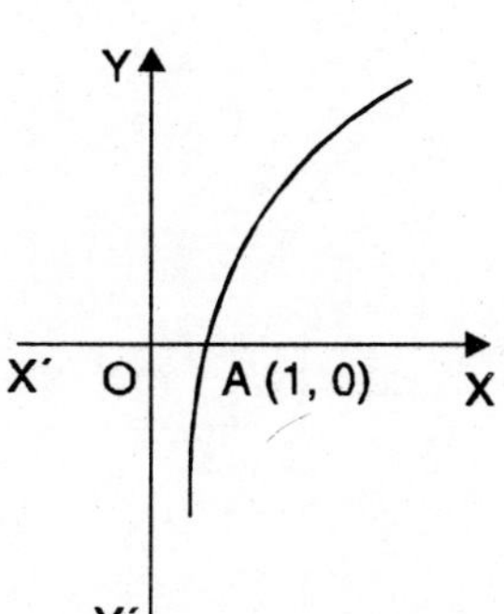
Y
X′
O
A (1, 0)
X
Y′

10.

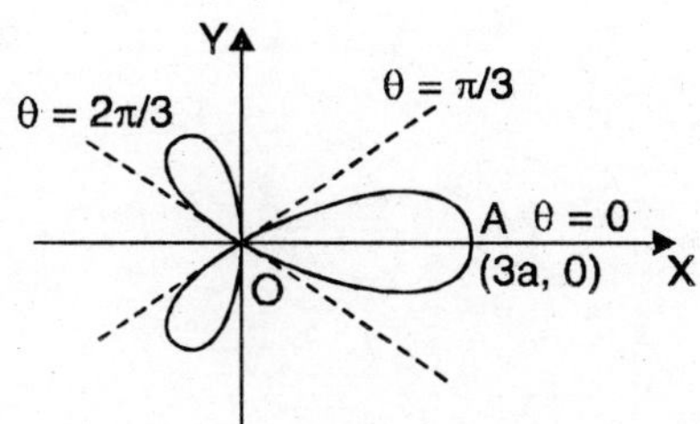
Y
θ = π/3
θ = 2π/3
A θ = 0
(3a, 0)
O
X

11.

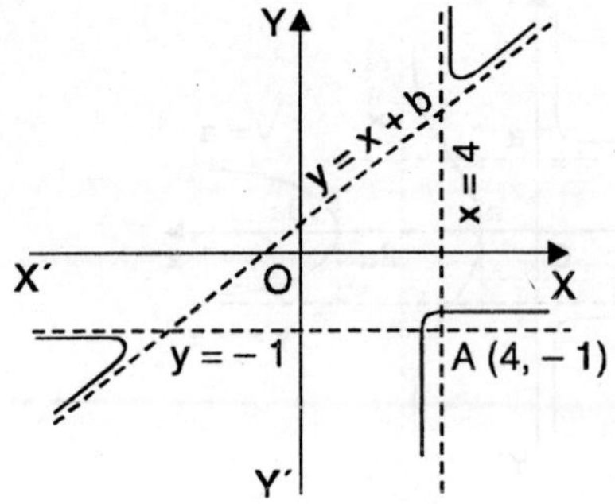

12.

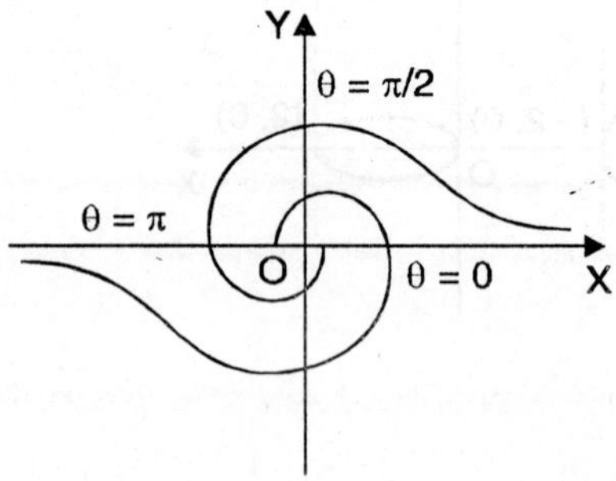

SECTION II

10

Reduction Formulae

1. Reduction Formula

In Integral Calculus, reduction formula is a formula which connects an integral with another integral which is of the same type but of lower degree or lower order and can be easily integrated. The reduction formulae are generally obtained by applying the rule of integration by the parts. The method is known as integration by *successive reduction.*

2. Find a reduction formula for $\int \tan^n x\, dx$ and hence find $\int \tan^5 x\, dx$

$$\int \tan^n x\, dx = \int \tan^{n-2} x \,.\, \tan^2 x\, dx$$

$$= \int \tan^{n-2} x\, (\sec^2 x - 1)\, dx$$

$$= \int \tan^{n-2} x \sec^2 x\, dx - \int \tan^{n-2} dx$$

$$= \frac{\tan^{n-1} x}{n-1} - \int \tan^{n-2} x\, dx \qquad \left(\because \int (f(x))^n f'(x)\, dx = \frac{(f(x))^{n+1}}{n+1}\right)$$

which is the required reduction formula.

To *evaluate* $\int tan^5 x\, dx$.

The above reduction formula is $\int \tan^n x\, dx = \frac{\tan^{n-1} x}{n-1} - \int \tan^{n-2} x\, dx$ ...(1)

Putting $n = 5$ in (1), $\int \tan^5 x\, dx = \frac{\tan^4 x}{4} - \int \tan^3 x\, dx$ ...(2)

Putting $n = 3$ in (1), $\int \tan^3 x\, dx = \frac{\tan^2 x}{2} - \int \tan x\, dx = \frac{\tan^2 x}{2} - \log \sec x$

Putting this value of $\int \tan^3 x\, dx$ in (2),

$$\int \tan^5 x\, dx = \frac{\tan^4 x}{4} - \left(\frac{\tan^2 x}{2} - \log \sec x\right) = \frac{\tan^4 x}{4} - \frac{\tan^2 x}{2} + \log \sec x.$$

3. To find a reduction formula for $\int \cot^n x\, dx$

$$\int \cot^n x\, dx = \int \cot^{n-2} x \cot^2 x\, dx$$

$$= \int \cot^{n-2} x\, (\text{cosec}^2 x - 1)\, dx = \int \cot^{n-2} x\, \text{cosec}^2 x\, dx - \int \cot^{n-2} x\, dx$$

$$= -\int \cot^{n-2} x\, (-\text{cosec}^2 x)\, dx - \int \cot^{n-2} x\, dx$$

$$= -\frac{\cot^{n-1} x}{n-1} - \int \cot^{n-2} x\, dx \qquad \left|\because \int (f(x))^n f'(x)\, dx = \frac{(f(x))^{n+1}}{n+1}\right.$$

which is the required reduction formula.

4. Find a reduction formula for $\int \sec^n x\,dx$ and hence evaluate $\int \sec^5 x\,dx$.

$\int \sec^n x\,dx = \int \sec^{n-2} x \sec^2 x\,dx$ (Integrate by parts)

$$= \sec^{n-2} x \tan x - \int (n-2)\sec^{n-3} x \,.\, \sec x \tan x \,.\, \tan x\,dx$$
$$= \sec^{n-2} x \tan x - (n-2)\int \sec^{n-2} x \,.\, \tan^2 x\,dx$$
$$= \sec^{n-2} x \tan x - (n-2)\int \sec^{n-2} x\,(\sec^2 x - 1)\,dx$$
$$= \sec^{n-2} x \tan x - (n-2)\int \sec^n x\,dx + (n-2)\int \sec^{n-2} x\,dx.$$

Transposing,

$$(1 + n - 2)\int \sec^n x\,dx = \sec^{n-2} x \tan x + (n-2)\int \sec^{n-2} x\,dx$$

$$\therefore \quad \int \sec^n x\,dx = \frac{\sec^{n-2} x \tan x}{n-1} + \frac{n-2}{n-1}\int \sec^{n-2} x\,dx \qquad \text{...(1)}$$

which is the required reduction formula.

Alternative Form of Art. 4

Use integration by parts to derive the reduction formula :

$$\int \sec^m x\,dx = \frac{1}{m-1}\sec^{m-2} x \tan x + \frac{m-2}{m-1}\int \sec^{m-2} x\,dx.$$

To Evaluate $\int \sec^5 x\,dx$.

Sol. Putting $n = 5$ in (1), we have

$$\int \sec^5 x\,dx = \frac{\sec^3 x \tan x}{4} + \frac{3}{4}\int \sec^3 x\,dx \qquad \text{...(2)}$$

Putting $n = 3$ in (1),

$$\int \sec^3 x\,dx = \frac{\sec x \tan x}{2} + \frac{1}{2}\int \sec x\,dx$$
$$= \frac{\sec x \tan x}{2} + \frac{1}{2}\log(\sec x + \tan x)$$

$\therefore$ From (2), $$\int \sec^5 x\,dx = \frac{\sec^3 x \tan x}{4} + \frac{3}{8}\sec x \tan x + \frac{3}{8}\log(\sec x + \tan x).$$

5. To find a reduction formula for $\int \operatorname{cosec}^n x\,dx$

$\int \operatorname{cosec}^n x\,dx = \int \operatorname{cosec}^{n-2} x \,.\, \operatorname{cosec}^2 x\,dx$ (Integrate by parts)

$$= \operatorname{cosec}^{n-2} x\,(-\cot x) - \int (n-2)\operatorname{cosec}^{n-3} x\,(-\operatorname{cosec} x \cot x)(-\cot x)\,dx$$
$$= -\operatorname{cosec}^{n-2} x \cot x - (n-2)\int \operatorname{cosec}^{n-2} x \cot^2 x\,dx$$
$$= -\operatorname{cosec}^{n-2} x \cot x - (n-2)\int \operatorname{cosec}^{n-2} x\,(\operatorname{cosec}^2 x - 1)\,dx$$
$$= -\operatorname{cosec}^{n-2} x \cot x - (n-2)\int \operatorname{cosec}^n x\,dx + (n-2)\int \operatorname{cosec}^{n-2} x\,dx$$

Transposing,

$$(1 + n - 2)\int \operatorname{cosec}^n x\,dx = -\operatorname{cosec}^{n-2} x \cot x + (n-2)\int \operatorname{cosec}^{n-2} x\,dx$$

$$\therefore \quad \int \operatorname{cosec}^n x\,dx = -\frac{\operatorname{cosec}^{n-2} x \cot x}{n-1} + \frac{n-2}{n-1}\int \operatorname{cosec}^{n-2} x\,dx$$

which of the required reduction formula.

Alternative form of Art. 5

Use integration by parts to derive the reduction formula :

$$\int cosec^m x\, dx = \frac{-1}{m-1} cosec^{m-2} x \cot x + \frac{m-2}{m-1} \int cosec^{m-2} x\, dx$$

6. Find a reduction formula for $\int x^m (\log x)^n dx$ and hence evaluate $\int x^m (\log x)^3 dx$

$\int x^m (\log x)^n dx = \int (\log x)^n . x^m dx$ (Integrate by parts)

$$= (\log x)^n . \frac{x^{m+1}}{m+1} - \int n(\log x)^{n-1} . \frac{1}{x} . \frac{x^{m+1}}{m+1} dx$$

$$= \frac{x^m + 1}{m+1} (\log x)^n - \frac{n}{m+1} \int x^m (\log x)^{n-1} dx \quad ...(1)$$

which is the required reduction formula.

To Evaluate $\int x^m (\log x)^3 dx$.

Sol. Putting $n = 3$ in (1),

$$\int x^m (\log x)^3 dx = \frac{x^{m+1}}{m+1} (\log x)^3 - \frac{3}{m+1} \int x^m (\log x)^2 dx \quad ...(2)$$

Putting $n = 2$ in (1),

$$\int x^m (\log x)^2 dx = \frac{x^{m+1}}{m+1} (\log x)^2 - \frac{2}{m+1} \int x^m \log x\, dx \quad ...(3)$$

Putting $n = 1$ in (1),

$$\int x^m \log x\, dx = \frac{x^{m+1}}{m+1} \log x - \frac{1}{m+1} \int x^m (\log x)^0 dx$$

$$= \frac{x^{m+1}}{m+1} \log x - \frac{1}{m+1} \int x^m dx = \frac{x^{m+1}}{m+1} \log x - \frac{x^{m+1}}{(m+1)^2} .$$

Substituting this value in (3), we have

$$\int x^m (\log x)^2 dx = \frac{x^{m+1}}{m+1} (\log x)^2 - \frac{2}{m+1} \left[\frac{x^{m+1}}{m+1} \log x - \frac{x^{m+1}}{(m+1)^2} \right]$$

$$= \frac{x^{m+1}}{m+1} (\log x)^2 - \frac{2}{(m+1)^2} x^{m+1} \log x + \frac{2}{(m+1)^3} x^{m+1}.$$

Substituting this value in (2),

$$\int x^m (\log x)^3 dx = \frac{x^{m+1}}{m+1} (\log x)^3$$

$$- \frac{3}{m+1} \left(\frac{x^{m+1}}{m+1} (\log x)^2 - \frac{2}{(m+1)^2} x^{m+1} \log x + \frac{2}{(m+1)^3} x^{m+1} \right)$$

$$= \frac{x^{m+1}}{m+1} \left((\log x)^3 - \frac{3}{m+1} (\log x)^2 + \frac{6}{(m+1)^2} \log x - \frac{6}{(m+1)^3} \right).$$

Example 1. *If $\mu_n = \int (\log x)^n dx$, prove that $\mu_n + n\, \mu_{n-1} = x (\log x)^n$.*

Sol. $\mu_n = \int (\log x)^n dx = \int (\log x)^n . 1\, dx$ (Integrate by parts)

$$= (\log x)^n x - \int n (\log x)^{n-1} \cdot \frac{1}{x} x\, dx$$
$$= x (\log x)^n - n \int (\log x)^{n-1} dx = x (\log x)^n - n\mu_{n-1}$$

$\therefore \qquad \mu_n + n\mu_{n-1} = x (\log x)^n.$

7. To find a reduction formula for $\int x^n e^{ax} dx$

Integrating by parts, we have

$$\int x^n e^{ax} dx = x^n \frac{e^{ax}}{a} - \int nx^{n-1} \cdot \frac{e^{ax}}{a} dx = \frac{x^n e^{ax}}{a} - \frac{n}{a} \int x^{n-1} e^{ax} dx$$

which is the required reduction formula.

Example 2. *Evaluate* $\int_0^\infty x^n e^{-x} dx$, *where n is a positive integer.*

Sol. Step I. Let $\quad I_n = \int_0^\infty \underset{\text{I}}{x^n} \underset{\text{II}}{e^{-x}} dx \qquad \text{...(1)}$

Integrating by product Rule $= \left[\frac{x^n e^{-x}}{-1}\right]_0^\infty - \int_0^\infty nx^{n-1} \frac{e^{-x}}{-1} dx$

$$= n \int_0^\infty x^{n-1} e^{-x} dx \qquad \left[\because \underset{x \to \infty}{\text{Lt}}\ x^n e^{-x} = 0\right]$$

or $\qquad \mathbf{I_n = n\, I_{n-1}} \qquad \text{...(2)}$

Step II. Changing n to $(n-1)$ in (2)*, $I_{n-1} = (n-1) I_{n-2}$ ...(3)

Step III. Putting the value of I_{n-1} from (3) in (2), $I_n = n(n-1) I_{n-2}$.

Step IV. Generalising $I_n = [n(n-1)(n-2) \ldots\ldots n \text{ factors}]\ I_{n-n} = n!\ I_0$ ...(4)

Putting $n = 0$ in (1), $\quad I_0 = \int_0^\infty x^0 e^{-x} dx = \int_0^\infty e^{-x} dx$

or $\qquad I_0 = \left[\frac{e^{-x}}{-1}\right]_0^\infty = -[e^{-\infty} - e^0] = -(0-1) = 1 \qquad (\because \quad e^{-\infty} = 0)$

Putting $I_0 = 1$ in (4), $\quad I_n = n!$

EXERCISE 1

1. (*a*) If $U_n = \int_0^{\pi/4} \tan^n x\, dx$, show that $U_n + U_{n-2} = \frac{1}{n-1}$ and deduce the value of U_5.

   (*b*) If $I_n = \int_0^{\pi/4} \tan^n \theta\, d\theta$.

   Prove that $n(I_{n-1} + I_{n+1}) = 1$ where n is a positive integer.

2. If $I_m = \int_0^1 x^n (\log x)^m dx$ ($m \geq 0$ and n being a positive integer), show that $I_m = -\frac{m}{n+1} I_{m-1}$ and hence evaluate I_3. *(M.D.U. 2005 A, 1996)*

---

*This step is being done on comparing the L.H.S and R.H.S. of equation (2).

3. Find a reduction formula for $\int \frac{x^m}{(\log x)^n}\, dx$.

$$\left[\textbf{Hint. } \frac{x^m}{(\log x)^n}\, dx = \int x^{m+1}\left(\frac{(\log x)^{-n}}{x}\right) dx\right].$$

ANSWERS

1. (*a*) $\frac{1}{2}\log 2 - \frac{1}{4}$ 2. $I_3 = \frac{-6}{(n+1)^4}$ 3. $\frac{-x^{m+1}}{(n-1)(\log x)^{n-1}} + \frac{m+1}{n-1}\int \frac{x^m}{(\log x)^{n-1}}\, dx.$

8. To find a reduction formula for $\int x^m \sin nx\, dx$

Integrating by parts,

i.e., using the formula $\int \text{I} . \text{II}\, dx = \text{I}\int \text{II}\, dx - \int \frac{d}{dx}(\text{I})\int \text{II}\, dx\, dx$

$$\int x^m \sin nx\, dx = x^m . \frac{-\cos nx}{n} - \int mx^{m-1} . \frac{-\cos nx}{n}\, dx$$

$$= -\frac{x^m \cos nx}{n} + \frac{m}{n}\int x^{m-1}\cos nx\, dx \qquad \text{(Integrating again by parts)}$$

$$= -\frac{x^m \cos nx}{n} + \frac{m}{n}\left(x^{m-1}\frac{\sin nx}{n} - \int (m-1)\, x^{m-2}\frac{\sin nx}{n}\, dx\right)$$

$$= -\frac{x^m \cos nx}{n} + \frac{mx^{m-1}\sin nx}{n^2} - \frac{m(m-1)}{n^2}\int x^{m-2}\sin nx\, dx$$

which is the required reduction formula.

9. To find a reduction formula for $\int e^{ax}\sin^n bx\, dx$

$$\int e^{ax}\sin^n bx\, dx = \int \sin^n bx\, e^{ax}\, dx \qquad \text{(Integrate by parts)}$$

$$= \sin^n bx . \frac{e^{ax}}{a} - \int (n\sin^{n-1} bx \cos bx . b)\frac{e^{ax}}{a}\, dx$$

$$= \frac{e^{ax}\sin^n bx}{a} - \frac{nb}{a}\int (\sin^{n-1} bx\cos bx)\, e^{ax}\, dx \qquad \text{(Integrate again by parts)}$$

$$= \frac{e^{ax}\sin^n bx}{a} - \frac{nb}{a}\left[(\sin^{n-1} bx\cos bx)\frac{e^{ax}}{a} - \int\left\{(n-1)\sin^{n-2} bx\cos bx . b(\cos bx) + \sin^{n-1} bx(-\sin bx . b)\right\}\frac{e^{ax}}{a}\, dx\right]$$

$$= \frac{e^{ax}\sin^n bx}{a} - \frac{nb}{a}\left[\frac{e^{ax}\sin^{n-1} bx\cos bx}{a} - \int\left\{(n-1)\, b\sin^{n-2} bx\cos^2 bx - b\sin^n bx\right\}\frac{e^{ax}}{a}\, dx\right]$$

$$= \frac{e^{ax}\sin^n bx}{a} - \frac{nb}{a}\left[\frac{e^{ax}\sin^{n-1} bx\cos bx}{a} - \int\left\{(n-1)b\sin^{n-2} bx(1-\sin^2 bx) - b\sin^n bx\right\}\frac{e^{ax}}{a}\, dx\right]$$

$$= \frac{e^{ax}\sin^n bx}{a} - \frac{nb}{a}\left[\frac{e^{ax}\sin^{n-1} bx\cos bx}{a} - \int\left((n-1)b\sin^{n-2} bx - nb\sin^n bx\right)\frac{e^{ax}}{a}\,dx\right]$$

$$= \frac{e^{ax}\sin^n bx}{a} - \frac{nb}{a^2}e^{ax}\sin^{n-1} bx\cos bx$$

$$+ \frac{n(n-1)b^2}{a^2}\int e^{ax}\sin^{n-2} bx\,dx - \frac{n^2b^2}{a^2}\int e^{ax}\sin^n bx\,dx$$

Transposing, $\left(1+\frac{n^2b^2}{a^2}\right)\int e^{ax}\sin^n bx\,dx = \frac{ae^{ax}\sin^n bx - nb\,e^{ax}\sin^{n-1} bx\cos bx}{a^2}$

$$+ \frac{n(n-1)b^2}{a^2}\int e^{ax}\sin^{n-2} bx\,dx.$$

$$\therefore \int e^{ax}\sin^n bx\,dx = \frac{e^{ax}\sin^{n-1} bx\,[a\sin bx - nb\cos bx]}{a^2+n^2b^2} + \frac{n(n-1)b^2}{a^2+n^2b^2}\int e^{ax}\sin^{n-2} bx\,dx.$$

10. To find a reduction formula for $\int \cos^m x \sin nx\,dx$

Integrating by parts, (Taking $\sin nx$ as second function)

$$\int \cos^m x\sin nx\,dx = \cos^m x\left(-\frac{\cos nx}{n}\right) - \int m\cos^{m-1}x(-\sin x)\left(-\frac{\cos nx}{n}\right)dx$$

$$= -\frac{\cos^m x\cos nx}{n} - \frac{m}{n}\int \cos^{m-1}x(\sin x\cos nx)\,dx \quad \text{...(1)}$$

Now $\sin(n-1)x = \sin nx\cos x - \cos nx\sin x$ **(Note this step)**

or $\sin x\cos nx = \sin nx\cos x - \sin(n-1)x.$

Substituting this value of $\sin x\cos nx$ in (1), $\int \cos^m x\sin nx\,dx$

$$= -\frac{\cos^m x\cos nx}{n} - \frac{m}{n}\int \cos^{m-1}x\,[\sin nx\cos x - \sin(n-1)x]\,dx$$

$$= -\frac{\cos^m x\cos nx}{n} - \frac{m}{n}\int \cos^m x\sin nx\,dx + \frac{m}{n}\int \cos^{m-1}x\sin(n-1)x\,dx.$$

Transposing, $\left(1+\frac{m}{n}\right)\int \cos^m x\sin nx\,dx$

$$= -\frac{\cos^m x\cos nx}{n} + \frac{m}{n}\int \cos^{m-1}x\sin(n-1)x\,dx$$

$$\therefore \int \cos^m x\sin nx\,dx = -\frac{\cos^m x\cos nx}{m+n} + \frac{m}{m+n}\int \cos^{m-1}x\sin(n-1)x\,dx.$$

11. Find a reduction formula for $\int \frac{\sin n\theta}{\cos\theta}\,d\theta$

Let us connect $\int \frac{\sin n\theta}{\cos\theta}\,d\theta$ with $\int \frac{\sin(n-2)\theta}{\cos\theta}\,d\theta.$

Consider $\frac{\sin n\theta}{\cos\theta} + \frac{\sin(n-2)\theta}{\cos\theta} = \frac{\sin n\theta + \sin(n-2)\theta}{\cos\theta}$

$$= \frac{2 \sin (n-1) \theta \cos \theta}{\cos \theta} = 2 \sin (n-1) \theta$$

$$\left(\because \sin C + \sin D = 2 \sin \frac{C+D}{2} \cos \frac{C-D}{2} \right)$$

Integrating both sides w.r.t. θ,

$$\int \frac{\sin n\theta}{\cos \theta} d\theta + \int \frac{\sin (n-2)\theta}{\cos \theta} d\theta = 2 \int \sin (n-1) \theta \, d\theta = -\frac{2 \cos (n-1)\theta}{n-1}$$

$$\therefore \quad \int \frac{\sin n\theta}{\cos \theta} d\theta = -\frac{2 \cos (n-1)\theta}{n-1} - \int \frac{\sin (n-2)\theta}{\cos \theta} d\theta.$$

EXERCISE 2

1. Find a reduction formula for $\int x^m \cos nx \, dx$.

2. If $u_n = \int_0^{\pi/2} x^n \sin x \, dx$ and $n > 1$, prove that

   $u_n + n(n-1) u_{n-2} = n \left(\frac{\pi}{2}\right)^{n-1}$. Hence evaluate u_5. *(M.D.U. 2003 ; K.U. 1998 S)*

3. Find a reduction formula for $\int e^{ax} \cos^n x \, dx$.

4. If $I_{m,n} = \int_0^{\pi/2} \cos^m x \sin nx \, dx$; show that

   $$I_{m,n} = \frac{1}{m+n} + \frac{m}{m+n} I_{m-1, n-1}.$$ *(K.U. 2000, 2005, 2006 ; M.D.U. 1999, 2006)*

5. (*a*) If $I_{m,n} = \int \cos^m x \cos nx \, dx$, prove that

   $$(m+n) I_{m,n} = \cos^m x \sin nx + m I_{m-1, n-1}.$$ *(M.D.U. 1998)*

   (*b*) If $I_{m,n} = \int_0^{\pi/2} \cos^m x \cos nx \, dx$, show that $I_{m,n} = \frac{m}{m+n} I_{m-1, n-1}$. *(K.U. 1999)*

6. (*a*) If $u_n = \int \cos n\theta \operatorname{cosec} \theta \, d\theta$, prove that $u_n - u_{n-2} = \frac{2 \cos (n-1) \theta}{n-1}$.

   (M.D.U. 2000 ; K.U. 1996 S, 2004 S, 2007)

   (*b*) If $I_n = \int \sin n\theta \sec \theta \, d\theta$, show that $I_n = -\frac{2 \cos (n-1)\theta}{n-1} - I_{n-2}$.

   [**Hint.** It is Art. 11.]

7. (*a*) Find a reduction formula for $\int \frac{\sin nx}{\sin x} dx$.

   (*b*) Prove that $\int_0^{\pi} \frac{\sin n\theta}{\sin \theta} d\theta$ is equal to 0 or π according as *n* is an even or odd positive integer.

8. Prove that $\int (\sin^{-1} x)^n dx = x (\sin^{-1} x)^n + n \sqrt{1-x^2} (\sin^{-1} x)^{n-1} - n(n-1) \int (\sin^{-1} x)^{n-2} dx$

   [Integrate (twice) by Product rule taking 1 as second function.]

ANSWERS

1. $\int x^m \cos nx \, dx = \frac{x^m}{n} \sin nx + \frac{m}{n^2} x^{m-1} \cos nx - \frac{m(m-1)}{n^2} \int x^{m-2} \cos nx \, dx.$

2. $\dfrac{5\pi^4}{16} - 15\pi^2 + 120.$

3. $\displaystyle\int e^{ax} \cos^n x\, dx = \frac{a \cos x + n \sin x}{a^2 + n^2} e^{ax} \cos^{n-1} x + \frac{n(n-1)}{a^2+n^2} \int e^{ax} \cos^{n-2} x\, dx.$

7. (a) $\displaystyle\int \frac{\sin nx}{\sin x}\, dx = \frac{2 \sin (n-1)x}{n-1} + \int \frac{\sin (n-2)x}{\sin x}\, dx.$

---

12. To find a reduction formula for $\int \sin^n x\, dx$ and hence evaluate $\int \sin^6 x\, dx$

(M.D.U. 2006)

$\int \sin^n x\, dx = \int \sin^{n-1} x \,.\, \sin x\, dx$ [Integrate by parts]

$$= \sin^{n-1} x(-\cos x) - \int (n-1) \sin^{n-2} x \cos x\, (-\cos x)\, dx$$
$$= -\sin^{n-1} x \cos x + (n-1) \int \sin^{n-2} x \cos^2 x\, dx$$
$$= -\sin^{n-1} x \cos x + (n-1) \int \sin^{n-2} x\, (1 - \sin^2 x)\, dx$$
$$= -\sin^{n-1} x \cos x + (n-1) \int \sin^{n-2} x\, dx - (n-1) \int \sin^n x\, dx$$

Transposing,

$$(1 + n - 1) \int \sin^n x\, dx = -\sin^{n-1} x \cos x + (n-1) \int \sin^{n-2} x\, dx.$$

$$\therefore \quad \int \sin^n x\, dx = -\frac{\sin^{n-1} x \cos x}{n} + \frac{n-1}{n} \int \sin^{n-2} x\, dx \qquad ...(1)$$

which is the required reduction formula.

To Evaluate $\int \sin^6 x\, dx$

Putting $n = 6$ in (1), $\displaystyle\int \sin^6 x\, dx = -\frac{\sin^5 x \,.\, \cos x}{6} + \frac{5}{6} \int \sin^4 x\, dx$ ...(2)

Again, putting $n = 4$ in (1), $\displaystyle\int \sin^4 x\, dx = -\frac{\sin^3 x \cos x}{4} + \frac{3}{4} \int \sin^2 x\, dx$ ...(3)

Again, putting $n = 2$ in (1), $\displaystyle\int \sin^2 x\, dx = -\frac{\sin x \cos x}{2} + \frac{1}{2} \int \sin^0 x\, dx$

$$= -\frac{1}{2} \sin x \cos x + \frac{1}{2} \int 1\, dx = -\frac{1}{2} \sin x \cos x + \frac{x}{2}$$

Substituting this value in (3), $\displaystyle\int \sin^4 x\, dx = -\frac{1}{4} \sin^3 x \cos x - \frac{3}{8} \sin x \cos x + \frac{3}{8} x$

Substituting this value in (2),

$$\int \sin^6 x\, dx = -\frac{1}{6} \sin^5 x \cos x - \frac{5}{24} \sin^3 x \cos x - \frac{5}{16} \sin x \cos x + \frac{5}{16} x.$$

13. To find a reduction formula for $\int \cos^n x\, dx$

$\int \cos^n x\, dx = \int \cos^{n-1} x \cos x\, dx$ [Integrate by parts]

$$= \cos^{n-1} x \,.\, \sin x - \int (n-1) \cos^{n-2} x\, (-\sin x) \sin x\, dx$$
$$= \cos^{n-1} x \,.\, \sin x + (n-1) \int \cos^{n-2} x \,.\, \sin^2 x\, dx$$
$$= \cos^{n-1} x \,.\, \sin x + (n-1) \int \cos^{n-2} x\, (1 - \cos^2 x)\, dx$$
$$= \cos^{n-1} x \,.\, \sin x + (n-1) \int \cos^{n-2} x\, dx - (n-1) \int \cos^n x\, dx$$

Transposing, $(1 + n - 1) \int \cos^n x\, dx = \cos^{n-1} x \,.\, \sin x + (n-1) \int \cos^{n-2} x\, dx$

$$\therefore \quad \int \cos^n x\, dx = \frac{\cos^{n-1} x \sin x}{n} + \frac{n-1}{n} \int \cos^{n-2} x\, dx$$

which is the required reduction formula.

14. To evaluate $\int_0^{\pi/2} \sin^n x\,dx$, where n is a positive integer greater than 1.

Reproducing Art. 12, we have

$$\int \sin^n x\,dx = -\frac{\sin^{n-1} x \cdot \cos x}{n} + \frac{n-1}{n}\int \sin^{n-2} x\,dx.$$

$$\therefore \quad \int_0^{\pi/2} \sin^n x\,dx = \left[-\frac{\sin^{n-1} x \cdot \cos x}{n}\right]_0^{\pi/2} + \frac{n-1}{n}\int_0^{\pi/2} \sin^{n-2} x\,dx$$

$\left[\text{But } \sin^{n-1} x \cos x = 0, \text{ when } x = \frac{\pi}{2} \text{ or } x = 0\right]$

$$\therefore \quad \int_0^{\pi/2} \sin^n x\,dx = \frac{n-1}{n}\int_0^{\pi/2} \sin^{n-2} x\,dx \qquad \text{...(1)}$$

Changing n to $n - 2$*, in (1)

$$\int_0^{\pi/2} \sin^{n-2} x\,dx = \frac{n-3}{n-2}\int_0^{\pi/2} \sin^{n-4} x\,dx.$$

Substituting this value in (1), $\int_0^{\pi/2} \sin^n x\,dx = \frac{(n-1)(n-3)}{n(n-2)}\int_0^{\pi/2} \sin^{n-4} x\,dx$ ...(2)

Generalizing from (1) and (2), two cases arise.

Case I. *When n is a + ve odd integer, then*

$$\int_0^{\pi/2} \sin^n x\,dx = \frac{(n-1)(n-3)\ldots\ldots 2}{n(n-2)\ldots\ldots 3}\int_0^{\pi/2} \sin^1 x\,dx$$

$$= \frac{(n-1)(n-3)\ldots\ldots 2}{n(n-2)\ldots\ldots 3} \qquad \left[\because \int_0^{\pi/2} \sin x\,dx = \Big[-\cos x\Big]_0^{\pi/2} = 1\right]$$

Case II. *When n is a + ve even integer, then*

$$\int_0^{\pi/2} \sin^n x\,dx = \frac{(n-1)(n-3)\ldots\ldots 1}{n(n-2)\ldots\ldots 2}\int_0^{\pi/2} \sin^0 x\,dx = \frac{(n-1)(n-3)\ldots\ldots 1}{n(n-2)\ldots\ldots 2}\cdot\frac{\pi}{2}$$

$$\left[\because \int_0^{\pi/2} \sin^0 x\,dx = \int_0^{\pi/2} 1\,dx = \Big(x\Big)_0^{\pi/2} = \frac{\pi}{2}\right]$$

Note. $\int_0^{\pi/2} \sin^n x\,dx = \frac{(n-1)\times \text{go on diminishing by 2}}{n \times \text{go on diminishing by 2}} \times \boxed{\frac{\pi}{2}}$ **only if n is + ve even integer** $\left(\textbf{otherwise no } \frac{\pi}{2}\right)$. **Above formula is called Walli's Formula.**

15. To evaluate $\int_0^{\pi/2} \cos^n x\,dx$, where n is a positive integer greater than 1.

Reproducing Art. 13, we have

$$\int \cos^n x\,dx = \frac{\cos^{n-1} x \sin x}{n} + \frac{n-1}{n}\int \cos^{n-2} x\,dx$$

*This step is being done on comparing the L.H.S. and R.H.S of equation (1).

$$\therefore \quad \int_0^{\pi/2} \cos^n x\,dx = \left[\frac{\cos^{n-1} x \sin x}{n}\right]_0^{\pi/2} + \frac{n-1}{n}\int_0^{\pi/2} \cos^{n-2} x\,dx$$

$$= \frac{n-1}{n}\int_0^{\pi/2} \cos^{n-2} x\,dx \quad ...(1) \quad \left[\because \cos^{n-1} x \sin x = 0, \text{ when } x = \frac{\pi}{2} \text{ or } 0\right]$$

Changing n to $n-2$, in (1) $\displaystyle\int_0^{\pi/2} \cos^{n-2} x\,dx = \frac{n-3}{n-2}\int_0^{\pi/2} \cos^{n-4} x\,dx.$

Substituting this value in (1), $\displaystyle\int_0^{\pi/2} \cos^n x\,dx = \frac{(n-1)(n-3)}{n(n-2)}\int_0^{\pi/2} \cos^{n-4} x\,dx$ ...(2)

Generalizing from (1) and (2), two cases arise.

Case I. *When n is a + ve odd integer*, then

$$\int_0^{\pi/2} \cos^n x\,dx = \frac{(n-1)(n-3)\,......\,2}{n(n-2)\,......\,3}\int_0^{\pi/2} \cos^1 x\,dx$$

$$= \frac{(n-1)(n-3)\,......\,2}{n(n-2)\,......\,3} \qquad \left[\because \int_0^{\pi/2} \cos x\,dx = \left(\sin x\right)_0^{\pi/2} = 1\right]$$

Case II. *When n is a + ve even integer*, then

$$\int_0^{\pi/2} \cos^n x\,dx = \frac{(n-1)(n-3)\,......\,1}{n(n-2)\,......\,2}\int_0^{\pi/2} \cos^0 x\,dx = \frac{(n-1)(n-3)\,......\,1}{n(n-2)\,......\,2}\cdot\frac{\pi}{2}$$

$$\left[\because \int_0^{\pi/2} \cos^0 x\,dx = \int_0^{\pi/2} 1\,dx = \left(x\right)_0^{\pi/2} = \frac{\pi}{2}\right]$$

Note. $\displaystyle\int_0^{\pi/2} \cos^n x\,dx = \frac{(n-1)\times \text{go on decreasing by 2}}{n \times \text{go on decreasing by 2}} \times \boxed{\frac{\pi}{2}}$ **only if n is a positive even integer. Above formula is also called Walli's formula.**

Example. *Evaluate*

(*i*) $\displaystyle\int_0^{\pi/2} \sin^6 x\,dx,$ (*ii*) $\displaystyle\int_0^{\pi/2} \cos^9 x\,dx.$

Sol. (*i*) $\displaystyle\int_0^{\pi/2} \sin^6 x\,dx = \frac{5.3.1}{6.4.2}\cdot\frac{\pi}{2} = \frac{5}{32}\pi.$

(*ii*) $\displaystyle\int_0^{\pi/2} \cos^9 x\,dx = \frac{8.6.4.2}{9.7.5.3} = \frac{128}{315}.$

EXERCISE 3

1. If $C_n = \displaystyle\int_0^{\pi/2} \cos^n x\,dx$, show that $C_n = \dfrac{n-1}{n}C_{n-2}$ where n is any positive integer.

Hence evaluate $\displaystyle\int_0^{\pi/2} \cos^n x\,dx.$

[Hint. It is Art 15.]

2. If $u_n = \int_0^{\pi/2} \theta \sin^n \theta \, d\theta$ and $n > 1$, prove that $u_n = \frac{n-1}{n} u_{n-2} + \frac{1}{n^2}$.

Deduce that $u_5 = \frac{149}{225}$. *(M.D.U. 2002, 1995 ; K.U. 1995)*

[Hint. $\int \theta \sin^n \theta \, d\theta = \int (\theta \sin^{n-1} \theta) \sin \theta \, d\theta$ and now apply product rule, taking $\theta \sin^{n-1} \theta$ as first function and $\sin \theta$ as second function]

3. If $I_n = \int_0^{\pi/2} x \cos^n x \, dx$, prove that $I_n = \frac{n-1}{n} I_{n-2} - \frac{1}{n^2}$. *(K.U. 1998, M.D.U. 2003 S)*

---

16. Reduction Formulae for $\int \sin^p x \cos^q x \, dx$

$\int \sin^p x \cos^q x \, dx$ can be connected with any one of the following six integrals :

(i) $\int \sin^{p-2} x \cos^q x \, dx$ *(ii)* $\int \sin^p x \cos^{q-2} x \, dx$

(iii) $\int \sin^{p+2} x \cos^q x \, dx$ *(iv)* $\int \sin^p x \cos^{q+2} x \, dx$

(v) $\int \sin^{p-2} x \cos^{q+2} x \, dx$ *(vi)* $\int \sin^{p+2} x \cos^{q-2} x \, dx$.

Thus in finding a reduction formula for $\int \sin^p x \cos^q x \, dx$, we may

(i) Decrease or increase by 2 the index of $\sin x$, leaving that of $\cos x$ unchanged *(i)*, *(iii)*.

(ii) Decrease or increase by 2 the index of $\cos x$, leaving that of $\sin x$ unchanged *(ii)*, *(iv)*.

(iii) Decrease the index of $\sin x$ by 2 and increase that of $\cos x$ by (2) *(v)*.

(iv) Increase the index of $\sin x$ by 2 and decrease that of $\cos x$ by 2 *(vi)*.

But we cannot increase or decrease by 2 the indices of both $\sin x$ and $\cos x$ in the same formula.

17. Rule for connecting $\int \sin^p x \cos^q x \, dx$ with any one of the above six integrals

Step I. Take $P = \sin^{\lambda+1} x \cos^{\mu+1} x$ where λ is smaller of the two indices of $\sin x$ and μ is smaller of the two indices of $\cos x$ in the two integrals which are to be connected.

Step II. Find $\frac{dP}{dx}$ and express it as a linear function of the two integrands whose integrals are being connected.

Step III. Integrate both sides w.r.t. x, transpose and solve for the given integral.

18. Connect $\int \sin^p x \cos^q x \, dx$ with $\int \sin^{p-2} x \cos^q x \, dx$ *(K.U. 1997, 2006 S ; M.D.U. 2007)*

Hence integrate $\int \sin^4 x \cos^2 x \, dx$.

Sol. *(a)* Let $P = \sin^{p-2+1} x \cos^{q+1} x = \sin^{p-1} x \cos^{q+1} x$

Differentiating both sides w.r.t. x

$$\frac{dP}{dx} = (p-1) \sin^{p-2} x \,.\, \cos x \,.\, \cos^{q+1} x + \sin^{p-1} x \times (q+1) \cos^q x (-\sin x)$$

$$= (p-1) \sin^{p-2} x \cos^{q+2} x - (q+1) \sin^p x \cos^q x$$

$$= (p-1) \sin^p x \, \mathbf{\cos^q x \cos^2 x}^* - (q+1) \sin^p x \cos^q x$$

$$= (p-1) \sin^{p-2} x \cos^q x . (1 - \sin^2 x) - (q+1) \sin^p x \cos^q x$$

$$= (p-1) \sin^{p-2} x \cos^q x - (p-1) \sin^p x \cos^q x - (q+1) \sin^p x \cos^q x$$

$$= (p-1) \sin^{p-2} x \cos^q x - (p+q) \sin^p x \cos^q x$$

Integrating both sides w.r.t. x

$$P = (p-1) \int \sin^{p-2} x \cos^q x \, dx - (p+q) \int \sin^p x \cos^q x \, dx$$

---

*[**Note.** Now we write $\cos^{q+2} x$ as $\cos^q x \cos^2 x$ and replace $\cos^2 x$ by $1 - \sin^2 x$, $\because$ $\cos^{q+2} x$ is not required while $\cos^q x$ is required.]

or $\qquad (p+q)\int \sin^p x \cos^q x\, dx = -\text{P} + (p-1)\int \sin^{p-2} x \cos^q x\, dx$

$$\therefore \qquad \int \sin^p x \cos^q x\, dx = -\frac{\sin^{p-1} x \,.\, \cos^{q+1} x}{p+q} + \frac{p-1}{p+q}\int \sin^{p-2} x \cos^q x\, dx \qquad ...(i)$$

which is the required reduction formula.

(*b*) To integrate $\int \sin^4 x \cos^2 x\, dx$, put $p = 4$, $q = 2$ in (*i*)

$$\int \sin^4 x \cos^2 x\, dx = -\frac{\sin^3 x \cos^3 x}{6} + \frac{3}{6}\int \sin^2 x \cos^2 x\, dx \qquad ...(ii)$$

Again, putting $p = 2$, $q = 2$ in (*i*)

$$\int \sin^2 x \cos^2 x\, dx = -\frac{\sin x \cos^3 x}{4} + \frac{1}{4}\int \sin^0 x \cos^2 x\, dx$$

$$= -\frac{\sin x \cos^3 x}{4} + \frac{1}{4}\int \frac{1+\cos 2x}{2}\, dx$$

$$= -\frac{\sin x \cos^3 x}{4} + \frac{1}{8}\left(x + \frac{\sin 2x}{2}\right).$$

Putting this value of $\int \sin^2 x \cos^2 x\, dx$ in (*ii*), we get

$$\int \sin^4 x \cos^2 x\, dx = -\frac{\sin^3 x \cos^3 x}{4} - \frac{\sin x \cos^3 x}{8} + \frac{1}{16}\left(x + \frac{\sin 2x}{2}\right).$$

Alternative Form of Art. 18. Find the reduction formula for $\int \sin^p x \cos^q x\, dx$ and hence evaluate $\int \sin^4 x \cos^2 x\, dx$.

Example. *Show that* $\int \sin^m x \cos^n x\, dx = \dfrac{\sin^{m+1} x \cos^{n-1} x}{m+n} + \dfrac{n-1}{m+n}\displaystyle\int \sin^m x \cos^{n-2} x\, dx.$

Sol. Let $\text{P} = \sin^{m+1} x \cos^{n-2+1} x = \sin^{m+1} x \cos^{n-1} x$

$$\therefore \qquad \frac{d\text{P}}{dx} = (m+1)\sin^m x \,.\, \cos x \,.\, \cos^{n-1} x + \sin^{m+1} x\,(n-1)\cos^{n-2} x\,(-\sin x)$$

$$= (m+1)\sin^m x \cos^n x - (n-1)\sin^{m+2} x \cos^{n-2} x$$

$$= (m+1)\sin^m x \cos^n x - (n-1)\sin^m x \,.\, \sin^2 x \cos^{n-2} x$$

$$= (m+1)\sin^m x \cos^n x - (n-1)\sin^m x\,(1-\cos^2 x)\cos^{n-2} x$$

$$= (m+1)\sin^m x \cos^n x - (n-1)\sin^m x \cos^{n-2} x + (n-1)\sin^m x \cos^n x$$

$$= (m+n)\sin^m x \cos^n x - (n-1)\sin^m x \cos^{n-2} x$$

Integrating both sides,

$$\text{P} = (m+n)\int \sin^m x \cos^n x\, dx - (n-1)\int \sin^m x \cos^{n-2} dx$$

or $\qquad (m+n)\int \sin^m x \cos^n x\, dx = \text{P} + (n-1)\int \sin^m x \cos^{n-2} x\, dx$

$$\therefore \qquad \int \sin^m x \cos^n x\, dx = \frac{\sin^{m+1} x \cos^{n-1} x}{m+n} + \frac{n-1}{m+n}\int \sin^m x \cos^{n-2} x\, dx$$

which is the required reduction formula.

19. To evaluate $\int_0^{\pi/2} \sin^p x \cos^q x\, dx$, where p and q are both positive integers greater than 1. *(M.D.U. 2004, 2007, 1997 ; K.U. 1996)*

Reproducing Art. 18, we have

$$\int \sin^p x \cos^q x\, dx = -\frac{\sin^{p-1} x \,.\, \cos^{q+1} x}{p+q} + \frac{p-1}{p+q}\int \sin^{p-2} x \cos^q x\, dx$$

$$\therefore \int_0^{\pi/2} \sin^p x \cos^q x \, dx = -\left[\frac{\sin^{p-1} x \cos^{q+1} x}{p+q}\right]_0^{\pi/2} + \frac{p-1}{p+q}\int_0^{\pi/2} \sin^{p-2} x \cos^q x \, dx$$

$$= \frac{p-1}{p+q}\int_0^{\pi/2} \sin^{p-2} x \cos^q x \, dx \quad ...(1)$$

$$\left[\because \sin^{p-1} x \cos^{q+1} x = 0, \text{ when } x = \frac{\pi}{2} \text{ or } 0\right]$$

Changing p to $p - 2$

$$\int_0^{\pi/2} \sin^{p-2} x \cos^q x \, dx = \frac{p-3}{p+q-2}\int_0^{\pi/2} \sin^{p-4} x \cos^q x \, dx$$

Putting this value in (1),

$$\int_0^{\pi/2} \sin^p x \cos^q x \, dx = \frac{(p-1)(p-3)}{(p+q)(p+q-2)}\int_0^{\pi/2} \sin^{p-4} x \cos^q x \, dx \quad ...(2)$$

Generalising from (1) and (2),

Case I. When p is a + ve odd integer.

$$\int_0^{\pi/2} \sin^p x \cos^q x \, dx = \frac{(p-1)(p-3) \ldots\ldots 2}{(p+q)(p+q-2) \ldots\ldots (q+3)}\int_0^{\pi/2} \sin^1 x \cos^q x \, dx \quad ...(3)$$

But $\displaystyle\int_0^{\pi/2} \sin x \cos^q x \, dx = -\int_0^{\pi/2} \cos^q x \, (-\sin x) \, dx$

$$= \left[-\frac{\cos^{q+1} x}{q+1}\right]_0^{\pi/2} = -\frac{1}{q+1}[0-1] = \frac{1}{q+1}$$

$\therefore$ from (3),

$$\int_0^{\pi/2} \sin^p x \cos^q x \, dx = \frac{(p-1)(p-3) \ldots\ldots 2}{(p+q)(p+q-2) \ldots\ldots (q+3)(q+1)} \quad ...(4)$$

Now multiplying both numerator and denominator by

$(q-1) \,.\, (q-3) \ldots..$, we have

$$\int_0^{\pi/2} \sin^p x \cos^q x \, dx = \frac{[(p-1)(p-3) \ldots\ldots 2][(q-1)(q-3) \ldots\ldots]}{(p+q)(p+q-2) \ldots\ldots (q+3)(q+1)(q-1)(q-3) \ldots\ldots}.$$

Case II. When p is a + ve even integer.

$$\int_0^{\pi/2} \sin^p x \cos^q x \, dx = \frac{(p-1)(p-3) \ldots\ldots 1}{(p+q)(p+q-2) \ldots\ldots (q+2)}\int_0^{\pi/2} \sin^0 x \cos^q x \, dx$$

$$= \frac{(p-1)(p-3) \ldots\ldots 1}{(p+q)(p+q-2) \ldots\ldots (q+2)}\int_0^{\pi/2} \cos^q x \, dx \quad ...(5)$$

Sub-case (*i*) When q is a + ve odd integer.

$$I_{0,q} = \int_0^{\pi/2} \cos^q x \, dx = \frac{(q-1)(q-3) \ldots\ldots 2}{q(q-2) \ldots\ldots 3}$$

$\therefore$ from (5),

$$\int_0^{\pi/2} \sin^p x \cos^q x \, dx = \frac{(p-1)(p-3) \ldots\ldots 1}{(p+q)(p+q-2) \ldots\ldots (q+2)} \cdot \frac{(q-1)(q-3) \ldots\ldots 2}{q(q-2) \ldots\ldots 3}$$

Sub-case (*ii*) When q is a + ve even integer.

$$\int_0^{\pi/2} \cos^q x\, dx = \frac{(q-1)(q-3)\ldots\ldots 1}{q(q-2)\ldots\ldots 2} \cdot \frac{\pi}{2}$$

∴ from (5),

$$\int_0^{\pi/2} \sin^p x \cos^q x\, dx = \frac{(p-1)(p-3)\ldots\ldots 1}{(p+q)(p+q-2)\ldots\ldots(q+2)} \cdot \frac{(q-1)(q-3)\ldots\ldots 1}{q(q-2)\ldots\ldots 2} \times \frac{\pi}{2}.$$

Note. $\int_0^{\pi/2} \sin^p x \cos^q x\, dx$

$$= \frac{[(p-1)\times \text{go on diminishing by 2}]\times[(q-1)\times \text{go on diminishing by 2}]}{(p+q)\times \text{go on diminishing by 2}} \times \boxed{\frac{\pi}{2}}$$

only if both p and q are + ve even integers $\left(\textbf{otherwise no } \frac{\pi}{2}\right)$.

Example 1. (*i*) *Evaluate* $\int_0^{\pi/2} \sin^4 x \cos^2 x\, dx$

(*ii*) $\int_0^{\pi/2} \sin^8 x \cos^4 x\, dx$

(*iii*) $\int_0^{\pi/2} \sin^5 x \cos^4 x\, dx$ (M.D.U. 1997, 2007)

(*iv*) $\int_0^{\pi/2} \cos^7 \theta \sin\theta\, d\theta$.

Sol. (*i*) $\int_0^{\pi/2} \sin^4 x \cos^2 x\, dx$ [Here $p = 4$, $q = 2$ are both even, $p + q = 6$]

$$= \frac{3\times 1.1}{6\times 4\times 2} \times \frac{\pi}{2} = \frac{\pi}{32}.$$

(*ii*) $$\int_0^{\pi/2} \sin^8 x \cos^4 x\, dx = \frac{7.5.3.1\times 3.1}{12.10.8.6.4.2} \times \frac{\pi}{2} = \frac{7\pi}{2048}.$$

(*iii*) $$\int_0^{\pi/2} \sin^5 x \cos^4 x\, dx = \frac{4.2.3.1}{9.7.5.3.1}$$ (Here $p = 5$, $q = 4$ and $p + q = 5 + 4 = 9$)

$$= \frac{8}{315}.$$

(*iv*) $\int_0^{\pi/2} \cos^7 \theta \sin\theta\, d\theta$

Here $p = 1$.

∴ Formula of Art. 19 is not applicable.

$$\therefore \int_0^{\pi/2} \cos^7 \theta \sin\theta\, d\theta = -\int_0^{\pi/2} (\cos\theta)^7 (-\sin\theta\, d\theta)$$

$$= -\left[\frac{(\cos\theta)^8}{8}\right]_0^{\pi/2} \qquad \left|\because \int (\mathbf{f}(\theta))^n\, \mathbf{f}'(\theta)\, d\theta = \frac{(\mathbf{f}(\theta))^{n+1}}{n+1}\right.$$

$$= \frac{-1}{8}\left[\left(\cos\frac{\pi}{2}\right)^8 - (\cos 0)^8\right] = \frac{-1}{8}(0-1) = \frac{1}{8}.$$

Example 2. *Evaluate* $\int_0^1 x^6 \sin^{-1} x \, dx.$

Sol. Put $\sin^{-1} x = \theta$

$\therefore \quad x = \sin\theta$

$\therefore \quad dx = \cos\theta \, d\theta$

To change the limits

When $x = 0$, $\sin\theta = 0 \quad \therefore \quad \theta = 0$

When $x = 1$, $\sin\theta = 1 \quad \therefore \quad \theta = \frac{\pi}{2}$

$$\therefore \quad \int_0^1 x^6 \sin^{-1} x \, dx = \int_0^{\pi/2} \sin^6\theta \,.\, \theta \,.\, \cos\theta \, d\theta$$

$$= \int_0^{\pi/2} \underset{\text{I}}{\theta} \,.\, \sin^6\theta \underset{\text{II}}{\cos}\theta \, d\theta \qquad \text{(Product Rule)}$$

$$= \left(\theta \int_0^{\pi/2} \sin^6\theta \cos\theta \, d\theta - \int_0^{\pi/2} \frac{d}{d\theta}(\theta) \int \sin^6\theta \cos\theta \, d\theta \, d\theta\right.$$

$$= \left(\theta \frac{\sin^7\theta}{7}\right)_0^{\pi/2} - \int_0^{\pi/2} \frac{\sin^7\theta \, d\theta}{7}$$

(See Example 1 (*iv*) above)

$$\frac{\pi}{2} \cdot \frac{1}{7} - \frac{1}{7} \cdot \frac{6.4.2}{7.5.3} = \frac{\pi}{14} - \frac{16}{245}.$$

Example 3. *Prove that* $\int_0^{\pi/2} \sin^{2n} x \, dx = \frac{2n\,!}{(2^n (n)\,!)^2} \frac{\pi}{2}.$

Sol. We know by Art. 14 that

$$\int_0^{\pi/2} \sin^{2n} x \, dx = \frac{(2n-1)(2n-3) \ldots\ldots 1}{2n(2n-2) \ldots\ldots 2} \times \frac{\pi}{2}.$$

Multiplying both numerator and denominator by denominator *i.e.*, by $2n(2n-2) \ldots\ldots 2$,

$$= \frac{2n(2n-1)(2n-2)(2n-3) \ldots\ldots 2\,.\,1}{[2n(2n-2) \ldots\ldots 2]^2} \cdot \frac{\pi}{2}$$

$$= \frac{2n\,!}{[2\,.\,n\,.\,2(n-1) \ldots\ldots 2\,.\,1]^2} \frac{\pi}{2}$$

$$= \frac{2n\,!}{[2^n \, n(n-1) \ldots\ldots 1]^2} \frac{\pi}{2} = \frac{2n\,!}{(2^n \, (n)\,!)^2} \frac{\pi}{2}.$$

Example 4. *Prove that*

$$\int_0^{2a} x^n \sqrt{2ax - x^2} \, dx = \pi a^{n+2} \frac{(2n+1)(2n-1) \ldots\ldots 3\,.\,1}{(n+2)(n+1)\, n \ldots\ldots 2\,.\,1}.$$

where n is a natural number. (M.D.U. 2004 S)

Sol.

$$\text{L.H.S.} = \int_0^{2a} x^n \sqrt{2ax - x^2} \, dx = \int_0^{2a} x^n \sqrt{x(2a-x)} \, dx$$

$$= \int_0^{2a} x^n \sqrt{x} \sqrt{2a - x} \, dx \qquad \ldots(1)$$

Put $\mathbf{x = 2a \sin^2 \theta}$

$\therefore$ $$dx = 2a \,.\, 2 \sin\theta \cos\theta \, d\theta = 4a \sin\theta \cos\theta \, d\theta$$

To change the limits

when $x = 0$, $2a \sin^2\theta = 0$. But $2a \neq 0$ $\therefore$ $\sin\theta = 0$

$\therefore$ $\theta = 0$

when $x = 2a$, $2a \sin^2\theta = 2a$ or $\sin^2\theta = 1 = \sin^2 \dfrac{\pi}{2}$

$\therefore$ $\theta = \dfrac{\pi}{2}$.

Putting these values in (1),

$$\text{L.H.S.} = \int_0^{\pi/2} (2a \sin^2\theta)^n \sqrt{2a \sin^2\theta}\, \sqrt{2a - 2a\sin^2\theta} \,.\, 4a \sin\theta \cos\theta \, d\theta$$

$$= \int_0^{\pi/2} (2a)^n \sin^{2n}\theta \,.\, \sqrt{2a}\sin\theta \,.\, \sqrt{2a(1-\sin^2\theta)}\; 4a \sin\theta\cos\theta\, d\theta$$

$$= \int_0^{\pi/2} (2a)^n \sin^{2n}\theta\, \sqrt{2a}\sin\theta\, \sqrt{2a}\cos\theta \,.\, 4a\sin\theta\cos\theta\, d\theta$$

$$= (2a)^n \sqrt{2a}\,\sqrt{2a}\; 4a \int_0^{\pi/2} \sin^{2n+2}\theta \cos^2\theta\, d\theta$$

$$= (2a)^{n+1} \,.\, 2 \,.\, 2a \,.\, \frac{(2n+1)(2n-1)\ldots\ldots 3.1.1}{(2n+4)(2n+2)\,2n \ldots\ldots 2} \,.\, \frac{\pi}{2} \left[\int_0^{\pi/2} \sin^p\theta \cos^q\theta\, d\theta\right]$$

$$= 2^{n+2} \,.\, a^{n+2} \frac{(2n+1)(2n-1)\ldots\ldots 3.1}{2(n+2)\,2\,(n+1)\ldots\ldots 2.n\ldots\ldots 2.1}\pi$$

$$= 2^{n+2} a^{n+2} \frac{(2n+1)(2n-1)\ldots\ldots 3.1}{2^{n+2}(n+2)(n+1).n.1}\pi$$

$$= \pi a^{n+2} \frac{(2n+1)(2n-1)\ldots\ldots 3.1}{(n+2)(n+1).n\ldots\ldots 1}.$$

EXERCISE 4

1. Prove that $\displaystyle\int_0^{\pi/2} \sin^{10} x\, dx = \int_0^{\pi/2} \cos^{10} x\, dx = \frac{63\pi}{512}$.

2. Prove that (*i*) $\displaystyle\int_0^{\pi/2} \sin^5 x \cos^5 x\, dx = \frac{1}{60}$

(*ii*) $\displaystyle\int_0^{\pi/2} \sin^4\theta \cos^8\theta\, d\theta = \frac{7\pi}{2048}$ (*iii*) $\displaystyle\int_0^{\pi/2} \sin^9\theta\cos\theta\, d\theta = \frac{1}{10}$

(*iv*) $\displaystyle\int_0^{\pi/2} \sin^5 x \cos^6 x\, dx = \frac{8}{693}$. (*M.D.U. 2004 A*)

3. Prove that $\int_0^{\pi} \sin^6 \frac{x}{2} \cos^8 \frac{x}{2}\, dx = \frac{5\pi}{2^{11}}$.

$\left[\textbf{Hint.} \text{ Put } \frac{x}{2} = t.\right]$

4. Prove that (*i*) $\int_0^{a} x^4 \sqrt{a^2 - x^2}\, dx = \frac{\pi a^6}{32}$. (*ii*) $\int_0^{a} \frac{x^4}{\sqrt{a^2 - x^2}}\, dx = \frac{3a^4\pi}{16}$.

[**Hint.** Put $x = a \sin \theta$.]

5. (*i*) $\int_0^{2a} x^2 \sqrt{2ax - x^2}\, dx$ [**Hint.** put $x = 2a \sin^2 \theta$]

(*ii*) Evaluate $\int_0^{2} x^{5/2} \sqrt{2 - x}\, dx$ (*D.U., B.Sc. (H)*)

(*iii*) Prove that $\int_0^{4} x^3 \sqrt{4x - x^2}\, dx = 28\pi$

(*iv*) Prove that $\int_0^{2a} x^{9/2} (2a - x)^{-1/2}\, dx = \frac{63\pi a^5}{8}$.

6. (*i*) Evaluate $\int_0^{\infty} \frac{dx}{(1 + x^2)^n}$.

(*ii*) $\int_0^{1} \frac{x^{2n}}{\sqrt{1 - x^2}}\, dx$ and hence find the sum of the series

$$\frac{1}{2n+1} + \frac{1}{2} \cdot \frac{1}{2n+3} + \frac{1}{2} \cdot \frac{3}{4} \cdot \frac{1}{2n+5} + \ldots\ldots \text{ to } \infty.$$

[**Hint.** $\int_0^{1} x^{2n} (1 - x^2)^{-1/2}\, dx$

$$= \int_0^{1} x^{2n} \left[1 + \frac{1}{2} x^2 + \frac{1}{2} \cdot \frac{3}{4} x^4 + \ldots\ldots \infty\right] dx$$

$$= \int_0^{1} \left[x^{2n} + \frac{1}{2} x^{2n+2} + \frac{1}{2} \cdot \frac{3}{4} x^{2n+4} + \ldots\ldots \infty\right] dx$$

$$= \left[\frac{x^{2n+1}}{2n+1} + \frac{1}{2} \cdot \frac{x^{2n+3}}{2n+3} + \frac{1}{2} \cdot \frac{3}{4} \frac{x^{2n+5}}{2n+5} + \ldots\ldots \infty\right]_0^1$$

$$= \frac{1}{2n+1} + \frac{1}{2} \cdot \frac{1}{2n+3} + \frac{1}{2} \cdot \frac{3}{4} \cdot \frac{1}{2n+5} + \ldots\ldots \Bigg].$$

7. Evaluate $\int_0^{1} x^5 \sqrt{\frac{1 + x^2}{1 - x^2}}\, dx$. (*D.U., B.C.A. Feb. 1998*)

[**Hint.** Put $x^2 = \cos \theta$.]

8. Evaluate $\int_0^{1} x^5 \sin^{-1} x\, dx$.

9. If $I_{p,q} = \int_0^{\pi/2} \sin^p x \cos^q x \, dx$, prove that

$$I_{p,q} = \frac{(p-1)}{(p+q)} \frac{(p-3)(p-5)\ldots\ldots 2.1}{(p+q-2)(p+q-4)\ldots\ldots(q+3)(q+1)}$$

where p **is an odd positive integer and** q is a positive integer, even or odd.
[**Hint.** Reproduce Art. 19 upto Eqn. (4).]

ANSWERS

5. (*i*) $\frac{5\pi}{8} a^4$ (*ii*) $\frac{5\pi}{8}$.

6. (*i*) $\frac{(2n-3)(2n-5)\ldots\ldots 3.1}{(2n-2)(2n-4)\ldots. 4.2} \cdot \frac{\pi}{2}$ (*ii*) $\frac{(2n-1)(2n-3)\ldots\ldots 3.1}{2^n . n!} \cdot \frac{\pi}{2}$.

7. $\frac{3\pi+8}{24}$ **8.** $\frac{11\pi}{192}$.

20. Reduction formula for $\int x^m (a + bx^n)^p \, dx$

$\int x^m (a + bx^n)^p \, dx$ can be connected with any one of the following six integrals :

(*i*) $\int x^m (a + bx^n)^{p-1} dx$ (*ii*) $\int x^{m-n} (a + bx^n)^p dx$

(*iii*) $\int x^m (a + bx^n)^{p+1} dx$ (*iv*) $\int x^{m+n} (a + bx^n)^p dx$

(*v*) $\int x^{m+n} (a + bx^n)^{p-1} dx$ (*vi*) $\int x^{m-n} (a + bx^n)^{p+1} dx$.

Thus the index m of the monomial x^m can only be increased or decreased by the index n of x^n in the binomial $a + bx^n$ and the index p of the binomial $a + bx^n$ can be increased or decreased by one only. The indices of binomial $a + bx^n$ and the monomial x^m cannot both be decreased or increased in the same formula.

Example 1. *Connect* $\int x^m (a + bx^n)^p dx$ *with* $\int x^m (a + bx^n)^{p-1} dx$.

Sol. Let $P = x^{m+1} (a + bx^n)^{p-1+1} = x^{m+1}(a + bx^n)^p$

$$\therefore \frac{dP}{dx} = (m+1) x^m (a + bx^n)^p + x^{m+1} . p (a + bx^n)^{p-1} . bn x^{n-1}$$

$$= (m+1) x^m (a + bx^n)^p + pn bx^{m+n} (a + bx^n)^{p-1}$$

$$= (m+1) x^m (a + bx^n)^p + pnx^m . bx^n(a + bx^n)^{p-1}$$

$$= (m+1) x^m (a + bx^n)^p + pnx^m . (\overline{a + bx^n} * - a)(a + bx^n)^{p-1}$$

$$= (m+1) x^m (a + bx^n)^p + pnx^m (a + bx^n)^p - pnax^m (a + bx^n)^{p-1}$$

$$= (m+1+pn) x^m (a + bx^n)^p - pnax^m (a + bx^n)^{p-1}.$$

Integrating both sides,

$$P = (m+1+pn) \int x^m (a + bx^n)^p dx - pna \int x^m (a + bx^n)^{p-1} dx$$

$$\therefore \quad (m+1+pn) \int x^m (a + bx^n)^p dx = P + pna \int x^m (a + bx^n)^{p-1} dx$$

$$\therefore \quad \int x^m (a + bx^n)^p dx = \frac{x^{m+1} (a + bx^n)^p}{m+1+pn} + \frac{pna}{m+1+pn} \times \int x^m (a + bx^n)^{p-1} dx$$

which is the required reduction formula.

---

*[**Note.** Now x^{m+n} is not required $\therefore$ we write $bx^{m+n} = x^m . x^n . b$.
x^n is also not required but powers of $a + bx^n$ are required.
$\therefore$ We write $bx^{m+n} = x^m . bx^n = x^m (a + bx^n - a)$].

Example 2. *Prove that*

$$\int \frac{dx}{(a^2+x^2)^n} = \frac{x}{2a^2(n-1)(a^2+x^2)^{n-1}} + \frac{(2n-3)}{2a^2(n-1)} \int \frac{dx}{(a^2+x^2)^{n-1}}.$$

Sol. Let us connect $\int \frac{dx}{(a^2+x^2)^n} = \int x^0 (a^2+x^2)^{-n}\, dx$

with $$\int \frac{dx}{(a^2+x^2)^{n-1}} = \int x^0 (a^2+x^2)^{-n+1}\, dx.$$

Let $$P = x^{0+1} (a^2+x^2)^{-n+1}$$ | "Smaller Index + 1" Method

or $$P = x(a^2+x^2)^{-n+1} = \frac{x}{(a^2+x^2)^{n-1}}$$

Diff. both sides w.r.t. x,

$$\frac{dP}{dx} = (a^2+x^2)^{-n+1} + x(-n+1)(a^2+x^2)^{-n}\,.\,2x$$

$$= \frac{1}{(a^2+x^2)^{n-1}} + 2(1-n)\frac{x^2}{(a^2+x^2)^n}$$

$$= \frac{1}{(a^2+x^2)^{n-1}} + 2(1-n)\frac{[(a^2+x^2)-a^2]}{(a^2+x^2)^n}$$

or $$\frac{dP}{dx} = \frac{1}{(a^2+x^2)^{n-1}} + \frac{2(1-n)}{(a^2+x^2)^{n-1}} - \frac{2a^2(1-n)}{(a^2+x^2)^n}$$

$$= \frac{3-2n}{(a^2+x^2)^{n-1}} - \frac{2a^2(1-n)}{(a^2+x^2)^n}$$

Integrating both sides w.r.t. x, we have

$$P = (3-2n)\int \frac{1}{(a^2+x^2)^{n-1}}\, dx - 2a^2(1-n)\int \frac{1}{(a^2+x^2)^n}\, dx$$

Transposing,

$$2a^2(1-n)\int \frac{1}{(a^2+x^2)^n}\, dx = -P + (3-2n)\int \frac{1}{(a^2+x^2)^{n-1}}\, dx$$

Putting the value of P and dividing both sides by $2a^2(1-n)$

$$\int \frac{1}{(a^2+x^2)^n}\, dx = -\frac{x}{2a^2(1-n)(a^2+x^2)^{n-1}} - \frac{(2n-3)}{2a^2(1-n)}\int \frac{1}{(a^2+x^2)^{n-1}}\, dx$$

$$= \frac{x}{2a^2(n-1)(a^2+x^2)^{n-1}} + \frac{(2n-3)}{2a^2(n-1)}\int \frac{1}{(a^2+x^2)^{n-1}}\, dx.$$

Example 3. *Obtain a reduction formula for* $\int \frac{x^n\, dx}{(x^3-1)^{1/3}}$ *and find the value of* $\int x^8 (x^3-1)^{-1/3}\, dx$. (M.D.U. 2005 ; K.U. 2005 S)

Sol. Let us connect the given integral

$\int x^n (x^3-1)^{-1/3}\, dx$ with the integral $\int x^{n-3}(x^3-1)^{-1/3}\, dx$

[**Note.** In the second integral we have taken x^{n-3} because of the presence of x^3 in the bracket]

Let $\quad P = x^{n-3+1}(x^3-1)^{-\frac{1}{3}+1}$ | "smaller index +1" method

or $\quad P = x^{n-2}(x^3-1)^{2/3}$

Diff. both sides w.r.t x, we have

$$\frac{dP}{dx} = x^{n-2}\cdot\frac{2}{3}(x^3-1)^{-1/3}\cdot 3x^2 + (n-2)x^{n-3}(x^3-1)^{2/3}$$

or $$\frac{dP}{dx} = 2x^n(x^3-1)^{-1/3} + (n-2)x^{n-3}(x^3-1)(x^3-1)^{2/3-1} \qquad \textbf{(Note)}$$

or $$\frac{dP}{dx} = 2x^n(x^3-1)^{-1/3} + (n-2)(x^n - x^{n-3})(x^3-1)^{-1/3}$$

or $$\frac{dP}{dx} = 2x^n(x^3-1)^{-1/3} + (n-2)x^n(x^3-1)^{-1/3} - (n-2)x^{n-3}(x^3-1)^{-1/3}$$

$$= (2+n-2)x^n(x^3-1)^{-1/3} - (n-2)x^{n-3}(x^3-1)^{-1/3}$$

Integrating both sides w.r.t x, we have

$$P = n\int x^n(x^3-1)^{-1/3}\,dx - (n-2)\int x^{n-3}(x^3-1)^{-1/3}\,dx$$

$$\therefore \quad n\int x^n(x^3-1)^{-1/3}\,dx = P + (n-2)\int x^{n-3}(x^3-1)^{-1/3}\,dx$$

Putting the value of P and dividing both sides by n,

$$\int x^n(x^3-1)^{-1/3}\,dx = \frac{x^{n-2}(x^3-1)^{2/3}}{n} + \frac{(n-2)}{n}\int x^{n-3}(x^3-1)^{-1/3}\,dx \qquad ...(1)$$

which is the required reduction formula.

To evaluate $\int x^8(x^3-1)^{-1/3}\,dx$

Putting $n = 8$ in (1), $$\int x^8(x^3-1)^{-1/3}\,dx = \frac{x^6(x^3-1)^{2/3}}{8} + \frac{6}{8}\int x^5(x^3-1)^{-1/3}\,dx \qquad ...(2)$$

Putting $n = 5$ in (1), $$\int x^5(x^3-1)^{-1/3}\,dx = \frac{x^3(x^3-1)^{2/3}}{5} + \frac{3}{5}\int x^2(x^3-1)^{-1/3}\,dx \qquad ...(3)$$

Now $$\int x^2(x^3-1)^{-1/3}\,dx = \frac{1}{3}\int (x^3-1)^{-1/3}(3x^2)\,dx$$

$$= \frac{1}{3}\frac{(x^3-1)^{2/3}}{\frac{2}{3}} + c \qquad \left[\because \int (f(x)f'(x)\,dx = \frac{(f(x))^{n+1}}{n+1} \text{ if } n \neq -1\right]$$

$$= \frac{1}{2}(x^3-1)^{2/3} + C$$

Putting this value in eqn. (3), we have

$$\int x^5(x^3-1)^{-1/3}\,dx = \frac{x^3(x^3-1)^{2/3}}{5} + \frac{3}{10}(x^3-1)^{2/3} + C$$

Putting this value in eqn. (2), we have

$$\int x^8(x^3-1)^{-1/3}\,dx = \frac{x^6(x^3-1)^{2/3}}{8} + \frac{3}{4}\left[\frac{x^3(x^3-1)^{2/3}}{5} + \frac{3}{10}(x^3-1)^{2/3}\right] + C$$

$$= \frac{1}{8}x^6(x^3-1)^{2/3} + \frac{3}{20}x^3(x^3-1)^{2/3} + \frac{9}{40}(x^3-1)^{2/3} + C$$

$$= (x^3-1)^{2/3}\left[\frac{x^6}{8} + \frac{3x^3}{20} + \frac{9}{40}\right] + C.$$

EXERCISE 5

1. Connect $\int x^m (a+bx^n)^p\,dx$ with $\int x^{m+n}(a+bx^n)^p\,dx$. *(K.U. 1997 S)*
2. If n is a positive integer, prove that

$$\int (a^2+x^2)^{n/2}\,dx = \frac{x(a^2+x^2)^{n/2}}{n+1} + \frac{na^2}{n+1}\int (a^2+x^2)^{n/2-1}\,dx.$$

3. If $I_n = \int x^n\sqrt{a^2-x^2}\,dx$, show that $I_n = \frac{-x^{n-1}(a^2-x^2)^{3/2}}{n+2} + \left(\frac{n-1}{n+2}\right)a^2 I_{n-2}$.
4. If $I_n = \int x^n\sqrt{a-x}\,dx$, prove that

$(2n+3)\,I_n = 2an\,I_{n-1} - 2x^n(a-x)^{3/2}$ and hence evaluate $\int_0^a x^2\sqrt{ax-x^2}\,dx$.

5. If I_n denotes $\int_0^a (a^2-x^2)^n\,dx$ and $n>0$, prove that $I_n = \frac{2na^2}{2n+1}I_{n-1}$.

Hence evaluate $\int_0^a (a^2-x^2)^3\,dx$.

6. If $I_{m,n} = \int_0^1 x^m(1-x)^n\,dx$, prove that $(m+n+1)\,I_{m,n} = n\,I_{m,n-1}$. *(M.D.U. 2006)*

Deduce the value of $\int_0^1 x^4(1-x)^3\,dx$.

7. If m is a positive integer, find a reduction formula for $\int x^m\sqrt{2ax-x^2}\,dx$.

Hence find the value of $\int_0^{2a} x^3\sqrt{2ax-x^2}\,dx$. *(K.U. 2003)*

$\left[\textbf{Hint. } x^m\sqrt{2ax-x^2} = x^m\sqrt{x}\sqrt{2a-x} = x^{m+\frac{1}{2}}\sqrt{2a-x}.\right]$

ANSWERS

1. $\int x^m(a+bx^n)^p\,dx = \frac{x^{m+1}(a+bx^n)^{p+1}}{(m+1)a} - \frac{b(pn+n+m+1)}{(m+1)a}\int x^{m+n}(a+bx^n)^p\,dx.$

4. $\frac{5\pi a^4}{128}$ 5. $\frac{16a^7}{35}$ 6. $\frac{1}{280}$.

7. $\int x^m\sqrt{2ax-x^2}\,dx = -x^{m-1}\frac{(2ax-x^2)^{3/2}}{m+2} + a\frac{(2m+1)}{m+2}\int x^{m-1}\sqrt{2ax-x^2}\,dx\,;\ \frac{7}{8}\pi a^5.$

11

Rectification

1. Rectification

Definition. The process of finding the length of an arc of a curve between two given points on it is called *rectification.*

2. Length of the Curve when its Cartesian Equation is given

Prove that the length of the arc of the curve $\mathbf{y = f(x)}$, *between the two points whose abscissae are a and b is given by*

$$\int_a^b \sqrt{1+\left(\frac{dy}{dx}\right)^2}\, dx$$

where y and dy/dx are *continuous* and *single-valued functions in the interval* $[a, b]$ *and the integrand does not change sign in this interval.*

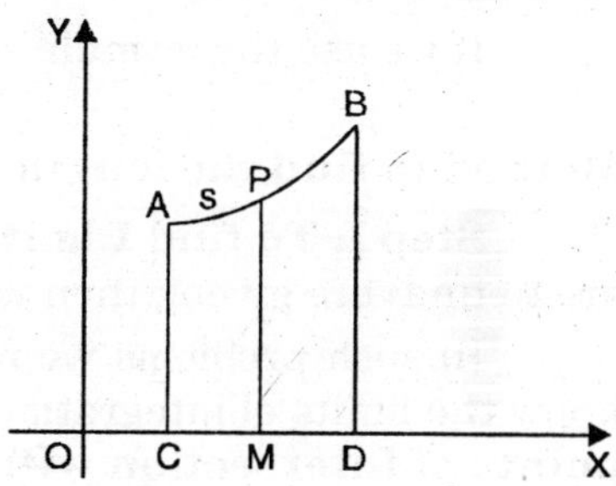

Let AB be the curve $y = f(x)$ and A, B are the points whose abscissae are a and b respectively and CA, DB are their ordinates.

Let $P(x, y)$ be *any* point on the curve and draw $MP \perp$ x-axis.

If s denotes the length of the arc AP measured from the fixed point A to the variable point P, then s is clearly a function of x.

From Differential Calculus, we know that

$$\frac{ds}{dx} = \sqrt{1+\left(\frac{dy}{dx}\right)^2} \qquad ...(1)$$

$$\therefore \quad \int_a^b \sqrt{1+\left(\frac{dy}{dx}\right)^2}\, dx = \int_a^b \frac{ds}{dx}\, dx = \Big[\, s \,\Big]_a^b$$

$$= (\text{value of } s \text{ at } x = b) - (\text{value of } s \text{ at } x = a)$$

$$= \text{arc AB} - 0 = \text{arc AB}.$$

Hence arc AB $= \displaystyle\int_a^b \sqrt{1+\left(\frac{dy}{dx}\right)^2}\, dx.$

Remark. It must be noted that in fact, $\dfrac{ds}{dx} = \pm\sqrt{1+\left(\dfrac{dy}{dx}\right)^2}$

We take the + ve sign with the radical under the assumption that s increases with x. The sign is – ve if s decreases with the increase of x.

If $\frac{ds}{dx}$ or the integrand changes sign at some value c within the range of integration $[a, b]$, then the definite integral from a to b must be broken into the sum of two definite integrals, one from a to c and other from c to b and the +ve value of the integrand taken in each. Otherwise, the result will be the difference of the lengths of two arcs.

Cor. 1. From (1), we have $s = \int \sqrt{1+\left(\frac{dy}{dx}\right)^2}\, dx.$

Cor. 2. *If the equation of the curve is of the form* **x = f(y),** *then the length of the arc between the points whose ordinates are c and d is*

$$\int_c^d \sqrt{1+\left(\frac{dx}{dy}\right)^2}\, dy.$$

[**Hint.** Proceed as in Art. 2 and instead of (1) use $\frac{ds}{dy} = \sqrt{1+\left(\frac{dx}{dy}\right)^2}$.]

Caution. To find the length of an arc of a cartesian curve, we must express either **y as an explicit function of x or x as an explicit function of y.**

Secondly, if we use the formula, $s = \int \sqrt{1+\left(\frac{dy}{dx}\right)^2}\, dx$, then $\frac{dy}{dx}$ must be a function of x alone.

If we use the formula $s = \int \sqrt{1+\left(\frac{dx}{dy}\right)^2}\, dy$, then $\frac{dx}{dy}$ must be a function of y alone.

Method to find the length of an arc of a cartesian curve

Step I. To find Limits of Integration. If the end points of the arc (whose length we are to find) are given ; then we know the limits of integration.

In such problems we need not trace the curve. Otherwise, trace the curve roughly to know the limits of integration. (While tracing the curve here, **Symmetry** of the curve and its **points of Intersection with the axes** are the two main points which should be discussed).

Step II. Express y as an explicit function of x or x as an explicit function of y.

Step III. Find $\frac{dy}{dx}$ or $\frac{dx}{dy}$.

Step IV. Use the formula, $s = \int_a^b \sqrt{1+\left(\frac{dy}{dx}\right)^2}\, dx$ or $\int_c^d \sqrt{1+\left(\frac{dx}{dy}\right)^2}\, dy$.

Example 1. *Show that in the catenary $y = c \cosh x/c$, the length of the arc from the vertex $(0, c)$ to any point (x, y) is given by (i) $s = c \sinh x/c$ and (ii) $s^2 = y^2 - c^2$.*

Sol. The equation of the curve is $y = c \cosh \frac{x}{c}$

$$\therefore \quad \frac{dy}{dx} = c \sinh \frac{x}{c} \cdot \frac{1}{c} = \sinh \frac{x}{c}.$$

(*i*) ∴ Reqd. length of arc from $x = 0$ to $x = x$ is

$$s = \int_0^x \sqrt{1+\left(\frac{dy}{dx}\right)^2}\, dx = \int_0^x \sqrt{1+\sinh^2 \frac{x}{c}}\, dx$$

$$= \int_0^x \cosh \frac{x}{c}\, dx = c\left[\sinh \frac{x}{c}\right]_0^x = c\left(\sinh \frac{x}{c} - 0\right) = c \sinh \frac{x}{c}.$$

(*ii*) Now $s^2 = c^2 \sinh^2 \frac{x}{c} = c^2\left[\cosh^2 \frac{x}{c} - 1\right] = c^2\left[\left(\frac{y}{c}\right)^2 - 1\right] = y^2 - c^2$

which proves the required result.

Example 2. *Find the length of arc of the parabola $y^2 = 4ax$*

(*i*) *from the vertex to an extremity of the latus rectum.* (M.D.U. 1997, 2007)

(*ii*) *cut off by the latus rectum.*

Sol. The equation of the parabola is $y^2 = 4ax$ ...(1)

Differentiating, $2y \frac{dy}{dx} = 4a$ or $\frac{dx}{dy} = \frac{y}{2a}$

Let O be the vertex and L be an extremity of latus rectum (a line through the focus and ⊥ to axis).

Extremities of latus rectum of the parabola $y^2 = 4ax$ are $(a, 2a)$ and $(a, -2a)$.

At O, we have $y = 0$ and at L, we have $y = 2a$.

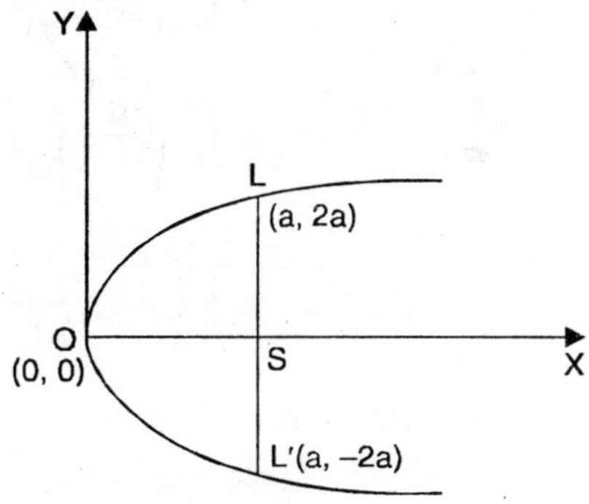

(*i*) ∴ Length of arc OL

$$= \int_0^{2a} \sqrt{1+\left(\frac{dx}{dy}\right)^2}\, dy = \int_0^{2a} \sqrt{1+\frac{y^2}{4a^2}}\, dy$$

$$= \frac{1}{2a}\int_0^{2a} \sqrt{y^2+4a^2}\, dy$$

$$= \frac{1}{2a}\left[\frac{1}{2} y\sqrt{y^2+4a^2} + 2a^2.\sinh^{-1}\frac{y}{2a}\right]_0^{2a}$$

$$\left(\because \int \sqrt{\mathbf{x}^2+\mathbf{a}^2}\,\mathbf{dx} = \frac{\mathbf{x}\sqrt{\mathbf{x}^2+\mathbf{a}^2}}{2} + \frac{\mathbf{a}^2}{2}\sinh^{-1}\frac{\mathbf{x}}{\mathbf{a}}\right)$$

$$= \frac{1}{2a}\left[a\sqrt{8a^2} + 2a^2 \sinh^{-1} 1 - 0\right] = a\,[\sqrt{2} + \sinh^{-1} 1]$$

$$= a\,[\sqrt{2} + \log(1+\sqrt{1+1})] \qquad \left[\because \sinh^{-1} x = \log(x+\sqrt{x^2+1})\right]$$

$$= a\,[\sqrt{2} + \log(1+\sqrt{2})].$$

(*ii*) Length of arc L′OL cut off by the latus rectum = 2 times arc OL = $2a\,[\sqrt{2} + \log(1+\sqrt{2})]$

Note 1. $\int \sqrt{\mathbf{a}^2-\mathbf{x}^2}\,\mathbf{dx} = \frac{\mathbf{x}\sqrt{\mathbf{a}^2-\mathbf{x}^2}}{2} + \frac{\mathbf{a}^2}{2}\sin^{-1}\frac{\mathbf{x}}{\mathbf{a}}.$

2. $\int \sqrt{\mathbf{x}^2-\mathbf{a}^2}\,\mathbf{dx} = \frac{\mathbf{x}\sqrt{\mathbf{x}^2-\mathbf{a}^2}}{2} - \frac{\mathbf{a}^2}{2}\cosh^{-1}\frac{\mathbf{x}}{\mathbf{a}}.$

3. $\int \sqrt{\mathbf{a}^2+\mathbf{x}^2}\,\mathbf{dx} = \frac{\mathbf{x}\sqrt{\mathbf{a}^2+\mathbf{x}^2}}{2} + \frac{\mathbf{a}^2}{2}\sinh^{-1}\frac{\mathbf{x}}{\mathbf{a}}.$

Example 3. *Show that the length of an arc of the curve $x^2 = a^2(1 - e^{y/a})$ measured from (0, 0) to (x, y) is $a \log \frac{a+x}{a-x} - x$.*

Sol. The given curve is $x^2 = a^2(1 - e^{y/a})$.

Let us solve for y in terms of x.

$\therefore \quad x^2 = a^2 - a^2 e^{y/a}$ or $a^2 e^{y/a} = a^2 - x^2 \quad \therefore \quad e^{y/a} = \frac{a^2 - x^2}{a^2}$

$\therefore \quad \frac{y}{a} = \log\left(\frac{a^2 - x^2}{a^2}\right)$ or $y = a \log \frac{a^2 - x^2}{a^2}$

$\therefore \quad \frac{dy}{dx} = a \cdot \frac{a^2}{a^2 - x^2}\left(-\frac{2x}{a^2}\right) = -\frac{2ax}{a^2 - x^2}$

$\therefore$ Required length of the arc from (0, 0) to (x, y) is

$$= \int_0^x \sqrt{1 + \left(\frac{dy}{dx}\right)^2}\, dx = \int_0^x \sqrt{1 + \frac{4a^2x^2}{(a^2 - x^2)^2}}\, dx$$

$$= \int_0^x \sqrt{\frac{(a^2 - x^2)^2 + 4a^2x^2}{(a^2 - x^2)^2}}\, dx = \int_0^x \frac{a^2 + x^2}{a^2 - x^2}\, dx$$

$$= \int_0^x \left(-1 + \frac{2a^2}{a^2 - x^2}\right) dx$$

$$-x^2 + a^2 \overline{)\, x^2 + a^2 \,}\ (-1$$
$$\underline{x^2 - a^2}$$
$$2a^2$$

$$= \left[-x + 2a^2 \cdot \frac{1}{2a} \log \frac{a+x}{a-x}\right]_0^x \qquad \left|\ \because \int \frac{1}{\mathbf{a^2 - x^2}}\, \mathbf{dx} = \frac{\mathbf{1}}{\mathbf{2a}} \mathbf{\log} \frac{\mathbf{a + x}}{\mathbf{a - x}}\right.$$

$$= \left(-x + a \log \frac{a+x}{a-x}\right) - (-0 + a \log 1) = a \log \frac{a+x}{a-x} - x. \qquad [\because \log 1 = 0]$$

Note. $\int \frac{\mathbf{1}}{\mathbf{x^2 - a^2}}\, \mathbf{dx} = \frac{\mathbf{1}}{\mathbf{2a}} \mathbf{\log} \frac{\mathbf{x - a}}{\mathbf{x + a}}$.

Example 4. *Show that the whole length of the curve $x^2(a^2 - x^2) = 8a^2y^2$ is $\pi a\sqrt{2}$.*

(K.U. 1996 S, 2006 S ; M.D.U. 2000)

Sol. Equation of the curve is $8a^2y^2 = x^2(a^2 - x^2)$ ...(1)

Let us trace the curve roughly to find the limits of integration.

(*i*) The curve is symmetrical about both the axes.

($\because$ There are only even powers of both x and y)

(*ii*) The curve passes through the origin and the tangents at the origin are given by $a^2x^2 = 8a^2y^2$ or $y = \pm \frac{1}{2\sqrt{2}} x$ showing that origin is a node.

(*iii*) The curve has no asymptotes.

(*iv*) To find points of intersection with the x-axis *i.e.* with the line $y = 0$ [Putting $y = 0$ in (1)], we have $x^2 (a^2 - x^2) = 0$ or $x = 0, \pm a$.

Hence the curve meets the x-axis at $(0, 0)$, $(a, 0)$ and $(-a, 0)$.

The curve meets the y-axis at the origin only.

(*v*) *Region*

From (1), $$y = \frac{1}{\sqrt{8a^2}} x \sqrt{a^2 - x^2} \quad \text{...(2)}$$

For y to be real, $a^2 - x^2 \geq 0$ *i.e.* $x^2 \leq a^2$.

$$\therefore \quad -a \leq x \leq a.$$

Thus, the whole curve lies between the lines $x = -a$ and $x = a$.

The shape of the curve is as shown in the figure. The curve consists of two symmetrical loops.

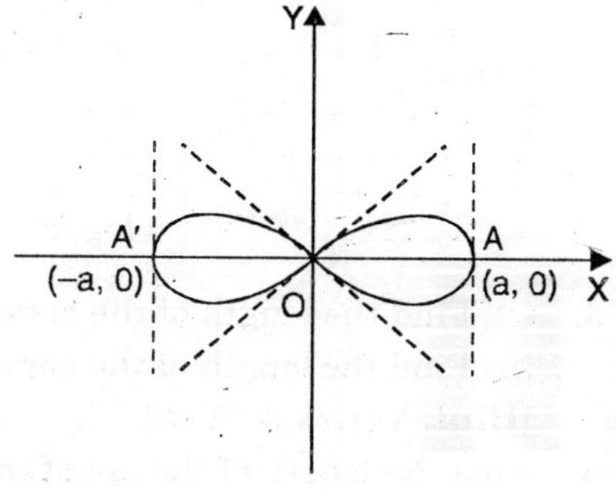

From (1), $$y = \frac{1}{\sqrt{8a^2}} x \sqrt{a^2 - x^2}$$

$$\therefore \quad \frac{dy}{dx} = \frac{1}{\sqrt{8a^2}} \left[x \left(\frac{-2x}{2\sqrt{a^2 - x^2}} \right) + \sqrt{a^2 - x^2} \right]$$

$$= \frac{1}{\sqrt{8a^2}} \left[\frac{a^2 - 2x^2}{\sqrt{a^2 - x^2}} \right]$$

$\therefore$ Required whole length of the curve = 4 (Arc OA)

$$= 4 \int_0^a \sqrt{1 + \left(\frac{dy}{dx}\right)^2}\, dx = 4 \int_0^a \sqrt{1 + \frac{(a^2 - 2x^2)^2}{8a^2(a^2 - x^2)}}\, dx$$

$$= 4 \int_0^a \sqrt{\frac{8a^4 - 8a^2x^2 + a^4 + 4x^4 - 4a^2x^2}{8a^2(a^2 - x^2)}}\, dx$$

$$= \frac{4}{\sqrt{8a^2}} \int_0^a \sqrt{\frac{9a^4 - 12a^2x^2 + 4x^4}{a^2 - x^2}}\, dx = \frac{4}{2a\sqrt{2}} \int_0^a \frac{3a^2 - 2x^2}{\sqrt{a^2 - x^2}}\, dx$$

$$= \frac{\sqrt{2}}{a} \int_0^a \frac{2(a^2 - x^2) + a^2}{\sqrt{a^2 - x^2}}\, dx \qquad \textit{(Note this step)}$$

$$= \frac{\sqrt{2}}{a} \int_0^a \left(2\sqrt{a^2 - x^2} + \frac{a^2}{\sqrt{a^2 - x^2}} \right) dx$$

$$= \frac{\sqrt{2}}{a} \left[2\left(x \frac{\sqrt{a^2 - x^2}}{2} + \frac{a^2}{2} \sin^{-1} \frac{x}{a} \right) + a^2 \sin^{-1} \frac{x}{a} \right]_0^a$$

$$= \frac{\sqrt{2}}{a} \left[x\sqrt{a^2 - x^2} + 2a^2 \sin^{-1} \frac{x}{a} \right]_0^a$$

$$= \frac{\sqrt{2}}{a}\,[2a^2 \sin^{-1} 1] = \frac{\sqrt{2}}{a}\,.\,2a^2\,.\,\frac{\pi}{2} = \pi a\,\sqrt{2}\,.$$

Note. It is a common mistake not to take into account the symmetry of the curve. So the student is suggested to make use of the symmetry of the curve while writing limit of integration.

EXERCISE 1

1. Find the length of the arc of the curve $y = \log \sec x$ from $x = 0$ to $x = \frac{\pi}{3}$.
2. Find the length of the arc of the curve $y = \log \frac{e^x - 1}{e^x + 1}$ from $x = 1$ to $x = 2$.

$\left[\textbf{Hint.} \int \frac{e^{2x}+1}{e^{2x}-1}\,dx = \int \frac{e^x + e^{-x}}{e^x - e^{-x}}\,dx\right.$ (Multiplying every term by e^{-x})

$= \log\,(e^x - e^{-x})\,.$ $\left.\left(\because \int \frac{f'(x)}{f(x)}\,dx = \log f(x)\right)\right]$

3. (*i*) Find the length of the curve $3ay^2 = 2x^3$ between the points (0, 0) and (x, y).
(*ii*) Find the length of the curve $ay^2 = x^3$ from the vertex to the point (a, a).
[**Hint.** Vertex is (0, 0)].
4. Find the length of the arc of the parabola $x^2 = 4ay$:
(*i*) from the vertex to an extremity of the latus rectum.
(*ii*) cut off by latus rectum. (*M.D.U 2005*)
5. (*i*) Find the arc length of the curve

$$y = \frac{1}{2}x^2 - \frac{1}{4}\log x \text{ from } x = 1 \text{ to } x = 2.$$

(*ii*) Find the length of the arc of the curve $y = x\,(2 - x)$ as x varies from 0 to 2.
6. (*i*) Find the perimeter of the loop of the curve $9ay^2 = (x - 2a)\,(x - 5a)^2$.
(*M.D.U. 2003, 2007 ; K.U. 2006*)

$\left[\textbf{Hint.} \int \frac{x-a}{\sqrt{x-2a}}\,dx = \int \frac{x - 2a + a}{\sqrt{x-2a}}\,dx.\right]$

(*ii*) Find the perimeter of the loop of the curve $9y^2 = x^3 - 12x^2 + 45x - 50$. (*K.U. 2004 S*)
[**Hint.** $x^3 - 12x^2 + 45x - 50$
$= x^3 - 2x^2 - 10x^2 + 20x + 25\,x - 50$
$= x^2\,(x - 2) - 10x\,(x - 2) + 25\,(x - 2) = (x - 2)\,(x^2 - 10\,x + 25)$
$= (x - 2)\,(x - 5)^2.$
Now it is part (*i*) with $a = 1$].
7. Find the whole length of the loop of the curve $3ay^2 = x(x - a)^2$. (*K.U. 1997*)
8. Rectify the loop of the curve $3ay^2 = x^2(a - x)$.
9. Find the whole length of the loop of the curve $3ax^2 = y^2\,(a - y)$.
10. Find the whole length of the loop of the curve $3ax^2 = y(y - a)^2$.
11. Find the length of the arc of the curve $x^2 + y^2 - 2ax = 0$ in first quadrant.

ANSWERS

1. $\log(2+\sqrt{3})$
2. $\log(e+e^{-1})$
3. (i) $\frac{4a}{9}\left[\left(1+\frac{3x}{2a}\right)^{3/2}-1\right]$ (ii) $\frac{a}{27}[13\sqrt{13}-8]$
4. (i) $a[\sqrt{2}+\log(\sqrt{2}+1)]$ (ii) $2a[\sqrt{2}+\log(\sqrt{2}+1)]$
5. (i) $\frac{3}{2}+\frac{1}{4}\log 2$ (ii) $\frac{1}{2}\log(2+\sqrt{5})+\sqrt{5}$
6. (i) $4\sqrt{3}a$ (ii) $4\sqrt{3}$
7. $\frac{4a}{\sqrt{3}}$
8. $\frac{4a}{\sqrt{3}}$
9. $\frac{4a}{\sqrt{3}}$
10. $\frac{4a}{\sqrt{3}}$
11. πa.

3. Length of the Curve when Parametric Equations are given

To prove that the length of the arc of the curve $x = f(t)$, $y = \phi(t)$ between the points where $t = t_1$ to $t = t_2$ is given by

$$\int_{t_1}^{t_2}\sqrt{\left(\frac{dx}{dt}\right)^2+\left(\frac{dy}{dt}\right)^2}\,dt$$

where x and y are continuous and single-valued functions of t, in the given interval $[t_1, t_2]$.

Let AB be the curve $x = f(t)$, $y = \phi(t)$ where A, B are two points where $t = t_1$ and $t = t_2$ and let CA and DB be their ordinates.

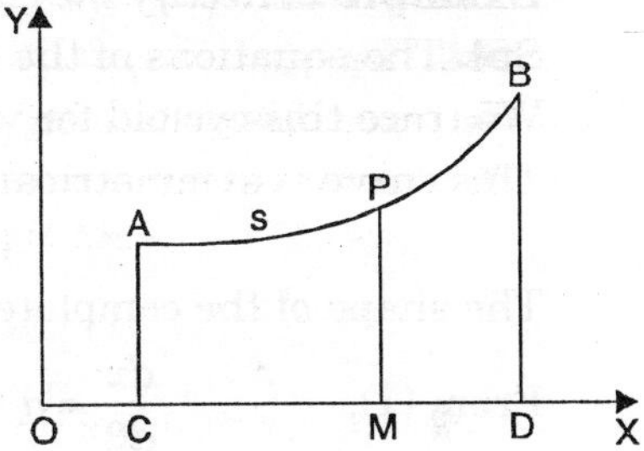

Let P (x, y) be any point on the curve and PM be its ordinate.

If s is the length of the arc from the fixed point A to the variable point P, then clearly s is a function of t.

From Differential Calculus, we know that

$$\frac{ds}{dt}=\sqrt{\left(\frac{dx}{dt}\right)^2+\left(\frac{dy}{dt}\right)^2} \quad \text{...(1)}$$

$$\therefore \int_{t_1}^{t_2}\sqrt{\left(\frac{dx}{dt}\right)^2+\left(\frac{dy}{dt}\right)^2}\,dt=\int_{t_1}^{t_2}\frac{ds}{dt}\,.\,dt=\Big[s\Big]_{t_1}^{t_2}$$

$$= (\text{value of } s \text{ when } t = t_2) - (\text{value of } s \text{ when } t = t_1)$$

$$= \text{arc AB} - 0 = \text{arc AB}.$$

$$\text{Hence arc AB} = \int_{t_1}^{t_2}\sqrt{\left(\frac{dx}{dt}\right)^2+\left(\frac{dy}{dt}\right)^2}\,dt.$$

Note. Here again it is assumed that the integrand does not change sign in the range of integration $[t_1, t_2]$.

Cor. From (1), $s=\sqrt{\left(\frac{dx}{dt}\right)^2+\left(\frac{dy}{dt}\right)^2}\,dt.$

Example 1. *Find the length of the curve*

$$x = e^{\theta}\left(\sin\frac{\theta}{2} + 2\cos\frac{\theta}{2}\right), y = e^{\theta}\left(\cos\frac{\theta}{2} - 2\sin\frac{\theta}{2}\right)$$

measured from θ = 0 to θ = π. (M.D.U. 1998)

Sol. The given curve is

$$x = e^{\theta}\left(\sin\frac{\theta}{2} + 2\cos\frac{\theta}{2}\right), \quad y = e^{\theta}\left(\cos\frac{\theta}{2} - 2\sin\frac{\theta}{2}\right)$$

$$\therefore \quad \frac{dx}{d\theta} = e^{\theta}\left(\frac{1}{2}\cos\frac{\theta}{2} - \sin\frac{\theta}{2}\right) + e^{\theta}\left(\sin\frac{\theta}{2} + 2\cos\frac{\theta}{2}\right) = \frac{5}{2}e^{\theta}\cos\frac{\theta}{2}$$

and

$$\frac{dy}{d\theta} = e^{\theta}\left(-\frac{1}{2}\sin\frac{\theta}{2} - \cos\frac{\theta}{2}\right) + e^{\theta}\left(\cos\frac{\theta}{2} - 2\sin\frac{\theta}{2}\right) = -\frac{5}{2}e^{\theta}\sin\frac{\theta}{2}$$

$$\therefore \quad \left(\frac{dx}{d\theta}\right)^2 + \left(\frac{dy}{d\theta}\right)^2 = \frac{25}{4}e^{2\theta}\left(\cos^2\frac{\theta}{2} + \sin^2\frac{\theta}{2}\right) = \frac{25}{4}e^{2\theta} \quad \text{...(1)}$$

$$\therefore \quad \text{Required Length} = \int_0^{\pi}\sqrt{\left(\frac{dx}{d\theta}\right)^2 + \left(\frac{dy}{d\theta}\right)^2}\,d\theta = \int_0^{\pi}\frac{5}{2}e^{\theta}\,d\theta \quad \text{| Using (1)}$$

$$= \frac{5}{2}\Big[e^{\theta}\Big]_0^{\pi} = \frac{5}{2}[e^{\pi} - 1].$$

Example 2. *Rectify the curve x = a (θ + sin θ), y = a (1 + cos θ).*

Sol. The equations of the cycloid are $x = a(\theta + \sin\theta), y = a(1 + \cos\theta)$ ...(1)

We trace this cycloid for values of θ in [– π, π] to get one **arch** of the cycloid.

This curve is symmetrical about *y*-axis.

(∵ On changing θ to – θ, *x* changes to – *x* and *y* is unchanged)

The shape of the complete cycloid is being shown below :

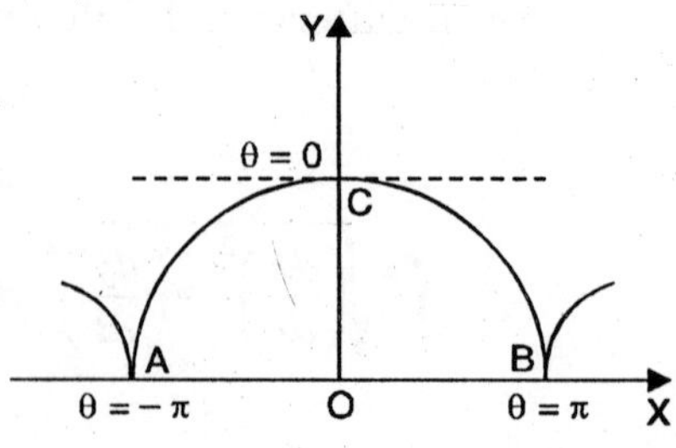

From (1), $\frac{dx}{d\theta} = a(1 + \cos\theta)$

and $\frac{dy}{d\theta} = -a\sin\theta$

$$\therefore \quad \left(\frac{dx}{d\theta}\right)^2 + \left(\frac{dy}{d\theta}\right)^2 = a^2(1 + \cos\theta)^2 + a^2\sin^2\theta$$

$$= a^2[1 + \cos^2\theta + 2\cos\theta + \sin^2\theta]$$

$$= 2a^2(1 + \cos\theta) = 4a^2\cos^2\frac{\theta}{2} \quad \text{...(2)}$$

∴ Length of the curve (*i.e.* length of *one arch of cycloid* or *complete cycloid*).

$$= 2\int_0^{\pi}\sqrt{\left(\frac{dx}{d\theta}\right)^2 + \left(\frac{dy}{d\theta}\right)^2}\,d\theta = 2\int_0^{\pi}\sqrt{4a^2\cos^2\frac{\theta}{2}}\,d\theta \quad \text{| From (2)}$$

$$= 2\int_0^{\pi} 2a\cos\frac{\theta}{2}\,d\theta = 4a\left(\frac{\sin\frac{\theta}{2}}{\frac{1}{2}}\right)_0^{\pi} = 8a\left[\sin\frac{\pi}{2} - \sin 0\right] = 8a.$$

Note. We have taken $\sqrt{4a^2 \cos^2 \theta/2} = 2a \cos \theta/2$ and not $-2a \cos \theta/2$ as $\cos \theta/2$ remains positive when θ increases 0 to π.

Example 3. *Rectify the ellipse $x = a \cos \theta$ and $y = b \sin \theta$.*

or $\left(\textit{Find the whole length of the ellipse } \frac{x^2}{a^2} + \frac{y^2}{b^2} = 1\right)$.

Sol. The ellipse $x = a \cos \theta, y = b \sin \theta, \left(\frac{x^2}{a^2} + \frac{y^2}{b^2} = 1\right)$ is symmetrical about both the axes.

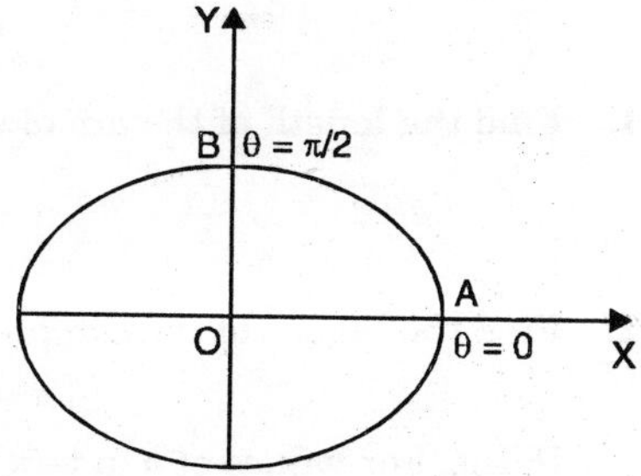

Diff. w.r.t. θ, $\frac{dx}{d\theta} = -a \sin \theta$ and $\frac{dy}{d\theta} = b \cos \theta$

$\therefore$ Perimeter of the ellipse

$$= 4 \text{ (Arc AB)}$$

$$= 4 \int_0^{\pi/2} \sqrt{\left(\frac{dx}{d\theta}\right)^2 + \left(\frac{dy}{d\theta}\right)^2}\, d\theta$$

$$= 4 \int_0^{\pi/2} \sqrt{a^2 \sin^2 \theta + b^2 \cos^2 \theta}\, d\theta$$

$$= 4 \int_0^{\pi/2} \sqrt{a^2 \sin^2 \theta + a^2(1-e^2) \cos^2 \theta}\, d\theta$$

[$\because$ For an ellipse $b^2 = a^2(1 - e^2)$, where e is the eccentricity of the ellipse]

$$= 4a \int_0^{\pi/2} \sqrt{(\sin^2 \theta + \cos^2 \theta) - e^2 \cos^2 \theta}\, d\theta$$

$$= 4a \int_0^{\pi/2} (1 - e^2 \cos^2 \theta)^{1/2}\, d\theta$$

$$= 4a \int_0^{\pi/2} \left[1 + \frac{1}{2} - (e^2 \cos^2 \theta) + \frac{\frac{1}{2}\left(\frac{1}{2} - 1\right)}{2!} - (e^2 \cos^2 \theta)^2 \right.$$

$$\left. + \frac{\frac{1}{2}\left(\frac{1}{2} - 1\right)\left(\frac{1}{2} - 2\right)}{3!} (-e^2 \cos^2 \theta)^3 + \ldots \right] d\theta$$

$\left[\because \text{ By Binomial Theorem, } (1+x)^n = 1 + nx + \frac{n(n-1)}{2!} x^2 + \frac{n(n-1)(n-2)}{3!} x^3 + \ldots\ldots\right]$

$$= 4a \int_0^{\pi/2} \left[1 - \frac{1}{2} e^2 \cos^2 \theta - \frac{1}{2.4} e^4 \cos^4 \theta - \frac{1.3}{2.4.6} e^6 \cos^6 \theta - \ldots\right] d\theta$$

$$= 4a \left[\int_0^{\pi/2} 1\, d\theta - \frac{1}{2} e^2 \int_0^{\pi/2} \cos^2 \theta\, d\theta - \frac{1}{2.4} e^4 \int_0^{\pi/2} \cos^4 \theta\, d\theta \right.$$

$$\left. - \frac{1.3}{2.4.6} e^6 \int_0^{\pi/2} \cos^6 \theta\, d\theta - \ldots \right]$$

$$= 4a\left[\frac{\pi}{2} - \frac{1}{2}e^2 \cdot \frac{1}{2}\frac{\pi}{2}^* - \frac{1}{2.4}e^4 \cdot \frac{3.1}{4.2}\frac{\pi}{2} - \frac{1.3}{2.4.6}e^6 \cdot \frac{5.3.1}{6.4.2}\frac{\pi}{2} - \ldots\right]$$

$$= 2a\pi\left[1 - \left(\frac{1}{2}\right)^2 e^2 - \left(\frac{1.3}{2.4}\right)^2 \frac{e^4}{3} - \left(\frac{1.3.5}{2.4.6}\right)^2 \frac{e^6}{5} - \ldots\right]$$

EXERCISE 2

1. Find the length of the arc of the curve $x = e^\theta \sin\theta$, $y = e^\theta \cos\theta$ from $\theta = 0$ to be $\theta = \frac{\pi}{2}$.

(*M.D.U. 1999*)

2. Find the length of the complete cycloid given by

$$x = a(\theta + \sin\theta), \quad y = a(1 - \cos\theta).$$ (*K.U. 2007*)

[**Hint.** For values of θ in $[-\pi, \pi]$, we get the complete cycloid. Length of complete cycloid is

$$2\int_0^\pi \sqrt{\left(\frac{dx}{d\theta}\right)^2 + \left(\frac{dy}{d\theta}\right)^2}\, d\theta.]$$

3. (*i*) Find the length of one arch of the cycloid $x = a(\theta - \sin\theta)$, $y = a(1 - \cos\theta)$.

(*ii*) and show that $\theta = \frac{2\pi}{3}$ divides it in the ratio 1 : 3. (*K.U. 1999*)

[**Hint.** For values of θ in $[0, 2\pi]$, we get the complete cycloid.

Length of complete cycloid is $2\int_0^\pi \sqrt{\left(\frac{dx}{d\theta}\right)^2 + \left(\frac{dy}{d\theta}\right)^2}\, d\theta.$]

4. Rectify the cycloid $x = a(\theta - \sin\theta)$, $y = a(1 + \cos\theta)$. (*K.U. 2005 ; M.D.U. 2006*)

[**Hint.** Same as for question 3.]

5. Show that the entire length of the astroid $x = a\cos^3 t$, $y = a\sin^3 t$ (or $x^{2/3} + y^{2/3} = a^{2/3}$ or $(x/a)^{2/3} + (y/a)^{2/3} = 1$) is $6a$. (*M.D.U. 2004 S, 2005 ; K.U. 2005 S*)

$$\left[\textbf{Hint. } \text{Entire Length} = 4\int_0^{\pi/2} \sqrt{\left(\frac{dx}{dt}\right)^2 + \left(\frac{dy}{dt}\right)^2}\, dt.\right]$$

6. Find the length of the arc in the first quadrant of the curve

$\left(\frac{x}{a}\right)^{2/3} + \left(\frac{y}{b}\right)^{2/3} = 1$ or ($x = a\cos^3 t$, $y = b\sin^3 t$). (*M.D.U. 1996, 2004*)

$$* \int_0^{\pi/2} \cos^n\theta\, d\theta = \frac{(n-1)\times \text{go on decreasing by 2}}{n \times \text{go on decreasing by 2}} \times \left(\frac{\pi}{2} \text{ if } n \text{ is an even positive integer}\right)$$

[**Hint.** Length of arc in **first** quadrant

$$= \int_0^{\pi/2} \sqrt{\left(\frac{dx}{dt}\right)^2 + \left(\frac{dy}{dt}\right)^2}\, dt = \int_0^{\pi/2} 3 \sin t \cos t \sqrt{a^2 \cos^2 t + b^2 \sin^2 t}\, dt.$$

[Put $a^2 \cos^2 t + b^2 \sin^2 t = z$.]

7. Prove that the loop of the curve $x = t^2, y = t - \frac{1}{3} t^3$ is of length $4\sqrt{3}$. *(K.U. 1995)*

[**Hint.** Trace this curve. Limits of integration are 0 to 3. The curve is symmetrical about x-axis.]

8. Prove that in the cycloid $x = a\,(\theta + \sin \theta), y = a\,(1 - \cos \theta)$; the length of the arc from the vertex to the point (x, y) is $\sqrt{8ay}$. Hence find the whole length of the curve.

9. (*i*) Find the length of the arc of the curve.

$x = a\,(\cos t + t \sin t), y = a\,(\sin t - t \cos t)$ from $t = 0$ to $t = 2$.

(*ii*) A curve is given by the equation $x = a\,(\cos \theta + \theta \sin \theta), y = a\,(\sin \theta - \theta \cos \theta)$.

Find the length of the arc from $\theta = 0$ to $\theta = \alpha$.

10. Show that the length of an arc of the curve $x \sin \theta + y \cos \theta = f'(\theta), x \cos \theta - y \sin \theta = f''(\theta)$ is given by $s = f(\theta) + f''(\theta) + c$. *(M.D.U. 2003 S)*

[**Hint.** Solving the two equations for x and y, we have $x = f'(\theta) \sin \theta + f''(\theta) \cos \theta, y = f'(\theta) \cos \theta - f''(\theta) \sin \theta$.]

ANSWERS

1. $\sqrt{2}\,(e^{\pi/2} - 1)$	**2.** $8a$	**3.** $8a$	**4.** $8a$
6. $\dfrac{a^2 + ab + b^2}{a + b}$	**8.** $8a$	**9.** (*i*) $2a$	(*ii*) $\dfrac{a\alpha^2}{2}$

Remark. For the sake of convenience of the readers, we are giving below the shapes of the curves in questions **2, 3, 4, 5, 6, 7.**

2.

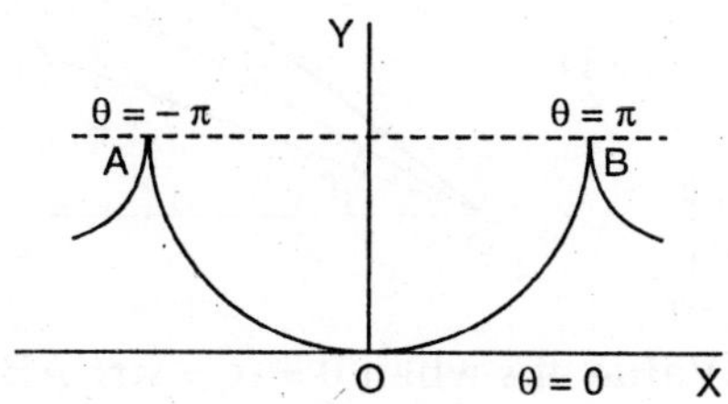

$x = a\,(\theta + \sin \theta)$

$y = a\,(1 - \cos \theta)$

3.

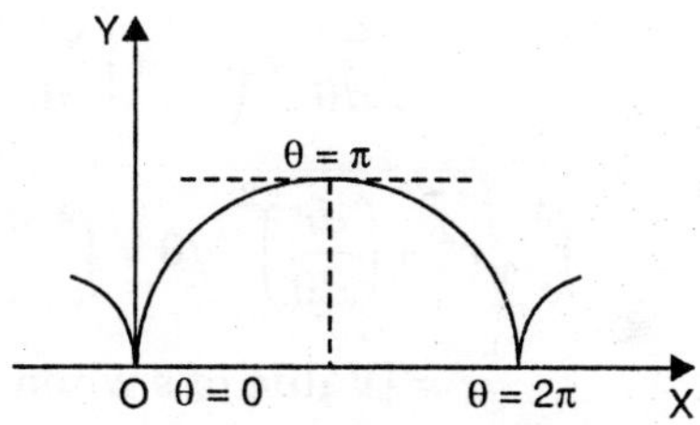

$x = a\,(\theta - \sin \theta)$

$y = a\,(1 - \cos \theta)$

4.

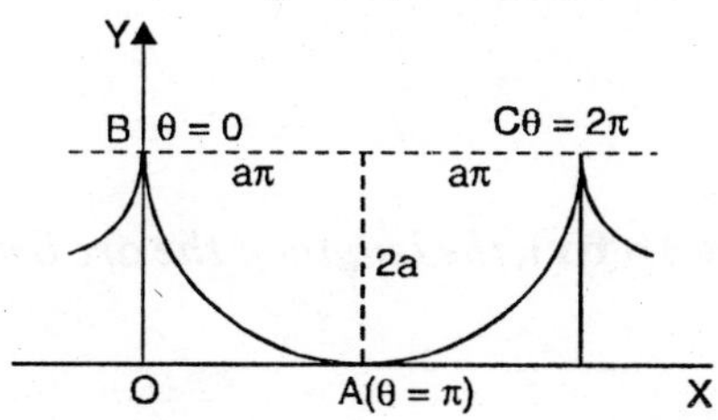

5.

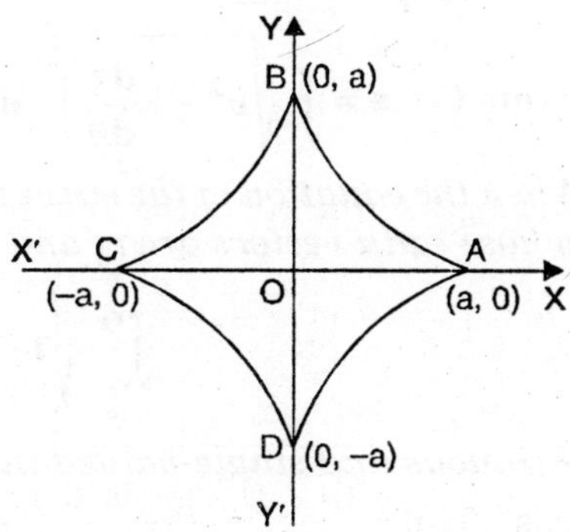

$x = a\,(\theta - \sin \theta)$

$y = a\,(1 + \cos \theta)$

6.

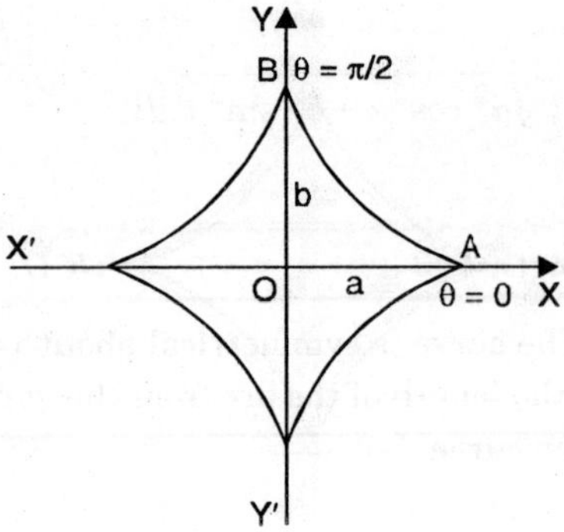

7.

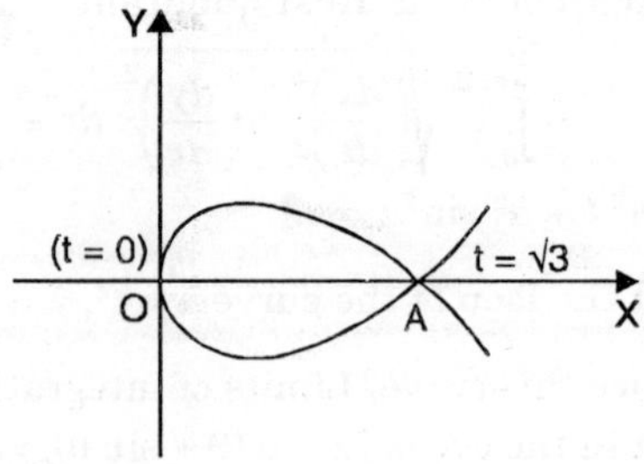

4. Length of the Polar Curves

To prove that the length of the arc of the curve $r = f(\theta)$ between points whose vectorial angles are α and β, is

$$\int_{\alpha}^{\beta} \sqrt{\mathbf{r}^2 + \left(\frac{\mathbf{dr}}{\mathbf{d\theta}}\right)^2}\, \mathbf{d\theta}$$

where $\frac{dr}{d\theta}$ *is continuous and single valued in* $[\alpha, \beta]$.

Let AB be the curve $r = f(\theta)$ and A, B the points where $\theta = \alpha$ and $\theta = \beta$.

Let P(r, θ) be *any point* on the curve and let s denotes the length of arc AP which is clearly a function of θ.

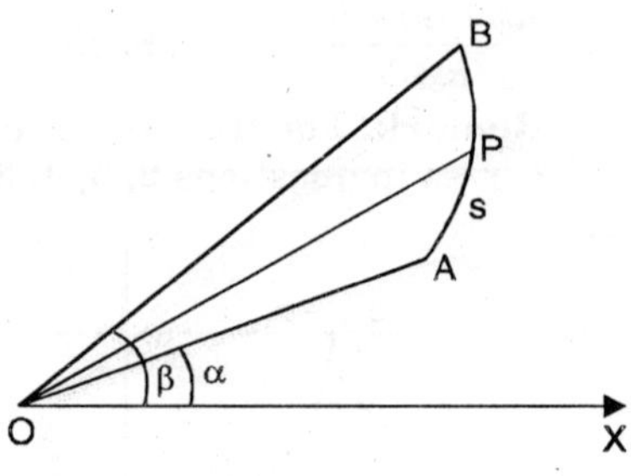

From Differential Calculus, we know that

$$\frac{ds}{d\theta} = \sqrt{r^2 + \left(\frac{dr}{d\theta}\right)^2} \qquad ...(1)$$

$$\therefore \quad \int_{\alpha}^{\beta} \sqrt{r^2 + \left(\frac{dr}{d\theta}\right)^2}\, d\theta = \int_{\alpha}^{\beta} \frac{ds}{d\theta} . d\theta = \Big[s \Big]_{\alpha}^{\beta}$$

= [Value of s when $\theta = \beta$] – [Value of s when $\theta = \alpha$] = arc AB – 0.

$$\therefore \quad \text{arc AB} = \int_{\alpha}^{\beta} \sqrt{r^2 + \left(\frac{dr}{d\theta}\right)^2}\, d\theta.$$

Cor. 1. From (1), $\mathbf{s} = \int \sqrt{\mathbf{r}^2 + \left(\frac{\mathbf{dr}}{\mathbf{d\theta}}\right)^2}\, \mathbf{d\theta}$.

Cor. 2. *When the equation of the curve is of the form* $\theta = \mathbf{f(r)}$, *the length of the arc between the two points whose radii vectors are* r_1 *and* r_2 *is*

$$\int_{r_1}^{r_2} \sqrt{1 + \mathbf{r}^2 \left(\frac{\mathbf{d\theta}}{\mathbf{dr}}\right)^2}\, \mathbf{dr}.$$

where $\frac{d\theta}{dr}$ *is continuous and single-valued in* $[r_1, r_2]$.

Proceed as in Art. 4, above and use $\frac{ds}{dr} = \sqrt{1 + r^2 \left(\frac{d\theta}{dr}\right)^2}$ instead of (1).

Example 1. *Find the entire length of the cardioid $r = a(1 + \cos\theta)$, and show that the arc of the upper half is bisected by $\theta = \pi/3$.* (M.D.U. 2006)

Sol. The equation of the curve is $r = a(1 + \cos\theta)$. ...(1)

The shape of the cardioid is as shown in the adjoining figure. It is symmetrical about the initial line and the upper half of the curve is traced when θ varies from 0 to π.

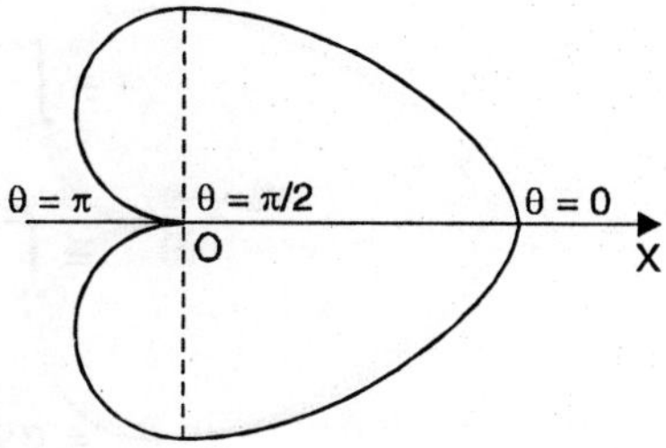

From (1), $$\frac{dr}{d\theta} = -a\sin\theta$$

$\therefore$ Length of whole arc = 2 × length of upper half

$$= 2\int_0^{\pi}\sqrt{r^2 + \left(\frac{dr}{d\theta}\right)^2}\,d\theta$$

$$= 2\int_0^{\pi}\sqrt{a^2(1+\cos\theta)^2 + a^2 \dot{+} \sin^2\theta\,d\theta}$$

$$= 2a\int_0^{\pi}\sqrt{2(1+\cos\theta)}\,d\theta = 2a\int_0^{\pi}2\cos\theta/2\,d\theta$$

$$= 4a\int_0^{\pi}\cos\theta/2\,d\theta = 4a\left[2\sin\frac{\theta}{2}\right]_0^{\pi} = 8a(1-0) = 8a.$$

$\therefore$ Entire length of cardioid = $8a$ and length of the upper half = $4a$.

Now the length of arc from $\theta = 0$ to $\theta^* = \pi/3$

$$= \int_0^{\pi/3}\sqrt{r^2 + \left(\frac{dr}{d\theta}\right)^2}\,d\theta = \int_0^{\pi/3}2a\cos\theta/2\,d\theta$$

$$= 2a\left|2\sin\frac{\theta}{2}\right|_0^{\pi/3} = 4a\left|\frac{1}{2} - 0\right| = 2a$$

$$= \frac{1}{2} \times \text{length of the upper half arc.}$$

Hence the length of the upper half arc is bisected at $\theta/\pi/3$.

Example 2. *Show that the arc of the hyperbolic spiral $r\,\theta = a$ taken from the point*

$$r = a \text{ to } r = 2a \text{ is } a\left[\sqrt{5} - \sqrt{2} + \log\left(\frac{2+\sqrt{8}}{1+\sqrt{5}}\right)\right].$$ (K.U. 2004 S)

Sol. The equation of the given curve is $r\theta = a$.

$$\therefore \quad \theta = \frac{a}{r} \quad \therefore \quad \frac{d\theta}{dr} = -\frac{a}{r^2}$$

$\therefore$ Required length from $r = a$ to $r = 2a$ is given by

$$s = \int_a^{2a}\sqrt{1 + r^2\left(\frac{d\theta}{dr}\right)^2}\,dr$$ (Cor. 2, Art. 4)

$$= \int_a^{2a}\sqrt{1 + r^2\cdot\frac{a^2}{r^4}}\,dr = \int_a^{2a}\frac{\sqrt{r^2+a^2}}{r}\,dr$$

---

*$\theta = \alpha$ is the equation of a half ray passing through the pole and making an angle α with the initial line.

Put $\sqrt{\mathbf{r^2 + a^2}} = \mathbf{t}$ $\therefore$ $r^2 + a^2 = t^2$

$\therefore$ $2r\,dr = 2t\,dt$ or $r\,dr = t\,dt$.

When $r = a$, $t = a\sqrt{2}$ and when $r = 2a$, $t = a\sqrt{5}$

$$\therefore \quad s = \int_{a\sqrt{2}}^{a\sqrt{5}} \frac{t}{t^2 - a^2} \cdot t\,dt = \int_{a\sqrt{2}}^{a\sqrt{5}} \frac{t^2}{t^2 - a^2}\,dt$$

$$= \int_{a\sqrt{2}}^{a\sqrt{5}} \left(1 + \frac{a^2}{t^2 - a^2}\right) dt = \left[t + a^2 \cdot \frac{1}{2a} \log \frac{t - a}{t + a}\right]_{a\sqrt{2}}^{a\sqrt{5}}$$

$$= a\sqrt{5} + \frac{a}{2} \log \frac{a\sqrt{5} - a}{a\sqrt{5} + a} - \left(a\sqrt{2} + \frac{a}{2} \log \frac{a\sqrt{2} - a}{a\sqrt{2} + a}\right)$$

$$= a\left[\sqrt{5} + \frac{1}{2} \log \frac{\sqrt{5} - 1}{\sqrt{5} + 1} - \sqrt{2} - \frac{1}{2} \log \frac{\sqrt{2} - 1}{\sqrt{2} + 1}\right]$$

$$= a\left[\sqrt{5} - \sqrt{2} + \frac{1}{2}\left(\log \frac{\sqrt{5} - 1}{\sqrt{5} + 1} - \log \frac{\sqrt{2} - 1}{\sqrt{2} + 1}\right)\right]$$

$$= a\left[\sqrt{5} - \sqrt{2} + \frac{1}{2} \log\left(\frac{(\sqrt{5} - 1)}{(\sqrt{5} + 1)} \frac{\sqrt{2} + 1}{\sqrt{2} - 1}\right)\right]$$

$$= a\left[\sqrt{5} - \sqrt{2} + \frac{1}{2} \log\left(\frac{(\sqrt{5} - 1)(\sqrt{5} + 1)}{(\sqrt{5} + 1)^2} \frac{(\sqrt{2} + 1)^2}{(\sqrt{2} + 1)(\sqrt{2} - 1)}\right)\right]$$

$$= a\left[\sqrt{5} - \sqrt{2} + \frac{1}{2} \log \frac{4(\sqrt{2} + 1)^2}{(\sqrt{5} + 1)^2}\right]$$

$$= a\left[\sqrt{5} - \sqrt{2} + \frac{1}{2} \log\left(\frac{2(\sqrt{2} + 1)}{\sqrt{5} + 1}\right)^2\right]$$

$$= a\left[\sqrt{5} - \sqrt{2} + 2 \cdot \frac{1}{2} \log \frac{\sqrt{8} + 2}{\sqrt{5} + 1}\right]$$

$$= a\left[\sqrt{5} - \sqrt{2} + \log\left(\frac{\sqrt{8} + 2}{\sqrt{5} + 1}\right)\right].$$

EXERCISE 3

1. Find the length of the arc of the cardioid $r = a\,(1 - \cos\theta)$ between the points whose vectorial angles are α and β.
2. Find the perimeter of the curve $r = a\,(1 - \cos\theta)$ and show that the arc of the upper half of the curve $r = a\,(1 - \cos\theta)$ is bisected by $\theta = \dfrac{2\pi}{3}$. *(K.U. 1998 S)*

3. Find the perimeter of the curve (circle) $r = a \cos \theta$.

$$\left[\textbf{Hint. } s = 2\int_0^{\pi/2} \sqrt{r^2 + \left(\frac{dr}{d\theta}\right)^2}\, d\theta.\right]$$

4. Find the length of a loop of the curve $r = a(\theta^2 - 1)$.

[**Hint.** To find the limits of integration for a loop, we generally put $r = 0$ and find two consecutive values of θ. Here putting $r = 0$, $\theta = \pm 1$.

$\therefore$ Limits of integration are -1 and 1.]

5. Find the length of the arc of the equiangular spiral $r = ae^{\theta \cot \alpha}$ between the points for which radii vectors are r_1 and r_2.

6. Find the length of the arc of the spiral $r = a\theta$ between the points whose radii vectors are r_1 and r_2.

7. (*a*) Prove that the cardioide $r = a(1 + \cos \theta)$ is divided by the line $4r \cos \theta = 3a$ into two parts such that the lengths of the arcs on either side of this line are equal.

[**Hint.** Solve $r = a(1 + \cos \theta)$ and $4r \cos \theta = 3a$ for r and θ to find the points of intersection.]

(*b*) Show that the length of the arc of that part of the cardioid $r = a(1 + \cos \theta)$ which lies on the side of the line $4r = 3a \sec \theta$ remote from the pole, is equal to $4a$.

8. Find the length of the arc of the parabola $\frac{2a}{r} = 1 + \cos \theta$ cut off by its latus rectum.

(M.D.U. 2004 S, 2003 S ; K.U. 2000 A)

$$\left[\textbf{Hint. } \text{The equation of latus rectum is } \theta = \frac{\pi}{2}.\right]$$

9. Prove that the perimeter of the limicon $r = a + b \cos \theta$ if $\frac{b}{a}$ is small, is approximately

$$2\pi a \left[1 + \frac{1}{4} \cdot \frac{b^2}{a^2}\right].$$

10. Show that the whole length of the limicon $r = a + b \cos \theta$ $(a > b)$ is equal to that of an ellipse whose semi-axes are equal in length to the maximum and minimum radii vectors of the limicon.

$$\left[\textbf{Hint. } \text{Length of the limicon} = 2\int_0^{\pi} \sqrt{r^2 + \left(\frac{dr}{d\theta}\right)^2}\, d\theta.\right.$$

Maximum and minimum radii vectors of limicon are $a + b$ and $(a - b)$.

$\therefore$ Equations of the ellipse are $x = (a + b) \cos t$, $y = (a - b) \sin t$.

$$\left.\text{Whole length of the ellipse is } 4\int_0^{\pi/2} \sqrt{\left(\frac{dx}{dt}\right)^2 + \left(\frac{dy}{dt}\right)^2}\, dt.\right]$$

ANSWERS

1. $4a\left(\cos\frac{\alpha}{2} - \cos\frac{\beta}{2}\right)$ **2.** $8a$ **3.** πa

4. $\frac{8a}{3}$ **5.** $(r_2 - r_1) \sec \alpha$

6. $\frac{1}{2a}[f(r_2) - f(r_1)]$ where $f(r) = r\sqrt{r^2 + a^2} + a^2 \sinh^{-1}\frac{r}{a}$ **8.** $2a[\sqrt{2} + \log(\sqrt{2} + 1)]$.

Remark. For the sake of convenience of the readers, we are giving below the shapes of the curves in Questions **2, 3, 8, 9** and **10.**

2.

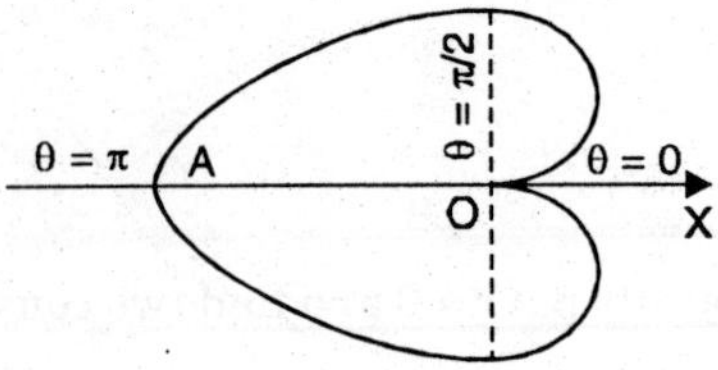

3.

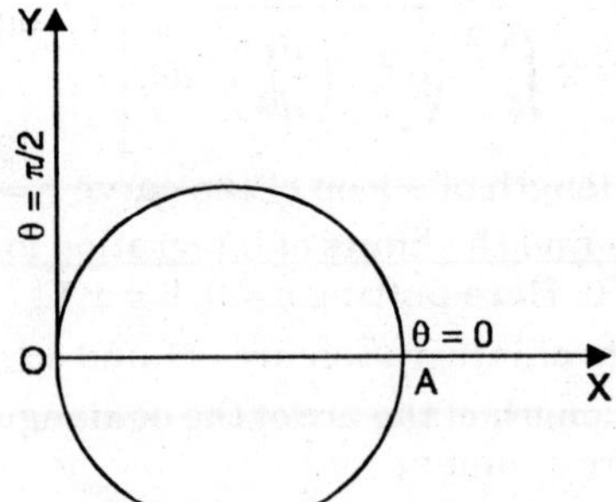

8.

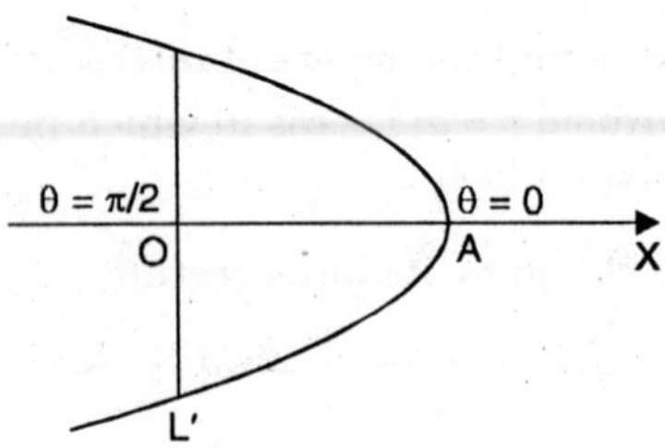

9. and 10.

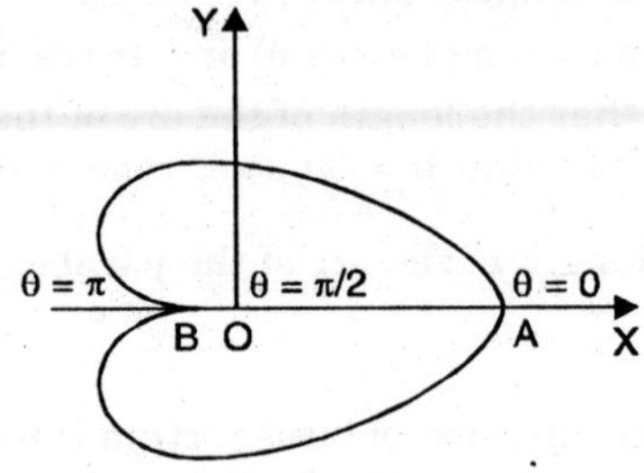

$r = a(1 - \cos\theta)$

---

5. To prove that the length of the arc of the curve p = f(r) between the points where r = a, r = b is

$$\int_a^b \frac{r}{\sqrt{r^2 - p^2}}\, dr.$$

From Cor. 2. Art. 4,

$$\text{Length of arc} = \int_a^b \sqrt{1 + r^2\left(\frac{d\theta}{dr}\right)^2}\, dr = \int_a^b \sqrt{1 + \tan^2\phi}\, dr \qquad \left(\because \tan\phi = r\frac{d\theta}{dr}\right)$$

$$= \int_a^b \sec\phi\, dr = \int_a^b \frac{r}{\sqrt{r^2 - p^2}}\, dr$$

$$\left(\because\ p = r\sin\phi\ \therefore\ \sec\phi = \frac{1}{\cos\phi} = \frac{1}{\sqrt{1 - \sin^2\phi}} = \frac{1}{\sqrt{1 - \frac{p^2}{r^2}}} = \frac{r}{\sqrt{r^2 - p^2}}\right)$$

[**Note. p** mentioned in the above Article is the length of the perpendicular from the pole on **any** tangent and ϕ is the angle between the tangent and radius vector at any point.]

EXERCISE 4

1. Show that the length of the arc of the equiangular spiral $p = r \sin\alpha$ between the points at which radii vectors are r_1 and r_2 is $(r_2 - r_1)\sec\alpha$.

2. Show that the length of the arc of the hyperbola $xy = a^2$ between the limits $x = b$ and $x = c$ is equal to the arc of the curve $p^2 (a^4 + r^4) = a^4 r^2$ between the limits $r = b$, $r = c$.

[**Hint**. Equations of two curves are $y = \dfrac{a^2}{x}$ and $p^2 = \dfrac{a^4 r^2}{a^4 + r^4}$.

Do not evaluate the two integrals. Show that they are equal.]

6. Intrinsic Equation of a Curve

A relation between the **variables** s and ψ is called **intrinsic equation** of a curve.

s is the length of the arc of a curve measured from a fixed point A on it to **ANY** point P and ψ is the angle which the tangent at P makes with the tangent at A (or with any other fixed line generally x-axis in the plane of the curve).

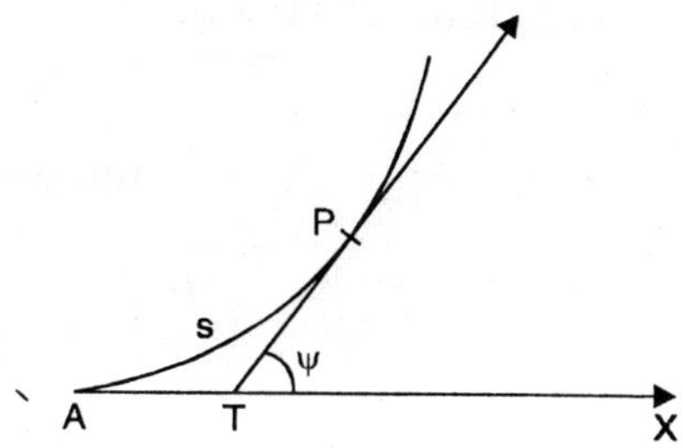

[**Note.** A relation between x and y is called *Cartesian Equation*.

A relation between r and θ is called *Polar Equation*.

A relation between p and r is called *Pedal Equation*.]

7. To find the intrinsic equation of a curve from the cartesian equation

Let $y = f(x)$ be *cartesian equation* of curve, where x-axis is the tangent at the origin O, the fixed point from which arc $s(=\text{OP})$ is measured. We know that

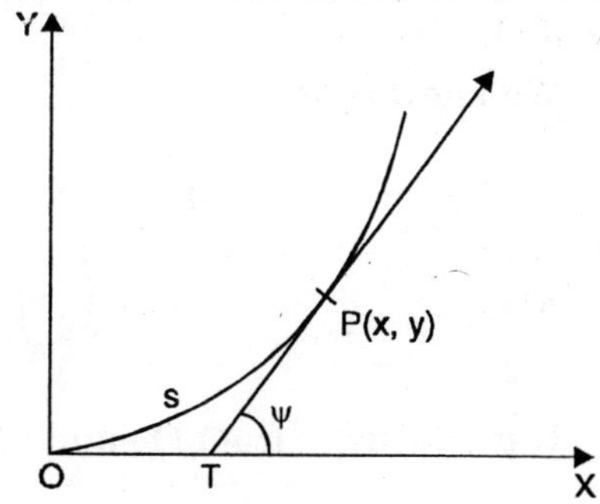

$$\tan\psi = \frac{dy}{dx} = f'(x) \qquad \text{...(1)}$$

and $$s = \int_0^x \sqrt{1+\left(\frac{dy}{dx}\right)^2}\, dx = \text{F}(x)\ (say) \qquad \text{...(2)}$$

Eliminating x between (1) and (2), we get a relation between s and ψ which is the intrinsic equation of the curve.

8. To find the intrinsic equation of the curve from the parametric equations

Let $x = f(t)$, $y = \text{F}(t)$, be the parametric equations of the curve and let the x-axis be the fixed straight line (tangent at the origin from which the arc is measured).

Let P(x, y) be any point on the curve and PT the tangent at P meeting the x-axis in T. Let arc OP $= s$ and $\angle$XTP $= \psi$.

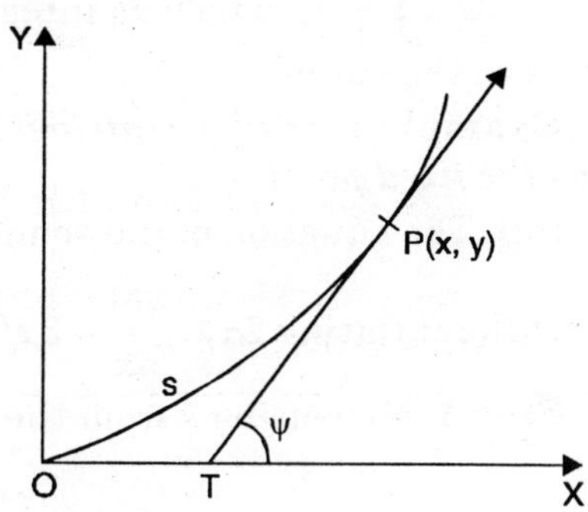

$$\text{Now } \tan\psi = \frac{dy}{dx} = \frac{dy/dt}{dx/dt} = \frac{\text{F}'(t)}{f'(t)} \qquad \text{...(1)}$$

and $$s = \int_0^t \sqrt{\left(\frac{dx}{dt}\right)^2 + \left(\frac{dy}{dt}\right)^2}\, dt$$

$$= \int_0^t \sqrt{[f'(t)]^2 + [\text{F}'(t)]^2}\, dt = \phi(t),\ (say) \qquad \text{...(2)}$$

Eliminating t between (1) and (2), we obtain a relation between s and ψ which is the intrinsic equation of the curve.

9. To find the intrinsic equation of the curve from polar equation

Let $r = f(\theta)$ be the polar equation of the curve, the initial line being the tangent to the curve at the pole O, the fixed point, from which arc is measured. Let P(r, θ) be any point on the curve, so that arc OP $= s$. Let tangent at P meet the initial line in T, so that $\angle$XTP $= \psi$.

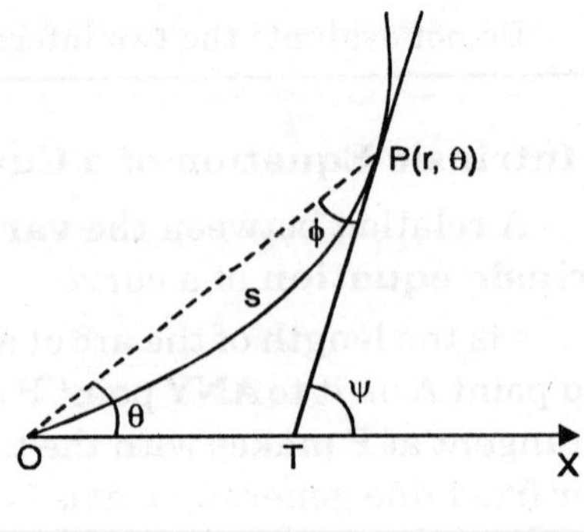

Now $$\psi = \theta + \phi \quad ...(1)$$

$$\tan\phi = r\frac{d\theta}{dr} = \frac{f(\theta)}{f'(\theta)} \quad ...(2)$$

and $$s = \int_0^\theta \sqrt{r^2 + \left(\frac{dr}{d\theta}\right)^2}\, d\theta = \int_0^\theta \sqrt{[f(\theta)]^2 + [f'(\theta)]^2}\, d\theta$$

$$= F(\theta) \quad (say) \quad ...(3)$$

Eliminating θ and ϕ between (1), (2) and (3), we get a relation between s and ψ, which is the intrinsic equation of the curve.

10. To find the intrinsic equation of the curve from the pedal equation

Let $p = f(r)$ be the pedal equation of the curve, the pole being at O, *a* fixed point on the curve and the initial line coinciding with the tangent at O, so here p and r vanish simultaneously at O.

We know that

$$\frac{ds}{d\psi} = \rho = r\frac{dr}{dp} = \frac{r}{dp/dr} = \frac{r}{f'(r)} \quad ...(1)$$

Also $$s = \int_0^r \frac{r\,dr}{\sqrt{r^2 - p^2}} = \int_0^r \frac{r\,dr}{\sqrt{r^2 - [f(r)]^2}} = f_1(r) \quad (say) \quad ...(2)$$

Eliminating r from (1) and (2), we get a relation between s and $\frac{ds}{d\psi}$ say $\frac{ds}{d\psi} = F(s)$ or $\frac{d\psi}{ds} = \frac{1}{F(s)}$.

$\therefore \quad \psi = \int \frac{ds}{F(s)}$ which on integration gives us a relation between ψ and s *i.e.* the intrinsic equation of the curve.

Example 1. *Find the intrinsic equation of the semi-cubical parabola, $ay^2 = x^3$, taking the cusp as the fixed point.*

Sol. The equation of the semi-cubical parabola is $ay^2 = x^3$ ...(1)

Differentiating $2ay \cdot \frac{dy}{dx} = 3x^2$ or $\frac{dy}{dx} = \frac{3x^2}{2ay}$.

Step I. Measuring s from the cusp (*at the origin, where $x = 0$*),

we have $$s = \int_0^x \sqrt{1 + \left(\frac{dy}{dx}\right)^2}\, dx = \int_0^x \sqrt{1 + \frac{9x^4}{4a^2y^2}}\, dx$$

$$= \int_0^x \sqrt{1+\frac{9x^4}{4ax^3}}\,dx = \int_0^x \sqrt{1+\frac{9x}{4a}}\,dx$$

$$= \left| \frac{\left(1+\frac{9x}{4a}\right)^{3/2}}{\frac{3}{2}\cdot\frac{9}{4a}} \right|_0^x = \frac{8a}{27}\left[\left(1+\frac{9x}{4a}\right)^{3/2} - 1\right] \quad ...(2)$$

Step II. $\tan\psi = \frac{dy}{dx} = \frac{3x^2}{2ay}$

$$\therefore \quad \tan^2\psi = \frac{9x^4}{4a^2y^2} = \frac{9x^4}{4ax^3} = \frac{9x}{4a} \quad ...(3)$$

Step III. Eliminating x from (2) and (3), we get

$$s = \frac{8a}{27}\,[(1+\tan^2\psi)^{3/2} - 1] = \frac{8a}{27}\,[\sec^3\psi - 1]$$

or $\quad 27s = 8a\,(\sec^3\psi - 1).$

this is the required intrinsic equation.

Example 2. *Find the intrinsic equation of the cycloid $x = a\,(t + \sin t)$, $y = a\,(1 - \cos t)$ and prove that $s^2 + \rho^2 = 16\,a^2$.* (M.D.U. 2003, 2006, 2006 S)

Sol. The equations of the cycloid are $x = a(t + \sin t)$, $y = a(1 - \cos t)$.

$$\therefore \quad \frac{dx}{dt} = a\,(1 + \cos t), \quad \frac{dy}{dt} = a\sin t$$

Step I. Measuring s from the point where $t = 0$,

$$s = \int_0^t \sqrt{\left(\frac{dx}{dt}\right)^2 + \left(\frac{dy}{dt}\right)^2}\,dt$$

$$= \int_0^t \sqrt{a^2(1+\cos t)^2 + a^2\sin^2 t}\,dt$$

$$= a\int_0^t \sqrt{1 + 2\cos t + \cos^2 t + \sin^2 t}\,dt$$

$$= a\int_0^t \sqrt{2(1+\cos t)}\,dt = a\int_0^t \sqrt{2.2\cos^2\frac{t}{2}}\,dt = 2a\int_0^t \cos\frac{t}{2}\,dt$$

$$= 2a\left[\frac{\sin t/2}{\frac{1}{2}}\right]_0^t = 4a\left[\sin\frac{t}{2}\right] \quad ...(1)$$

Step II. $\tan\psi = \frac{dy}{dx} = \frac{dy/dt}{dx/dt} = \frac{a\sin t}{a(1+\cos t)} = \frac{2\sin t/2\,.\cos t/2}{2\cos^2 t/2} = \tan t/2$

$$\therefore \quad \psi = t/2 \quad ...(2)$$

Step III. Eliminating t from (1) and (2) [by putting $t/2 = \psi$ from (2) in (1)], we get $s = 4a\sin\psi$. This is the required intrinsic equation.

Diff. w.r.t. ψ, $\frac{ds}{d\psi} = 4a\cos\psi$ or $\rho = 4a\cos\psi$

$\therefore \qquad s^2 + \rho^2 = 16a^2 \sin^2 \psi + 16a^2 \cos^2 \psi$
$= 16a^2 (\sin^2 \psi + \cos^2 \psi) = 16a^2.$

Example 3. *Find the intrinsic equation of the cardioid $r = a(1 - \cos\theta)$.* (M.D.U. 2005 ; K.U. 2006)

Sol. Suppose s is measured from the pole, where (putting $r = 0$), we have $\theta = 0$. Now

$$\frac{dr}{d\theta} = a \sin\theta.$$

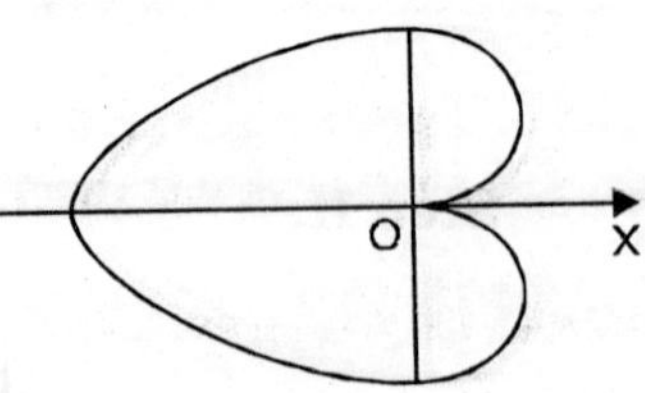

Step I.

$$s = \int_0^\theta \sqrt{r^2 + \left(\frac{dr}{d\theta}\right)^2}\, d\theta$$
$$= \int_0^\theta \sqrt{a^2(1 - \cos\theta)^2 + a^2 \sin^2\theta}\, d\theta$$
$$= \int_0^\theta \sqrt{2a^2(1 - \cos\theta)}\, d\theta = \int_0^\theta \sqrt{2a^2 . 2\sin^2\frac{\theta}{2}}\, d\theta$$
$$= \int_0^\theta 2a^2 \sin\theta/2\, d\theta = -4a\left[\cos\frac{\theta}{2}\right]_0^\theta$$
$$= -4a[\cos\theta/2 - 1] = 4a(1 - \cos\theta/2)$$
$$= 4a . 2\sin^2\theta/4 = 8a\sin^2\theta/4 \qquad ...(1)$$

Step II. Now $\tan\phi = r\dfrac{d\theta}{dr} = \dfrac{a(1-\cos\theta)}{a\sin\theta} = \tan\theta/2$

$\therefore \qquad \phi = \dfrac{\theta}{2}$ ...(2)

Step III. Again $\psi = \theta + \phi = \theta + \theta/2 = 3\theta/2$ [Using (2)]

$\therefore \qquad \theta = 2\psi/3$ or $\theta/4 = \psi/6$ ...(3)

Step IV. Eliminating θ from (1) and (3), we get

$$s = 8a\sin^2\psi/6.$$

This is the required intrinsic equation of the curve.

Example 4. *Find the intrinsic equation of the curve whose pedal equation is*

$$p^2 = r^2 - a^2.$$

Sol. The given equation of the curve is $p^2 = r^2 - a^2$. ...(1)

$\therefore \qquad 2p\dfrac{dp}{dr} = 2r$ or $p\dfrac{dp}{dr} = r$ or $\dfrac{p}{r} = \dfrac{dr}{dp}$

$\therefore \qquad \dfrac{\mathbf{ds}}{\mathbf{d\psi}} = \rho = \mathbf{r}\dfrac{\mathbf{dr}}{\mathbf{dp}} = r . \dfrac{p}{r} = p$ ...(2)

Step I. Let s be measured from the point **where r = a.** (*i.e.* $s = 0$)

[$\because$ For $r = 0$, $p^2 = -a^2$ from (1) and hence p is Imaginary]

Step II. $s = \displaystyle\int_a^r \frac{r\,dr}{\sqrt{r^2 - p^2}} = \int_a^r \frac{r}{a}\, dr$ [$\because$ from (1), $r^2 - p^2 = a^2$]

$$= \frac{1}{2a}\Big| r^2 \Big|_a^r = \frac{1}{2}(r^2 - a^2) = \frac{p^2}{2a} \qquad ...(3) \quad \text{[Using (1)]}$$

Step III. From (2) and (3), eliminating p, we get

[From (3), $p^2 = 2as$ $\therefore$ $p = \sqrt{2as}$. Putting this value of p in (2)]

$$\frac{ds}{d\psi} = \sqrt{2as}.$$

$$\therefore \quad \frac{ds}{\sqrt{s}} = \sqrt{2a}\, d\psi$$

Step IV. Integrating, we have $\int s^{-1/2}\, ds = \sqrt{2a} \int 1\, d\psi + k$

i.e. $$2\sqrt{s} = \sqrt{2a}\,\psi + k, \qquad \text{...(4)}$$

where k is some constant of integration.

Let when $\psi = 0, s = 0$ $\therefore$ From (4), $k = 0$.

$\therefore$ From (4), $2\sqrt{s} = \sqrt{2a}\,\psi$

Squaring both sides, $4s = 2a\psi^2$ or $s = \dfrac{a}{2}\psi^2$.

This is the required intrinsic equation.

EXERCISE 5

1. Find the intrinsic equation of the catenary $y = c \cosh \dfrac{x}{c}$, s being measured from the vertex $(0, c)$ of the catenary and show that $c\rho = c^2 + s^2$, ρ being the radius of curvature. *(K.U. 2003, 2007)*
2. Show that the intrinsic equation of the parabola $y^2 = 4ax$ is
$$s = a \cot\psi \operatorname{cosec}\psi + a \log(\cot\psi + \operatorname{cosec}\psi).$$
3. Show that the intrinsic equation of the parabola $x^2 = 4ay$ is $s = a \tan\psi \sec\psi + a \log(\tan\psi + \sec\psi)$.
4. Find the intrinsic equation of the curve $y = a \log \sec \dfrac{x}{a}$, s being measured from the origin.
5. Find the intrinsic equation of the parabola $x = at^2$, $y = 2at$, these are being measured from the vertex.
6. Find the intrinsic equation of the curve $x = a(1 + \sin t)$, $y = a(1 + \cos t)$.
7. In the four-cusped hypocycloid $x^{2/3} + y^{2/3} = a^{2/3}$, show that

   (*i*) $s = \dfrac{3a}{4} \cos 2\psi$, s being measured from the vertex.

   [**Hint.** Parametric equations of this curve are $x = a\cos^3\theta$, $y = a\sin^3\theta$. At the vertex (the point of intersection of the given curve with its line of symmetry $y = x$), $\theta = \pi/4$].

   (*ii*) $s = \dfrac{3a}{2} \sin^2\psi$ when $(a, 0)$ is taken as the fixed point.
8. Find the intrinsic equation of the cardioid $r = a(1 + \cos\theta)$ and show that $s^2 + 9\rho^2 = 16a^2$. *(K.U. 2005 ; M.D.U. 2007)*
9. (*i*) Show that the intrinsic equation of the equiangular spiral $r = a\, e^{\theta \cot\alpha}$, measured s from the point $(a, 0)$ is $s = a \sec\alpha\, [e^{\psi \cot\alpha} - 1]$.

   (*ii*) Show that intrinsic equation of the curve $r = a\, e^{m\theta}$ is $s = \dfrac{a\sqrt{1+m^2}}{m}\, [e^{m\psi} - 1]$.
10. Find the intrinsic equation of the curve whose pedal equation is $p = r \sin\alpha$.

ANSWERS

1. $s = c \tan \psi$ **4.** $s = a \log (\tan \psi + \sec \psi)$

5. $s = a \cot \psi \operatorname{cosec} \psi + a \log (\cot \psi + \operatorname{cosec} \psi)$

6. $s + a\psi = 0$ **8.** $s = 4a \sin \left(\frac{\psi}{3} - \frac{\pi}{6}\right)$ **10.** $s = e^{\psi \cot \alpha}$.

REVISION EXERCISE

1. Show that the length of the arc of the parabola $y^2 = 4ax$ cut off by the line $3y = 8x$ is $a\left(\log_e 2 + \frac{15}{16}\right)$.

2. Show that the length of the curve given by $x = a(3 \sin \theta - \sin^3 \theta)$ and $y = a \cos^3 \theta$ measured from $(0, a)$ to any point (x, y) is $\frac{3a}{2}(\theta + \sin \theta \cos \theta)$.

3. Prove that the length of the arc of the curve $x = a \sin 2\theta (1 + \cos 2\theta)$, $y = a \cos 2\theta (1 - \cos 2\theta)$ measured from $(0, 0)$ to (x, y) is $\frac{4a}{3} \sin 3\theta$. *(M.D.U. 2005 A)*

Find the intrinsic equation of the following curves : (**4 to 8**).

4. $x = a (2 \cos t - \cos 2t)$, $y = a (2 \sin t - \sin 2t)$.

5. $x = a (3 \cos \theta - \cos 3\theta)$, $y = a (3 \sin \theta - \sin 3\theta)$; s being measured from $\theta = \frac{\pi}{2}$.

6. $x = a \sin 2t (1 + \cos 2t)$, $y = a \cos 2t (1 - \cos 2t)$; the fixed point being the origin.

7. $r = a\theta$, s being measured from the pole.

8. $x = a (3 \sin t - \sin^3 t)$, $y = a \cos^3 t$, where s is measured from the point $(0, a)$.

9. Show that the length s of the curve $x^{2/3} + y^{2/3} = a^{2/3}$ measured from $(0, a)$ to the point (x, y) is given by $s = \frac{3}{2} \sqrt[3]{ax^2}$.

10. Show that in the parabola $\frac{2a}{r} = 1 + \cos \theta$, $\frac{ds}{d\psi} = \frac{2a}{\sin^3 \psi}$.

Hence show that the arc intercepted between the vertex and the extremity of latus rectum is $a \{\sqrt{2} + \log (1 + \sqrt{2})\}$.

11. Find the length of any arc of the curve $y^2 (a - x) = x^3$ or $r = \frac{a \sin^2 \theta}{\cos \theta}$.

ANSWERS

4. $s = 16a \sin^2 \frac{\psi}{6}$ **5.** $s = -6a \cos \frac{\psi}{2}$ **6.** $3s = 4a \sin 3\psi$

7. $s = \frac{a'}{2} [\theta \sqrt{\theta^2 + 1} + \log (\theta + \sqrt{\theta^2 + 1})]$ where $\psi = \theta + \tan^{-1} \theta$ **8.** $4s + 3a(2\psi + \sin 2\psi) = 0$

11. $\frac{a}{2}\left[2\sqrt{\frac{4a - 3x}{a - x}} + \sqrt{3} \cosh^{-1} \frac{6x - 7a}{a}\right]_{x_1}^{x_2}$.

12

Quadrature (Areas of Curves)

1. Area Formulae for Cartesian Equations

We know that the area bounded by the curve $y = f(x)$, the x-axis and the ordinates $x = a$, $x = b$ is $\int_a^b$ ***y dx***, *where $y = f(x)$, is a continuous single valued function and y does not change sign in the interval [a, b].*

Thus, if AB is the curve $y = f(x)$ and CA, DB, the ordinates $x = a$, $x = b$, then

$$\text{Area ACDB} = \int_a^b y\,dx.$$

Note 1. We have supposed that y is a strictly increasing function of x in the interval $[a, b]$. But this condition may be removed. If $f(x)$ goes on decreasing as we go from C to D, the above result is still true.

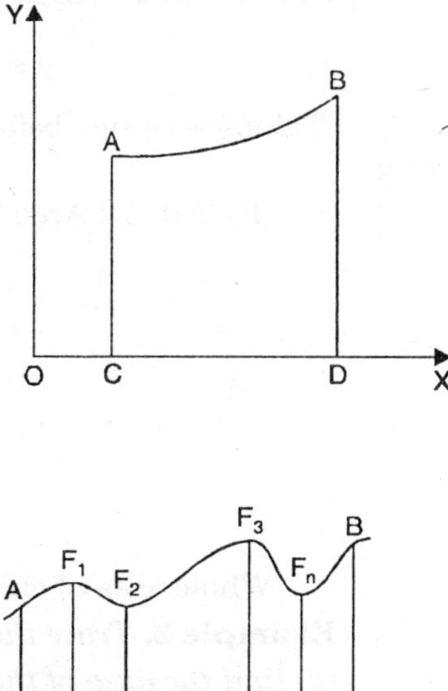

If however, $y = f(x)$ increases in certain parts of the interval. $[a, b]$ and decreases in other parts, as in the adjoining figure, then the area ACDB = the area ACE_1F_1 + the area $F_1E_1E_2F_2 + \ldots +$ the area $E_n\ DBF_n$ ...(1)

where E_1F_1, E_2F_2 ..... $E_n\ F_n$ are the maximum and minimum ordinates of the curve (n being finite), in the interval $[a, b]$.

If $c_1, c_2, \ldots\ldots c_n$ are the abscissae of the ordinates E_1F_1, E_2F_2 ..., E_nF_n respectively, then it follows from (1) that

$$\text{Area ACDB} = \int_a^{c_1} y\,dx + \int_{c_1}^{c_2} y\,dx + \ldots\ldots$$

$$+ \int_{c_{n-1}}^{c_n} y\,dx + \int_{c_n}^{b} y\,dx = \int_a^b y\,dx.$$

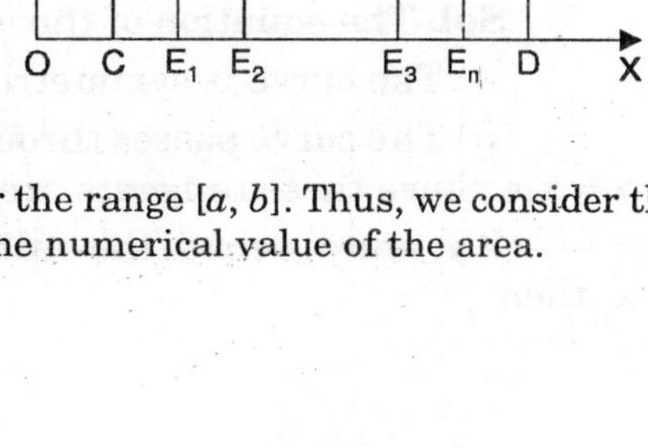

Note 2. The area bounded by the curve AB, the ordinates at A and B, and the x-axis is often called *the area under the curve AB.*

Note 3. *The process of finding the area bounded by a given portion of a curve is called* **quadrature.**

Note 4. Sign of the area. We know that if $y = f(x) > 0$ over the range $a \le x < b$, then the area $\int_a^b f(x)\,dx$ is + ve and if y is – ve in the range $a \le x \le b$, then the area $\int_a^b f(x)\,dx$ is also negative. The curve $y = f(x)$ in this case lies below the x-axis over the range $[a, b]$. Thus, we consider the areas below x-axis as – ve. By the area in such case, we mean the numerical value of the area.

If as x changes from a to b, y = f(x) changes sign at some intermediate point c (say), then the areas from a to c and c to b are, calculated separately and their numerical values are added. Similarly, this result can be extended if y changes sign at more than one intermediate place in the interval $[a, b]$.

2. *To prove that the area of the curve $x = f(y)$ between the y-axis and the lines $y = c$, to $y = d$ is given by* $\int_c^d \mathbf{x\,dy}$.

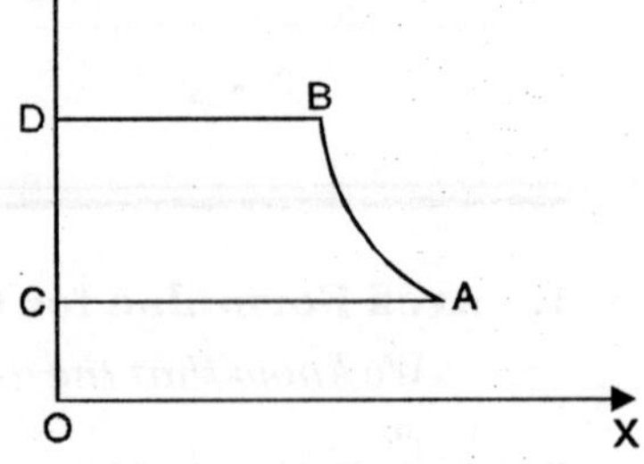

The result follows easily from Art. 1, on interchanging x and y with the proper limits of integration.

Thus, area ACDB = $\int_c^d x\,dy$, where OC = c, OD = d.

Example 1. *Calculate the area of the ellipse $x^2/a^2 + y^2/b^2 = 1$.*

Sol. The equation of the ellipse is $\dfrac{x^2}{a^2} + \dfrac{y^2}{b^2} = 1$ ...(*i*)

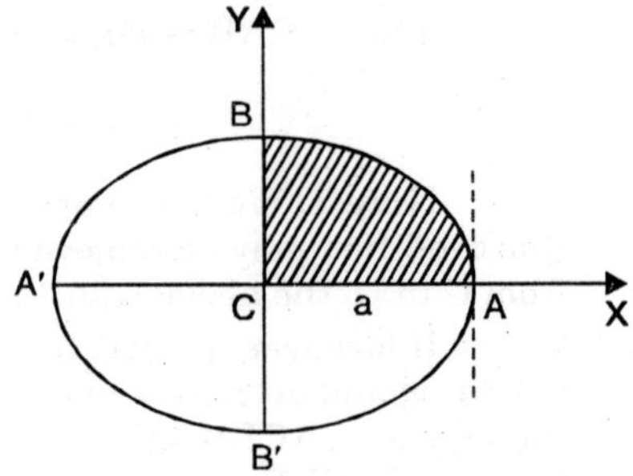

Since the ellipse is symmetrical about both the axes, therefore the area of the whole ellipse is four times the area under the curve in first quadrant.

i.e., area of ellipse = 4 × area CAB.

Now for area CAB, x varies from 0 to a and from (*i*),

$$y = \frac{b}{a}\sqrt{a^2 - x^2}$$

[Taking + ve sign before radical sign, in the first quadrant]

∴ (By Art. 1.) Area CAB

$$= \int_0^a y\,dx = \int_0^a \frac{b}{a}\sqrt{a^2 - x^2}\,dx$$

$$= \frac{b}{a}\int_0^a \sqrt{a^2 - x^2}\,dx = \frac{b}{a}\left|\frac{x}{2}\sqrt{a^2 - x^2} + \frac{a^2}{2}\sin^{-1}\frac{x}{a}\right|_0^a$$

$$= \frac{b}{a}\left[0 + \frac{a^2}{2}\sin^{-1}1 - 0\right] = \frac{b^2}{a}\cdot\frac{a^2}{2}\cdot\frac{\pi}{2} = \frac{\pi ab}{4}$$

∴ Whole area of ellipse = 4 × area CAB = 4 × $\pi ab/4 = \pi ab$.

Example 2. *Trace the curve $y^2(a - x) = x^2(a + x)$ and*

(*a*) *find the area of the loop.* (M.D.U. 1999 A)

(*b*) *find the area of the portion bounded by the curve and its asymptote.* (D.U. 1997)

Sol. The equation of the curve is $y^2(a - x) = x^2(a + x)$ ...(1)

(*i*) The curve is symmetrical about x-axis, since (1) contains only even powers of y.

(*ii*) The curve passes through the origin and tangents at the origin are given by $y^2 = x^2$ or $y = \pm x$. Since these tangents are real and different, ∴ origin is a node.

If y_c and y_t denote the ordinates of the curve and the tangent $y = x$, for the same value of x, then

$$y_c^2 - y_t^2 = \frac{x^2(a+x)}{a-x} - x^2 = \frac{2x^2}{a-x}$$

which is > 0 for small + ve values of x and is < 0 for $x < 0$. Thus, the curve lies above the tangent $y = x$ in the first quadrant and below the tangent in the second quadrant.

(*iii*) The asymptote parallel to y-axis is given by $a - x = 0$ *i.e.*, $x = a$. The curve has no other asymptote.

(*iv*) The curve meets x-axis, in the point $(-a, 0)$ and the y-axis in $(0, 0)$.

(*v*) From (1), we have $y^2 = \dfrac{x^2(a+x)}{a-x}$

$$\therefore \quad 2y\frac{dx}{dx} = \frac{(a-x)(2ax+3x^2) - x^2(a+x)(-1)}{(a-x)^2}$$

$$= \frac{2a^2x + ax^2 - 3x^2 + ax^2 + x^3}{(a-x)^2} = \frac{2x(a^2+ax-x^2)}{(a-x)^2}$$

or
$$\frac{dy}{dx} = \frac{x(a^2+ax-x^2)}{(a-x)^2} \cdot \frac{\sqrt{a-x}}{x\sqrt{a+x}} = \frac{a^2+ax-x^2}{(a-x)^{3/2}\sqrt{a+x}}$$

$$\therefore \quad \frac{dy}{dx} = 0 \text{ when } x^2 - ax - a^2 = 0$$

or
$$x = \frac{a \pm \sqrt{a^2+4a^2}}{2} = \frac{a \pm 2.2a}{2} = 1.6a \text{ or } -.6a$$

Thus, the tangent is || to x-axis at $x = -.6a$. The value of $x = 1.6a$ is rejected as we shall see in the next step that y is imaginary when $x = 1.6\,a$.

Again, $dy/dx = \infty$ when $x = \pm a$. Thus, at $(-a, 0)$, the tangent is parallel to y-axis and as $x \to a$, $y \to \infty$.

$\therefore \quad x = a$ is an asymptote, a fact already established.

(*vi*) From (1), we have (taking + ve root), $y = x\sqrt{\dfrac{a+x}{a-x}}$ ...(2)

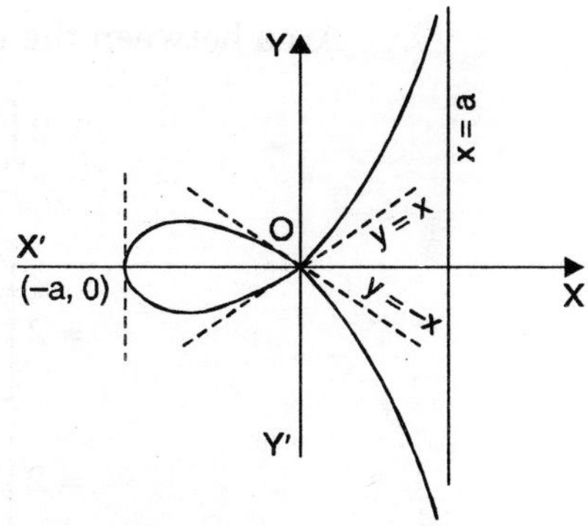

When $x < -a$ or when $x > a$, y is imaginary. Hence no portion of the curve lies beyond the lines $x = \pm a$. Thus, the shape of the curve is as shown in the figure.

Now for the loop, x varies from $-a$ to 0 and the loop is symmetrical about x-axis.

(*a*) $\therefore$ Area of the loop = 2 × area of the upper half of the loop

$$= 2\int_{-a}^{0} y\,dx = 2\int_{-a}^{0} x\sqrt{\frac{a+x}{a-x}}\,dx \quad ...(3)$$

Now
$$\int x\sqrt{\frac{a+x}{a-x}}\,dx.$$

Rationalising the numerator by multiplying the numerator and denominator by $\sqrt{a+x}$.

$$= \int x \frac{(a+x)}{\sqrt{a^2-x^2}} dx = \int \frac{ax}{\sqrt{a^2-x^2}} dx + \int \frac{x^2}{\sqrt{a^2-x^2}} dx$$

$$= a \int x (a^2 - x^2)^{-1/2} dx + \int \frac{a^2 - (a^2 - x^2)}{\sqrt{a^2 - x^2}} dx$$

$$= -\frac{a}{2} \int (a^2 - x^2)^{-1/2} (-2x) \, dx + a^2 \int \frac{1}{\sqrt{a^2-x^2}} dx - \int \sqrt{a^2 - x^2} \, dx$$

$$= -\frac{a}{2} \frac{(a^2-x^2)^{1/2}}{\frac{1}{2}} + a^2 \sin^{-1}\frac{x}{a} - \left[\frac{x\sqrt{a^2-x^2}}{2} + \frac{a^2}{2}\sin^{-1}\frac{x}{a}\right]$$

$$= -a\sqrt{a^2-x^2} + \frac{a^2}{2}\sin^{-1}\frac{x}{a} - \frac{x\sqrt{a^2-x^2}}{2} \qquad \text{...(4)}$$

$\therefore$ From (3), area of the loop

$$= 2\left[-a\sqrt{a^2-x^2} + \frac{a^2}{2}\sin^{-1}\frac{x}{a} - \frac{x\sqrt{a^2-x^2}}{2}\right]_{-a}^{0}$$

$$= 2\left[-a^2 - \frac{a^2}{2}\sin^{-1}(-1)\right] = 2\left[\frac{\pi a^2}{4} - a^2\right]$$

$$= \frac{2a^2}{4}(\pi - 4) = \frac{a^2}{2}(4 - \pi) \text{ numerically} \qquad (\because \pi < 4)$$

(*b*) For half of the area bounded by the curve and its asymptote x varies from 0 to a.

$\therefore$ Area between the curve and its asymptote

$$= 2\int_0^a y\,dx = 2\int_0^a x\sqrt{\frac{a+x}{a-x}}\,dx$$

[Using (4)]

$$= 2\left[-a\sqrt{a^2-x^2} + \frac{a^2}{2}\sin^{-1}\frac{x}{a} - \frac{x\sqrt{a^2-x^2}}{2}\right]_0^a$$

$$= 2\left[\frac{a^2}{2}\sin^{-1}1 - (-a \cdot a)\right] = 2\left[\frac{a^2}{2}\frac{\pi}{2} + a^2\right]$$

$$= 2a^2\left(\frac{\pi+4}{4}\right) = \frac{a^2}{2}(\pi + 4).$$

Example 3. *Find the area included between the curve $xy^2 = 4a^2(2a - x)$ and its asymptote.*

Sol. The equation of the curve is $xy^2 = 4a^2(2a - x)$ ...(1)

Let us first trace the curve roughly to find limits of integration for the area.

(*i*) The curve is symmetrical about x-axis.

(*ii*) The curve does not pass through the origin.

(*iii*) The curve meets the x-axis in the point $(2a, 0)$. Shifting the origin to point $(2a, 0)$, equation (1) transforms to $(X + 2a)Y^2 = -4aX$, and tangent to the curve at the new origin is $X = 0$, *i.e.*, the new y-axis. Hence at the point $(2a, 0)$ tangent to the curve is || to y-axis.

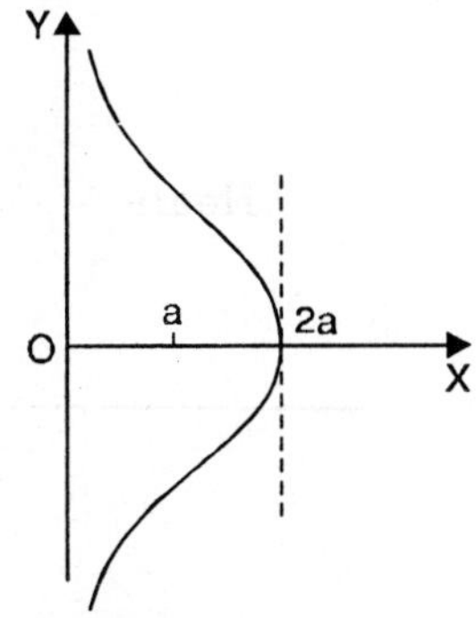

The curve does not meet the y-axis.

(*iv*) $x = 0$ *i.e.*, y-axis is the only asymptote of the curve.

(*v*) From (1), $$y = \pm\, 2a\sqrt{\frac{2a - x}{x}} \qquad \ldots(2)$$

When x is negative and $x > 2a$, y is imaginary. Hence no portion of curve lies to the left of y-axis and to right of line $x = 2a$.

From (1), $$x = \frac{8a^3}{y^2 + 4a^2}.$$

Now area required

$= 2 \times$ area between the upper half of the curve, y-axis and x-axis

$$= 2\int_0^\infty x\,dy = 2\int_0^\infty \frac{8a^3}{y^2 + 4a^2}\,dy$$

$$= 16a^3 \cdot \frac{1}{2a} \cdot \left| \tan^{-1}\frac{y}{2a} \right|_0^\infty$$

$$= 8a^2\left(\frac{\pi}{2} - 0\right) = 4\pi a^2.$$

Example 4. *If A is the vertex, O the centre and P any point on the hyperbola* $\frac{x^2}{a^2} - \frac{y^2}{b^2} = 1$, *prove that* $x = a\cosh\frac{2S}{ab}$, $y = b\sinh\frac{2S}{ab}$, *where S is the sectorial area AOP.*

(M.D.U. 2007)

Sol. Draw PM $\perp$ on OX.

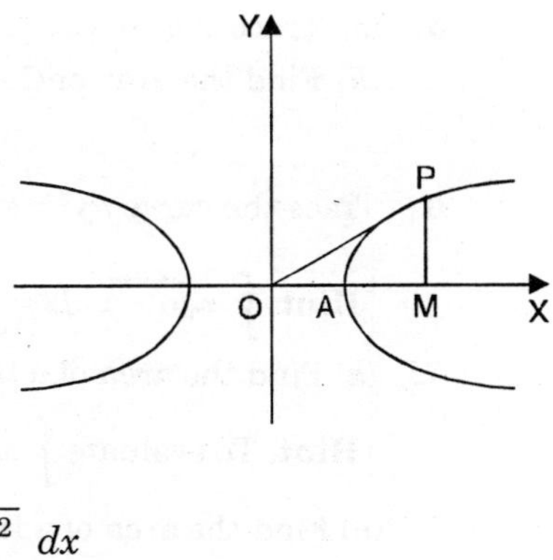

$$S = \text{Area OAP}$$

$$= \text{Area of } \Delta\text{OMP} - \text{area AMP}$$

$$= \tfrac{1}{2}xy - \int_a^x y\,dx \text{ where}$$

$$y = \frac{b}{a}\sqrt{x^2 - a^2}$$

$$\left(\because \text{ From the hyperbola } \frac{x^2}{a^2} - \frac{y^2}{b^2} = 1,\ y = \frac{b}{a}\sqrt{x^2 - a^2}\right)$$

$$= \frac{x}{2}\cdot\frac{b}{a}\sqrt{x^2 - a^2} - \frac{b}{a}\int_a^x \sqrt{x^2 - a^2}\,dx$$

$$= \frac{bx\sqrt{x^2 - a^2}}{2} - \frac{b}{a}\left[\frac{x\sqrt{x^2 - a^2}}{2} - \frac{a^2}{2}\cosh^{-1}\frac{x}{a}\right]_a^x$$

or $$S = \frac{bx\sqrt{x^2 - a^2}}{2a} - \frac{b}{a}\left[\frac{x\sqrt{x^2 - a^2}}{2} - \frac{a^2}{2}\cosh^{-1}\frac{x}{a}\right] \qquad \left|\ \because\ \cosh^{-1} 1 = 0\right.$$

or $$S = \frac{ab}{2}\cosh^{-1}\frac{x}{a}$$

$$\therefore \qquad \frac{2S}{ab} = \cosh^{-1}\frac{x}{a} \quad \therefore \quad x = a\cosh\frac{S}{ab}$$

Hence
$$y = \frac{b}{a}\sqrt{x^2 - a^2} = \frac{b}{a}\sqrt{a^2\cosh^2\frac{S}{ab} - a^2}$$

or
$$y = b\sinh\frac{2S}{ab}.$$

EXERCISE 1

1. Find the area bounded by each of the following curves, the x-axis and the ordinates $x = 0, x = h$:

(*i*) $y = e^x$ (*ii*) $y = c\cosh\frac{x}{c}$.

2. (*a*) Find the areas bounded by each of the following curves, the x-axis and the ordinates $x = a$, $x = b$:

(*i*) $xy = c^2$ (*ii*) $y = \log x$.

$\left[\textbf{Hint.} \int \log x\,dx = \int \underset{\text{I}}{\log x}\,.\,\underset{\text{II}}{1}\,dx = (\log x)x - \int \frac{1}{x}x\,dx = x\log x - x = x(\log x - 1).\right]$

(*b*) In the catenary $y = a\cosh\frac{x}{a}$, prove that the area between the curve, the x-axis and the ordinates of two points on the curve varies as the length of the intervening arc.

3. Show that the area cut off a parabola by any double ordinate is two-thirds of the corresponding rectangle contained by the double ordinate and its distance from the vertex.

4. Obtain the area bounded by the parabola $y^2 = 4ax$ and its latus rectum.

5. (*a*) Trace the curve $a^2y^2 = a^2x^2 - x^4$. Find the whole area within it.

(*b*) Find the area enclosed by the curve.

$$y^2 = x^2 - x^4.$$

6. Trace the curve $ay^2 = x^2(a - x)$ and show that the area of its loop is $\frac{8}{15}a^2$.

$\left[\textbf{Hint.} \int x\sqrt{a-x}\,dx = \int [a - (a - x)]\sqrt{a-x}\,dx = a\int (a-x)^{1/2}dx - \int (a-x)^{3/2}dx.\right]$

7. (*a*) Find the area of a loop of the curve $a^3y^2 = x^4(b + x)$.

$\left[\textbf{Hint.} \text{ To evaluate } \int x^2\sqrt{b+x}\,dx, \text{ put } \sqrt{b+x} = t.\right]$

(*b*) Find the area of a loop of the curve $3ay^2 = x(x - a)^2$

(*c*) Show that the area of a loop of the curve $y^2 = x(4 - x^2)$ is $\frac{16}{3}$.

8. Trace the curve $a^4y^2 = x^5(2a - x)$ and prove that its area is to that of the circle whose radius is a as 5 to 4.

9. Prove that the area of the curve $a^2x^2 = y^3(2a - y)$ is equal to that of the circle whose radius is a. [**Hint.** Area = $\int x\,dy$.]

10. Find the area enclosed by the curve $xy^2 = 4(2 - x)$ and y-axis.

11. Find the area included between the curve $x^2y = 4a^2(2a - y)$ and its asymptote.

12. (*i*) Show that the area of the loop of the curve

$$ay^2 = (x - a)(x - 5a)^2 \text{ is } \frac{256}{15} a^2.$$

(*ii*) Show that the area of the loop of the curve $y^2 = x^3 - 11x^2 + 35x - 25$ is $\frac{256}{15}$.

[**Hint.** Equation of the curve is $y^2 = x^3 - 11x^2 + 35x - 25$.

$$\therefore \quad y = x^3 - x^2 - 10x^2 + 10x + 25x - 25$$
$$= x^2 (x - 1) - 10x (x - 1) + 25 (x - 1)$$
$$= (x - 1)(x^2 - 10x + 25)$$
$$= (x - 1)(x - 5)^2.$$

It is part (*i*) with $a = 1$.]

13. (*a*) Find the whole area of the curve $y^2 = x^2 \left(\frac{a^2 - x^2}{a^2 + x^2}\right)$.

$$\left[\textbf{Hint.} \text{ It will be found that area} = 4 \int_0^a x \sqrt{\frac{a^2 - x^2}{a^2 + x^2}}\, dx. \text{ Put } \sqrt{a^2 + x^2} = t.\right]$$

(*b*) Find the whole area of the curve $x^2 (x^2 + y^2) = a^2 (x^2 - y^2)$.

[**Hint.** It is same as part (*i*).]

14. Find the area of a loop of the curve $x(x^2 + y^2) = a(x^2 - y^2)$. Also find the area between this curve and its asymptote.

15. Find the area between the curve $x^2y^2 = a^2 (y^2 - x^2)$ and its asymptote. (*K.U. 2005 S*)

16. Find the area between the curve $y^2 = \frac{x^3}{2a - x}$ and its asymptote.

17. Show that the area of the infinite region enclosed between the curve $x^2 (1 - y)y = 1$ and its asymptote is 2π.

18. In the ellipse $\frac{x^2}{a^2} + \frac{y^2}{b^2} = 1$, prove that $x = a \cos \frac{2S}{ab}$, $y = b \sin \frac{2S}{ab}$, where S is the sectorial area bounded by the ellipse, x-axis and the line joining (0, 0) to (x, y).

ANSWERS

1. (*i*) $e^h - 1$ (*ii*) $c^2 \sinh \frac{h}{c}$

2. (*a*) (*i*) $c^2 \log \frac{b}{a}$ (*ii*) $b \log b - a \log a - (b - a)$ **4.** $\frac{8}{3} a^2$

5. (*a*) $\frac{4}{3} a^2$ (*b*) $\frac{4}{3}$

7. (*a*) $\frac{32}{105} b^{7/2} a^{-3/2}$ (*b*) $\frac{8a^2}{5\sqrt{3}}$

10. 4π **11.** $4\pi a^2$

13. (*a*) $a^2 (\pi - 2)$ (*b*) $a^2 (\pi - 2)$

14. $\frac{a^2}{2} (4 - \pi)$, $\frac{a^2}{2} (\pi + 4)$ **15.** $4a^2$

16. $3\pi a^2$.

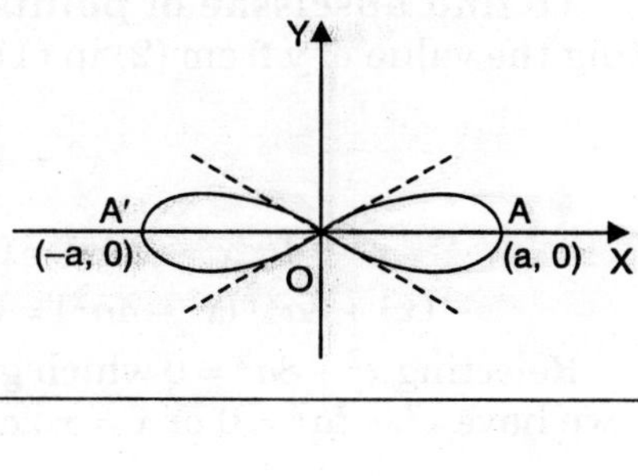

3. Area between two Curves

To prove that the area bounded by the curves $y = f(x)$, $y = \phi(x)$ and the ordinates $x = a$, $x = b$ is $\int_a^b (y_1 - y_2)\, dx$

where y_1 is the 'y' of the upper curve and y_2 that of the lower curve.

Let AB, CD be the curves $y = f(x)$, $y = \phi(x)$ and MCA, NDB be the ordinates $x = a$, $x = b$. Then area ACDB = area AMNB – area CMND

$$= \int_a^b f(x)\, dx - \int_a^b \phi(x)\, dx$$

$$= \int_a^b [f(x) - \phi(x)]\, dx$$

$$= \int_a^b (y_1 - y_2)\, dx$$

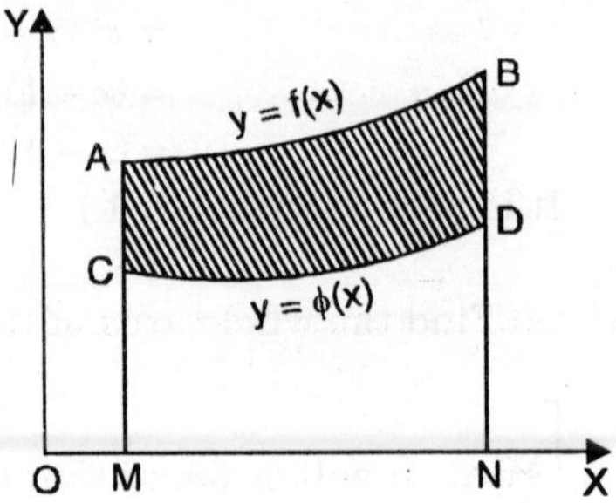

where $y_1 = f(x)$ is the 'y' of the upper curve AB and $y_2 = \phi(x)$, is the 'y' of lower curve CD.

Cor. The area bounded by the curves $x = f(y)$, $x = F(y)$ and the abscissae $y = c$, $y = d$ is given by

$$\int_c^d [(x \text{ of outer curve}) - (x \text{ of inner curve})]\, dy.$$

Caution. *While applying formula of Art. 3 one must see the figure of the problem very carefully. In some problems we may have to use $\int (y_1 + y_2)\, dx$ (see example 2).*

Example 1. *Find the area included between the parabola $x^2 = 4ay$ and the curve*

$$y(x^2 + 4a^2) = 8a^3.$$

Sol. The equation of the parabola is $x^2 = 4ay$ ...(1)

It is an upward parabola with y-axis as its axis and the origin as its vertex.

The equation of the curve is $y = \dfrac{8a^3}{x^2 + 4a^2}$ ...(2)

(*i*) The curve (2) is symmetrical about y-axis.

(*ii*) The curve does not pass through the origin. The curve does not meet x-axis and it meets y-axis in the point $(0, 2a)$.

(*iii*) $y = 0$ is an asymptote of the curve (2).

From (2), $y > 0$.

$\therefore$ The curve given by equation (2) lies above the axis of x.

Thus, the shape of the curve (2) is as shown in the figure (along with the parabola (1)).

To find abscissae of points of intersection eliminating y between (1) and (2), [*i.e.*, putting the value of y from (2) in (1)], we get

$$x^2 = 4a \cdot \frac{8a^3}{x^2 + 4a^2}$$

or $$x^4 + 4a^2x^2 - 32a^4 = 0$$

or $$(x^2 + 8a^2)(x^2 - 4a^2) = 0$$

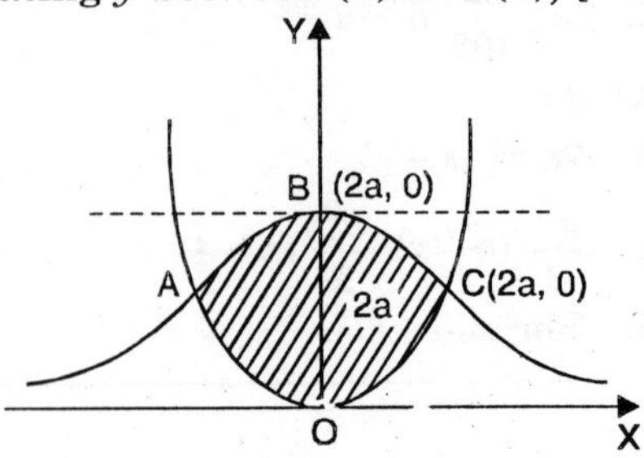

Rejecting $x^2 + 8a^2 = 0$ which gives imaginary values of x, we have $x^2 - 4a^2 = 0$ or $x = \pm 2a$.

When $x = \pm 2a, y = a$ [from (1)].

Hence the points of intersection are A(– 2a, a), C(2a, a).

$\therefore$ Required area OABC = 2 × area OBC

$$= 2 \int_0^{2a} (y_1 - y_2)\, dx = 2 \int_0^{2a} \left(\frac{8a^3}{x^2 + 4a^2} - \frac{x^2}{4a} \right) dx$$

$$= 2 \left| 8a^3 \cdot \frac{1}{2a} \tan^{-1} \frac{x}{2a} - \frac{1}{4a} \cdot \frac{x^3}{3} \right|_0^{2a}$$

$$= 2 \left[4a^2 (\pi/4 - 0) - \frac{1}{12a} (8a^3 - 0) \right]$$

$$= 2 \left(\pi a^2 - \frac{2}{3} a^2 \right) = \frac{2}{3} a^2 (3\pi - 2).$$

Remark. We know that **the equation of the circle whose centre is (α, β) and radius a is $(x - \alpha)^2 + (y - \beta)^2 = a^2$.**

If centre (α, β) is (0, 0) ; then equation of circle is $x^2 + y^2 = a^2$.

Example 2. *Find the area common to the parabola $y^2 = 4x$ and the circle $4x^2 + 4y^2 = 9$.*

Sol. The required area is the area common to the interiors of the parabola

$$y^2 = 4x \qquad ...(1)$$

[Parabola (1) is a rightward parabola and is symmetrical about x-axis.]

and the circle $4x^2 + 4y^2 = 9$ ...(2)

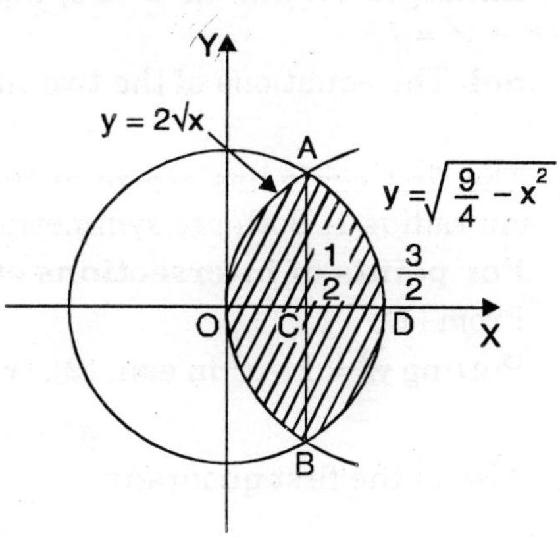

Dividing every term of eqn. (2) by 4,

$$x^2 + y^2 = \frac{9}{4} = \left(\frac{3}{2}\right)^2$$

which is a circle whose centre is origin and radius is $\frac{3}{2}$.

Putting $y^2 = 4x$ from (1) in (2), $4x^2 + 16x - 9 = 0$

$$\therefore \quad x = \frac{-16 \pm \sqrt{256 + 144}}{8} = \frac{-16 \pm 20}{8} = \frac{1}{2}, -\frac{9}{2}.$$

When $x = -\frac{9}{2}$, from (1), $y^2 = -18$ is negative and hence y is imaginary and hence impossible.

When $x = \frac{1}{2}$, from (1), $y^2 = 4x = 4 \times \frac{1}{2} = 2$

$\therefore$ The two points of intersection of parabola (1) and circle (2) are

$$A\left(\frac{1}{2}, \sqrt{2}\right) \text{ and } B\left(\frac{1}{2}, -\sqrt{2}\right).$$

Both the curves are symmetrical about x-axis.

For the parabola (1), $y = 2\sqrt{x}$ in the first quadrant.

For the circle (2), $4y^2 = 9 - 4x^2$ or $y^2 = \frac{9}{4} - x^2$

or $$y = \sqrt{\frac{9}{4} - x^2} \text{ in first quadrant.}$$

Required area OADBO (shaded)

$= 2 \times$ Area OADO $= 2$ [Area OAC + Area CAD]

$$= 2\left[{}^*\int_0^{1/2} 2\sqrt{x}\, dx + \int_{1/2}^{3/2} \sqrt{\frac{9}{4} - x^2}\, dx\right] \qquad [\text{Area} = \int y\, dx]$$

$$= 2\left[\left\{2 \cdot \frac{x^{3/2}}{3/2}\right\}_0^{1/2} + \left\{\frac{x\sqrt{\frac{9}{4} - x^2}}{2} + \frac{\frac{9}{4}}{2}\sin^{-1}\frac{x}{3/2}\right\}_{1/2}^{3/2}\right]$$

$$= 2\left[\frac{4}{3} \times \frac{1}{2\sqrt{2}} + \frac{9}{8}\sin^{-1} 1 - \frac{\frac{1}{2}\sqrt{2}}{2} - \frac{9}{8}\sin^{-1}\frac{1}{3}\right]$$

$$= 2\left[\frac{\sqrt{2}}{3} + \frac{9}{8} \cdot \frac{\pi}{2} - \frac{\sqrt{2}}{4} - \frac{9}{8}\sin^{-1}\frac{1}{3}\right] = \frac{9\pi}{8} - \frac{9}{4}\sin^{-1}\frac{1}{3} + \frac{\sqrt{2}}{6}.$$

Example 3. *Find the area of the region enclosed between the two circles $x^2 + y^2 = 1$ and $(x - 1)^2 + y^2 = 1$.* (M.D.U. 2006)

Sol. The equations of the two circles are $x^2 + y^2 = 1$ ...(1)

and $(x - 1)^2 + y^2 = 1$ ...(2)

The first circle has centre at the origin and radius 1. The second circle has centre at (1, 0) and radius 1. Both are symmetrical about the x-axis.

For points of intersections of circles (1) and (2)

From (1), $y^2 = 1 - x^2$

Putting $y^2 = 1 - x^2$ in eqn. (2), $(x - 1)^2 + 1 - x^2 = 1$

or $x^2 + 1 - 2x + 1 - x^2 = 1$ or $-2x + 1 = 0$ $\therefore$ $x = \frac{1}{2}$

Also in the first quadrant,

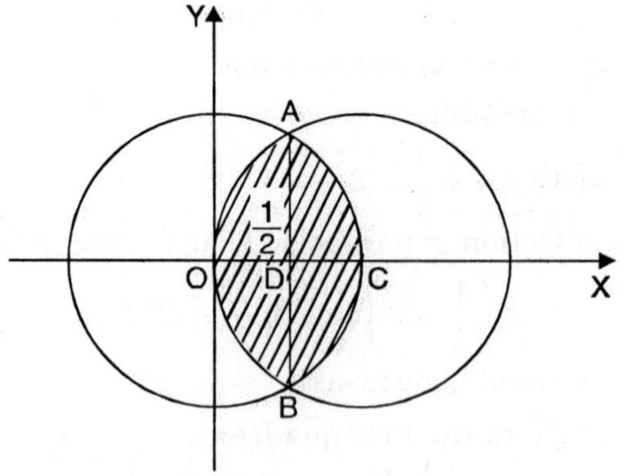

*Limits of integration for parabola are $x = 0$ to x of point of intersection and for circle are x of point of intersection to x = radius of circle.

From (1), $y = \sqrt{1-x^2}$

From (2), $y = \sqrt{1-(x-1)^2}$.

Required area OACBO (shaded)

$= 2 \times$ Area OAC

$= 2$ [Area OAD + Area DAC]

$$= 2\left[\int_0^{1/2} y \text{ of circle (2) } dx + \int_{1/2}^1 y \text{ of circle (1) } dx\right]$$

(Note)

$$= 2\left[\int_0^{1/2} \sqrt{1-(x-1)^2}\, dx + \int_{1/2}^1 \sqrt{1-x^2}\, dx\right]$$

$$= 2\left[\left\{\frac{(x-1)\sqrt{1-(x-1)^2}}{2} + \frac{1}{2}\sin^{-1}(x-1)\right\}_0^{1/2} + \left\{\frac{x\sqrt{1-x^2}}{2} + \frac{1}{2}\sin^{-1}x\right\}_{1/2}^1\right]$$

$$= \left\{-\frac{1}{2}\sqrt{\frac{3}{4}} + \sin^{-1}\left(-\frac{1}{2}\right)\right\} - \{\sin^{-1}(-1)\} + \sin^{-1}1 - \left\{\frac{1}{2}\sqrt{\frac{3}{4}} + \sin^{-1}\frac{1}{2}\right\}$$

$$= -\frac{\sqrt{3}}{4} - \frac{\pi}{6} + \frac{\pi}{2} + \frac{\pi}{2} - \frac{\sqrt{3}}{4} - \frac{\pi}{6} = \frac{2\pi}{3} - \frac{\sqrt{3}}{2}.$$

Example 4. *Find the area of the region lying above x-axis and included between the circle $x^2 + y^2 = 2ax$ and parabola $y^2 = ax$.*

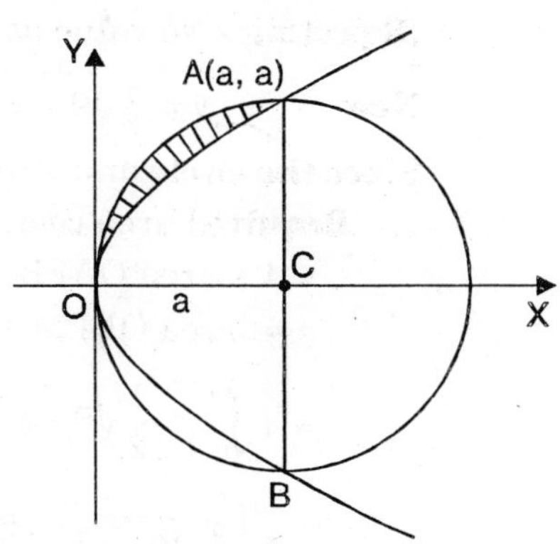

Sol. Equation of circle is $x^2 + y^2 = 2ax$ ...(1)

Equation of parabola is $y^2 = ax$ ...(2)

Adding a^2 to both sides of eqn. (1),

$$x^2 - 2ax + a^2 + y^2 = a^2$$

or $$(x-a)^2 + y^2 = a^2$$

$\therefore$ $$y = \sqrt{a^2 - (x-a)^2} \quad ...(3)$$

which represents a circle whose centre is $(a, 0)$ and radius is a. Also circle (1) passes through the origin.

To find points of intersections of circle (1) and parabola (2).

Putting $y^2 = ax$ from (2) in (1), we have

$$x^2 + ax = 2ax \quad \text{or} \quad x^2 - ax = 0 \text{ or } x(x-a) = 0 \quad \therefore \quad x = 0, a$$

When $x = 0, y = 0$; when $x = a, y^2 = a^2$ so that $y = a$ (in 1st quadrant)

$\therefore$ The two points of intersection are O(0, 0) and A(a, a).

Required area (shaded)

$$= \int_0^a (y \text{ of circle (1) } i.e., (3))\, dx - \int_0^a (y \text{ of parabola (2)})\, dx$$

$$= \int_0^a \sqrt{a^2 - (x-a)^2}\, dx - \int_0^a \sqrt{a}\sqrt{x}\, dx$$

$$= \left[\frac{(x-a)\sqrt{a^2-(x-a)^2}}{2} + \frac{a^2}{2}\sin^{-1}\frac{x-a}{a}\right]_0^a - \sqrt{a}\left[\frac{x^{3/2}}{\frac{3}{2}}\right]_0^a$$

$$= (0+0) - \left[0 + \frac{a^2}{2}\sin^{-1}(-1)\right] - \frac{2}{3}\sqrt{a}\,a^{3/2}$$

$$= \frac{a^2}{2}\cdot\frac{\pi}{2} - \frac{2a^2}{3} = \frac{a^2}{12}(3\pi - 8).$$

Example 2. *Find the area common to the circle $x^2 + y^2 = 4$ and the ellipse $x^2 + 4y^2 = 9$.*

(M.D.U. 2003, 2005 ,K.U. 1996 S, 2006 S)

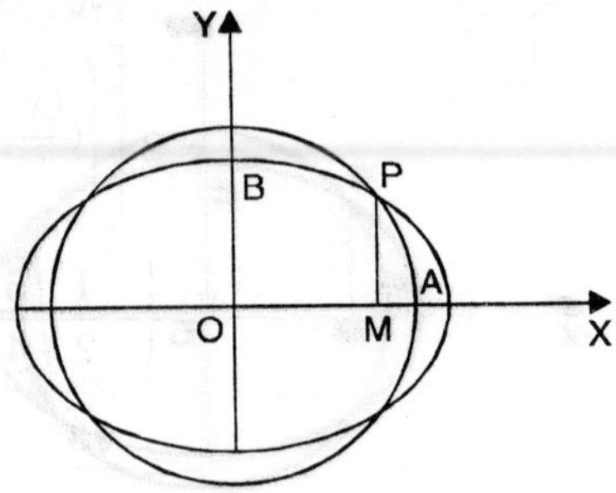

Sol. Let P be the point of intersection in the first quadrant, of the circle and the ellipse, as shown in the figure.

Draw MP$\perp$ x-axis.

To find abscissae of the points of intersection, eliminating y^2 from the two equations of the curves, we have (Putting $y^2 = 4 - x^2$ from first equation in the second equation)

$x^2 + 4(4 - x^2) = 9$ or $3x^2 = 7$

which gives $x = \sqrt{7/3}$ for P.

[Rejecting – ve value as P lies in first quadrant]

Now $y = \frac{1}{2}\sqrt{9-x^2}$, for the ellipse and $y = \sqrt{4-x^2}$, for the circle.

Since the circle and ellipse are symmetrical about the axes.

$\therefore$ Required area common to circle and ellipse

= 4 × area OAPB

= 4 [area OBPM + area PMA]

$$= 4\left[\int_0^{\sqrt{7/3}} \frac{1}{2}\sqrt{9-x^2}\,dx + \int_{\sqrt{7/3}}^{2} \sqrt{4-x^2}\,dx\right]$$

$$= 2\left[\frac{x}{2}\sqrt{9-x^2} + \frac{9}{2}\sin^{-1}\frac{x}{3}\right]_0^{\sqrt{7/3}} + 4\left[\frac{x}{2}\sqrt{4-x^2} + \frac{4}{2}\sin^{-1}\frac{x}{2}\right]_{\sqrt{7/3}}^{2}$$

$$\left[\because \int\sqrt{a^2-x^2}\,dx = \frac{x}{2}\sqrt{a^2-x^2} + \frac{a^2}{2}\sin^{-1}\frac{x}{a}\right]$$

$$= \left[\sqrt{\frac{7}{3}}\cdot\sqrt{9-\frac{7}{3}} + 9\sin^{-1}\sqrt{\frac{7}{27}} - 0\right] + 2\left[0 + 4\sin^{-1}1 - \sqrt{\frac{7}{3}}\sqrt{4-\frac{7}{3}} - 4\,.\sin^{-1}\sqrt{\frac{7}{12}}\right]$$

$$= \sqrt{\frac{7}{3}}\left[\sqrt{\frac{20}{3}} - 2\sqrt{\frac{5}{3}}\right] + 9\sin^{-1}\sqrt{\frac{7}{27}} + \frac{8\pi}{2} - 8\sin^{-1}\sqrt{\frac{7}{12}}$$

$$= 4\pi + 9\sin^{-1}\sqrt{\frac{7}{27}} - 8\sin^{-1}\sqrt{\frac{7}{12}}.$$

EXERCISE 2

1. (*a*) Find the area bounded by the parabola $x^2 = 8y$ and the line $x - 2y + 8 = 0$.

(*b*) Show that the area of the segment cut off from the parabola $y^2 = 2x$ by the straight line $y = 4x - 1$ is 9/32.

2. (*a*) Find the area of the segment cut off from the parabola $y^2 = 4x$ by the straight line
$$2x - 3y + 4 = 0.$$

(*b*) Find the area bounded by the curve $y^2 = x^3$ and the line $y = 2x$.

3. Find the area bounded by the parabola $y = 2 - x^2$ and the straight line $y = -x$.

4. (*a*) Find the area common to the parabolas $y^2 = 4ax$ and $x^2 = 4ay$. (*K.U. 2000*)

(*b*) Find the area included between curves $y^2 = 4bx$ and $x^2 = 4ay$.

(*c*) Find the area enclosed between the parabola $y^2 = 8x$ and $x^2 = 12y$.

5. Find the area common to the parabola $y^2 = ax$ and the circle $x^2 + y^2 = 4ax$. (*M.D.U. 2003 S ; K.U. 2005*)

6. Show that the larger of the two areas into which the circle $x^2 + y^2 = 64a^2$ is divided by the parabola $y^2 = 12ax$ is $\dfrac{16}{3}a^2(8\pi - \sqrt{3})$.

7. Find the area of the smaller portion enclosed by the curves $y^2 = 8x$ and $x^2 + y^2 = 9$.

8. Show that the area enclosed between the parabolas
$$y^2 = 4a(x + a) \text{ and } y^2 = -4a(x - a) \text{ is } \frac{16a^2}{3}.$$

9. Find the area enclosed between the parabolas
$$y^2 = 4a\,(x + a) \text{ and } y^2 = 4b(b - x).$$
[**Hint.** The points of intersection of the two parabolas are
$$x = b - a,\ y = \pm 2\sqrt{ab}.$$
$$\text{Required Area} = 2\left[\int_{-a}^{b-a} \sqrt{4a\,(x+a)}\,dx + \int_{b-a}^{b} \sqrt{4b(b-x)}\,dx\right].$$

10. Find the area of the region lying above x-axis and included between the circle $x^2 + y^2 = 4x$ and the parabola $y^2 = 2x$. (*D.U., B.C.A. 1998*)
[It is Example 4 with $a = 2$]

ANSWERS

1. (*a*) 36 Numerically **2.** (*a*) $\dfrac{1}{3}$ (*b*) $\dfrac{16}{5}$

4. (*a*) $\dfrac{16}{3}a^2$ (*b*) $\dfrac{16ab}{3}$ (*c*) 32

5. $a^3\left(3\sqrt{3} + \dfrac{4\pi}{3}\right)$ **7.** $\dfrac{2\sqrt{2}}{3} + \dfrac{9\pi}{2} - 9\sin^{-1}\dfrac{1}{3}$

9. $\dfrac{8}{3}(a + b)\sqrt{ab}$ **10.** $\dfrac{1}{3}(3\pi - 8)$.

4. Area Formulae for Curves given by Parametric Equations

(i) To prove that the area bounded by the curves $x = f(t)$, $y = \phi(t)$, the x-axis and the ordinates at the points where, $t = a$, $t = b$ is given by

$$\int_a^b y\frac{dx}{dt}\,dt \qquad ...(1)$$

The parametric equations of the curve are $x = f(t), y = \phi(t)$.

From Art. 1, we know that

$$\text{Required area} = \int_{t=a}^{t=b} y\,dx = \int_a^b y\frac{dx}{dt}\,dt.$$

(ii) Similarly, the area bounded by the curve $x = f(t)$, $y = \phi(t)$, the y-axis and the abscissae at the points where $t = c$, $t = d$ is given by

$$\int_c^d x\frac{dy}{dt}\,dt \qquad ...(2)$$

Example 1. *Find the area included between the cycloid $x = a(\theta - \sin\theta)$, $y = a(1 - \cos\theta)$ and its base.*

Sol. For the first half of the cycloid, θ varies from 0 to π.

$\therefore$ Required area bounded by the cycloid.

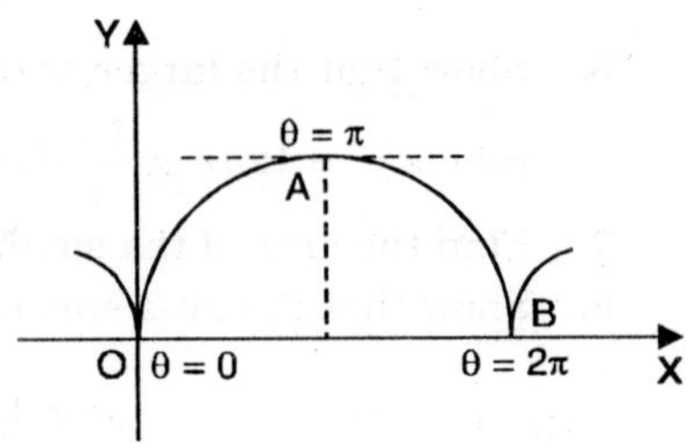

$$= 2\int_0^\pi y\,.\frac{dx}{d\theta}\,d\theta \qquad \text{[By Art. 4 }(i)]$$

$$= 2\int_0^\pi a(1 - \cos\theta)\,.\,a(1 - \cos\theta)\,d\theta$$

$$[\because \; y = a(1 - \cos\theta) \text{ and } x = a(\theta - \sin\theta)]$$

$$= 2a^2\int_0^\pi (1 - \cos\theta)^2\,d\theta = 2a^2\int_0^\pi\left(2\sin^2\frac{\theta}{2}\right)^2 d\theta$$

$$= 2a^2\int_0^\pi 4\sin^4\frac{\theta}{2}\,d\theta$$

Put $\frac{\theta}{2} = t \quad \therefore \quad \theta = 2t \quad \therefore \quad d\theta = 2dt$

When $\theta = 0, t = 0$; when $\theta = \pi, t = \frac{\pi}{2}$

$$= 8a^2\int_0^{\pi/2}\sin^4 t\,.\,2dt = 16a^2\int_0^{\pi/2}\sin^4 t\,dt = 16a^2\,.\,\frac{3.1}{4.2}\,.\,\frac{\pi}{2} = 3\pi a^2.$$

EXERCISE 3

1. (*a*) Find the area of the curve $x = a\cos^3 t$, $y = b\sin^3 t$ or the curve $\left(\frac{x}{a}\right)^{2/3} + \left(\frac{y}{b}\right)^{2/3} = 1$.

$$\left[\textbf{Hint. } \text{Area of curve} = 4\int_0^{\pi/2} y\frac{dx}{dt}\,.\,dt = 4\int_0^{\pi/2} b\sin^3 t(-3a\cos^2 t\sin t)\,dt\right]$$

(*b*) Find the area of the curve $x^{2/3} + y^{2/3} = a^{2/3}$

or $\qquad x = a\cos^3 t, y = a\sin^3 t.$

(*c*) Show that the area of the ellipse $x = a\cos t$, $y = b\sin t$ is πab.

2. Prove that the whole area between the four infinite branches of the tractrix.

$$x = a\cos t + \frac{a}{2}\log\tan^2\frac{t}{2}\ .\ y = a\sin t \text{ is } \pi a^2.$$

$\left[\textbf{Hint.}\ \text{Area} = 4\int_0^{\pi/2} y\frac{dx}{dt}dt.\right]$

3. Find the whole area of the curve $x = a\left(\frac{1-t^2}{1+t^2}\right), y = \frac{2at}{1+t^2}$.

[**Hint.** Eliminating t (by squaring and adding), we have $x^2 + y^2 = a^2$ which is a circle of radius a. Hence area $= \pi a^2$.]

4. Find the area included between the cycloid $x = a(\theta + \sin\theta)$, $y = a(1 - \cos\theta)$ and its base.

5. Find the area of the loop of the curve $x = a(1 - t^2)$, $y = at(1 - t^2)$. *(M.D.U. 2002)*

[**Hint.** Dividing $t = \frac{y}{x}$, putting this value of t in $x = a(1 - t^2)$ we have $ay^2 = x^2(a - x)$

$\therefore$ Area of loop $= 2\int_0^a y\,dx$. Also See Q. 6, Exercise 6 (*a*).]

6. Show that the area bounded by the cissoid $x = a\sin^2 t$, $y = a\frac{\sin^3 t}{\cos t}$ and its asymptote is $\frac{3\pi a^2}{4}$.

(M.D.U. 2004 A)

[**Hint.** Eliminating t, the equation of the curve is $y^2(a - x) = x^3$. Now it is Q. 16, Page 287]

7. Show that the area of a loop of the curve $x = a\sin 2t$, $y = a\sin t$ is $\frac{4}{3}a^2$.

[**Hint.** Change to cartesians for tracing.]

ANSWERS

1. (*a*) $\frac{3\pi}{8}ab$ (Numerically) (*b*) $\frac{3}{8}\pi a^2$

3. πa^2

4. $3\pi a^2$ **5.** $\frac{8}{15}a^2$ (Numerically).

Note. For sake of convenience of the reader, the shape of the curve in question (2) is being given here.

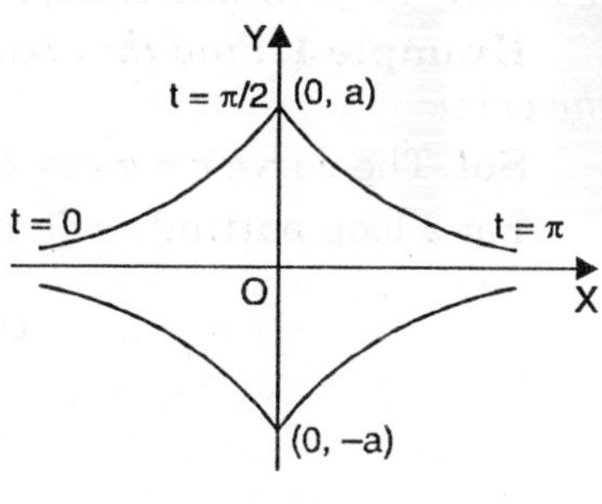

5. Area Formula for Curves given by Polar Equations

To prove that the area bounded by the curve $r = f(\theta)$ *and the radii vectors* $\theta = \alpha$, $\theta = \beta$ *is* $\int_\alpha^\beta \frac{1}{2}\mathbf{r}^2\,\mathbf{d}\theta$, *where* $r = f(\theta)$ *is a continuous and single valued function of* θ *in* $[\alpha, \beta]$.

Let AB be the curve $r = f(\theta)$, and OA, OB the radii vectors

$$\theta = \alpha, \theta = \beta.$$

Let P(r, θ) be *any point* on the curve. Let the *sectorial area* OAP be denoted by A, which clearly is a function of θ.

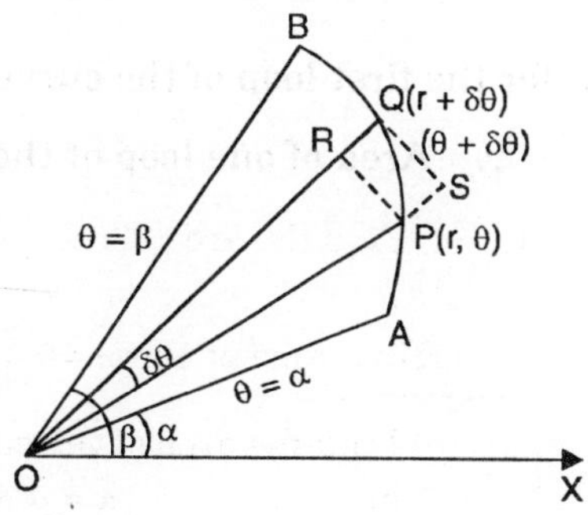

Let Q $(r + \delta r, \theta + \delta\theta)$ be a point on the curve close to P.

Then the area OPQ = δA.

With O as centre and OP, OQ as radii, draw circular arcs to meet OQ, OP (produced) in R and S respectively.

Then the area OPQ lies between areas of two circular sectors OPR and OSQ.

i.e., δA lies between $\frac{1}{2} r^2 \delta\theta$ and $\frac{1}{2}(r + \delta r)^2 \delta\theta$.

[$\because$ area of a circular sector = $\frac{1}{2}$ (radius)^2 × circular measure of angle of the sector]

$\therefore$ $\frac{\delta A}{\delta\theta}$ lies between $\frac{1}{2} r^2$ and $\frac{1}{2}(r + \delta r)^2$

Proceeding to limits as $\delta\theta \to 0$ and $\therefore \delta r \to 0$, we get

$\frac{dA}{d\theta}$ lies between $\frac{1}{2} r^2$ and a quantity which $\to \frac{1}{2} r^2$ $\quad\therefore\quad \frac{dA}{d\theta} = \frac{1}{2} r^2$.

Integrating both sides within the limits α and β.

$$\therefore \quad \int_\alpha^\beta \frac{1}{2} r^2 \, d\theta = \int_\alpha^\beta \frac{dA}{d\theta} d\theta = [A]_\alpha^\beta$$

$$= (\text{value of A when } \theta = \beta) - (\text{value of A when } \theta = \alpha)$$

$$= \text{area OAB} - 0$$

Hence area **OAB** = $\displaystyle\int_\alpha^\beta \frac{1}{2} \mathbf{r^2 \, d\theta}$.

Note 1. In the above result, we have supposed that r is an increasing function of θ in the interval $[\alpha, \beta]$. The result is still true even if the radius vector r decreases as θ varies from α to β.

Note 2. In some cases, it is more convenient to transform a given cartesian equation into polars than to solve for y. In such cases, the formula of the present article is applied after changing to polars.

Example 1. *Find the area of a loop of the curve $r = a \cos 2\theta$ and hence find the total area of the curve.*

Sol. The curve $r = a \cos 2\theta$ has **four** equal loops (see Note 1, the end of this example).

For a loop putting $r = 0$, we get

$$\cos 2\theta = 0 \quad \therefore \quad 2\theta = \frac{-\pi}{2}, \frac{\pi}{2} \qquad \text{(two consecutive values)}$$

(See Note 2 at the end of this example)

$$\therefore \qquad \theta = \frac{-\pi}{4}, \frac{\pi}{4}$$

i.e., for the first loop of the curve θ varies from $\frac{-\pi}{4}$ to $\frac{\pi}{4}$

$\therefore$ Area of one loop of the curve

$$= \int_{-\pi/4}^{\pi/4} \frac{1}{2} r^2 d\theta = \frac{1}{2} \int_{-\pi/4}^{\pi/4} a^2 \cos^2 2\theta \, d\theta$$

$$= \frac{a^2}{2} \times 2 \int_0^{\pi/4} \cos^2 2\theta \, d\theta$$

[$\because$ $\cos^2 2\theta$ is an even function of θ and for an even function $\int_{-a}^{a} f(x) \, dx = 2 \int_0^a f(x) \, dx$]

$$= a^2 \int_0^{\pi/2} \cos^2 t \frac{dt}{2} \qquad \text{(Putting } 2\theta = t\text{)}$$

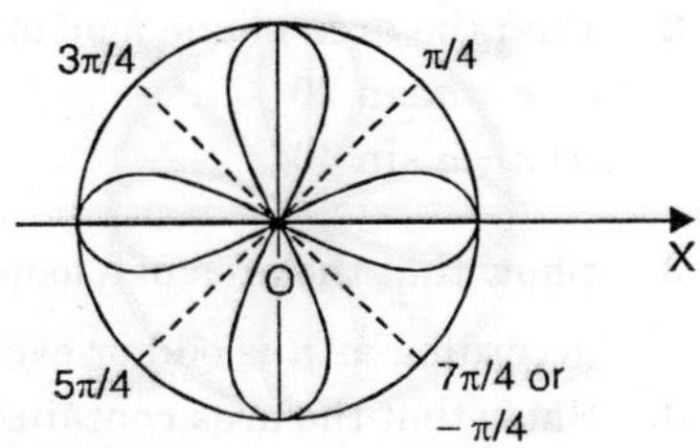

$$\frac{a^2}{2} \cdot \frac{1}{2} \cdot \frac{\pi}{2} = \frac{\pi a^2}{8}$$

$\therefore$ Total area of the curve = 4 × (Area of one loop)

$$= 4 \times \frac{\pi a^2}{8} = \frac{\pi a^2}{2}.$$

Note. 1. It should be carefully remembered that the curves $r = a \sin n\theta$ or $r = a \cos n\theta$ have n *equal loops if n is odd* and $2n$ *equal loops if n is even.*

Note. 2. To find the limits of integration for a loop we generally put $r = 0$ and find *two consecutive* values of θ.

Example 2. *Find the area of the loop of the folium of Descrates, $x^3 + y^3 = 3axy$.*

(M.D.U. 2005 ; K.U. 2005 S, 2006 A)

Sol. The equation of the curve is $x^3 + y^3 = 3axy$

Changing to polars, (see Note 2, Art. 5) by putting $x = r \cos \theta$, $y = r \sin \theta$; (*i*) reduces to

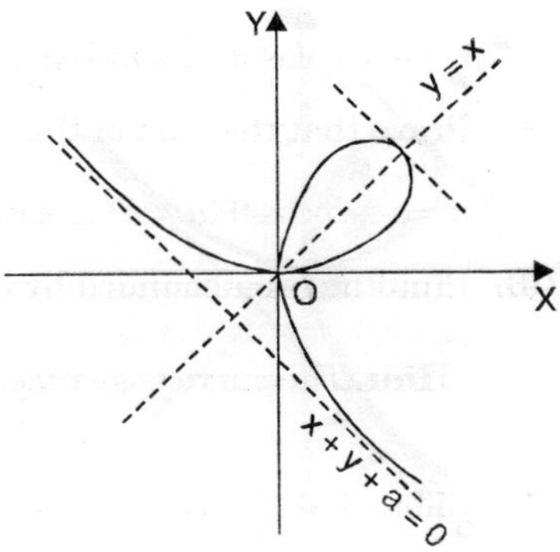

$$r^3 (\sin^3 \theta + \cos^3 \theta) = 3ar^2 \sin \theta \cos \theta$$

or
$$r = \frac{3a \sin \theta \cos \theta}{\sin^3 \theta + \cos^3 \theta}$$

For the loop, putting $r = 0$, we get $\sin \theta \cos \theta = 0$ *i.e.,*

$\sin \theta = 0$ and $\cos \theta = 0$ or $\theta = 0, \frac{\pi}{2}$.

Hence for the loop θ varies from 0 to $\frac{\pi}{2}$.

$\therefore$ Area of the loop

$$= \int_0^{\pi/2} \frac{1}{2} r^2 d\theta = \frac{9a^2}{2} \int_0^{\pi/2} \frac{\sin^2 \theta \cos^2 \theta \, d\theta}{(\sin^3 \theta + \cos^3 \theta)^2}$$

$$= \frac{9a^2}{2} \int_0^{\pi/2} \frac{\tan^2 \theta \sec^2 \theta \, d\theta}{(1 + \tan^3 \theta)^2} \qquad \text{(On dividing the num. and den. by } \cos^6 \theta\text{)}$$

(Putting $\tan^3 \theta = t$, so that $3 \tan^2 \theta \sec^2 \theta \, d\theta = dt$)

$$= \frac{3a^2}{2} \int_0^{\infty} \frac{dt}{(1+t)^2}$$

$$= \frac{3a^2}{2} \left[-\frac{1}{1+t} \right]_0^{\infty} = \frac{3a^2}{2} [0 - (-1)] = \frac{3a^2}{2}.$$

EXERCISE 4

1. Find the area of any sector of each of the following curves :

(*i*) $r\theta = a$ ($\theta = \alpha$ to $\theta = \beta$)

(*ii*) $r = a\, e^{\theta \cot \alpha}$ ($\theta = \beta$ to $\theta = \beta + \gamma$, (γ being $< 2\pi$)).

2. Find the area of one loop of the following curves :
(i) $r = a \sin 2\theta$ (ii) $r = a \sin 3\theta$ (K.U. 1996)
(iii) $r = a \sin 4\theta$ (iv) $r = a \sin n\theta$. Also state the total area.

3. Show that the area of a loop of $r = a \cos n\theta$ is $\frac{\pi a^2}{4n}$. Also prove that the whole area is $\frac{\pi a^2}{4}$ or $\frac{\pi a^2}{2}$ according as n is odd or even.

4. Show that the area contained between the circle $r = a$ and the curve $r = a \cos 5\theta$ is equal to three-fourths of the area of the circle.

5. Find the area of a loop of the curve $r = a\theta \cos \theta$ between $\theta = 0$ and $\theta = \frac{\pi}{2}$.

6. Trace the curve $r^2 = a^2 \cos 2\theta$ and find its area.

7. Find the areas bounded by :
(i) the cardioid $r = a(1 - \cos \theta)$
(ii) the cardioid $r = a(1 + \cos \theta)$.

8. Show that the whole area of the curve represented by the equation
$r = a + b \cos \theta$, assuming $a > b$ is $\frac{\pi}{2}(2a^2 + b^2)$.

9. Prove that the sum of the area of the two loops of the limacon
$r = a + b \cos \theta$ $(a < b)$ is equal to $\frac{\pi}{2}(2a^2 + b^2)$.

10. Find the area bounded by $r^2 = a^2 \cos^2 \theta + b^2 \sin^2 \theta$.

[**Hint.** The curve is symmetrical about the lines $\theta = 0$ and $\theta = \frac{\pi}{2}$. $\therefore$ Area $= 4 \int_0^{\pi/2} \frac{1}{2} r^2 \, d\theta$.]

11. Show that the area of *a* loop of the curve $r = \sqrt{3} \cos 3\theta + \sin 3\theta$ is $\frac{\pi}{3}$. (K.U. 2003 ; M.D.U. 2007)

12. Find the area of the loop of the following curves :
(a) $(x^2 + y^2)^2 = 4axy^2$ (b) $x^4 + y^4 = 4a^2 xy$
(c) $x^5 + y^5 = 5ax^2y^2$. (K.U. 2005)

13. Prove that the area of the loop of the curve $x^3 + y^3 = 3axy$ is three times the area of a loop of the curve $r^2 = a^2 \cos 2\theta$.

14. Show that the area bounded by the curve $p = f(r)$ and the two radii vectors $r = a$,
$r = b$ is $\frac{1}{2} \int_a^b \sqrt{\frac{pr\,dr}{r^2 - p^2}}$.

[**Hint.** Area $= \int \frac{1}{2} r^2 \, d\theta = \int \frac{1}{2} r^2 \frac{d\theta}{dr} dr = \int \frac{1}{2} r \tan \phi \, dr \left(\because \tan \phi = r \frac{d\theta}{dr} \right)$.

Now use $p = r \sin \phi$.]

ANSWERS

1. (i) $a^2 \left(\frac{1}{\alpha} - \frac{1}{\beta} \right)$ (ii) $\frac{a^2}{4} \tan \alpha \, e^{2 \beta \cot \alpha} [e^{2 \gamma \cot \alpha} - 1]$

2. (i) $\frac{\pi a^2}{8}$ (ii) $\frac{\pi a^2}{12}$ (iii) $\frac{\pi a^2}{16}$

(*iv*) $\frac{\pi a^2}{4n}$; total area $= \frac{\pi a^2}{4}$ or $\frac{\pi a^2}{2}$ according as n is an odd or even positive integer

5. $\frac{\pi a^2}{16}\left(\frac{\pi^2}{6}-1\right)$

6.

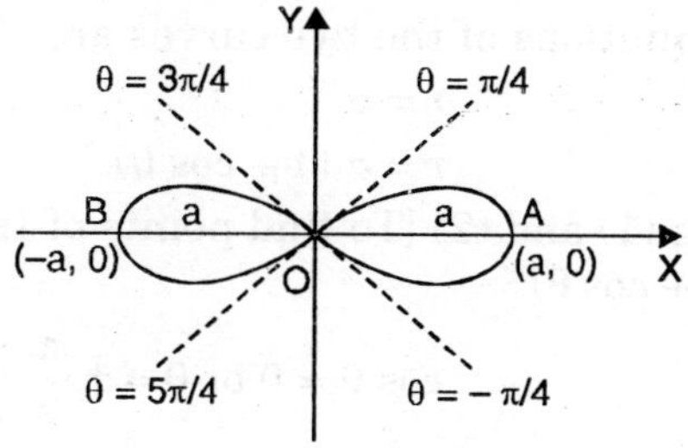

7. (*i*) $\frac{3}{2}\pi a^2$ (*ii*) $\frac{3}{2}\pi a^2$ **10.** $\frac{\pi}{2}(a^2+b^2)$

12. (*a*) $\frac{\pi a^2}{4}$ (*b*) $\frac{\pi a^2}{2}$ (*c*) $\frac{5a^2}{2}$.

Note. For the sake of convenience of the reader the shapes of the curves in question 2 (*ii*) and question 4 are being given below.

2. (*ii*)

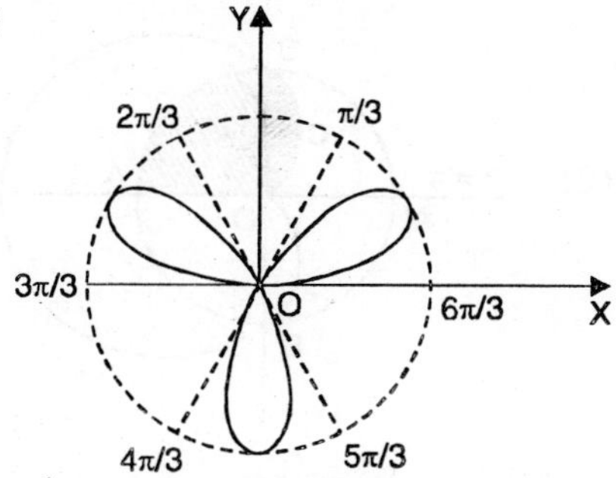

4.

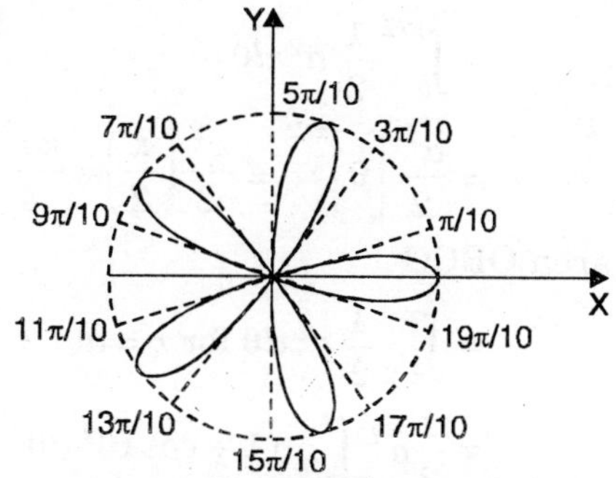

---

6. Area between two Polar Curves

To prove that the area bounded by the curves $r = f(\theta)$, $r = F(\theta)$ and the radii vectors $\theta = \alpha$, $\theta = \beta$ is

$$\int_\alpha^\beta \tfrac{1}{2}\,(\mathbf{r_1}^2 - \mathbf{r_2}^2)\mathbf{d\theta}$$

where r_1 is the 'r' of the outer curve and r_2 that of the inner curve.

Let AB, CD be the curves $r = f(\theta)$, $r = F(\theta)$ and OCA, ODB the radii vectors $\theta = \alpha$, $\theta = \beta$ respectively.

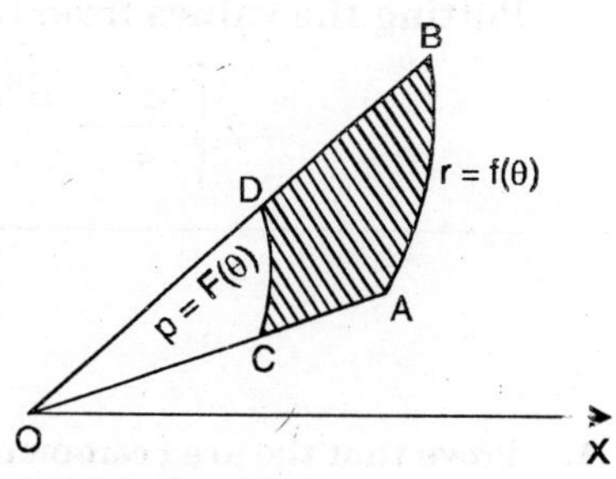

Then area CABD

$$= \text{area OAB} - \text{area OCD}$$

$$= \int_\alpha^\beta \tfrac{1}{2}\,[f(\theta)]^2\,d\theta - \int_\alpha^\beta \tfrac{1}{2}\,[F(\theta)]^2\,d\theta$$

$$= \int_\alpha^\beta \tfrac{1}{2}\,[[f(\theta)]^2 - [F(\theta)]^2]\,d\theta$$

$$= \int_\alpha^\beta \tfrac{1}{2}\,(r_1^2 - r_2^2)\,d\theta$$

where $r_1 = [f(\theta)]$ is the value of r of the outer curve and $r_2 = [F(\theta)]$ is that of the inner curve.

Example 1. *Find the area common to the circle r = a and the cardioid r = a (1 + cos θ).* (M.D.U. 2007)

Sol. Equations of the two curves are

$$r = a \quad \text{...(1) (circle with pole as centre and radius } a)$$

and $$r = a(1 + \cos\theta) \quad \text{...(2) (cardioid)}$$

Solving (1) and (2) [To find points of Intersection], by putting $r = a$, from (1) in (2), we have $a = a(1 + \cos\theta)$

or $$\cos\theta = 0 \text{ or } \theta = \pm\frac{\pi}{2}$$

i.e., two curves cut each other at points where $\theta = \pm\frac{\pi}{2}$.

∴ Required area = 2 (the shaded area) = 2 [Area OABO + Area OBCO] ...(1)

Now area OABO

$$= \int_0^{\pi/2} \frac{1}{2} r^2\, d\theta \text{ for } r = a$$

$$= \int_0^{\pi/2} \frac{1}{2} a^2\, d\theta$$

$$= \frac{a^2}{2}\Big[\theta\Big]_0^{\pi/2} = \frac{a^2}{2}\left(\frac{\pi}{2}\right) = \frac{\pi a^2}{4} \quad \text{...(2)}$$

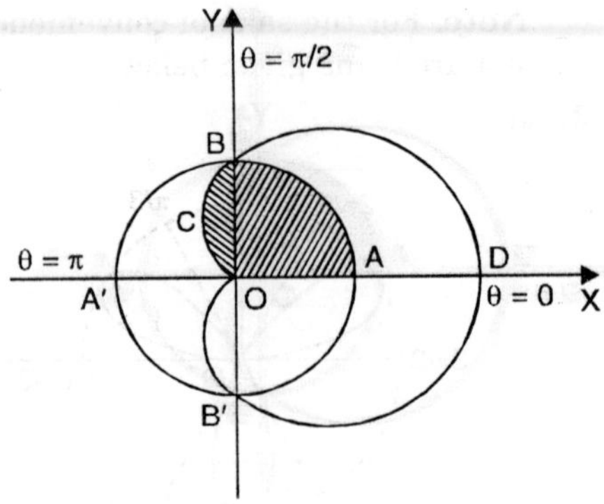

And area OBCO

$$= \int_{\pi/2}^{\pi} \frac{1}{2} r^2\, d\theta \text{ for } r = a(1 + \cos\theta)$$

$$= \frac{1}{2}a^2 \int_{\pi/2}^{\pi} (1 + \cos\theta)^2\, d\theta$$

$$= \frac{1}{2}a^2 \int_{\pi/2}^{\pi} (1 + 2\cos\theta + \cos^2\theta)\, d\theta$$

$$= \frac{a^2}{2}\int_{\pi/2}^{\pi}\left(1 + 2\cos\theta + \frac{1+\cos 2\theta}{2}\right) d\theta = \frac{a^2}{2}\int_{\pi/2}^{\pi}\left(\frac{3}{2} + 2\cos\theta + \frac{1}{2}\cos 2\theta\right) d\theta$$

$$= \frac{a^2}{2}\left[\frac{3}{2}\theta + 2\sin\theta + \frac{1}{4}\sin 2\theta\right]_{\pi/2}^{\pi} = \frac{a^2}{2}\left[\frac{3\pi}{2} - \left(\frac{3\pi}{4} + 2\right)\right] = \frac{a^2}{8}(3\pi - 8) \quad \text{...(3)}$$

Putting the values from (2) and (3) in (1) the required area

$$= 2\left[\frac{\pi a^2}{4} + \frac{a^2}{8}(3\pi - 8)\right] = a^2\left(\frac{5\pi}{4} - 2\right).$$

EXERCISE 5

1. Prove that the area common to the circles $r = a\sqrt{2}$ and $r = 2a\cos\theta$ is $a^2(\pi - 1)$.
2. Prove that the area of the region included between the cardioids $r = a(1 + \cos\theta)$ and $r = a(1 - \cos\theta)$ is $\frac{a^2}{2}(3\pi - 8)$.

3. Find the area outside the circle $r = 2a \cos \theta$ and inside the cardioid $r = a(1 + \cos \theta)$.

4. Find the area inside the circle $r = \sin \theta$ and outside the cardioid $r = 1 - \cos \theta$. *(K.U. 2004 S)*

ANSWERS

3. $\frac{\pi a^2}{2}$ **4.** $1 - \frac{\pi}{4}$.

MISCELLANEOUS EXERCISE

1. (*a*) Find the area of the segment cut off from the parabola $y^2 = 4x$ by the straight line
$$2x - 3y + 4 = 0.$$
(*b*) Find the area bounded by the curve $y^2 = x^3$ and the line $y = 2x$.

2. Show that the area included between one of the branches of the curve $x^2y^2 = a^2(x^2 + y^2)$ and its asymptotes is equal to that of the square whose side is a.

$$\left[\textbf{Hint. } \text{Area in first quadrant} = \int (y_1 - y_2)\, dx = \int_a^\infty \left(\frac{ax}{\sqrt{x^2 - a^2}} - a \right) dx. \right]$$

3. Show that the area of the loop of the curve $a^2y^2 = x^2(2a - x)(x - a)$ is $\frac{3}{8} a^2 \pi$.

4. Show that the area of the loop of the curve $y^2 = 3x^3 - x^4 - 2x^2$ is $\frac{3\pi}{8}$.

[**Hint.** Equation of the curve is $y^2 = 3x^3 - x^4 - 2x^2$
$$= x^2(-x^2 + 3x - 2)$$
$$= x^2(-x^2 + x + 2x - 2) = x^2[-x(x - 1) + 2(x - 1)]$$
or
$$y^2 = x^2(x - 1)(2 - x)$$
Now it is Q. No. 3 with $a = 1$].

5. Show that the area enclosed by the curves $xy^2 = a^2(a - x)$ and $(a - x)y^2 = a^2x$ is $(\pi - 2)a^2$.

[**Hint.** The point of intersection is $x = \frac{a}{2}$].

6. Find the area of the loop of the curve $x = \frac{a \sin 3\theta}{\sin \theta}$, $y = \frac{a \sin 3\theta}{\cos \theta}$.

[**Hint.** Dividing $\tan \theta = \frac{y}{x}$ $\therefore$ $\sin \theta = \frac{y}{\sqrt{x^2 + y^2}}$

Eliminating θ, equation of the curve is $y^2(a + x) = x^2(3a - x)$].

7. Find the area of a loop of the curve $y^2x + (x + a)^2(x + 2a) = 0$.

$$\left[\textbf{Hint. } \text{Area of loop} = 2\int_{-2a}^{-a} y\, dx = 2\int_{-2a}^{-a} (x + a)\sqrt{\frac{x + 2a}{-x}}\, dx. \text{ Put } x = -2a \sin^2 \theta. \right]$$

8. Prove that the area included between the folium $x^3 + y^3 = 3axy$ and its asymptote is equal to the area of its loop.

9. Show that the area of a loop of the curve $r = \sqrt{3} \cos 3\theta + \sin 3\theta$ is $\frac{\pi}{3}$. *(K.U. 2007)*

10. Find the area of the ellipse $\frac{l}{r} = 1 + e \cos \theta$.

11. Find the ratio of the two parts into which the parabola $2a = r(1 + \cos \theta)$ divides the area of the cardioid $r = 2a(1 + \cos \theta)$.

12. Show that the ratio of the area of the larger to the area of the smaller loop of the curve $r = \frac{1}{2} + \cos 2\theta$ is $\frac{(4\pi + 3\sqrt{3})}{(2\pi - 3\sqrt{3})}$.

13. Show that the area of the region enclosed between the two loops of the curve $r = a(1 + 2\cos\theta)$ is $a^2(\pi + 3\sqrt{3})$.

14. Show that the area of the segment cut off from the parabola $y^2 = 2x$ by the straight line $y = 4x - 1$ is $\frac{9}{32}$.

15. Find the area bounded by the parabola $x^2 = 8y$ and the line $x - 2y + 8 = 0$.

16. Prove that the area of a sector of the ellipse of semi-axes a and b between the major axis and a radius vector from the focus is $\frac{1}{2} ab(\theta - e\sin\theta)$ where e is the eccentricity of the ellipse and θ is the eccentric angle of the point to which the radius vector is drawn.

17. If A is the area contained between the catenary $y = c\cosh\frac{x}{c}$, the two axes and an ordinate at the extremity of the arc s, show that $A = cs$; s being measured from the vertex.

18. Prove that the area of a loop of the curve $r = a\cos 3\theta + b\sin 3\theta$ is $\frac{1}{12}\pi(a^2 + b^2)$.

19. Find the area of the loop of the curve $x^4 + 3x^2y^2 + 2y^4 = a^2xy$.

ANSWERS

1. (a) $\frac{1}{3}$ (b) $\frac{16}{5}$ 6. $3\sqrt{3}\,a^2$

7. $2a^2\left(1 - \frac{\pi}{4}\right)$ 10. $\frac{\pi l^2}{(1 - e^2)^{3/2}}$ 11. $\frac{9\pi - 16}{9\pi + 16}$

15. $\frac{76}{3}$ 19. $\frac{a^2}{4}\log 2$.

13

Volumes and Surfaces of Solids of Revolution

1. Introduction

A plane area when made to revolve about a fixed straight line lying in its own plane, generates **a solid of revolution** and its boundary generates a **surface of revolution.** The straight line about which the plane area is rotated is called the **axis of revolution.**

Every section of such a solid by a plane perpendicular to the axis of revolution is a circle having its centre on the axis.

For example (*a*) A right circular cylinder is generated by the revolution of a rectangle about one of its sides.

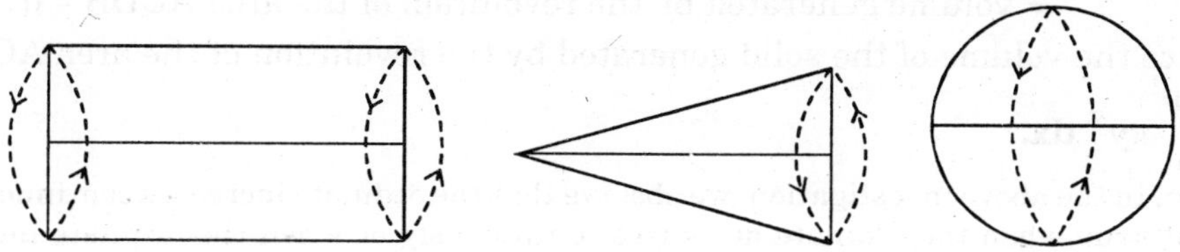

(a) (b) (c)

(*b*) A right circular cone is generated by the revolution of a right-angled triangle about its base.

(*c*) A sphere is generated when a semi-circle is rotated about its bounding diameter.

2. Volume formulae for the Cartesian Equations

If $y = f(x)$ be continuous, finite and a single valued function of x in the interval $[a, b]$ *where a and b are finite and $a < b$, then the volume of solid generated by revolution about x-axis, of the area bounded by the curve $y = f(x)$, the x-axis and the bounded ordinates $x = a$, $x = b$ is* $\int_a^b \pi y^2 \, dx$.

Let AB be the curve $y = f(x)$ and CA, DB be the two ordinates $x = a$, $x = b$ respectively.

Let P (x, y) be *any* point on the curve AB.

Draw PM $\perp$ OX. $\therefore$ OM $= x$, PM $= y$.

Let V denote the volume of the solid generated by the revolution about x-axis of the area ACMP.

As x increases *i.e.* PM moves towards the right, V also increases.

$\therefore$ V is a function of x.

Let Q$(x + \delta x, y + \delta y)$ be another point on the curve in the immediate neighbourhood of P and NQ its ordinate.

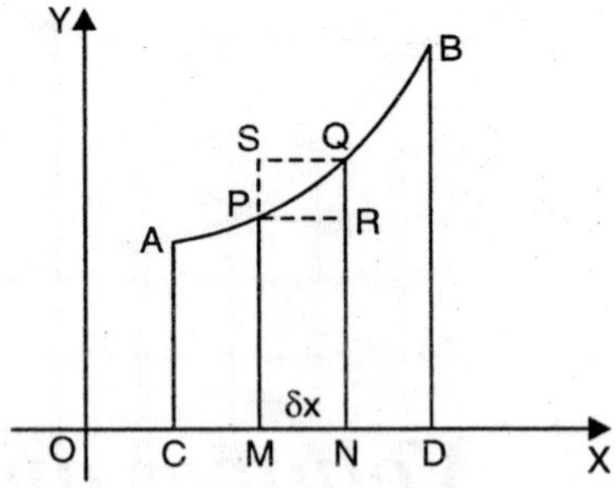

Then the volume of the solid generated by the revolution about the x-axis of the area PMNQ is δV.

Complete the rectangle PRQS.

Then the volume of the solid generated by the revolution of the area PMNQ lies between the volumes of the circular cylinders generated by the rectangles PMNR and SMNQ *i.e.* δV lies between $\pi y^2\ \delta x$ and $\pi(y + \delta y)^2\ \delta x$.

or $$\pi y^2\ \delta x < \delta V < \pi(y + \delta y)^2\ \delta x$$

or $$\pi y^2 < \delta V/\delta x < \pi(y + \delta y)^2$$

Since y is a continuous function. $\therefore$ $\delta y \to 0$ as $\delta x \to 0$.

$\therefore$ Taking limits as $\delta x \to 0$

$$\underset{\delta x \to 0}{\text{Lt}} \frac{\delta V}{\delta x} \text{ lies between } \pi y^2 \text{ and a quantity which} \to \pi y^2$$

$$\therefore \quad \frac{dV}{dx} = \pi y^2$$

$$\therefore \quad \int_a^b \pi y^2\ dx = \int_a^b \frac{dV}{dx}\ dx = [V]_a^b$$

$$= (\text{volume V when } x = b) - (\text{volume V when } x = a)$$

$$= \text{volume generated by the revolution of the area ACDB} - 0$$

Hence the volume of the solid generated by the revolution of the area ACDB about the x-axis is $\int_a^b \pi y^2\ dx$.

Note. In the above investigation, we observe that the ordinate increases continuously. The above result is still true when the ordinate decreases continuously or when the ordinate increases in some parts and decreases in the other parts. It is also assumed that curve does not cross the x axis, *i.e.* the axis of revolution.

Cor. 1. Revolution about y-axis. *The volume of the solid generated by revolution about the* **y-axis** *of the area bounded by the curve $x = f(y)$, the y-axis and the abscissae $y = a$, $y = b$ is*

$$\int_a^b \pi x^2\ dy.$$

The result follows immediately on interchanging x and y in the above proposition.

Cor. 2. Volume formulae for parametric equations. *If $x = f(t)$ and $y = \phi(t)$ are the parametric equations of a curve, then the volume of the solid generated by revolving the area about x-axis is*

$$\int \pi y^2 \frac{dx}{dt} . dt,$$

the limits of integration being so taken as to cover the whole area revolved and volume of the solid formed by revolving the area about y-axis is

$$\int \pi x^2 \frac{dy}{dt} . dt,$$

limits of integration being so taken as to cover the whole area rotated.

Note. An important observation. (*a*) If the generating curve *is symmetrical about the* x-axis, then the volume generated by the revolution of the area about the x-axis is the same as the volume generated by the revolution of its upper (or lower) half area.

Thus, the volume generated by the area DCOAB = vol. generated by the area OABM or OCDM.

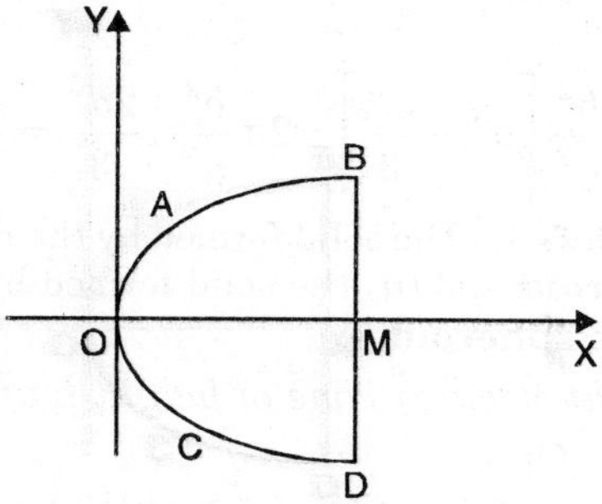

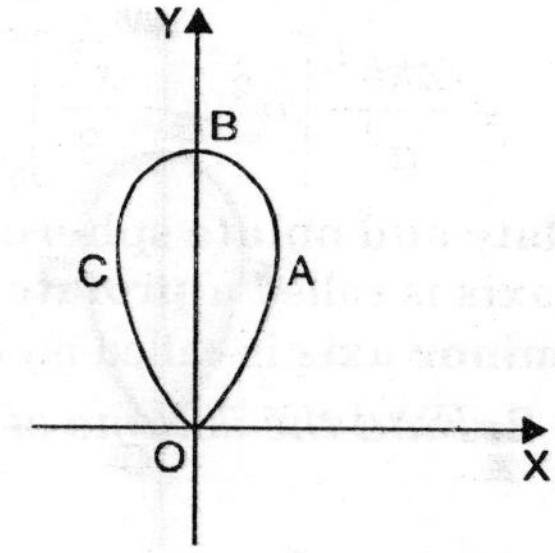

(*b*) If the curve is *symmetrical about the y-axis* and the same curve be made to *revolve about the x-axis* (the curve lying on one side of x-axis), then the volume generated

= 2 × volume generated by the portion OAB lying in the first quadrant.

3. Volume formulae for Polar Equations

The volume of the solid generated by the revolution of the area bounded by the curve $r = f(\theta)$ and the radii vectors $\theta = \alpha$, $\theta = \beta$:

(*i*) **about the initial line OX** $(\theta = 0) = \displaystyle\int_{\alpha}^{\beta} \frac{2}{3}\pi r^3 \sin\theta \, d\theta.$

(*ii*) **about the line OY** $\left(\theta = \frac{\pi}{2}\right) = \displaystyle\int_{\alpha}^{\beta} \frac{2}{3}\pi r^3 \cos\theta \, d\theta.$

The proofs of these formulae are beyond the scope of this book.

Note. The volume formulae of cartesian curves can also be used for polar curves by converting the polar equation to cartesian equation by using the relations $x = r\cos\theta$ and $y = r\sin\theta$.

$$\left(i.e.\ r^2 = x^2 + y^2,\ \cos\theta = \frac{x}{r} \text{ and } \sin\theta = \frac{y}{r}\right).$$

Example 1. *Find the volume of the prolate spheroid generated by the revolution of the ellipse* $\frac{x^2}{a^2} + \frac{y^2}{b^2} = 1$ *about the x-axis.* (M.D.U. 2003 A)

Sol. The equation of the ellipse is $\frac{x^2}{a^2} + \frac{y^2}{b^2} = 1$

or $$\frac{y^2}{b^2} = 1 - \frac{x^2}{a^2} = \frac{a^2 - x^2}{a^2} \quad \therefore \quad y^2 = \frac{b^2}{a^2}(a^2 - x^2) \qquad ...(1)$$

Since the ellipse is symmetrical about the x-axis

∴ the required volume = 2 × volume generated by revolving the area OABO about x-axis.

[We are multiplying by 2 because of symmetry about y-axis. See Note Cor. 2, Art. 2.]

Now for the area OABO, x varies from 0 to a.

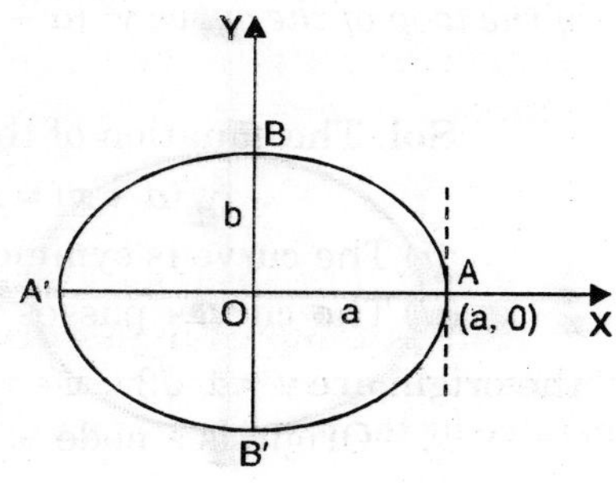

∴ Required volume $= 2\displaystyle\int_0^a \pi y^2 \, dx$

$$= 2\pi \int_0^a \frac{b^2}{a^2}(a^2 - x^2)\, dx \qquad [\because \text{ of } (1)]$$

$$= \frac{2\pi b^2}{a^2}\left[a^2 x - \frac{x^3}{3}\right]_0^a = 2\pi \frac{b^2}{a^2}\left[a^3 - \frac{a^3}{3}\right] = 2\pi \frac{b^2}{a^2} \cdot \frac{2a^3}{3} = \frac{4}{3}\pi a b^2.$$

Note. Prolate and oblate spheroids. Defs. (*i*) The solid formed by the revolution of the ellipse about the **major axis** is called a **prolate spheroid** and (*ii*) the solid formed by the revolution of the ellipse about the **minor axis** is called an **oblate spheroid.**

Example 2. *Find the volume of a right circular cone of height h and radius a.*

Or

Find the volume of the right circular cone formed by revolution of a right angled triangle about a side which contains the right angle.

Sol. We know by example (*b*) Art. 1, that a right circular cone is generated by the revolution of a right angled triangle ΔOCA about is base OC (taken as x-axis here)

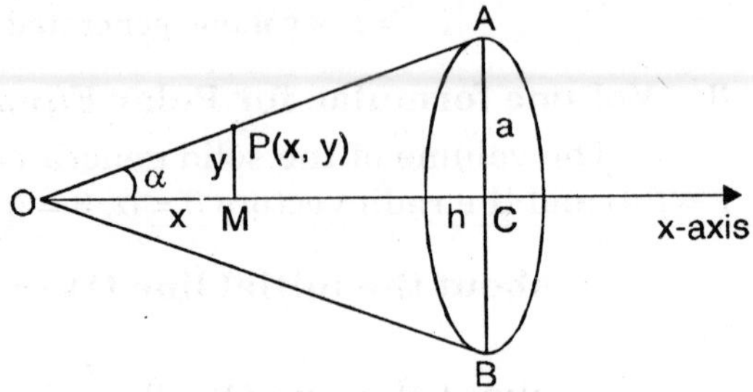

Let h be the height of the cone and a be the radius of its circular base.

In ΔOMP, $\tan \alpha = \dfrac{\text{MP}}{\text{OM}} = \dfrac{y}{x}$

$\therefore \qquad y = x \tan \alpha \qquad ...(1)$

Required volume of the cone $= \displaystyle\int_{x=0}^{h} \pi y^2 dx$

Putting $y = x \tan \alpha$ from (1),

$$= \int_0^h \pi x^2 \tan^2 \alpha\, dx = \pi \tan^2 \alpha \int_0^h x^2\, dx$$

$$= \pi \tan^2 \alpha \left(\frac{x^3}{3}\right)_0^h = \frac{1}{3}\pi h^3 \tan^2 \alpha \qquad ...(1)$$

Again in right angled ΔOCA, $\tan \alpha = \dfrac{\text{AC}}{\text{OA}} = \dfrac{a}{h} \qquad ...(2)$

Putting this value of $\tan \alpha$ from (2) in (1),

Required volume of the cone $= \dfrac{1}{3}\pi h^3 \cdot \dfrac{a^2}{h^2} = \dfrac{1}{3}\pi a^2 h.$

Example 3. *Find the volumes formed by the revolution of the loop of the curve $y^2 (a + x) = x^2 (3a - x)$, about the x-axis.*

(K.U. 2004 S ; M.D.U. 2007)

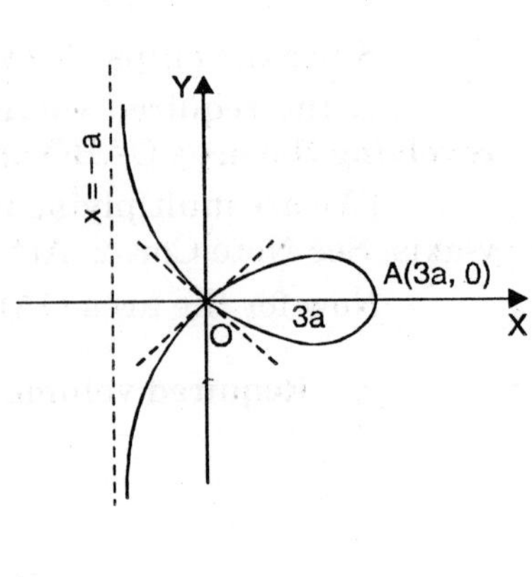

Sol. The equation of the curve is

$$y^2(a + x) = x^2(3a - x) \qquad ...(1)$$

(*i*) The curve is symmetrical about the x-axis.

(*ii*) The curves passes the origin and the tangents at the origin are $y = \pm \sqrt{3} \,.\, x$

$\therefore$ Origin is a node.

(*iii*) The curve meets the x-axis in points (0, 0) and $(3a, 0)$. It meets the y-axis at the origin only.

(*iv*) The curve has an asymptote $x + a = 0$, || to y-axis and there is no other asymptote.

(*v*) From (1), $y = x \cdot \sqrt{\dfrac{3a - x}{a + x}}$, when $x < -a$, or $x > 3a$, y is imaginary, $\therefore$ the entire curve lies between the lines $x = -a$ and $x = 3a$.

Thus, the shape of curve as shown in the figure and for the upper half of its loop x varies from 0 to $3a$.

Required volume formed by the revolution of loop

$$= \int_0^{3a} \pi y^2 \, dx = \pi \int_0^{3a} \frac{x^2 (3a - x)}{a + x} \, dx$$

$$= \pi \int_0^{3a} \frac{-x^3 + 3ax^2}{x + a} \, dx$$

$$= \pi \int_0^{3a} \left(-x^2 + 4ax - 4a^2 + \frac{4a^3}{x + a} \right) dx$$

[Dividing the Num. by the denominator]

$$= \pi \left[\frac{-x^3}{3} + 2ax^2 - 4a^2x + 4a^3 \log (x + a) \right]_0^{3a}$$

$$= \pi \, [-9a^3 + 18a^3 - 12a^3 + 4a^3 \log 4a - 4a^3 \log a]$$

$$= \pi a^3 (-3 + 4 \log 4) = \pi a^3 (8 \log 2 - 3).$$

Example 4. *A basin is formed by the revolution of the curve $x^3 = 64y$ $(y > 0)$ about the axis of y. If the depth of the basin is 8 inches, how many cubic inches of water will it hold ?*

Sol. The equation of the generating curve is $x^3 = 64y$ ...(1)

The curve is symmetrical in opposite quadrants.

The shape of the curve is as shown in the figure by thick lines.

The height of basin is given to be 8 inches, so that when $y = 8$, from (1),

$$x^3 = 64 \times 8 \quad \therefore \quad x = 8.$$

Hence A(8, 8) is point of the curve (1) at a height of 8 inches.

Thus, the basin is formed by the revolution of the arc OA about the y-axis where A is (8, 8).

$\therefore$ Required volume

$$= \int_0^8 \pi x^2 dy = \int_0^8 \pi (64\,y)^{2/3} \, dy$$

$$= 16\pi \int_0^8 y^{2/3} \, dy$$

$$= 16\pi \cdot \left[\frac{y^{5/3}}{\frac{5}{3}} \right]_0^8 = \frac{48}{5} \pi \; [(8)^{5/3} - 0]$$

$$= \frac{48\pi}{5} \times 32 = \frac{1536\pi}{5} \text{ cubic inches.}$$

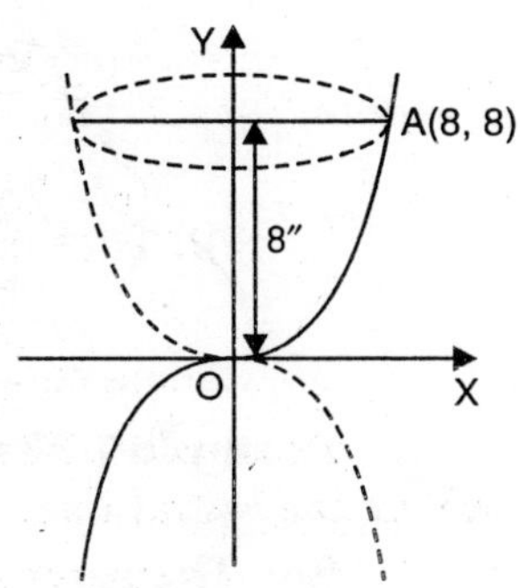

Example 5. *Find the volume generated by revolving one arch of the cycloid* $x = a(\theta - \sin\theta)$, $y = a(1 - \cos\theta)$ *about its base.* (K.U. 2006, 2006 S)

Sol. The equations of the cycloid are $x = a(\theta - \sin\theta)$, $y = a(1 - \cos\theta)$.

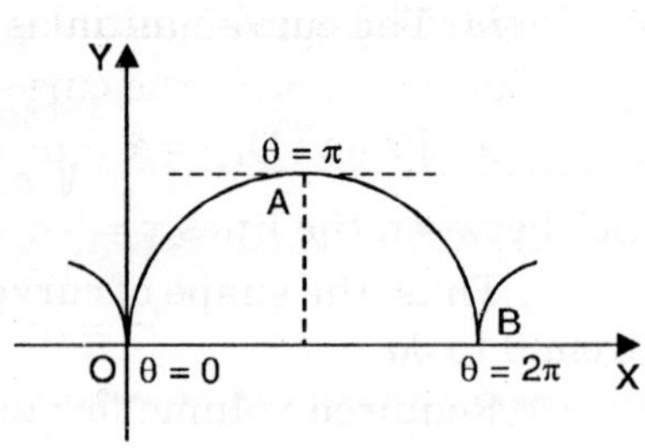

For the first half of the cycloid in the first quadrant, θ varies from 0 to π.

∴ By Cor 2, Art. 2 required volume

$$= 2\int_{\theta=0}^{\pi} \pi y^2\, dx = 2\int_0^{\pi} \pi y^2 \frac{dx}{d\theta}\, d\theta$$

$$= 2\pi \int_0^{\pi} a^2 (1 - \cos\theta)^2 \,.\, a(1 - \cos\theta)\, d\theta$$

$$= 2\pi a^3 \int_0^{\pi} (1 - \cos\theta)^3\, d\theta = 2\theta a^3 \int_0^{\pi} (2\sin^2 \theta/2)^3\, d\theta \qquad ...(1)$$

Put $\frac{\theta}{2} = t$ ∴ $\theta = 2t$ and $d\theta = 2dt$

When $\theta = 0, t = 0$

When $\theta = \pi, t = \frac{\pi}{2}$

∴ Using (1)

$$\text{Required volume} = 2\pi a^3 \int_0^{\pi/3} (2\sin^2 t)^3\, 2dt$$

$$= 32\,\pi a^3 \int_0^{\pi/2} \sin^6 t\, dt = 32\,\pi a^3 \,.\, \frac{5\,.\,3\,.\,1}{6\,.\,4\,.\,2} \,.\, \frac{\pi}{2} = 5\pi^2 a^3.$$

Example 6. *Find the volume of the solid obtained by revolving the cardioid* $r = a(1 - \cos\theta)$ *about the initial line.*

Sol. The equation of the curve is $r = a(1 - \cos\theta)$.

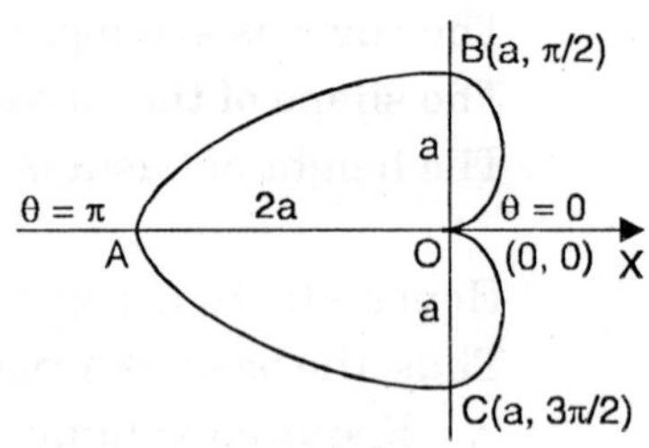

The curve is symmetrical about the initial line and for its upper half θ varies from 0 to π.

∴ By Art. 3, required volume

$$= \int_0^{\pi} \frac{2}{3} \pi \mathbf{r}^3 \sin\theta\, \mathbf{d}\theta$$

$$= \frac{2}{3}\pi \int_0^{\pi} a^3 (1 - \cos\theta)^3 \sin\theta\, d\theta$$

$$= \frac{2\pi a^3}{3} \left[\frac{(1 - \cos\theta)^4}{4}\right]_0^{\pi} \qquad \Bigg|\; \because \int (f(x))^n f'(x)\, dx = \frac{(f(x))^{n+1}}{n+1} \text{ if } n \neq -1$$

$$= \frac{2}{12} \pi a^3 [(1 - \cos\pi)^4 - (1 - \cos 0)^4]$$

$$= \frac{2}{12} \pi a^3 [2^4 - 0] = \frac{8}{3} \pi a^3.$$

Example 7. *Find the volume of the solid obtained by revolving the lemniscate* $r^2 = a^2 \cos 2\theta$ *about the initial line.*

Sol. The curve $r^2 = a^2 \cos 2\theta$ is symmetrical about the initial line.

The curve consists of two equal loops.

The required volume, therefore is twice the volume generated by the revolution of the half-loop OA about the initial line.

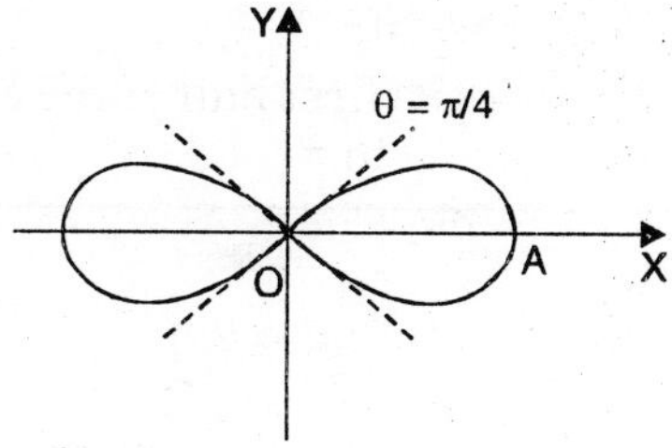

The equation of the curve is

$$r^2 = a^2 \cos 2\theta$$

or $$r^2 = a^2 (\cos^2 \theta - \sin^2 \theta)$$

or $$r^2 = a^2 \left(\frac{x^2}{r^2} - \frac{y^2}{r^2} \right)$$

i.e., $$r^4 = a^2 (x^2 - y^2)$$

or $$(x^2 + y^2)^2 = a^2 (x^2 - y^2)$$

or $$y^4 + (2x^2 + a^2) y^2 + (x^4 - a^2 x^2) = 0$$

It is a quadratic in y^2

Solving for y^2,

$$y^2 = \frac{-(2x^2 + a^2) \pm \sqrt{(2x^2 + a^2)^2 - 4(x^4 - a^2x^2)}}{2}$$

or $$y^2 = \tfrac{1}{2} [-(2x^2 + a^2) \pm \sqrt{8a^2x^4 + a^4}]$$

or $$y^2 = \tfrac{1}{2} [-(2x^2 + a^2) + \sqrt{8a^2x^2 + a^4}] \qquad \text{...(1)}$$

Negative sign before the radical on the right is rejected, because otherwise y^2 is negative and hence y is imaginary.

For the half-loop OA, x varies from 0 to a.

∴ Required volume

$$= 2 \int_0^a \pi y^2 \, dx$$

$$= \pi \int_0^a \left[-(2x^2 + a^2) + 2\sqrt{2}a \sqrt{x^2 + \frac{a^2}{8}} \, dx \right] \qquad \text{[From (1)]}$$

$$= \pi \left[-\frac{2}{3}x^3 - a^2x + 2a\sqrt{2} \left\{ \frac{x\sqrt{x^2 + \frac{a^2}{8}}}{2} + \frac{\frac{a^2}{8}}{2} \sinh^{-1} \frac{x}{\left(\frac{a}{2\sqrt{2}}\right)} \right\} \right]_0^a$$

$$= \pi \left[-\frac{2}{3}a^3 - a^3 + 2a\sqrt{2} \left\{ \frac{a \cdot 3a}{2 \cdot 2\sqrt{2}} + \frac{a^2}{16} \sinh^{-1} 2\sqrt{2} \right\} \right]$$

$$= \pi \left[-\frac{2}{3}a^3 - a^3 + \frac{3a^3}{2} + \frac{\sqrt{2}a^3}{8} \log (2\sqrt{2} + 3) \right]$$

$$\left[\because \sinh^{-1} x = \log (x + \sqrt{x^2 + 1}) \right]$$

$$= \pi \left[-\frac{a^3}{6} + \frac{\sqrt{2}a^3}{8} \log (\sqrt{2} + 1)^2 \right]$$

$$= \frac{\pi a^3}{2}\left[\frac{-1}{3} + \frac{\sqrt{2}}{2}\log(\sqrt{2}+1)\right]$$

$$= \frac{\pi a^3}{2}\left[\frac{1}{\sqrt{2}}\log(\sqrt{2}+1) - \frac{1}{3}\right].$$

EXERCISE 1

1. Find the volume generated by the revolution of an arc of the catenary $y = c\cosh\frac{x}{c}$ about the axis of x.
2. Find the volume generated by rotating about the y-axis the area bounded by the coordinate axes and the graph of the curve $y = \cos x$ from $x = 0$ to $x = \pi/2$.

   $\left[\textbf{Hint.} \text{ Volume} = \int \pi x^2\, dy = \int_0^{\pi/2} \pi x^2 \frac{dy}{dx}\, dx.\right]$
3. Let B be a number > 1. What is the volume of the solid generated by the area under the curve $y = e^{-x}$ between 1 and B (the axis of revolution being the x-axis) ? Does the volume approach a limit as B becomes large ? If so, what limit ?
4. The area of the parabola $y^2 = 4ax$ lying between the vertex and the latus rectum is revolved about the x-axis. Find the volume generated.

   Or

   Find the volume of the spindle formed by the revolution of a parabolic arc about the line joining the vertex to one extremity of the latus rectum.

   [**Hint.** We know that equation of latus rectum of the parabola $y^2 = 4ax$ is $x = a$.

   $\therefore$ Required volume $= \int_0^a \pi y^2\, dx = \int_0^a \pi \cdot 4ax\, dx = 4a\pi\int_0^a x\, dx = 4a\pi\left(\frac{x^2}{2}\right)_0^a = 2\pi a^2.$]
5. The part of the parabola $y^2 = 4ax$ cut off by the latus rectum revolves about the tangent at the vertex. Find the volume of the reel thus formed.

   [**Hint.** Tangent at the vertex is y-axis.]
6. Find the volume of a sphere of radius a. *(M.D.U. 2006)*

   [**Hint.** See Art. 1 example (c) page 352. The equation of circle is $x^2 + y^2 = a^2$.

   Volume of sphere $= 2\int_0^a \pi y^2\, dx = 2\int_0^a \pi(a^2 - x^2)\, dx.$]
7. (*i*) Find the volume of the oblate spheroid formed by revolving the ellipse about minor axis.

   (*ii*) Find the volume of the solid generated by revolving the ellipse. $x^2 + 4y^2 = 4$ about y-axis.
8. Prove that the volume of the solid generated by the revolution of an ellipse round its minor axis, is a mean proportional (*i.e.* Geometric Mean) between those generated by the revolution of the ellipse and the auxiliary circle round the major axis.

   [**Hint.** Equation of auxiliary circle (a circle drawn on major axis as diameter) of the ellipse

   $\frac{x^2}{a^2} + \frac{y^2}{b^2} = 1$ is $x^2 + y^2 = a^2$.]

9. The loop of the curve $2ay^2 = x(x-a)^2$ revolves about x-axis, find the volume of the solid so generated. *(M.D.U. 2005)*

10. Find the volume of the solid obtained by revolving the loop of the curve $a^2y^2 = x^2(2a - x)(x - a)$ about x-axis.

11. (*a*) Find the volume formed by revolution of the loop of the curve $y^2 = \dfrac{x^2(a-x)}{a+x}$ about the x-axis.

[**Hint.** See Example 3, Page 312]

(*b*) Find the volume of the solid produced by the revolution of the loop of the curve $y^2 = \dfrac{x^2(a+x)}{a-x}$ about the axis of x.

12. Find the volume of the solid generated by the revolution of the area between the curve $xy^2 = 4a^2(2a - x)$ and its asymptote about the asymptote.

13. (*a*) Find the volume of the solid formed by the revolution of the cycloid $x = a(\theta + \sin\theta)$, $y = a(1 + \cos\theta)$, about its base. *(M.D.U. 2004)*

[**Hint.** Base of the cycloid is x-axis.]

(*b*) Find the volume of the solid formed by revolution of the cycloid $x = a(\theta + \sin\theta)$, $y = a(1 - \cos\theta)$ about the tangent at the vertex. *(M.D.U. 2005, 2007)*

[**Hint.** Tangent at the vertex is x-axis.]

14. Find the volume of the solid generated by revolving the curve $x^{2/3} + y^{2/3} = a^{2/3}$ (*i.e.* $x = a\cos^3 t$, $y = a\sin^3 t$) about x-axis.

15. Prove that the volume of the solid generated by the revolution of the tractrix $x = a\cos t + \dfrac{a}{2}\log\tan^2\dfrac{t}{2}$, $y = a\sin t$ about its asymptote equals half that of a sphere of radius a.

16. Find the volume of the solid generated by revolving the loop of the curve $x = t^2$, $y = t - \frac{1}{3}t^3$ about x-axis.

17. Find the volume of the solid generated by revolution of $r = 2a\cos\theta$ about the initial line.

18. The cardioid $r = a(1 + \cos\theta)$ revolves about the initial line. Find the volume of the solid generated. *(K.U. 2003, 2007)*

19. The arc of the cardioid $r = a(1 + \cos\theta)$ specified by $-\dfrac{\pi}{2} \le \theta \le \dfrac{\pi}{2}$ is rotated about the line $\theta = 0$, prove that the volume generated is $\dfrac{5}{2}\pi a^3$.

20. Show that the volume of the solid formed by the revolution of the curve $r = a + b\cos\theta$ $(a > b)$ about the initial line is $\dfrac{4}{3}\pi a(a^2 + b^2)$. *(K.U. 2005 S ; M.D.U. 2006)*

21. Find the volume of the solid generated by revolving the curve $r^2 = a^2\cos 2\theta$ about the line $\theta = \pi/2$. *(M.D.U. 2004 S)*

[**Hint.** $V = 2\displaystyle\int_0^{\pi/4} \frac{2}{3}\pi r^3 \cos\theta\, d\theta$, where $r = \sqrt{a^2\cos 2\theta} = a\sqrt{1 - 2\sin^2\theta}$.

Now to integrate, put $\sqrt{2}\sin\theta = \sin\phi$.]

ANSWERS

1. $\left[\dfrac{\pi c^2}{4}\left(2x + c\sinh\dfrac{2x}{c}\right)\right]_a^b$ **2.** $\pi(\pi - 2)$ **3.** $-\dfrac{\pi}{2}[e^{-2B} - e^{-2}]$, $\dfrac{\pi}{2e^2}$

4. $2\pi a^3$ 5. $\frac{4}{5}\pi a^3$ 6. $\frac{4}{3}\pi a^3$

7. (i) $\frac{4}{3}\pi a^2 b$ (ii) $\frac{16\pi}{3}$ 9. $\frac{\pi a^3}{24}$

10. $\frac{23\pi a^3}{60}$ 11. (a) $2\pi a^3\left[\log 2 - \frac{2}{3}\right]$ (b) $2\pi a^3\left[\log 2 - \frac{2}{3}\right]$

12. $4\pi^2 a^3$ 13. (a) $5\pi^2 a^3$ (b) $\pi^2 a^3$

14. $\frac{32}{105}\pi a^3$ 16. $\frac{3\pi}{4}$ 17. $\frac{4}{3}\pi a^3$

18. $\frac{8}{3}\pi a^3$ 21. $\frac{\pi^2 a^3}{4\sqrt{2}}$.

---

4. Surfaces of the Solids of Revolution

The area of the surface of the solid generated by the revolution about x-axis of the area bounded by the curve $y = f(x)$, the x-axis and the ordinates

$x = a$, $x = b$, is $\int_{x=a}^{x=b} 2\pi y \, ds$

where s is the length of the arc of curve measured from a fixed point on it to any point (x, y).

Let AB be the curve $y = f(x)$ and CA, BD the ordinates $x = a$, $x = b$. Let P(x, y) be *any* point on the curve and let the arc AP be s. Draw PM ⊥ OX.

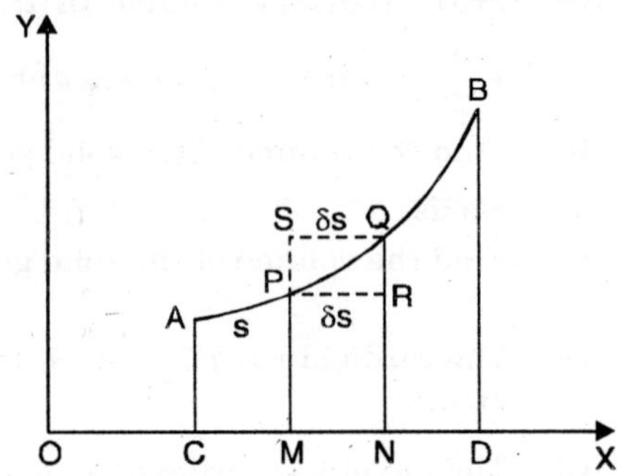

If S denotes the curved surface of the solid generated by the revolution of the area ACMP about the x-axis, then clearly S is a function of s.

Let Q$(x + \delta x, y + \delta y)$ be a point on the curve in the neighbourhood of P. Draw QN ⊥ OX, and let the arc AQ be $s + \delta s$, so that arc PQ $= \delta s$.

∴ The curved surface of the solid generated by the revolution of the area PMNQ about the x-axis is δS. Draw PR and QS parallel to x-axis and each equal in length to the arc PQ *i.e.*, δs.

Then it may be taken as an axiom that the curved surface of the solid generated by the revolution of the area PMNQ lies between the curved surfaces of the right circular cylinders whose base radii are MP, NQ and heights are PR, QS.

∴ δS lies between $2\pi y\, \delta s$ and $2\pi(y + \delta y)\, \delta s$

or $$2\pi y\, \delta s < \delta S < 2\pi\, (y + \delta y)\, \delta s \quad \text{or} \quad 2\pi y < \frac{\delta S}{\delta s} < 2\pi\, (y + \delta y).$$

Since $y = f(x)$ is continuous, ∴ as $\delta x \to 0$, δy also $\to 0$.

∴ Proceeding to the limit as $\delta x \to 0$ (∴ $\delta y \to 0$),

$$\frac{dS}{ds} = 2\pi y \qquad \text{...(1)}$$

∴ $$\int_{x=a}^{x=b} 2\pi y\, ds = \int_{x=a}^{x=b} \frac{dS}{ds}\, ds = [S]_{x=a}^{x=b}$$

= (value of S when $x = b$) – (value of S when $x = a$)

= curved surface of the solid generated by the revolution of the area ABCD – 0.

$\therefore$ Surface area of the solid generated by the revolution of the area ACDB

$$= \int_{x=a}^{x=b} 2\pi y \, ds.$$

Note 1. The above result has been proved for the case when the ordinate increases continuously. It is equally true in case, where the ordinate decreases continuously. It is also assumed that the curve does not cross the x-axis *i.e.*, the axis of revolution.

Cor. From (1), $S = \int 2\pi y \, ds$.

Practical forms of the surface formula

(*i*) **Cartesian form [for the curve y = f(x)]**

$$S = \int 2\pi y \frac{ds}{dx} dx, \text{ where } \frac{ds}{dx} = \sqrt{1 + \left(\frac{dy}{dx}\right)^2}.$$

(*ii*) **Parametric form [for the curve x = f(t), y = F(t)]**

$$S = \int 2\pi y \frac{ds}{dt} dt, \text{ where } \frac{ds}{dt} = \sqrt{\left(\frac{dx}{dt}\right)^2 + \left(\frac{dy}{dt}\right)^2}.$$

(*iii*) **Polar form [for the curve r = f(θ)]**

$$S = \int 2\pi y \frac{ds}{d\theta} d\theta, \text{ where}$$

$$y = r \sin\theta \text{ and } \frac{ds}{d\theta} = \sqrt{r^2 + \left(\frac{dr}{d\theta}\right)^2}.$$

Note 2. *Interchanging* x and y in the above formula, we see that the curved surface of the solid generated by the revolution about y-axis of the area bounded by the curve $x = f(y)$, y-axis, and abscissae $y = c$, $y = d$ is $\int_{y=c}^{y=d} 2\pi x \, ds$.

Example 1. *Find the area of the surface formed by the revolution of* $y^2 = 4ax$, *about the* x-*axis, by the arc from the vertex to one end of latus rectum.* (M.D.U. 2005, 2007 ; K.U. 2006)

Sol. The equation of the parabola is $y^2 = 4ax$.

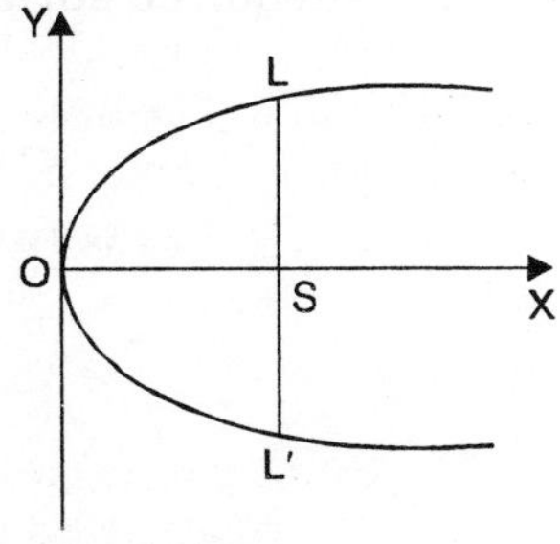

Differentiating, we get $2y\frac{dy}{dx} = 4a$ or $\frac{dy}{dx} = \frac{2a}{y}$

$$\therefore \quad \frac{ds}{dx} = \sqrt{1 + \left(\frac{dy}{dx}\right)^2} = \sqrt{1 + \frac{4a^2}{y^2}}$$

$$= \sqrt{1 + \frac{4a^2}{4ax}} \qquad (\because \; y^2 = 4ax)$$

$$= \sqrt{\frac{x+a}{x}}$$

For the arc from the vertex O to L, the end of the latus rectum x varies from 0 to a.

$$\therefore \quad \text{Required surface} = \int_0^a 2\pi y \frac{ds}{dx} dx \qquad [\textit{Practical form (i)}]$$

$$= 2\pi \int_0^a \sqrt{4ax} \cdot \sqrt{\frac{x+a}{x}}\, dx = 4\pi \sqrt{a} \int_0^a (x+a)^{1/2}\, dx$$

$$= 4\pi \sqrt{a} \left| \frac{(x+a)^{3/2}}{3/2} \right|_0^a = \frac{8\pi}{3} \sqrt{a}\, [(2a)^{3/2} - a^{3/2}]$$

$$= \frac{8}{3} \pi a^2 (2\sqrt{2} - 1).$$

Example 2. *Prove that the surface generated by revolution of the tractrix $x = a \cos t + \frac{a}{2} \log \tan^2 \frac{t}{2}$, $y = a \sin t$ about its asymptote is equal to the surface of a sphere of radius a.*

(K.U. 2007)

Sol. The equations of the curve are

$$\left. \begin{aligned} x &= a \cos t + \frac{a}{2} \log \tan^2 \frac{t}{2} = a \cos t + \log \tan^2 \frac{t}{2} \\ y &= a \sin t. \end{aligned} \right\} \quad ...(1)$$

The curve (1) is symmetrical about both the axes and its asymptote is $y = 0$, *i.e.*, x-axis.

From (1),
$$\frac{dx}{dt} = -a \sin t + a \cdot \frac{1}{\tan \frac{t}{2}} \sec^2 \frac{t}{2} \cdot \frac{1}{2} = -a \sin t + \frac{a}{2 \sin \frac{t}{2} \cos \frac{t}{2}}$$

$$= -a \sin t + \frac{a}{\sin t} = a \frac{(1 - \sin^2 t)}{\sin t} = a \frac{\cos^2 t}{\sin t}$$

and
$$\frac{dy}{dt} = a \cos t$$

$\therefore$
$$\frac{ds}{dt} = \sqrt{\left(\frac{dx}{dt}\right)^2 + \left(\frac{dy}{dt}\right)^2} = \sqrt{\frac{a^2 \cos^4 t}{\sin^2 t} + a^2 \cos^2 t}$$

$$= \sqrt{\frac{a^2 \cos^2 t (\cos^2 t + \sin^2 t)}{\sin^2 t}} = \frac{a \cos t}{\sin t}$$

For the curve in the second quadrant t varies from 0 to $\pi/2$.

$\therefore$ Required surface $= 2 \int_0^{\pi/2} 2\pi y \frac{ds}{dt}\, dt$ [*Practical Form (ii)*]

$$= 4\pi \int_0^{\pi/2} a \sin t \cdot \frac{a \cos t}{\sin t}\, dt$$

$$= 4\pi a^2 \int_0^{\pi/2} \cos t\, dt$$

$$= 4\pi a^2 [\sin t]_0^{\pi/2}$$

$$= 4\pi a^2 (1 - 0) = 4\pi a^2$$

= Surface of a sphere of radius a.

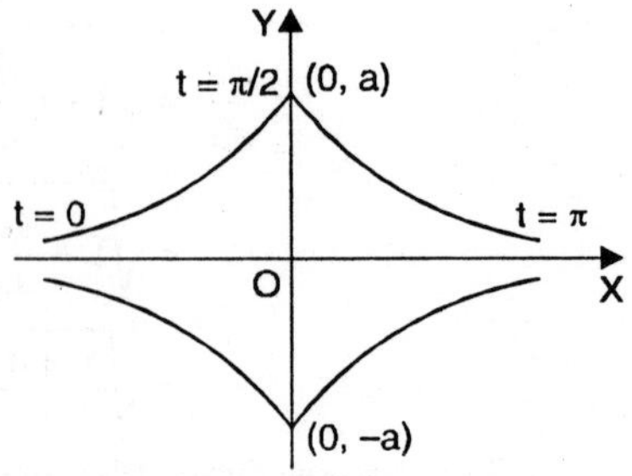

Example 3. *The arc of the cardioid $r = a(1 + \cos \theta)$ included between $-\pi/2 \le \theta \le \pi/2$ is rotated about line $\theta = \pi/2$. Find the area of the surface generated.*

Sol. Equation of cardioid is $r = a\,(1 + \cos \theta)$ ...(1)

The area OCAB (shown shaded in the figure) revolves about the line $\theta = \pi/2$ *i.e.*, y-axis. Also the curve is symmetrical about the initial line or x- axis.

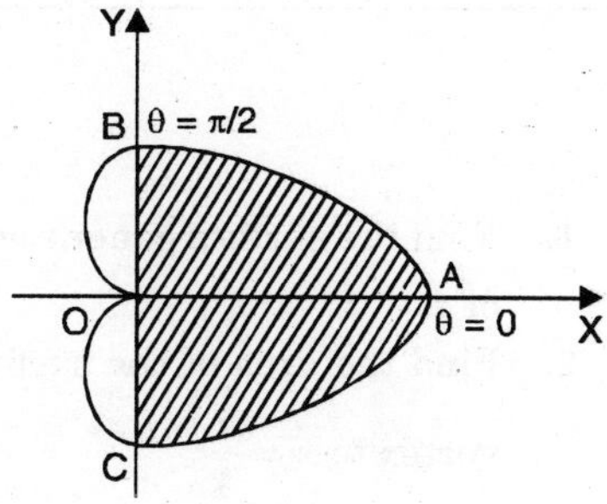

$\therefore$ Required surface area

$= 2 \times$ surface generated by revolution of arc AB about y-axis.

$$= 2 \int_0^{\pi/2} 2\pi x \frac{ds}{d\theta}\, d\theta \qquad \ldots(2)$$

(See Note 2, Art. 4)

From (1), $\dfrac{dr}{d\theta} = -a \sin\theta$

$$\therefore \quad \frac{ds}{d\theta} = \sqrt{r^2 + \left(\frac{dr}{d\theta}\right)^2} = \sqrt{a^2 (1+\cos\theta)^2 + a^2 \sin^2\theta}$$

$$= a\sqrt{2(1+\cos\theta)} = a \,.\, 2\cos\theta/2 = 2a\cos\theta/2 \quad \text{and} \quad x = r\cos\theta.$$

Now from (2), we have required surface area $= 4\pi \displaystyle\int_0^{\pi/2} x \frac{ds}{d\theta}\, d\theta$

$$= 4\pi \int_0^{\pi/2} r\cos\theta \,.\, 2a\cos\theta/2 \, d\theta$$

$$= 4\pi \int_0^{\pi/2} a(1+\cos\theta)\,.\,\cos\theta\,.\,2a\cos\theta/2\, d\theta$$

$$= 8\pi a^2 \int_0^{\pi/2} (1+\cos\theta)\cos\theta\,.\,\cos\theta/2\, d\theta$$

$$= 8\pi^2 \int_0^{\pi/2} (2 - 2\sin^2\theta/2)(1 - 2\sin^2\theta/2)\,.\,\cos\theta/2\, d\theta$$

$$\left(\because \cos\theta = 1 - 2\sin^2\frac{\theta}{2}\right)$$

$$= 16\pi a^2 \int_0^{\pi/2} (1 - \sin^2\theta/2)(1 - 2\sin^2\theta/2)\,.\,\cos\theta/2\, d\theta$$

$$= 16\pi a^2 \int_0^{\pi/2} [1 - 3\sin^2\theta/2 + 2\sin^4\theta/2] \cos\theta/2\, d\theta$$

$$\left[\text{Put } \sin\frac{\theta}{2} = t, \text{ so that } \frac{1}{2}\cos\frac{\theta}{2}\, d\theta = dt \text{ and when } \theta = 0, t = 0\ ;\ \text{when } \theta = \frac{\pi}{2}; t = \frac{1}{\sqrt{2}}\right]$$

$$= 16\pi a^2 \int_0^{1/\sqrt{2}} (1 - 3t^2 + 2t^4)\, 2dt = 32\pi a^2 \left[t - t^3 + \frac{2t^5}{5}\right]_0^{1/\sqrt{2}}$$

$$= 32\pi a^2 \left[\frac{1}{\sqrt{2}} - \frac{1}{2\sqrt{2}} + \frac{2}{5}\cdot\frac{1}{4\sqrt{2}}\right]$$

$$= 32\pi a^2 \left[\frac{20 - 10 + 2}{20\sqrt{2}}\right] = 32\pi a^2 \times \frac{12}{20\sqrt{2}} = \frac{96\pi a^2}{5\sqrt{2}}.$$

EXERCISE 2

1. Find the surface generated by the revolution of an arc of the catenary $y = c \cosh \frac{x}{c}$ about the axis of x. (*M.D.U. 2003*)
2. Find the area of the surface formed by revolution of $y^2 = 4ax$ about y-axis by the arc from the vertex to $x = \frac{a}{4}$. (*M.D.U. 1995*)
3. Find the surface of a sphere of radius a.

   Or

   The circle $x^2 + y^2 = a^2$ is revolved about x-axis. Find the area of the sphere generated.
4. Find the area of the surface generated by rotating about x-axis the arc of the curve $y = x^3$ between $x = 0$ and $x = 1$.
5. Find the surface of the right circular cone formed by the revolution of a right-angled triangle about a side which contains the right angle. (*M.D.U. 2004, 2006*)
6. Find the surface of the solid generated by the revolution of the astroid $x^{2/3} + y^{2/3} = a^{2/3}$ or $x = a \cos^3 t$, $y = a \sin^3 t$ about the x-axis. (*K.U. 2005 ; M.D.U. 2007*)
7. Find the surface area of the solid generated by revolving the cycloid :

   (*a*) $x = a(\theta - \sin \theta)$, $y = a(1 - \cos \theta)$ about the x-axis. (*K.U. 2005 S*)

   (*b*) $x = a(\theta + \sin \theta)$, $y = a(1 - \cos \theta)$ about the tangent at the vertex. (*M.D.U. 2006*)

   [**Hint.** Tangent at the vertex is x-axis.]

   (*c*) Find the area of the curved surface generated by the revolution of the cycloid $x = a(\theta + \sin \theta)$, $y = a(1 - \cos \theta)$ about its base. (*K.U. 1998 A*)
8. Find the surface of the solid generated by revolving the loop of the curve $x = t^2$, $y = t - \frac{1}{3}t^3$ about x-axis.
9. Find the area of the surface of revolution formed by revolving the curve $r = 2a \cos \theta$ about the initial line.
10. (*a*) The curve $r = a(1 + \cos \theta)$ revolves about the initial line. Find the surface of the figure so formed. (*M.D.U. 2004 S ; K.U. 1998 S*)

    (*b*) Find the surface of the solid generated by the revolution of the cardioid $r = a(1 - \cos \theta)$ about the initial line. (*K.U. 1997 S, 2004 S, 2006 S*)
11. Find the surface of the solid generated by the revolution of the leminscate $r^2 = a^2 \cos 2\theta$ about the initial line.
12. The part of the parabola $y^2 = 4ax$ cut off by the latus rectum revolves about the tangent at the vertex. Find the curved surface of the reel thus generated.
13. Prove that the surface of the solid obtained by revolving the ellipse $b^2x^2 + a^2y^2 = a^2b^2$ about x-axis is $2\pi ab \left[\sqrt{1 - e^2} + \frac{1}{e} \sin^{-1} e\right]$, where e is the eccentricity of the ellipse. (*K.U. 2003*)
14. Show that the surface of the spherical zone contained between two parallel planes is $2\pi ah$, where a is the radius of the sphere and h is the distance between parallel planes. (*M.D.U. 2003 S*)
15. A quadrant of a circle of radius a revolves round its chord. Find the area of the surface of the spindle generated.

ANSWERS

1. $\pi c\left[b-a+\frac{c}{2}\sinh\frac{2b}{c}-\frac{c}{2}\sinh\frac{2a}{c}\right]$ 3. $4\pi a^2$ 4. $\frac{\pi}{27}(10\sqrt{10}-1)$

5. πrl 6. $\frac{12}{5}\pi a^2$ 7. (a) $\frac{64}{3}\pi a^2$ (b) $\frac{32}{3}\pi a^2$ (c) $\frac{32}{3}\pi a^2$

8. 3π 9. $4\pi a^2$ 10. (a) $\frac{32}{5}\pi a^2$ (b) $\frac{32}{5}\pi a^2$

11. $4\pi a^2\left(1-\frac{1}{\sqrt{2}}\right)$ 12. $\pi a^2[3\sqrt{2}-\log(\sqrt{2}+1)]$.

---

5. Revolution about any axis

The **volume** *of the solid generated by the revolution, about any axis CD, of the area bounded by the curve AB, the axis CD and the perpendiculars AC, BD, on the axis is*

$$\int_{OC}^{OD}\pi\,(PM)^2\,d(OM)$$

where O is a fixed point on the axis CD, and PM is perpendicular from any point P of the curve AB on the axis CD.

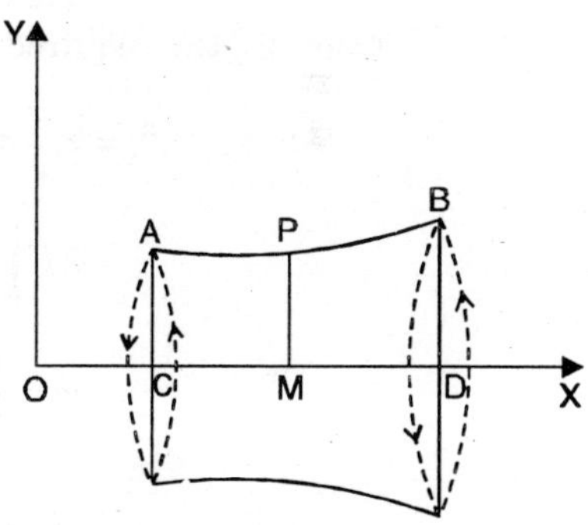

Take the fixed point O on CD as origin ; and OCD the axis of revolution as the x-axis, and OY, perpendicular to it as the y-axis.

Let the co-ordinates of P be (x, y), so that OM $= x$, PM $= y$.

If OC $= a$ and OD $= b$, then the required volume

$$=\int_a^b \pi y^2\,dx=\int_{OC}^{OD}\pi(PM)^2\,d(OM).$$

Example 1. *Show that the volume of the solid generated by the revolution of the curve $(a-x)y^2=a^2x$ about its asymptote is $\frac{1}{2}\pi^2a^3$.*

Sol. The equation of the curve is $(a-x)y^2=a^2x$ ...(1)

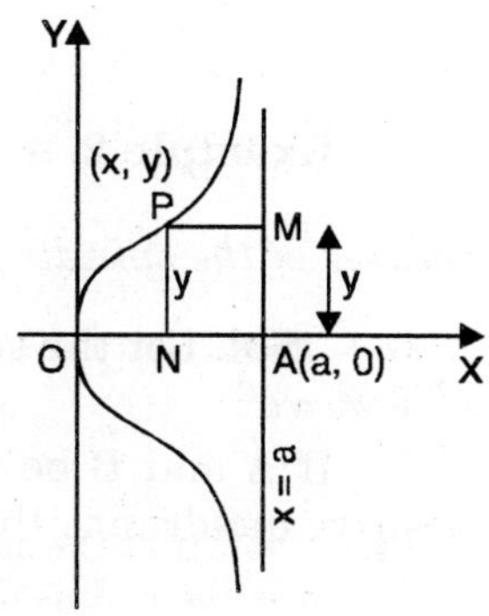

The curve is symmetrical about the x-axis and by equating to zero the co-efficient of the highest power of y the asymptote parallel to y-axis is

$$a-x=0 \text{ or } x=a$$

Let P(x, y) be any point on the curve and PM, the perpendicular on the asymptote (the axis of revolution), the PM $= a-x$, and AM $= y$, ON $= x$; A being the point of intersection of the asymptote with the x-axis.

$\therefore$ Required volume $= 2\int\pi(PM)^2\,d(AM)$

$$=2\int_0^\infty\pi(a-x)^2\,dy \qquad \left[\because \text{ for upper half of curve } y \text{ varies from } 0 \text{ to } \infty\right] \quad ...(2)$$

$$=2\pi\int_0^\infty\left[a-\frac{ay^2}{a^2+y^2}\right]^4 dy \qquad \left[\because \text{ from (1) } x(a^2+y^2)=ay^2 \text{ so that } x=\frac{ay^2}{a^2+y^2}\right]$$

$$= 2\pi \int_0^{\infty} \frac{a^6 dy}{(y^2 + a^2)^2}$$

$\left\{\begin{array}{l}\text{Put } y = a \tan \theta \; \therefore dy = a \sec^2 \theta \, d\theta \\ \text{when } v = 0, \theta = 0 \text{ and when } y = \infty, \\ \theta = \pi/2.\end{array}\right\}$

$$= 2\pi \int_0^{\pi/2} \frac{a^6 . a \sec^2 \theta \, d\theta}{a^4 \sec^4 \theta}$$

$$= 2\pi a^3 \int_0^{\pi/2} \cos^2 \theta \, d\theta = 2\pi a^3 . \frac{1}{2} . \frac{\pi}{2} = \frac{1}{2} \pi^2 a^3.$$

Note. From (2), we can also proceed as follows :

We have AM $= y = \dfrac{a\sqrt{x}}{\sqrt{a - x}}$

$$\therefore \quad d(\text{AM}) = dy = a . \frac{\sqrt{a-x} . \dfrac{1}{2\sqrt{x}} - \sqrt{x} . \dfrac{1}{2\sqrt{a-x}} (-1)}{a - x} dx$$

$$= \frac{a^2 \, dx}{2\sqrt{x} . (a - x)^{3/2}}$$

$\therefore$ From (2), the required volume

$$= 2 \int \pi \, (\text{PM})^2 \, d(\text{AM}) = 2\pi \int_0^a (a - x)^2 \, dy$$

$$= 2\pi \int_0^a (a - x)^2 . \frac{a^2 \, dx}{2\sqrt{x} . (a - x)^{3/2}}$$

$$= \pi a^2 \int_0^a \frac{\sqrt{a - x}}{\sqrt{x}} dx$$

$\left[\begin{array}{l}\text{Put } x = a \sin^2 \theta \\ \therefore \; dx = 2a \sin \theta \cos \theta \, d\theta \\ \text{When } x = 0, \theta = 0, \text{ and when } x = a, \theta = \dfrac{\pi}{2}\end{array}\right]$

$$= \pi a^2 \int_0^{\pi/2} \frac{\sqrt{a \cos^2 \theta}}{\sqrt{a \sin^2 \theta}} . 2a \sin \theta \cos \theta \, d\theta$$

$$= 2\pi a^3 \int_0^{\pi/2} \cos^2 \theta \, d\theta = 2\pi a^3 . \frac{1}{2} . \frac{\pi}{2} = \frac{1}{2} \pi^2 a^3.$$

Example 2. *A quadrant of a circle of radius a, revolves about its chord. Show that the volume of the spindle generated is* $\dfrac{\pi}{6\sqrt{2}} (10 - 3\pi) a^3$.

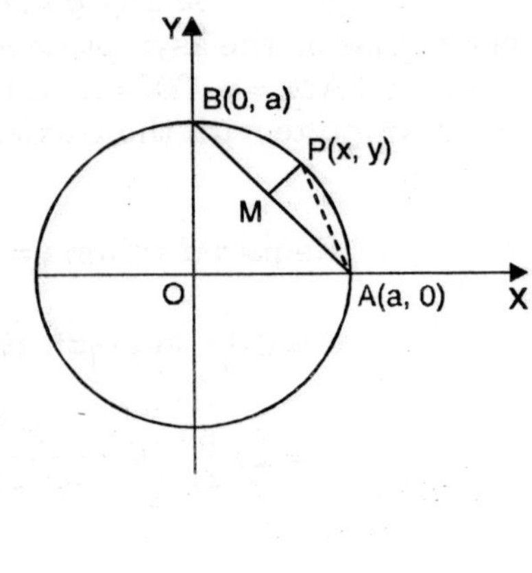

Sol. Let the equation of the generating circle be

$$x^2 + y^2 = a^2 \quad ...(1)$$

If A and B be the extremities of the arc (in the positive quadrant), then the equation of line AB is

$$y - 0 = \frac{0 - a}{a - 0} (x - a)$$

or $\quad x + y - a = 0.$

If P(x, y) be any point on the arc AB. Draw PM $\perp$ on chord AB and join AP.

Then $\quad \text{PM} = \dfrac{x+y-a}{\sqrt{2}} = \dfrac{x-a+\sqrt{a^2-x^2}}{\sqrt{2}}$ **[From (1)]**

and $\quad \text{PM}^2 = \dfrac{x^2+a^2-2ax+a^2-x^2+2(x-a)\sqrt{a^2-x^2}}{2}$

$$= (a-x)\,(a-\sqrt{a^2-x^2})$$

Now $\quad \text{AM}^2 = \text{AP}^2 - \text{PM}^2$

$$= (x-a^2) + y^2 - \frac{(x+y-a)^2}{2}$$

$$= \frac{1}{2}\,[2(x-a)^2 + 2y^2 - (x-a)^2 - y^2 - 2y(x-a)]$$

$$= \frac{1}{2}\,[(x-a)^2 + y^2 - 2y(x-a)] = \frac{1}{2}\,(x-a-y)^2$$

$\therefore \quad \text{AM} = \dfrac{1}{\sqrt{2}}\,(x-y-a) = \dfrac{1}{\sqrt{2}}\,(x-a-\sqrt{a^2-x^2})$ [From (1)]

$$\therefore \quad d(\text{AM}) = \frac{1}{\sqrt{2}}\left[1+\frac{x}{\sqrt{a^2-x^2}}\right]dx = \frac{\sqrt{a^2-x^2}+x}{\sqrt{2}\,\sqrt{a^2-x^2}}\,dx$$

Also for the arc AB, x varies from 0 to a.

$\therefore$ **Required volume** $= \displaystyle\int \pi\,(\text{PM})^2\,d(\text{AM})$

$$= \int_0^a \pi(a-x)(a-\sqrt{a^2-x^2})\,.\,\frac{\sqrt{a^2-x^2}+x}{\sqrt{2}\,\sqrt{a^2-x^2}}\,dx$$

[Put $x = a\sin\theta$, so that $dx = a\cos\theta\,d\theta$, and when $x = 0, \theta = 0$, when $x = a, \theta = \pi/2$]

$$= \int_0^{\pi/2} \pi\,.\,a(1-\sin\theta)\,.\,a(1-\cos\theta)\,.\,\frac{a\cos\theta + a\sin\theta}{\sqrt{2}\,.\,a\cos\theta}\,.\,a\cos\theta\,d\theta$$

$$= \frac{\pi a^3}{\sqrt{2}}\int_0^{\pi/2} (1-\sin\theta)(1-\cos\theta)\,(\sin\theta+\cos\theta)\,d\theta$$

$$= \frac{\pi a^3}{\sqrt{2}}\int_0^{\pi/2} (\sin\theta + \cos\theta - 2\sin\theta\cos\theta - 1 + \sin^2\theta\cos\theta + \cos^2\theta\sin\theta)\,d\theta$$

$$= \frac{\pi a^3}{\sqrt{2}}\left[-\cos\theta + \sin\theta + \cos^2\theta - \theta + \frac{1}{3}\sin^3\theta - \frac{1}{3}\cos^3\theta\right]_0^{\pi/2}$$

$$= \frac{\pi a^3}{\sqrt{2}}\left[\left(-0+1+0-\frac{\pi}{2}+\frac{1}{3}-0\right)-\left(-1+0+1-0+0-\frac{1}{3}\right)\right]$$

$$= \frac{\pi a^3}{\sqrt{2}}\left(\frac{5}{3}-\frac{\pi}{2}\right) = \frac{\pi a^3}{6\sqrt{2}}\,(10-3\pi).$$

6. Volume between two solids.

The volume of the solid generated by the revolution about the x-axis, of the area bounded by the curves $y = f(x)$, $y = \phi(x)$ and the ordinates $x = a$, $x = b$ is

$$\int_a^b \pi[(\text{y of upper curve})^2 - (\text{y of lower curve})^2]\, dx.$$

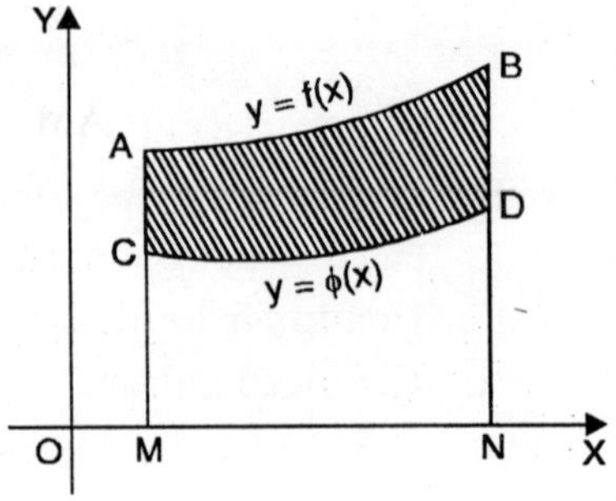

Let AB be the curve $y = f(x)$ and CD be the curve $y = \phi(x)$ both between the ordinates MCA ($x = a$) and NDB ($x = b$).

$\therefore$ Volume of the solid generated by the revolution about x-axis, of the shaded area ACDB.

= volume of the solid generated by revolving the area AMNB about x-axis – volume of the solid generated by revolution about x-axis, of the area CMND.

$$= \int_a^b \pi\, [f(x)]^2\, dx - \int_a^b \pi[\phi(x)]^2\, dx.$$

Example 1. *The figure bounded by a parabola and the tangents at the extremities of its latus rectum revolves about the axis of the parabola. Find the volume of the solid thus obtained.*

Sol. Let the equation of the parabola be $y^2 = 4ax$, and let LSL′ be its latus rectum and LT, L′T the tangents at L and L′. The equation of the tangent at L $(a, 2a)$ is $y.2a = 2a(x + a)$ or $y = x + a$.

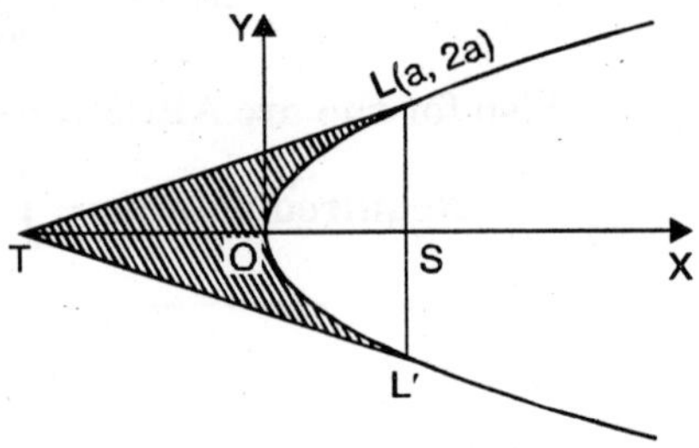

This meets x-axis in T$(-a, 0)$. The curve and the two tangents TL, TL′, are symmetrical about x-axis.

$\therefore$ Required volume of the solid formed by revolving the area TOL about the axis of parabola (*i.e.*, x-axis) = volume of the solid generated by revolving the area TSL about x-axis –volume of the solid generated by the revolution about x-axis of the area OLS.

$$= \int_{-a}^{a} \pi y^2\, dx - \int_0^a \pi y^2\, dx$$

(for tangent) (for curve)

$$= \int_{-a}^{a} \pi(x + a)^2\, dx - \int_0^a \pi\, .\, 4ax\, dx$$

$$= \pi\left[\frac{(x+a)^3}{3}\right]_{-a}^{a} - 2a\pi\left[x^2\right]_0^a$$

$$= \frac{\pi}{3}(8a^3 - 0) - 2a\pi\, .\, a^2 = \frac{2}{3}\pi a^3.$$

Example 2. *Show that if the area lying within the cardioid $r = 2a(1 + \cos\theta)$ and without the parabola $r(1 + \cos\theta) = 2a$, revolves about the initial line, the volume generated is $18\pi a^3$.*

Sol. The equation of the cardioid is $r = 2a(1 + \cos\theta)$ ...(1)

and the equation of the parabola is $r = \dfrac{2a}{1+\cos\theta}$ ...(2)

Both the curves are symmetrical about the initial line.

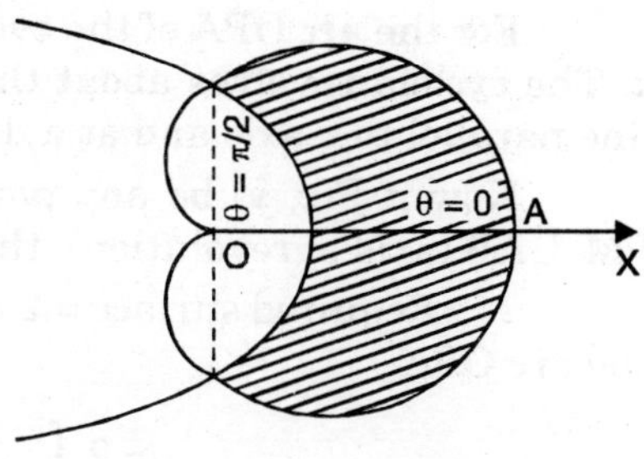

The upper half of the shaded area revolves about the initial line and it will generate the required volume.

The two curves intersect where solving (1) and (2). (Eliminating r) [*i.e.* Putting the value of r from (1) in (2)]

$$2a(1+\cos\theta) = \frac{2a}{1+\cos\theta}$$

or $\quad (1+\cos\theta)^2 = 1 \quad$ or $\quad 1 + 2\cos\theta + \cos^2\theta = 1$

or $\quad \cos\theta(2+\cos\theta) = 0 \quad \therefore \quad \cos\theta = 0$ or -2

But $\cos\theta \neq -2$, [$\because$ $\cos\theta$ can never be numerically > 1]

$\therefore \quad \cos\theta = 0$ or $\theta = \pm\pi/2$.

For the upper half, θ varies from 0 to $\pi/2$.

$\therefore$ Required volume

$$= \int_0^{\pi/2} \frac{2\pi}{3} \, [(r \text{ of outer curve})^3 - (r \text{ of inner curve})^3] \sin\theta \, d\theta$$

$$= \frac{2\pi}{3}\int_0^{\pi/2}\left[[2a(1+\cos\theta)]^3 - \left(\frac{2a}{1+\cos\theta}\right)^3\right]\sin\theta\,d\theta$$

$$= \frac{2\pi}{3}.\,8a^3\left[-\frac{(1+\cos\theta)^4}{4} + \frac{(1+\cos\theta)^{-2}}{-2}\right]_0^{\pi/2}$$

$$= \frac{16\pi a^3}{3}\left[-\left(\frac{1-16}{4}\right) - \frac{1}{2}\left(1-\frac{1}{4}\right)\right]$$

$$= \frac{16\pi a^3}{3}\left(\frac{15}{4} - \frac{3}{8}\right) = \frac{16\pi a^3}{3} \times \frac{27}{8} = 18\,\pi a^3.$$

7. Revolution about any axis

The **surface** *area of the solid generated by the revolution about any axis CD of the arc AB is*

$$\int_{x=OC}^{x=OD} 2\pi\,(PM)\,.\,ds$$

where PM, AC, BD are perpendiculars from any point P, A and B of the arc AB on CD and arc PA = s and O is any fixed point on the axis CD.

Take the fixed point O as origin, the axis of revolution OCD as the x-axis and OY perpendicular to it as y-axis (*Refer to figure of Art. 5, Page 366*).

Let the co-ordinates of P be (x, y) so the OM $= x$ and MP $= \bar{y}$.

Then the curved surface

$$= \int_{x=OC}^{x=OD} 2\pi y\,ds = \int_{x=OC}^{x=OD} 2\pi\,(PM)\,ds.$$

Example 1. *Find the area of the curved surface generated by the revolution of the cycloid $x = a(\theta + \sin\theta)$, $y = a(1-\cos\theta)$ about its base.*

Sol. The equations of the cycloid are $x = a(\theta + \sin\theta)$, $y = a(1-\cos\theta)$.

The cycloid is symmetrical about the y-axis.

For the arc OPA of the cycloid θ varies from 0 to π. The cycloid revolves about the base BA, which is a line parallel to x-axis and at a distance $2a$ from it.

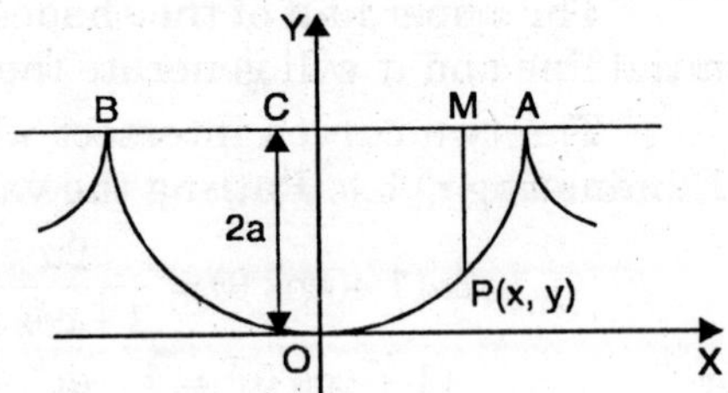

Now if P(x, y) be any point on the cycloid and PM ⊥ AB (axis of revolution), then PM $= 2a - y$.

∴ Required surface = 2 × surface generated by the arc OPA

$$= 2\int_0^{\pi} 2\pi\,(\text{PM})\,\frac{ds}{d\theta}\,d\theta$$

$$= 4\pi\int_0^{\pi} (2a - y)\sqrt{\left(\frac{dx}{d\theta}\right)^2 + \left(\frac{dy}{d\theta}\right)^2}\,d\theta$$

$$= 4\pi\int_0^{\pi} (2a - a + a\cos\theta)\sqrt{a^2(1+\cos\theta)^2 + a^2\sin^2\theta}\,d\theta$$

$$= 4\pi a^2\int_0^{\pi} (1+\cos\theta)\sqrt{2(1+\cos\theta)}\,d\theta$$

$$= 4\pi a^2\int_0^{\pi} (2\cos^2\theta/2)(2\cos\theta/2)\,d\theta$$

$$= 16\pi a^2\int_0^{\pi/2} \cos^3 t\,.\,2\,dt, \text{ where } t = \theta/2$$

$$= 32\,\pi a^2\,.\,\frac{2}{3} = \frac{64\pi a^2}{3}.$$

EXERCISE 3

1. (a) Show that the volume of the solid formed by the revolution of the cissoid $y^2(2a - x) = x^3$ about its asymptote is $2\pi^2a^3$. *(M.D.U. 2003 S)*

   (b) Find the volume of the solid generated by the arc of the cissoid $x = 2a\sin^2 t,\ y = \dfrac{2a\sin^3 t}{\cos t}$ about its asymptote.

2. Find the volume of the solid of revolution obtained by rotating the area included between the curves $y^2 = x^3$ and $x^2 = y^3$ about x-axis.
3. Find the area of the surface generated if an arch of the cycloid $x = a(\theta - \sin\theta),\ y = a(1 - \cos\theta)$ revolves about the line $y = 2a$.

ANSWERS

1. (a) $2\pi^2a^3$ (b) $2\pi^2a^3$ 2. $\dfrac{5}{28}\pi$. 3. $\dfrac{32}{3}\pi a^2$.

MISCELLANEOUS EXERCISE

1. Find the volumes of the oblate and prolate spheroids generated by an ellipse whose major and minor axis are $(24\pi)^{1/3}$ and $(3\pi)^{1/3}$.
2. Find the volume of a spherical cap of height h, cut off from a sphere of radius a.

[**Hint.** If we rotate the part of the circle $x^2 + y^2 = a^2$ lying in first quadrant from $y = a - h$ to $y = a$, we get a spherical cap of height h.

Thus the required volume $= \int_{y=a-h}^{a} \pi x^2\, dy = \pi \int_{a-h}^{a} (a^2 - y^2)\, dy = \pi \left(a^2 y - \frac{y^3}{3}\right)_{a-h}^{a}$.]

3. Show that the volume of the solid obtained by revolving the area included between the curves

$$y = x^2 \text{ and } y = 4 - x^2 \text{ about } x\text{-axis is } \frac{64\sqrt{2}\,\pi}{3}.$$

[**Hint.** The given curves are symmetrical about y-axis and intersect at $x = \pm\sqrt{2}$].

By Art. 6 required volume $= 2\left[\int_0^{\sqrt{2}} \pi y^2\, dx \text{ for curve } y = (4 - x^2) - \int_0^{\sqrt{2}} \pi y^2\, dx \text{ for curve } y = x^2\right]$.

4. Show that the volume of the solid generated by the revolution of the cycloid $x = a(\theta + \sin\theta)$, $y = a(1 - \cos\theta)$ about the y-axis is $\pi a^3 \left(\frac{3}{2}\pi^2 - \frac{8}{3}\right)$.

5. The area of the ellipse $\frac{x^2}{a^2} + \frac{y^2}{b^2} = 1$ lying in the first quadrant revolves about the line $y = b$. Show that the volume of the solid thus obtained is $\frac{10 - 3\pi}{6}\pi ab^2$.

6. Find the volume of the solid formed by the revolution of the area enclosed between the curve $27ay^2 = 4(x - 2a)^3$, the x-axis and the parabola $y^2 = 4ax$ about the x-axis.

7. Find the area of the surface formed by the revolution of $x^2 + 4y^2 = 16$ about its major axis.

8. A circular arc revolves about its chord. Prove that the area of the surface generated is $4\pi a^2 (\sin\alpha - \alpha\cos\alpha)$, where a is the radius and 2α is the angle subtended by the arc at the centre.

9. The area of the inner loop of the curve $r = 1 + 2\cos\theta$ is rotated through two right angles about the initial line. Show that the volume of the solid so formed is $\frac{\pi}{12}$.

10. A segment is cut off from a sphere of radius a by a plane at a distance $\frac{a}{2}$ from the centre. Show that the volume of the segment is $\frac{5}{32}$ of the volume of the sphere.

11. A $(0, a)$, P(x, y) are two points on the curve whose equation is $y = a\cosh\frac{x}{a}$ and s is the length of the arc AP. If the curve makes a complete revolution about the x-axis, prove that the area S of the curved surface bounded by planes through A and P perpendicular to the x-axis and the corresponding volume V are connected by $aS = 2V = \pi a(ax + sy)$.

12. The hyperbola $\frac{x^2}{a^2} - \frac{y^2}{b^2} = 1$ revolves about the axis of x. Show that the volume cut off from one of the solids thus formed by a plane perpendicular to x-axis and distant h from the vertex is $\frac{\pi b^2 h^2 (3a + h)}{3a^2}$.

ANSWERS

1. $2\pi^2, \pi^2$. **2.** $\pi h^2 \left(a - \frac{h}{3}\right)$. **4.** $\frac{10 - 3\pi}{6}\pi ab^2$.

6. $80\pi a^3$. **7.** $8\pi\left(1 + \frac{4\pi}{3\sqrt{3}}\right)$.

PART B : ORDINARY DIFFERENTIAL EQUATIONS

SECTION III

14

Geometrical Meaning of a Differential Equation

1. Geometrical Meaning of a Differential Equation of First Order and First Degree

Let $$\frac{dy}{dx} = f(x, y) \qquad ...(1)$$

be a differential equation of first order and first degree.

We know that $\frac{dy}{dx}$ represents the slope of the tangent to the curve of the function $y(x)$.

Let (1) has a solution $y(x)$ passing through a point $A_1(x_1, y_1)$ of the xy-plane. Putting the coordinates of the point $A_1(x_1, y_1)$ in (1), we get the value of $\frac{dy}{dx}$ at $A_1(x_1, y_1)$. Let it be denoted by m_1. Thus m_1 is the slope of the tangent to the curve at $A_1(x_1, y_1)$.

Now a point which moves, subject to the restriction imposed by (1), on passing through $A_1(x_1, y_1)$ must go in the direction m_1. Let the point move from $A_1(x_1, y_1)$ in the direction m_1 for an infinitesimal small distance, to a point $A_2(x_2, y_2)$. Let the slope of the tangent at $A_2(x_2, y_2)$ be m_2. Now let the point move in the direction m_2. Let the point move through an infinitesimal small distance in the direction m_2 and reach the position $A_3(x_3, y_3)$. In this way, let the point go to a point $A_4(x_4, y_4)$ and so on to the successive points. Hence through the point $A_1(x_1, y_1)$ will pass a curve.

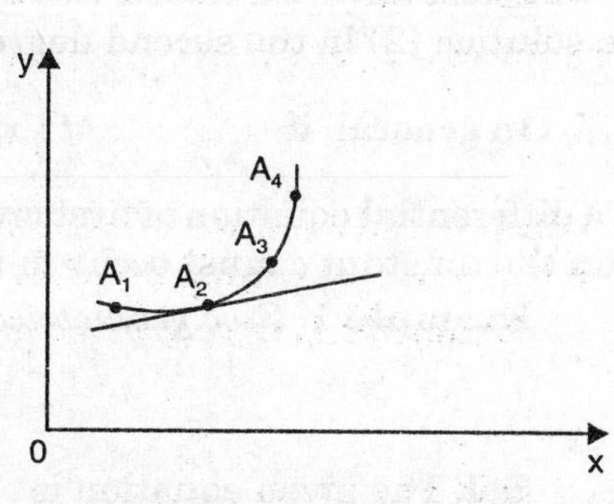

Similarly through every point of the xy-plane, there passes a curve, the coordinates of every point of which and the direction of the tangent there at will satisfy the differential equation (1).

Hence equation (1) represents a family of curves such that through every point of the xy-plane there passes one curve of the family.

2. Geometrical Meaning of the General Solution of a Differential Equation of First Order and First Degree

Let $$\frac{dy}{dx} = f(x, y) \qquad ...(1)$$

be a differential equation of first order and first degree.

Since the differential equation (1) is of the first order, therefore its general solution will contain one arbitrary constant. Let the general solution be

$$\phi(x, y, c) = 0 \qquad ...(2)$$

It is evident that (2) represents one-parameter family of curves such that through each point of xy-plane, there passes one member of the family. If (x_1, y_1) be a point of the xy-plane, then putting $x = x_1$ and $y = y_1$ in (2), we have

$$\phi(x_1, y_1, c) = 0.$$

This equation will give one value of c say c_1.

Thus through the point (x_1, y_1) in the plane, there will pass one member $\phi(x, y, c_1) = 0$ of the family of curves given by (2). Hence if we put $c = c_1$ in (2), we have $\phi(x, y, c_1) = 0$ which evidently represents a particular member of the family (2) and is a particular solution of the differential equation (1).

Note. It may be noted that the general solution (2) contains one parameter c for the first order differential equation (1). Thus the solution of a differential equation of the first order will represent infinitely many curves for different values of the parameter.

3. Geometrical Meaning of a Differential Equation of First Order and Degree Higher than One

Let $$f\left(x, y, \frac{dy}{dx}\right) = 0 \qquad ...(1)$$

be a differential equation of first order and of second degree. Let us take a particular point (x_1, y_1) in the xy-plane, then putting $x = x_1$ and $y = y_1$ in (1), there will be two values of $\frac{dy}{dx}$ at this point (x_1, y_1). Therefore, the moving point can pass through each point in two directions and hence two curves of the system (which is the locus of the general solution of (1)) pass through each point of the xy-plane.

Let the general solution of equation (1) be

$$\phi(x, y, c) = 0 \qquad ...(2)$$

$\therefore$ (2) must have, therefore, two different values of c for each point and hence c must occur in the solution (2) in the second degree.

In general, if $$f\left(x, y, \frac{dy}{dx}\right) = 0$$

be a differential equation of first order and of degree n and if its general solution is $\phi(x, y, c) = 0$, then the constant c must occur in the nth degree.

Example 1. *Give geometrical significance of the solution of the differential equation*

$$\frac{dy}{dx} = -\frac{x}{y}.$$

Sol. The given equation is

$$\frac{dy}{dx} = -\frac{x}{y}. \qquad ...(1)$$

(1) $\Rightarrow$ $$x\,dx + y\,dy = 0$$

Integrating, we have

$$\frac{x^2}{2} + \frac{y^2}{2} = c_1 \quad \text{or} \quad x^2 + y^2 = c \qquad ...(2)$$

(2) represents a family of concentric circles. (Putting $c = 2c_1$)

Let (h, k) be any point in the xy-plane.

$\therefore$ The circle passing through the point (h, k) is $x^2 + y^2 = h^2 + k^2$, which is a particular solution.

Thus the locus of the solution of the differential equation (1) is made of all circles, infinite in number, that have their centres at the origin.

$\therefore$ The geometrical meaning of the solution of equation (1) is that a point (x, y) moving so as to satisfy the equation, moves in a direction perpendicular to the line joining the point (x, y) to the origin, that is, the point moving under the condition (1) describes a circle about the origin as centre.

Example 2. *Give geometrical meaning of the solution of the differential equation*

$$\frac{d^2y}{dx^2} = 0.$$

Sol. The differential equation is

$$\frac{d^2y}{dx^2} = 0 \qquad \text{...(1)}$$

Integrating (1), we have

$$\frac{dy}{dx} = m, \quad m \text{ being an arbitrary constant.}$$

Again integrating, we have $y = mx + c$ ...(2)

where c is another arbitrary constant.

(2) is the general solution of differential equation (1).

Now a line through any point $(0, c)$ drawn in any direction m is the locus of a particular integral of (1). If we give a particular value c_1 to c, there will be infinity of lines corresponding to the infinity of values that m can have, and all these lines will be the loci of the integrals. Since to each of the infinity of values that can be given to c, there corresponds an infinity of lines (because of the infinity of values that can be given to m), therefore, the general solution of the equation represents a doubly infinite system of lines. In other words, the locus of the differential equation (1) consists of a doubly infinite system of lines.

EXERCISE 1

Give geometrical meaning of the solution of the following differential equations :

1. $\dfrac{dy}{dx} = -\dfrac{y}{x}$
2. $\dfrac{dy}{dx} = m.$

Answers

1. Infinite number of hyperbolas having coordinate axes as asymptotes.
2. Infinite number of straight lines each having slope m.

15

Exact Differential Equations

1. Definition

The differential equation Mdx + Ndy = 0, where M and N are functions of x and y, is called an **exact differential equation** if Mdx + Ndy is the differential of some function u of x and y *i.e.* if, Mdx + Ndy = du. In such a case, the equation Mdx + Ndy = 0 reduces to $du = 0$ and so its general solution is $u = c$ *i.e.* $u(x, y) = c$.

Examples of exact differential equations :

(*i*) The equation $xdy + ydx = 0$ is exact ; for $xdy + ydx = d(xy)$. Its general solution is $xy = c$.

(*ii*) The equation $\frac{1}{x}\,dy - \frac{y}{x^2}\,dx = 0$ is exact ; for

$$\frac{1}{x}\,dy - \frac{y}{x^2}\,dx = \frac{xdy - ydx}{x^2} = d\left(\frac{y}{x}\right).$$

Its general solution is $y/x = c$.

2. Necessary and Sufficient Conditions for the Equation Mdx + Ndy = 0 to be Exact

(*K.U. 2004, 2007 ; M.D.U. 2005*)

Necessary condition. Let Mdx + Ndy = 0 be an exact differential equation.

∴ Mdx + Ndy is the differential of some function, say, u of x and y.

$$\therefore \quad Mdx + Ndy = \frac{\partial u}{\partial x}\,dx + \frac{\partial u}{\partial y}\,dy$$

$$\therefore \quad M = \frac{\partial u}{\partial x} \quad \text{and} \quad N = \frac{\partial u}{\partial y}$$

$$\therefore \quad \frac{\partial M}{\partial y} = \frac{\partial}{\partial y}\left(\frac{\partial u}{\partial x}\right) = \frac{\partial}{\partial x}\left(\frac{\partial u}{\partial y}\right) = \frac{\partial N}{\partial x} \qquad \left(\because \frac{\partial^2 u}{\partial y \partial x} = \frac{\partial^2 u}{\partial x \partial y}\right)$$

$$\therefore \quad \frac{\partial M}{\partial y} = \frac{\partial N}{\partial x}.$$

Sufficient condition. Let $\frac{\partial M}{\partial y} = \frac{\partial N}{\partial x}$.

Let $v = \int^x M\,dx$, where $\int^x$ denotes integration w.r.t. x treating y constant.

$$\therefore \quad M = \frac{\partial v}{\partial x}$$

Now $$\frac{\partial}{\partial x}\left(N - \frac{\partial v}{\partial y}\right) = \frac{\partial N}{\partial x} - \frac{\partial^2 v}{\partial x \partial y} = \frac{\partial N}{\partial x} - \frac{\partial^2 v}{\partial y \partial x} \qquad \left(\because \frac{\partial^2 v}{\partial x \partial y} = \frac{\partial^2 v}{\partial y \partial x}\right)$$

$$= \frac{\partial N}{\partial x} - \frac{\partial M}{\partial y} = \frac{\partial N}{\partial x} - \frac{\partial N}{\partial x} = 0 \qquad \left(\because \frac{\partial M}{\partial y} = \frac{\partial N}{\partial x}\right)$$

$\therefore$ $N - \dfrac{\partial v}{\partial y}$ is a function of y only.

Let $u = v + \int \left(N - \dfrac{\partial v}{\partial y}\right) dy$

$$\therefore \qquad du = dv + \left(N - \frac{\partial v}{\partial y}\right) dy = \left(\frac{\partial v}{\partial x} dx + \frac{\partial v}{\partial y} dy\right) + \left(N - \frac{\partial v}{\partial y}\right) dy$$

$$= M dx + \frac{\partial v}{\partial y} dy + N\, dy - \frac{\partial v}{\partial y} dy = M\, dx + N\, dy$$

$\therefore$ $M\, dx + N\, dy$ is the differential of the function u.

$\therefore$ The equation $M\, dx + N\, dy = 0$ is exact.

3. Inspection Method of Solving an Exact Differential Equation

At times an exact differential equation may be solved by inspection by regrouping the terms of the given equation, by recognizing a certain group of terms of an exact differential. For example :

(*i*) $xdy + ydx = d(xy)$ (*ii*) $xdx + ydy = d\left(\dfrac{x^2 + y^2}{2}\right)$

(*iii*) $\cos ydx - x\sin ydy = d(x \cos y)$ etc.

Example 1. *Solve* $xdx + ydy + x^4\, dx + \sin y\, dy = 0$.

Sol. We have

$$y\, dx + x\, dy + x^4\, dx + \sin y dy = 0 \qquad ...(1)$$

$$\Rightarrow \qquad (y + x^4)\, dx + (x + \sin y)\, dy = 0$$

Here $\qquad M = y + x^4$ and $N = x + \sin y$

$$\therefore \qquad \frac{\partial M}{\partial y} = 1 \text{ and } \frac{\partial N}{\partial x} = 1$$

$$\therefore \qquad \frac{\partial M}{\partial y} = \frac{\partial N}{\partial x}.$$

$\therefore$ The equation (1) is exact.

$$(1) \Rightarrow \qquad d(xy) + d\left(\frac{x^5}{5}\right) + d(-\cos y) = 0 \qquad \Rightarrow \qquad d\left(xy + \frac{x^5}{5} - \cos y\right) = 0$$

$\therefore$ The general solution is $\mathbf{xy + \dfrac{x^5}{5} - \cos y = c}$.

Example 2. *Solve* $(4x^3y^3 - 2xy)\, dx + (3x^4y^2 - x^2)dy = 0$.

Sol. We have $(4x^3y^3 - 2xy)\, dx + (3x^4y^2 - x^2)\, dy = 0$

Here $\qquad M = 4x^3y^3 - 2xy$ and $N = 3x^4y^2 - x^2$

$$\therefore \qquad \frac{\partial M}{\partial y} = 12x^3y^2 - 2x \text{ and } \frac{\partial N}{\partial x} = 12x^3y^2 - 2x$$

$$\therefore \qquad \frac{\partial M}{\partial y} = \frac{\partial N}{\partial x}. \qquad \therefore \text{ The given equation is exact.}$$

(1) $\Rightarrow$ $(4x^3y^3dx + 3x^4y^2dy) - (2xydx + x^2dy) = 0$

$\Rightarrow$ $d(x^4y^3) - d(x^2y) = 0$ $\Rightarrow$ $d(x^4y^3 - x^2y) = 0$

$\therefore$ The general solution is $\mathbf{x^4y^3 - x^2y = c}$.

Example 3. *Solve (cos y + y cos x) dx + (sin x – x sin y) dy = 0.*

Sol. We have $(\cos y + y \cos x)\, dx + (\sin x - x \sin y)\, dy = 0$ ...(1)

Here $M = \cos y + y \cos x$ and $N = \sin x - x \sin y$

$\therefore$ $\dfrac{\partial M}{\partial y} = -\sin y + \cos x$ and $\dfrac{\partial N}{\partial x} = \cos x - \sin y$

$\therefore$ $\dfrac{\partial M}{\partial y} = \dfrac{\partial N}{\partial x}$ $\therefore$ The given equation is exact.

(1) $\Rightarrow$ $(\cos y\, dx - x \sin y dy) + (y \cos x dx + \sin x dy) = 0$

$\Rightarrow$ $d(x\cos y) + d(y\sin x) = 0$ $\Rightarrow$ $d(x\cos y + y\sin x) = 0$

$\therefore$ The general solution is $\mathbf{x\cos y + y\sin x = c}$.

EXERCISE 1

Solve the following differential equations :

1. $xdx + ydy + x^3dx + y^4dy = 0$
2. $xdy + ydx + \sin x dx = 0$
3. $\cos ydx - x\sin ydy + 5y^4\, dy = 0$
4. $e^y\, dx + (xe^y + 2y)dy = 0$
5. $e^{2y}\, dx + 2(xe^{2y} - y)\, dy = 0$
6. $2x(ye^{x^2} - 1)dx + e^{x^2} dy = 0$
7. $(6x^5y^3 + 4x^3y^5)dx + (3x^6y^2 + 5x^4y^4)\, dy = 0$
8. $(2x^3 + 3y)dx + (3x + y - 1)\, dy = 0.$

Answers

1. $\dfrac{x^2}{2} + \dfrac{y^2}{2} + \dfrac{x^4}{4} + \dfrac{y^5}{5} = c$
2. $xy - \cos x = c$
3. $x\cos y + y^5 = c$
4. $xe^y + y^2 = c$
5. $xe^{2y} - y^2 = c$
6. $ye^{x^2} - x^2 = c$
7. $x^6y^3 + x^4y^5 = c$
8. $\dfrac{1}{2}x^4 + \dfrac{1}{2}y^2 - y + 3xy = c.$

4. General Method of Solving an Exact Differential Equation

Let $Mdx + Ndy = 0$ be an exact differential equation

$\therefore$ $Mdx + Ndy$ is the differential of some function, say, u of x and y

$\therefore$ $$Mdx + Ndy = \frac{\partial u}{\partial x}dx + \frac{\partial u}{\partial y}dy$$

$\therefore$ $$M = \frac{\partial u}{\partial x} \quad ...(1) \quad \text{and} \quad N = \frac{\partial u}{\partial y} \quad ...(2)$$

$\therefore$ $$\frac{\partial M}{\partial y} = \frac{\partial}{\partial y}\left(\frac{\partial u}{\partial x}\right) = \frac{\partial}{\partial x}\left(\frac{\partial u}{\partial y}\right) = \frac{\partial N}{\partial x} \quad \textit{i.e.,} \quad \frac{\partial M}{\partial y} = \frac{\partial N}{\partial x}$$

Also, (1) $\Rightarrow$ $u = \int^x M\, dx + \phi(y)$, where $\phi(y)$ is some function of y only

$$(2) \quad \Rightarrow \quad N = \frac{\partial u}{\partial y} = \frac{\partial}{\partial y}\left(\int^x M\, dx\right) + \frac{d\phi}{dy}$$

$$\Rightarrow \quad \frac{d\phi}{dy} = N - \frac{\partial}{\partial y}\left(\int^x M\, dx\right) = N - \int^x \frac{\partial M}{\partial y}\, dx = N - \int^x \frac{\partial N}{\partial x}\, dx \qquad \left(\because \frac{\partial M}{\partial y} = \frac{\partial N}{\partial x}\right)$$

$$\therefore \quad u = \int^x M\, dx + \int \left(N - \int^x \frac{\partial N}{\partial x}\, dx\right) dy$$

$$\Rightarrow \quad u = \int^x M\, dx + \int \begin{pmatrix}\text{terms of N not}\\ \text{containing } x\end{pmatrix} dy$$

∴ The general solution of the given exact equation is $u = c$ *i.e.*,

$$\int^{\mathbf{x}} \mathbf{M\, dx} + \int \begin{pmatrix}\textbf{terms of N not}\\ \textbf{containing x}\end{pmatrix} \mathbf{dy = c}$$, **where c is an arbitrary constant.**

Remark. If each term of N contain x, then the general solution reduces to $\int^{\mathbf{x}} \mathbf{M\, dx = c}$.

Example 1. *Show that the equation*

$$(1 + e^{x/y})\, dx + e^{x/y}\, (1 - x/y)\, dy = 0$$

is exact, and hence solve it. (KU 1995)

Sol. We have

$$(1 + e^{x/y})\, dx + e^{x/y}\, (1 - x/y)\, dy = 0.$$

Here $\quad M = 1 + e^{x/y} \quad$ and $\quad N = e^{x/y}\, (1 - x/y)$

$$\therefore \quad \frac{\partial M}{\partial y} = e^{x/y} \cdot \left(\frac{-x}{y^2}\right) = -\frac{x}{y^2}\, e^{x/y} \qquad ...(1)$$

and

$$\frac{\partial N}{\partial x} = e^{x/y}\, \frac{1}{y}\left(1 - \frac{x}{y}\right) + e^{x/y}\left(-\frac{1}{y}\right) = -\frac{x}{y^2}\, e^{x/y} \qquad ...(2)$$

$$\therefore \quad \frac{\partial M}{\partial y} = \frac{\partial N}{\partial x}$$

Hence the given equation is exact and its general solution is

$$\int^x M dx + \int (\text{terms of N not containing } x)\, dy = c.$$

$$\Rightarrow \quad \int^x (1 + e^{x/y})\, dx = c$$

$$\Rightarrow \quad \mathbf{x + e^{x/y}\, y = c}.$$ This is the required solution.

Example 2. *Solve*

$$(x^4 - 2xy^2 + y^4)\, dx - (2x^2y - 4xy^3 + \sin y)\, dy = 0. \qquad \text{(MDU 2000)}$$

Sol. We have

$$(x^4 - 2xy^2 + y^4)\, dx - (2x^2y - 4xy^3 + \sin y)\, dy = 0. \qquad ...(1)$$

Comparing it with $Mdx + Ndy = 0$,

we have $\quad M = x^4 - 2xy^2 + y^4 \quad$ and $\quad N = -(2x^2y - 4xy^3 + \sin y)$.

$$\therefore \quad \frac{\partial M}{\partial y} = -4xy + 4y^3 \quad \text{and} \quad \frac{\partial N}{\partial x} = -4xy + 4y^3$$

$\therefore \quad \dfrac{\partial M}{\partial y} = \dfrac{\partial N}{\partial x}$. ∴ The equation is exact.

∴ The general solution is

$$\int^x M dx + \int (\text{terms of N not containing } x)\, dy = c.$$

$$\Rightarrow \quad \int^x (x^4 - 2xy^2 + y^4)\, dx + \int (-\sin y)\, dy = c$$

$$\Rightarrow \quad \mathbf{\frac{x^5}{5} - x^2y^2 + y^4x + \cos y = c.}$$

Example 3. *Solve $(ax + hy + g)\, dx + (hx + by + f)\, dy = 0$ and show that this differential equation represents a family of conics.* (K.U. 1995 C)

Sol. We have

$$(ax + hy + g)\, dx + (hx + by + f)\, dy = 0.$$

Comparing it with $\quad M dx + N dy = 0,$

we get $\quad M = ax + hy + g \quad$ and $\quad N = hx + by + f$

$$\therefore \quad \frac{\partial M}{\partial y} = h \quad \text{and} \quad \frac{\partial N}{\partial x} = h$$

$\therefore \quad \dfrac{\partial M}{\partial y} = \dfrac{\partial N}{\partial x}$. ∴ The given equation is exact.

∴ The general solution is

$$\int^x M dx + \int (\text{terms of N not containing } x)\, dy = c.$$

$$\int^x (ax + hy + g)\, dx + \int (by + f)\, dy = c_1$$

$$\Rightarrow \quad a\,\frac{x^2}{2} + hyx + gx + \frac{by^2}{2} + fy = c_1$$

$$\Rightarrow \quad ax^2 + by^2 + 2hxy + 2gx + 2fy - 2c_1 = 0$$

$\Rightarrow \quad \mathbf{ax^2 + by^2 + 2hxy + 2gx + 2fy + c = 0,}$ where $c = -2c_1$.

This being a second degree in x and y represents a family of conics.

Example 4. *Solve $x\, dx + y\, dy = a^2 \left(\dfrac{x\, dy - y\, dx}{x^2 + y^2}\right)$.* (K.U. 2006 S)

Sol. We have $\quad x\, dx + y\, dy = \dfrac{a^2 x}{x^2 + y^2}\, dy - \dfrac{a^2 y\, dx}{x^2 + y^2}$

$$\Rightarrow \quad \left(x + \frac{a^2 y}{x^2 + y^2}\right) dx + \left(y - \frac{a^2 x}{x^2 + y^2}\right) dy = 0$$

Comparing it with $M dx + N dy = 0$, we have

$$M = x + \frac{a^2 y}{x^2 + y^2} \quad \text{and} \quad N = y - \frac{a^2 x}{x^2 + y^2}$$

$$\therefore \quad \frac{\partial M}{\partial y} = 0 + \left[\frac{(x^2+y^2)a^2 - a^2 y \cdot 2y}{(x^2+y^2)^2}\right] = \frac{a^2(x^2-y^2)}{(x^2+y^2)^2}$$

$$\frac{\partial N}{\partial x} = 0 - \left[\frac{(x^2+y^2)a^2 - a^2 x \cdot 2x}{(x^2+y^2)^2}\right] = \frac{a^2(x^2-y^2)}{(x^2+y^2)^2}$$

$\therefore \quad \frac{\partial M}{\partial y} = \frac{\partial N}{\partial x}$. $\quad \therefore$ The given equation is exact.

$\therefore$ The general solution is

$$\int^x M dx + \int \text{(terms of N not containing } x) \, dy = c.$$

$$\Rightarrow \quad \int \left(x + \frac{a^2 y}{x^2+y^2}\right) dx + \int y \, dy = c$$

$$\Rightarrow \quad \int x \, dx + a^2 y \int \frac{1}{x^2+y^2} dx + \frac{y^2}{2} = c$$

$$\Rightarrow \quad \mathbf{\frac{x^2}{2} + a^2 \tan^{-1} \frac{x}{y} + \frac{y^2}{2} = c.}$$

Example 5. *Solve* $(x^3 + 3xy^2)\, dx + (3x^2y + y^3)\, dy = 0.$ (M.D.U. 2006)

Sol. We have $(x^3 + 3xy^2)\, dx + (3x^2y + y^3)\, dy = 0$

Comparing with M dx + N $dy = 0$, we get

$$M = x^3 + 3xy^2 \text{ and } N = 3x^2y + y^3$$

$$\therefore \quad \frac{\partial M}{\partial y} = 6xy \quad \text{and} \quad \frac{\partial N}{\partial x} = 6xy$$

$\therefore \quad \frac{\partial M}{\partial y} = \frac{\partial N}{\partial x} \quad \therefore$ The given equation is exact.

$\therefore$ The general solution is $\int^x M \, dx + \int$ (terms of N not containing x) $dy = c$.

$$\Rightarrow \quad \int^x (x^3 + 3xy^2)\, dx + \int y^3 \, dy = c$$

$$\Rightarrow \quad \frac{x^4}{4} + 3y^2 \cdot \frac{x^2}{2} + \frac{y^4}{4} = c$$

$$\Rightarrow \quad \mathbf{x^4 + 6x^2y^2 + y^4 = c_1.} \qquad \text{(Putting } c_1 = 4c)$$

EXERCISE 2

Solve the following differential equations :

1. $(a^2 - 2xy - y^2)\, dx - (x+y)^2\, dy = 0$ *(M.D.U. 2001)*
2. $(2xy + y - \tan y)\, dx + (x^2 - x \tan^2 y + \sec^2 y)\, dy = 0$
3. $\left[y\left(1 + \frac{1}{x}\right) + \cos y\right] dx + (x + \log x - x \sin y)\, dy = 0$
4. $\cos x\,(\cos x - \sin \alpha \sin y)\, dx + \cos y\,(\cos y - \sin \alpha \sin x)\, dy = 0.$
5. $(e^y + 1) \cos x \, dx + e^y \sin x \, dy = 0$
6. $x\, dx + y\, dy + \frac{x\, dy - y\, dx}{x^2 + y^2} = 0$

7. $y \sin 2x\, dx - (y^2 + \cos^2 x)\, dy = 0$

8. $(x^2 + y^2 + e^x)\, dx + 2xy\, dy = 0$

9. $(x^2 - 2xy + 3y^2)\, dx + (4y^3 + 6xy - x^2)\, dy = 0$

10. $(x^2 - ay)\, dx = (ax - y^2)\, dy$

11. $\sec^2 x \tan y\, dx + \sec^2 y \tan x\, dy = 0$

12. $(x^3 + 3xy^2)\, dx + (3x^2y + y^3)\, dy = 0$. (*M.D.U. 2006*)

Answers

1. $a^2x - x^2y - xy^2 - \dfrac{y^3}{3} = c$

2. $x^2y + xy - x \tan y + \tan y = c$

3. $y(x + \log x) + x \cos y = c$

4. $\dfrac{1}{2}\left(x + \dfrac{\sin 2x}{2}\right) - \sin \alpha \sin x \sin y + \dfrac{1}{2}\left(y + \dfrac{\sin 2y}{2}\right) = c$

5. $(e^y + 1) \sin x = c$

6. $x^2 + y^2 - 2 \tan^{-1} \dfrac{x}{y} = c$

7. $y \cos^2 x + \dfrac{y^3}{3} + c = 0$

8. $\dfrac{x^3}{3} + xy^2 + e^x = c$

9. $\dfrac{x^3}{3} - x^2y + 3xy^2 + y^4 + c = 0$

10. $x^3 - 3axy + y^3 = c$

11. $\tan x \tan y = c$

12. $x^4 + 6x^2y^2 + y^4 = c$.

5. Integrating Factor

In the foregoing section, we have seen that the equation $Mdx + Ndy = 0$ is exact if $\dfrac{\partial M}{\partial y} = \dfrac{\partial N}{\partial x}$ and hence the equation can be solved. Evidently, if $\dfrac{\partial M}{\partial y} \neq \dfrac{\partial N}{\partial x}$, the equation is not exact and therefore cannot be solved. But sometimes, the equation becomes exact if it is multiplied by some function of x and y. Such a function of x and y is called an *integrating factor*.

An **integrating factor** of a differential equation is a function of x and y which is such that if the differential equation is multiplied by it, the resulting equation becomes exact. An integrating factor is briefly written as I.F.

We give below some rules for finding integrating factors.

6. Rule I for Finding Integrating Factor

At times an integrating factor may be found by inspection, after regrouping the terms of the given equation, by recognizing a certain group of terms as being a part of an exact differential. For this purpose, the following list of exact differentials would be found very useful.

S. No.	Group of terms	Integrating factor	Exact differential
		$\dfrac{1}{x^2}$	$\dfrac{xdy - ydx}{x^2} = d\left(\dfrac{y}{x}\right)$
		$\dfrac{1}{y^2}$	$-\dfrac{ydx - xdy}{y^2} = d\left(-\dfrac{x}{y}\right)$
1.	$xdy - ydx$	$\dfrac{1}{xy}$	$\dfrac{dy}{y} - \dfrac{dx}{x} = d\left(\log \dfrac{y}{x}\right)$

		$\dfrac{1}{x^2+y^2}$	$\dfrac{xdy-ydx}{x^2+y^2}=\dfrac{\dfrac{xdy-ydx}{x^2}}{1+\left(\dfrac{y}{x}\right)^2}=d\left(\tan^{-1}\dfrac{y}{x}\right)$
2.	$x\,dy+y\,dx$	$\dfrac{1}{xy}$	$\dfrac{dy}{y}+\dfrac{dx}{x}=d(\log xy)$
		$\dfrac{1}{(xy)^n},\ n\neq 1$	$\dfrac{x\,dy+y\,dx}{(xy)^n}=d\left(\dfrac{(xy)^{-n+1}}{-n+1}\right)=d\left(\dfrac{-1}{(n-1)(xy)^{n-1}}\right)$
3.	$x\,dx+y\,dy$	$\dfrac{1}{x^2+y^2}$	$\dfrac{x\,dx+y\,dy}{x^2+y^2}=\dfrac{1}{2}\cdot\dfrac{d(x^2+y^2)}{x^2+y^2}=d\left(\dfrac{1}{2}\log(x^2+y^2)\right)$
		$\dfrac{1}{(x^2+y^2)^n},\ n\neq 1$	$\dfrac{x\,dx+y\,dy}{(x^2+y^2)^n}=\dfrac{1}{2}(x^2+y^2)^{-n}\,d(x^2+y^2)$ $=d\left(\dfrac{-1}{2(n-1)(x^2+y^2)^{n-1}}\right)$

Example 1. *Solve* $xdx+ydy+(x^2+y^2)dy=0.$

Sol. We have $xdx+ydy+(x^2+y^2)dy=0$

The last term suggests to have $\dfrac{1}{x^2+y^2}$ as an I.F.

Dividing by x^2+y^2, we get, $\dfrac{xdx+ydy}{x^2+y^2}+dy=0$

$$\Rightarrow\quad \frac{1}{x^2+y^2}\cdot\frac{1}{2}\,d(x^2+y^2)+dy=0 \quad\Rightarrow\quad \frac{1}{2}\,d(\log(x^2+y^2))+dy=0$$

$$\Rightarrow\quad d\left(\frac{1}{2}\log(x^2+y^2)+y\right)=0.$$

$\therefore$ The general solution is $\mathbf{\dfrac{1}{2}\log(x^2+y^2)+y=c.}$

Example 2. *Solve* $xdy-ydx-(1-x^2)dx=0.$

Sol. We have $xdy-ydx-(1-x^2)\,dx=0$ ...(1)

The last term suggests to have $\dfrac{1}{x^2}$ as an I.F.

Dividing (1) by x^2, we get $\dfrac{xdy-ydx}{x^2}-\left(\dfrac{1}{x^2}-1\right)dx=0$

$$\Rightarrow\quad d\left(\frac{y}{x}\right)+d\left(x+\frac{1}{x}\right)=0 \quad\Rightarrow\quad d\left(\frac{y}{x}+x+\frac{1}{x}\right)=0$$

$\therefore$ The general solution is $\dfrac{y}{x}+x+\dfrac{1}{x}=c$ or $\mathbf{y+x^2+1=cx.}$

Example 3. *Solve* $\frac{dy}{dx} = \frac{y - xy^2 - x^3}{x + x^2y + y^3}$.

Sol. We have $\frac{dy}{dx} = \frac{y - xy^2 - x^3}{x + x^2y + y^3}$

$$\Rightarrow \quad (x + x^2y + y^3)dy - (y - xy^2 - x^3)dx = 0$$

$$\Rightarrow \quad (x^3 + xy^2 - y)dx + (y^3 + x^2y + x)dy = 0$$

$$\Rightarrow \quad (x^3 + xy^2)\, dx + (y^3 + x^2y)\, dy + (xdy - ydx) = 0$$

$$\Rightarrow \quad (x^2 + y^2)(xdx + ydy) + (xdy - ydx) = 0$$

The first term suggests to have $\frac{1}{x^2 + y^2}$ as an I.F.

Dividing by $x^2 + y^2$, we get, $xdx + ydy + \frac{xdy - ydx}{x^2 + y^2} = 0$

$$\Rightarrow \quad d\left(\frac{x^2}{2}\right) + d\left(\frac{y^2}{2}\right) + \frac{\frac{xdy - ydx}{x^2}}{1 + \left(\frac{y}{x}\right)^2} = 0 \quad \Rightarrow \quad d\left(\frac{x^2 + y^2}{2}\right) + \frac{1}{1 + \left(\frac{y}{x}\right)^2} d\left(\frac{y}{x}\right) = 0$$

$$\Rightarrow \quad d\left(\frac{x^2 + y^2}{2}\right) + d\left(\tan^{-1}\frac{y}{x}\right) = 0 \quad \Rightarrow \quad d\left(\frac{x^2 + y^2}{2} + \tan^{-1}\frac{y}{x}\right) = 0.$$

$\therefore$ The general solution is $\frac{x^2 + y^2}{2} + \tan^{-1}\frac{y}{x} = c$

or $\quad \mathbf{x^2 + y^2 + 2\tan^{-1}\frac{y}{x} = c_1.}$ (Putting $c_1 = 2c$)

EXERCISE 3

Solve the following differential equations :

1. $ydx - xdy + (x^2 + y^2)\, dx = 0$
2. $xdx + ydy + 4y^3 (x^2 + y^2)dy = 0$
3. $(x + x^4 + 2x^2y^2 + y^4)\, dx + ydy = 0$
4. $(x^3 + xy^2 + a^2y)dx + (y^3 + yx^2 - a^2x)\, dy = 0$
5. $ydx - xdy + (1 + x^2)dx + x^2 \sin y\, dy = 0$
6. $x^2 \frac{dy}{dx} + xy = \sqrt{1 - x^2y^2}$
7. $xdx + ydy = m(xdy - ydx)$
8. $(1 + xy)y\, dx - x(1 - xy)dy = 0.$

Answers

1. $\tan^{-1}\frac{y}{x} = x + c$
2. $\frac{1}{2}\log (x^2 + y^2) + y^4 = c$
3. $(x^2 + y^2)(2x + c) = 1$
4. $x^2 + y^2 + 2a^2 \tan^{-1}\frac{x}{y} = c$
5. $x^2 - y - 1 - x\cos y = cx$
6. $\sin^{-1}(xy) - \log x = c$
7. $\log (x^2 + y^2) - 2m \tan^{-1}\frac{y}{x} = c$
8. $\log \frac{x}{y} - \frac{1}{xy} = c.$

$$\leq |x|^2 \qquad \left(\because \left|\sin \frac{1}{x}\right| \leq 1 \text{ as } |\sin\theta| \leq 1\right)$$

Let us take $\delta = {}^*\sqrt{\varepsilon}$

$\therefore$ For $0 < |x - c| < \delta$ *i.e.,* $0 < |x| < \delta$ $(\because\ x \to 0 \text{ means } c = 0)$

$$|f(x) - l| \leq |x|^2 < \delta^2 < \varepsilon \qquad (\because\ \delta = \sqrt{\varepsilon})$$

$$\therefore \quad \underset{x\to 0}{\text{Lt}}\ f(x) = l \text{ i.e., } \underset{x\to 0}{\text{Lt}}\ x^2 \sin\frac{1}{x} = 0.$$

Example 3. *Using $\varepsilon - \delta$ definition of limit, prove that* $\underset{x\to c}{Lt}\ \sin x = \sin c$

Sol. Let $\varepsilon > 0$, However small.

Let us choose a positive number $\delta (= \varepsilon)$ such that

$$0 < |x - c| < \delta \quad ...(1) \quad (x \neq c \Rightarrow 0 < |x - c|)$$

Let $f(x) = \sin x \quad \therefore \quad f(c) = \sin c$

$$|f(x) - f(c)| = |\sin x - \sin c|$$

$$= \left|2\cos\frac{x+c}{2}\sin\frac{x-c}{2}\right| \qquad \text{(C – D formula)}$$

$$= 2\left|\cos\frac{x+c}{2}\right|\left|\sin\frac{x-c}{2}\right| \qquad (\because\ |\alpha\beta| = |\alpha||\beta|)$$

$$\leq 2\,(1)\left|\frac{x-c}{2}\right| \qquad \left|\ \because\ \cos\theta \leq 1 \text{ and } |\sin\theta| \leq |\theta|\right.$$

or $|f(x) - f(c)| \leq |x - c| < \delta\ (= \varepsilon)$ (By (1)) whenever $0 < |x - c| < \delta$

$\therefore$ By Def. $\underset{x\to c}{\text{Lt}}\ f(x) = f(c)$

i.e., $\underset{x\to c}{\text{Lt}}\ \sin x = \sin c.$

Example 4. *Using ε-δ definition, prove that* $\underset{x\to 1}{Lt}\dfrac{x^3-1}{x^2-1} = \dfrac{3}{2}.$

Sol. Here $f(x) = \dfrac{x^3-1}{x^2-1}$ when $x \to 1$ *i.e.* $x \neq 1$ and $l = \dfrac{3}{2}$ (given)

Let $\varepsilon > 0$, However small.

$$|f(x) - l| = \left|\frac{x^3-1}{x^2-1} - \frac{3}{2}\right| = \left|\frac{(x-1)(x^2+x+1)}{(x-1)(x+1)} - \frac{3}{2}\right|$$

$$(\because\ a^3 - b^3 = (a-b)(a^2 + b^2 + ab))$$

$$= \left|\frac{x^2+x+1}{x+1} - \frac{3}{2}\right| = \left|\frac{2x^2+2x+2-3x-3}{2(x+1)}\right| = \left|\frac{2x^2-x-1}{2(x+1)}\right|$$

$$= \left|\frac{(x-1)(2x+1)}{2x+2}\right|$$

$$\left|\ \because\ 2x^2 - x - 1 = 2x^2 - 2x + x - 1 = 2x(x-1) + (x-1) = (x-1)(2x+1)\right.$$

---

*Because we are free to take any value of δ depending on ε.

Here on looking at $|x|^2$, we have taken $\delta^2 = \varepsilon$ *i.e.,* $\delta = \sqrt{\varepsilon}$.

$$= |x-1|\left|\frac{2x+1}{2x+2}\right|$$

$$< |x-1| \qquad \left[\because |2x+1|<|2x+2| \text{ for all real } x \;\therefore \left|\frac{2x+1}{2x+2}\right| < 1\right.$$

Let us take $\delta = \varepsilon$ (See footnotes examples 1 and 2 on page 8)

$\therefore$ For $0 < |x-1| < \delta\,(=\varepsilon)$,

we have $|f(x) - l| \left(= \left|f(x) - \frac{3}{2}\right|\right) < |x-1| < \delta\,(=\varepsilon)$

$\therefore$ By Def. of limit, $\underset{x\to1}{\text{Lt}}\ f(x) = \frac{3}{2}$.

8. Next we consider the behaviour of a function $f(x)$ which increases continuously as x tends to c. If the value of $f(x)$ becomes and remains greater than any positive number (however large) when $x \to c$, then $f(x)$ increases beyond limit and we say that it approaches to the limit $+\infty$.

Def. 1. *A function $f(x)$ is said to tend to plus infinity $(+\infty)$, if corresponding to any positive number G (however large), we can find a positive number δ, such that*

$$f(x) < G \text{ for } 0 < |x-c| < \delta. \quad i.e., \quad \underset{x\to c}{\text{Lt}}\ f(x) = +\infty.$$

Def. 2. *A function $f(x)$ is said to tend to minus infinity $(-\infty)$, if corresponding to any positive number G (however large), we can find a positive number δ, such that*

$$f(x) < -G \text{ for } 0 < |x-c| < \delta. \quad i.e., \quad \underset{x\to c}{\text{Lt}}\ f(x) = -\infty.$$

Def. 3. *If $f(x)$ does not tend to a unique limit (finite or infinite) as $x \to c$, $f(x)$ is said to* **oscillate.**

If for all values of x in the neighbourhood of c, $f(x)$ remains numerically less than a finite constant k, we say that $f(x)$ oscillates finitely, otherwise it is said to oscillate infinitely.

Def. 4. *A function $f(x)$ is said to tend to the limit L as $x \to \infty$, if corresponding to any given positive number ε [however small], we can find a positive number N, such that,*

$$|f(x) - L| < \varepsilon \quad \text{for} \quad x > N. \quad i.e., \quad \underset{x\to+\infty}{\text{Lt}}\ f(x) = L.$$

Def. 5. *A function $f(x)$ is said to tend to plus infinity as $x \to \infty$, if corresponding to any given positive number G [however large], we can find a positive N, such that*

$$f(x) > G \quad \text{for} \quad x > N \quad i.e., \quad \underset{x\to+\infty}{\text{Lt}}\ f(x) = +\infty.$$

9. *To prove that $\underset{x\to0}{\text{Lt}}\ x^m = 0$, where m is a positive integer.*

Let $\varepsilon > 0$, However small. Let us choose a positive number $\delta\,(=\varepsilon)$ such that $|x-0| = |x| < \delta$.

Sol. Case I. Let $x \to 0$ through positive values.

$\because$ $x \to 0$, we may take $x < 1$

$\therefore$ $|x^m - 0| = x^m$

$\leq x\ (\because\ x < 1) = |x|\ (\because\ x > 0)$

$< \delta = \varepsilon$

Hence given $\varepsilon > 0$, we can find a positive number δ [equal to ε in this case], such that

$$|x^m| < \varepsilon \quad \text{for} \quad |x| < \varepsilon \quad \therefore \quad \underset{x\to 0+}{\text{Lt}} \; x^m = 0.$$

Case II. Let x be negative

Put $x = -y, \quad \therefore \quad y$ is positive

$$|x^m - 0| = |(-1)^m y^m| = y^m \leq y \qquad \text{(By Case I)}$$

$$< \varepsilon \text{ for } y < \varepsilon$$

$$\therefore \quad |x^m| < \varepsilon \text{ for } -\delta < x < 0 \text{ or } |x| < \delta \text{ (and } x < 0) \quad \therefore \quad \underset{x\to 0-}{\text{Lt}} \; x^m = 0.$$

Hence, right and left hand limits being equal, $x^m \to$ to the unique limit zero as $x \to 0$.

EXERCISE 1

Using the $\varepsilon - \delta$ definition of limit, prove the following :

1. $\underset{x\to 2}{\text{Lt}} \; (4x-5) = 3$
2. $\underset{x\to 1}{\text{Lt}} \; \dfrac{x^2-1}{x-1} = 2$
3. $\underset{x\to 2}{\text{Lt}} \; \dfrac{x^2-4}{x-2} = 4$
4. $\underset{x\to 3}{\text{Lt}} \; (1-3x) = -8$
5. $\underset{x\to -1}{\text{Lt}} \; (2-3x) = 5$
6. $\underset{x\to 2}{\text{Lt}} \; \dfrac{x^3-8}{x^2-4} = 3$ *(M.D.U. 2005 A)*
7. $\underset{x\to 0}{\text{Lt}} \; x \sin\dfrac{1}{x} = 0$ *(K.U. 2003)*
8. $\underset{x\to 0}{\text{Lt}} \; x^2 \cos\dfrac{1}{x} = 0$
9. $\underset{x\to c}{\text{Lt}} \; \cos x = \cos c.$

10. Some Standard Results on Limits

Here we remind the reader of some of the standard results, he or she has learnt in class XII and these results are

(*i*) $\underset{x\to 0}{\text{Lt}} \; \dfrac{\sin x}{x} = 1$ and $\underset{x\to 0}{\text{Lt}} \; \dfrac{\tan x}{x} = 1$ where x is the angle measured in Radians.

(*ii*) $\underset{x\to 0}{\text{Lt}} \; \dfrac{a^x - 1}{x} = \log a$ (*iii*) $\underset{x\to 0}{\text{Lt}} \; (1+x)^{\frac{1}{x}} = e$

(*iv*) $\underset{x\to \infty}{\text{Lt}} \left(1+\dfrac{1}{x}\right)^x = e$ (*v*) $\underset{x\to 0}{\text{Lt}} \; \dfrac{\sin^{-1} x}{x} = 1$ and $\underset{x\to 0}{\text{Lt}} \; \dfrac{\tan^{-1} x}{x} = 1.$

Here we also remind the reader about the left hand limit and right hand limit as learnt in class XII.

Left hand limit = $\underset{x\to c-}{\text{Lt}} \; f(x).$

$x \to c- \;\; \Rightarrow \;\; x$ is slightly less than c.

Put $x = c - h$ where $h \to 0^+$.

Right hand limit = $\underset{x\to c+}{\text{Lt}} \; f(x).$

$x \to c+ \;\; \Rightarrow \;\; x$ is slightly $> c$.

Put $x = c + h$ where $h \to 0^+$.

If Left Hand Limit $\underset{x\to c-}{\text{Lt}} \; f(x)$ = Right Hand Limit $\underset{x\to c+}{\text{Lt}} \; f(x)$, then we say that $\underset{x\to c}{\text{Lt}} \; f(x)$ exists.

Note. In problems on modulus function, greatest integer function, functions with partitioned domain *i.e.* $f(x)$ is defined differently for $x < c$ (say) and for $x > c$ and $f(x)$ involving $e^{\frac{1}{x}}$, we are to find left hand limit and right hand limit seperately.

Here, we explain the results of this Art. 10 by the help of some examples.

11. Algebra of Limits

If $f(x)$ and $\phi(x)$ are two functions of x, such that

$$\underset{x \to c}{\text{Lt}} \; f(x) = l \text{ and } \underset{x \to c}{\text{Lt}} \; \phi(x) = m.$$

Then

(i) $\underset{x \to c}{\text{Lt}} \; [f(x) \pm \phi(x)] = \underset{x \to c}{\text{Lt}} \; f(x) \pm \underset{x \to c}{\text{Lt}} \; \phi(x) = l \pm m$

(ii) $\underset{x \to c}{\text{Lt}} \; af(x) = a \underset{x \to c}{\text{Lt}} \; f(x) = a.l.$

(iii) $\underset{x \to c}{\text{Lt}} \; [f(x).\phi(x)] = \underset{x \to c}{\text{Lt}} \; f(x).\underset{x \to c}{\text{Lt}} \; \phi(x) = l.m$

(iv) $\underset{x \to c}{\text{Lt}} \left[\dfrac{f(x)}{\phi(x)}\right] = \underset{x \to c}{\text{Lt}} \; f(x) \cdot \underset{x \to c}{\text{Lt}} \; \dfrac{1}{\phi(x)} = \dfrac{l}{m}.$

12. Absolute Function and Greatest Integer Function

(a) **Absolute function** $|x| = x$ if $x \ge 0$ and $= -x$ if $x < 0$

i.e., $|x|$ is the Numerical value of x.

Note. $|x|$ is also defined as $\sqrt{x^2}$.

(b) **Greatest integer function**

If x is any real number,

then the greatest integer $\le x$ is denoted by $[x]$.

e.g., if $x = 3.9$; $[x] = 3$

if $x = 3.1$, $[x] = 3$

if $x = 5$, $[x] = 5$

if $x = -4.3$, $[x] = -5$.

Let $x = c$ be any real number.

Case I. c is an integer i.e., c ∈ Z or I

As $h \to 0^+$, $[c - h] = c - 1$ and $[c + h] = c$.

Case II. c ∈ R – Z i.e., c is a real number not an integer.

As $h \to 0^+$, then $[c - h] = [c]$ and $[c + h] = [c]$.

Note. $x - 1 < [x] \le x$.

Example 1. *Evaluate the following limits :*

(i) $\underset{x \to 2}{Lt} \dfrac{x^3 - 8}{x^2 - 4}$ (ii) $\underset{x \to 2}{Lt} \dfrac{x^2 - 4}{\sqrt{3x - 2} - \sqrt{x + 2}}$ (iii) $\underset{x \to 0}{Lt} \dfrac{(1 + x)^n - 1}{x}$.

Sol. (i) $\lim\limits_{x \to 2} \dfrac{x^3 - 8}{x^2 - 4} = \lim\limits_{x \to 2} \dfrac{x^3 - 2^3}{x^2 - 2^2}$ | Form $\dfrac{0}{0}$

$= \lim\limits_{x \to 2} \dfrac{(x - 2)(x^2 + 2x + 4)}{(x - 2)(x + 2)}$ | $\because a^3 - b^3 = (a - b)(a^2 + ab + b^2)$

$= \lim\limits_{x \to 2} \dfrac{x^2 + 2x + 4}{x + 2}$ [$\because x \to 2 \Rightarrow x \ne 2$]

$= \dfrac{2^2 + 2 \times 2 + 4}{2 + 2} = \dfrac{12}{4} = 3$

(ii) $\lim\limits_{x \to 2} \dfrac{x^2 - 4}{\sqrt{3x-2} - \sqrt{x+2}}$ $\quad \Big| \text{ Form } \dfrac{0}{0}$

Rationalising, $= \lim\limits_{x \to 2} \dfrac{x^2 - 4}{\sqrt{3x-2} - \sqrt{x+2}} \times \dfrac{\sqrt{3x-2} + \sqrt{x+2}}{\sqrt{3x-2} + \sqrt{x+2}}$

$$= \lim_{x \to 2} \frac{(x^2 - 4)(\sqrt{3x-2} + \sqrt{x+2})}{(3x-2) - (x+2)} = \lim_{x \to 2} \frac{(x^2 - 4)(\sqrt{3x-2} + \sqrt{x+2})}{2x - 4}$$

$$= \lim_{x \to 2} \frac{(x+2)(x-2)(\sqrt{3x-2} + \sqrt{x+2})}{2(x-2)}$$

Cancelling $(x - 2)$,

$$= \lim_{x \to 2} \frac{(x+2)(\sqrt{3x-2} + \sqrt{x+2})}{2} = \frac{(2+2)(\sqrt{3 \times 2 - 2} + \sqrt{2+2})}{2}$$

$$= \frac{(4)(2+2)}{2} = 8.$$

(iii) $\mathop{\text{Lt}}\limits_{x \to 0} \dfrac{(1+x)^n - 1}{x}$ $\quad \Big| \text{ Form } \dfrac{0}{0}$

Expanding by Binomial Theorem

$$= \mathop{\text{Lt}}_{x \to 0} \frac{1 + nx + \dfrac{n(n-1)}{2!} x^2 + \dfrac{n(n-1)(n-2)}{3!} x^3 + \ldots - 1}{x}$$

$$= \mathop{\text{Lt}}_{x \to 0} \frac{nx + \dfrac{n(n-1)}{2!} x^2 + \dfrac{n(n-1)(n-2)}{3!} x^3 + \ldots}{x}$$

$$= \mathop{\text{Lt}}_{x \to 0} \frac{nx \left(1 + \dfrac{(n-1)}{2!} x + \dfrac{(n-1)(n-2)}{3!} x^2 + \ldots\right)}{x}$$

$$= \mathop{\text{Lt}}_{x \to 0} n \left(1 + \frac{(n-1)}{2!} x + \frac{(n-1)(n-2)}{3!} x^2 + \ldots\right)$$

$$= n(1 + 0) = n.$$

Example 2. *Evaluate the following limits*

(i) $\mathop{Lt}\limits_{\theta \to 0} \dfrac{1 - \cos\theta}{\theta^2}$ *(ii)* $\mathop{Lt}\limits_{x \to 0} \dfrac{\sin x^\circ}{x}$ *(iii)* $\mathop{Lt}\limits_{\theta \to 0} \dfrac{1 - \cos m\theta}{1 - \cos n\theta}$

where m, n are fixed non-zero real numbers.

Sol. (i) $\mathop{\text{Lt}}\limits_{\theta \to 0} \dfrac{1 - \cos\theta}{\theta^2} = \mathop{\text{Lt}}\limits_{\theta \to 0} \dfrac{2\sin^2 \dfrac{\theta}{2}}{\theta^2} = 2 \mathop{\text{Lt}}\limits_{\theta \to 0} \left(\dfrac{\sin \dfrac{\theta}{2}}{2 \cdot \dfrac{\theta}{2}}\right)^2$

$$= \frac{2}{4} \mathop{\text{Lt}}_{\theta \to 0} \left(\frac{\sin \dfrac{\theta}{2}}{\dfrac{\theta}{2}}\right)^2 = \frac{1}{2} \left[\mathop{\text{Lt}}_{\frac{\theta}{2} \to 0} \frac{\sin \dfrac{\theta}{2}}{\dfrac{\theta}{2}}\right]^2 = \frac{1}{2}(1)^2 = \frac{1}{2}.$$

$$(ii)\quad \underset{x\to 0}{\text{Lt}} \frac{\sin x^0}{x} = \underset{x\to 0}{\text{Lt}} \frac{\sin \frac{\pi x}{180}}{x} \qquad \left[\begin{array}{l} \because\ 180 \text{ degrees} = \pi \text{ radians} \\ \therefore\ x \text{ degrees} = \frac{\pi x}{180} \text{ radians}\end{array}\right]$$

$$= \underset{x\to 0}{\text{Lt}} \frac{\sin \frac{\pi x}{180}}{\frac{\pi x}{180}} \times \frac{\pi}{180} = \frac{\pi}{180}\left[\underset{\pi x/180\to 0}{\text{Lt}} \frac{\sin \frac{\pi x}{180}}{\frac{\pi x}{180}}\right] = \frac{\pi}{180}(1) = \frac{\pi}{180}.$$

$$(iii)\quad \underset{\theta\to 0}{\text{Lt}} \frac{1-\cos m\theta}{1-\cos n\theta} = \underset{\theta\to 0}{\text{Lt}} \frac{2\sin^2 \frac{m\theta}{2}}{2\sin^2 \frac{n\theta}{2}} = \underset{\theta\to 0}{\text{Lt}} \left[\frac{\sin \frac{m\theta}{2}}{\frac{m\theta}{2}} \cdot \frac{\frac{m\theta}{2}}{\frac{n\theta}{2}} \cdot \frac{\frac{n\theta}{2}}{\sin \frac{n\theta}{2}}\right]^2$$

$$= \left[\underset{\frac{m\theta}{2}\to 0}{\text{Lt}} \frac{\sin \frac{m\theta}{2}}{\frac{m\theta}{2}}\right]^2 \left[\frac{m}{n}\right]^2 \left[\underset{\frac{n\theta}{2}\to 0}{\text{Lt}} \frac{\frac{n\theta}{2}}{\sin \frac{n\theta}{2}}\right]^2 = (1)^2\left(\frac{m^2}{n^2}\right)(1)^2 = \frac{m^2}{n^2}.$$

Example 3. *Evaluate the following limits :*

$$(i)\ \underset{x\to 0}{Lt} \frac{\sqrt{1+x}-\sqrt{1-x}}{\sin^{-1} x} \qquad (ii)\ \underset{x\to 1}{Lt} \frac{1-\sqrt{x}}{(\cos^{-1} x)^2}.$$

Sol. (i) $\underset{x\to 0}{\text{Lt}} \frac{\sqrt{1+x}-\sqrt{1-x}}{\sin^{-1} x}$

Rationalising, $= \underset{x\to 0}{\text{Lt}} \frac{(\sqrt{1+x}-\sqrt{1-x})(\sqrt{1+x}+\sqrt{1-x})}{\sin^{-1} x(\sqrt{1+x}+\sqrt{1-x})}$

$$= \underset{x\to 0}{\text{Lt}}\ 2\frac{x}{\sin^{-1} x} \cdot \frac{1}{\sqrt{1+x}+\sqrt{1-x}} \qquad [\because\ 1+x-(1-x) = 2x]$$

$$= 2(1)\frac{1}{1+1} = 1. \qquad \left|\ \because\ \underset{x\to 0}{\text{Lt}} \frac{\sin^{-1} x}{x} = 1\right.$$

(ii) Put $\cos^{-1} x = \theta$, then $x = \cos\theta$ so that as $x \to 1$, $\theta \to 0$

$$\therefore\ \underset{x\to 1}{\text{Lt}} \frac{1-\sqrt{x}}{(\cos^{-1} x)^2} = \underset{\theta\to 0}{\text{Lt}} \frac{1-\sqrt{\cos\theta}}{\theta^2}$$

Rationalising, $= \underset{\theta\to 0}{\text{Lt}} \frac{(1-\sqrt{\cos\theta})(1+\sqrt{\cos\theta})}{\theta^2(1+\sqrt{\cos\theta})} = \underset{\theta\to 0}{\text{Lt}} \frac{1-\cos\theta}{\theta^2(1+\sqrt{\cos\theta})}$

$$= \underset{\theta\to 0}{\text{Lt}} \frac{2\sin^2 \frac{\theta}{2}}{\theta^2(1+\sqrt{\cos\theta})} = \underset{\theta\to 0}{\text{Lt}} \frac{2\left(\frac{\sin \frac{\theta}{2}}{2\cdot\frac{\theta}{2}}\right)^2}{1+\sqrt{\cos\theta}} = \frac{2\cdot\left(\frac{1}{2}\right)^2}{1+1} = \frac{1}{4}$$

Remark. In problems on limits of Inverse Trigonometric functions, you can always put inverse function equal to θ.

Example 4. *Evaluate the following limits :*

$$(i)\ \underset{x\to\infty}{Lt} \frac{x}{\sqrt{4x^2+1}-1}$$

(ii) $\underset{n \to \infty}{Lt} \dfrac{1^2 + 2^2 + 3^2 + \ldots\ldots + n^2}{n^3}$ or $\underset{n \to \infty}{Lt} \dfrac{\Sigma n^2}{n^3}$.

Sol. (i) $\underset{x \to \infty}{\text{Lt}} \dfrac{x}{\sqrt{4x^2 + 1} - 1}$ Dividing the num. and denom. by x,

$$= \underset{x \to \infty}{\text{Lt}} \frac{1}{\sqrt{4 + \frac{1}{x^2}} - \frac{1}{x}} = \frac{1}{\sqrt{4 + 0} - 0} = \frac{1}{2}.$$

(ii) $$\underset{n \to \infty}{\text{Lt}} \frac{1^2 + 2^2 + 3^2 + \ldots\ldots + n^2}{n^3} = \underset{n \to \infty}{\text{Lt}} \frac{\frac{n(n+1)(2n+1)}{6}}{n^3}$$

$$= \underset{n \to \infty}{\text{Lt}} \frac{(n+1)(2n+1)}{6n^2}$$ Dividing the num. and denom. by n^2,

$$= \underset{n \to \infty}{\text{Lt}} \frac{\left(1 + \frac{1}{n}\right)\left(2 + \frac{1}{n}\right)}{6} = \frac{(1+0)(2+0)}{6} = \frac{1}{3}.$$

Example 5. *Evaluate the following limits :*

(i) $\underset{x \to 0}{Lt} \dfrac{3^x - 2^x}{x}$ (ii) $\underset{x \to 0}{Lt} \dfrac{(ab)^x - a^x - b^x + 1}{x^2}$

(iii) $\underset{x \to 1}{Lt} \dfrac{\log x}{x - 1}$ (iv) $\underset{x \to \infty}{Lt} \left(\dfrac{x+6}{x+1}\right)^{x+4}$.

Sol. (i) $\underset{x \to 0}{\text{Lt}} \dfrac{3^x - 2^x}{x}$ Subtract and add 1,

$$= \underset{x \to 0}{\text{Lt}} \frac{3^x - 1 - 2^x + 1}{x} = \underset{x \to 0}{\text{Lt}} \frac{(3^x - 1) - (2^x - 1)}{x} = \underset{x \to 0}{\text{Lt}} \left[\frac{3^x - 1}{x} - \frac{(2^x - 1)}{x}\right]$$

$$= \underset{x \to 0}{\text{Lt}} \left(\frac{3^x - 1}{x}\right) - \underset{x \to 0}{\text{Lt}} \left(\frac{2^x - 1}{x}\right) = \log 3 - \log 2 = \log \frac{3}{2}.$$

(ii) $(ab)^x - a^x - b^x + 1 = a^x b^x - a^x - b^x + 1 = a^x (b^x - 1) - (b^x - 1) = (a^x - 1)(b^x - 1)$

$\therefore$ Required Limit $= \underset{x \to 0}{\text{Lt}} \dfrac{(ab)^x - a^x - b^x + 1}{x^2} = \underset{x \to 0}{\text{Lt}} \dfrac{(a^x - 1)(b^x - 1)}{x \cdot x}$

$$= \underset{x \to 0}{\text{Lt}} \frac{a^x - 1}{x} \cdot \underset{x \to 0}{\text{Lt}} \frac{b^x - 1}{x} = (\log a)(\log b).$$

(iii) Put $x = 1 + h$ so that as $x \to 1, h \to 0$

$$\therefore \underset{x \to 1}{\text{Lt}} \frac{\log x}{x - 1} = \underset{h \to 0}{\text{Lt}} \frac{\log(1+h)}{1 + h - 1} = \underset{h \to 0}{\text{Lt}} \frac{\log(1+h)}{h} = \underset{h \to 0}{\text{Lt}} \frac{1}{h} \log(1+h)$$

$$= \underset{h \to 0}{\text{Lt}} \log(1+h)^{1/h} = \log e = 1. \quad \left[\because \underset{x \to 0}{\text{Lt}} (1+x)^{1/x} = e\right]$$

(iv) $$\underset{x \to \infty}{\text{Lt}} \left(\frac{x+6}{x+1}\right)^{x+4} = \underset{x \to \infty}{\text{Lt}} \left(\frac{x+1+5}{x+1}\right)^{x+4}$$

$$= \underset{x \to \infty}{\text{Lt}} \left(1 + \frac{5}{x+1}\right)^{x+4} = \underset{x \to \infty}{\text{Lt}} \left[\left(1 + \frac{5}{x+1}\right)^{\frac{x+1}{5}}\right]^{(x+4)\cdot\frac{5}{x+1}}$$

$$= e^5. \quad \left[\because \underset{x \to \infty}{\text{Lt}} \left(1 + \frac{1}{x}\right)^x = e\right.$$

and $$\underset{x \to \infty}{\text{Lt}} \frac{5(x+4)}{x+1} = \underset{x \to \infty}{\text{Lt}} \frac{5x\left(1+\frac{4}{x}\right)}{x\left(1+\frac{1}{x}\right)} = \underset{x \to \infty}{\text{Lt}} \frac{5\left(1+\frac{4}{x}\right)}{1+\frac{1}{x}} = 5\right]$$

Example 6. *Show that* $\underset{x \to 2}{Lt} \dfrac{|x-2|}{x-2}$ *does not exist.*

Sol. $$\text{L.H.L.} = \underset{x \to 2-}{\text{Lt}} \frac{|x-2|}{x-2} = \underset{h \to 0+}{\text{Lt}} \frac{|2-h-2|}{2-h-2} \quad [\text{by putting } x = 2-h,\ h > 0]$$

$$= \underset{h \to 0+}{\text{Lt}} \frac{|-h|}{-h} = \underset{h \to 0+}{\text{Lt}} \frac{h}{-h} = \underset{h \to 0+}{\text{Lt}} (-1) = -1$$

$$\text{R.H.L.} = \underset{x \to 2+}{\text{Lt}} \frac{|x-2|}{x-2} = \underset{h \to 0+}{\text{Lt}} \frac{|2+h-2|}{2+h-2} \quad [\text{by putting } x = 2+h,\ h > 0]$$

$$= \underset{h \to 0+}{\text{Lt}} \frac{|h|}{h} = \underset{h \to 0+}{\text{Lt}} \frac{h}{h} = \underset{h \to 0+}{\text{Lt}} 1 = 1.$$

Since L.H.L. ≠ R.H.L.

∴ $\underset{x \to 2}{\text{Lt}} \dfrac{|x-2|}{x-2}$ does not exist.

Example 7. *Find* $\underset{x \to 0}{Lt} \dfrac{x}{|x|+x^2}$.

Sol. $$\text{L.H.L.} = \underset{x \to 0-}{\text{Lt}} \frac{x}{|x|+x^2} = \underset{x \to 0-}{\text{Lt}} \frac{x}{-x+x^2} \quad [\because x \to 0- \Rightarrow x < 0 \Rightarrow |x| = -x]$$

$$= \underset{x \to 0-}{\text{Lt}} \frac{x}{x(-1+x)} = \underset{x \to 0-}{\text{Lt}} \frac{1}{-1+x} = \frac{1}{-1+0} = -1$$

$$\text{R.H.L.} = \underset{x \to 0+}{\text{Lt}} \frac{x}{|x|+x^2} = \underset{x \to 0+}{\text{Lt}} \frac{x}{x+x^2}$$

$$= \underset{x \to 0+}{\text{Lt}} \frac{x}{x(1+x)} = \underset{x \to 0+}{\text{Lt}} \frac{1}{1+x} = \frac{1}{1+0} = 1$$

Since L.H.L. ≠ R.H.L.

∴ $\underset{x \to 0}{\text{Lt}} \dfrac{x}{|x|+x^2}$ does not exist.

Example 8. *For what value of p does* $\underset{x \to 1}{Lt} f(x)$ *exist where f is defined by*

$$f(x) = \begin{cases} 2px + 3 & \text{if } x < 1 \\ 1 - px^2 & \text{if } x > 1. \end{cases}$$

Sol. $$\underset{x \to 1-}{\text{Lt}} f(x) = \underset{x \to 1-}{\text{Lt}} (2px + 3) \quad [\because \text{ when } x < 1, f(x) = 2px + 3]$$

$$= 2p(1) + 3 = 2p + 3$$

$$\underset{x \to 1+}{\text{Lt}}\ f(x) = \underset{x \to 1+}{\text{Lt}}\ (1 - px^2) \qquad [\because \text{ when } x > 1, f(x) = 1 - px^2]$$
$$= 1 - p(1)^2 = 1 - p.$$

Now, for the existence of $\underset{x \to 1}{\text{Lt}}\ f(x)$, we must have

$$\underset{x \to 1-}{\text{Lt}}\ f(x) = \underset{x \to 1+}{\text{Lt}}\ f(x)$$

i.e.,
$$2p + 3 = 1 - p \quad \text{or} \quad 3p = -2$$

Hence,
$$p = -\frac{2}{3}.$$

Example 9. *Show that* $\underset{x \to c}{Lt}\ [x]$ *where* $[x]$ *denotes the greatest integer* $\le x$ *does not exist if* c *is an integer.*

Sol.
$$\underset{x \to c-}{\text{Lt}}\ [x] = \underset{h \to 0+}{\text{Lt}}\ [c - h] \qquad [\text{Putting } x = c - h, h > 0]$$
$$= \underset{h \to 0+}{\text{Lt}}\ (c - 1) \qquad [\because \ c - 1 < c - h < c]$$
$$= c - 1 \qquad (\text{Also see Art. } 12(b))$$
$$\underset{x \to c+}{\text{Lt}}\ [x] = \underset{h \to 0+}{\text{Lt}}\ [c + h] \qquad [\text{Putting } x = c + h, h > 0]$$
$$= \underset{h \to 0+}{\text{Lt}}\ c \qquad [\because c < c + h < c + 1]$$
$$= c \qquad (\text{Also see Art. } 12(b))$$

Since
$$\underset{x \to c-}{\text{Lt}}\ [x] \ne \underset{x \to c+}{\text{Lt}}\ [x]$$

$\therefore \quad \underset{x \to c}{\text{Lt}}\ [x]$ does not exist.

Example 10. *Prove that* $\underset{x \to c}{Lt}\ [x]$ *where* $[x]$ *denotes the greatest integer* $\le x = [c]$, *if* c *is not an integer.*

Sol. Since c is not an integer, let n be the greatest integer less than c. *i.e.,* let $[c] = n$

Then $[c] = n$ where $n < c < n + 1$

$$\underset{x \to c-}{\text{Lt}}\ [x] = \underset{h \to 0+}{\text{Lt}}\ [c - h] \qquad [\text{Putting } x = c - h, h > 0]$$
$$= \underset{h \to 0+}{\text{Lt}}\ [c] = [c] = n \qquad [\text{By Art. } 12(b)]$$
$$\underset{x \to c+}{\lim}\ [x] = \underset{h \to 0+}{\lim}\ [c + h] \qquad [\text{Putting } x = c + h, h > 0]$$
$$= \underset{h \to 0+}{\text{Lt}}\ [c] = [c] = n \qquad [\text{By Art. } 12(b)]$$

Since
$$\underset{x \to c-}{\text{Lt}}\ [x] = \underset{x \to c+}{\text{Lt}}\ [x] = n \quad \therefore \quad \underset{x \to c}{\text{Lt}}\ [x] = n = [c].$$

EXERCISE 2

Evaluate the following limits : (1 – 9)

1. (*i*) $\underset{x \to 4}{\text{Lt}}\ \dfrac{x^2 - 16}{x^3 - 64}$ (*ii*) $\underset{x \to a}{\text{Lt}}\ \dfrac{\sqrt{x} - \sqrt{a}}{x - a}$

(*iii*) $\underset{x \to 1}{\text{Lt}}\ \dfrac{\sqrt{x^2 - 1} + \sqrt{x - 1}}{\sqrt{x^2 - 1}}$ (*iv*) $\underset{x \to 3}{\text{Lt}}\ \dfrac{x^2 - 9}{\sqrt{3x - 2} - \sqrt{x + 4}}$.

2. (i) $\underset{h \to 0}{\text{Lt}} \frac{1}{h}\left(\frac{1}{x+h} - \frac{1}{x}\right)$ (ii) $\underset{h \to 0}{\text{Lt}} \frac{1}{h}\left(\frac{1}{\sqrt{x+h}} - \frac{1}{\sqrt{x}}\right)$

3. (i) $\underset{\theta \to 0}{\text{Lt}} \frac{1-\cos\theta}{2\theta^2}$ (ii) $\underset{t \to 0}{\text{Lt}} \frac{1-\cos 6t}{t}$

(iii) $\underset{x \to 0}{\text{Lt}} \frac{\tan x - \sin x}{\sin^3 x}$ (iv) $\underset{x \to 0}{\text{Lt}} \frac{\tan x - \sin x}{x^3}$

4. (i) $\underset{x \to \frac{\pi}{2}}{\text{Lt}} \frac{1+\cos 2x}{(\pi - 2x)^2}$ (ii) $\underset{x \to \frac{\pi}{4}}{\text{Lt}} \frac{1-\tan x}{x - \frac{\pi}{4}}$ (iii) $\underset{x \to \frac{\pi}{2}}{\text{Lt}} (\sec x - \tan x)$

5. (i) $\underset{x \to 1}{\text{Lt}} \frac{1-x}{(\cos^{-1} x)^2}$ (ii) $\underset{x \to 1}{\text{Lt}} \frac{1-x}{\pi - 2\sin^{-1} x}$

6. (i) $\underset{n \to \infty}{\text{Lt}} \frac{1+2+3+\ldots+n}{1-n^2}$ (ii) $\underset{n \to \infty}{\text{Lt}} \frac{\Sigma n^3}{n^4}$

7. (i) $\underset{x \to 0}{\text{Lt}} \frac{a^x - 1}{b^x - 1}, a > 0, b > 0$ (ii) $\underset{x \to 0}{\text{Lt}} \frac{2^x - 1}{\sqrt{1+x} - 1}$ (*M.D.U. 2006*)

8. (i) $\underset{x \to 0}{\text{Lt}} \frac{\log(3+x) - \log(3-x)}{x}$ (ii) $\underset{x \to 0^+}{\text{Lt}} \frac{\log(1+x^3)}{\sin^3 x}$

9. (i) $\underset{x \to 0}{\text{Lt}} (1+3x)^{\frac{1}{x}}$ (ii) $\underset{x \to \infty}{\text{Lt}} \left(\frac{x+2}{x+3}\right)^x$.

10. Show that $\underset{x \to c}{\text{Lt}} \frac{|x-c|}{x-c}$ does not exist.

11. Show that the following limits do not exist :

(i) $\underset{x \to 3}{\text{Lt}} [x]$ (ii) $\underset{x \to 2}{\text{Lt}} (x - [x])$

where $[x]$ denotes the greatest integer $\le x$.

12. Find $\underset{x \to 3}{\text{Lt}} f(x)$ where $f(x) = \begin{cases} 4 & \text{if } x \ge 3 \\ x+1 & \text{if } x \le 3 \end{cases}$.

13. Show that $\underset{x \to 0}{\text{Lt}} e^{-1/x}$ does not exist.

ANSWERS

1. (i) $\frac{1}{6}$ (ii) $\frac{1}{2\sqrt{a}}$ (iii) $1+\frac{1}{\sqrt{2}}$ (iv) $6\sqrt{7}$

2. (i) $-\frac{1}{x^2}$ (ii) $-\frac{1}{2x^{3/2}}$

3. (i) $\frac{1}{4}$ (ii) 0 (iii) $\frac{1}{2}$ (iv) $\frac{1}{2}$

4. (i) $\frac{1}{2}$ (ii) -2 (iii) 0

5. (i) $\frac{1}{2}$ (ii) 0

6. (i) $\frac{-1}{2}$ (ii) $\frac{1}{4}$ 7. (i) $\log_b a$ (ii) $2 \log 2$

8. (i) $\frac{2}{3}$ (ii) 1

9. (i) e^3 (ii) $\frac{1}{e}$ 12. 4.

SECTION B
CONTINUITY

13. (*i*) Definition of Continuity at a Point

A function f (x) is said to be continuous at a point x = c if for $\varepsilon > 0$, however small, there exists a positive number δ depending on ε such that $|f(x) - f(c)| < \varepsilon$ for all x s.t. $|x - c| < \delta$.

Note. Comparing with the Definition of limit in Art. 5, we have $l = f(c)$ and instead of a deleted neighbourhood of c ($0 < |x - c| < \delta$) we have neighbourhood of c (*i.e.*, $|x - c| < \delta$).

From the above definition of continuity, we can say that

A function f (x) is said to be continuous at a point x = c, if and only if

$$\operatorname*{Lt}_{x \to c} f(x) = f(c)$$

i.e., the limit of the function = value of the function at that point.

This definition is based on the assumption that the function $f(x)$ has a well-defined value at the point, and the limit of $f(x)$ as $x \to c$ in any way is equal to its value.

Further $\operatorname*{Lt}_{x \to c} f(x)$ exists, if and only if both the limits of $f(x)$ as $x \to c$ from the right and $x \to c$ from the left exist and are equal.

Hence a function is continuous, if

$$\underset{\text{L.H. Lt}}{\operatorname*{Lt}_{x \to c-}} f(x) = f(c) = \underset{\text{R.H. Lt}}{\operatorname*{Lt}_{x \to c+}} f(x)$$

Cor. If $\operatorname*{Lt}_{x \to c-} f(x) = f(c)$, while either $\operatorname*{Lt}_{x \to c+} f(x)$ does not exist or is not equal to $f(c)$, then $f(x)$ is said to be **continuous on the left of c.**

If $f(c) = \operatorname*{Lt}_{x \to c+} f(x)$, while $\operatorname*{Lt}_{x \to c-} f(x)$ either does not exist or is not equal to $f(c)$, then $f(x)$ is said to be **continuous on the right of c.**

(*ii*) Definition of Continuity in an Interval. *A function $f(x)$ is said to be continuous in the closed interval $[a, b]$, if it is continuous for every value of x lying between a and b, continuous to the right of a, (i.e., $\operatorname*{Lt}_{x \to a+} f(x) = f(a)$) and continuous to the left of b (i.e., $\operatorname*{Lt}_{x \to b-} f(x) = f(b)$).*

14. Algebra of Continuous Functions

Theorem 1. The algebraic sum and product of a finite number of functions which are continuous at x = c, are continuous at x = c.

Proof. Let $f(x)$, $\phi(x)$, $\psi(x)$, ..., be finite number of functions which are continuous at $x = c$

$$\therefore \quad \operatorname*{Lt}_{x \to c} f(x) = f(c), \quad \operatorname*{Lt}_{x \to c} \phi(x) = \phi(c),$$

$$\operatorname*{Lt}_{x \to c} \psi(x) = \psi(c), \text{ and so on.}$$

From limits, we know that

$$\operatorname*{Lt}_{x \to c} [f(x) + \phi(x) - \psi(x) + \ldots\ldots]$$

$$= \underset{x \to c}{\text{Lt}} f(x) + \underset{x \to c}{\text{Lt}} \phi(x) - \underset{x \to c}{\text{Lt}} \psi(x) + \ldots\ldots$$

$$= f(c) + \phi(c) - \psi(c) + \ldots\ldots$$

$= \text{value of the function } [f(x) + \phi(x) - \psi(x) + \ldots\ldots]$

at $x = c$.

$\therefore$ $[f(x) + \phi(x) - \psi(x) + \ldots]$ is also continuous at $x = c$.

Hence the theorem.

Similarly, the theorem can be proved for the product.

Theorem 2. The quotient of two functions which are continuous at x = c is continuous at x = c, provided the denominator is not zero.

Proof. Let $f(x)$ and $\phi(x)$ be any two functions which are continuous at $x = c$.

$$\therefore \qquad \underset{x \to c}{\text{Lt}} f(x) = f(c) \text{ and } \underset{x \to c}{\text{Lt}} \phi(x) = \phi(c)$$

From limits, we know that

$$\underset{x \to c}{\text{Lt}} \frac{f(x)}{\phi(x)} = \frac{\underset{x \to c}{\text{Lt}} f(x)}{\underset{x \to c}{\text{Lt}} \phi(x)} \text{ provided } \underset{x \to c}{\text{Lt}} \phi(x) \neq 0 = \frac{f(c)}{\phi(c)}$$

$$= \text{value of the function } \frac{f(x)}{\phi(x)} \text{ at } x = c.$$

$$\therefore \qquad \frac{f(x)}{\phi(x)} \text{ is also continuous at } x = c.$$

Hence the theorem.

Theorem 3. If f is continuous, | f | is also continuous,

Proof. Let f *i.e.*, $f(x)$ be continuous at $x = c$.

$\therefore$ By Def. Art. 13(*i*), for $\varepsilon > 0$, (however small), there exists a positive real number δ (depending on ε) s.t.

$$|f(x) - f(c)| < \varepsilon \text{ whenever } |x - c| < \delta \qquad \ldots(1)$$

We know that $\quad ||f(x)| - |f(c)|| \leq |f(x) - f(c)| \qquad (\because \ ||\alpha| - |\beta|| \leq |\alpha - \beta|)$

$< \varepsilon$ whenever $|x - c| < \delta$ *by (1)*

$$\therefore \qquad \underset{x \to c}{\text{Lt}} |f(x)| = |f(c)|$$

$\therefore$ $\quad |f|$ is continuous at $x = c$.

Remark. But the converse is not true.

Note. (*i*) $a + bx$ is a continuous function of x, where a and b are real numbers.

(*ii*) Polynomials in x are continuous functions of x.

(*iii*) A rational function in x is a continuous function of x at all those values of x which do not make the denominator zero.

(*iv*) $\dfrac{1}{x}$ is a continuous function of x, at all $x \neq 0$.

(*v*) e^x, e^{-x}, $\sin x$ and $\cos x$ are continuous functions of x for all real values of x.

(*vi*) $\tan x$ is a continuous function except at $x = \pm\dfrac{\pi}{2}, \pm\dfrac{3\pi}{2}, \ldots\ldots$

Example 1. *Using the $\varepsilon - \delta$ definition of continuity, show that sin x is continuous for every value of x.* (K.U. 2007)

Sol. Let $f(x) = \sin x$

Let c be any real number.

$$\therefore \quad f(c) = \sin c.$$

Let $\varepsilon > 0$ be any positive number, however small. For $f(x)$ to be continuous at $x = c$,

$$| f(x) - f(c) | < \varepsilon \text{ whenever } | x - c | < \delta. \quad \text{(By Def. Art. 13}(i))$$

We have $\quad | f(x) - f(c) | = | \sin x - \sin c |$

$$= \left| 2 \cos \frac{x+c}{2} \sin \frac{x-c}{2} \right|$$

$$= 2 \left| \cos \frac{x+c}{2} \right| \left| \sin \frac{x-c}{2} \right| \quad [\because \ | ab | = | a | \ | b | \ ]$$

$$\leq 2 \left| \sin \frac{x-c}{2} \right| \quad \left(\because \left| \cos \frac{x+c}{2} \right| \leq 1 \text{ as } |\cos \theta| \leq 1 \right)$$

$$\leq 2 \left| \frac{x-c}{2} \right|. \quad [\because \ | \sin \theta | \leq | \theta |]$$

$$\therefore \quad | f(x) - f(c) | \leq | x - c | \quad \ldots(1)$$

Let us take $\delta = \varepsilon$.

$\therefore$ For $| x - c | < \delta \, (= \varepsilon)$, we have from (1) that $| f(x) - f(c) | \leq | x - c | < \delta \, (= \varepsilon)$

i.e., $\quad | f(x) - f(c) | < \varepsilon$ whenever $| x - c | < \delta$.

$\therefore$ $f(x)$ is continuous at $x = c$.

But c is any real value of x.

$\therefore$ $f(x)$ is continuous for all values of x.

Example 2. *Using the $\varepsilon - \delta$ definition of continuity, show that $\cos^2 x$ is continuous for all real values of x.* (M.D.U. 2003 S ; K.U. 2005, 2006)

Sol. Let $\quad f(x) = \cos^2 x$

Let c be any real number.

$$\therefore \quad f(c) = \cos^2 c$$

Let $\varepsilon > 0$ be any positive number, however small.

$$| f(x) - f(c) | = | \cos^2 x - \cos^2 c |$$
$$= | (1 - \sin^2 x) - (1 - \sin^2 c) | = | - \sin^2 x + \sin^2 c |$$
$$= | - (\sin^2 x - \sin^2 c) | = | \sin^2 x - \sin^2 c |$$
$$= | \sin (x + c) \sin (x - c) | \quad (\because \ \sin^2 A - \sin^2 B = \sin (A + B) \sin (A - B))$$
$$= | \sin (x + c) | \ | \sin (x - c) |$$
$$\leq 1 \, . \, | x - c | \quad | \because \ | \sin (x + c) | \leq 1 \text{ and } | \sin \theta | \leq | \theta |$$

or $\quad | f(x) - f(c) | \leq | x - c | \quad \ldots(1)$

Let us take $\quad \delta = \varepsilon$.

$\therefore$ For $| x - c | < \delta \, (= \varepsilon)$, we have from (1) that $| f(x) - f(c) | \leq | x - c | < \delta \, (= \varepsilon)$

i.e., $\quad | f(x) - f(c) | < \varepsilon$ whenever $| x - c | < \delta$.

$\therefore$ $\quad f(x)$ is continuous at $x = c$.

But c is any real value of x.

$\therefore$ $f(x)$ is continuous for all real values of x.

Example 3. $f(x) = x \sin \frac{1}{x}$; $x \neq 0$.

$= 0$ when $x = 0$

Using the $\varepsilon - \delta$ *definition of continuity, show that* $f(x)$ *is continuous at* $x = 0$.

Sol. Let $\varepsilon > 0$ be any positive number, however small.

$$| f(x) - f(0) | = \left| x \sin \frac{1}{x} - 0 \right| = \left| x \sin \frac{1}{x} \right| = |x| \left| \sin \frac{1}{x} \right|$$

$$\leq |x| (1) \qquad | \because \ | \sin \theta | \leq 1 \text{ for all real } \theta$$

or $$| f(x) - f(0) | \leq | x | \qquad ...(1)$$

Let us choose a positive number $\delta\ (= \varepsilon)$

s.t. $| x - 0 | < \delta$ *i.e.,* $| x | < \delta$

$\therefore$ From (1) $| f(x) - f(0) | \leq | x | < \delta\ (= \varepsilon)$

whenever $| x - 0 | < \delta$.

$\therefore$ $f(x)$ is continuous at $x = 0$.

15. Discontinuous Function

(M.D.U. 2003)

A function which is not continuous is said to be **discontinuous at the point.**

A function is discontinuous :

(*i*) If a function is not defined at the point.

(*ii*) If one or both the limits [L.H. Lt. and R.H. Lt.] do not exist as x approaches the point.

(*iii*) If both these limits exist and are either unequal or if equal, not equal to the value of the function at the point.

(*iv*) If either or both the limits are infinite.

16. Kinds of Discontinuity

(M.D.U. 2003)

(*i*) **Discontinuity of the First Kind. Def.** If $\underset{x \to a+}{\text{Lt}} f(x)$ and $\underset{x \to a-}{\text{Lt}} f(x)$ exist but are not equal to one another, then $f(x)$ is said to have a discontinuity of the first kind at $x = a$.

(*ii*) **Discontinuity of the Second Kind. Def.** If neither $\underset{x \to a+}{\text{Lt}} f(x)$ nor $\underset{x \to a-}{\text{Lt}} f(x)$ exists, then the function $f(x)$ is said to have discontinuity of the second kind at $x = a$.

(*iii*) **Mixed Discontinuity. Def.** If only one of the limits $\underset{x \to a^+}{\text{Lt}} f(x)$ or $\underset{x \to a-}{\text{Lt}} f(x)$ exists, then the function $f(x)$ is said to have a mixed discontinuity at $x = a$.

(*iv*) **Removable Discontinuity. Def.** If $\underset{x \to a+}{\text{Lt}} f(x) = \underset{x \to c-}{\text{Lt}} f(x)$ *i.e.,* $\underset{x \to a}{\text{Lt}} f(x)$ exists and $\neq f(a)$, then $f(x)$ is said to have a removable discontinuity at $x = a$.

(*v*) **Infinite Discontinuity.** If either or both limits $\underset{x \to a+}{\text{Lt}} f(x)$ and $\underset{x \to c-}{\text{Lt}} f(x)$ are infinite, the function $f(x)$ is said to have an infinite discontinuity at $x = a$.

Examples on each of the five types of discontinuities. (*M.D.U. 2003*)

(*i*) Let $f(x)=\dfrac{|x|}{x}$ when $x \neq 0$

$= 0$ when $x = 0$

Left Hand Limit $= \underset{x \to c-}{\text{Lt}} f(x) = \underset{x \to 0+}{\text{Lt}} \dfrac{|x|}{x}$

put $x = 0 - h = -h$ as $h \to 0+ = \underset{h \to 0+}{\text{Lt}} \dfrac{|-h|}{-h} = \underset{h \to 0+}{\text{Lt}} \dfrac{h}{-h} = \underset{h \to 0+}{\text{Lt}} -1 = -1$ ...(1)

Right Hand Limit $= \underset{x \to 0+}{\text{Lt}} f(x) = \underset{x \to 0+}{\text{Lt}} \dfrac{|x|}{x}$

put $x = 0 + h = h$ as $h \to 0+ = \underset{h \to 0+}{\text{Lt}} \dfrac{|h|}{h} = \underset{h \to 0+}{\text{Lt}} \dfrac{h}{h} = \underset{h \to 0+}{\text{Lt}} 1 = 1$ ...(2)

From (1) and (2) we can say that $\underset{x \to c-}{\text{Lt}} f(x)$ and $\underset{x \to a+}{\text{Lt}} f(x)$ both exist but are not equal.

$\therefore$ $f(x)$ is discontinuous at $x = 0$ and *discontinuity is of first kind* (By Art. 16)

(*ii*) Let $f(x) = \sin\dfrac{1}{x}$ when $x \neq 0$ and $= 0$ when $x = 0$.

Here neither $\underset{x \to c-}{\text{Lt}} f(x)$ nor $\underset{x \to a+}{\text{Lt}} f(x)$ exists (For proof see example 9, page 29)

$\therefore$ $f(x)$ is discontinuous at $x = 0$ and *discontinuity is of second kind.*

(*iii*) Let $f(x) = 1$ when $x \leq 0 = \sin\dfrac{1}{x}$ when $x > 0$

$\underset{x \to 0-}{\text{Lt}} f(x) = \underset{x \to 0-}{\text{Lt}} 1 = 1$ exists

$\underset{x \to c-}{\text{Lt}} f(x) = \underset{x \to a+}{\text{Lt}} \sin\dfrac{1}{x}$ does not exist

(See example 9, page 29)

$\therefore$ $f(x)$ is discontinuous at $x = 0$ and discontinuity is a **mixed discontinuity** (By Art. 16)

(*iv*) Let $f(x) = \dfrac{x + \sin x}{x}$ ...(1)

$\underset{x \to 0}{\text{Lt}} f(x) = \underset{x \to 0}{\text{Lt}} \dfrac{x + \sin x}{x} = \underset{x \to 0}{\text{Lt}} \left(1 + \dfrac{\sin x}{x}\right) = 1 + 1 = 2$ ...(2)

Putting $x = 0$ in (1), $f(0) = \dfrac{0}{0}$ which is undefined. ...(3)

From (2) and (3), we can say that $\underset{x \to 0}{\text{Lt}} f(x) \neq f(0)$.

$\therefore$ $f(x)$ is discontinuous at $x = 0$ and $x = 0$ is a point of **removable discontinuity** (By Art. 16)

To remove this discontinuity, let us define $f(x)$ as below :

$f(x) = \dfrac{x + \sin x}{x}$ if $x \neq 0$ and $= 2$ if $x = 0$.

$\therefore$ $f(x)$ is continuous at $x = 0$ [$\because$ By (2) above $\underset{x \to 0}{\text{Lt}} f(x) = 2 = f(0)$]

(*v*) Let $f(x) = \dfrac{1}{x}$

$\text{Lt}\ f(x) = \underset{x \to 0+}{\text{Lt}} \dfrac{1}{x} = \infty$ and hence does not exist.

$$\underset{x \to 0-}{\text{Lt}} f(x) = \underset{x \to 0-}{\text{Lt}} \frac{1}{x} = -\infty \text{ and hence does not exist.}$$

$\therefore$ $x = 0$ is a point of *infinite discontinuity* of $f(x) = \dfrac{1}{x}$

17. Graphical Representation of Continuity

Let $y = f(x)$ be continuous at $x = a$, and let $P[a, f(a)]$, and $Q[a + h, f(a + h)]$ be two points on the graph.

Since $f(x)$ is continuous at $x = a$; $\therefore$ $\underset{x \to a}{\text{Lt}} f(x) = f(a)$

or $$f(a + h) \to f(a) \text{ as } h \to 0.$$

i.e., as $h \to 0$ the abscissa and ordinate of the points Q tend to the abscissa and ordinate respectively of the point *P*.

i.e., as $h \to 0$, point Q $\to$ P

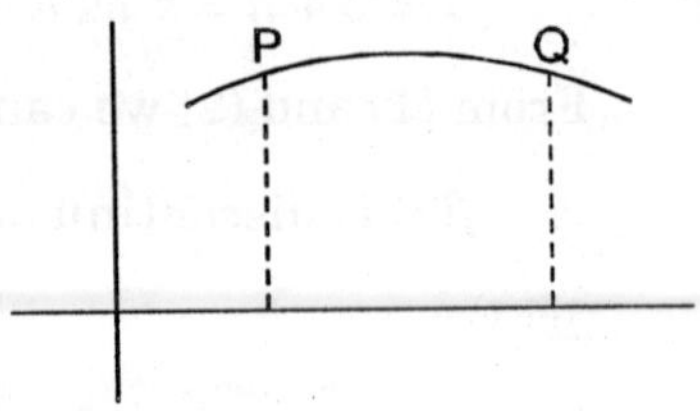

Thus the graph of the function $y = f(x)$ is **without break** at *P*.

Hence if a function is continuous throughout in an interval (a, b), the graph of the function in this interval is without break, *e.g.,* graphs of $\sin x$, $\cos x$.

If a function is discontinuous at a point $x = a$, then there is necessarily a break in its graph at the point corresponding to $x = a$.

If the function is not defined at $x = a$, then there is a break in the graph as there is no point on the graph corresponding to $x = a$, *e.g.,* graphs of $\tan x$, $\cot x$, $\sec x$, $\text{cosec } x$.

Note. If the value of the function at the point is not given in question the function may be said as **not defined** and evidently, the function will be discontinuous function.

Example 1. *Discuss the continuity of* $f(x) = x \sin \dfrac{1}{x}$, *when* $x \neq 0$ *and* $f(0) = 0$ *at* $x = 0$.

Sol. Here, we have $f(x) = x \sin \dfrac{1}{x}$, for $x \neq 0$ and $f(0) = 0$.

Continuity at x = 0.

$$\underset{x \to 0}{\text{Lt}} f(x) = \underset{x \to 0}{\text{Lt}} x \sin \frac{1}{x}$$

Put $x = 0 + h = h$ as $h \to 0 = \underset{h \to 0}{\text{Lt}} h \sin \dfrac{1}{h}$

$= 0 \times$ [A finite quantity between -1 and $+1$]

$= 0$ $\therefore$ $\underset{x \to 0}{\text{Lt}} f(x)$ exists and $= 0$.

Also $f(0) = 0$ (given)

$\therefore$ $\underset{x \to 0}{\text{Lt}} f(x)$ exists and $= f(0)$.

Hence the function is continuous at $x = 0$.

Remark. $\underset{x \to 0}{\text{Lt}} x \sin \dfrac{1}{x} = \underset{x \to 0}{\text{Lt}} \dfrac{\sin \frac{1}{x}}{\frac{1}{x}} \neq 1.$

$$\Rightarrow \qquad x = \frac{3p^3}{2a} - \frac{\lambda}{a}p + \frac{\lambda^2}{a}\log(p+\lambda) - c.$$

Putting the value of x in (1), we get

$$y = \frac{p^3}{a} + \frac{\lambda p^2}{a} - \frac{3\lambda p^2}{2a} + \frac{\lambda^2 p}{a} - \frac{\lambda^3}{a}\log(p+\lambda) + c\lambda$$

$$\Rightarrow \qquad y = \frac{p^3}{a} - \frac{\lambda}{2a}p^2 + \frac{\lambda^2 p}{a} - \frac{\lambda^3}{a}\log(p+\lambda) + c\lambda.$$

$\therefore$ The general solution of the given equation is

$$\mathbf{x = \frac{3p^2}{2a} - \frac{\lambda}{a}p + \frac{\lambda^2}{a}\log(p+\lambda) - c, \quad y = \frac{p^3}{a} - \frac{\lambda}{2a}p^2 + \frac{\lambda^2}{a}p - \frac{\lambda^3}{a}\log(p+\lambda) + c\lambda,}$$

where p is a parameter.

Example 6. *Solve* $x = yp + p^2$.

Sol. We have $x = yp + p^2 \Rightarrow y = \dfrac{x}{p} - p$

Differentiating w.r.t. x, we get

$$\frac{dy}{dx} = \frac{1}{p}.1 + x\left(-\frac{1}{p^2}\right)\frac{dp}{dx} - \frac{dp}{dx} \qquad \Rightarrow \qquad p - \frac{1}{p} = -\left(\frac{x}{p^2}+1\right)\frac{dp}{dx}$$

$$\Rightarrow \qquad p^3 - p = -(x+p^2)\frac{dp}{dx} \qquad \Rightarrow \qquad (p^3-p)\frac{dx}{dp} + x = -p^2$$

[Note this step]

$$\Rightarrow \qquad \frac{dx}{dp} + \frac{x}{p^3-p} = \frac{-p}{p^2-1} \qquad \text{...(1)}$$

(1) is a linear equation.

$$\text{I.F.} = e^{\int \frac{1}{p^3-p}dp}$$

Now
$$\int \frac{dp}{p^3-p} = \int \frac{dp}{p(p-1)(p+1)}$$

$$= \int \left[\frac{1}{p(-1)(1)} + \frac{1}{(1)(p-1)(2)} + \frac{1}{(-1)(-2)(p+1)}\right]dp$$

$$= \int \left[-\frac{1}{p} + \frac{1}{2(p-1)} + \frac{1}{2(p+1)}\right]dp$$

$$= -\log p + \frac{1}{2}\log(p-1) + \frac{1}{2}\log(p+1) = \log\frac{\sqrt{p^2-1}}{p}.$$

$$\therefore \qquad \text{I.F.} = e^{\log\frac{\sqrt{p^2-1}}{p}} = \frac{\sqrt{p^2-1}}{p}.$$

$\therefore$ The solution of (1) is

$$x \cdot \frac{\sqrt{p^2-1}}{p} = \int -\frac{p}{p^2-1} \cdot \frac{\sqrt{p^2-1}}{p}\, dp + c$$

$$\Rightarrow \quad \frac{x\sqrt{p^2-1}}{p} = -\log(p+\sqrt{p^2-1}) + c$$

$$\Rightarrow \quad x = -\frac{p}{\sqrt{p^2-1}} \log(p+\sqrt{p^2-1}) + \frac{cp}{\sqrt{p^2-1}}$$

Putting the value of x in the given equation, we get

$$y = -\frac{1}{\sqrt{p^2-1}} \log(p+\sqrt{p^2-1}) + \frac{c}{\sqrt{p^2-1}} - p.$$

$\therefore$ The general solution of the given equation is

$$\mathbf{x = -\frac{p}{\sqrt{p^2-1}} \log(p+\sqrt{p^2-1}) + \frac{cp}{\sqrt{p^2-1}},}$$

$$\mathbf{y = -\frac{1}{\sqrt{p^2-1}} \log(p+\sqrt{p^2-1}) + \frac{c}{\sqrt{p^2-1}} - p,}$$

where p is a parameter.

Example 7. *Solve* $y = xp^2 - \dfrac{1}{p}$. (M.D.U. 2006)

Sol. We have $\quad y = xp^2 - \dfrac{1}{p}$. ...(1)

Differentiating (1) w.r.t. x, we get

$$\frac{dy}{dx} = x \cdot 2p\frac{dp}{dx} + p^2 \cdot 1 + \frac{1}{p^2}\frac{dp}{dx}$$

$$\Rightarrow \quad p = \left(2xp + \frac{1}{p^2}\right)\frac{dp}{dx} + p^2$$

$$\Rightarrow \quad p - p^2 = \frac{2xp^3+1}{p^2}\frac{dp}{dx}$$

$$\Rightarrow \quad p^3 - p^4 = (2xp^3+1)\frac{dp}{dx}$$

$$\Rightarrow \quad \frac{dx}{dp} = \frac{2xp^3+1}{p^3(1-p)} = \frac{2x}{1-p} + \frac{1}{p^3(1-p)}$$

$$\Rightarrow \quad \frac{dx}{dp} + \left(\frac{2}{p-1}\right)x = \frac{1}{p^3(1-p)} \qquad ...(2)$$

(2) is a linear equation.

$$\text{I.F.} = e^{\int \frac{2}{p-1}\, dp} = e^{2 \log (p-1)} = e^{\log (p-1)^2} = (p-1)^2$$

$\therefore$ The solution of (2) is

$$x \,.\, (p-1)^2 = \int \frac{(p-1)^2}{p^3 (1-p)}\, dp + c$$

$$\Rightarrow \qquad x (p-1)^2 = \int \left(\frac{1}{p^3} - \frac{1}{p^2} \right) dp + c$$

$$\Rightarrow \qquad x (p-1)^2 = -\frac{1}{2p^2} + \frac{1}{p} + c$$

$$\Rightarrow \qquad x (p-1)^2 = \frac{2p-1}{2p^2} + c$$

$$\Rightarrow \qquad x = \frac{2p-1}{2p^2 (p-1)^2} + \frac{c}{(p-1)^2}$$

Putting the value of x in (1), we get

$$y = p^2 \left[\frac{2p-1}{2p^2 (p-1)^2} + \frac{c}{(p-1)^2} \right] - \frac{1}{p} = \frac{2p-1}{2(p-1)^2} + \frac{cp^2}{(p-1)^2} - \frac{1}{p}$$

$\therefore$ The general solution of the given equation is

$$x = \frac{2p-1}{2p^2 (p-1)^2} + \frac{c}{(p-1)^2},$$

$$y = \frac{2p-1}{2(p-1)^2} + \frac{cp^2}{(p-1)^2} - \frac{1}{p},$$

where p is a parameter.

EXERCISE 2

Solve the following differential equations :

1. $xp^2 - 2yp + ax = 0$

2. $y - 2px = \tan^{-1} (xp^2)$

3. $16x^2 + 2p^2y - p^3x = 0$

4. $y = x + 2 \tan^{-1} p$

5. $y = 3x + \log p$ *(M.D.U. 2006)*

6. $x - yp = ap^2$ *(M.D.U. 2007)*

7. $x^2 \left(\frac{dy}{dx} \right)^4 + 2x \frac{dy}{dx} - y = 0$

8. $y = 1 + xp^3$

9. $xp^2 - 2yp - ax = 0$

10. $y = 2p + \sqrt{1+p^2}$

11. $x^2 + p^2x = yp$

12. $y = p \tan p + \log \cos p$

13. $y = x(1 + p) + p^2$

14. $y = 2px + \phi(xp^2)$.

Answers

1. $2y = cx^2 + \dfrac{a}{c}$ 2. $y = 2\sqrt{cx} + \tan^{-1} c$ 3. $16 + 2c^2y - c^3x^2 = 0$

4. $x + c = \log \dfrac{p-1}{\sqrt{p^2+1}} - \tan^{-1} p,\ y = x + 2\tan^{-1} p$ 5. $y = 3x + \log \dfrac{3}{1 - ce^{3x}}$

6. $x = \dfrac{p}{\sqrt{p^2-1}}(c - a\log(p + \sqrt{p^2-1})),\ y = \dfrac{1}{\sqrt{p^2-1}}(c - a\log(p+\sqrt{p^2-1})) - ap$

7. $y = c^2 + 2\sqrt{cx}$ 8. $x^{2/3} = (y-1)^{2/3} + c$

9. $2cy = c^2x^2 + a$

10. $x = 2\log p + \log\left(p + \sqrt{1+p^2}\right) + c,\ y = 2p + \sqrt{1+p^2}$

11. $x = c\sqrt{p} - \dfrac{1}{3}p^2,\ y = \dfrac{c}{3}p^{3/2} - \dfrac{2}{9}p^3 + c^2$ 12. $x = \tan p + c,\ y = p\tan p + \log\cos p$

13. $x = 2(1-p) + ce^{-p},\ y = 2 + p^2 + (1+p)ce^{-p}$ 14. $y = 2\sqrt{cx} + \phi(c)$.

4. Equations Solvable for x

If the given equation is solvable for x, we can express x explicitly in terms of y and p. Thus, the equations of this type can be put as $x = f(y, p)$. ...(1)

Differentiating (1) w.r.t. y, we get

$$\frac{dx}{dy} = \frac{1}{p} = \frac{\partial f}{\partial y} + \frac{\partial f}{\partial p}\frac{dp}{dy} = F\left(y, p, \frac{dp}{dy}\right), \text{ say} \quad \Rightarrow \quad \frac{1}{p} = F\left(x, p, \frac{dp}{dy}\right) \qquad ...(2)$$

Equation (2) is a differential equation of first order in p and y.

Suppose the solution of (2) is $\phi(y, p, c) = 0$ ...(3)

The elimination of p from (1) and (3) gives the general solution of (1).

Note. If p cannot be easily eliminated, then we solve equations (1) and (3) for x and y to get

$$\mathbf{x = \phi_1(p, c),\ y = \phi_2(p, c).}$$

These two relations together constitute the general solution of the given equation with p as a parameter.

Example 1. *Solve* $x = y + a\log p.$

Sol. We have $x = y + a\log p.$ ...(1)

Diff. w.r.t. y, we get $\dfrac{dx}{dy} = 1 + \dfrac{a}{p}\dfrac{dp}{dy}$.

$$\Rightarrow \quad \frac{1}{p} - 1 = \frac{a}{p}\frac{dp}{dy} \quad \Rightarrow \quad 1 - p = a\frac{dp}{dy} \quad \Rightarrow \quad dy = \frac{a}{1-p}dp$$

Integrating, we get $y = -a\log(1-p) + c$

Putting the value of y in (1), we get

$$x = -a\log(1-p) + c + a\log p \quad \text{or} \quad x = a\log\frac{p}{1-p} + c.$$

$\therefore$ The general solution of the given equation is

$$\mathbf{x = a \log \frac{p}{1-p} + c, \; y = -a \log (1-p) + c,}$$

where p is a parameter.

Example 2. *Solve* $xp^3 = a + bp$.

Sol. We have $x = \dfrac{a}{p^3} + \dfrac{b}{p^2}$.

Diff. w.r.t. y, we get $\dfrac{dx}{dy} = \left(-\dfrac{3a}{p^4} - \dfrac{2b}{p^3}\right)\dfrac{dp}{dy}$

$$\Rightarrow \quad \frac{1}{p} = -\frac{1}{p^4}(3a + 2bp)\frac{dp}{dy} \quad \Rightarrow \quad dy = -\left(\frac{3a}{p^3} + \frac{2b}{p^2}\right)dp$$

Integrating, we get $y = -\dfrac{3ap^{-2}}{-2} - \dfrac{2bp^{-1}}{-1} + c \quad \Rightarrow \quad y = \dfrac{3a}{2p^2} + \dfrac{2b}{p} + c$ ...(1)

Using given equation and (1), the general solution of the given equation is

$$\mathbf{x = \frac{a}{p^3} + \frac{b}{p^2}, \; y = \frac{3a}{2p^2} + \frac{2b}{p} + c,}$$

where p is a parameter.

Example 3. *Solve* $y = 2px + y^2p^3$.

Sol. We have $y = 2px + y^2p^3$ ...(1)

$$\Rightarrow \quad 2px = y - y^2p^3 \quad \Rightarrow \quad x = \frac{1}{2}\left(\frac{y}{p} - y^2p^2\right) \qquad ...(2)$$

Differentiating w.r.t. y, we get

$$\frac{dx}{dy} = \frac{1}{2}\left(\frac{1}{p} - \frac{y}{p^2}\frac{dp}{dy} - 2yp^2 - 2y^2p\frac{dp}{dy}\right)$$

$$\Rightarrow \quad \frac{1}{p} = \frac{1}{2p} - yp^2 - \left(\frac{y}{2p^2} + y^2p\right)\frac{dp}{dy}$$

$$\Rightarrow \quad \frac{1}{2p} + yp^2 + \left(\frac{y}{2p^2} + y^2p\right)\frac{dp}{dy} = 0$$

$$\Rightarrow \quad \frac{p}{y}\left(\frac{y}{2p^2} + y^2p\right) + \left(\frac{y}{2p^2} + y^2p\right)\frac{dp}{dy} = 0$$

$$\Rightarrow \quad \left(\frac{y}{2p^2} + y^2p\right)\left(\frac{p}{y} + \frac{dp}{dy}\right) = 0$$

$$\Rightarrow \quad \frac{p}{y} + \frac{dp}{dy} = 0 \quad \Rightarrow \quad \frac{dy}{y} + \frac{dp}{p} = 0$$

Integrating $\log y + \log p = \log c$ or $py = c$ or $p = \dfrac{c}{y}$

Putting this value of p in (1), we have

$$y = 2\left(\frac{c}{y}\right)x + y^2\left(\frac{c}{y}\right)^3 \quad \text{or} \quad \mathbf{y^2 = 2cx + c^3.}$$

This is the general solution of the given equation.

Example 4. *Solve* $p = \tan\left(x - \dfrac{p}{1+p^2}\right)$.

Sol. We have $$p = \tan\left(x - \frac{p}{1+p^2}\right) \quad \text{...(1)}$$

$$\Rightarrow \quad \tan^{-1} p = x - \frac{p}{1+p^2} \quad \Rightarrow \quad x = \tan^{-1} p + \frac{p}{1+p^2}. \quad \text{...(2)}$$

Differentiating w.r.t. y, we get

$$\frac{dx}{dy} = \frac{1}{1+p^2}\frac{dp}{dy} + \frac{(1+p^2).1 - p.2p}{(1+p^2)^2}\frac{dp}{dy}$$

$$\Rightarrow \quad \frac{1}{p} = \left(\frac{1}{1+p^2} + \frac{1-p^2}{(1+p^2)^2}\right)\frac{dp}{dy}$$

$$\Rightarrow \quad \frac{1}{p} = \frac{2}{(1+p^2)^2}\frac{dp}{dy} \quad \Rightarrow \quad dy = \frac{2p}{(1+p^2)^2}\,dp$$

Integrating, $$y = \int (1+p^2)^{-2}\, 2p\, dp + c = \frac{(1+p^2)^{-1}}{-1} + c$$

or $$y = c - \frac{1}{1+p^2}. \quad \text{...(3)}$$

Using given equation and (3), the general solution of the given equation is

$$\mathbf{x = \tan^{-1}p + \frac{p}{1+p^2}, \qquad y = c - \frac{1}{1+p^2},}$$

where p is a parameter.

EXERCISE 3

Solve the following differential equations :

1. $y = 3px + 6p^2y^2$
2. $y = 2px + p^2y$
3. $p^3 - 4xyp + 8y^2 = 0$
4. $y^2 \log y = xyp + p^2$
5. $x = y + p^2$
6. $yp^2 - 2xp + y = 0$
7. $p^3 - 2xyp + 4y^2 = 0$
8. $4x = py(p^2 - 3)$.

Answers

1. $y^3 = 3cx + 6c^2$
2. $y^2 = 2cx + c^2$
3. $64y = c(4x - c)^2$
4. $\log y = cx + c^2$
5. $x = c - 2p - 2 \log (p - 1), y = c - p^2 - 2p - 2 \log (p - 1)$
6. $y^2 = 2cx - c^2$
7. $2y = c(x - c)^2$
8. $x = \dfrac{cp(p^2 - 3)}{4(p^2 - 4)^{9/10} (p^2 + 1)^{3/5}}, y = \dfrac{c}{(p^2 - 4)^{9/10} (p^2 + 1)^{3/5}}$.

---

5. Clairaut's Equation

An equation of the form

$$y = px + f(p) \qquad ...(1)$$

is known as **Clairaut's equation.**

Differentiating (1) w.r.t. x, we get

$$p = \left(p + x \frac{dp}{dx}\right) + f'(p) \frac{dp}{dx} \quad \text{or} \quad [x + f'(p)] \frac{dp}{dx} = 0$$

Discarding the factor $[x + f'(p)]$ because it does not contain $\frac{dp}{dx}$, we have $\frac{dp}{dx} = 0$.

Integrating $\quad p = c$

Putting $\quad p = c$ in (1), the required solution is $\mathbf{y = cx + f(c)}$.

Thus, the solution of Clairaut's equation is obtained by writing c for p in the equation.

Example 1. *Solve* $(y - px)(p - 1) = p$.

Sol. We have $\quad (y - px)(p - 1) = p \qquad ...(1)$

$$\Rightarrow \quad y - px = \frac{p}{p - 1} \quad \Rightarrow \quad y = px + \frac{p}{p - 1} \qquad ...(2)$$

(2) is of the form $y = px + f(p)$ with $f(p) = \dfrac{p}{p - 1}$.

$\therefore$ (2) is a Clairaut's equation.

$\therefore$ The general solution of the given equation is $\mathbf{y = cx + \dfrac{c}{c - 1}}$. (Replacing p by c)

Example 2. *Solve* $\sin px \cos y = \cos px \sin y + p$.

Sol. We have $\quad \sin px \cos y - \cos px \sin y = p$.

$$\Rightarrow \quad \sin (px - y) = p \Rightarrow px - y = \sin^{-1} p \Rightarrow y = px - \sin^{-1} p.$$

This is a Clairaut's equation.

$\therefore$ Its general solution is $\mathbf{y = cx - \sin^{-1} c}$. (By replacing p by c)

Example 3. *Solve* $(y - px)^2 = 1 + p^2$.

Sol. We have $\quad (y - px)^2 = 1 + p^2$.

$$\Rightarrow \quad y - px = \pm\sqrt{1 + p^2} \Rightarrow y = px \pm \sqrt{1 + p^2} \qquad ...(1)$$

Taking +ve sign, (1) $\Rightarrow$ $y = px + \sqrt{1+p^2}$.

This is a Clairaut's equation and its solution is $y = cx + \sqrt{1+c^2}$. (By replacing p by c)

$$\Rightarrow \qquad y - cx - \sqrt{1+c^2} = 0. \qquad ...(2)$$

Taking –ve sign, (1) $\Rightarrow$ $y = px - \sqrt{1+p^2}$

This is a Clairaut's equation and its solution is $y = cx - \sqrt{1+c^2}$.

$$\Rightarrow \qquad y - cx + \sqrt{1+c^2} = 0. \qquad ...(3)$$

Using (2) and (3), the general solution of the given equation is

$$(y - cx - \sqrt{1+c^2})(y - cx + \sqrt{1+c^2}) = 0 \quad \text{or} \quad \mathbf{(y - cx)^2 = 1 + c^2.}$$

Example 4. *Solve $p^2x(x-2) + p(2y - 2xy - x + 2) + y^2 + y = 0$.*

Sol. We have

$$p^2x(x-2) + p(2y - 2xy - x + 2) + y^2 + y = 0.$$

$$\therefore \qquad p = \frac{-2y + 2xy + x - 2 \pm \sqrt{(2y - 2xy - x + 2)^2 - 4x(x-2)(y^2+y)}}{2x(x-2)}$$

$$= \frac{-2y + 2xy + x - 2 \pm (2y - x + 2)}{2x(x-2)} = \frac{y}{x-2}, \frac{y+1}{x}.$$

$$p = \frac{y}{x-2} \qquad \Rightarrow \quad y = px - 2p \qquad ...(1)$$

$$p = \frac{y+1}{x} \qquad \Rightarrow \quad y = px - 1 \qquad ...(2)$$

(1) is a Clairaut's equation.

$\therefore$ Its solution is $y = cx - 2c$ or $y - cx + 2c = 0$.

(2) is also a Clairaut's equation.

$\therefore$ Its solution is $y = cx - 1$ or $y - cx + 1 = 0$.

$\therefore$ The general solution of the given equation is

$$\mathbf{(y - cx + 2c)(y - cx + 1) = 0.}$$

Note. Many differential equations can be reduced to Clairaut's form by suitably changing the variables.

Example 5. *Solve $e^{4x}(p-1) + e^{2y}p^2 = 0$.*

Sol. We have $\qquad e^{4x}(p-1) + e^{2y}p^2 = 0 \qquad ...(1)$

Let $\qquad X = e^{2x}$ and $Y = e^{2y}$*

***Why this step.** In problems involving e^{lx} and e^{my}, use the substitutions $X = e^{kx}$ and $y = e^{ky}$, where k is the H.C.F. of l and m.

$$\therefore \quad \frac{dX}{dx} = 2e^{2x} \quad \text{and} \quad \frac{dY}{dx} = 2e^{2y}\frac{dy}{dx} = 2e^{2y}p$$

$$\therefore \quad \frac{dY}{dX} = \frac{dY}{dx}\Big/\frac{dX}{dx} = \frac{2e^{2y}\,p}{2e^{2x}} = \frac{e^{2y}}{e^{2x}}\,p = \frac{Y}{X}\,p$$

Let $P = \dfrac{dY}{dX}$. $\therefore$ $P = \dfrac{Y}{X}\,p$ or $p = \dfrac{X}{Y}\,P$.

$$\therefore \ (1) \Rightarrow X^2\left(\frac{X}{Y}P - 1\right) + Y\,.\,\frac{X^2}{Y^2}\,P^2 = 0 \quad \text{or} \quad XP - Y + P^2 = 0$$

or $Y = PX + P^2$

This is a Clairaut's equation.

$\therefore$ Its solution is $Y = cX + c^2$ (By replacing P by c)

$$\Rightarrow \quad \mathbf{e^{2y} = ce^{2x} + c^2.}$$

This is the general solution of the given equation.

Example 6. *Solve $(px - y)(py + x) = 2p$.*

Sol. We have $(px - y)(py + x) = 2p$ ...(1)

Let $X = x^2$ and $Y = y^2$.

$$\therefore \quad \frac{dX}{dx} = 2x \quad \text{and} \quad \frac{dY}{dx} = 2yp$$

$$\therefore \quad \frac{dY}{dX} = \frac{2yp}{2x} = \frac{y}{x}\,p$$

Let $P = \dfrac{dY}{dX}$

$$\therefore \quad P = \frac{y}{x}\,p \quad \text{or} \quad p = \frac{\sqrt{X}}{\sqrt{Y}}\,P.$$

$$\therefore \ (1) \Rightarrow \left(\frac{\sqrt{X}}{\sqrt{Y}}P\,.\,\sqrt{X} - \sqrt{Y}\right)\left(\frac{\sqrt{X}}{\sqrt{Y}}P\,.\,\sqrt{Y} + \sqrt{X}\right) = 2\,\frac{\sqrt{X}}{\sqrt{Y}}\,P$$

or $(PX - Y)(P + 1) = 2P$ or $PX - Y = \dfrac{2P}{P+1}$ or $Y = PX - \dfrac{2P}{P+1}$

This is a Clairaut's equation.

$\therefore$ Its solution is $Y = cX - \dfrac{2c}{c+1}$ (By replacing P by c)

$$\Rightarrow \quad \mathbf{y^2 = cx^2 - \frac{2c}{c+1}}.$$

This is the general solution of the given equation.

Example 7. *Solve $(px^2 + y^2)(px + y) = (p + 1)^2$ by using the substitutions $X = x + y$ and $Y = xy$.*

Sol. Given equation is $(px^2 + y^2)(px + y) = (p + 1)^2$ ...(1)

We have $X = x + y$ and $Y = xy$

$\therefore \quad \frac{dX}{dx} = 1 + p$ and $\frac{dY}{dx} = y + xp$

$\therefore \quad \frac{dY}{dX} = \frac{y + xp}{1 + p}$

Let $P = \frac{dY}{dX}$. $\therefore \quad P = \frac{xp + y}{1 + p}$.

(1) $\Rightarrow \quad (px^2 + y^2 + xyp + xy - xyp - xy)(px + y) = (p + 1)^2$ [Note this step]

$\Rightarrow \quad [(px + y)(x + y) - (p + 1)xy](px + y) = (p + 1)^2$

$\Rightarrow \quad (px + y)^2 (x + y) - (p + 1)xy(px + y) = (p + 1)^2$

$$\Rightarrow \quad \frac{(px + y)^2 (x + y)}{(p + 1)^2} - \frac{xy(px + y)}{p + 1} = 1$$

$$\Rightarrow \quad (x + y)\left(\frac{px + y}{p + 1}\right)^2 - xy\left(\frac{px + y}{p + 1}\right) = 1$$

$$\Rightarrow \quad XP^2 - YP = 1 \Rightarrow YP = XP^2 - 1 \Rightarrow Y = PX - \frac{1}{P}.$$

This is a Clairaut's equation.

$\therefore$ Its solution is $\quad Y = cX - \frac{1}{c}$. (By replacing P by c)

$$\Rightarrow \quad \mathbf{xy = c(x + y) - \frac{1}{c}}$$

This is the general solution of the given equation.

EXERCISE 4

Solve the following differential equations (Q. No. 1—10) :

1. $y = xp + \frac{a}{p}$ **2.** $y = px + \sqrt{a^2p^2 + b^2}$ **3.** $p = \log(px - y)$

4. $p = \sin(y - px)$ **5.** $p^2(x^2 - 1) - 2pxy + y^2 - 1 = 0$ **6.** $e^{3x}(p - 1) + p^3e^{2y} = 0$

7. $x^2(y - px) = yp^2$ **8.** $(y + px)^2 = x^2p$ **9.** $y = 3px + 6y^2p^2$

10. $p^2 \cos^2 y + p \sin x \cos x \cos y - \sin y \cos^2 x = 0$.

11. Reduce the equation $y^2(y - xp) = x^4p^2$ to Clairaut's equation by using the substitutions $x = \frac{1}{u}$, $y = \frac{1}{v}$ and hence solve the equation.

Answers

1. $y = cx + \dfrac{a}{c}$
2. $y = cx + \sqrt{a^2c^2 + b^2}$
3. $y = cx - e^c$
4. $y = cx + \sin^{-1} c$
5. $(y - cx)^2 = 1 + c^2$
6. $e^y = ce^x + c^2$
7. $y^2 = cx^2 + c^2$
8. $xy = cx - c^2$
9. $y^3 = cx + \dfrac{2}{3}c^2$
10. $\sin y = c \sin x + c^2$
11. $x = cy + c^2xy.$

Hints

7. Try $X = x^2$, $Y = y^2$
8. Try $X = x$, $Y = xy$
9. Try $X = x$, $Y = y^3$
10. Try $X = \sin x$, $Y = \sin y$.

17

Singular Solutions

1. Introduction

We know that an ordinary differential equation of order one admitting solution has one arbitrary constant in its general solution. We can also find its particular solutions by giving different values to the arbitrary constant in its general solution. If a differential equation of order one is of degree greater than one then there may exist a solution of this equation which is neither general nor particular. Such a solution cannot be derived from the general solution by giving some particular value to the arbitrary constant.

In the present chapter, we shall discuss the possibility of such solutions.

2. Discriminant Relation

Let $f(X) = 0$ be an equation in the variable X. The relation obtained by eliminating X from the equations $f(X) = 0$ and $\frac{df}{dX} = 0$ is called the **X-discriminant relation** of the equation $f(X) = 0$. This relation gives the required condition for the given equation $f(X) = 0$ to have multiple roots.

For example, let $f(X) = aX^2 + bX + c \quad \therefore \quad \frac{df}{dX} = 2aX + b.$

$$\frac{df}{dX} = 0 \quad \Rightarrow \quad 2aX + b = 0 \quad \Rightarrow \quad X = -\frac{b}{2a}$$

$\therefore$ Putting $X = -\frac{b}{2a}$ in $f(X) = 0$, we get

$$a\left(-\frac{b}{2a}\right)^2 + b\left(-\frac{b}{2a}\right) + c = 0 \quad \text{or} \quad b^2 - 2b^2 + 4ac = 0 \quad \text{or} \quad b^2 - 4ac = 0.$$

$\therefore$ The X-discriminant relation of the equation $aX^2 + bX + c = 0$ is $b^2 - 4ac = 0$.

Verification : $f(X) = 0 \Rightarrow X = \frac{-b \pm \sqrt{b^2 - 4ac}}{2a}$

$\therefore$ If $b^2 - 4ac = 0$, then the equation $f(X) = 0$ has multiple roots *i.e.*, equal roots $-\frac{b}{2a}, -\frac{b}{2a}$.

3. The p-Discriminant Relation

Let $$f(x, y, p) = 0 \qquad \text{...(1)}$$

be a differential equation of order one and degree greater than one.

By definition, *p–discriminant relation* of (1) is obtained by eliminating p between the equations

$$f(x, y, p) = 0 \quad \text{and} \quad \frac{\partial f}{\partial p} = 0.$$

Thus, the *p-discriminant relation* represents the locus, for each point of which $f(x, y, p) = 0$ has equal values of p.

4. The c-Discriminant Relation

Let $\phi(x, y, c) = 0$ ...(1)

be a one parameter family of curves and of at least second degree in c.

By definition, *c-discriminant relation* of (1) is obtained by eliminating c between the equations

$$\phi(x, y, c) = 0 \quad \text{and} \quad \frac{\partial \phi}{\partial c} = 0.$$

Thus, the *c-discriminant relation* represents the locus, for each point of which $\phi(x, y, c) = 0$ has equal values of c.

5. The Envelope of a Family of Curves

Let $\phi(x, y, c) = 0$ be a one parameter family of curves and of at least second degree in c. This family represents infinitely many curves of same kind. Let the values of c be arranged in order of magnitude . The successive c's thus differing by infinitesimal amounts. Suppose that all these curves are drawn. Curves corresponding to two consecutive values of c are called **consecutive curves** and their point of intersection is called an **ultimate point of intersection.** The limiting position of these points of intersection includes the envelope of the given family of curves. Since in the limit the c's for two consecutive curves become equal and the c-discriminant relation represents the locus of points for which $\phi(x, y, c)$ will have equal values of c, the envelope is a part of the locus of the c-discriminant relation. In differential calculus, we prove that at any point of the envelope, the latter is touched by some curve of the family $\phi(x, y, c) = 0$.

$\therefore$ At every point of the envelope, x, y and p have the same value for the envelope and the curve of the family that touches it there.

6. Singular Solution

Let $f(x, y, p) = 0$ be a differential equation of order one and of degree greater than one.

Let $\phi(x, y, c) = 0$ be the general solution of this differential equation. Let $E(x, y) = 0$ be the equation of the envelope of the family $\phi(x, y, c) = 0$. We have seen that $E(x, y) = 0$ is a part of the c-discriminant relation of the family $\phi(x, y, c) = 0$. Also, at every point of the envelope, x, y and p have the same value for the envelope and the curve of the family $\phi(x, y, c) = 0$ that touches it there.

$\therefore$ The equation $E(x, y) = 0$ of the envelope represents a solution of the differential equation $f(x, y, p) = 0$.

This solution is called the **singular solution** of the differential equation $f(x, y, p) = 0$. This solution does not contain any arbitrary constant and is not deducible from the general solution $\phi(x, y, c) = 0$ by giving a particular value to the arbitrary constant in it.

We can also show that the equation of the envelope $E(x, y) = 0$ *i.e.*, the singular solution of the equation $f(x, y, p) = 0$ is also a part of the p-discriminant relation of the equation $f(x, y, p) = 0$. This is because at the ultimate points of intersection of consecutive curves of the members of the family $\phi(x, y, c) = 0$, the p's for the intersecting curves become equal. Thus the locus of the points where p's have equal roots will include the envelope.

$\therefore$ The p-discriminant relation of $f(x, y, p) = 0$ contains the equation of the envelope *i.e.*, the singular solution of the equation $f(x, y, p) = 0$.

$\therefore$ The singular solution is contained in both p and c discriminant relations.

Remark. The p and c discriminant relations may sometimes represent other loci beside the envelope *i.e.*, they may contain other equations beside the equation of the singular solution. These loci are called **Tac locus, Nodal locus** and **Cuspidal locus.** Generally the equations of the tac locus, nodal locus and cuspidal locus do not satisfy the equation of the differential equation. This is why, these are called **extraneous loci**. The detail regarding these loci is beyond the scope of this book.

Working rules for finding singular solution

Step I. Find the general solution of the given differential equation.

Step II. Find the p-discriminant relation of the given differential equation and write it in the form $ET^2C = 0$, where E, T, C are some expressions in x and y.

Step III. Find the c-discriminant relation of the general solution and write it in the form $EN^2C^3 = 0$, where N is some expression in x and y.

Step IV. Consider the equation $E = 0$. If this equation satisfies the given differential equation then this constitutes the singular solution of the given differential equation. $T = 0$, $N = 0$ and $C = 0$ respectively gives the equation of Tac locus, Nodal locus and Cuspidal locus.

Remarks 1. In case $f(x, y, p) = 0$ is quadratic in p then it is advisable to find its p-discriminant relation by equating the discriminant of $f(x, y, p)$ to zero. Similar argument also work for c-discriminant relation.

2. If the p-discriminant relation and the c-discriminant relation are same, then the common equation gives $E = 0$.

Example 1. *Solve the differential equation* $y = px + \frac{a}{p}$ *and obtain its singular solution.* (MDU 2003)

Sol. We have $y = px + \dfrac{a}{p}$. ...(1)

This is a Clairaut's equation.

$\therefore$ By replacing p by c, its general solution is $y = cx + \dfrac{a}{c}$ *i.e.* $\mathbf{cy = c^2x + a}$.

(1) $\Rightarrow$ $py = p^2x + a \Rightarrow xp^2 - yp + a = 0$

Let $f(x, y, p) = xp^2 - yp + a$

This is quadratic in p.

$\therefore$ Disc. $= 0 \Rightarrow y^2 - 4ax = 0$ ...(2)

Let $\phi(x, y, c) = c^2x + a - cy \Rightarrow \phi(x, y, c) = xc^2 - yc + a$

This is quadratic in c.

$\therefore$ Disc. $= 0 \Rightarrow y^2 - 4ax = 0$ ...(3)

Using (2), the p-discriminant relation ($ET^2C = 0$) can be written as $(y^2 - 4ax) \cdot 1^2 \cdot 1 = 0$.

Using (3), the c-discriminant relation ($EN^2C^3 = 0$) can be written as $(y^2 - 4ax)\, 1^2 \cdot 1^3 = 0$.

$\therefore$ $E = 0 \Rightarrow y^2 - 4ax = 0$ ...(4)

(4) $\Rightarrow \quad 2y\dfrac{dy}{dx} - 4a = 0 \Rightarrow \quad 2yp = 4a \quad \Rightarrow \quad p = \dfrac{2a}{y}$

Also, $y^2 - 4ax = 0 \qquad \Rightarrow \quad x = \dfrac{y^2}{4a}$.

Putting the values of p and x in (1), we get $y = \dfrac{2a}{y} \cdot \dfrac{y^2}{4a} + a \cdot \dfrac{y}{2a}$ or $y = y$, which is true.

$\therefore \quad y^2 - 4ax = 0$ satisfies the given equation.

$\therefore \quad \mathbf{y^2 - 4ax = 0}$ is the singular solution of the given equation.

Example 2. *Investigate the equation $y = px - 2p^2$ for singular solution.* (KU 2005)

Sol. We have $y = px - 2p^2$ ...(1)

This is a Clairaut's equation.

Replacing p by c, the general solution of (1) is $y = cx - 2c^2$ *i.e.,* $y - cx + 2c^2 = 0$ ...(2)

Let $f(x, y, p) = y - px + 2p^2$. $\quad \therefore \quad \dfrac{\partial f}{\partial p} = -x + 4p$

$\therefore \qquad f(x, y, p) = 0 \quad \Rightarrow \quad y - px + 2p^2 = 0$ ...(3)

and $\qquad \dfrac{\partial f}{\partial p} = 0 \quad \Rightarrow \quad -x + 4p = 0$ ...(4)

(4) $\Rightarrow \qquad p = \dfrac{x}{4}$. Putting the value of p in (3), we get $y - \left(\dfrac{x}{4}\right)x + 2\left(\dfrac{x}{4}\right)^2 = 0$

$\Rightarrow \qquad y - \dfrac{x^2}{4} + \dfrac{x^2}{8} = 0 \quad \Rightarrow \quad y - \dfrac{x^2}{8} = 0 \quad \Rightarrow \quad x^2 - 8y = 0$ ...(5)

Let $\phi(x, y, c) = y - cx + 2c^2$.

Eliminating c between $\phi(x, y, c) = 0$ and $\dfrac{\partial \phi}{\partial c} = 0$, we get $x^2 - 8y = 0$ ...(6)

Using (5), the p-discriminant relation ($ET^2C = 0$) can be written as $(x^2 - 8y)\ 1^2 \cdot 1 = 0$.

Using (6), the c-discriminant relation ($EN^2C^3 = 0$) can be written as $(x^2 - 8y). 1^2 . 1^3 = 0$.

$\therefore \qquad E = 0 \quad \Rightarrow \quad x^2 - 8y = 0$ ...(7)

(7) $\Rightarrow \quad y = \dfrac{x^2}{8}$ and thus $p = \dfrac{x}{4}$. Putting the values of y and p in (1), we get

$\dfrac{x^2}{8} = \left(\dfrac{x}{4}\right)x - 2\left(\dfrac{x}{4}\right)^2$, which is true.

$\therefore \qquad x^2 - 8y = 0$ satisfies the given equation.

$\therefore \quad \mathbf{x^2 - 8y = 0}$ is the singular solution of the given equation.

Example 3. *Obtain the singular solution of the equation $p^2 + y^2 = 1, p \geq 0$. Also interpret the result geometrically.*

Sol. We have $\qquad p^2 + y^2 = 1$. ...(1)

$\Rightarrow \qquad p = \sqrt{1 - y^2}$ $\qquad (\because \quad p \geq 0)$

$\Rightarrow \qquad \dfrac{dy}{\sqrt{1 - y^2}} = dx$

Integrating, we get $\sin^{-1} y = x + c$. $\Rightarrow$ $y = \sin(x + c)$ or $y - \sin(x + c) = 0$. ...(2)
This is the general solution.

Let $f(x, y, p) = p^2 + y^2 - 1$ $\therefore$ $\dfrac{\partial f}{\partial p} = 2p$

$\therefore$ $f(x, y, p) = 0$ $\Rightarrow$ $p^2 + y^2 - 1 = 0$ ...(3)

and $\dfrac{\partial f}{\partial p} = 0$ $\Rightarrow$ $2p = 0$ ...(4)

Eliminating p between (3) and (4), we get $0 + y^2 - 1 = 0$ *i.e.,* $y^2 - 1 = 0$. ...(5)

Let $\phi(x, y, c) = y - \sin(x + c)$. $\therefore$ $\dfrac{\partial \phi}{\partial c} = -\cos(x + c)$

$\therefore$ $\phi(x, y, c) = 0$ $\Rightarrow$ $y - \sin(x + c) = 0$ ...(6)

and $\dfrac{\partial \phi}{\partial c} = 0$ $\Rightarrow$ $-\cos(x + c) = 0$ ...(7)

We have $\sin^2(x + c) + \cos^2(x + c) = 1$

$\therefore$ $y^2 + 0^2 = 1$ *i.e.,* $y^2 - 1 = 0$ ...(8)

Using (5), the p-discriminant relation ($ET^2C = 0$) can be written as $(y^2 - 1)\, 1^2 \,.\, 1 = 0$.
Using (8), the c-discriminant relation ($EN^2C^3 = 0$) can be written as $(y^2 - 1)\, 1^2 \,.\, 1^3 = 0$.

$\therefore$ $E = 0$ $\Rightarrow$ $y^2 - 1 = 0$ ...(9)

(9) $\Rightarrow$ $y = \pm 1$ and thus $p = 0$.

Putting the values of y and p in (1), we get $0 + 1 = 1$, which is true.

$\therefore$ $y^2 - 1 = 0$ satisfies the given equation.

$\therefore$ $\mathbf{y^2 - 1 = 0}$ is the singular solution of the given equation. The equation $y = \sin(x + c)$ represents the family of sine curves lying between $y = 1$ and $y = -1$. The equation $y^2 - 1 = 0$ represents the lines $y = 1$ and $y = -1$. The lines $y = 1$, $y = -1$ represent the envelope of the family of sine curves.

Example 4. *Find the general solution and singular solution of the differential equation* $4p^2(x - 2) = 1$.

Sol. We have $4p^2(x - 2) = 1$. ...(1)

$\Rightarrow$ $p^2 = \dfrac{1}{4(x - 2)}$ $\Rightarrow$ $p = \pm \dfrac{1}{2\sqrt{x - 2}}$

$\therefore$ $p = \dfrac{1}{2\sqrt{x - 2}}$ ...(2) and $p = -\dfrac{1}{2\sqrt{x - 2}}$ ...(3)

Integrating (2), we get

$$y = \sqrt{x - 2} + c \quad \text{or} \quad y - \sqrt{x - 2} - c = 0.$$

Integrating (3), we get

$$y = -\sqrt{x - 2} + c \quad \text{or} \quad y + \sqrt{x - 2} - c = 0$$

$\therefore$ The general solution of (1) is

$$(y - \sqrt{x - 2} - c)(y + \sqrt{x - 2} - c) = 0 \quad \text{or} \quad \mathbf{(y - c)^2 - (x - 2) = 0.}$$

Let $f(x, y, p) = 4p^2(x - 2) - 1$, $f(x, y, p)$ is quadratic in p.

$\therefore$ Disc. $= 0$ $\Rightarrow$ $(0)^2 - 4 \,.\, 4(x - 2)(-1) = 0$ $\Rightarrow$ $x - 2 = 0$ ...(4)

Let $\phi(x, y, c) = (y - c)^2 - (x - 2)$

$\Rightarrow \qquad \phi(x, y, c) = c^2 - 2yc + (y^2 - x + 2)$

$\phi(x, y, c)$ is quadratic in c.

$\therefore$ Disc. = 0 $\Rightarrow$ $4y^2 - 4.1(y^2 - x + 2) = 0 \Rightarrow 4(x-2) = 0 \Rightarrow x - 2 = 0$ ...(5)

Using (4), the p-discriminant relation ($ET^2C = 0$) can be written as $(x - 2).1^2 . 1 = 0$.

Using (5), the c-discriminant relation ($EN^2C^3 = 0$) can be written as $(x - 2) . 1^2 . 1^3 = 0$.

$\therefore \qquad E = 0 \Rightarrow x - 2 = 0$ ...(6)

$$x - 2 = 0 \Rightarrow x = 2 \Rightarrow \frac{dx}{dy} = 0 \Rightarrow \frac{1}{p} = 0$$

$$\therefore \quad x - 2 = 0 \Rightarrow 4(x-2) = \frac{1}{p^2} \Rightarrow 4p^2(x-2) = 0$$

$\therefore$ $x - 2 = 0$ satisfies the given equation

$\therefore$ $\boldsymbol{x - 2 = 0}$ is the singular solution of the given equation.

Example 5. *Find the general and singular solutions of the equation $xp^2 - 2yp + 4x = 0$.* (MDU 2004)

Sol. We have $\qquad xp^2 - 2yp + 4x = 0.$

$$\Rightarrow \quad y = \frac{xp^2 + 4x}{2p} \Rightarrow y = \frac{px}{2} + \frac{2x}{p} \qquad ...(1)$$

Differentiating w.r.t. x, we get

$$\frac{dy}{dx} = \frac{1}{2}\left(p.1 + x\frac{dp}{dx}\right) + 2\left(\frac{1}{p}.1 - \frac{x}{p^2}\frac{dp}{dx}\right)$$

$$\Rightarrow \quad p = \frac{p}{2} + \frac{x}{2}\frac{dp}{dx} + \frac{2}{p} - \frac{2x}{p^2}\frac{dp}{dx}$$

$$\Rightarrow \quad \frac{p}{2} - \frac{2}{p} = \left(\frac{x}{2} - \frac{2x}{p^2}\right)\frac{dp}{dx} \Rightarrow \frac{p^2 - 4}{2p} = x\left(\frac{p^2 - 4}{2p^2}\right)\frac{dp}{dx}$$

$$\Rightarrow \quad \frac{p^2 - 4}{2p} = \frac{x}{p}\left(\frac{p^2 - 4}{2p}\right)\frac{dp}{dx} \Rightarrow 1 = \frac{x}{p}\frac{dp}{dx}$$

$$\Rightarrow \quad \frac{dp}{p} = \frac{dx}{x}$$

Integrating, we get

$$\log p = \log x + \log c \Rightarrow p = cx.$$

Putting the value of p in (1), we get $\quad y = \dfrac{cx . x}{2} + \dfrac{2x}{cx}$

$$\Rightarrow \quad y = \frac{cx^2}{2} + \frac{2}{c} \Rightarrow 2cy = c^2x^2 + 4 \Rightarrow \mathbf{2cy - c^2x^2 - 4 = 0.}$$

This represents the general solution of (1).

Let $\quad f(x, y, p) = xp^2 - 2yp + 4x$

$f(x, y, p)$ is quadratic in p.

$\therefore \qquad$ Disc. = 0 $\Rightarrow 4y^2 - 4 . x. 4x = 0 \Rightarrow y^2 - 4x^2 = 0$ ...(2)

Let $\phi(x, y, c) = 2cy - c^2x^2 - 4 \quad \Rightarrow \quad \phi(x, y, c) = -c^2x^2 + 2cy - 4$

$\therefore$ $\phi(x, y, c)$ is quadratic in c.

$\therefore$ Disc. = 0 $\Rightarrow 4y^2 - 4(-x^2)(-4) = 0 \Rightarrow y^2 - 4x^2 = 0$ ...(3)

Using (2), the p-discriminant relation ($ET^2C = 0$) can be written as $(y^2 - 4x^2) \cdot 1^2 \cdot 1 = 0$.

Using (3), the c-discriminant relation ($EN^2C^3 = 0$) can be written as $(y^2 - 4x^2) \cdot 1^2 \cdot 1^3 = 0$

$\therefore$ $E = 0 \Rightarrow y^2 - 4x^2 = 0.$ ...(4)

$$(4) \Rightarrow \quad 2y\frac{dy}{dx} - 8x = 0 \quad \Rightarrow p = \frac{4x}{y}$$

Putting the value of p in (1), we get

$$x \cdot \frac{16x^2}{y^2} - 2y \cdot \frac{4x}{y} + 4x = 0 \quad \text{or} \quad 16x^3 - 4xy = 0 \quad \text{or} \quad 4x(4x^2 - y^2) = 0,$$

which is true, because $y^2 - 4x^2 = 0$.

$\therefore$ $y^2 - 4x^2 = 0$ satisfies the given equation.

$\therefore$ **$y^2 - 4x^2 = 0$ is the singular solution of the given equation.**

Example 6. *Find the singular solution of the equation $xp^2 - (x - a)^2 = 0$.* (KU 2003 C)

Sol. We have $xp^2 - (x - a)^2 = 0.$ ...(1)

$$(1) \Rightarrow \quad p^2 = \frac{(x-a)^2}{x} \quad \Rightarrow \quad p = \pm\frac{x-a}{\sqrt{x}} \quad \Rightarrow \quad \frac{dy}{dx} = \pm\left(\sqrt{x} - \frac{a}{\sqrt{x}}\right)$$

$$\frac{dy}{dx} = \sqrt{x} - \frac{a}{\sqrt{x}} \quad \Rightarrow \quad y = \frac{2x^{3/2}}{3} - 2a\sqrt{x} + c$$

$$\frac{dy}{dx} = -\left(\sqrt{x} - \frac{a}{\sqrt{x}}\right) \quad \Rightarrow \quad y = -\frac{2x^{3/2}}{3} + 2a\sqrt{x} + c.$$

$\therefore$ General solution of (1) is

$$\left(y - \frac{2x^{3/2}}{3} + 2a\sqrt{x} - c\right)\left(y + \frac{2x^{3/2}}{3} - 2a\sqrt{x} - c\right) = 0$$

$$\Rightarrow \quad (y - c)^2 - \left(\frac{2x^{3/2}}{3} - 2a\sqrt{x}\right)^2 = 0$$

$$\Rightarrow \quad (y - c)^2 - \frac{4x}{9}(x - 3a)^2 = 0.$$

Let $f(x, y, p) = xp^2 - (x - a)^2$

$f(x, y, p)$ is quadratic in p.

$\therefore$ Disc. = 0 $\Rightarrow 0^2 - 4 \cdot x \cdot -(x - a)^2 = 0 \Rightarrow x(x - a)^2 = 0$ ...(2)

Let $\phi(x, y, c) = (y - c)^2 - \dfrac{4x}{9}(x - 3a)^2$

$$\Rightarrow \quad \phi(x, y, c) = c^2 - 2cy + \left(y^2 - \frac{4x}{9}(x - 3a)^2\right)$$

$\therefore$ $\phi(x, y, c)$ is quadratic in c.

$$\text{Disc.} = 0 \quad \Rightarrow \quad 4y^2 - 4.1.\left(y^2 - \frac{4x}{9}(x - 3a)^2\right)$$

$$= \frac{4x}{9}(x - 3a)^2 = 0 \quad \Rightarrow \quad x(x - 3a)^2 = 0 \qquad ...(3)$$

Using (2), the p-discriminant relation (ET^{2}C = 0) can be written as $x.(x - a)^2 \,.\, 1 = 0$.

Using (3), the c-discriminant relation (EN^{2}C^3 = 0) can be written as $x.(x - 3a)^2 \,.\, 1^3 = 0$.

$$\therefore \qquad E = 0 \quad \Rightarrow \quad x = 0. \qquad ...(4)$$

$x = 0$ satisfies the given equation when written in the form $\left(\dfrac{dx}{dy}\right)^2 = \dfrac{x}{(x - a)^2}$.

$\therefore$ **x = 0** is the singular solution of the given equation.

Remark. $T = 0 \Rightarrow x - a = 0$ and $N = 0 \Rightarrow x - 3a = 0$.

$\therefore$ The equations of Tac locus and Nodal locus are respectively $x - a = 0$ and $x - 3a = 0$.

Example 7. *Find the singular solution of the differential equation* $yp^2 - 2xp + y = 0$.

(K.U. 2004, 2006)

Sol. We have $yp^2 - 2xp + y = 0$.

$$\Rightarrow \qquad x = \frac{yp}{2} + \frac{y}{2p} \qquad ...(1)$$

Differentiating w.r.t. y, we get

$$\frac{dx}{dy} = \frac{1}{2}\left(p.1 + y\frac{dp}{dy}\right) + \frac{1}{2}\left(\frac{1}{p} - \frac{y}{p^2}\frac{dp}{dy}\right)$$

$$\Rightarrow \qquad \frac{1}{p} = \frac{p}{2} + \frac{y}{2}\frac{dp}{dy} + \frac{1}{2p} - \frac{y}{2p^2}\frac{dp}{dy}$$

$$\Rightarrow \qquad \frac{1}{2p} - \frac{p}{2} = \left(\frac{y}{2} - \frac{y}{2p^2}\right)\frac{dp}{dy} \qquad \Rightarrow \qquad -\left(\frac{p}{2} - \frac{1}{2p}\right) = \frac{y}{p}\left(\frac{p}{2} - \frac{1}{2p}\right)\frac{dp}{dy}$$

$$\Rightarrow \qquad -1 = \frac{y}{p}\frac{dp}{dy} \qquad \Rightarrow \qquad \frac{dy}{y} + \frac{dp}{p} = 0$$

Integrating, we get $\log y + \log p = \log c$.

$$\Rightarrow \qquad yp = c \qquad \Rightarrow \qquad p = \frac{c}{y}$$

Putting the value of p in (1), we get

$$x = \frac{y}{2}.\frac{c}{y} + \frac{y}{2}.\frac{y}{c} \quad \Rightarrow \quad x = \frac{c}{2} + \frac{y^2}{2c} \quad \Rightarrow \quad y^2 - 2cx + c^2 = 0.$$

This is the general solution of (1)

Let $f(x, y, p) = yp^2 - 2xp + y$.

This is quadratic in p.

$$\therefore \qquad \text{Disc.} = 0 \quad \Rightarrow \quad 4x^2 - 4.y.\,y = 0 \quad \Rightarrow \quad x^2 - y^2 = 0 \qquad ...(2)$$

Let $\phi(x, y, c) = y^2 - 2cx + c^2$

$\Rightarrow \quad \phi(x, y, c) = c^2 - 2xc + y^2$

This is quadratic in c.

$$\therefore \qquad \text{Disc.} = 0 \quad \Rightarrow \quad 4x^2 - 4.1.y^2 = 0 \quad \Rightarrow \quad x^2 - y^2 = 0 \qquad ...(3)$$

Using (2), the p-discriminant relation (ET^2 C = 0) can be written as $(x^2 - y^2) \,.\, 1^2 \,.1 = 0$.

Using (3), the c-discriminant relation ($EN^2C^3 = 0$) can be written as $(x^2 - y^2) \cdot 1^2 \cdot 1^3 = 0$.

$\therefore \quad E = 0 \quad \Rightarrow \quad x^2 - y^2 = 0 \quad ...(4)$

$(3) \Rightarrow \quad 2x - 2y \dfrac{dy}{dx} = 0 \quad \Rightarrow \quad p = \dfrac{x}{y}$.

Putting the value of p in (1), we get

$$x = \frac{y}{2} \cdot \frac{x}{y} + \frac{y}{2} \cdot \frac{y}{x} \quad \text{or} \quad x = \frac{x}{2} + \frac{x^2}{2x}, \text{ which is true.} \quad [\because \; x^2 - y^2 = 0]$$

$\therefore$ $x^2 - y^2 = 0$ satisfies the given equation.

$\therefore$ $\mathbf{x^2 - y^2 = 0}$ is the singular solution of the given equation.

Example 8. *Examine and investigate for singular solution of the differential equation :*
$(8p^3 - 27)\, x = 12p^2 y.$

Sol. We have $(8p^3 - 27)x = 12p^2y. \quad ...(1)$

$$(1) \Rightarrow \quad y = \frac{8p^3x - 27x}{12p^2} \quad \Rightarrow \quad y = \frac{2}{3}px - \frac{9x}{4p^2} \quad ...(2)$$

Differentiating w.r.t. x, we get

$$\frac{dy}{dx} = \frac{2}{3}\left(p \cdot 1 + x\frac{dp}{dx}\right) - \frac{9}{4}\left(\frac{1}{p^2} - \frac{2x}{p^3}\frac{dp}{dx}\right)$$

$$\Rightarrow \quad p = \frac{2}{3}p + \frac{2x}{3}\frac{dp}{dx} - \frac{9}{4p^2} + \frac{9x}{2p^3}\frac{dp}{dx}$$

$$\Rightarrow \quad \frac{p}{3} + \frac{9}{4p^2} = \left(\frac{2x}{3} + \frac{9x}{2p^3}\right)\frac{dp}{dx}$$

$$\Rightarrow \quad p\left(\frac{1}{3} + \frac{9}{4p^3}\right) = 2x\left(\frac{1}{3} + \frac{9}{4p^3}\right)\frac{dp}{dx} \quad \Rightarrow \quad p = 2x\frac{dp}{dx} \quad \Rightarrow \quad \frac{2dp}{p} = \frac{dx}{x}$$

Integrating, we get $2 \log p = \log x + \log c \quad \Rightarrow \quad p^2 = cx \quad \Rightarrow \quad p = \sqrt{cx}$

Putting the value of p in (1), we get $(8\, cx\sqrt{cx} - 27)x = 12\, cxy$

$$\Rightarrow \quad 8\, cx\sqrt{cx} - 27 = 12\, cy \quad \Rightarrow \quad 8\, cx\sqrt{cx} = 12cy + 27$$

$$\Rightarrow \quad 64\, c^3x^3 = (12\, cy + 27)^2 \quad \Rightarrow \quad x^3 = \frac{9}{64c^3}(4cy + 9)^2$$

$$\Rightarrow \quad x^3 = \frac{9(16c^2)}{64c^3}\left(y + \frac{9}{4c}\right)^2 \quad \Rightarrow \quad x^3 = \frac{9}{4c}\left(y + \frac{9}{4c}\right)^2$$

$$\Rightarrow \quad x^3 = c_1(y + c_1)^2, \text{ where } c_1 = \frac{9}{4c}.$$

This represents the general solution of the given equation.

Let $f(x, y, p) = (8p^3 - 27)x - 12p^2y$

$$\therefore \quad \frac{\partial f}{\partial p} = 24p^2x - 24py$$

$\therefore \quad f(x, y, p) = 0 \quad \Rightarrow \quad (8p^3 - 27)x - 12p^2y = 0 \quad ...(3)$

and $\quad \dfrac{\partial f}{\partial p} = 0 \quad \Rightarrow \quad 24p^2x - 24\, py = 0 \quad ...(4)$

(4) $\Rightarrow$ $p(px - y) = 0$ $\Rightarrow$ $p = 0, \frac{y}{x}$

$p = 0$ in (3) $\Rightarrow$ $(0 - 27)x - 0 = 0 \Rightarrow x = 0$

$p = \frac{y}{x}$ in (3) $\Rightarrow$ $\left(\frac{8y^3}{x^3} - 27\right)x - 12\left(\frac{y^2}{x^2}\right)y = 0 \Rightarrow 8y^3 - 27x^3 - 12y^3 = 0$

$\Rightarrow 4y^3 + 27x^3 = 0$

$\therefore$ The eliminant of $f(x, y, p) = 0$ and $\frac{\partial f}{\partial p} = 0$ is $x(4y^3 + 27x^3) = 0$ ...(5)

Let $\phi(x, y, c_1) = x^3 - c_1(y + c_1)^2$

$\therefore$ $\frac{\partial \phi}{\partial c_1} = 0 - 1.(y + c_1)^2 - c_1 . 2(y + c_1) . 1$

$\therefore$ $\phi(x, y, c_1) = 0 \Rightarrow x^3 - c_1(y + c_1)^2 = 0$ ...(6)

and $\frac{\partial \phi}{\partial c_1} = 0 \Rightarrow -(y + c_1)(y + c_1 + 2c_1) = 0 \Rightarrow (y + c_1)(y + 3c_3) = 0$ ...(7)

(7) $\Rightarrow$ $c_1 = -y, -\frac{y}{3}$

Putting $c_1 = -y$ in (6) $\Rightarrow$ $x^3 + y(y - y)^2 = 0 \Rightarrow x^3 = 0$

And $c_1 = -\frac{y}{3}$ in (6) $\Rightarrow$ $x^3 + \frac{y}{3}\left(y - \frac{y}{3}\right)^2 = 0 \Rightarrow 4y^3 + 27x^3 = 0$

$\therefore$ The eliminant of $\phi(x, y, c_1) = 0$ and $\frac{\partial \phi}{\partial c} = 0$ is $x^3(4y^3 + 27x^3) = 0$ ...(8)

Using (5), the p-discriminant relation ($ET^2 C = 0$) can be written as

$$(4y^3 + 27x^3) . 1^2 . x = 0$$

Using (8), the c-discriminant relation ($EN^2C^3 = 0$) can be written as

$$(4y^3 + 27x^3) . 1^2 . x^3 = 0$$

$\therefore$ $E = 0 \Rightarrow 4y^3 + 27x^3 = 0$ ...(9)

(9) $\Rightarrow$ $y = -\frac{3}{4^{1/3}}x$ and $p = -\frac{3}{4^{1/3}}$.

Putting the values of y and p in (1), we get

$$\left[8\left(-\frac{27}{4}\right) - 27\right]x = 12 . \frac{9}{4^{2/3}}\left(-\frac{3}{4^{1/3}}\right)x$$

$\Rightarrow$ $-81x = -81x$, which is true.

$\therefore$ $4y^3 + 27x^3 = 0$ satisfies the given equation

$\therefore$ $\mathbf{4y^3 + 27x^3 = 0}$ **is the singular solution of the given equation.**

Example 9. *Find the general and singular solutions of the differential equation $p^2y^2 \cos^2 \alpha - 2pxy \sin^2 \alpha + y^2 - x^2 \sin^2 \alpha = 0$. Also interpret the result geometrically.*

Sol. We have $p^2y^2 \cos^2 \alpha - 2pxy \sin^2 \alpha + y^2 - x^2 \sin^2 \alpha = 0.$ ...(1)

Dividing by $\cos^2 \alpha$, we get $p^2y^2 - 2pxy \tan^2 \alpha + y^2 \sec^2 \alpha - x^2 \tan^2 \alpha = 0$

This is quadratic in p.

$$\therefore \quad p = \frac{2xy \tan^2 \alpha \pm \sqrt{4x^2y^2 \tan^4 \alpha - 4y^4 \sec^2 \alpha + 4x^2y^2 \tan^2 \alpha}}{2y^2}$$

$$= \frac{2xy \tan^2 \alpha \pm 2y\sqrt{x^2 \tan^4 \alpha - y^2 \sec^2 \alpha + x^2 \tan^2 \alpha}}{2y^2}$$

$$= \frac{x \tan^2 \alpha \pm \sqrt{x^2 \tan^2 \alpha (1 + \tan^2 \alpha) - y^2 \sec^2 \alpha}}{y}$$

$$= \frac{x \tan^2 \alpha \pm \sec \alpha \sqrt{x^2 \tan^2 \alpha - y^2}}{y}$$

$$\Rightarrow \quad yp = x \tan^2 \alpha \pm \sec \alpha \sqrt{x^2 \tan^2 \alpha - y^2}$$

$$\Rightarrow \quad y \frac{dy}{dx} - x \tan^2 \alpha = \pm \sec \alpha \sqrt{x^2 \tan^2 \alpha - y^2}$$

$$\Rightarrow \quad \pm \frac{y\, dy - x \tan^2 \alpha\, dx}{\sqrt{x^2 \tan^2 \alpha - y^2}} = \sec \alpha\, dx$$

or
$$\pm \frac{x \tan^2 \alpha\, dx - y dy}{\sqrt{x^2 \tan^2 \alpha - y^2}} = - \sec \alpha\, dx$$

or
$$\pm \frac{\frac{1}{2} d(x^2 \tan^2 \alpha - y^2)}{\sqrt{x^2 \tan^2 \alpha - y^2}} = - \sec \alpha\, dx$$

Integrating, we get

$$\pm \sqrt{x^2 \tan^2 \alpha - y^2} = - \sec \alpha \,.\, x + c$$

Squaring, $\quad x^2 \tan^2 \alpha - y^2 = \sec^2 \alpha\, x^2 + c^2 - 2cx \sec \alpha$

or $\quad \mathbf{x^2 + y^2 - 2cx \sec \alpha + c^2 = 0.}$

This is the general solution of the given equation.

Let $f(x, y, p) = p^2y^2 \cos^2 \alpha - 2pxy \sin^2 \alpha + y^2 - x^2\sin^2 \alpha$

This is quadratic in p.

$\therefore$ Disc. = 0 $\Rightarrow$ $(- 2xy \sin^2 \alpha)^2 - 4y^2 \cos^2 \alpha (y^2 - x^2 \sin^2 \alpha) = 0$

$\Rightarrow$ $4y^2[x^2 \sin^4 \alpha - y^2 \cos^2 \alpha + x^2 \sin^2 \alpha \cos^2 \alpha] = 0$

$\Rightarrow$ $y^2 [x^2 \sin^2 \alpha (\sin^2 \alpha + \cos^2 \alpha) - y^2 \cos^2 \alpha] = 0$

$\Rightarrow$ $y^2 (x^2 \sin^2 \alpha - y^2 \cos^2 \alpha) = 0$

$\Rightarrow$ $y^2 \cos^2 \alpha (x^2 \tan^2 \alpha - y^2) = 0$

$\Rightarrow$ $y^2(x^2 \tan^2 \alpha - y^2) = 0.$ ...(2)

Let $\phi(x, y, c) = x^2 + y^2 - 2cx \sec \alpha + c^2$

$\Rightarrow$ $\phi(x, y, c) = c^2 - (2x \sec \alpha)\, c + (x^2 + y^2)$

This is quadratic in c.

$\therefore$ Disc. = 0 $\Rightarrow$ $4x^2 \sec^2 \alpha - 4.\ 1.(x^2 + y^2) = 0.$

$\Rightarrow$ $4x^2(\sec^2 \alpha - 1) - 4y^2 = 0$ $\Rightarrow$ $x^2 \tan^2 \alpha - y^2 = 0.$ ...(3)

Using (2), the p-discriminant relation ($ET^2C = 0$) can be written as

$$(x^2 \tan^2 \alpha - y^2)\,(y)^2 \,.\, 1 = 0.$$

Using (3), the c-discriminant relation ($EN^2C^3 = 0$) can be written as

$$(x^2 \tan^2 \alpha - y^2)\,.\,1^2\,.\,1^3 = 0.$$

$$\therefore \quad E = 0 \quad \Rightarrow \quad x^2 \tan^2 \alpha - y^2 = 0. \qquad ...(4)$$

$$(4) \Rightarrow 2x \tan^2 \alpha = 2yp \quad \Rightarrow \quad p = \frac{x \tan^2 \alpha}{y}$$

Putting the value of p in (1), we get

$$\left(\frac{x^2 \tan^4 \alpha}{y^2}\right) y^2 \cos^2 \alpha - 2\left(\frac{x \tan^2 \alpha}{y}\right) xy \sin^2 \alpha + y^2 - x^2 \sin^2 \alpha = 0$$

$$\Rightarrow \quad \frac{x^2 \sin^4 \alpha}{\cos^2 \alpha} - \frac{2x^2 \sin^4 \alpha}{\cos^2 \alpha} + x^2 \tan^2 \alpha - x^2 \sin^2 \alpha = 0$$

$$\Rightarrow \quad x^2 \sin^4 \alpha - 2x^2 \sin^4 \alpha + x^2 \sin^2 \alpha - x^2 \sin^2 \alpha \cos^2 \alpha = 0$$

$$\Rightarrow \quad x^2 \sin^2 \alpha\,(\sin^2 \alpha - 2 \sin^2 \alpha + 1 - \cos^2 \alpha) = 0$$

$\Rightarrow$ $0 = 0$, which is true.

$\therefore$ $x^2 \tan^2 \alpha - y^2 = 0$ satisfies the given equation.

$\therefore$ $\mathbf{x^2 \tan^2 \alpha - y^2 = 0}$ is the singular solution of the given equation. The singular solution represents a pair of straight lines $y = \pm\, x \tan \alpha$. The general solution represents a family of circles whose envelope is the pair of lines $y = \pm\, x \tan \alpha$.

Example 10. *Reduce the equation $xyp^2 - (x^2 + y^2 - 1)p + xy = 0$ to Clairaut's form by substituting $x^2 = u$ and $y^2 = v$. Hence show that the general solution of the equation represents a family of conics touching the four sides of a square.* (KU 2004 S)

Sol. Given equation is $xyp^2 - (x^2 + y^2 - 1)\,p + xy = 0.$ ...(1)

We have $u = x^2$ and $v = y^2$

$$\therefore \quad \frac{du}{dx} = 2x \quad \text{and} \quad \frac{dv}{dx} = 2y\,\frac{dy}{dx} = 2yp$$

$$\therefore \quad \frac{dv}{du} = \frac{dv}{dx}\Big/\frac{du}{dx} = \frac{2yp}{2x} = \frac{y}{x}\,p$$

$$\text{Let} \quad P = \frac{dv}{du}. \quad \therefore \quad P = \frac{y}{x}\,p \quad \text{or} \quad p = \frac{x}{y}\,P \quad \text{or} \quad p = \frac{\sqrt{u}}{\sqrt{v}}\,P$$

$$\therefore \quad (1) \Rightarrow \sqrt{u}\,\sqrt{v}\,.\,\frac{u}{v}\,P^2 - (u + v - 1)\,\frac{\sqrt{u}}{\sqrt{v}}\,P + \sqrt{u}\,\sqrt{v} = 0$$

$$\Rightarrow \quad \frac{\sqrt{u}}{\sqrt{v}}\,(uP^2 - (u + v - 1)\,P + v) = 0$$

$$\Rightarrow \quad uP^2 - (u + v - 1)\,P + v = 0 \quad \Rightarrow \quad u(P^2 - P) - v(P - 1) + P = 0$$

$$\Rightarrow \quad uP - v + \frac{P}{P-1} = 0 \quad \Rightarrow \quad v = Pu + \frac{P}{P-1} \qquad ...(2)$$

This is a Clairaut's equation.

By replacing P by c, the general solution of (2) is $\quad v = cu + \dfrac{c}{c-1}.$

$\Rightarrow \qquad y^2 = cx^2 + \dfrac{c}{c-1} \qquad \Rightarrow \qquad (y^2 - cx^2)(c-1) = 0$

$\Rightarrow \qquad (c - c^2)\,x^2 + (c-1)\,y^2 - c = 0.$

This is the general solution of (1).

This represents a family of conics.

Let $f(x, y, p) = xyp^2 - (x^2 + y^2 - 1)p + xy.$

This is quadratic in p.

$\therefore$ Disc. = 0 $\Rightarrow \qquad (x^2 + y^2 - 1)^2 - 4\,xy\,.\,xy = 0$

$\Rightarrow \qquad (x^2 + y^2 - 1 - 2xy)(x^2 + y^2 - 1 + 2xy) = 0$

$\Rightarrow \qquad ((x-y)^2 - 1)((x+y)^2 - 1) = 0$

$\Rightarrow \qquad (x - y - 1)(x - y + 1)(x + y - 1)(x + y + 1) = 0$

Let $\phi(x, y, c) = (c - c^2)x^2 + (c-1)\,y^2 - c$

$\Rightarrow \quad \phi(x, y, c) = -\,c^2x^2 + (x^2 + y^2 - 1)\,c - y^2$

This is quadratic in c.

$\therefore$ Disc. = 0 $\Rightarrow \quad (x^2 + y^2 - 1)^2 - 4(-\,x^2)(-\,y^2) = 0$

$\Rightarrow \quad (x^2 + y^2 - 1)^2 - 4x^2y^2 = 0 \qquad \Rightarrow \qquad (x^2 + y^2 - 1 - 2xy)(x^2 + y^2 - 1 + 2xy) = 0$

$\Rightarrow \quad ((x-y)^2 - 1)\,((x+y)^2 - 1) = 0 \qquad \Rightarrow \qquad (x - y - 1)(x - y + 1)(x + y - 1)(x + y + 1) = 0.$

Since p-discriminant relation and c-discriminant relation are same,

we have $\qquad E = 0 \quad \Rightarrow \quad (x - y - 1)(x - y + 1)(x + y - 1)(x + y + 1) = 0 \qquad ...(3)$

(3) $\Rightarrow \quad x - y - 1 = 0 \qquad ...(4) \qquad x - y + 1 = 0 \qquad ...(5)$

$x + y - 1 = 0 \qquad ...(6) \qquad$ and $\qquad x + y + 1 = 0 \qquad ...(7)$

(4) $\Rightarrow \quad 1 - \dfrac{dy}{dx} - 0 = 0 \qquad \Rightarrow \qquad p = 1.$ Also, (4) $\Rightarrow \quad y = x - 1.$

Putting the values of p and y in (1), we get

$$x(x-1).1^2 - (x^2 + (x-1)^2 - 1)(1) + x(x-1) = 0$$

$\Rightarrow \quad x^2 - x - 2x^2 + 2x + x^2 - x = 0,$ which is true.

$\therefore$ (4) satisfies the given equation.

Similarly, (5), (6), (7) satisfy the given equation.

$\therefore$ The singular solution of the given equation is

$$(x - y - 1)(x - y + 1)(x + y - 1)(x + y + 1) = 0.$$

This represents four lines $x - y - 1 = 0, x - y + 1 = 0, x + y - 1 = 0$ and $x + y + 1 = 0$ covering a square.

Since the singular solution is the envelope of the curves given by the general solution, the conics given by the general solution are touched by the lines given by the singular solution.

EXERCISE 1

Investigate the following differential equations for singular solutions and extraneous loci (Q. No. 1 — 12) :

1. $p^3 + px - y = 0$

2. $4xp^2 - (3x - 1)^2 = 0$

3. $y = 2xp - yp^2$

4. $6p^2y^2 + 3px - y = 0$

5. $9yp^2 + 4 = 0$

6. $xp^2 - 2yp + x + 2y = 0$

7. $p^3 - 4xyp + 8y^2 = 0$
8. $x^3p^2 + x^2yp + a^3 = 0$
9. $2p^2 - 2px^2 + 3xy = 0$
10. $(a^2 - x^2)p^2 + 2xyp + b^2 - y^2 = 0$
11. $\sin px \cos y = \cos px \sin y + p$ *(MDU 2005)*
12. $p = \log (px - y)$.
13. Reduce the equation $y^2(y - xp) = x^4 p^2$ to Clairaut's form by using the substitutions $x = \frac{1}{u}, y = \frac{1}{v}$ and hence find its singular solution and equations of extraneous loci.
14. Transform $xp^2 - 2yp + x + 2y = 0$ to Clairaut's form by the transformations $x^2 = u, y - x = v$. Hence obtain and interpret the general solution and singular solution of the given equation. Show that the general solution represents a family of parabolas touching a pair of straight lines.
15. Reduce the differential equation $x^2p^2 + yp(2x + y) + y^2 = 0$, where $p = \frac{dy}{dx}$, to Clairaut's form by the substitutions $u = y, v = xy$. Hence or otherwise solve it. Also prove that $y + 4x = 0$ is singular solution of the given equation. *(KU 2003)*
16. Reduce $(px - y)(x - py) = 2p$ to Clairaut's form by using the substitutions $x^2 = u$ and $y^2 = v$ and find the general solution and its singular solution, if any. *(CDLU 2004)*

Answers

1. S.S. $27y^2 + 4x^3 = 0$
2. S.S. $x = 0$, T.L. $3x - 1 = 0$, N.L. $x - 1 = 0$
3. S.S. $x^2 - y^2 = 0$
4. S.S. $3x^2 + 8y^3 = 0$
5. C.L. $y = 0$
6. S.S. $x^2 + 2xy - y^2 = 0$
7. S.S. $y(27y - 4x^3) = 0$, T.L. $y = 0$
8. S.S. $x(xy^2 - 4a^3) = 0$, T.L. $x = 0$
9. S.S. $x^3 - 6y = 0$, C.L. $x = 0$
10. S.S. $\frac{x^2}{a^2} + \frac{y^2}{b^2} = 1$
11. S.S. $y = \sqrt{x^2 - 1} - \sin^{-1} \frac{\sqrt{x^2 - 1}}{x}$
12. S.S. $x + y - x \log x = 0$
13. S.S. $y(y + 4x^2) = 0$, T.L. $xy = 0$
14. G.S. $2c^2x^2 - 2c(y - x) + 1 = 0$. S.S. $y = (\sqrt{2} + 1)x, y = (-\sqrt{2} + 1)x$
15. G.S. $xy = cy + c^2$
16. G.S. $(1 - c)y^2 - c(1 - c)x^2 + 2c = 0$, S.S. $(x^2 + y^2 - 2)^2 - 4x^2y^2 = 0$.

Hints

6. Try $u = x^2$ and $v = y - x$.
10. Given equation is expressible as $y = px \pm \sqrt{a^2p^2 + b^2}$.
13. The reduced equation is $v = Pu + P^2$, where $P = \frac{dv}{du}$.
14. $u = x^2 \Rightarrow \frac{du}{dx} = 2x, \quad v = y - x \Rightarrow \frac{dv}{dx} = p - 1$

$$\therefore \quad P = \frac{dv}{du} = \frac{p-1}{2x} \quad \text{or} \quad p = 2xP + 1.$$

The given equation reduces to $v = Pu + \frac{1}{2P}$.

15. $u = y \Rightarrow \frac{du}{dx} = p, \quad v = xy \Rightarrow \frac{dv}{dx} = y + xp$

$$\therefore \quad P = \frac{dv}{du} = \frac{y + xp}{p} \quad \text{or} \quad p = \frac{y}{P - x}.$$

The given equation reduces to $v = Pu + P^2$.

16. $u = x^2 \Rightarrow \dfrac{du}{dx} = 2x, \quad v = y^2 \Rightarrow \dfrac{dv}{dx} = 2yp$

$$\therefore \quad P = \frac{dv}{du} = \frac{y}{x} p \quad \text{or} \quad p = \frac{x}{y} P.$$

The given equation reduces to $v = Pu - \dfrac{2P}{1 - P}$.

18

Orthogonal Trajectories

1. Orthogonal Trajectories

(*i*) **Trajectory.** A curve which cuts every member of a given family of curves according to some definite law is called a **trajectory** of the family.

(*ii*) **Orthogonal trajectory.** A curve which cuts every member of a given family of curves at right angle is called an **orthogonal trajectory** of the family.

(*iii*) **Self orthogonal family.** A family of curves is called a **self orthogonal family** if the family of orthogonal trajectories of this family coincides with the given family.

2. Working Rules to Find the Equation of Orthogonal Trajectories

(*a*) **Let the family of cartesian curves be f(x, y, c) = 0.** ...(1)

Step I. Differentiate (1) w.r.t. x.

Step II. Eliminate the arbitrary constant c between (1) and the resulting equation. This gives the differential equation of the family (1).

Let it be $$F\left(x, y, \frac{dy}{dx}\right) = 0. \qquad ...(2)$$

Step III. Replace $\dfrac{dy}{dx}$ by $-\dfrac{dx}{dy}$ in (2).

The differential equation of the orthogonal trajectories is

$$F\left(x, y, -\frac{dx}{dy}\right) = 0. \qquad ...(3)$$

Step IV. Solve (3) to get the equation of the required orthogonal trajectories.

(*b*) **Let the family of polar curves be f (r, θ, c) = 0.** ...(1)

Step I. Differentiate (1) w.r.t. θ.

Step II. Eliminate the arbitrary constant c between (1) and the resulting equation. This gives the differential equation of the family (1).

Let it be $$F\left(r, \theta, \frac{dr}{d\theta}\right) = 0. \qquad ...(2)$$

Step III. Replace $\dfrac{dr}{d\theta}$ by $-r^2\dfrac{d\theta}{dr}$ in (2).

The differential equation of the orthogonal trajectories is

$$F\left(r, \theta, -r^2\frac{d\theta}{dr}\right) = 0. \qquad ...(3)$$

Step IV. Solve (3) to get the equation of the required orthogonal trajectories.

Example 1. *Find the equation of orthogonal trajectories of the family of parabolas* $y^2 = 4ax$.

Sol. The equation of the family of parabolas is

$$y^2 = 4ax. \qquad ...(1)$$

Step I. Differentiating (1) with respect to x, we get

$$2y\frac{dy}{dx} = 4a \quad i.e., \quad y\frac{dy}{dx} = 2a \qquad ...(2)$$

Step II. Eliminating a between (1) and (2), we get

$$y^2 = 2y\frac{dy}{dx}x \quad \text{or} \quad y = 2x\frac{dy}{dx}. \qquad ...(3)$$

Step III. Replacing $\frac{dy}{dx}$ by $-\frac{dx}{dy}$ in (3), the differential equation of orthogonal trajectories is

$$y = -2x\frac{dx}{dy}.$$

$$\Rightarrow \quad y\,dy = -2xdx \qquad ...(4)$$

Step IV. Integrating (4), we have

$$\int y\,dy = -2\int xdx + c_1 \quad \Rightarrow \quad \frac{y^2}{2} = -x^2 + c_1$$

$$\Rightarrow \quad \mathbf{2x^2 + y^2 = k.} \qquad ...(k = 2c_1)$$

This is the equation of the orthogonal trajectories.

Example 2. *Find the orthogonal trajectories of the family of curves* $\frac{x^2}{a^2} + \frac{y^2}{b^2 + \lambda} = 1$, *where* λ *is a parameter.* (MDU 2005)

Sol. The equation of the given family is

$$\frac{x^2}{a^2} + \frac{y^2}{b^2 + \lambda} = 1. \qquad ...(1)$$

Step I. Differentiating (1) w.r.t. x, we get

$$\frac{2x}{a^2} + \frac{2y}{b^2 + \lambda}\frac{dy}{dx} = 0 \quad \text{or} \quad \frac{x}{a^2} + \frac{y}{b^2 + \lambda}\frac{dy}{dx} = 0 \qquad ...(2)$$

Step II. To eliminate the parameter λ, we equate the values of $b^2 + \lambda$ from (1) and (2).

From (1), $$\frac{y^2}{b^2 + \lambda} = 1 - \frac{x^2}{a^2} = \frac{a^2 - x^2}{a^2} \quad \text{or} \quad b^2 + \lambda = \frac{a^2y^2}{a^2 - x^2}$$

From (2) $$b^2 + \lambda = -\frac{a^2y}{x}\frac{dy}{dx}$$

$$\therefore \quad \frac{a^2y^2}{a^2 - x^2} = -\frac{a^2y}{x}\frac{dy}{dx} \quad \text{or} \quad \frac{xy}{a^2 - x^2} + \frac{dy}{dx} = 0. \qquad ...(3)$$

This is the differential equation of the given family (1).

Step III. Replacing $\frac{dy}{dx}$ by $-\frac{dx}{dy}$ in (3), we get

$$\frac{xy}{a^2 - x^2} - \frac{dx}{dy} = 0 \quad \text{or} \quad y\,dy - \left(\frac{a^2 - x^2}{x}\right) dx = 0. \qquad ...(4)$$

This is the differential equation of the orthogonal trajectories.

Step IV. Integrating (4), we get

$$\int y\,dy - \int \left(\frac{a^2}{x} - x\right) dx = c \quad \text{or} \quad \frac{y^2}{2} - a^2 \log x + \frac{x^2}{2} = c$$

or $$\mathbf{x^2 + y^2 = 2a^2 \log x + C.} \qquad \mathbf{(C = 2c)}$$

This is the equation of the orthogonal trajectories of (1).

Example 3. *Prove that the system of confocal and coaxial parabolas $y^2 = 4a(x + a)$ is self orthogonal.*

Sol. The equation of the given family of parabolas is

$$y^2 = 4a(x + a). \qquad ...(1)$$

Step I. Differentiating (1) w.r.t. x, we get

$$2y\frac{dy}{dx} = 4a \quad \text{or} \quad y\frac{dy}{dx} = 2a \qquad ...(2)$$

Step II. Eliminating a between (1) and (2), we have

$$y^2 = 4 \,.\, \frac{y}{2}\frac{dy}{dx}\left[x + \frac{y}{2} \,.\, \frac{dy}{dx}\right] = 2xy\frac{dy}{dx} + y^2\left(\frac{dy}{dx}\right)^2$$

or $$y\left(\frac{dy}{dx}\right)^2 + 2x\frac{dy}{dx} - y = 0. \qquad ...(3)$$

This is the differential equation of the given family (1).

Step. III. Replacing $\frac{dy}{dx}$ by $-\frac{dx}{dy}$ in (3), we get $y\left(-\frac{dx}{dy}\right)^2 - 2x\frac{dx}{dy} - y = 0$

or $$y - 2x\frac{dy}{dx} - y\left(\frac{dy}{dx}\right)^2 = 0 \quad \text{or} \quad y\left(\frac{dy}{dx}\right)^2 + 2x\frac{dy}{dx} - y = 0. \qquad ...(4)$$

This is the differential equation of the orthogonal trajectories.

Since (4) is the same as (3), the orthogonal trajectories of the given family (1) are the curves of this family itself.

$\therefore$ The given family of parabolas is self orthogonal.

Example 4. *Prove that the system of confocal conics*

$$\frac{x^2}{a^2 + \lambda} + \frac{y^2}{b^2 + \lambda} = 1 \textit{ is self orthogonal.}$$

Sol. The system of confocal conics is

$$\frac{x^2}{a^2 + \lambda} + \frac{y^2}{b^2 + \lambda} = 1. \qquad ...(1)$$

Step I. Differentiating (1) w.r.t. x, we get

$$\frac{2x}{a^2 + \lambda} + \frac{2y}{b^2 + \lambda}\frac{dy}{dx} = 0$$

or $$\frac{x}{a^2+\lambda}+\frac{y}{b^2+\lambda}\,p=0, \quad \text{where } p=\frac{dy}{dx} \qquad \text{...(2)}$$

Step II. (2) $\Rightarrow$ $(b^2+\lambda)\,x+(a^2+\lambda)\,yp=0$ or $\lambda(x+yp)=-(a^2yp+b^2x)$

$$\therefore \quad \lambda=-\frac{a^2yp+b^2x}{x+yp}$$

$$\therefore \quad a^2+\lambda=\frac{(a^2-b^2)\,x}{x+yp}, \quad b^2+\lambda=-\frac{(a^2-b^2)\,yp}{x+yp} \qquad \text{...(3)}$$

$\therefore$ From (1) and (2), eliminating λ, we get

$$\frac{x^2(x+yp)}{(a^2-b^2)x}-\frac{y^2(x+yp)}{(a^2-b^2)yp}=1 \quad \text{or} \quad (x+yp)\left(x-\frac{y}{p}\right)=a^2-b^2. \qquad \text{...(4)}$$

This is the differential equation of the family of curves (1).

Step III. Replacing p by $-\frac{1}{p}$, the differential equation of the orthogonal trajectories is

$$\left(x-\frac{y}{p}\right)(x+yp)=a^2-b^2. \qquad \text{...(5)}$$

This is the same as (4).

Since the differential equation of the family of curves (1) and that of the orthogonal trajectories are same, therefore, the orthogonal trajectories of the given family are the curves of this family itself.

$\therefore$ The family of curves (1) is self orthogonal.

Example 5. *Find the equation of orthogonal trajectories of the family of cardioids*

$$r=a(1-\cos\theta). \qquad \text{(M.D.U. 2006, 2007)}$$

Sol. The equation of the given family of cardioids is

$$r=a(1-\cos\theta). \qquad \text{...(1)}$$

Step I. Differentiating (1) w.r.t. θ, we get $\frac{dr}{d\theta}=a\sin\theta$ ...(2)

Step II. Dividing (2) by (1) [to eliminate a], we get

$$\frac{1}{r}\cdot\frac{dr}{d\theta}=\frac{\sin\theta}{1-\cos\theta}. \qquad \text{...(3)}$$

Step III. Replacing $\frac{dr}{d\theta}$ *by* $-r^2\frac{d\theta}{dr}$ in (3), we get

$$\frac{1}{r}\left(-r^2\frac{d\theta}{dr}\right)=\frac{\sin\theta}{1-\cos\theta} \quad \Rightarrow \quad \frac{1-\cos\theta}{\sin\theta}\,d\theta=-\frac{dr}{r}. \qquad \text{...(4)}$$

This is the differential equation of the orthogonal trajectories.

Step IV. Integrating (4), we get

$$\int\frac{1-\cos\theta}{\sin\theta}\,d\theta=-\int\frac{dr}{r}+\log k$$

$$\Rightarrow \quad \int\frac{\sin\theta}{1+\cos\theta}\,d\theta=-\log r+\log k$$

$\Rightarrow \quad -\log(1 + \cos\theta) = -\log r + \log k \quad \Rightarrow \quad \mathbf{r = k(1 + \cos\theta).}$

This is the equation of the orthogonal trajectories of (1).

Example 6. *Find the equation of orthogonal trajectories of the system of curves*

$r^n \sin n\theta = a^n$, where a is a parameter. (MDU 2004)

Sol. The given system of curves is $r^n \sin n\theta = a^n$. ...(1)

Step I. Taking 'logs' of both sides, we get

$$n \log r + \log \sin n\theta = n \log a.$$

Differentiating w.r.t. θ, we have

$$\frac{n}{r}\frac{dr}{d\theta} + \frac{n \cos n\theta}{\sin n\theta} = 0 \quad \text{or} \quad \frac{dr}{d\theta} = -r \cot n\theta. \qquad ...(2)$$

This is the differential equation of the family (1) and it does not contain 'a'.

Step II. Replacing $\frac{dr}{d\theta}$ by $-r^2 \frac{d\theta}{dr}$ in (2), we get

$$-r^2 \frac{d\theta}{dr} = -r \cot n\theta \textit{ i.e.,} \quad r \frac{d\theta}{dr} = \cot n\theta$$

$$\therefore \quad \frac{dr}{r} = \tan n\theta \, d\theta \qquad ...(3)$$

Step III. Integrating (3), we get

$$\int \frac{dr}{r} = \int \tan n\theta \, d\theta + c'$$

$$\therefore \quad \log r = \frac{1}{n} \log \sec n\theta + \log c, \quad \text{where } c' = \log c$$

or $\quad n \log r = \log \sec n\theta + n \log c \quad \text{or} \quad \log r^n = \log \sec n\theta + \log c^n$

i.e., $\quad r^n = c^n \sec n\theta \quad \text{or} \quad \mathbf{r^n \cos n\theta = c^n.}$

This is the equation of the orthogonal trajectories of (1).

Example 7. *Find the equation of orthogonal trajectories of the family of curves $r = a(\sec\theta + \tan\theta)$ where a is a parameter.*

Sol. Given family of curves is $r = a(\sec\theta + \tan\theta)$. ...(1)

Step I. Differentiating (1) w.r.t. θ we get

$$\frac{dr}{d\theta} = a\,(\sec\theta \tan\theta + \sec^2\theta) \quad \Rightarrow \quad \frac{dr}{d\theta} = a \sec\theta\,(\tan\theta + \sec\theta)$$

Step II. Putting $a(\tan\theta + \sec\theta)$ equal to r, we get

$$\frac{dr}{d\theta} = r \sec\theta \qquad ...(2)$$

Step III. Replacing $\frac{dr}{d\theta}$ by $-r^2 \frac{d\theta}{dr}$ in (2), we get

$$-r^2 \frac{d\theta}{dr} = r \sec\theta \quad \Rightarrow \quad -\cos\theta \, d\theta = \frac{dr}{r}$$

Step IV. Integrating, we get $\quad -\sin\theta = \log r + \log k$

$\Rightarrow \quad rk = e^{-\sin\theta} \quad \Rightarrow \quad \mathbf{r = \frac{1}{k} e^{-\sin\theta}}$

This is the equation of the orthogonal trajectories of (1).

EXERCISE 1

1. Find the orthogonal trajectories of the following families of curves :
   (*i*) $xy = c$ (*ii*) $y = ax^2$ (*M.D.U. 2006*) (*iii*) $y = ax^3$
   (*iv*) $y = ax^n$ (*v*) $ay^2 = x^3$ (*vi*) $x^2 + y^2 = a^2$.
2. Find the orthogonal trajectories of the following families of curves :
   (*i*) $x + 2y = c$ (*ii*) $x^2 + 2y^2 = c$ (*iii*) $y = ce^{-2x}$
   (*iv*) $y = x - 1 + ce^{-x}$ (*v*) $y^2 = 2x^2(1 - cx)$ (*vi*) $y^2 = \dfrac{x^3}{c - x}$
   (*vii*) $px^2 + qy^2 = a^2$, where p and q are fixed constants. (*K.U. 2003 ; M.D.U. 2005*)
3. Find the orthogonal trajectories of the family of ellipses $\dfrac{x^2}{a^2} + \dfrac{y^2}{a^2 + \lambda} = 1$, where λ is a parameter.
4. Show that the family of parabolas $x^2 = 4a(y + a)$ is self-orthogonal. (*K.U. 2004 S*)
5. Find the orthogonal trajectories of the family of curves $x^{2/3} + y^{2/3} = a^{2/3}$.
6. (*i*) Find the orthogonal trajectories of the family of circles $x^2 + y^2 + 2gx + c = 0$, c being a parameter. (*K.U. 2007*)
   (*ii*) Find the orthogonal trajectories of the family of circles $x^2 + y^2 + 2gx + c = 0$, g being a parameter. (*K.U. 2003 C ; M.D.U. 2007*)
7. Find the orthogonal trajectories of the following families of curves :
   (*i*) $r = a \cos \theta$ (*ii*) $r = a(1 + \cos \theta)$ (*iii*) $r = a + \sin 5\theta$ (*iv*) $r = a(1 + \sin \theta)$.
8. Find the orthogonal trajectories of the family of curves $px^2 + qy^2 = a^2$, where p and q are fixed constants. (*K.U. 2003*)
9. Find the orthogonal trajectories of the following families of curves :
   (*i*) $r^2 = a^2 \cos 2\theta$ (*ii*) $r^n = a^n \cos n\theta$ (*iii*) $r = \dfrac{2a}{1 + \cos \theta}$ (*iv*) $r = a^{\theta}$.
10. Find the orthogonal trajectories of the family of curves $\left(r + \dfrac{k^2}{r}\right) \cos \theta = a$, where a is a parameter.

Answers

1. (*i*) $x^2 - y^2 = k$ (*ii*) $x^2 + 2y^2 = c$ (*iii*) $x^2 + 3y^2 = c$
   (*iv*) $x^2 + ny^2 = c$ (*v*) $2x^2 + 3y^2 = c$ (*vi*) $y = cx$
2. (*i*) $y - 2x = k$ (*ii*) $y = kx^2$ (*iii*) $y^2 = x + k$
   (*iv*) $x = y - 1 + ke^{-y}$ (*v*) $x^2 + 3y^2 \log (ky) = 0$ (*vi*) $(x^2 + y^2)^2 = k(2x^2 + y^2)$
   (*vii*) $y^p = cx^q$
3. $x^2 + y^2 = 2a^2 \log x + c$ **5.** $x^{4/3} - y^{4/3} = c$ **6.** (*i*) $x - ky + g = 0$
   (*ii*) $x^2 + y^2 - ky - c = 0$ **7.** (*i*) $r = c \sin \theta$ (*ii*) $r = c(1 - \cos \theta)$
   (*iii*) $r \log (\sec 5\theta + \tan 5\theta) + cr = 25$ (*iv*) $r = c(1 - \sin \theta)$
8. $y^p = cx^q$ **9.** (*i*) $r^2 = c^2 \sin 2\theta$ (*ii*) $r^n = c^n \sin n\theta$
   (*iii*) $r = \dfrac{c}{1 - \cos \theta}$ (*iv*) $r = e^{\sqrt{k - \theta^2}}$ **10.** $r^2 = k^2 + cr \operatorname{cosec} \theta$.

Hints

2. (vi) $y^2 = \dfrac{x^3}{c-x} \Rightarrow y^2(c-x) = x^3 \Rightarrow 2yy'(c-x) + y^2(-1) = 3x^2$

$$\Rightarrow \quad 2yy' . \frac{x^3}{y^2} - y^2 = 3x^2 \Rightarrow y' = \frac{3x^2y + y^3}{2x^3}.$$

Replacing y' by $-\dfrac{dx}{dy}$, we get

$$-\frac{dx}{dy} = \frac{3x^2y + y^3}{2x^3} \quad \text{or} \quad \frac{dy}{dx} = -\frac{2x^3}{3x^2y + y^3}.$$

This is a homogeneous differential equation.

6. (i) $x^2 + y^2 + 2gx + c = 0 \Rightarrow 2x + 2yy' + 2g = 0 \Rightarrow y' = -\dfrac{x+g}{y}$

Replacing y' by $-\dfrac{dx}{dy}$, we get $-\dfrac{dx}{dy} = -\dfrac{x+g}{y}$.

$$\Rightarrow \quad \frac{dx}{x+g} = \frac{dy}{y} \quad \Rightarrow \log(x+g) = \log y + \log k$$

(ii) $x^2 + y^2 + 2gx + c = 0 \quad \Rightarrow 2x + 2yy' + 2g = 0$

$\therefore \quad x^2 + y^2 + (-2x - 2yy')x + c = 0 \Rightarrow y^2 - x^2 - 2xyy' + c = 0$

Replacing y' by $-\dfrac{dx}{dy}$, we get $y^2 - x^2 + 2xy\dfrac{dx}{dy} + c = 0$ or $2x\dfrac{dx}{dy} - \dfrac{x^2}{y} = -y - \dfrac{c}{y}$

$$z = x^2 \Rightarrow \frac{dz}{dy} = 2x\frac{dx}{dy} \quad \therefore \quad \frac{dz}{dy} + z\left(-\frac{1}{y}\right) = -y - \frac{c}{y}.$$

This is a linear equation.

19

Linear Differential Equations with Constant Coefficients

1. Introduction

Till now, we have been discussing ordinary differential equations of order first. These differential equations were of degree one or more than one. Now, we shall consider the solution of differential equations of order more than one. A differential equation of order more than one may or may not be linear. In the present chapter, we shall consider linear differential equations of order more than one and with constant coefficients.

2. Linear Differential Equation

A differential equation in which the dependent variable and its differential coefficients occur only in the first degree and are not multiplied together is called a **linear differential equation.**

A linear differential equation of order n is of the form

$$P_0 \frac{d^n y}{dx^n} + P_1 \frac{d^{n-1} y}{dx^{n-1}} + P_2 \frac{d^{n-2} y}{dx^{n-2}} + \ldots\ldots + P_n y = Q \qquad \ldots(1)$$

where $P_0 \neq 0, P_1, P_2, \ldots\ldots, P_n$, Q are functions of x.

For example,

$$x^2 \frac{d^3 y}{dx^3} + \sin^2 x \frac{d^2 y}{dx^2} + 6y = 1 + \cos x$$

is a linear differential equation of order 3.

In particular, if $P_0, P_1, P_2, \ldots\ldots, P_n$ are all constants then the equation (1) is called a **linear differential equation with constant coefficients.**

For example,

$$2\frac{d^4 y}{dx^4} + 7\frac{d^3 y}{dx^3} + 9\frac{dy}{dx} + 15y = e^{4x}$$

is a linear differential equation with constant coefficients. The order of this differential equation is 4.

3. Differential Operator

In calculus the symbols $\frac{d}{dx}, \frac{d^2}{dx^2}, \frac{d^3}{dx^3}, \ldots\ldots$ stand respectively for first derivative w.r.t. x, second derivative w.r.t. x, third derivative w.r.t. x, ... We denote the symbols

$\frac{d}{dx}, \frac{d^2}{dx^2}, \frac{d^3}{dx^3}, ...$ respectively by D, D^2, D^3, ... The symbols D, D^2, D^3, ... are called **operators.** It may be verified that these operators satisfy the following laws :

(*i*) $D^m + D^n = D^n + D^m$ (*ii*) $D^m D^n = D^n D^m = D^{m+n}$

(*iii*) $D(u + v) = Du + Dv$ (*iv*) $(D - \alpha)(D - \beta) = (D - \beta)(D - \alpha)$.

Let us verify law (*iv*)

$$(D - \alpha)(D - \beta)\, y = (D - \alpha)(D\beta - \beta y)$$

$$= (D - \alpha)\left(\frac{dy}{dx} - \beta y\right) = D\left(\frac{dy}{dx} - \beta y\right) - \alpha\left(\frac{dy}{dx} - \beta y\right)$$

$$= \frac{d}{dx}\left(\frac{dy}{dx} - \beta y\right) - \alpha\frac{dy}{dx} + \alpha\beta y = \frac{d^2y}{dx^2} - \beta\frac{dy}{dx} - \alpha\frac{dy}{dx} + \alpha\beta y$$

$$= \frac{d^2y}{dx^2} - (\beta + \alpha)\frac{dy}{dx} + \alpha\beta y.$$

Similarly, we can show that

$$(D - \beta)(D - \alpha)\, y = \frac{d^2y}{dx^2} - (\beta + \alpha)\frac{dy}{dx} + \alpha\beta\, .$$

$$\therefore \qquad (D - \alpha)(D - \beta)y = (D - \beta)(D - \alpha)y$$

$$\Rightarrow \qquad (D - \alpha)(D - \beta) = (D - \beta)(D - \alpha).$$

Remarks 1. We have shown that $(D - \alpha)(D - \beta)y = (D^2 - (\beta + \alpha)\, D + \alpha\beta)y$.

$$\therefore \qquad (D - \alpha)(D - \beta) = D^2 - (\beta + \alpha)D + \alpha\beta$$

Thus the factors $D - \alpha$ and $D - \beta$ can be multiplied algebrically.

2. In terms of operator D, the differential equation (1) can be written as

$$(P_0 D^n + P_1 D^{n-1} + P_2 D^{n-2} + ... + P_n)\, y = Q.$$

or as $f(D)y = Q$ where $f(D) = P_0 D^n + P_1 D^{n-1} + P_2 D^{n-2} + ... + P_n$. $f(D)$ is a polynomial in D.

4. Linearly Independent Functions

A set of functions, $y_1, y_2, ..., y_n$ is said to be **linearly independent** if the equality $c_1 y_1 + c_2 y_2 + ... + c_n y_n = 0$ holds only when $c_1, c_2, ..., c_n$ are all zero.

Thus, if $y_1, y_2, ..., y_n$ are linearly independent then any one of these functions cannot be expressed as a linear combination of remaining functions.

For example, the functions e^{3x}, e^{4x}, $5e^{4x}$ are not linearly independent, because we have

$$c_1 e^{3x} + c_2 e^{4x} + c_3 (5e^{4x}) = 0,$$

where $\qquad c_1 = 0, c_2 = 10, c_3 = -2.$

A necessary and sufficient condition for the functions $y_1, y_2, ..., y_n$ to be linearly independent is

$$\begin{vmatrix} y_1 & y_2 & \cdots\cdots & y_n \\ y_1' & y_2' & \cdots\cdots & y_n' \\ \cdots\cdots & \cdots\cdots & \cdots\cdots & \cdots\cdots \\ \cdots\cdots & \cdots\cdots & \cdots\cdots & \cdots\cdots \\ y_1^{n-1} & y_2^{n-1} & \cdots\cdots & y_n^{n-1} \end{vmatrix} \neq 0.$$

Theorem I. *If $y = y_1, y = y_2, ..., y = y_n$ are n linearly independent solutions of the linear differential equation*

$$P_0\frac{d^n y}{dx^n}+P_1\frac{d^{n-1}y}{dx^{n-1}}+P_2\frac{d^{n-2}y}{dx^{n-2}}+\ldots\ldots+P_n y=0$$

of order n, where $P_0 \neq 0, P_1, P_2, \ldots\ldots, P_n$ are functions of x then $y = c_1y_1 + c_2y_2 + \ldots\ldots + c_ny_n$ is the general solution of this differential equation, where $c_1, c_2, \ldots\ldots, c_n$ are arbitrary constants.

Proof. Given equation is

$$P_0\frac{d^n y}{dx^n}+P_1\frac{d^{n-1}y}{dx^{n-1}}+P_2\frac{d^{n-2}y}{dx^{n-2}}+\ldots\ldots+P_n y=0. \quad \ldots(1)$$

$$\Rightarrow \left(P_0\frac{d^n}{dx^n}+P_1\frac{d^{n-1}}{dx^{n-1}}+P_2\frac{d^{n-2}}{dx^{n-2}}+\ldots\ldots+P_n\right)y=0$$

$$\Rightarrow (P_0D^n + P_1D^{n-1} + P_2D^{n-2} + \ldots\ldots + P_n)\,y = 0$$

$$\Rightarrow f(D)\,y = 0 \quad \ldots(2)$$

where $f(D) = P_0D^n + P_1D^{n-1} + P_2D^{n-2} + \ldots\ldots + P_n$.

$\therefore$ $y = y_1, y = y_2, \ldots\ldots, y = y_n$ are solutions of (2).

$\therefore$ $f(D)\,y_1 = 0,\ f(D)\,y_2 = 0, \ldots\ldots, f(D)\,y_n = 0$.

Now $f(D)(c_1y_1 + c_2y_2 + \ldots\ldots + c_ny_n)$

$$= f(D)\,c_1y_1 + f(D)c_2y_2 + \ldots\ldots + f(D)c_ny_n$$
$$= c_1f(D)\,y_1 + c_2f(D)\,y_2 + \ldots\ldots + c_nf(D)\,y_n$$
$$= c_1 . 0 + c_2 . 0 + \ldots\ldots + c_n . 0 = 0.$$

$\therefore$ $y = c_1y_1 + c_2y_2 + \ldots\ldots + c_ny_n$

is also a solution of the given differential equation.

Since the functions $y_1, y_2, \ldots\ldots, y_n$ are given to be linearly independent, no function can be expressed as a linear combination of remaining functions.

$\therefore$ No term in $c_1y_1 + c_2y_2 + \ldots\ldots + c_ny_n$ can be eliminated.

$\therefore$ The solution $y = c_1y_1 + c_2y_2 + \ldots\ldots + c_ny_n$ contains n arbitrary constants $c_1, c_2, \ldots\ldots, c_n$. Also the order of given differential equation is n and the solution $y = c_1y_1 + c_2y_2 + \ldots\ldots + c_ny_n$ contains n arbitrary constant.

$\therefore$ The solution $y = c_1y_1 + c_2y_2 + \ldots\ldots + c_ny_n$ represents the general solution of the given differential equation.

5. General Solution of Linear Differential Equation with Constant Coefficients and Second Member Zero

Let
$$P_0\frac{d^n y}{dx^n}+P_1\frac{d^{n-1}y}{dx^{n-1}}+P_2\frac{d^{n-2}y}{dx^{n-2}}+\ldots\ldots+P_n y=0 \quad \ldots(1)$$

where $P_0 \neq 0, P_1, P_2, \ldots\ldots, P_n$ are constants be a linear differential equation of order n with constant coefficients and second member zero.

(1) can also be written as

$$(P_0D^n + P_1D^{n-1} + P_2D^{n-2} + \ldots\ldots + P_n)\,y = 0 \quad \text{or} \quad f(D)\,y = 0, \quad \ldots(2)$$

where $f(D) = P_0D^n + P_1D^{n-1} + P_2D^{n-2} + \ldots\ldots + P_n$. $\quad \ldots(3)$

Let $y = e^{mx}$ be a solution of (1).

$$\therefore \quad \frac{d^k y}{dx^k} = m^k e^{mx} \quad \forall \quad k = 1, 2, 3, \ldots\ldots$$

$\therefore$ (1) $\Rightarrow$ $P_0 m^n e^{mx} + P_1 m^{n-1} e^{mx} + P_2 m^{n-2} e^{mx} + \ldots\ldots + P_n e^{mx} = 0$

$\Rightarrow$ $(P_0 m^n + P_1 m^{n-1} + P_2 m^{n-2} + \ldots\ldots + P_n) e^{mx} = 0$

$\Rightarrow$ $P_0 m^n + P_1 m^{n-1} + P_2 m^{n-2} + \ldots\ldots + P_n = 0.$ ...(4)

$(\because \; e^{mx} \neq 0)$

This is a polynomial equation of degree n in m and is called the **auxiliary equation** of the differential equation (1).

On comparing (4) and $f(D) = 0$, we see that the roots of the equation $f(D) = 0$ will coincide with the roots of (4). In practice, we call $f(D) = 0$ as the auxiliary equation and solve this equation to find the roots of the equation (4).

Let the roots of the auxiliary equation of (1) be $m_1, m_2, m_3, \ldots\ldots, m_n$.

$\therefore$ $y = e^{m_1 x}$, $y = e^{m_2 x}$, $y = e^{m_3 x}$, ..., $y = e^{m_n x}$ are solutions of the given differential equation.

The general solution of the given differential equation will depend upon the nature of the roots $m_1, m_2, m_3, \ldots\ldots, m_n$.

Case I. Roots of real and distinct

In this case, the solutions

$$y = e^{m_1 x}, y = e^{m_2 x}, y = e^{m_3 x}, \ldots\ldots, y = e^{m_n x} \text{ are linearly independent.}$$

$\therefore$ The general solution of the given equation is

$$\mathbf{y = c_1 e^{m_1 x} + c_2 e^{m_2 x} + c_3 e^{m_3 x} + \ldots\ldots + c_n e^{m_n x}},$$

where $c_1, c_2, c_3, \ldots\ldots, c_n$ are n arbitrary constants.

Case II. Roots are distinct and not all real

The coefficients of the auxiliary equation are real.

$\therefore$ The imaginary roots occurs in conjugate pairs.

Let $m_1 = \alpha + i\beta$, $m_2 = \alpha - i\beta$ and all other roots be real and distinct.

$\therefore$ The solutions $y = e^{m_1 x}, y = e^{m_2 x}, y = e^{m_3 x}, \ldots\ldots, y = e^{m_n x}$ are linearly independent.

$\therefore$ The general solution of the given equation is

$$y = c_1 e^{m_1 x} + c_2 e^{m_2 x} + c_3 e^{m_3 x} + \ldots\ldots + c_n e^{m_n x}$$

$$\Rightarrow \quad y = c_1 e^{(\alpha + i\beta)x} + c_2 e^{(\alpha - i\beta)x} + c_3 e^{m_3 x} + \ldots\ldots + c_n e^{m_n x}$$

$$= e^{\alpha x}[c_1 e^{i\beta x} + c_2 e^{-i\beta x}] + c_3 e^{m_3 x} + \ldots\ldots + c_n e^{m_n x}$$

$$= e^{\alpha x}[c_1 (\cos \beta x + i \sin \beta x) + c_2 (\cos \beta x - i \sin \beta x)] + c_3 e^{m_3 x} + \ldots\ldots + c_n e^{m_n x}$$

$$\therefore \quad \mathbf{y = e^{\alpha x}(A \cos \beta x + B \sin \beta x) + c_3 e^{m_3 x} + \ldots\ldots + c_n e^{m_n x}},$$

where $A = c_1 + c_2$, $B = i(c_1 - c_2)$.

This is the general solution of the given equation.

Case III. Roots are real and distinct except two roots

Let $m_1 = m_2 \neq m_3 \neq \ldots\ldots \neq m_n$

The equation $f(D)\, y = 0$ can be written as

$$P_0 (D - m_1)^2 (D - m_3) \ldots\ldots (D - m_n)\, y = 0 \qquad (\because \; m_1 = m_2)$$

The factors on the left side are commutative.

$\therefore$ The solution of each of the equations

$$(D - m_1)^2 y = 0, (D - m_3) y = 0, \ldots\ldots, (D - m_n) y = 0$$

is a solution of the given equation.

We consider the solution of the equation $(D - m_1)^2 y = 0$. ...(5)

(5) $\Rightarrow (D - m_1)(D - m_1)y = 0$

$\Rightarrow \quad (D - m_1)u = 0$, where $u = (D - m_1)y$

$$\Rightarrow \frac{du}{dx} - m_1 u = 0 \quad \Rightarrow \quad \frac{du}{u} = m_1\, du \quad \Rightarrow \quad \log u = m_1 x + \log c_1$$

$$\Rightarrow \quad \frac{u}{c_1} = e^{m_1 x} \quad \Rightarrow \quad u = c_1 e^{m_1 x}$$

$$\therefore \quad (D - m_1) y = c_1 e^{m_1 x} \quad \Rightarrow \quad \frac{dy}{dx} - m_1 y = c_1 e^{m_1 x} \qquad \ldots(6)$$

This is a linear differential equation.

$$\text{I.F.} = e^{\int - m_1\, dx} = e^{-m_1 x}$$

$\therefore$ The solution of (6) is

$$y\,(e^{-m_1 x}) = \int (c_1 e^{m_1 x})(e^{-m_1 x})\, dx + c_2$$

$$\Rightarrow \quad y e^{-m_1 x} = c_1 x + c_2 \quad \text{or} \quad y = (c_1 x + c_2)\, e^{m_1 x}$$

The general solution of the given equation is

$$\mathbf{y = (c_1 x + c_2) e^{m_1 x} + c_3 e^{m_3 x} + \ldots\ldots + c_n e^{m_n x}.}$$

Remark. If the root m_1 is repeated r times and all other roots are distinct then the general solution takes the form.

$$y = (c_1 x^{r-1} + c_2 x^{r-2} + \ldots\ldots + c_r)\, e^{m_1 x} + c_{r+1} e^{m_{r+1} x} + \ldots\ldots + c_n e^{m_n x}.$$

In practical problems, it is convenient to write the general solution in the form

$$y = (c_1 + c_2 x + \ldots\ldots + c_r x^{r-1})\, e^{m_1 x} + c_{r+1} e^{m_{r+1} x} + \ldots\ldots + c_n e^{m_n x}.$$

Case IV. Two equal pairs of imaginary roots and all other roots are real and distinct

Let $m_1 = m_2 = \alpha + i\beta$, $m_3 = m_4 = \alpha - i\beta$ and all other roots are real and distinct.

$\therefore$ The general solution is

$$y = (c_1 + c_2 x)\, e^{(\alpha + i\beta)x} + (c_3 + c_4 x)\, e^{(\alpha - i\beta)x} + c_5 e^{m_5 x} + \ldots\ldots + c_n e^{m_n x}$$

$$= e^{\alpha x}[(c_1 + c_2 x)\, e^{i\beta x} + (c_3 + c_4 x)\, e^{-i\beta x}] + \ldots\ldots + c_n e^{m_n x}$$

$$= e^{\alpha x}[(c_1 + c_2 x)(\cos \beta x + i \sin \beta x) + (c_3 + c_4 x)(\cos \beta x - i \sin \beta x)] + \ldots\ldots + c_n e^{m_n x}$$

$$\therefore \quad \mathbf{y = e^{\alpha x}[(A + Bx) \cos \beta x + (C + Dx) \sin \beta x] + \ldots\ldots + c_n e^{m_n x},}$$

where A, B, C, D, ... , c_n are arbitrary constants.

6. Working Rules

Let the given differential equation be

$$P_0 \frac{d^n y}{dx^n} + P_1 \frac{d^{n-1} y}{dx^{n-1}} + P_2 \frac{d^{n-2} y}{dx^{n-2}} + \ldots\ldots + P_n y = 0, \qquad \ldots(1)$$

where $P_0 \neq 0, P_1, P_2, \ldots\ldots, P_n$ are constants.

The auxiliary equation is

$$P_0 D^n + P_1 D^{n-1} + P_2 D^{n-2} + \ldots\ldots + P_n = 0.$$

Let the roots of this equation be $m_1, m_2, \ldots\ldots, m_n$.

Case I. Roots are real and distinct

The general solution of (1) is

$$y = c_1 e^{m_1 x} + c_2 e^{m_2 x} + \ldots\ldots + c_n e^{m_n x}.$$

Case II. $m_1 = \alpha + i\beta$, $m_2 = \alpha - i\beta$ and all other are real and distinct.

The general solution is

$$y = e^{\alpha x}(c_1 \cos \beta x + c_2 \sin \beta x) + c_3 e^{m_3 x} + \ldots\ldots + c_n e^{m_n x}.$$

Case III. Roots are real and $m_1 = m_2 \neq m_3 \neq \ldots\ldots \neq m_n$

The general solution is

$$y = (c_1 + c_2 x)e^{m_1 x} + c_3 e^{m_3 x} + \ldots\ldots + c_n e^{m_n x}.$$

Case IV. $m_1 = m_2 = \alpha + i\beta$, $m_3 = m_4 = \alpha - i\beta$ and all other are real and distinct.

The general solution is

$$y = e^{\alpha x}[(c_1 + c_2 x) \cos \beta x + (c_3 + c_4 x) \sin \beta x] + c_5 e^{m_5 x} + \ldots\ldots + c_n e^{m_n x}.$$

Example 1. *Solve* $\dfrac{d^2y}{dx^2} + 4\dfrac{dy}{dx} + 3y = 0.$

Sol. We have $\dfrac{d^2y}{dx^2} + 4\dfrac{dy}{dx} + 3y = 0$ ...(1)

$\Rightarrow$ $(D^2 + 4D + 3)\,y = 0.$

$\therefore$ The A.E. is $D^2 + 4D + 3 = 0 \Rightarrow (D + 1)(D + 3) = 0$

$\Rightarrow$ $D = -1, -3.$

The roots are real and distinct.

$\therefore$ The general solution of (1) is $y = c_1 e^{-x} + c_2 e^{-3x}.$

Example 2. *Solve* $(D^4 + 5D^2 - 36)\,y = 0.$

Sol. We have $(D^4 + 5D^2 - 36)\,y = 0.$ ...(1)

$\therefore$ The A.E. is $D^4 + 5D^2 - 36 = 0$

$\Rightarrow$ $D^2 = 4, -9.$ $\therefore$ $D = \pm 2, \pm 3i.$

$\therefore$ The general solution of (1) is

$$y = c_1 e^{2x} + c_2 e^{-2x} + e^{0x}(c_3 \cos 3x + c_4 \sin 3x)$$

or

$$y = c_1 e^{2x} + c_2 e^{-3x} + c_3 \cos 3x + c_4 \sin 3x.$$

Example 3. *Solve* $\dfrac{d^4y}{dx^4} + 5\dfrac{d^3y}{dx^3} + 6\dfrac{d^2y}{dx^2} - 4\dfrac{dy}{dx} - 8y = 0.$

Sol. The symbolic form of the given equation is

$(D^4 + 5D^3 + 6D^2 - 4D - 8)\,y = 0.$ ...(1)

$\therefore$ The A.E. is $D^4 + 5D^3 + 6D^2 - 4D - 8 = 0.$ ...(2)

Putting $D = 1$, we see that (2) is satisfied.

$\therefore$ $D = 1$ is a solution of (2) by inspection.

$\therefore$ D – 1 is a factor of $D^4 + 5D^3 + 6D^2 - 4D - 8$.

The synthetic division by D – 1 gives ;

1	1	5	6	– 4	– 8
		1	6	12	8
	1	6	12	8	0

$\therefore$ Auxiliary equation (2) can be written as

$$(D - 1)(D^3 + 6D^2 + 12D + 8) = 0 \quad \text{or} \quad (D - 1)(D + 2)^3 = 0.$$

$\therefore$ $D = 1, -2, -2, -2.$

$\therefore$ The general solution of (1) is

$$\mathbf{y = c_1e^x + (c_2 + c_3x + c_4x^2)e^{-2x}.}$$

Example 4. *Solve* $\dfrac{d^4y}{dx^4} + m^4y = 0.$

Sol. The symbolic form of the given equation is

$$(D^4 + m^4)y = 0. \qquad ...(1)$$

$\therefore$ The A.E. is $D^4 + m^4 = 0 \Rightarrow (D^2)^2 + (m^2)^2 = 0$

Completing square by adding and subtracting $2D^2m^2$, we get

$$D^4 + 2m^2D^2 + m^4 - 2m^2D^2 = 0$$

$$\Rightarrow \quad (D^2 + m^2)^2 - 2m^2D^2 = 0$$

$$\Rightarrow \quad [(D^2 + m^2) - \sqrt{2}mD]\,[D^2 + m^2 + \sqrt{2}mD] = 0.$$

$$\therefore \quad D^2 + m^2 - \sqrt{2}mD = 0 \quad \text{or} \quad D^2 + m^2 + \sqrt{2}mD = 0$$

$$\Rightarrow \quad D = \frac{m\sqrt{2} \pm \sqrt{2m^2 - 4m^2}}{2}, \frac{-m\sqrt{2} \pm \sqrt{2m^2 - 4m^2}}{2}$$

$$= \frac{m\sqrt{2} \pm im\sqrt{2}}{2}, \frac{-m\sqrt{2} \pm im\sqrt{2}}{2} = \frac{m}{\sqrt{2}} \pm i\frac{m}{\sqrt{2}}, \; -\frac{m}{\sqrt{2}} \pm i\frac{m}{\sqrt{2}}.$$

We have two pairs of unequal imaginary roots.

$\therefore$ The general solution of (1) is

$$\mathbf{y = e^{\frac{m}{\sqrt{2}}x}\left[c_1 \cos \frac{m}{\sqrt{2}}x + c_2 \sin \frac{m}{\sqrt{2}}x\right] + e^{-\frac{m}{\sqrt{2}}x}\left[c_3 \cos \frac{m}{\sqrt{2}}x + c_4 \sin \frac{m}{\sqrt{2}}x\right].}$$

Example 5. *Solve* $(D^2 - 8D + 65)^2\, y = 0.$

Sol. We have $\qquad (D^2 - 8D + 65)^2\, y = 0. \qquad ...(1)$

$\therefore$ The A.E. is $\qquad (D^2 - 8D + 65)^2 = 0$

$$D^2 - 8D + 65 = 0 \Rightarrow D = \frac{8 \pm \sqrt{64 - 260}}{2} = 4 + 7i, \; 4 - 7i$$

$\therefore$ The roots of the A.E. are $4 + 7i, 4 + 7i, 4 - 7i, 4 - 7i$.

$\therefore$ The general solution of (1) is

$$\mathbf{y = e^{4x}\,[(c_1 + c_2x) \cos 7x + (c_3 + c_4x) \sin 7x].}$$

Example 6. *If* λ_1, λ_2 *are real and distinct roots of the equation* $\lambda^2 + a_1\lambda + a_2 = 0$ *and* $y_1 = e^{\lambda_1 x}, y_2 = e^{\lambda_2 x}$, *then prove that* $y = c_1y_1 + c_2y_2$ *is a solution of the differential equation* $\dfrac{d^2y}{dx^2} + a_1\dfrac{dy}{dx} + a_2y = 0$. *Is this solution general ?* (MDU 1998)

Sol. λ_1 and λ_2 are roots of $\quad \lambda^2 + a_1\lambda + a_2 = 0.$ ...(1)

Given equation is $\quad \dfrac{d^2y}{dx^2} + a_1 \dfrac{dy}{dx} + a_2 y = 0.$

$\Rightarrow \quad (D^2 + a_1 D + a_2)\, y = 0$ ...(2)

Let $y = e^{mx}$ be a solution of (2).

$\therefore \quad (D^2 + a_1 D + a_2)\, e^{mx} = 0$

$\Rightarrow \quad (m^2 + a_1 m + a_2)\, e^{mx} = 0$

$\Rightarrow \quad m^2 + a_1 m + a_2 = 0.$ ...(3)

Comparing (1) and (3), we see that λ_1, λ_2 are also roots of (3).

$\therefore \quad y_1 = e^{\lambda_1 x}, y = e^{\lambda_2 x}$ are solutions of (2).

Let $y = c_1 y_1 + c_2 y_2$, where c_1 and c_2 are arbitrary constants.

Now $\quad (D^2 + a_1 D + a_2)(c_1 y_1 + c_2 y_2)$

$$= c_1 (D^2 + a_1 D + a_2) y_1 + c_2 (D^2 + a_1 D + a_2) y_2$$

$$= c_1 (\lambda_1^2 + a_1 \lambda_1 + a_2) e^{\lambda_1 x} + c_2 (\lambda_2^2 + a_1 \lambda_2 + a_2) e^{\lambda_2 x}$$

$$= c_1 . 0 . e^{\lambda_1 x} + c_2 . 0 . e^{\lambda_2 x} = 0.$$

$\therefore \quad y = c_1 y_1 + c_2 y_2$ is a solution of (2).

Also, $\quad \begin{vmatrix} y_1 & y_2 \\ y_1' & y_2' \end{vmatrix} = \begin{vmatrix} e^{\lambda_1 x} & e^{\lambda_2 x} \\ \lambda_1 e^{\lambda_1 x} & \lambda_2 e^{\lambda_2 x} \end{vmatrix}$

$$= e^{\lambda_1 x} e^{\lambda_2 x} \begin{vmatrix} 1 & 1 \\ \lambda_1 & \lambda_2 \end{vmatrix} = e^{(\lambda_1 + \lambda_2)x} (\lambda_2 - \lambda_1) \neq 0. \qquad (\because \lambda_1 \neq \lambda_2)$$

$\therefore \quad y_1$ and y_2 are linearly independent.

$\therefore \quad y = c_1 y_1 + c_2 y_2$ is the general solution of the given differential equation.

Example 7. *If λ_1, λ_1 are real and equal roots of the equation $\lambda^2 + a_1\lambda + a_2 = 0$ then prove that $y = (c_1 + c_2 x)\, e^{\lambda_1 x}$ is a solution of the differential equation $\dfrac{d^2y}{dx^2} + a_1 \dfrac{dy}{dx} + a_2 y = 0$.*

Is this solution general ? (KU 1997)

Sol. λ_1, λ_1 are roots of $\quad \lambda^2 + a_1\lambda + a_2 = 0.$ ...(1)

$\therefore \quad$ (1) is same as $\quad (\lambda - \lambda_1)^2 = 0.$

Given equation is $\quad \dfrac{d^2y}{dx^2} + a_1 \dfrac{dy}{dx} + a_2 y = 0 \quad$ *i.e.,* $\quad (D^2 + a_1 D + a_2)\, y = 0$ ...(2)

(1) show that the roots of $D^2 + a_1 D + a_2 = 0$ are also λ_1, λ_1 and $D^2 + a_1 D + a_2 = (D - \lambda_1)^2$.

$\therefore \quad (2) \Rightarrow (D - \lambda_1)^2 y = 0 \Rightarrow (D - \lambda_1)(D - \lambda_1)\, y = 0$

$\Rightarrow \quad (D - \lambda_1)\, u = 0$, where $u = (D - \lambda_1) y$

$\Rightarrow \quad u = k_1 e^{\lambda_1 x}$

$\therefore \quad (D - \lambda_1)\, y = k_1 e^{\lambda_1 x} \quad \therefore \quad \dfrac{dy}{dx} - \lambda_1 y = k_1 e^{\lambda_1 x}$ ...(3)

This is a linear differential equation.

$$\text{I.F.} = e^{-\int \lambda_1 \, dx} = e^{-\lambda_1 x}$$

$\therefore$ The solution of (3) is $y\,(e^{-\lambda_1 x}) = \int (k_1 e^{\lambda_1 x})(e^{-\lambda_1 x})\,dx + k_2$

$\Rightarrow$ $y\,e^{-\lambda_1 x} = k_1 x + k_2$ or $y = (k_1 x + k_2)e^{\lambda_1 x}$.

Let $k_1 = c_2,\ k_2 = c_1$

$\therefore$ $y = (c_1 + c_2 x)e^{\lambda_1 x}$

$\therefore$ $y = c_1 e^{\lambda_1 x} + c_2 x e^{\lambda_1 x}$ is a solution of (2).

Let $y_1 = e^{\lambda_1 x},\ y_2 = x e^{\lambda_1 x}$

$$\therefore \quad \begin{vmatrix} y_1 & y_2 \\ y_1' & y_2' \end{vmatrix} = \begin{vmatrix} e^{\lambda_1 x} & x e^{\lambda_1 x} \\ \lambda_1 e^{\lambda_1 x} & (1 + \lambda_1 x) e^{\lambda_1 x} \end{vmatrix}$$

$$= e^{\lambda_1 x} e^{\lambda_1 x} \begin{vmatrix} 1 & x \\ \lambda_1 & 1 + \lambda_1 x \end{vmatrix} = e^{2\lambda_1 x}.1 = e^{2\lambda_1 x} \neq 0.$$

$\therefore$ y_1 and y_2 are linearly independent.

$$\therefore \quad y = c_1 y_1 + c_2 y_2 = c_1 e^{\lambda_1 x} + c_2\, x e^{\lambda_2 x} = (c_1 + c_2 x)\, e^{\lambda_1 x}$$

is the general solution of the given differential equation.

EXERCISE 1

Find the general solution of the following differential equations (Q. No. 1—4) :

1. (*i*) $\dfrac{d^2y}{dx^2} - 5\dfrac{dy}{dx} + 6y = 0$ (*ii*) $\dfrac{d^2y}{dx^2} - 4\dfrac{dy}{dx} + y = 0$

   (*iii*) $\dfrac{d^3y}{dx^3} - \dfrac{d^2y}{dx^2} - 12\dfrac{dy}{dx} = 0$ (*iv*) $\dfrac{d^3y}{dx^3} + 2\dfrac{d^2y}{dx^2} - 5\dfrac{dy}{dx} - 6y = 0$.

2. (*i*) $\dfrac{d^2y}{dx^2} + b^2 y = 0$ (*ii*) $(D^2 - 2D + 10)\,y = 0$

   (*iii*) $(D^3 - D^2 + 9D - 9)y = 0$ (*iv*) $\dfrac{d^4y}{dx^4} - a^4 y = 0$

   (*v*) $\dfrac{d^4y}{dx^4} + 4y = 0$ (*vi*) $\dfrac{d^4y}{dx^4} + 13\dfrac{d^2y}{dx^2} + 36y = 0$.

3. (*i*) $\dfrac{d^2y}{dx^2} - 6\dfrac{dy}{dx} + 9y = 0$ (*ii*) $(D + 2)(D^2 + 6D + 9)y = 0$

   (*iii*) $(D^3 - 3D^2 + 3D - 1)y = 0$ (*iv*) $\dfrac{d^4y}{dx^4} + 5\dfrac{d^3y}{dx^3} + 6\dfrac{d^2y}{dx^2} - 4\dfrac{dy}{dx} - 8y = 0$

   (*v*) $(D^4 + 6D^3 + 5D^2 - 24D - 36)y = 0$.

4. (*i*) $(D^2 - 2D + 5)^2 y = 0$ (*ii*) $\dfrac{d^4y}{dx^4} + 6\dfrac{d^2y}{dx^2} + 9y = 0$

(*iii*) $(D^2 - 4)(D^2 + 25)^2 y = 0$ (*iv*) $(D^6 + 9D^4 + 24D^2 + 16)y = 0$.

5. (*i*) Solve $\dfrac{d^2x}{dt^2} + 4\dfrac{dx}{dt} + 3x = 0$ given that for $t = 0, x = 0$ and $\dfrac{dx}{dt} = 12$.

(*ii*) Solve $\dfrac{d^2y}{dx^2} + y = 0$ given that $y = 2$ for $x = 0$ and $y = -2$ for $x = \dfrac{\pi}{2}$.

6. Show that the solution of the differential equation $\dfrac{d^2x}{dt^2} + k\dfrac{dx}{dt} + \mu x = 0$ is

$x = e^{(-1/2)kt}$ (A cos nt + B sin nt), where $n^2 = \mu - \dfrac{1}{4}k^2$ and n is real.

7. Prove that the general solution of $(D - m)^2 y = 0$ is $y = (c_1 + c_2 x)e^{mx}$, where $D \equiv \dfrac{d}{dx}$ and c_1, c_2 are arbitrary constants.

Answers

1. (*i*) $y = c_1 e^{2x} + c_2 e^{3x}$ (*ii*) $y = c_1 e^{(2+\sqrt{3})x} + c_2 e^{(2-\sqrt{3})x}$
(*iii*) $y = c_1 + c_2 e^{4x} + c_3 e^{-3x}$ (*iv*) $y = c_1 e^{2x} + c_2 e^{-x} + c_3 e^{-3x}$
2. (*i*) $y = c_1 \cos bx + c_2 \sin bx$ (*ii*) $y = e^x (c_1 \cos 3x + c_2 \sin 3x)$
(*iii*) $y = c_1 e^x + c_2 \cos 3x + c_3 \sin 3x$ (*iv*) $y = c_1 \cos ax + c_2 \sin ax + c_3 e^{ax} + c_4 e^{-ax}$
(*v*) $y = e^{-x}(c_1 \cos x + c_2 \sin x) + e^x (c_3 \cos x + c_4 \sin x)$
(*vi*) $y = c_1 \cos 2x + c_2 \sin 2x + c_3 \cos 3x + c_4 \sin 3x$
3. (*i*) $y = (c_1 + c_2 x)e^{3x}$ (*ii*) $y = c_1 e^{-2x} + (c_2 + c_3 x) e^{-3x}$
(*iii*) $y = (c_1 + c_2 x + c_3 x^2)e^x$ (*iv*) $y = c_1 e^x + (c_2 + c_3 x + c_4 x^2)e^{-2x}$
(*v*) $y = c_1 e^{2x} + c_2 e^{-2x} + (c_3 + c_4 x)e^{-3x}$
4. (*i*) $y = e^x [(c_1 + c_2 x) \cos 2x + (c_3 + c_4 x) \sin 2x]$
(*ii*) $y = (c_1 + c_2 x) \cos \sqrt{3}\, x + (c_3 + c_4 x) \sin \sqrt{3}\, x$
(*iii*) $y = c_1 e^{2x} + c_2 e^{-2x} + (c_3 + c_4 x) \cos 5x + (c_5 + c_6 x) \sin 5x$
(*iv*) $y = c_1 \cos x + c_2 \sin x + (c_3 + c_4 x) \cos 2x + (c_5 + c_6 x) \sin 2x$
5. (*i*) $x = -6e^{-3t} + 6e^{-t}$ (*ii*) $y = 2(\cos x - \sin x)$.

---

7. General Solution of Linear Differential Equation with Constant Coefficients

Theorem. If $y = u$ is any particular solution of the linear differential equation

$$P_0 \frac{d^ny}{dx^n} + P_1 \frac{d^{n-1}y}{dx^{n-1}} + P_2 \frac{d^{n-2}y}{dx^{n-2}} + \ldots\ldots + P_n y = Q$$

of order n, where $P_0 \neq 0, P_1, P_2, \ldots\ldots, P_n$ and Q are functions of the variable x and $y = Y$ is the general solution of the equation

$$P_0 \frac{d^ny}{dx^n} + P_1 \frac{d^{n-1}y}{dx^{n-1}} + P_2 \frac{d^{n-2}y}{dx^{n-2}} + \ldots\ldots + P_n y = 0$$

then prove that $y = Y + u$ is the general solution of the equation

$$P_0 \frac{d^n y}{dx^n} + P_1 \frac{d^{n-1} y}{dx^{n-1}} + P_2 \frac{d^{n-2} y}{dx^{n-2}} + \ldots\ldots + P_n y = Q.$$

Proof. $y = u$ is a particular solution of the equation

$$P_0 \frac{d^n y}{dx^n} + P_1 \frac{d^{n-1} y}{dx^{n-1}} + P_2 \frac{d^{n-2} y}{dx^{n-2}} + \ldots\ldots + P_n y = Q$$

or $$(P_0 D^n + P_1 D^{n-1} + \ldots\ldots + P_n) y = Q$$

or $$f(D) y = Q, \text{ say.}$$

$$\therefore \quad f(D)\, u = Q \qquad \ldots(1)$$

Also, $y = Y$ is a solution of $f(D)\, y = 0$.

$$\therefore \quad f(D) Y = 0 \qquad \ldots(2)$$

Now $$f(D)(Y + u) = f(D) Y + f(D) u = 0 + Q = Q \qquad \text{[Using (1) and (2)]}$$

$$\therefore \quad f(D)(Y + u) = Q.$$

$\therefore$ $y = Y + u$ is a solution of the equation $f(D)\, y = Q$.

Since Y is the general solution of the equation $f(D) y = 0$, it must contain n arbitrary constants.

$\therefore$ The solution $y = Y + u$ of the equation $f(D)\, y = Q$ also contain n arbitrary constants.

$\therefore$ $y = Y + u$ is the general solution of the equation $f(D)\, y = Q$.

8. Method of Solving Linear Differential Equation with Constant Coefficients

Let $$f(D) y = Q \qquad \ldots(1)$$

be a linear differential equation of order n with constant coefficients, where

$$f(D) = P_0 D^n + P_1 D^{n-1} + \ldots\ldots + P_n$$

and $P_0 \neq 0, P_1, \ldots\ldots, P_n$ are constants and Q is some function of x.

We have proved that if $y = Y$ is the general solution of the equation $f(D)\, y = 0$ and u is any particular solution of the equation $f(D) y = Q$, then $y = Y + u$ is the general solution of the given equation $f(D) y = Q$.

The function Y is called the **complementary function** (C.F.) of the general solution of the equation $f(D) y = Q$. We have already studied the method of finding the complementary function Y for the equation $f(D) y = Q$, which is nothing but the general solution of the equation $f(D) y = 0$.

Now we shall take up the method of finding a particular integral (u) of the equation $f(D) y = Q$. We define the operator '$\frac{1}{f(D)}$'.

$\frac{1}{f(D)} Q$ is defined to be that function of x which, when operated upon by $f(D)$, gives Q.

The operator $\frac{1}{f(D)}$, according to this definition, is the inverse of the operator $f(D)$. It can be shown that the operator $\frac{1}{f(D)}$ can be broken up into factors which may be taken in any order or into partial fractions.

By the definition of the operator '$\frac{1}{f(D)}$', we have

$$f(D)\left(\frac{1}{f(D)}Q\right) = Q$$

$\therefore \quad y = \frac{1}{f(D)}Q$ is a particular integral of the equation $f(D)y = Q$.

Let the operator $\frac{1}{f(D)}$ may be factored as $\frac{1}{D-m_1}\cdot\frac{1}{D-m_2}\cdots\cdots\frac{1}{D-m_n}$, then the particular integral may be written as $\frac{1}{D-m_1}\cdot\frac{1}{D-m_2}\cdots\cdots\frac{1}{D-m_n}Q$.

On operating with the first symbolic factor, beginning at the right, there is obtained $\frac{1}{D-m_1}\cdot\frac{1}{D-m_2}\cdots\cdots\frac{1}{D-m_{n-1}}e^{m_n x}\int Qe^{-m_n x}\,dx$ then, on operating with the second and remaining factors in succession, taking them from right to left, the value of the particular integral is obtained.

Alternatively, the operator $\frac{1}{f(D)}$ may be decomposed into its partial fractions as

$\frac{N_1}{D-m_1}+\frac{N_2}{D-m_2}+\cdots\cdots+\frac{N_n}{D-m_n}$ and then the particular integral will have the form

$N_1 e^{m_1 x}\int Q\,e^{-m_1 x}\,dx + N_2\,e^{m_2 x}\int Q\,e^{-m_2 x}\,dx + \cdots\cdots + N_n e^{m_n x}\int Q\,e^{-m_n x}\,dx$.

Of the above two methods, the latter is generally preferred.

Remark 1. In calculating the particular integral of the equation $f(D)y = Q$, no arbitrary constant is introduced.

Remark 2. In finding the particular integral, we have used the result $\frac{1}{D-m}Q = e^{mx}\int Q\,e^{-mx}\,dx$. This result is very important. For the sake of completeness, we prove this result. (*K.U. 2001*)

Let $$y = \frac{1}{D-m}Q.$$

$\therefore$ By definition of the operator $\frac{1}{D-m}Q$, we have $(D-m)\left[\frac{1}{D-m}Q\right] = Q$.

$$\Rightarrow \quad (D-m)y = Q \quad \Rightarrow \quad \frac{dy}{dx} - my = Q \qquad ...(1)$$

(1) is a linear differential equation of first order.

$$\text{I.F.} = e^{\int -m\,dx} = e^{-mx}$$

$\therefore$ The solution of (1) is

$$y(e^{-mx}) = \int Q\,e^{-mx}\,dx \quad \text{or} \quad y = e^{mx}\int Q\,e^{-mx}\,dx.$$

$$\therefore \quad \mathbf{\frac{1}{D-m}Q = e^{mx}\int Q\,e^{-mx}\,dx.}$$ (*M.D.U. 2007*)

9. Particular Integral of f(D)y = Q when Q is in Some Standard Form

When Q is in one of the following standard forms, then the particular integral of the equation $f(D)y = Q$ can be obtained by using shorter methods than by using the general methods given above. We shall discuss the following standard forms of the function Q of the equation $f(D)\,y = Q$:

(i) e^{ax}, where a is any constant

(ii) $\sin ax$, $\cos ax$, where a is any constant

(iii) x^m, where m is any positive integer

(iv) $e^{ax}V$, where V is any function of x

(v) xV, where V is any function of x.

10. Q is of the form e^{ax}, where a is any constant

Let the linear differential equation be $f(D)y = Q$, where $f(D)$ is a polynomial in D with constant coefficients and $Q = e^{ax}$.

$$\therefore \qquad \text{P.I.} = \frac{1}{f(D)} e^{ax} \qquad \text{...(1)}$$

Successive differentiation gives $D^k e^{ax} = a^k e^{ax}$.

Since, $f(D)$ is a polynomial in D with constant coefficients, we have

$$f(D)\, e^{ax} = f(a)\, e^{ax}$$

Operating on both sides with $\frac{1}{f(D)}$, we get $\frac{1}{f(D)} f(D)\, e^{ax} = \frac{1}{f(D)} f(a)\, e^{ax}$

Since $\frac{1}{f(D)}$ and $f(D)$ are inverse operators and $f(a)$ is an algebraic multiplier, we get

$$e^{ax} = f(a) \frac{1}{f(D)} e^{ax} \quad \text{or} \quad \mathbf{\frac{1}{f(D)} e^{ax} = \frac{1}{f(a)} e^{ax}}.$$

This method fails if $f(a) = 0$.

Let $f(a) = 0$. $\therefore$ a is a root of $f(D) = 0$.

Case I. a is a non-repeated root of f(D) = 0.

$\therefore$ $D - a$ is a factor of $f(D)$.

Let $\qquad f(D) = (D - a)\, \phi(D)$ and $\phi(a) \neq 0$

$$\therefore \qquad \frac{1}{f(D)} e^{ax} = \frac{1}{D-a} \cdot \frac{1}{\phi(D)} e^{ax} = \frac{1}{D-a} \cdot \frac{1}{\phi(a)} e^{ax}$$

$$= e^{ax} \int \left(\frac{1}{\phi(a)} e^{ax} \right) e^{-ax}\, dx = \frac{xe^{ax}}{\phi(a)}.$$

Also, $\qquad f'(D) = 1 \,.\, \phi(D) + (D - a)\, \phi'(D)$

$\therefore \qquad f'(a) = \phi(a) + (a - a)\, \phi'(a) = \phi(a)$.

$$\therefore \qquad \mathbf{\frac{1}{f(D)} e^{ax} = \frac{x\, e^{ax}}{f'(a)}}.$$

Case II. a is a repeated root of f(D) = 0

Let a is a twice repeated root of $f(D) = 0$.

$\therefore$ $(D - a)^2$ is a factor of $f(D)$.

Let $f(D) = (D - a)^2 \psi(D)$ and $\psi(a) \neq 0$.

$$\therefore \quad \frac{1}{f(D)} e^{ax} = \frac{1}{(D-a)^2} \cdot \frac{1}{\psi(D)} e^{ax} = \frac{1}{(D-a)^2} \cdot \frac{1}{\psi(a)} e^{ax}$$

$$= \frac{1}{D-a} \cdot e^{ax} \int \left(\frac{1}{\psi(a)} e^{ax} \right) e^{-ax} \, dx = \frac{1}{\psi(a)} \cdot \frac{1}{D-a} x e^{ax}$$

$$= \frac{1}{\psi(a)} \cdot e^{ax} \int (x\, e^{ax}) e^{-ax} \, dx = \frac{1}{\psi(a)} \cdot e^{ax} \frac{x^2}{2} = \frac{x^2 e^{ax}}{2\,\psi(a)}.$$

Also $f'(D) = 2(D - a)\, \psi(D) + (D - a)^2 \psi'(D)$

and $f''(D) = 2\psi(D) + 2(D - a)\, \psi'(D) + 2(D - a)\, \psi'(D) + (D - a)^2 \psi''(D)$.

$\therefore$ $f''(a) = 2\psi(a)$.

$$\therefore \quad \mathbf{\frac{1}{f(D)} e^{ax} = \frac{x^2 e^{ax}}{f''(a)}}.$$

Similarly, if a is a root repeated r times, then

$$\mathbf{\frac{1}{f(D)} e^{ax} = \frac{x^r e^{ax}}{f^{(r)}(a)}}.$$

Example 1. *Solve* $4 \dfrac{d^2y}{dx^2} + 4 \dfrac{dy}{dx} - 3y = e^{2x}$.

Sol. We have

$$4 \frac{d^2y}{dx^2} + 4 \frac{dy}{dx} - 3y = e^{2x}. \qquad \text{...(1)}$$

$\Rightarrow$ $(4D^2 + 4D - 3)y = e^{2x}$

$\therefore$ The A.E. is $4D^2 + 4D - 3 = 0$.

$$\therefore \quad D = \frac{-4 \pm \sqrt{16 + 48}}{8} = \frac{-4 \pm \sqrt{64}}{8} = \frac{-4 \pm 8}{8} = \frac{1}{2}, -\frac{3}{2}$$

$\therefore$ C.F. $= c_1 e^{x/2} + c_2 e^{-3x/2}$.

$$\text{P.I.} = \frac{1}{4D^2 + 4D - 3} e^{2x} = \frac{1}{4(2)^2 + 4(2) - 3} e^{2x} = \frac{e^{2x}}{21}.$$

$\therefore$ The general solution, of (1) is y = C.F. + P.I.

$$\therefore \quad \mathbf{y = c_1 e^{x/2} + c_2 e^{-3x/2} + \frac{e^{2x}}{21}}.$$

Example 2. *Solve*

$$\frac{d^2y}{dx^2} + 2 \frac{dy}{dx} + y = 2e^{2x} \textit{ given that } x = 0, \textit{ when } y = 0.$$

Sol. The equation in the symbolic form is

$$(D^2 + 2D + 1)y = 2e^{2x}. \qquad \text{...(1)}$$

$\therefore$ The A.E. is $D^2 + 2D + 1 = 0$ *i.e.,* $(D + 1)^2 = 0$

$\therefore \quad D = -1, -1$

$\therefore \quad$ C.F. $= (c_1 + c_2 x)e^{-x}$

$$\text{P.I.} = \frac{1}{(D+1)^2} 2e^{2x} = 2\frac{1}{(D+1)^2} e^{2x} = 2\frac{1}{(2+1)^2} e^{2x} = \frac{2}{9} e^{2x}$$

$\therefore$ The general solution of (1) is y = C.F. + P.I.

$$\therefore \quad y = (c_1 + c_2 x)e^{-x} + \frac{2}{9} e^{2x}$$

It is given that $\quad y = 0$, when $x = 0$

$$\therefore \quad 0 = c_1 + \frac{2}{9} \quad \text{or} \quad c_1 = -\frac{2}{9}$$

$$\therefore \quad y = \left(-\frac{2}{9} + c_2 x\right)e^{-x} + \frac{2}{9} e^{2x} \quad \text{or} \quad \mathbf{y = \frac{2}{9}(e^{2x} - e^{-x}) + c_2 xe^{-x}}.$$

This is the required solution.

Example 3. *Solve* $\dfrac{d^2y}{dx^2} + 6\dfrac{dy}{dx} + 25y = 104\, e^{3x}$. (K.U. 2003 ; M.D.U. 2007)

Sol. We have

$$\frac{d^2y}{dx^2} + 6\frac{dy}{dx} + 25y = 104\, e^{3x} \quad \text{or} \quad (D^2 + 6D + 25)y = 104\, e^{3x}. \qquad ...(1)$$

$\therefore$ The A.E. is $\quad D^2 + 6D + 25 = 0.$

$$\therefore \quad D = \frac{-6 \pm \sqrt{36 - 100}}{2} = -3 \pm 4i.$$

$\therefore \quad$ C.F. $= e^{-3x}(c_1 \cos 4x + c_2 \sin 4x)$

$$\text{P.I.} = \frac{1}{D^2 + 6D + 25} 104\, e^{3x} = 104 \frac{1}{D^2 + 6D + 25} e^{3x}$$

$$= 104 \frac{1}{(3)^2 + 6(3) + 25} e^{3x} = 2e^{3x}.$$

Hence the general solution of (1) is

$$\mathbf{y = e^{-3x}(c_1 \cos 4x + c_2 \sin 4x) + 2e^{3x}}.$$

Remark. $\cosh x = \dfrac{e^x + e^{-x}}{2}$ and $\sinh x = \dfrac{e^x - e^{-x}}{2}$.

Example 4. *Solve* $(D^2 - 3D + 2)\, y = \cosh x$. (K.U. 1995 C)

Sol. We have $\quad (D^2 - 3D + 2)\, y = \cosh x. \qquad ...(1)$

$\therefore$ The A.E. is $\quad D^2 - 3D + 2 = 0. \quad \therefore \quad D = 1, 2$

$\therefore \quad$ C.F. $= c_1 e^x + c_2 e^{2x}$

$$\text{P.I.} = \frac{1}{D^2 - 3D + 2} \cosh x = \frac{1}{(D^2 - 3D + 2)}\left(\frac{e^x + e^{-x}}{2}\right).$$

$$= \frac{1}{2}\left(\frac{1}{D^2 - 3D + 2} e^x\right) + \frac{1}{2}\left(\frac{1}{D^2 - 3D + 2} e^{-x}\right) \qquad \text{(case of failure)}$$

$$= \frac{1}{2}\left(\frac{x}{(2D-3)_{D=1}}e^{x}\right) + \frac{1}{2}\left(\frac{1}{(-1)^2 - 3(-1) + 2}e^{-x}\right) \qquad \text{(Case of failure)}$$

$$= -\frac{1}{2}xe^{x} + \frac{1}{2}\cdot\frac{1}{6}e^{-x} = -\frac{1}{2}xe^{x} + \frac{1}{12}e^{-x}.$$

$\therefore$ The general solution of (1) is y = C.F. + P.I.

$\therefore$ $$\mathbf{y = c_1e^{x} + c_2e^{2x} - \frac{1}{2}xe^{x} + \frac{1}{12}e^{-x}.}$$

EXERCISE 2

Solve the following differential equations (Q. No. 1—15) :

1. $\frac{d^2y}{dx^2} - 5\frac{dy}{dx} + 6y = e^{4x}$

2. $\frac{d^2y}{dx^2} - 3\frac{dy}{dx} + 2y = e^{5x}$

3. $\frac{d^2y}{dx^2} - 2k\frac{dy}{dx} + k^2y = e^{x}$, $k \neq 1$

4. $(D^3 - 3D^2 + 4)y = e^{3x}$

5. $(D^3 + 2D^2 + D)y = e^{2x}$

6. $\frac{d^3y}{dx^3} + y = 3 + 5e^{x}$

7. $\frac{d^2y}{dx^2} + 2p\frac{dy}{dx} + (p^2 + q^2)y = e^{ax}$

8. $\frac{d^3y}{dx^3} + y = 3 + e^{-x} + 5e^{2x}$

9. $(D^2 + 4D + 4)y = e^{2x} - e^{-2x}$ *(KU 2000)*

10. $(D^2 - a^2)y = e^{ax} + e^{nx}$

11. $\frac{d^3y}{dx^3} - \frac{d^2y}{dx^2} - \frac{dy}{dx} + y = e^{x}$

12. $\frac{d^2y}{dx^2} + 4\frac{dy}{dx} + 4y = 2\sinh x$

13. $\frac{d^2y}{dx^2} - y = \cosh x$

14. $\frac{d^2y}{dx^2} + 4\frac{dy}{dx} + 4y = 2\sinh 2x$

15. $\frac{d^3y}{dx^3} - 5\frac{d^2y}{dx^2} + 7\frac{dy}{dx} - 3y = e^{2x}\cosh x.$

16. If $\frac{d^2x}{dt^2} + \frac{g}{b}(x - a) = 0$; a, b and g being positive numbers and $x = a'$ and $\frac{dx}{dt} = 0$ when $t = 0$; show that $x = a + (a' - a)\cos\left(\sqrt{\frac{g}{b}}\,t\right)$.

Answers

1. $y = c_1e^{2x} + c_2e^{3x} + \frac{1}{2}e^{4x}$

2. $y = c_1e^{x} + c_2e^{2x} + \frac{1}{12}e^{5x}$

3. $y = (c_1 + c_2x)e^{kx} + \frac{1}{(1-k)^2}e^{x}$

4. $y = c_1e^{-x} + (c_2 + c_3x)e^{2x} + \frac{1}{4}e^{3x}$

5. $y = c_1 + (c_2 + c_3x)e^{-x} + \frac{1}{18}e^{2x}$

6. $y = c_1 e^{-x} + e^{x/2}\left(c_2 \cos \frac{\sqrt{3}}{2} x + c_3 \sin \frac{\sqrt{3}}{2} x\right) + 3 + \frac{5}{2} e^x$

7. $y = e^{-px}(c_1 \cos qx + c_2 \sin qx) + \frac{1}{(a+p)^2 + q^2} e^{ax}$

8. $y = c_1 e^{-x} + e^{x/2}\left(c_2 \cos \frac{\sqrt{3}}{2} x + c_2 \sin \frac{\sqrt{3}}{2} x\right) + 3 + x \frac{e^{-x}}{3} + \frac{5}{9} e^{2x}$

9. $y = (c_1 + c_2 x) e^{-2x} + \frac{e^{2x}}{16} - \frac{x^2}{2} e^{-2x}$

10. $y = c_1 e^{ax} + c_2 e^{-ax} + \frac{xe^{ax}}{2a} + \frac{1}{n^2 - a^2} e^{nx}$

11. $y = (c_1 + c_2 x) e^x + c_3 e^{-x} + \frac{1}{4} x^2 e^x$

12. $y = (c_1 + c_2 x) e^{-2x} + \frac{1}{9} e^x - e^{-x}$

13. $y = c_1 e^x + c_2 e^{-x} + \frac{x}{2} \sinh x$

14. $y = (c_1 + c_2 x) e^{-2x} + \frac{1}{16} e^{2x} - \frac{1}{2} x^2 e^{-2x}$

15. $y = (c_1 + c_2 x) e^x + c_3 e^{3x} + \frac{x}{8} e^{3x} - \frac{1}{8} x^2 e^x.$

---

11. Q is of the Form sin ax or cos ax

Let the linear differential equation be $f(D)y = Q$, where $f(D)$ is a polynomial in D with constant coefficients and $Q = \sin ax$.

$$\therefore \qquad \text{P.I.} = \frac{1}{f(D)} \sin ax \qquad ...(1)$$

Successive differentiation of $\sin ax$ gives

$$D \sin ax = a \cos ax, \qquad D^2 \sin ax = -a^2 \sin ax$$

$$D^3 \sin ax = -a^3 \cos ax, \quad D^4 \sin ax = a^4 \sin ax = (-a^2)^2 \sin ax$$

....

$\therefore$ In general, $(D^2)^k \sin ax = (-a^2)^k \sin ax.$

$f(D)$ may or may not be a function of D^2.

Let $f(D)$ be a function of D^2 and $f(D) = \phi(D^2)$, say.

Since, $\phi(D^2)$ is a polynomial in D^2 with constant coefficients, we have

$$\therefore \qquad \phi(D^2) \sin ax = \phi(-a^2) \sin ax \Rightarrow \frac{1}{\phi(D^2)} \phi(D^2) \sin ax = \frac{1}{\phi(D^2)} \phi(-a^2) \sin ax$$

$$\Rightarrow \qquad \sin ax = \phi(-a^2) \frac{1}{\phi(D^2)} \sin ax \quad \text{or} \quad \mathbf{\frac{1}{\phi(D^2)} \sin ax = \frac{1}{\phi(-a^2)} \sin ax.}$$

(*MDU 2005*)

Similarly, if Q is of the form $\cos ax$, then

$$\mathbf{\frac{1}{\phi(D^2)} \cos ax = \frac{1}{\phi(-a^2)} \cos ax.}$$

These results fail if $\phi(-a^2) = 0$.

Let $\qquad \phi(-a^2) = 0$

We have $\qquad e^{iax} = \cos ax + i \sin ax$

$$\therefore \qquad \frac{1}{\phi(D^2)} (\cos ax + i \sin ax) = \frac{1}{\phi(D^2)} e^{iax}$$

Now $\quad D = ia \Rightarrow \phi(D^2) = \phi((ia)^2) = \phi(-a^2) = 0$

$$\therefore \quad \frac{1}{\phi(D^2)} e^{iax} = x \frac{1}{\frac{d}{dD}(\phi(D^2))} e^{iax} = x \frac{1}{\frac{d}{dD}(\phi(D^2))} (\cos ax + i \sin ax)$$

Equating real and imaginary parts, we get

$$\mathbf{\frac{1}{\phi(D^2)} \cos ax = x \frac{1}{\frac{d}{dD}(\phi(D^2))} \cos ax}$$

and
$$\mathbf{\frac{1}{\phi(D^2)} \sin ax = x \frac{1}{\frac{d}{dD}(\phi(D^2))} \sin ax.}$$

If by using these rules, the denominator again vanishes, we repeat the rule once again.

Now we consider the possibility that $f(D)$ is not a function of D^2. In this case, D^2 is replaced by $-a^2$ every where in $\frac{1}{f(D)} \sin ax$. In the next step, the operator is rationalised and again D^2 is put equal to $-a^2$. This is simplified to get the required particular integral. Same procedure is also followed in case of $\frac{1}{f(D)} \cos ax$.

Remark. The following results of Trigonometry are very useful in solving problems :

(*i*) $\sin^2 x = \frac{1-\cos 2x}{2}$ (*ii*) $\cos^2 x = \frac{1+\cos 2x}{2}$

(*iii*) $2 \sin A \cos B = \sin (A + B) + \sin (A - B)$

(*iv*) $2 \cos A \sin B = \sin (A + B) - \sin (A - B)$

(*v*) $2 \cos A \cos B = \cos (A + B) + \cos (A - B)$

(*vi*) $2 \sin A \sin B = \cos (A - B) - \cos (A + B)$.

Example 1. *Solve* $\frac{d^3y}{dx^3} + \frac{d^2y}{dx^2} - \frac{dy}{dx} - y = \cos 2x.$ (MDU 1998)

Sol. Given equation in the symbolic form is

$$(D^3 + D^2 - D - 1)y = \cos 2x. \quad \text{...(1)}$$

$\therefore$ The A.E. is $\quad D^3 + D^2 - D - 1 = 0.$

$\Rightarrow \quad D^2(D + 1) - (D + 1) = 0 \quad \Rightarrow \quad (D^2 - 1)(D + 1) = 0.$

$\therefore \quad D = 1, -1, -1.$

$\therefore \quad \text{C.F.} = c_1 e^x + (c_2 + c_3x) e^{-x}$

$$\text{P.I.} = \frac{1}{D^3 + D^2 - D - 1} \cos 2x = \frac{1}{DD^2 + D^2 - D - 1} \cos 2x$$

$$= \frac{1}{D(-4) + (-4) - D - 1} \cos 2x = \frac{1}{-5D - 5} \cos 2x$$

$$= -\frac{1}{5(D+1)} \cos 2x = -\frac{1.(D-1)}{5(D+1)(D-1)} \cos 2x$$

$$= -\frac{1}{5(D^2-1)} (D-1) \cos 2x = -\frac{(D-1)}{5} \left(\frac{1}{D^2-1} \cos 2x \right)$$

$$= -\frac{1}{5}(D-1)\left(\frac{1}{-4-1}\cos 2x\right)$$

$$= \frac{1}{25}(D-1)\cos 2x = \frac{1}{25}(D\cos 2x - \cos 2x)$$

$$= \frac{1}{25}(-2\sin 2x - \cos 2x) = -\frac{1}{25}(2\sin 2x + \cos 2x).$$

$\therefore$ The general solution of (1) is y = C.F. + P.I.

$$\therefore \qquad \mathbf{y = c_1e^x + (c_2 + c_3x)e^{-x} - \frac{1}{25}(2\sin 2x + \cos 2x).}$$

Example 2. *Solve* $\dfrac{d^2y}{dx^2} - 4y = e^x + \sin 2x.$

Sol. Given equation in the symbolic form is

$$(D^2 - 4)y = e^x + \sin 2x. \qquad \text{...(1)}$$

$\therefore$ The A.E. is $\quad D^2 - 4 = 0. \quad \therefore \quad D = \pm 2$

$\therefore \qquad$ C.F. $= c_1e^{2x} + c_2e^{-2x}$

$$\text{P.I.} = \frac{1}{D^2-4}(e^x + \sin 2x) = \left(\frac{1}{D^2-4}e^x\right) + \left(\frac{1}{D^2-4}\sin 2x\right)$$

$$= \left(\frac{1}{1^2-4}e^x\right) + \left(\frac{1}{-4-4}\sin 2x\right) = -\frac{1}{3}e^x - \frac{1}{8}\sin 2x$$

$\therefore$ The general solution of (1) is $\quad y$ = C.F. + P.I.

$$\therefore \qquad \mathbf{y = c_1e^{2x} + c_2e^{-2x} - \frac{1}{3}e^x - \frac{1}{8}\sin 2x.}$$

Example 3. *Solve* $(D^2 - 4D + 3)y = \sin 2x \cos x.$

Sol. Given equation is $\quad (D^2 - 4D + 3)y = 0. \qquad$ ...(1)

$\therefore$ The A.E. is $\quad D^2 - 4D + 3 = 0. \quad \therefore \quad D = 3, 1$

$\therefore \qquad$ C.F. $= c_1e^{3x} + c_2e^x.$

$$\text{P.I.} = \frac{1}{D^2-4D+3}[\sin 2x\cos x] = \frac{1}{D^2-4D+3}\left[\frac{1}{2}(2\sin 2x\cos x)\right]$$

$$= \frac{1}{D^2-4D+3}\left[\frac{1}{2}(\sin 3x + \sin x)\right] = \frac{1}{2}\left(\frac{1}{D^2-4D+3}\sin 3x\right) + \frac{1}{2}\left(\frac{1}{D^2-4D+3}\sin x\right)$$

$$= \frac{1}{2}\left[\frac{1}{-9-4D+3}\sin 3x + \frac{1}{-1-4D+3}\sin x\right]$$

$$= \frac{1}{2}\left[\frac{1}{-6-4D}\sin 3x + \frac{1}{2-4D}\sin x\right]$$

$$= \frac{1}{2}\left[-\frac{1}{2(3+2D)}\sin 3x + \frac{1}{2(1-2D)}\sin x\right]$$

$$= \frac{1}{4}\left[-\frac{3-2D}{(3+2D)(3-2D)}\sin 3x + \frac{1+2D}{(1-2D)(1+2D)}\sin x\right]$$

$$= \frac{1}{4}\left[\frac{-3+2D}{9-4D^2}\sin 3x + \frac{1+2D}{1-4D^2}\sin x\right]$$

$$= \frac{1}{4}\left[\frac{-3+2D}{9-4(-9)}\sin 3x + \frac{(1+2D)\sin x}{1-4(-1)}\right]$$

$$= \frac{1}{4}\left[\frac{(-3+2D)}{45}\sin 3x + \left(\frac{1+2D}{5}\right)\sin x\right]$$

$$= \frac{1}{4}\left[\frac{1}{45}(-3\sin 3x + 2D\sin 3x) + \frac{1}{5}\sin x + \frac{2}{5}D(\sin x)\right]$$

$$= \frac{1}{4}\left[\frac{1}{45}(-3\sin 3x + 6\cos 3x) + \frac{1}{5}\sin x + \frac{2}{5}\cos x\right]$$

$$= \frac{1}{4}\left[-\frac{1}{15}\sin 3x + \frac{2}{15}\cos 3x + \frac{1}{5}\sin x + \frac{2}{5}\cos x\right]$$

$$= \frac{1}{60}[-\sin 3x + 2\cos 3x + 3\sin x + 6\cos x]$$

$\therefore$ The general solution of (1) is $y =$ C.F. + P.I.

$$\therefore \qquad \mathbf{y = c_1e^{3x} + c_2e^{x} + \frac{1}{60}[-\sin 3x + 2\cos 3x + 3\sin x + 6\cos x].}$$

Example 4. *Solve* $\dfrac{d^2y}{dx^2} + 4y = e^x + \sin 2x.$

Sol. Given equation is

$$\frac{d^2y}{dx^2} + 4y = e^x + \sin 2x \quad \text{or} \quad (D^2 + 4)\,y = e^x + \sin 2x. \qquad ...(1)$$

$\therefore$ The A.E. is $D^2 + 4 = 0$. $\therefore$ $D = \pm 2i$

$$\therefore \qquad \text{C.F.} = e^{0x}(c_1\cos 2x + c_2\sin 2x) = c_1\cos 2x + c_2\sin 2x$$

$$\text{P.I.} = \frac{1}{D^2+4}(e^x + \sin 2x) = \frac{1}{D^2+4}e^x + \frac{1}{D^2+4}\sin 2x$$

$$= \frac{1}{1+4}e^x + x\frac{1}{\frac{d}{dD}(D^2+4)}\sin 2x \qquad \text{[case of failure]}$$

$$= \frac{1}{5}e^x + x\frac{1}{2D}\sin 2x = \frac{1}{5}e^x + x\frac{1}{2D^2}\,.\,D\sin 2x$$

$$= \frac{1}{5}e^x + \frac{x}{2}D\left(\frac{1}{D^2}\sin 2x\right) = \frac{1}{5}e^x + \frac{x}{2}D\left(-\frac{1}{4}\sin 2x\right)$$

$$= \frac{e^x}{5} - \frac{x}{8}D(\sin 2x) = \frac{e^x}{5} - \frac{x}{8}2\cos 2x = \frac{e^x}{5} - \frac{x}{4}\cos 2x.$$

$\therefore$ The general solution of (1) is $y =$ C.F. + P.I.

$$\therefore \qquad \mathbf{y = c_1\cos 2x + c_2\sin 2x + \frac{1}{5}e^x - \frac{x}{4}\cos 2x.}$$

Example 5. *Solve* $\dfrac{d^4y}{dx^4} + 2n^2\dfrac{d^2y}{dx^2} + n^4y = \cos mx.$ (KU 1995)

Sol. The given equation in symbolic form is

$$(D^4 + 2n^2 D^2 + n^4)\, y = \cos mx. \quad ...(1)$$

$\therefore$ The A.E. is $D^4 + 2n^2D^2 + n^4 = 0.$

$\Rightarrow$ $(D^2 + n^2)^2 = 0 \Rightarrow D = \pm ni, \pm ni$

$\therefore$ C.F. $= e^{ox}((c_1 + c_2x)\cos nx + (c_3 + c_4x)\sin nx)$
$= (c_1 + c_2x)\cos nx + (c_3 + c_4x)\sin nx$

$$\text{P.I.} = \frac{1}{(D^2+n^2)^2}\cos mx = \frac{1}{((-m^2)+n^2)^2}\cos mx = \frac{\cos mx}{(m^2-n^2)^2}$$

$\therefore$ The general solution of (1) is y = C.F. + P.I.

$$\therefore \quad \mathbf{y = (c_1 + c_2x)\cos mx + (c_3 + c_4x)\sin mx + \frac{\cos mx}{(m^2-n^2)^2}}.$$

EXERCISE 3

Solve the following differential equations (Q. No. 1—14) :

1. $\dfrac{d^3y}{dx^3} + y = \cos 2x$

2. $\dfrac{d^2y}{dx^2} - 2\dfrac{dy}{dx} + 5y = \sin 3x$

3. $\dfrac{d^2y}{dx^2} - 4\dfrac{dy}{dx} + y = a\cos 2x$

4. $(D^4 - 1)\, y = \sin 2x$ *(M.D.U. 2006)*

5. $(D^2 - 3D + 2)\, y = 6e^{-3x} + \sin 2x$

6. $\dfrac{d^2y}{dx^2} - 4\dfrac{dy}{dx} + 3y = \sin 3x \cos 2x$ *(K.U. 2000)*

7. $(D^2 + 1)\, y = \sin x \sin 2x$

8. $\dfrac{d^2y}{dx^2} + a^2y = \sin ax$

9. $\dfrac{d^2y}{dx^2} + \dfrac{dy}{dx} + y = \sin 2x$ *(K.U. 2001)*

10. $(D^2 - 4D + 4)\, y = e^{-4x} + 5\cos 3x$ *(K.U. 1996 C)*

11. $\dfrac{d^2y}{dx^2} + 2a\dfrac{dy}{dx} + a^2y = b\cos px$

12. $\dfrac{d^2y}{dx^2} - 4\dfrac{dy}{dx} + 3y = \sin^2 x - e^{3x}$ *(K.U. 1996)*

13. $\dfrac{d^3y}{dx^3} + y = \sin 3x - \cos^2 \dfrac{x}{2}$

14. $\dfrac{d^3y}{dx^3} + 6\dfrac{d^2y}{dx^2} + 11\dfrac{dy}{dx} + 6y = 2\sin x$

15. Solve $\dfrac{d^2s}{dt^2} + b^2s = k\cos bt$, given that $s = 0, \dfrac{ds}{dt} = 0$ when $t = 0$ *(K.U. 1995 C)*

16. Solve $(D^4 - m^4)\, y = \sin mx$. *(M.D.U. 2006)*

Answers

1. $y = c_1 e^{-x} + e^{x/2}\left[c_2 \cos\dfrac{\sqrt{3}}{2}x + c_3 \cos\dfrac{\sqrt{3}}{2}x\right] + \dfrac{1}{65}(\cos 2x - 8\sin 2x)$

2. $y = e^x[c_1\cos 2x + c_2\sin 2x] + \dfrac{1}{26}[3\cos 3x - 2\sin 3x]$

3. $y = e^{2x}\left[c_1 e^{\sqrt{3}x} + c_2 e^{-\sqrt{3}x}\right] - \dfrac{a}{73}[8\sin 2x + 3\cos 2x]$

4. $y = (c_1 e^x + c_2 e^{-x} + c_3 \cos x + c_4 \sin x) + \frac{1}{15} \sin 2x$

5. $y = c_1 e^x + c_2 e^{2x} + \frac{3}{10} e^{-3x} + \frac{1}{20} [3 \cos 2x - \sin 2x]$

6. $y = c_1 e^x + c_2 e^{3x} + \frac{1}{884} [10 \cos 5x - 11 \sin 5x] + \frac{1}{20} [\sin x + 2 \cos x]$

7. $y = c_1 \cos x + c_2 \sin x + \frac{x}{4} \sin x + \frac{1}{16} \cos 3x$

8. $y = c_1 \cos ax + c_2 \sin ax - \frac{x \cos ax}{2a}$

9. $y = e^{-\frac{1}{2}x} \left(c_1 \cos \frac{\sqrt{3}}{2} x + c_2 \sin \frac{\sqrt{3}}{2} x \right) - \frac{1}{13} (2 \cos 2x + 3 \sin 2x)$

10. $y = (c_1 + c_2 x) e^{2x} + \frac{1}{36} e^{-4x} - \frac{5}{169} (12 \sin 3x + 5 \cos 3x)$

11. $y = (c_1 + c_2 x) e^{-ax} + \frac{b}{(a^2 + p^2)^2} [2ap \sin px + (a^2 - p^2) \cos px]$

12. $y = c_1 e^x + c_2 e^{3x} + \frac{1}{6} + \frac{1}{130} (\cos 2x + 8 \sin 2x) - \frac{1}{2} x e^{3x}$

13. $y = c_1 e^{-x} + e^{x/2} \left[c_2 \cos \frac{\sqrt{3}}{2} x + c_2 \sin \frac{\sqrt{3}}{2} x \right] + \frac{1}{730} [\sin 3x + 27 \cos 3x] - \frac{1}{2} - \frac{1}{4} (\cos x - \sin x)$

14. $y = c_1 e^{-x} + c_2 e^{-2x} + c_3 e^{-3x} - \frac{1}{5} \cos x$

15. $s = \frac{kt}{2b} \sin bt$

16. $y = c_1 e^{mx} + c_2 e^{-mx} + c_3 \cos mx + c_4 \sin mx + \frac{x}{4m^3} \cos mx.$

---

12. Q is of the Form x^m, where m is any Positive Integer

Let the linear differential equation be $f(D)\, y = Q$, where $f(D)$ is a polynomial in D with constant coefficients and $Q = x^m$.

$$\therefore \qquad \text{P.I.} = \frac{1}{f(D)} x^m$$

In the first step, take out the lowest degree term from $f(D)$, so that it is reduced to the form $(1 \pm \phi(D))^n$ where n may be 1, 2, ... In the second step, we take it to the numerator and get $(1 \pm \phi(D))^{-n}$. Now we expand this by using Binomial theorem, namely

$$(1 + x)^n = 1 + nx + \frac{n(n-1)}{1.2} x^2 + \ldots\ldots .$$

If $Q = x^m$ or a polynomial of degree m, then the terms of the expansion beyond D^m need not be written because the result of their operation on x^m would be zero.

Example 1. *Solve* $\frac{d^2y}{dx^2} - 3\frac{dy}{dx} + 2y = x.$

Sol. The equation in the symbolic form is

$$(D^2 - 3D + 2)y = x. \qquad \ldots(1)$$

$\therefore$ The A.E. is $D^2 - 3D + 2 = 0. \quad \therefore \quad D = 1, 2.$

$\therefore$ $$\text{C.F.} = c_1 e^x + c_2 e^{2x}$$

$$\text{P.I.} = \frac{1}{D^2 - 3D + 2} x = \frac{1}{2\left[1 - \frac{3D}{2} + \frac{D^2}{2}\right]} x$$

$$= \frac{1}{2}\left[1 - \frac{3D}{2} + \frac{D^2}{2}\right]^{-1} x = \frac{1}{2}\left[1 - \left(\frac{3D}{2} - \frac{D^2}{2}\right)\right]^{-1} x$$

$$= \frac{1}{2}\left[1 + \left(\frac{3D}{2} - \frac{D^2}{2}\right) + \ldots\ldots\right] x = \frac{1}{2}\left[x + \frac{3}{2} Dx - \frac{1}{2} D^2 x + \ldots\ldots\right]$$

$$= \frac{1}{2}\left[x + \frac{3}{2}.1 - 0 + \ldots\ldots\right] = \frac{1}{2}\left[x + \frac{3}{2}\right] = \frac{x}{2} + \frac{3}{4}.$$

The general solution of (1) is y = C.F. + P.I.

$\therefore$ $$\mathbf{y = c_1 e^x + c_2 e^{2x} + \frac{x}{2} + \frac{3}{4}}.$$

Example 2. *Solve* $\dfrac{d^3y}{dx^3} - \dfrac{d^2y}{dx^2} - 6\dfrac{dy}{dx} = 1 + x^2$. (K.U. 2006 S)

Sol. The symbolic form of the given equation is

$$(D^3 - D^2 - 6D)\, y = 1 + x^2. \quad \ldots(1)$$

$\therefore$ The A.E. is $D^3 - D^2 - 6D = 0$. $\therefore$ $D(D^2 - D - 6) = 0$ or $D(D - 3)(D + 2) = 0$.

$\therefore$ $$D = 0, 3, -2$$

$\therefore$ $$\text{C.F.} = c_1 e^{0x} + c_2 e^{3x} + c_3 e^{-2x} = c_1 + c_2 e^{3x} + c_3 e^{-2x}$$

$$\text{P.I.} = \frac{1}{D^3 - D^2 - 6D}(1 + x^2) = \frac{1}{-6D\left[1 + \frac{D}{6} - \frac{D^2}{6}\right]}(1 + x^2)$$

$$= -\frac{1}{6D}\left[1 + \left(\frac{D}{6} - \frac{D^2}{6}\right)\right]^{-1}(1 + x^2)$$

$$= -\frac{1}{6D}\left[1 - \left(\frac{D}{6} - \frac{D^2}{6}\right) + \left(\frac{D}{6} - \frac{D^2}{6}\right)^2 + \ldots\ldots\right](1 + x^2)$$

$$[\because \ (1 + x)^{-1} = 1 - x + x^2 - \ldots\ldots]$$

$$= -\frac{1}{6D}\left[1 - \frac{D}{6} + \frac{D^2}{6} + \frac{D^2}{36} + \ldots\ldots\right](1 + x^2) \quad \text{[Retaining terms only upto } D^2]$$

$$= -\frac{1}{6D}\left[1 - \frac{D}{6} + \frac{7D^2}{36} + \ldots\ldots\right](1 + x^2)$$

$$= -\frac{1}{6D}\left[1 + x^2 - \frac{1}{6}D(1 + x^2) + \frac{7}{36}D^2(1 + x^2) + \ldots\ldots\right]$$

$$= -\frac{1}{6D}\left[1 + x^2 - \frac{1}{6}(0 + 2x) + \frac{7}{36}(0 + 2) + 0 + \ldots\ldots\right]$$

$$= -\frac{1}{6D}\left[1 + x^2 - \frac{x}{3} + \frac{7}{18}\right] = -\frac{1}{6D}\left[x^2 - \frac{x}{3} + \frac{25}{18}\right]$$

$$= -\frac{1}{6}\left[\int x^2\, dx - \frac{1}{3}\int x dx + \frac{25}{18}\int dx\right]$$

$$= -\frac{1}{6}\left[\frac{x^3}{3} - \frac{x^2}{6} + \frac{25}{18}x\right] = -\frac{1}{108}[6x^3 - 3x^2 + 25x]$$

The general solution of (1) is y = C.F. + P.I.

$$\therefore \quad \mathbf{y = c_1 + c_2e^{3x} + c_3e^{-2x} - \frac{1}{108}[6x^3 - 3x^2 + 25x].}$$

Example 3. *Solve* $\frac{d^2y}{dx^2} - 4\frac{dy}{dx} + 4y = x^2 + e^x + \cos 2x.$

Sol. Writing the given equation in the symbolic form, we have

$$(D^2 - 4D + 4)y = x^2 + e^x + \cos 2x \qquad ...(1)$$

∴ The A.E. is $D^2 - 4D + 4 = 0$. ∴ $D = 2, 2$

$$\therefore \quad \text{C.F.} = (c_1 + c_2x)e^{2x}$$

$$\text{P.I.} = \frac{1}{D^2 - 4D + 4}(x^2 + e^x + \cos 2x) = \frac{1}{(D-2)^2}(x^2 + e^x + \cos 2x)$$

$$= \frac{1}{(D-2)^2}x^2 + \frac{1}{(D-2)^2}e^x + \frac{1}{(D-2)^2}\cos 2x$$

$$= \frac{1}{4(1-D/2)^2}x^2 + \frac{1}{(D-2)^2}e^x + \frac{1}{D^2 - 4D + 4}\cos 2x$$

$$= \frac{1}{4}\left[\left(1 - \frac{D}{2}\right)^{-2}x^2\right] + \frac{e^x}{(1-2)^2} + \frac{1}{-(2)^2 - 4D + 4}\cos 2x$$

$$= \frac{1}{4}\left[1 + 2\left(\frac{D}{2}\right) + \frac{(-2)(-3)}{2}\left(\frac{-D}{2}\right)^2 + \ldots\ldots\right]x^2 + e^x - \frac{1}{4D}\cos 2x$$

$$= \frac{1}{4}\left[1 + D + \frac{3}{4}D^2 + \ldots\right]x^2 + e^x - \frac{1}{4}\frac{\sin 2x}{2}$$

$$= \frac{1}{4}\left[x^2 + 2x + \frac{3}{4}.2\right] + e^x - \frac{\sin 2x}{8} = \frac{1}{4}\left(x^2 + 2x + \frac{3}{2}\right) + e^x - \frac{1}{8}\sin 2x.$$

∴ The general solution of (1) is y = C.F. + P.I.

$$\therefore \quad \mathbf{y = (c_1 + c_2x)e^{2x} + \frac{1}{4}\left(x^2 + 2x + \frac{3}{2}\right) + e^x - \frac{1}{8}\sin 2x.}$$

EXERCISE 4

Solve the following differential equations :

1. $\frac{d^2y}{dx^2} - \frac{dy}{dx} - 6y = x$

2. $\frac{d^2y}{dx^2} + \frac{dy}{dx} - 6y = x + k$

3. $\frac{d^3y}{dx^3} + 3\frac{d^2y}{dx^2} + 2\frac{dy}{dx} = x^2$

4. $\frac{d^3y}{dx^3} + 2\frac{d^2y}{dx^2} + \frac{dy}{dx} = e^{2x} + x^2 + x$

5. $\frac{d^2y}{dx^2} + \frac{dy}{dx} - 2y = x + \sin x$ *(MDU 1997 ; KU 1998)*

6. $(D - 1)^2(D + 1)^2 y = \sin^2 \frac{x}{2} + e^x + x$

7. $(D^4 - a^4)\, y = x^4 + \sin kx$

8. $\frac{d^2y}{dx^2} + 4y = 3e^x + \sin 3x + x^2.$ *(MDU 1995)*

Answers

1. $y = c_1e^{3x} + c_2e^{-2x} - \frac{1}{6}\left(x - \frac{1}{6}\right)$

2. $y = c_1e^{2x} + c_2e^{-3x} - \frac{1}{6}(x + k + \frac{1}{6})$

3. $y = c_1 + c_2e^{-x} + c_3e^{-2x} + \frac{1}{2}\left[\frac{x^3}{3} - \frac{3x^2}{2} + \frac{7x}{2}\right]$

4. $y = c_1 + (c_2 + c_3x)e^{-x} + \frac{1}{18}e^{2x} + \frac{x^3}{3} - \frac{3x^2}{2} + 4x$

5. $y = c_1e^{-2x} + c_2e^{x} - \frac{1}{2}\left(x + \frac{1}{2}\right) - \frac{1}{10}(\cos x + 3 \sin x)$

6. $y = (c_1 + c_2x)e^x + (c_3 + c_4x)e^{-x} + \frac{1}{2}(1 + 2x) - \frac{1}{8}\cos x + \frac{x^2e^x}{8}$

7. $y = c_1e^{ax} + c_2e^{-ax} + c_3 \cos ax + c_4 \sin ax - \frac{1}{a^4}\left(x^4 + \frac{24}{a^4}\right) + \frac{\sin kx}{k^4 - a^4}$

8. $y = c_1 \cos 2x + c_2 \sin 2x + \frac{3}{5}e^x - \frac{1}{5}\sin 3x + \frac{1}{4}x^2 - \frac{1}{8}.$

---

13. Q is of the Form e^{ax} V, where V is any Function of x *(MDU 2000)*

Let the linear differential equation be $f(D)\, y = Q$, where $f(D)$ is a polynomial in D with constant coefficients and $Q = e^{ax}V$.

$$\therefore \qquad \text{P.I.} = \frac{1}{f(D)} e^{ax}V$$

For any function U of x, we have

$$De^{ax}U = e^{ax} . DU + ae^{ax} . U = e^{ax}(D + a)U$$

$$D^2e^{ax}U = e^{ax} . D(D + a)U + ae^{ax}(D + a)U = e^{ax}(D + a)^2 U$$

....

$\therefore$ In general $D^k e^{ax} U = e^{ax}(D + a)^k U$

Since $f(D)$ is a polynomial in D with constant coefficients, we have

$$f(D)e^{ax} U = e^{ax} f(D + a)\, U. \qquad \text{...(1)}$$

Let $\qquad f(D + a)U = V \quad i.e., \quad U = \frac{1}{f(D+a)} V$

$$\therefore \quad (1) \Rightarrow \quad f(D)\, e^{ax} \frac{1}{f(D+a)} V = e^{ax}\, V.$$

Operating with $\frac{1}{f(D)}$, we get $\quad e^{ax} \frac{1}{f(D+a)} V = \frac{1}{f(D)} e^{ax}\, V.$

$$\therefore \qquad \mathbf{\frac{1}{f(D)} e^{ax}\, V = e^{ax} \frac{1}{f(D+a)} V.}$$

Example 1. *Solve* $(D^2 - 4D + 4)\, y = e^{2x} \cos^2 x.$ (MDU 2001)

Sol. We have $(D^2 - 4D + 4)y = e^{2x} \cos^2 x.$ ...(1)

$\therefore$ The A.E. is $D^2 - 4D + 4 = 0$. $\therefore$ $D = 2, 2$.

$\therefore$ $$\text{C.F.} = (c_1 + c_2 x)\, e^{2x}$$

$$\text{P.I.} = \frac{1}{(D-2)^2} e^{2x} \cos^2 x = e^{2x} \frac{1}{(D+2-2)^2} \cos^2 x$$

$$= e^{2x} \frac{1}{D^2}\left(\frac{1+\cos 2x}{2}\right) = \frac{e^{2x}}{2}\left[\frac{1}{D^2}(1) + \frac{1}{D^2}\cos 2x\right]$$

$$= \frac{e^{2x}}{2}\left[\frac{x^2}{2} + \frac{1}{-(2)^2}\cos 2x\right] = \frac{e^{2x}}{8}[2x^2 - \cos 2x]$$

The general solution of (1) is y = C.F. + P.I.

$\therefore$ $$\mathbf{y = (c_1 + c_2\, x)e^{2x} + \frac{e^{2x}}{8}\,[2x^2 - \cos 2x].}$$

Example 2. *Solve* $\dfrac{d^3y}{dx^3} - 3\dfrac{dy}{dx} + 2y = x^2e^x$. (K.U. 2006)

Sol. The given equation in the symbolic form is

$$(D^3 - 3D + 2)\, y = x^2e^x. \qquad \text{...(1)}$$

$\therefore$ The A.E. is $D^3 - 3D + 2 = 0$ or $(D-1)(D^2 + D - 2) = 0$

or $(D-1)(D+2)(D-1) = 0$. $\therefore$ $D = 1, 1, -2$

$\therefore$ $$\text{C.F.} = (c_1 + c_2x)\, e^x + c_3e^{-2x}$$

$$\text{P.I.} = \frac{1}{D^3 - 3D + 2} x^2e^x = e^x \frac{1}{(D+1)^3 - 3(D+1) + 2} x^2$$

$$= e^x \frac{1}{D^3 + D^2 + 3D + 1 - 3D - 3 + 2}$$

$$= e^x \frac{1}{D^3 + 3D^2} x^2 = e^x \frac{1}{3D^2\left[1 + \frac{D}{3}\right]} x^2$$

$$= \frac{e^x}{3} \cdot \frac{1}{D^2}\left(1 + \frac{D}{3}\right)^{-1} x^2 = \frac{e^x}{3} \cdot \frac{1}{D^2}\left[1 - \frac{D}{3} + \left(\frac{D}{3}\right)^2 + \ldots\ldots\right] x^2$$

(Retaining terms only upto D^2)

$$= \frac{e^x}{3} \cdot \frac{1}{D^2}\left(x^2 - \frac{1}{3}Dx^2 + \frac{1}{9}D^2x^2 + 0 + \ldots\ldots\right)$$

$$= \frac{e^x}{3} \cdot \frac{1}{D^2}\left(x^2 - \frac{2x}{3} + \frac{2}{9}\right) = \frac{e^x}{3}\frac{1}{D}\left(\int x^2\, dx - \frac{2}{3}\int x dx + \frac{2}{9}\int dx\right)$$

$$= \frac{e^x}{3} \cdot \frac{1}{D}\left[\frac{x^3}{3} - \frac{x^2}{3} + \frac{2}{9}x\right] = \frac{e^x}{3}\left(\int \frac{x^3}{3}\, dx - \int \frac{x^2}{3}\, dx + \int \frac{2}{9}\, x dx\right)$$

$$= \frac{e^x}{3}\left[\frac{x^4}{12} - \frac{x^3}{9} + \frac{x^2}{9}\right] = \frac{e^x}{108}(3x^4 - 4x^3 + 4x^2) = \frac{e^x\, x^2}{108}(3x^2 - 4x + 4)$$

The general solution of (1) is y = C.F. + P.I.

$\therefore$ $$\mathbf{y = (c_1 + c_2x)\, e^x + c_3e^{-2x} + \frac{1}{108}\, x^2e^x(3x^2 - 4x + 4).}$$

Example 3. *Solve* $\frac{d^2y}{dx^2} + 2y = x^2 e^{3x} + e^x \cos 2x.$ (M.D.U. 2000 ; K.U. 2004 S, 2007)

Sol. The equation in the symbolic form is

$$(D^2 + 2)\, y = x^2 e^{3x} + e^x \cos 2x. \qquad ...(1)$$

$\therefore$ The A.E. is $D^2 + 2 = 0. \quad \therefore \quad D = \pm \sqrt{2}\, i$

$$\therefore \quad \text{C.F.} = e^{o.x}(c_1 \cos \sqrt{2}\,x + c_2 \sin \sqrt{2}\,x) = c_1 \cos \sqrt{2}\,x + c_2 \sin \sqrt{2}\,x$$

$$\text{P.I.} = \frac{1}{D^2+2}(x^2 e^{3x} + e^x \cos 2x)$$

$$= \frac{1}{D^2+2} x^2 e^{3x} + \frac{1}{D^2+2} e^x \cos 2x$$

$$= e^{3x} \frac{1}{(D+3)^2+2} x^2 + e^x \frac{1}{(D+1)^2+2} \cos 2x$$

$$= e^{3x} \frac{1}{D^2+6D+9+2} x^2 + e^x \frac{1}{D^2+2D+3} \cos 2x$$

$$= e^{3x} \frac{1}{11+6D+D^2} x^2 + e^x \frac{1}{-4+2D+3} \cos 2x$$

$$\left[\because \quad \frac{1}{f(D^2)} \cos ax = \frac{1}{f(-a^2)} \cos ax\right]$$

$$= e^{3x} \frac{1}{11\left[1+\frac{6D+D^2}{11}\right]} x^2 + e^x \frac{1}{-1+2D} \cos 2x$$

$$= e^{3x} \frac{1}{11}\left(1+\frac{6D+D^2}{11}\right)^{-1} x^2 + \frac{e^x (2D+1)}{(2D-1)(2D+1)} \cos 2x$$

$$= e^{3x} \frac{1}{11}\left[1-\frac{6D+D^2}{11}+\left(\frac{6D+D^2}{11}\right)^2 - \ldots\ldots\right] x^2 + \frac{e^x (2D+1)}{4D^2-1} \cos 2x$$

$$= e^{3x} \frac{1}{11}\left[1-\frac{6D+D^2}{11}+\frac{36}{121}D^2\right] x^2 + \frac{e^x (2D+1)}{4(-4)-1} \cos 2x$$

(Retaining terms only upto D^2)

$$= e^{3x} \frac{1}{11}\left[1-\frac{6D}{11}+\frac{25D^2}{121}\right] x^2 + \frac{e^x}{-17}(2D+1)\cos 2x$$

$$= e^{3x} \frac{1}{11}\left[x^2-\frac{6}{11}Dx^2+\frac{25}{121}D^2x^2\right] + \frac{e^x}{-17}(2D \cos 2x + \cos 2x)$$

$$= e^{3x} \cdot \frac{1}{11}\left[x^2-\frac{12}{11}x+\frac{50}{121}\right] - \frac{e^x}{17}(-4 \sin 2x + \cos 2x)$$

$$= \frac{1}{11} e^{3x}\left[x^2-\frac{12x}{11}+\frac{50}{121}\right] - \frac{e^x}{17}(-4 \sin 2x + \cos 2x).$$

The general solution of (1) is y = C.F. + P.I.

$$\therefore \quad \mathbf{y = c_1 \cos \sqrt{2}\, x + c_2 \sin \sqrt{2}\, x + \frac{e^{3x}}{11}\left[x^2 - \frac{12x}{11} + \frac{50}{121}\right] - \frac{e^x}{17}[-4 \sin 2x + \cos 2x].}$$

EXERCISE 5

Solve the following differential equations :

1. $\frac{d^2y}{dx^2} - 2\frac{dy}{dx} + y = x^2 e^{3x}$ (*KU 2000*)

2. $\frac{d^2y}{dx^2} + y = x e^{2x}$

3. $\frac{d^3y}{dx^3} - 7\frac{dy}{dx} - 6y = e^{2x}(1 + x)$ (*KU 1996*)

4. $\frac{d^2y}{dx^2} - 2\frac{dy}{dx} + 4y = e^x \cos x$ (*KU 2003 C*)

5. $\frac{d^4y}{dx^4} - y = e^x \cos x$

6. $\frac{d^2y}{dx^2} - 4\frac{dy}{dx} + 3y = e^x \cos 2x + \cos 3x$ (*KU 1997*)

7. $\frac{d^2y}{dx^2} + 2y = e^x x^2 + e^{2x} \sin x$

8. $4\frac{d^2y}{dx^2} + 12\frac{dy}{dx} + 9y = 144\, xe^{-3x/2}$

9. $(D^3 + 1)\, y = e^x \cos x$

10. $(D^3 - D^2 - 9D + 9)y = e^x(x^2 + \cos 2x)$

11. $(D^2 - 4D + 4)\, y = e^{2x} \cos^2 x$ (*MDU 2001*)

12. $\frac{d^2y}{dx^2} + 2\frac{dy}{dx} + y = e^x \sin 2x + x^2$ (*MDU 1996*)

13. $\frac{d^2y}{dx^2} - 2\frac{dy}{dx} + 4y = e^x \cos x + x$ (*KU 1996 C ; MDU 1995*)

14. $\frac{d^3y}{dx^3} - 2\frac{d^2y}{dx^2} + \frac{dy}{dx} - 2y = e^{2x} \cos x + k$, where k is a constant.

Answers

1. $y = (c_1 + c_2 x)\, e^x + \frac{1}{4} e^{3x}\left(x^2 - 2x + \frac{3}{2}\right)$

2. $y = c_1 \cos x + c_2 \sin x + \frac{1}{5} e^{2x}\left(x - \frac{4}{5}\right)$

3. $y = c_1 e^{3x} + c_2 e^{-x} + c_3 e^{-2x} - \frac{1}{12} e^{2x}\left(x + \frac{17}{12}\right)$

4. $y = e^x (c_1 \cos \sqrt{3}x + c_2 \sin \sqrt{3}x) + \frac{1}{2} e^x \cos x$

5. $y = c_1 e^x + c_2 e^{-x} + c_3 \cos x + c_4 \sin x - \frac{1}{5} e^x \cos x$

6. $y = c_1 e^x + c_2 e^{3x} - \frac{e^x}{8}(\cos 2x + \sin 2x) - \frac{1}{30}(\cos 3x + 2 \sin 3x)$

7. $y = c_1 \cos \sqrt{2}x + c_2 \sin \sqrt{2}x + \frac{e^x}{3}\left[x^2 - \frac{4}{3}x + \frac{2}{9}\right] + \frac{e^{2x}}{41}[5 \sin x - 4 \cos x]$

8. $y = (c_1 + c_2 x)\, e^{-3x/2} + 6x^3 e^{-3x/2}$

9. $y = c_1 e^{-x} + e^{x/2}\left(c_2 \cos \frac{\sqrt{3}}{2} x + c_3 \sin \frac{\sqrt{3}}{2} x\right) + \frac{1}{5} e^x (2 \sin x - \cos x)$

10. $y = c_1 e^x + c_2 e^{3x} + c_3 e^{-3x} - \frac{e^x}{8}\left(\frac{x^3}{3} + \frac{x^2}{4} + \frac{3x}{8}\right) - \frac{e^x}{80}(3 \sin 2x + \cos 2x)$

11. $y = (c_1 + c_2 x)\, e^{2x} + \frac{1}{8} e^{2x}(2x^2 - \cos 2x)$

12. $y = (c_1 + c_2 x)\, e^{-x} - \frac{1}{8} e^x \cos 2x + x^2 - 4x + 6$

13. $y = e^x (c_1 \cos \sqrt{3}\, x + c_2 \sin \sqrt{3}\, x) + \frac{1}{2} e^x \cos x + \frac{x}{4} + \frac{1}{8}$

14. $y = c_1 e^{2x} + c_2 \cos x + c_3 \sin x - \frac{e^{2x}}{8}(\cos x - \sin x) - \frac{k}{2}$.

14. Q is of the Form xV, where V is any Function of x

Let the linear differential equation be $f(D)\,y = Q$, where $f(D)$ is a polynomial in D with constant coefficients and $Q = xV$.

$$\therefore \qquad \text{P.I.} = \frac{1}{f(D)}\,xV$$

For any function U of x, we have

$$D\,xU = x\,DU + U$$

$$D^2\,xU = (x\,D^2U + DU) + DU = x\,D^2U + 2DU$$

....

In general, $\quad D^k\,xU = x\,D^k\,U + kD^{k-1}\,U \quad$ or $\quad D^k\,xU = x\,D^kU + \left(\dfrac{d}{dD}D^k\right)U$

Since $f(D)$ is a polynomial in D with constant coefficients, we have

$$\therefore \qquad f(D)\,xU = x\,f(D)U + f'(D)U. \qquad ...(1)$$

Let $\qquad f(D)U = V \quad$ *i.e.,* $\quad U = \dfrac{1}{f(D)}V$

$$\therefore \quad (1) \Rightarrow f(D)\,x\,\frac{1}{f(D)}V = xV + f'(D)\frac{1}{f(D)}V$$

Operating with $\dfrac{1}{f(D)}$, we get

$$x\,\frac{1}{f(D)}V = \frac{1}{f(D)}xV + \frac{1}{f(D)}f'(D)\frac{1}{f(D)}V$$

$$\Rightarrow \qquad \frac{1}{f(D)}xV = x\,\frac{1}{f(D)}V - \frac{1}{f(D)}f'(D)\frac{1}{f(D)}V$$

$$\therefore \qquad \mathbf{\frac{1}{f(D)}xV = \left(x - \frac{1}{f(D)}f'(D)\right)\frac{1}{f(D)}V.}$$

Corollary. We have $\quad \dfrac{1}{f(D)}xV = \left(x - \dfrac{1}{f(D)}f'(D)\right)\dfrac{1}{f(D)}V.$

$$\Rightarrow \qquad \frac{1}{f(D)}xV = x\,\frac{1}{f(D)}V - \frac{f'(D)}{(f(D))^2}V$$

$$\Rightarrow \qquad \mathbf{\frac{1}{f(D)}xV = x\,\frac{1}{f(D)}V + \left[\frac{d}{dD}\left(\frac{1}{f(D)}\right)\right]V.}$$

In practical problems, it is advisable to use this form.

Example 1. *Solve* $\dfrac{d^2y}{dx^2} + 4y = x\sin x.$

Sol. The equation in the symbolic form is

$$(D^2 + 4)\,y = x\sin x. \qquad ...(1)$$

$\therefore\;$ The A.E. is $\qquad D^2 + 4 = 0. \quad \therefore \quad D = \pm\,2i$

$\therefore \qquad \text{C.F.} = e^{0.x}(c_1 \cos 2x + c_2 \sin 2x) = c_1 \cos 2x + c_2 \sin 2x$

$$\text{P.I.} = \frac{1}{D^2+4}(x \sin x) = x\frac{1}{D^2+4}\sin x + \left\{\frac{d}{dD}\left(\frac{1}{D^2+4}\right)\right\}\sin x$$

$$= x\frac{1}{-1+4}\sin x + \left(\frac{-1}{(D^2+4)^2}2D\right)\sin x$$

$$= \frac{x \sin x}{3} - \frac{2D}{(-1+4)^2}\sin x = \frac{x\sin x}{3} - \frac{2}{9}D\sin x = \frac{x \sin x}{3} - \frac{2}{9}\cos x$$

$\therefore$ The general solution of (1) is $y = \text{C.F.} + \text{P.I.}$

$$\therefore \qquad \mathbf{y = c_1 \cos 2x + c_2 \sin 2x + \frac{x \sin x}{3} - \frac{2}{9}\cos x.}$$

Example 2. *Solve* $\dfrac{d^2y}{dx^2} + 3\dfrac{dy}{dx} + 2y = xe^x \sin x.$ (K.U. 2004, 2006 S)

Sol. The equation in the symbolic form is

$$(D^2 + 3D + 2)\, y = x\, e^x \sin x. \qquad ...(1)$$

$\therefore$ The A.E. is $\qquad D^2 + 3D + 2 = 0. \quad \therefore \quad D = -1, -2$

$\therefore \qquad \text{C.F.} = c_1 e^{-x} + c_2 e^{-2x}$

$$\text{P.I.} = \frac{1}{D^2+3D+2}(xe^x \sin x)$$

$$= e^x \frac{1}{(D+1)^2+3(D+1)+2} x \sin x = e^x \frac{1}{D^2+5D+6}(x \sin x)$$

$$= e^x\left[x\frac{1}{D^2+5D+6}\sin x + \left(\frac{d}{dD}\frac{1}{D^2+5D+6}\right)\sin x\right]$$

$$= e^x\left[x\frac{1}{-1+5D+6}\sin x + \frac{-1(2D+5)}{(D^2+5D+6)^2}\sin x\right]$$

$$= e^x\left[x\frac{1}{5(1+D)}\sin x - \frac{(2D+5)}{(-1+5D+6)^2}\sin x\right]$$

$$= e^x\left[x\frac{(1-D)\sin x}{5(1+D)(1-D)} - \frac{2D+5}{(5+5D)^2}\sin x\right]$$

$$= e^x\left[x\frac{1}{5(1-D^2)}(1-D)\sin x - \frac{2D+5}{25(D+1)^2}\sin x\right]$$

$$= e^x\left[x\frac{1}{5(1+1)}(1-D)\sin x - \frac{2D+5}{25(D^2+2D+1)}\sin x\right]$$

$$= e^x\left[\frac{x}{10}(\sin x - D\sin x) - \frac{1}{25(-1+2D+1)}(2D+5)\sin x\right]$$

$$= e^x\left[\frac{x}{10}(\sin x - \cos x) - \frac{1}{50}\left(2+\frac{5}{D}\right)\sin x\right]$$

$$= e^x\left[\frac{x}{10}(\sin x - \cos x) - \frac{2}{50}\sin x - \frac{1}{10}\frac{1}{D}\sin x\right]$$

$$= e^x \left[\frac{x}{10}(\sin x - \cos x) - \frac{1}{25}\sin x + \frac{1}{10}\cos x\right]$$

$\therefore$ The general solution of (1) is y = C.F. + P.I.

$$\therefore \quad \mathbf{y = c_1e^{-x} + c_2e^{-2x} + e^x\left[\frac{x}{10}(\sin x - \cos x) - \frac{1}{25}\sin x + \frac{1}{10}\cos x\right].}$$

EXERCISE 6

Solve the following differential equations :

1. $\dfrac{d^2y}{dx^2} + 4y = x\cos x$

2. $\dfrac{d^2y}{dx^2} - 2\dfrac{dy}{dx} + y = x\sin x$ *(K.U. 1995 C)*

3. $\dfrac{d^2y}{dx^2} - 2\dfrac{dy}{dx} + y = xe^x\sin x$ *(M.D.U. 1999, 2006 ; K.U. 1997, 2001)*

4. $\dfrac{d^2y}{dx^2} - 9y = x\cos 2x$

5. $\dfrac{d^2y}{dx^2} - y = x\sin x + e^x(1 + x^2)$

6. $\dfrac{d^2y}{dx^2} + 3\dfrac{dy}{dx} + 2y = xe^x\sin x$

7. $\dfrac{d^2y}{dx^2} - 2\dfrac{dy}{dx} + y = xe^x\sin 2x$ *(M.D.U. 1996)*

8. $\dfrac{d^2y}{dx^2} + y = xe^x\cos 2x.$

Answers

1. $y = c_1\cos 2x + c_2\sin 2x + \dfrac{1}{3}x\cos x + \dfrac{2}{9}\sin x$
2. $y = (c_1 + c_2x)e^x + \dfrac{x}{2}\cos x + \dfrac{1}{2}(\cos x - \sin x)$
3. $y = (c_1 + c_2x)e^x - e^x(x\sin x + 2\cos x)$
4. $y = c_1e^{3x} + c_2e^{-3x} - \dfrac{x}{13}\cos 2x + \dfrac{4}{169}\sin 2x$
5. $y = c_1e^x + ce^{-x} - \dfrac{1}{2}(x\sin x + \cos x) + \dfrac{1}{12}xe^x(2x^2 - 3x + 9)$
6. $y = c_1e^{-x} + c_2e^{-2x} + e^x\left[\dfrac{x}{10}(\sin x - \cos x) - \dfrac{1}{25}\sin x + \dfrac{1}{10}\cos x\right]$
7. $y = (c_1 + c_2x)e^x - \dfrac{e^x}{4}(x\sin 2x + \cos 2x)$
8. $y = c_1\cos x + c_2\sin x + e^x\left[\dfrac{5x-1}{25}\sin 2x + \dfrac{11-5x}{50}\cos 2x\right].$

---

15. General Method of Evaluating Particular Integral

In case any of the shorter method is not applicable, then we evaluate the particular integral by using the general method. For this purpose the following formula is used very frequently :

$$\mathbf{\frac{1}{D-\alpha}Q = e^{\alpha x}\int Qe^{-\alpha x}\,dx.}$$

Example 1. *Solve the differential equation*

$$(D^2 - 3D + 2)\, y = \sin e^{-x}.$$

Sol. We have $(D^2 - 3D + 2)y = \sin e^{-x}$. ...(1)

$\therefore$ The A.E. is $D^2 - 3D + 2 = 0$. $\therefore$ $D = 1, 2$

$\therefore$ C.F. $= c_1 e^x + c_2 e^{2x}$

$$\text{P.I.} = \frac{1}{D^2 - 3D + 2} \sin e^{-x} = \frac{1}{(D-2)(D-1)} \sin e^{-x} = \frac{1}{D-2}\left(\frac{1}{D-1} \sin e^{-x}\right)$$

$$= \frac{1}{D-2}\left(e^x \int \sin (e^{-x})\, e^{-x}\, dx\right) \qquad \left|\ \because \frac{1}{D-\alpha} Q = e^{\alpha x} \int Q e^{-\alpha x}\, dx \right. \quad \text{Here } \alpha = 1,\ Q = \sin e^{-x}$$

Let $e^{-x} = t$. $\therefore$ $-e^{-x}\, dx = dt$ or $e^{-x}\, dx = -\, dt$

$$\therefore \quad \text{P.I.} = \frac{1}{D-2}\left(e^x \int - \sin t\, dt\right) = \frac{1}{D-2} e^x \cos t = \frac{1}{D-2} e^x \cos e^{-x}$$

$$= e^{2x} \int e^x \cos e^{-x} . e^{-2x}\, dx \qquad [\text{Here } \alpha = 2]$$

$$= e^{2x} \int e^{-x} \cos e^{-x} dx = e^{2x} \int - \cos t\, dt, \text{ where } t = e^{-x}$$

$$= - e^{2x} \sin t = - e^{2x} \sin e^{-x}.$$

$\therefore$ The general solution of (1) is y = C.F. + P.I.

$$\therefore \quad \mathbf{y = c_1 e^x + c_2 e^{2x} - e^{2x} \sin e^{-x}.}$$

Remark. We know that $e^x = 1 + x + \dfrac{x^2}{2!} + \dfrac{x^3}{3!} + \dfrac{x^4}{4!} + \ldots\ldots$

Putting $x = i\theta$, $e^{i\theta} = 1 + i\theta + \dfrac{(i\theta)^2}{2!} + \dfrac{(i\theta)^3}{3!} + \dfrac{(i\theta)^4}{4!} + \ldots\ldots$

$$\Rightarrow \quad e^{i\theta} = 1 + i\theta - \frac{\theta^2}{2!} - i\frac{\theta^3}{3!} + \frac{\theta^4}{4!} + i\frac{\theta^5}{5!} \quad \Rightarrow \quad e^{i\theta} = \left(1 - \frac{\theta^2}{2!} + \frac{\theta^4}{4!} - \ldots\ldots\right) + i\left(\theta - \frac{\theta^3}{3!} + \frac{\theta^5}{5!} - \ldots\ldots\right)$$

$\Rightarrow$ $\mathbf{e^{i\theta} = \cos \theta + i \sin \theta}$. ...(1)

$\therefore$ $\cos \theta$ is Real part of $e^{i\theta}$ and briefly written as **R.P.** of $e^{i\theta}$

and $\sin \theta$ is Imaginary part of $e^{i\theta}$ and briefly written as **I.P.** of $e^{i\theta}$.

Changing i to $-i$ in equation (1), we get $e^{-i\theta} = \cos \theta - i \sin \theta$.

Example 2. *Solve* $\dfrac{d^4y}{dx^4} + 2\dfrac{d^2y}{dx^2} + y = x^2 \sin x.$

Sol. We have

$$(D^4 + 2D^2 + 1)\, y = x^2 \sin x. \qquad \text{...(1)}$$

$\therefore$ The A.E. is $D^4 + 2D^2 + 1 = 0$ or $(D^2 + 1)^2 = 0$.

$$\therefore \quad D = \pm i, \ \pm i$$

$$\therefore \quad \text{C.F.} = e^{0.x} ((c_1 + c_2x) \cos 1.x + (c_3 + c_4 x) \sin 1.x)$$

$$\therefore \quad \text{C.F.} = (c_1 + c_2x) \cos x + (c_3 + c_4x) \sin x$$

$$\text{P.I.} = \frac{1}{D^4 + 2D^2 + 1} x^2 \sin x$$

$$= \text{I.P. of } \frac{1}{D^4 + 2D^2 + 1} x^2 e^{ix} \qquad [\because \sin \theta \text{ is I.P. of } e^{i\theta}]$$

$$= \text{I.P. of } \frac{1}{(D^2+1)^2} x^2 e^{ix} = \text{I.P. of } e^{ix} \frac{1}{((D+i)^2+1)^2} x^2$$

$$= \text{I.P. of } e^{ix} \frac{1}{(D^2+2iD-1+1)^2} x^2 \qquad | \because \; i^2 = -1$$

$$= \text{I.P. of } e^{ix} \frac{1}{(2iD+D^2)^2} x^2 = \text{I.P. of } e^{ix} \frac{1}{4i^2D^2\left(1+\frac{D}{2i}\right)^2} x^2$$

$$= \text{I.P. of } \frac{-1}{4} e^{ix} \frac{1}{D^2}\left(1+\frac{D}{2i}\right)^{-2} x^2$$

$$= \text{I.P of } \frac{-1}{4} e^{ix} \frac{1}{D^2}\left(1-2\left(\frac{D}{2i}\right)+3\left(\frac{D}{2i}\right)^2\right) x^2 \qquad [\because \; D^3x^2 = 0, D^4x^2 = 0, \ldots]$$

$$= \text{I.P. of } \frac{-1}{4} e^{ix} \left[\frac{1}{D^2} + i\frac{1}{D} - \frac{3}{4}\right] x^2 \qquad \Big| \because \; \frac{1}{i} = -i$$

$$= \text{I.P. of } \frac{-1}{4} (\cos x + i \sin x)\left(\frac{x^4}{12} + \frac{ix^3}{3} - \frac{3}{4}x^2\right)$$

$$= -\frac{1}{4}\left[\frac{x^3}{3}\cos x + \frac{x^4}{12}\sin x - \frac{3}{4}x^2 \sin x\right]$$

∴ The general solution of (1) is $y = \text{C.F.} + \text{P.I.}$

$$\therefore \qquad \mathbf{y = (c_1 + c_2x)\cos x + (c_3 + c_4 x)\sin x - \frac{1}{4}\left[\frac{x^3}{3}\cos x + \frac{x^4}{12}\sin x - \frac{3}{4}x^2 \sin x\right].}$$

Example 3. *Solve* $\frac{d^2y}{dx^2} + a^2y = \sec ax.$ (KU 2005)

Sol. We have $(D^2 + a^2)\, y = \sec ax.$ ...(1)

∴ The A.E. is $D^2 + a^2 = 0.$ ∴ $D = \pm ai$

$$\therefore \qquad \text{C.F.} = e^{0.x}(c_1 \cos ax + c_2 \sin ax) = c_1 \cos ax + c_2 \sin ax$$

$$\text{P.I.} = \frac{1}{D^2+a^2}\sec ax = \frac{1}{D^2-(-a^2)}\sec ax$$

$$= \frac{1}{D^2-(ia)^2}\sec ax = \frac{1}{(D-ia)(D+ia)}\sec ax$$

$$= \frac{1}{2ia}\frac{(D+ia)-(D-ia)}{(D-ia)(D+ia)}\sec ax$$

or
$$\text{P.I.} = \frac{1}{2ia}\left[\frac{1}{D-ia} - \frac{1}{D+ia}\right]\sec ax$$

$$\therefore \quad \text{P.I.} = \frac{1}{2ia}\left[\frac{1}{D-ia}\sec ax - \frac{1}{D+ia}\sec ax\right] \qquad \ldots(2)$$

Now $\frac{1}{D-ia}\sec ax = e^{iax}\int \sec ax\, e^{-ix}\, dx$ $\left[\because \; \frac{1}{D-\alpha}Q = e^{\alpha x}\int Q\, e^{-\alpha x}\, dx \quad \text{Here } \alpha = ia\right]$

$$= e^{iax} \int \frac{\cos ax - i \sin ax}{\cos ax} dx \qquad (\because \quad e^{-i\theta} = \cos\theta - i \sin\theta)$$

$$= e^{iax} \int (1 - i \tan ax)\, dx$$

$$= (\cos ax + i \sin ax)\left(x - \frac{i \log \sec ax}{a}\right)$$

$$\therefore \quad \frac{1}{D - ia} \sec ax = x \cos ax - \frac{i}{a} \cos ax \log \sec ax + ix \sin ax + \frac{1}{a} \sin ax \log \sec ax \qquad ...(3)$$

Changing i to $-i$ in (3), we get

$$\frac{1}{D + ia} \sec ax = x \cos ax + \frac{i}{a} \cos ax \log \sec ax - ix \sin ax + \frac{1}{a} \sin ax \log \sec ax \qquad ...(4)$$

$$(3) - (4) \quad \Rightarrow \quad \frac{1}{D - ia} \sec ax - \frac{1}{D + ia} \sec ax = -\frac{2i}{a} \cos ax \log \sec ax + 2ix \sin ax$$

Putting this value in (2), we get

$$\text{P.I.} = \frac{1}{2ia} 2i\left(-\frac{1}{a} \cos ax \log \sec ax + x \sin ax\right)$$

$$= -\frac{1}{a^2} \cos ax \log \sec ax + \frac{x}{a} \sin ax.$$

$\therefore$ The general solution of (1) is y = C.F. + P.I.

$$\therefore \quad \mathbf{y = c_1 \cos ax + c_2 \sin ax - \frac{1}{a^2} \cos ax \log \sec ax + \frac{x}{a} \sin ax.}$$

EXERCISE 7

Solve the following differential equations :

1. $(D^2 - 3D + 2)\, y = \cos e^{-x}$ *(M.D.U. 2007)*
2. $\dfrac{d^2y}{dx^2} - y = x^2 \cos x$ *(K.U. 2005 S)*
3. $\dfrac{d^2y}{dx^2} + 16y = \sec 4x$ *(M.D.U. 2003)*
4. $\dfrac{d^2y}{dx^2} + y = \text{cosec}\, x$ *(M.D.U. 2005)*
5. $\dfrac{d^2y}{dx^2} + a^2y = \text{cosec}\, ax$
6. $\dfrac{d^4y}{dx^4} - y = x \sin x.$ *(M.D.U. 2003)*

Answers

1. $y = c_1 e^x + c_2 e^{2x} - e^{2x} \cos e^{-x}$
2. $y = c_1 e^x + c_2 e^{-x} - \frac{1}{2}(x^2 - 1) \cos x + x \sin x$
3. $y = c_1 \cos 4x + c_2 \sin 4x - \frac{1}{16} \cos 4x \log \sec 4x + \frac{x}{4} \sin 4x$
4. $y = c_1 \cos x + c_2 \sin x + \sin x \log \sin x - x \cos x$
5. $y = c_1 \cos ax + c_2 \sin ax + \frac{1}{a^2} \sin ax \log \sin ax - \frac{x}{a} \cos ax$
6. $y = c_1 \cos x + c_2 \sin x + c_3 e^x + c_4 e^{-x} + \frac{1}{8} x^2 \cos x - \frac{3}{8} x \sin x.$

MISCELLANEOUS QUESTIONS

Solve the following differential equations :

1. $\dfrac{d^2y}{dx^2} - y = (1 + x^2)\, e^x$
2. $(D^4 + 6D^3 + 11D^2 + 6D)y = 20\, e^{-2x} \sin x$
3. $(D^3 + D^2 + D + 1)\, y = 2e^{-x} + x^2$
4. $(D^4 - 2D^2 + 1)y = 40 \cosh x$
5. $\dfrac{d^2y}{dx^2} - 4\dfrac{dy}{dx} + 4y = 3x^2 e^{2x} \sin 2x$ (*K.U. 2005 S*)
6. $(D^2 + 1)y = 3 \cos^2 x + 2 \sin^3 x$
7. $\dfrac{d^2y}{dx^2} + y = \sec x$ (*K.U. 2004*)
8. $(D^2 - 4D + 3)\, y = e^x \cos 2x + \cos 3x.$ (*C.D.L.U. 2004*)

Answers

1. $y = c_1 e^x + c_2 e^{-x} + \dfrac{1}{12} xe^x (2x^2 - 3x + 9)$
2. $y = c_1 + c_2 e^{-x} + c_3 e^{-2x} + c_4 e^{-3x} + 2e^{-2x} (\sin x - 2 \cos x)$
3. $y = c_1 e^{-x} + c_2 \cos x + c_3 \sin x + xe^{-x} + x^2 - 2x$
4. $y = (c_1 + c_2 x)\, e^x + (c_3 + c_4 x)\, e^{-x} + 5x^2 \cosh x$
5. $y = (c_1 + c_2 x)\, e^{2x} + \dfrac{3}{8} e^{2x} (3 \sin 2x - 4x \cos 2x - 2x^2 \sin 2x)$
6. $y = c_1 \cos x + c_2 \sin x + \dfrac{3}{2} - \dfrac{1}{2} \cos 2x - \dfrac{3}{4} x \cos x + \dfrac{1}{16} \sin 3x$
7. $y = c_1 \cos x + c_2 \sin x - \cos x \log \sec x + x \sin x$
8. $y = c_1 e^x + c_2 e^{3x} - \dfrac{e^x}{8} (\sin 2x + \cos 2x) - \dfrac{1}{30} (2 \sin 3x + \cos 3x).$

20

Homogeneous Linear Ordinary Differential Equations

1. Introduction

In the preceding chapter, we learnt the method of solving linear differential equations with constant coefficients. In this chapter, we shall learn the method of solving linear differential equations with a particular type of variable coefficients, namely, the homogeneous linear equations.

2. Homogeneous Linear Equation

A differential equation of the form

$$P_0 x^n \frac{d^n y}{dx^n} + P_1 x^{n-1} \frac{d^{n-1} y}{dx^{n-1}} + P_2 x^{n-2} \frac{d^{n-2} y}{dx^{n-2}} + \ldots\ldots + P_n y = Q$$

where $P_0 \neq 0, P_1, P_2, \ldots\ldots, P_n$ are all constants and Q is a function of x is called a **homogeneous linear differential equation** of order n.

For example,

$$2x^3 \frac{d^3 y}{dx^3} + 6x^2 \frac{d^2 y}{dx^2} + x \frac{dy}{dx} + 7y = \sin x$$

is a homogeneous linear differential equation of order 3.

Remark. A homogeneous linear differential equation is also called a **Cauchy linear equation.**

3. General Solution of Homogeneous Linear Differential Equation

Let $$P_0 x^n \frac{d^n y}{dx^n} + P_1 x^{n-1} \frac{d^{n-1} y}{dx^{n-1}} + P_2 x^{n-2} \frac{d^{n-2} y}{dx^{n-2}} + \ldots\ldots + P_n y = Q \qquad \ldots(1)$$

where $P_0 \neq 0, P_1, P_2, \ldots\ldots, P_n$ are all constants and Q is a function of x be a homogeneous linear differential equation of order n.

Let $z = \log x$ *i.e.*, $x = e^z$.

Now, $$\frac{dy}{dx} = \frac{dy}{dz}\frac{dz}{dx} = \frac{dy}{dz} \cdot \frac{1}{x} = \frac{1}{x}\frac{dy}{dz}$$

$$\frac{d^2 y}{dx^2} = \frac{d}{dx}\left(\frac{1}{x}\frac{dy}{dz}\right) = \frac{1}{x}\frac{d^2 y}{dz^2}\frac{dz}{dx} - \frac{1}{x^2}\frac{dy}{dz} = \frac{1}{x^2}\left(\frac{d^2 y}{dz^2} - \frac{dy}{dz}\right)$$

$$\frac{d^3 y}{dx^3} = \frac{d}{dx}\left[\frac{1}{x^2}\left(\frac{d^2 y}{dz^2} - \frac{dy}{dz}\right)\right] = \frac{1}{x^2}\left(\frac{d^3 y}{dz^3}\frac{dz}{dx} - \frac{d^2 y}{dz^2}\frac{dz}{dx}\right) - \frac{2}{x^3}\left(\frac{d^2 y}{dz^2} - \frac{dy}{dz}\right)$$

$$= \frac{1}{x^3}\left(\frac{d^3y}{dz^3} - \frac{d^2y}{dz^2}\right) - \frac{2}{x^3}\left(\frac{d^2y}{dz^2} - \frac{dy}{dz}\right) = \frac{1}{x^3}\left(\frac{d^3y}{dz^3} - 3\frac{d^2y}{dz^2} + 2\frac{dy}{dz}\right)$$

....

....

On putting D for $\frac{d}{dz}$ and clearing the fractions, we get

$$x\frac{dy}{dx} = Dy$$

$$x^2\frac{d^2y}{dx^2} = (D^2 - D)y = D(D-1)y$$

$$x^3\frac{d^3y}{dx^3} = (D^3 - 3D^2 + 2D)y = D(D-1)(D-2)y$$

....

....

In general, $x^n \frac{d^ny}{dx^n} = D(D-1)(D-2) \ldots\ldots (D-(n-1))y.$

$\therefore$ (1) becomes

$$[P_0 D(D-1)(D-2) \ldots\ldots (D-(n-1)) + P_1 D(D-1)(D-2) \ldots\ldots (D-(n-2)) + \ldots\ldots + P_{n-1} D + P_n]y = Z \quad \ldots(2)$$

where Z is the function of z into which Q is changed.

(2) is a linear differential equation with constant coefficients. This equation can be solved. Let the general solution of equation (2) be $y = f(z)$.

$\therefore$ The general solution of the given equation (1) is $y = f(\log x)$.

Example 1. *Solve* $x^3 \frac{d^3y}{dx^3} + 6x^2 \frac{d^2y}{dx^2} + 4x\frac{dy}{dx} - 4y = 0.$ (M.D.U. 2007)

Sol. We have

$$x^3\frac{d^3y}{dx^3} + 6x^2\frac{d^2y}{dx^2} + 4x\frac{dy}{dx} - 4y = 0. \quad \ldots(1)$$

Let $z = \log x. \quad \therefore \quad x = e^z$

$\therefore \quad x\frac{d}{dx} = D,\ x^2\frac{d^2}{dx^2} = D(D-1)$ and $x^3\frac{d^3}{dx^3} = D(D-1)(D-2)$, where $D = \frac{d}{dz}$.

$\therefore$ (1) becomes, $[D(D-1)(D-2) + 6D(D-1) + 4D - 4]y = 0$

i.e., $$(D^3 + 3D^2 - 4)y = 0 \quad \ldots(2)$$

$\therefore$ The A.E. of (2) is $D^3 + 3D^2 - 4 = 0 \Rightarrow (D-1)(D^2+4D+4) = 0$

$\Rightarrow (D-1)(D+2)^2 = 0.$

$\therefore \quad D = 1, -2, -2$

$\therefore$ The general solution of (2) is $y = c_1e^z + (c_2 + c_3z)e^{-2z}$.

$\therefore$ The general solution of (1) is $\mathbf{y = c_1x + (c_2 + c_3 \log x)\, x^{-2}}$. $(\because e^{-2z} = (e^z)^{-2} = x^{-2})$

Example 2. *Solve* $x^2\frac{d^2y}{dx^2} - 2y = x^2 + \frac{1}{x}.$ (K.U. 2006)

Sol. We have $x^2 \dfrac{d^2y}{dx^2} - 2y = x^2 + \dfrac{1}{x}$. ...(1)

Let $z = \log x. \quad \therefore \quad x = e^z$

$\therefore \quad x \dfrac{d}{dx} = D$ and $x^2 \dfrac{d^2}{dx^2} = D(D-1)$, where $D = \dfrac{d}{dz}$.

$\therefore$ (1) $\Rightarrow \quad [D(D-1) - 2]y = e^{2z} + e^{-z}$

$\Rightarrow \quad (D^2 - D - 2)y = e^{2z} + e^{-z}$ ...(2)

$\therefore$ The A.E. of (2) is $D^2 - D - 2 = 0. \quad \therefore \quad D = -1, 2$

$\therefore \quad$ C.F. $= c_1e^{-z} + c_2e^{2z} = c_1x^{-1} + c_2x^2$

$$\text{P.I.} = \frac{1}{D^2 - D - 2}(e^{2z} + e^{-z}) = \frac{1}{D^2 - D - 2} e^{2z} + \frac{1}{D^2 - D - 2} e^{-z}$$

$$= z \frac{1}{2D - 1} e^{2z} + z \frac{1}{2D - 1} e^{-z} \qquad \text{(Case of failure)}$$

$$= z \frac{1}{2(2) - 1} e^{2z} + z \frac{1}{2(-1) - 1} e^{-z} = \frac{z}{3} e^{2z} + \frac{z}{-3} e^{-z}$$

$$= \frac{z}{3}(e^{2z} - e^{-z}) = \frac{(\log x)}{3}(x^2 - x^{-1}) = \frac{1}{3}\left(x^2 - \frac{1}{x}\right) \log x.$$

$\therefore$ The general solution of (1) is $y =$ C.F. + P.I.

$$\therefore \quad \mathbf{y = c_1x^{-1} + c_2x^2 + \frac{1}{3}\left(x^2 - \frac{1}{x}\right) \log x.}$$

Example 3. *Solve* $x^2 \dfrac{d^2y}{dx^2} - x \dfrac{dy}{dx} + y = 2 \log x$. (KU 2001 ; MDU 2001)

Sol. We have

$$x^2 \frac{d^2y}{dx^2} - x \frac{dy}{dx} + y = 2 \log x. \qquad ...(1)$$

Let $z = \log x. \quad \therefore \quad x = e^z$

$\therefore \quad x \dfrac{d}{dx} = D$ and $x^2 \dfrac{d^2}{dx^2} = D(D-1)$, where $D = \dfrac{d}{dz}$.

$\therefore$ (1) $\Rightarrow \quad [D(D-1) - D + 1]y = 2z$ *i.e.,* $(D^2 - 2D + 1)y = 2z$ ...(2)

$\therefore$ The A.E. of (2) is $(D - 1)^2 = 0. \quad \therefore \quad D = 1, 1$

C.F. $= (c_1 + c_2z)e^z = (c_1 + c_2 \log x)\, x$

$$\text{P.I.} = \frac{1}{(D-1)^2} 2z = 2 \frac{1}{(D-1)^2} z = 2(1 - D)^{-2} z = 2(1 + 2D + \ldots\ldots) z$$

$$= 2[z + 2 \,.\, 1] = 2z + 4 = 2 \log x + 4.$$

$\therefore$ The general solution of (1) is $y =$ C.F. + P.I.

$$\therefore \quad \mathbf{y = (c_1 + c_2 \log x)\, x + 2 \log x + 4.}$$

Example 4. *Solve* $x^2 \dfrac{d^2y}{dx^2} - 3x \dfrac{dy}{dx} + 5y = \sin(\log x)$. (MDU 1998 ; DLU 2004)

Sol. We have $x^2 \dfrac{d^2y}{dx^2} - 3x \dfrac{dy}{dx} + 5y = \sin \log x$. ...(1)

Let $$z = \log x. \quad \therefore \quad x = e^z$$

$$\therefore \quad x\frac{d}{dx} = D \quad \text{and} \quad x^2\frac{d^2}{dz^2} = D(D-1), \text{ where } D = \frac{d}{dz}.$$

$\therefore$ (1) becomes $[D(D-1) - 3D + 5]y = \sin z$

$$\Rightarrow \quad (D^2 - 4D + 5)y = \sin z \qquad ...(2)$$

$\therefore$ The A.E. of (2) is $D^2 - 4D + 5 = 0. \quad \therefore \quad D = 2 \pm i$

$$\therefore \quad \text{C.F.} = e^{2z}(c_1 \cos z + c_2 \sin z) = x^2[c_1 \cos(\log x) + c_2 \sin(\log x)]$$

$$\text{P.I.} = \frac{1}{D^2 - 4D + 5}\sin z = \frac{1}{-1 - 4D + 5}\sin z$$

$$= \frac{1}{4(1-D)}\sin z = \frac{1}{4}\frac{(1+D)}{1-D^2}\sin z = \frac{1}{4}\frac{(1+D)}{1-(-1)}\sin z = \frac{1}{8}(1+D)\sin z$$

$$= \frac{1}{8}[\sin z + \cos z] = \frac{1}{8}[\sin(\log x) + \cos(\log x)]$$

$\therefore$ The general solution of (1) is $y = \text{C.F.} + \text{P.I.}$

$$\therefore \quad \mathbf{y = x^2[c_1 \cos(\log x) + c_2 \sin(\log x)] + \frac{1}{8}[\sin(\log x) + \cos(\log x)].}$$

Example 5. *Solve* $x^2\dfrac{d^2y}{dx^2} + 3x\dfrac{dy}{dx} + y = \dfrac{1}{(1-x)^2}$.

Sol. We have $$x^2\frac{d^2y}{dx^2} + 3x\frac{dy}{dx} + y = \frac{1}{(1-x)^2}. \qquad ...(1)$$

Let $$z = \log x. \quad \therefore \quad x = e^z$$

$$\therefore \quad x\frac{d}{dx} = D \quad \text{and} \quad x^2\frac{d^2}{dx^2} = D^2, \text{ where } D = \frac{d}{dz}.$$

$$\therefore \quad (1) \Rightarrow \quad [D(D-1) + 3D + 1]y = \frac{1}{(1-e^z)^2} \quad i.e., \quad (D^2 + 2D + 1)y = \frac{1}{(1-e^z)^2}$$

or $$(D+1)^2 y = (1-e^z)^{-2} \qquad ...(2)$$

$\therefore$ The A.E. of (2) is $(D+1)^2 = 0. \quad \therefore \quad D = -1, -1$

$$\therefore \quad \text{C.F.} = (c_1 + c_2 z)e^{-z} = (c_1 + c_2 \log x)x^{-1}$$

$$\text{P.I.} = \frac{1}{(D+1)^2}(1-e^z)^{-2} = \frac{1}{(D+1)}\frac{1}{(D+1)}\frac{1}{(1-e^z)^2}$$

$$= \frac{1}{(D+1)}e^{-z}\int \frac{1}{(1-e^z)^2}e^z\,dz \qquad \left[\because \frac{1}{D-\alpha}Q = e^{\alpha x}\int Qe^{-\alpha x}\,dx\right]$$

$$= \frac{1}{D+1}e^{-z}\int \frac{1}{(1-t)^2}\,dt, \quad \text{where } e^z = t$$

$$= \frac{1}{D+1}e^{-z}\left[\frac{-(1-t)^{-2+1}}{-2+1}\right] = \frac{1}{D+1}e^{-z}(1-t)^{-1} = \frac{1}{D+1}e^{-z}(1-e^{-z})^{-1}$$

$$= e^{-z}\int e^{-z}(1-e^z)^{-1}\,dz \,.\, e^z\,dz \qquad \left[\because \frac{1}{D-\alpha}Q = e^{\alpha x}\int Qe^{-\alpha x}\,dx\right]$$

$$= e^{-z} \int \frac{1}{1-e^{z}}\, dz = e^{-z} \int \frac{e^{-z}\, dz}{e^{-z}-1} = e^{-z}\, [-\log (e^{-z} - 1)]$$

$$= -\, e^{-z} \log (e^{-z} - 1) = -\, x^{-1} \log (x^{-1} - 1) = -\, \frac{1}{x} \log \left(\frac{1}{x} - 1\right)$$

$\therefore$ The general solution of (1) is y = C.F. + P.I.

$$\therefore \qquad y = (c_1 + c_2 \log x)\, x^{-1} - \frac{1}{x} \log \left(\frac{1}{x} - 1\right)$$

or
$$\mathbf{y = \frac{1}{x}\left[c_1 + c_2 \log x - \log\left(\frac{1}{x} - 1\right)\right].}$$

EXERCISE 1

Solve the following differential equations :

1. $x^2 \dfrac{d^2y}{dx^2} + 9x \dfrac{dy}{dx} + 25y = 5$

2. $x^2 \dfrac{d^2y}{dx^2} - 3x \dfrac{dy}{dx} + 4y = 2x^2$ *(M.D.U. 1992)*

3. $x^2 \dfrac{d^2y}{dx^2} - 2x \dfrac{dy}{dx} - 4y = x^4$

4. $x^2 \dfrac{d^2y}{dx^2} + 5x \dfrac{dy}{dx} + 4y = x^3$

5. $(x^2D^2 + xD - 1)y = x^m$

6. $x^2 \dfrac{d^3y}{dx^3} - 4x \dfrac{d^2y}{dx^2} + 6 \dfrac{dy}{dx} = 4$

7. $x^2 \dfrac{d^2y}{dx^2} + 2x \dfrac{dy}{dx} - 20y = (x + 1)^2$

8. $\dfrac{d^3y}{dx^3} - \dfrac{4}{x} \dfrac{d^2y}{dx^2} + \dfrac{5}{x^2} \dfrac{dy}{dx} - \dfrac{2y}{x^3} = 1$ *(K.U. 2004 S)*

9. $x^3 \dfrac{d^3y}{dx^3} + 2x^2 \dfrac{d^2y}{dx^2} + 2y = 10\left(x + \dfrac{1}{x}\right)$ *(M.D.U. 1997 ; K.U. 1998)*

10. $x^2 \dfrac{d^2y}{dx^2} - x \dfrac{dy}{dx} + 2y = x \log x$ *(M.D.U. 1996, 99)*

11. $x^2 \dfrac{d^2y}{dx^2} - x \dfrac{dy}{dx} - 3y = x^2 \log x$ *(M.D.U. 2006 ; K.U. 2005 S)*

12. $x^2 \dfrac{d^2y}{dx^2} + 8x \dfrac{dy}{dx} + 13y = \log x.$ *(K.U. 1996, 96 C, 2000, 2005, 2007)*

13. $x^2 \dfrac{d^2y}{dx^2} + x \dfrac{dy}{dx} - y = x^2 e^x$

14. $x^2 \dfrac{d^2y}{dx^2} + 4x \dfrac{dy}{dx} + 2y = x + \sin x.$ *(K.U. 1995 C, 97 ; M.D.U. 2000)*

15. $(x^3D^3 + 3x^2D^2 + xD + 8)\, y = 65 \cos (\log x)$

16. $(x^4D^4 + 6x^3D^3 + 9x^2D^2 + 3xD + 1)\, y = (1 + \log x)^2$

17. $(x^2D^2 - xD + 4)y = \cos (\log x) + x \sin (\log x).$

Answers

1. $y = \dfrac{1}{5} + x^{-4}\, [c_1 \cos (\log x^3) + c_2 \sin (\log x^3)]$

2. $y = x^2 (c_1 + c_2 \log x) + x^2 (\log x)^2$

3. $y = c_1x^4 + c_2x^{-1} + \dfrac{1}{5}\, x^4 \log x$

4. $y = (c_1 + c_2 \log x)\, x^{-2} + \dfrac{x^3}{25}$

5. $y = c_1x + c_2x^{-1} + \dfrac{x^m}{m^2 - 1}$

6. $y = c_1 + c_2x^3 + c_3x^4 + \dfrac{2}{3}\, x$

7. $y = c_1x^{-5} + c_2x^4 - \frac{1}{14}x^2 - \frac{1}{9}x - \frac{1}{20}$ **8.** $y = c_1x^2 + x^{5/2}\left[c_2x^{\frac{1}{2}\sqrt{21}} + c_3x^{-\frac{1}{2}\sqrt{21}}\right] - \frac{x^3}{5}$

9. $y = c_1x^{-1} + x\,[c_2 \cos(\log x) + c_3 \sin(\log x)] + 5x + \frac{2}{x}\log x$

10. $y = x[c_1 \cos(\log x) + c_2 \sin(\log x)] + x \log x$

11. $y = c_1x^3 + \frac{c_2}{x} - \frac{x^2}{3}\left(\frac{2}{3} + \log x\right)$

12. $y = x^{-7/2}\left[c_1 \cos\left(\frac{\sqrt{3}}{2}\log x\right) + c_2 \sin\left(\frac{\sqrt{3}}{2}\log x\right)\right] + \frac{1}{169}(13\log x - 7)$

13. $y = c_1x + c_2x^{-1} + \left(1 - \frac{1}{x}\right)e^x$ **14.** $y = \frac{c_1}{x} + \frac{c_2}{x^2} + \frac{x}{6} - \frac{1}{x^2}\sin x$

15. $y = c_1x^{-2} + x[c_2 \cos(\sqrt{3}\log x) + c_3 \sin(\sqrt{3}\log x)] + 8\cos(\log x) - \sin(\log x)$

16. $y = (c_1 + c_2 \log x)\cos(\log x) + (c_3 + c_4 \log x)\sin(\log x) + (\log x)^2 + 2\log x - 3$

17. $y = x\,[c_1 \cos(\sqrt{3}\log x) + c_2 \sin(\sqrt{3}\log x)] + \frac{1}{13}[3\cos(\log x) - 2\sin(\log x)] + \frac{1}{2}x\sin(\log x).$

4. Equations Reducible to Homogeneous Linear Form

Consider the differential equation

$$P_0(a+bx)^n \frac{d^ny}{dx^n} + P_1(a+bx)^{n-1}\frac{d^{n-1}y}{dx^{n-1}} + P_2(a+bx)^{n-2}\frac{d^{n-2}y}{dx^{n-2}} + \ldots\ldots + P_n\, y = Q^* \quad \ldots(1)$$

where $P_0 \neq 0$, P_1, P_2, ... , P_n are all constants and Q is a function of x. This equation can be reduced to a homogeneous linear equation by putting $a + bx$ equal to some other variable.

We solve the equation (1).

Let $z = \log(a + bx)$ *i.e.,* $a + bx = e^z$

Now, $$\frac{dy}{dx} = \frac{dy}{dz}\frac{dz}{dx} = \frac{dy}{dz}\cdot\frac{b}{a+bx} = \frac{b}{a+bx}\frac{dy}{dz}$$

$$\frac{d^2y}{dx^2} = \frac{d}{dx}\left(\frac{b}{a+bx}\frac{dy}{dz}\right) = \frac{b}{a+bx}\frac{d^2y}{dz^2}\frac{dz}{dx} - \frac{b^2}{(a+bx)^2}\frac{dy}{dz}$$

$$= \frac{b^2}{(a+bx)^2}\left(\frac{d^2y}{dz^2} - \frac{dy}{dz}\right)$$

$$\frac{d^3y}{dx^3} = \frac{d}{dx}\left[\frac{b^2}{(a+bx)^2}\left(\frac{d^2y}{dz^2} - \frac{dy}{dz}\right)\right]$$

$$= \frac{b^2}{(a+bx)^2}\left[\frac{d^3y}{dz^3}\frac{dz}{dx} - \frac{d^2y}{dz^2}\frac{dz}{dx}\right] - \frac{2b^3}{(a+bx)^3}\left(\frac{d^2y}{dz^2} - \frac{dy}{dz}\right)$$

$$= \frac{b^3}{(a+bx)^3}\left[\frac{d^3y}{dz^3} - \frac{d^2y}{dz^2} - 2\frac{d^2y}{dz^2} + 2\frac{dy}{dz}\right]$$

*A differential equation of this type is called a **Legendre linear equation.**

$$= \frac{b^3}{(a+bx)^3}\left[\frac{d^3y}{dz^3} - 3\frac{d^2y}{dz^2} + 2\frac{dy}{dz}\right]$$

....

....

On putting D for $\frac{d}{dz}$ and clearing the fractions, we get

$$(a+bx)\frac{dy}{dx} = b\text{D}y$$

$$(a+bx)^2\frac{d^2y}{dx^2} = b^2(\text{D}^2 - \text{D})\,y = b^2\,\text{D}(\text{D}-1)y$$

$$(a+bx)^3\frac{d^3y}{dx^3} = b^3(\text{D}^3 - 3\text{D}^2 + 2\text{D})\,y = b^3\,\text{D}(\text{D}-1)(\text{D}-2)y$$

....

....

In general,

$$(a+bx)^n\frac{d^ny}{dx^n} = b^n\,\text{D}(\text{D}-1)(\text{D}-2)\,......\,(\text{D}-(n-1))y$$

$\therefore$ (1) becomes

$$[\text{P}_0 b^n\,\text{D}(\text{D}-1)(\text{D}-2)\,......\,(\text{D}-(n-1)) + \text{P}_1 b^{n-1}\,\text{D}(\text{D}-1)(\text{D}-2)\,......\,(\text{D}-(n-2)) + ... + \text{P}_{n-1}\,b\text{D} + \text{P}_n]\,y = \text{Z} \qquad ...(2)$$

where Z is the function of z into which Q is changed.

(2) is a linear differential equation with constant coefficients. This equation can be solved. Let the general solution of equation (2) be $y = f(z)$.

$\therefore$ The general solution of the given equation (1) is $y = f(\log(a+bx))$.

Example 1. *Solve* $(x+a)^2\frac{d^2y}{dx^2} - 4(x+a)\frac{dy}{dx} + 6y = x.$ *(MDU 2001)*

Sol. We have $\quad (x+a)^2\frac{d^2y}{dx^2} - 4(x+a)\frac{dy}{dx} + 6y = x. \qquad ...(1)$

Let $\quad z = \log(x+a). \quad \therefore \quad x + a = e^z$

$\therefore \quad (x+a)\frac{d}{dx} = 1.\text{D} = \text{D}, \quad (x+a)^2\frac{d^2}{dx^2} = 1^2.\text{D}(\text{D}-1) = \text{D}(\text{D}-1)$, where $\text{D} = \frac{d}{dz}$.

$\therefore$ (1) $\Rightarrow \quad [\text{D}(\text{D}-1) - 4\text{D} + 6]y = e^z - a$

$\Rightarrow \quad (\text{D}^2 - 5\text{D} + 6)y = e^z - a. \qquad ...(2)$

$\therefore$ The A.E. of (2) is $\quad \text{D}^2 - 5\text{D} + 6 = 0. \quad \therefore \quad \text{D} = 2, 3$

$\therefore \quad$ C.F. $= c_1 e^{2z} + c_2 e^{3z} = c_1(x+a)^2 + c_2(x+a)^3$

$$\text{P.I.} = \frac{1}{\text{D}^2 - 5\text{D} + 6}(e^z - a) = \frac{1}{\text{D}^2 - 5\text{D} + 6}e^z - a\frac{1}{\text{D}^2 - 5\text{D} + 6}e^0$$

$$= \frac{1}{1^2 - 5(1) + 6}e^z - a\frac{1}{0 - 0 + 6}e^{0z} = \frac{e^z}{2} - \frac{a}{6} = \frac{x+a}{2} - \frac{a}{6} = \frac{1}{6}(3x + 2a)$$

$\therefore$ The general solution of (1) is y = C.F. + P.I.

$\therefore$ $$\mathbf{y = c_1(x + a)^2 + c_2(x + a)^3 + \frac{1}{6}\,(3x + 2a).}$$

Example 2. *Solve* $(1 + x)^2 \dfrac{d^2y}{dx^2} + (1 + x)\dfrac{dy}{dx} + y = 4 \cos \log (1 + x).$

Sol. We have

$$(1 + x)^2 \frac{d^2y}{dx^2} + (1 + x)\frac{dy}{dx} + y = 4 \cos \log(1 + x). \qquad ...(1)$$

Let $z = \log (1 + x)$. $\therefore$ $1 + x = e^z$

$\therefore$ $(1 + x)\dfrac{d}{dx} = 1.\text{D} = \text{D}$, $(1 + x)^2 \dfrac{d^2}{dx^2} = 1^2.\text{D}(\text{D} - 1) = \text{D}(\text{D} - 1)$, where $\text{D} = \dfrac{d}{dz}$.

$\therefore$ (1) $\Rightarrow$ $(\text{D}(\text{D} - 1) + \text{D} + 1)\, y = 4 \cos z$ $\Rightarrow$ $(\text{D}^2 + 1)\, y = 4 \cos z$ ...(2)

$\therefore$ The A.E. of (2) is $\text{D}^2 + 1 = 0$. $\therefore$ $\text{D} = \pm i$

$\therefore$ $$\text{C.F.} = e^{oz}\,(c_1 \cos 1 \,.\, z + c_2 \sin 1 \,.\, z) = c_1 \cos z + c_2 \sin z$$
$$= c_1 \cos(\log (1 + x)) + c_2 \sin (\log (1 + x))$$

$$\text{P.I.} = \frac{1}{\text{D}^2 + 1}\, 4 \cos z = 4\,\frac{1}{\text{D}^2 + 1} \cos z = 4z \,.\, \frac{1}{2\text{D}} \cos z$$

$$= 2z \sin z = 2 \log (1 + x) \sin \log (1 + x) \qquad \left(\because \frac{1}{\text{D}} \cos z = \int \cos z\, dz = \sin z\right)$$

The general solution of (1) is y = C.F. + P.I.

$\therefore$ $$\mathbf{y = c_1 \cos \log (1 + x) + c_2 \sin \log (1 + x) + 2 \log (1 + x) \sin \log (1 + x).}$$

EXERCISE 2

Solve the following differential equations :

1. $(1 + 2x)^2 \dfrac{d^2y}{dx^2} - 6(1 + 2x)\dfrac{dy}{dx} + 16y = 8(1 + 2x)^2$ *(M.D.U. 1999 ; K.U. 2003)*

2. $(3x + 2)^2 \dfrac{d^2y}{dx^2} + 3(3x + 2)\dfrac{dy}{dx} - 36y = 3x^2 + 4x + 1$ *(K.U. 1992)*

3. $(5 + 2x)^2 \dfrac{d^2y}{dx^2} - 6(5 + 2x)\dfrac{dy}{dx} + 8y = 0$

4. $(2x - 1)^3 \dfrac{d^3y}{dx^3} + (2x - 1)\dfrac{dy}{dx} - 2y = 0$ *(K.U. 1995 ; M.D.U. 2007)*

5. $(x + 1)^2 \dfrac{d^2y}{dx^2} + (x + 1)\dfrac{dy}{dx} = 4x^2 + 14x + 12.$ *(M.D.U. 1998, 2000)*

Answers

1. $y = (1 + 2x)^2\, [(\log (1 + 2x))^2 + c_1 + c_2 \log (1 + 2x)]$

2. $y = c_1\, (3x + 2)^2 + c_2\, (3x + 2)^{-2} + \dfrac{1}{108}\, [(3x + 2)^2 \log (3x + 2) + 1]$

3. $y = c_1(5 + 2x)^{2+\sqrt{2}} + c_3(2x - 1)^{-\sqrt{3}/2}$

4. $y = (2x - 1)[c_1 + c_2\,(2x - 1)^{\sqrt{3}/2} + c_3\,(2x - 1)^{-\sqrt{3}/2}]$

5. $y = c_1 + c_2 \log(1 + x) + x^2 + 8x + [\log(x + 1)]^2.$

21*

Total Differential Equations

1. Introduction

Till now we have been discussing the methods of solving differential equations involving only two variables. In the present chapter, we shall study differential equations involving more than two variables. A differential equation involving more than two variables is either *ordinary* or *partial*. If only one independent variable is involved then the differential equation is *ordinary* and the differential equation is *partial* if more than one independent variable are involved in the differential equation. We shall be discussing ordinary differential equations containing more than two variables.

2. Total Differential Equation

Let $P_1, P_2, ..., P_n$ be n functions of n variables $x_1, x_2, ..., x_n$. The ordinary differential equation of the form

$$P_1\, dx_1 + P_2\, dx_2 + ... + P_n\, dx_n = 0$$

is called a **total differential equation** in n variables.

In particular, if P, Q and R be functions of three variables x, y and z then

$$P\, dx + Q\, dy + R\, dz = 0$$

is a total differential equation in three variables.

We shall be studying the methods of solving total differential equations involving only three variables.

3. Homogeneous Total Differential Equation

Let $Pdx + Qdy + Rdz = 0$ be a total differential equation in three variables x, y and z. Here each of P, Q and R is a function of x, y and z. This differential equation is called a **homogeneous total differential equation** or simply a **homogeneous equation** if the functions P, Q and R are homogeneous and of the same degree.

4. Integrable Total Differential Equation

Let $$P\, dx + Q\, dy + R\, dz = 0 \qquad ...(1)$$

be a total differential equation in three variables x, y and z. Here each of P, Q, R is a function of x, y and z.

The total differential equation (1) is said to be an **integrable total differential equation** if there exists a function $u(x, y, z)$ of x, y and z such that the equation (1) can be obtained by differentiating the relation $u(x, y, z) = c$, where c is an arbitrary constant.

---

*This chapter is in the syllabus of **M.D.U. Rohtak** and not in the syllabus of **K.U. Kurukshetra.**

In such a case, the relation $u(x, y, z) = c$ is called the **solution** of the total differential equation (1). The solution of the differential equation (1) is also called the **complete integral** of the total differential equation (1).

5. Integrability of a Total Differential Equation

Theorem. *Let $Pdx + Qdy + Rdz = 0$ be a total differential equation, where P, Q and R are functions of x, y and z. This equation is integrable if and only if*

$$P\left(\frac{\partial Q}{\partial z}-\frac{\partial R}{\partial y}\right)+Q\left(\frac{\partial R}{\partial x}-\frac{\partial P}{\partial z}\right)+R\left(\frac{\partial P}{\partial y}-\frac{\partial Q}{\partial x}\right)=0.$$

Proof. The given total differential equation is

$$\mathrm{P}\,dx + \mathrm{Q}\,dy + \mathrm{R}\,dz = 0 \qquad \text{...(1)}$$

Necessity. Let the equation (1) be integrable and its solution be

$$u(x, y, z) = c \qquad \text{...(2)}$$

Differentiating (2), we get

$$\frac{\partial u}{\partial x}\,dx+\frac{\partial u}{\partial y}\,dy+\frac{\partial u}{\partial z}\,dz=0 \qquad \text{...(3)}$$

Since (2) is a solution of (1), we have

$$\frac{\partial u}{\partial x}\,dx+\frac{\partial u}{\partial y}\,dy+\frac{\partial u}{\partial z}\,dz=\lambda(\mathrm{P}\,dx+\mathrm{Q}\,dy+\mathrm{R}\,dz),$$

where λ is some function of x, y and z. In particular, if no multiplication on the right side is required then we have $\lambda(x, y, z) = 1$.

$$\therefore \qquad \frac{\partial u}{\partial x}=\lambda\mathrm{P},\quad \frac{\partial u}{\partial y}=\lambda\mathrm{Q},\quad \frac{\partial u}{\partial z}=\lambda\mathrm{R}$$

Assuming $\dfrac{\partial^2 u}{\partial y\partial x}=\dfrac{\partial^2 u}{\partial x\partial y}$, we have $\dfrac{\partial}{\partial y}\left(\dfrac{\partial u}{\partial x}\right)=\dfrac{\partial}{\partial x}\left(\dfrac{\partial u}{\partial y}\right)$.

$$\Rightarrow \qquad \frac{\partial}{\partial y}(\lambda\mathrm{P})=\frac{\partial}{\partial x}(\lambda\mathrm{Q})$$

$$\Rightarrow \qquad \lambda\frac{\partial \mathrm{P}}{\partial y}+\mathrm{P}\frac{\partial \lambda}{\partial y}=\lambda\frac{\partial \mathrm{Q}}{\partial x}+\mathrm{Q}\frac{\partial \lambda}{\partial x}$$

$$\Rightarrow \qquad \lambda\left(\frac{\partial \mathrm{P}}{\partial y}-\frac{\partial \mathrm{Q}}{\partial x}\right)=\mathrm{Q}\frac{\partial \lambda}{\partial x}-\mathrm{P}\frac{\partial \lambda}{\partial y} \qquad \text{...(4)}$$

Similarly, $\dfrac{\partial^2 u}{\partial z\partial y}=\dfrac{\partial^2 u}{\partial y\partial z}$ implies

$$\lambda\left(\frac{\partial \mathrm{Q}}{\partial z}-\frac{\partial \mathrm{R}}{\partial y}\right)=\mathrm{R}\frac{\partial \lambda}{\partial y}-\mathrm{Q}\frac{\partial \lambda}{\partial z} \qquad \text{...(5)}$$

and $\dfrac{\partial^2 u}{\partial x\partial z}=\dfrac{\partial^2 u}{\partial z\partial x}$ implies

$$\lambda\left(\frac{\partial \mathrm{R}}{\partial x}-\frac{\partial \mathrm{P}}{\partial z}\right)=\mathrm{P}\frac{\partial \lambda}{\partial z}-\mathrm{R}\frac{\partial \lambda}{\partial x} \qquad \text{...(6)}$$

Multiplying (4), (5) and (6) by R, P and Q respectively and adding, we get

$$\lambda R\left(\frac{\partial P}{\partial y}-\frac{\partial Q}{\partial x}\right)+\lambda P\left(\frac{\partial Q}{\partial z}-\frac{\partial R}{\partial y}\right)+\lambda Q\left(\frac{\partial R}{\partial x}-\frac{\partial P}{\partial z}\right)=0$$

$$\Rightarrow \quad P\left(\frac{\partial Q}{\partial z}-\frac{\partial R}{\partial y}\right)+Q\left(\frac{\partial R}{\partial x}-\frac{\partial P}{\partial z}\right)+R\left(\frac{\partial P}{\partial y}-\frac{\partial Q}{\partial x}\right)=0 \qquad (\because \quad \lambda \neq 0)$$

Sufficiency. Let

$$P\left(\frac{\partial Q}{\partial z}-\frac{\partial R}{\partial y}\right)+Q\left(\frac{\partial R}{\partial x}-\frac{\partial P}{\partial z}\right)+R\left(\frac{\partial P}{\partial y}-\frac{\partial Q}{\partial x}\right)=0 \qquad \text{...(7)}$$

Let μ be any function of x, y and z.

Let $\quad P_1 = \mu P,\ Q_1 = \mu Q,\ R_1 = \mu R$

We shall show that (1) also hold if P, Q and R are replaced by P_1, Q_1 and R_1 respectively.

$$\frac{\partial Q_1}{\partial z}-\frac{\partial R_1}{\partial y}=\frac{\partial}{\partial z}(\mu Q)-\frac{\partial}{\partial y}(\mu R)$$

$$=\left(\mu\frac{\partial Q}{\partial z}+Q\frac{\partial \mu}{\partial z}\right)-\left(\mu\frac{\partial R}{\partial y}+R\frac{\partial \mu}{\partial y}\right)$$

$$=\mu\left(\frac{\partial Q}{\partial z}-\frac{\partial R}{\partial y}\right)+\left(Q\frac{\partial \mu}{\partial z}-R\frac{\partial \mu}{\partial y}\right)$$

Similarly, $\quad \dfrac{\partial R_1}{\partial x}-\dfrac{\partial P_1}{\partial z}=\mu\left(\dfrac{\partial R}{\partial x}-\dfrac{\partial P}{\partial z}\right)+\left(R\dfrac{\partial \mu}{\partial x}-P\dfrac{\partial \mu}{\partial z}\right)$

and $\quad \dfrac{\partial P_1}{\partial y}-\dfrac{\partial Q_1}{\partial x}=\mu\left(\dfrac{\partial P}{\partial y}-\dfrac{\partial Q}{\partial x}\right)+\left(P\dfrac{\partial \mu}{\partial y}-Q\dfrac{\partial \mu}{\partial x}\right)$

Multiplying these relations by P_1, Q_1 and R_1 and adding, we get

$$P_1\left(\frac{\partial Q_1}{\partial z}-\frac{\partial R_1}{\partial y}\right)+Q_1\left(\frac{\partial R_1}{\partial x}-\frac{\partial P_1}{\partial z}\right)+R_1\left(\frac{\partial P_1}{\partial y}-\frac{\partial Q_1}{\partial x}\right)$$

$$=\mu^2\left[P\left(\frac{\partial Q}{\partial z}-\frac{\partial R}{\partial y}\right)+Q\left(\frac{\partial R}{\partial x}-\frac{\partial P}{\partial z}\right)+R\left(\frac{\partial P}{\partial y}-\frac{\partial Q}{\partial x}\right)\right]$$

$$+\mu\left[P\left(Q\frac{\partial \mu}{\partial z}-R\frac{\partial \mu}{\partial y}\right)+Q\left(R\frac{\partial \mu}{\partial x}-P\frac{\partial \mu}{\partial z}\right)+R\left(P\frac{\partial \mu}{\partial y}-Q\frac{\partial \mu}{\partial x}\right)\right]=\mu^2(0)+\mu(0)=0$$

$$\therefore \quad P_1\left(\frac{\partial Q_1}{\partial z}-\frac{\partial R_1}{\partial x}\right)+Q_1\left(\frac{\partial R_1}{\partial x}-\frac{\partial P_1}{\partial z}\right)+R_1\left(\frac{\partial P_1}{\partial y}-\frac{\partial Q_1}{\partial x}\right)=0$$

Thus if relation (7) holds for the equation $P dx + Q dy + R dz = 0$ then a similar relation holds for the equation

$$\mu P\, dx + \mu Q\, dy + \mu R\, dz = 0 \qquad \text{...(8)}$$

where μ is any function of x, y and z.

Now $P\, dx + Q\, dy$ may or may not be an exact differential with respect to x and y. If $P\, dx + Q\, dy$ is not an exact differential with respect to x and y, an integrating factor μ can be found for it and equation (8) can then be taken as the equation to be considered. Hence there is no loss of generality in assuming $Pdx + Qdy$ as an axact differential with respect to x and y.

$\therefore$ We have $$\frac{\partial P}{\partial y} = \frac{\partial Q}{\partial x} \quad \text{and} \quad \int (P\,dx + Q\,dy) = V, \text{ say}$$

$$\therefore \quad P\,dx + Q\,dy = \frac{\partial V}{\partial x}dx + \frac{\partial V}{\partial y}dy$$

$$\Rightarrow \quad P = \frac{\partial V}{\partial x} \quad \text{and} \quad Q = \frac{\partial V}{\partial y}$$

$$\therefore \quad (7) \Rightarrow \frac{\partial V}{\partial x}\left(\frac{\partial}{\partial z}\left(\frac{\partial V}{\partial y}\right) - \frac{\partial R}{\partial y}\right) + \frac{\partial V}{\partial y}\left(\frac{\partial R}{\partial x} - \frac{\partial}{\partial z}\left(\frac{\partial V}{\partial x}\right)\right)$$
$$+ R\left(\frac{\partial}{\partial y}\left(\frac{\partial V}{\partial x}\right) - \frac{\partial}{\partial x}\left(\frac{\partial V}{\partial y}\right)\right) = 0$$

$$\Rightarrow \quad \frac{\partial V}{\partial x}\left(\frac{\partial^2 V}{\partial z \partial y} - \frac{\partial R}{\partial y}\right) + \frac{\partial V}{\partial y}\left(\frac{\partial R}{\partial x} - \frac{\partial^2 V}{\partial z \partial x}\right) + R(0) = 0$$

$$\Rightarrow \quad \frac{\partial V}{\partial x}\left(\frac{\partial}{\partial y}\left(\frac{\partial V}{\partial z} - R\right)\right) - \frac{\partial V}{\partial y}\left(\frac{\partial}{\partial x}\left(\frac{\partial V}{\partial z} - R\right)\right) = 0$$

$$\Rightarrow \quad \begin{vmatrix} \dfrac{\partial V}{\partial x} & \dfrac{\partial}{\partial x}\left(\dfrac{\partial V}{\partial z} - R\right) \\ \dfrac{\partial V}{\partial y} & \dfrac{\partial}{\partial y}\left(\dfrac{\partial V}{\partial z} - R\right) \end{vmatrix} = 0$$

$\therefore$ *There exists a relation between V and $\frac{\partial V}{\partial z} - R$ which is independent of x and y.

$\therefore$ $\frac{\partial V}{\partial z} - R$ can be expressed as a function of z and V alone.

Let $$\frac{\partial V}{\partial z} - R = \phi(z, V)$$

$\therefore$ The given equation $P\,dx + Q\,dy + R\,dz = 0$ reduces to

$$\frac{\partial V}{\partial x}dx + \frac{\partial V}{\partial y}dy + Rdz + \left(\frac{\partial V}{\partial z}dz - \frac{\partial V}{\partial z}dz\right) = 0$$

or $$\left(\frac{\partial V}{\partial x}dx + \frac{\partial V}{\partial y}dy + \frac{\partial V}{\partial z}dz\right) + \left(R - \frac{\partial V}{\partial z}\right)dz = 0$$

or $$dV - \phi(z, V)\,dz = 0 \qquad ...(9)$$

This is an equation in two variables and hence solvable. Let the solution of (9) be $f(z, V) = 0$. Putting the value of V in $f(z, V) = 0$, we get the solution of the given total differential equation $P\,dx + Q\,dy + R\,dz = 0$.

---

* This is a standard result.

$\therefore$ **The total differential equation Pdx + Qdy + Rdz = 0 is integrable i.e., solvable if and only if**

$$\mathbf{P}\left(\frac{\partial \mathbf{Q}}{\partial \mathbf{z}}-\frac{\partial \mathbf{R}}{\partial \mathbf{y}}\right)+\mathbf{Q}\left(\frac{\partial \mathbf{R}}{\partial \mathbf{x}}-\frac{\partial \mathbf{P}}{\partial \mathbf{z}}\right)+\mathbf{R}\left(\frac{\partial \mathbf{P}}{\partial \mathbf{y}}-\frac{\partial \mathbf{Q}}{\partial \mathbf{x}}\right)=\mathbf{0}.$$

Remark. The necessary and sufficient condition for the equation $P\,dx + Q\,dy + R\,dz = 0$ to be integrable can be easily remembered because P, Q, R, x, y, z appear in it in a regular cyclical order.

6. Exact Total Differential Equation

Let $$P\,dx + Q\,dy + R\,dz = 0 \qquad \text{...(1)}$$

be an exact total differential equation in three variables x, y and z.

Since (1) is exact, the expression $P\,dx + Q\,dy + R\,dz$ is equal to the differential of a function say, $u(x, y, z)$ of x, y and z.

$$\therefore \quad P\,dx + Q\,dy + R\,dz = \frac{\partial u}{\partial x}dx + \frac{\partial u}{\partial y}dy + \frac{\partial u}{\partial z}dz$$

$$\Rightarrow \quad P = \frac{\partial u}{\partial x},\ Q = \frac{\partial u}{\partial y},\ R = \frac{\partial u}{\partial z}$$

$$\therefore \quad \frac{\partial P}{\partial y} = \frac{\partial}{\partial y}\left(\frac{\partial u}{\partial x}\right) = \frac{\partial^2 u}{\partial y \partial x} = \frac{\partial^2 u}{\partial x \partial y} = \frac{\partial}{\partial x}\left(\frac{\partial u}{\partial y}\right) = \frac{\partial Q}{\partial x}$$

$$\frac{\partial Q}{\partial z} = \frac{\partial}{\partial z}\left(\frac{\partial u}{\partial y}\right) = \frac{\partial^2 u}{\partial z \partial y} = \frac{\partial^2 u}{\partial y \partial z} = \frac{\partial}{\partial y}\left(\frac{\partial u}{\partial z}\right) = \frac{\partial R}{\partial y}$$

and $$\frac{\partial R}{\partial x} = \frac{\partial}{\partial x}\left(\frac{\partial u}{\partial z}\right) = \frac{\partial^2 u}{\partial x \partial z} = \frac{\partial^2 u}{\partial z \partial x} = \frac{\partial}{\partial z}\left(\frac{\partial u}{\partial x}\right) = \frac{\partial P}{\partial z}$$

$$\therefore \quad \frac{\partial P}{\partial y} = \frac{\partial Q}{\partial x},\ \frac{\partial Q}{\partial z} = \frac{\partial R}{\partial y} \text{ and } \frac{\partial R}{\partial x} = \frac{\partial P}{\partial z}.$$

$\therefore$ **If the equation P dx + Q dy + R dz = 0 is exact then we have**

$$\frac{\partial \mathbf{P}}{\partial \mathbf{y}} = \frac{\partial \mathbf{Q}}{\partial \mathbf{x}},\ \frac{\partial \mathbf{Q}}{\partial \mathbf{z}} = \frac{\partial \mathbf{R}}{\partial \mathbf{y}} \textbf{ and } \frac{\partial \mathbf{R}}{\partial \mathbf{x}} = \frac{\partial \mathbf{P}}{\partial \mathbf{z}}.$$

Remark. If the equation $Pdx + Qdy + Rdz = 0$ is exact then

$$P\left(\frac{\partial Q}{\partial z}-\frac{\partial R}{\partial y}\right)+Q\left(\frac{\partial R}{\partial x}-\frac{\partial P}{\partial z}\right)+R\left(\frac{\partial P}{\partial y}-\frac{\partial Q}{\partial x}\right)$$

$= P(0) + Q(0) + R(0) = 0$ *i.e.*, the condition of integrability is trivially satisfied.

7. Solution of Total Differential Equation

Let $$P\,dx + Q\,dy + R\,dz = 0 \qquad \text{...(1)}$$

be a total differential equation in three variables x, y and z. The equation (1) is solvable only when the functions P, Q and R satisfy the relation

$$P\left(\frac{\partial Q}{\partial z}-\frac{\partial R}{\partial y}\right)+Q\left(\frac{\partial R}{\partial x}-\frac{\partial P}{\partial z}\right)+R\left(\frac{\partial P}{\partial y}-\frac{\partial Q}{\partial x}\right)=0$$

An equation of the form (1) and not satisfying above relation cannot be solved. There are number of methods of solving integrable total differential equations in there variables. Some of the methods depend upon the exactness of the given equation and some on the homogeneity of the functions P, Q and R.

8. (Method I) Solution of Exact Total Differential Equation

Let $P\,dx + Q\,dy + R\,dz = 0$ be an exact total differential equation.

$$\therefore \quad \frac{\partial P}{\partial y} = \frac{\partial Q}{\partial x}, \quad \frac{\partial Q}{\partial z} = \frac{\partial R}{\partial y}, \quad \frac{\partial R}{\partial x} = \frac{\partial P}{\partial z}$$

Since the given equation is exact, we would be able to solve this equation after some desired regrouping of terms.

The following is the list of some important exact differentials. These formulae are used to regroup the terms of the given equation. In the given formulae, we can replace x by y (or z) and so on as desired.

(*i*) $x\,dy + y\,dx = d(xy)$

(*ii*) $xy\,dz + xz\,dy + yz\,dx = d(xyz)$

(*iii*) $y^2dx + 2xy\,dy = d(xy^2)$

(*iv*) $3x^2y\,dx + x^3\,dy = d(x^3y)$

(*v*) $\dfrac{x\,dy - y\,dx}{x^2} = d\left(\dfrac{y}{x}\right)$

(*vi*) $\dfrac{y\,dx - x\,dy}{y^2} = d\left(\dfrac{x}{y}\right)$

(*vii*) $-\dfrac{x\,dy + y\,dx}{x^2y^2} = d\left(\dfrac{1}{xy}\right)$

(*viii*) $\dfrac{x\,dy - y\,dx}{x^2 + y^2} = d\left(\tan^{-1}\dfrac{y}{x}\right)$

(*ix*) $\dfrac{x\,dx + y\,dy}{x^2 + y^2} = d\left(\dfrac{1}{2}\log(x^2 + y^2)\right)$

(*x*) $\dfrac{d(f(x, y, z))}{f(x, y, z)} = d(\log(f(x, y, z)))$

(*xi*) $\dfrac{d(f(x, y, z))}{(f(x, y, z))^n} = d\left(\dfrac{(f(x, y, z))^{n+1}}{n+1}\right), \quad n \neq -1$

(*xii*) $\dfrac{ye^x\,dx - e^x\,dy}{y^2} = d\left(\dfrac{e^x}{y}\right)$.

Example 1. *Solve the following total differential equations :*

(*i*) $(x - y)dx - x\,dy + z\,dz = 0$

(*ii*) $(yz + 2x)dx + (zx - 2z)dy + (xy - 2y)dz = 0$

(*iii*) $(y - z)(y + z - 2x)dx + (z - x)(z + x - 2y)dy + (x - y)(x + y - 2z)dz = 0$

(*iv*) $(3x^2y^2 - e^xz)dx + (2x^3y + \sin z)dy + (y\cos z - e^x)dz = 0.$

Sol. (*i*) We have $(x - y)dx - x\,dy + z\,dz = 0$ ...(1)

Here $P = x - y, \quad Q = -x, \quad R = z.$

$$\therefore \quad \frac{\partial P}{\partial y} = -1, \quad \frac{\partial P}{\partial z} = 0, \quad \frac{\partial Q}{\partial x} = -1, \quad \frac{\partial Q}{\partial z} = 0, \quad \frac{\partial R}{\partial x} = 0, \quad \frac{\partial R}{\partial y} = 0$$

$$\therefore \quad \frac{\partial P}{\partial y} = \frac{\partial Q}{\partial x}, \quad \frac{\partial Q}{\partial z} = \frac{\partial R}{\partial y}, \quad \frac{\partial R}{\partial x} = \frac{\partial P}{\partial z}$$

$\therefore$ The equation (1) is exact.

On regrouping the terms,

(1) $\Rightarrow$ $x\,dx - (y\,dx + x\,dy) + z\,dz = 0$

$$\Rightarrow \quad d\left(\frac{x^2}{2}\right) - d(xy) + d\left(\frac{z^2}{2}\right) = 0 \Rightarrow d\left(\frac{x^2}{2} - xy + \frac{z^2}{2}\right) = 0$$

Integrating, we get

$$\frac{x^2}{2} - xy + \frac{z^2}{2} = k \Rightarrow \mathbf{x^2 - 2y + z^2 = c.} \qquad \text{(Putting } c = 2k\text{)}$$

This is the required solution.

(*ii*) We have $(yz + 2x)dx + (zx - 2z)dy + (xy - 2y)dz = 0$ ...(1)

Here $P = yz + 2x,\quad Q = zx - 2z,\quad R = xy - 2y$

$$\therefore \quad \frac{\partial P}{\partial y} = z,\ \frac{\partial P}{\partial z} = y,\ \frac{\partial Q}{\partial x} = z,\ \frac{\partial Q}{\partial z} = x - 2,\ \frac{\partial R}{\partial x} = y,\ \frac{\partial R}{\partial y} = x - 2$$

$$\therefore \quad \frac{\partial P}{\partial y} = \frac{\partial Q}{\partial x},\ \frac{\partial Q}{\partial z} = \frac{\partial R}{\partial y},\ \frac{\partial R}{\partial x} = \frac{\partial P}{\partial z}$$

$\therefore$ The equation (1) is exact.

On regrouping the terms,

(1) $\Rightarrow$ $(yz\,dx + zx\,dy + xy\,dz) + 2x\,dx - 2(z\,dy + y\,dz) = 0$

$\Rightarrow$ $d(xyz) + d(x^2) - 2d(yz) = 0$

$\Rightarrow$ $d(xyz + x^2 - 2yz) = 0$

Integrating, we get $\mathbf{xyz + x^2 - 2yz = c.}$

This is the required solution.

(*iii*) We have

$$(y - z)(y + z - 2x)dx + (z - x)(z + x - 2y)dy + (x - y)(x + y - 2z)dz = 0 \qquad ...(1)$$

Here
$$P = (y - z)(y + z - 2x) = y^2 - z^2 - 2x(y - z)$$
$$Q = (z - x)(z + x - 2y) = z^2 - x^2 - 2y(z - x)$$
$$R = (x - y)(x + y - 2z) = x^2 - y^2 - 2z(x - y)$$

$$\therefore \quad \frac{\partial P}{\partial y} = 2y - 2x,\quad \frac{\partial P}{\partial z} = -2z + 2x,$$

$$\frac{\partial Q}{\partial x} = -2x + 2y,\quad \frac{\partial Q}{\partial z} = 2z - 2y,$$

$$\frac{\partial R}{\partial x} = 2x - 2z,\quad \frac{\partial R}{\partial y} = -2y + 2z$$

$$\therefore \quad \frac{\partial P}{\partial y} = \frac{\partial Q}{\partial x},\ \frac{\partial Q}{\partial z} = \frac{\partial R}{\partial y},\ \frac{\partial R}{\partial x} = \frac{\partial P}{\partial z}$$

$\therefore$ The equation (1) is exact.

On regrouping the terms,

(1) $\Rightarrow$ $(y^2 - z^2 - 2xy + 2xz)\,dx + (z^2 - x^2 - 2yz + 2xy)\,dy + (x^2 - y^2 - 2zx + 2yz)\,dz = 0$

$\Rightarrow$ $(y^2\,dx + 2xy\,dy) - (z^2\,dx + 2zx\,dz) - (2xy\,dx + x^2\,dy) + (2xz\,dx + x^2\,dz)$
$+ (z^2\,dy + 2yz\,dz) - (2yz\,dy + y^2\,dz) = 0$

$\Rightarrow \qquad d(y^2x) - d(z^2x) - d(x^2y) + d(x^2z) + d(z^2y) - d(y^2z) = 0$

$\Rightarrow \qquad d(y^2x - z^2x - x^2y + x^2z + z^2y - y^2z) = 0$

Integrating, we get $\qquad y^2x - z^2x - x^2y + x^2z + z^2y - y^2z = c$

$\Rightarrow \qquad \mathbf{x(y^2 - z^2) + y(z^2 - x^2) + z(x^2 - y^2) = c.}$

This is the required solution.

(*iv*) We have

$$(3x^2y^2 - e^x z)dx + (2x^3y + \sin z)dy + (y \cos z - e^x)dz = 0 \qquad \ldots(1)$$

Here $P = 3x^2y^2 - e^xz$, $Q = 2x^3y - \sin z$ and $R = y \cos z - e^x$

$$\therefore \quad \frac{\partial P}{\partial y} = 6x^2y, \quad \frac{\partial P}{\partial z} = -e^x, \quad \frac{\partial Q}{\partial x} = 6x^2y, \quad \frac{\partial Q}{\partial z} = \cos z,$$

$$\frac{\partial R}{\partial x} = -e^x, \quad \frac{\partial R}{\partial y} = \cos z$$

$$\therefore \quad \frac{\partial P}{\partial y} = \frac{\partial Q}{\partial x}, \quad \frac{\partial Q}{\partial z} = \frac{\partial R}{\partial y}, \quad \frac{\partial R}{\partial x} = \frac{\partial P}{\partial z}$$

$\therefore$ The equation (1) is exact. On regrouping the terms

(1) $\Rightarrow \quad (3x^2y^2\, dx + 2x^3\, y\, dy) - (e^x\, z\, dx + e^x\, dz) + (\sin z\, dy + y \cos z\, dz) = 0$

$\Rightarrow \qquad d(x^3y^2) - d(e^x z) + d(y \sin z) = 0$

$\Rightarrow \qquad d(x^3y^2 - e^x z + y \sin z) = 0$

Integrating, we get $\qquad \mathbf{x^3y^2 - e^x z + y \sin z = c.}$

This is the required solution.

EXERCISE 1

Solve the following total differential equations :

1. $x\, dx + 6y^2\, dy + z^3\, dz = 0$
2. $\dfrac{1}{x}\, dx + y\, dz + z\, dy = 0$
3. $(y + 3z)dx + (x + 2z)dy + (3x + 2y)dz = 0$
4. $(2x + y^2 + 2xz)dx + 2xy\, dy + x^2\, dz = 0$
5. $(yz + 2x)dx + (zx - 2z)dy + (xy - 2y)dz = 0$
6. $(2x + y^2 + 2xz)dx + 2xy\, dy + x^2\, dz = 0$
7. $\dfrac{y}{x^2}\, dx - \dfrac{1}{x}\, dy + 2z\, dz = 0$
8. $(\cos x + e^x\, y)dx + (e^x + e^yz)dy + e^y\, dz = 0.$

Answers

1. $2x^2 + 8y^3 + z^4 = c$
2. $\log x + yz = c$
3. $xy + 2yz + 3xz = c$
4. $x^2 + xy^2 + x^2z = c$
5. $xyz + x^2 - 2yz = c$
6. $x^2 + xy^2 + zx^2 = c$
7. $\dfrac{y}{x} - z^2 = c$
8. $e^xy + e^yz + \sin x = c.$

---

9. (Method II) Solution of Exact and Homogeneous Total Differential Equation of Degree ($n \neq -1$)

Let $P\, dx + Q\, dy + R\, dz = 0$ be an exact and homogeneous total differential equation of degree n ($\neq -1$).

$$\therefore \qquad \frac{\partial P}{\partial y} = \frac{\partial Q}{\partial x}, \quad \frac{\partial Q}{\partial z} = \frac{\partial R}{\partial y}, \quad \frac{\partial R}{\partial x} = \frac{\partial P}{\partial z} \qquad ...(1)$$

Since the given equation is homogeneous, so by Euler's theorem, we have

$$x\frac{\partial P}{\partial x} + y\frac{\partial P}{\partial y} + z\frac{\partial P}{\partial z} = nP \qquad ...(2)$$

$$x\frac{\partial Q}{\partial x} + y\frac{\partial Q}{\partial y} + z\frac{\partial Q}{\partial z} = nQ \qquad ...(3)$$

$$x\frac{\partial R}{\partial x} + y\frac{\partial R}{\partial y} + z\frac{\partial R}{\partial z} = nR \qquad ...(4)$$

Let $\qquad xP + yQ + zR = c \qquad ...(5)$

Differentiating **(5)**, we get

$$\left(1.P + x\frac{\partial P}{\partial x} + y\frac{\partial Q}{\partial x} + z\frac{\partial R}{\partial x}\right)dx + \left(x\frac{\partial P}{\partial y} + 1.Q + y\frac{\partial Q}{\partial y} + z\frac{\partial R}{\partial y}\right)dy$$
$$+ \left(x\frac{\partial P}{\partial z} + y\frac{\partial Q}{\partial z} + 1.R + z\frac{\partial R}{\partial z}\right)dz = 0$$

$$\Rightarrow \left(P + x\frac{\partial P}{\partial x} + y\frac{\partial P}{\partial y} + z\frac{\partial P}{\partial z}\right)dx + \left(Q + x\frac{\partial Q}{\partial x} + y\frac{\partial Q}{\partial y} + z\frac{\partial Q}{\partial z}\right)dy$$
$$+ \left(R + x\frac{\partial R}{\partial x} + y\frac{\partial R}{\partial y} + z\frac{\partial R}{\partial z}\right)dz = 0$$

[By using (1)]

$$\Rightarrow \qquad (P + nP)\,dx + (Q + nQ)\,dy + (R + nR)\,dz = 0$$

[By using (2), (3) and (4)]

$$\Rightarrow \qquad (n + 1)(P\,dx + Q\,dy + R\,dz) = 0$$

$$\Rightarrow \qquad P\,dx + Q\,dy + R\,dz = 0 \qquad (\because \quad n \neq -1)$$

$\therefore$ $\qquad xP + yQ + zR = c$ is a solution of the given equation.

Example 1. *Solve the following total differential equations :*

(i) $(4y + 2z)dx + (4x - z)dy + (2x - y)dz = 0$

(ii) $(y^2 + z^2 + 2xy + 2xz)dx + (x^2 + z^2 + 2xy + 2yz)dy + (x^2 + y^2 + 2xz + 2yz)dz = 0.$

Sol. *(i)* We have $\quad (4y + 2z)dx + (4x - z)dy + (2x - y)dz = 0 \qquad ...(1)$

Here $\quad P = 4y + 2z, \quad Q = 4x - z, \quad R = 2x - y$

$\therefore$ Equation (1) is a homogeneous equation of degree $1(\neq -1)$.

Also
$$\frac{\partial P}{\partial y} = 4, \quad \frac{\partial P}{\partial z} = 2, \quad \frac{\partial Q}{\partial x} = 4, \quad \frac{\partial Q}{\partial z} = -1, \quad \frac{\partial R}{\partial x} = 2, \quad \frac{\partial R}{\partial y} = -1$$

$$\therefore \qquad \frac{\partial P}{\partial y} = \frac{\partial Q}{\partial x}, \quad \frac{\partial Q}{\partial z} = \frac{\partial R}{\partial y}, \quad \frac{\partial R}{\partial x} = \frac{\partial P}{\partial z}$$

$\therefore$ The equation (1) is also exact.

$\therefore$ The solution of (1) is

$$xP + yQ + zR = c$$

i.e., $\qquad x(4y + 2z) + y(4x - z) + z(2x - y) = c$

or $\qquad 8xy - 2yz + 4zx = c$

or $$\mathbf{4xy - yz + 2zx = k.}$$ (Putting $k = c/2$)

This is the required solution.

(*ii*) We have $(y^2 + z^2 + 2xy + 2xz)dx$

$$+ (x^2 + z^2 + 2xy + 2yz)dy + (x^2 + y^2 + 2xz + 2yz)dz = 0 \quad ...(1)$$

Here $\quad P = y^2 + z^2 + 2xy + 2xz$

$Q = x^2 + z^2 + 2xy + 2yz$

$R = x^2 + y^2 + 2xz + 2yz$

$\therefore$ Equation (1) is a homogeneous equation of degree $2(\neq -1)$.

Also $$\frac{\partial P}{\partial y} = 2y + 2x, \quad \frac{\partial P}{\partial z} = 2z + 2x, \quad \frac{\partial Q}{\partial x} = 2x + 2y, \quad \frac{\partial Q}{\partial z} = 2z + 2y,$$

$$\frac{\partial R}{\partial x} = 2x + 2z, \quad \frac{\partial R}{\partial y} = 2y + 2z$$

$$\therefore \quad \frac{\partial P}{\partial y} = \frac{\partial Q}{\partial x}, \quad \frac{\partial Q}{\partial z} = \frac{\partial R}{\partial y}, \quad \frac{\partial R}{\partial x} = \frac{\partial P}{\partial z}$$

$\therefore$ The equation (1) is also exact.

$\therefore$ The solution of (1) is $xP + yQ + zR = c$

i.e., $$\mathbf{x(y^2 + z^2 + 2xy + 2xz) + y(x^2 + z^2 + 2xy + 2yz) + z(x^2 + y^2 + 2xz + 2yz) = c.}$$

This is the required solution.

EXERCISE 2

Solve the following total differential equations :

1. $(x - y)dx - x\,dy + z\,dz = 0$
2. $(y + 3z)dx + (x + 2z)dy + (3x + 2y)dz = 0$
3. $(2y - z)dx + 2(x - z)dy - (x + 2y)dz = 0$
4. $(y - 7z)dx + (x - 3z)dy - (7x + 3y)dz = 0$
5. $(x - 3y - z)dx + (2y - 3x)dy + (z - x)dz = 0$
6. $(y - z)(y + z - 2x)dx + (z - x)(z + x - 2y)dy + (x - y)(x + y - 2z)dz = 0.$

Answers

1. $x^2 - 2xy + z^2 = c$
2. $xy + 2yz + 3xz = c$
3. $2xy - xz - 2yz = c$
4. $xy - 3yz - 7xz = c$
5. $x^2 + 2y^2 + z^2 - 6xy - 2xz = c$
6. $x(y^2 - z^2) + y(z^2 - x^2) + z(x^2 - y^2) = c.$

10. (Method III) Solution of Integrable Total Differentiable Equation by Finding Integrating Factor by Inspection

Let $Pdx + Qdy + Rdz = 0$ be an integrable total differential equation. If the given equation is not exact, we find an integrating factor $\lambda(x, y, z)$ by inspection. By multiplying the given equation by $\lambda(x, y, z)$, we get an exact differential equation and this equation is solved by using the method discussed earlier.

Example 1. *Solve the following total differential equations :*

(*i*) $y^2\,dx - z\,dy + y\,dz = 0$

(*ii*) $yz \log z\,dx - zx \log z\,dy + xy\,dz = 0$

(*iii*) $(2x^3y + 1)dx + x^4\,dy + x^2 \tan z\,dz = 0$

(*iv*) $(2x^2 + 2xy + 2xz^2 + 1)dx + dy + 2z\,dz = 0.$

Sol. (*i*) We have $y^2\,dx - z\,dy + y\,dz = 0$ ...(1)

Here $P = y^2$, $Q = -z$, $R = y$

$$\therefore \quad \frac{\partial P}{\partial y} = 2y, \quad \frac{\partial P}{\partial z} = 0, \quad \frac{\partial Q}{\partial x} = 0, \quad \frac{\partial Q}{\partial z} = -1, \quad \frac{\partial R}{\partial x} = 0, \quad \frac{\partial R}{\partial y} = 1$$

$$\frac{\partial P}{\partial y} \neq \frac{\partial Q}{\partial x} \Rightarrow \text{Equation (1) is not exact.}$$

Now $$P\left(\frac{\partial Q}{\partial z} - \frac{\partial R}{\partial y}\right) + Q\left(\frac{\partial R}{\partial x} - \frac{\partial P}{\partial z}\right) + R\left(\frac{\partial P}{\partial y} - \frac{\partial Q}{\partial x}\right)$$

$$= y^2(-1-1) + (-z)(0-0) + y(2y-0) = -2y^2 + 2y^2 = 0$$

$\therefore$ The equation (1) is integrable. We take $\dfrac{\text{'1'}}{y^2}$ as an integrating factor.

Multiplying (1) by $\dfrac{1}{y^2}$, we get $dx - \dfrac{z}{y^2}\,dy + \dfrac{1}{y}\,dz = 0$

$$\Rightarrow \quad dx + \frac{y\,dz - z\,dy}{y^2} = 0 \Rightarrow dx + d\left(\frac{z}{y}\right) = 0 \Rightarrow d\left(x + \frac{z}{y}\right) = 0$$

Integrating, we get $\mathbf{x + \dfrac{z}{y} = c.}$

This is the required solution.

(*ii*) We have $yz \log z\,dx - zx \log z\,dy + xy\,dz = 0$ ...(1)

Here $P = yz \log z$, $Q = -zx \log z$, $R = xy$

$$\therefore \quad \frac{\partial P}{\partial y} = z \log z, \quad \frac{\partial P}{\partial z} = y\left(z \cdot \frac{1}{z} + 1 \cdot \log z\right) = y + y \log z$$

$$\frac{\partial Q}{\partial x} = -z \log z, \quad \frac{\partial Q}{\partial z} = -x\left(z \cdot \frac{1}{z} + 1 \cdot \log z\right) = -x - x \log z$$

$$\frac{\partial R}{\partial x} = y, \quad \frac{\partial R}{\partial y} = x$$

$$\frac{\partial P}{\partial y} \neq \frac{\partial Q}{\partial x} \Rightarrow \text{Equation (1) is not exact.}$$

Now $P\left(\frac{\partial Q}{\partial z}-\frac{\partial R}{\partial y}\right)+Q\left(\frac{\partial R}{\partial x}-\frac{\partial P}{\partial z}\right)+R\left(\frac{\partial P}{\partial y}-\frac{\partial Q}{\partial x}\right)$

$$= yz \log z\,(-x - x \log z - x) - zx \log z(y - y - y \log z) + xy\,(z \log z + z \log z)$$
$$= -2xyz \log z - xyz\,(\log z)^2 + xyz\,(\log z)^2 + 2xyz \log z = 0$$

$\therefore$ The equation (1) is integrable. We take $\frac{1}{xyz \log z}$ as an integrating factor.

Multiplying (1) by $\frac{1}{xyz \log z}$, we get

$$\frac{1}{x}\,dx - \frac{1}{y}\,dy + \frac{1}{z \log z}\,dz = 0$$

$$\Rightarrow \quad d(\log x) - d(\log y) + d(\log \log z) = 0$$

$$\Rightarrow \quad d\,(\log x - \log y + \log \log z) = 0$$

$$\Rightarrow \quad d\left(\log\left(\frac{x}{y}\log z\right)\right) = 0$$

Integrating, we get $\log\left(\frac{x}{y}\log z\right) = c.$

$$\Rightarrow \quad \frac{x}{y}\log z = e^c \quad \Rightarrow \quad \frac{x}{y}\log z = k \qquad \text{(Putting } k = e^c\text{)}$$

$\Rightarrow$ **x log z = ky.**

This is the required solution.

(*iii*) We have $(2x^3y + 1)dx + x^4\,dy + x^2 \tan z\,dz = 0$ ...(1)

Here $P = 2x^3y + 1$, $Q = x^4$, $R = x^2 \tan z$

$$\therefore \quad \frac{\partial P}{\partial y} = 2x^3, \quad \frac{\partial P}{\partial z} = 0, \quad \frac{\partial Q}{\partial x} = 4x^3, \quad \frac{\partial Q}{\partial z} = 0, \quad \frac{\partial R}{\partial x} = 2x \tan z, \quad \frac{\partial R}{\partial y} = 0$$

$$\frac{\partial P}{\partial y} \neq \frac{\partial Q}{\partial x} \quad \Rightarrow \quad \text{Equation (1) is not exact.}$$

Now $P\left(\frac{\partial Q}{\partial z}-\frac{\partial R}{\partial y}\right)+Q\left(\frac{\partial R}{\partial x}-\frac{\partial P}{\partial z}\right)+R\left(\frac{\partial P}{\partial y}-\frac{\partial Q}{\partial x}\right)$

$$= 2x^3\,(0 - 0) + x^4\,(2x \tan z - 0) + x^2 \tan z\,(2x^3 - 4x^3) = 0 + 2x^5 \tan z - 2x^5 \tan z = 0$$

$\therefore$ The equation (1) is integrable. Taking integrating factor $\frac{1}{x^2}$, equation (1) becomes

$$\frac{2x^3y + 1}{x^2}\,dx + x^2dy + \tan z\,dz = 0$$

$$\Rightarrow \quad 2xy\,dx + \frac{1}{x^2}\,dx + x^2\,dy + \tan z\,dz = 0$$

$$\Rightarrow \qquad (2xy\,dx + x^2\,dy) + \frac{1}{x^2}\,dx + \tan z dz = 0$$

$$\Rightarrow \qquad d(x^2\,y) + d\left(-\frac{1}{x}\right) + d(\log \sec z) = 0$$

$$\Rightarrow \qquad d\left(x^2 y - \frac{1}{x} + \log \sec z\right) = 0$$

Integrating, we get $\mathbf{x^2y - \frac{1}{x} + \log\sec z = c.}$

This is the required solution.

(*iv*) We have $(2x^2 + 2xy + 2xz^2 + 1)dx + dy + 2z\,dz = 0$ ...(1)

Here $P = 2x^2 + 2xy + 2xz^2 + 1$, $Q = 1$, $R = 2z$

$$\therefore \qquad \frac{\partial P}{\partial y} = 2x, \quad \frac{\partial P}{\partial z} = 4xz, \quad \frac{\partial Q}{\partial x} = 0, \quad \frac{\partial Q}{\partial z} = 0, \quad \frac{\partial R}{\partial x} = 0, \quad \frac{\partial R}{\partial y} = 0$$

$$\frac{\partial P}{\partial y} \neq \frac{\partial Q}{\partial x} \quad \Rightarrow \quad \text{Equation (1) is not exact.}$$

Now $P\left(\frac{\partial Q}{\partial z} - \frac{\partial R}{\partial y}\right) + Q\left(\frac{\partial R}{\partial x} - \frac{\partial P}{\partial z}\right) + R\left(\frac{\partial P}{\partial y} - \frac{\partial Q}{\partial x}\right)$

$$= (2x^2 + 2xy + 2xz^2 + 1)(0 - 0) + 1\,.\,(0 - 4xz) + 2z(2x - 0) = -\,4xz + 4xz = 0$$

$\therefore$ The equation (1) is integrable.

(1) $\Rightarrow$ $(2x^2 + 2xy + 2xz^2)dx + (dx + dy + 2zdz) = 0$ **(Note this step)**

$\Rightarrow$ $2x(x + y + z^2)\,dx + d(x + y + z^2) = 0$ ...(2)

Taking integrating factor $\dfrac{1}{x + y + z^2}$, equation (2) becomes

$$2xdx + \frac{d(x + y + z^2)}{x + y + z^2} = 0$$

$$\Rightarrow \qquad d(x^2) + d(\log\,(x + y + z^2)) = 0$$

$$\Rightarrow \qquad d(x^2 + \log\,(x + y + z^2)) = 0$$

Integrating, we get $\mathbf{x^2 + \log\,(x + y + z^2) = c.}$

This is the required solution.

EXERCISE 3

Solve the following total differential equations :

1. $x\,dy - y\,dx - 2x^2z\,dz = 0$
2. $2yz\,dx + zx\,dy - xy\,(1 + z)dz = 0$
3. $(a - z)(ydx + xdy) + xy\,dz = 0$
4. $2xz\,dx + z\,dy - dz = 0$
5. $x(y^2 - a^2)dx + y(x^2 - z^2)dy - z(y^2 - a^2)dz = 0$
6. $(z + z^2)\cos x\,dx - (z + z^2)dy + (1 - z^2)(y - \sin x)dz = 0$
7. $2yz\,dx - 3zx\,dy - 4xy\,dz = 0$
8. $(2x^3 - z)z\,dx + 2x^2yz\,dy + x(z + x)dz = 0.$

Answers

1. $\frac{y}{x} - z^2 = c$ 2. $x^2y = cze^z$ 3. $xy = c(a - z)$
4. $x^2 + y - \log z = c$ 5. $(x^2 - z^2)(y^2 - a^2) = c$ 6. $y - \sin x = cze^{-z}$
7. $x^2 = cy^3z^4$ 8. $x^2 + y^2 + \log z + \frac{z}{x} = c.$

Hints

1. Use I.F. $= \frac{1}{x^2}$ 2. Use I.F. $= \frac{1}{xyz}$
3. Use I.F. $= \frac{1}{xy(a - z)}$. 4. Use I.F. $= \frac{1}{z}$
5. Use I.F. $= \frac{1}{(y^2 - a^2)(x^2 - z^2)}$ 6. Use I.F. $= \frac{1}{(y - \sin x)(z + z^2)}$
8. Use I.F. $= \frac{1}{x^2 z}$.

---

11. (Method IV) Solution of Integrable Homogeneous Total Differential Equation

Let $$P\,dx + Q\,dy + R\,dz = 0 \qquad ...(1)$$
be an integrable homogeneous total differential equation of degree n.

$\therefore$ The functions P, Q and R are homogeneous functions of degree n in the variables x, y and z.

Let $$x = uz \quad \text{and} \quad y = vz.$$
$\therefore$ $$dx = u\,dz + z\,du \quad \text{and} \quad dy = v dz + z dv$$
$\therefore$ (1) $\Rightarrow$ $$P(uz, vz, z)\,[u\,dz + z\,du] + Q(uz, vz, z)\,[v\,dz + z\,dv] + R(uz, vz, z)dz = 0 \qquad ...(2)$$

Since P, Q, R are homogeneous functions of degree n, we have
$$P(uz, vz, z) = z^n\,P(u, v, 1),$$
$$Q(uz, vz, z) = z^n\,Q(u, v, 1) \text{ and } R(uz, vz, z) = z^n\,R(u, v, 1).$$
$\therefore$ (1) $\Rightarrow$ $$z^n\,P(u, v, 1)\,[u\,dz + z\,du] + z^n\,Q(u, v, 1)\,[v\,dz + z\,dv] + z^n\,R(u, v, 1)dz = 0$$
Dividing by z^n, we get $$P(u, v, 1)\,[u\,dz + z\,du] + Q(u, v, 1)\,[vdz + zdv] + R(u, v, 1)dz = 0$$
$\Rightarrow$ $$P_1\,(u\,dz + z\,du) + Q_1\,(v\,dz + z\,dv) + R_1\,dz = 0$$
(Putting $P_1 = P(u, v, 1)$ etc.)
$\Rightarrow$ $$z(P_1\,du + Q_1\,dv) + (uP_1 + vQ_1 + R_1)dz = 0$$
$\Rightarrow$ $$\frac{P_1}{uP_1 + vQ_1 + R_1}\,du + \frac{Q_1}{uP_1 + vQ_1 + R_1}\,dv + \frac{1}{z}\,dz = 0 \qquad ...(3)$$

In equation (3), the variable z has been separated from the variables u and v and it is occurring only in the last term.

Since equation (1) is integrable, the equation (3) is also integrable.

$\therefore$ $$\frac{P_1}{uP_1 + vQ_1 + R_1}(0 - 0) + \frac{Q_1}{uP_1 + vQ_1 + R_1}(0 - 0) + \frac{1}{z}\left(\frac{\partial}{\partial v}\left(\frac{P_1}{uP_1 + vQ_1 + R_1}\right) - \frac{\partial}{\partial u}\left(\frac{Q_1}{uP_1 + vQ_1 + R_1}\right)\right) = 0$$

$$\Rightarrow \quad \frac{\partial}{\partial v}\left(\frac{P_1}{uP_1 + vQ_1 + R_1}\right) = \frac{\partial}{\partial u}\left(\frac{Q_1}{uP_1 + vQ_1 + R_1}\right)$$

$\Rightarrow \quad \dfrac{P_1}{uP_1 + vQ_1 + R_1}\, du + \dfrac{Q_1}{uP_1 + vQ_1 + R_1}\, dv$ is an exact differential. Also the third term of (3) is exact differential of $\log z$.

$\therefore$ The equation (3) is an exact equation and is thus solvable. In the solution of (3), we put $u = z/x$ and $v = z/y$ and get the required solution of the given equation (1).

Example 1. *Solve the following total differential equations :*

(*i*) $yz\, dx - z^2\, dy - xy\, dz = 0$

(*ii*) $(yz + z^2)dx - xz\, dy + xy\, dz = 0$

(*iii*) $(2xz - yz)dx + (2yz - xz)dy - (x^2 - xy + y^2)dz = 0$

(*iv*) $yz(y + z)dx + xz(x + z)\, dy + xy(x + y)dz = 0.$

Sol. (*i*) We have $\quad yz\, dx - z^2\, dy - xy\, dz = 0 \qquad ...(1)$

Here $\quad P = yz, \quad Q = -z^2, \quad R = -xy$

$\therefore$ (1) is a homogeneous equation of degree 2.

$$\frac{\partial P}{\partial y} = z, \quad \frac{\partial P}{\partial z} = y, \quad \frac{\partial Q}{\partial x} = 0, \quad \frac{\partial Q}{\partial z} = -2z, \quad \frac{\partial R}{\partial x} = -y, \quad \frac{\partial R}{\partial y} = -x$$

$$\therefore \quad P\left(\frac{\partial Q}{\partial z} - \frac{\partial R}{\partial y}\right) + Q\left(\frac{\partial R}{\partial x} - \frac{\partial P}{\partial z}\right) + R\left(\frac{\partial P}{\partial y} - \frac{\partial Q}{\partial x}\right)$$

$$= yz(-2z + x) - z^2(-y - y) - xy(z - 0)$$

$$= -2yz^2 + xyz + 2yz^2 - xyz = 0$$

$\therefore$ The equation (1) is integrable.

Let $\quad x = uz \quad$ and $\quad y = vz.$

$\therefore \quad dx = u\, dz + z\, du \quad$ and $\quad dy = v\, dz + z\, dv$

$\therefore$ (1) reduces to $\quad (vz)z[u\, dz + z\, du] - z^2[v\, dz + z\, dv] - (uz)(vz)dz = 0$

$\Rightarrow$ Dividing by z^2, we get $\quad v[u\, dz + z\, du] - [v\, dz + z\, dv] - uv\, dz = 0$

$\Rightarrow \quad vz\, du - z\, dv - v\, dz = 0$

$\Rightarrow \quad z(v\, du - dv) - v\, dz = 0$

$$\Rightarrow \quad \frac{v\, du - dv}{v} - \frac{1}{z}\, dz = 0$$

$$\Rightarrow \quad du - \frac{1}{v}\, dv - \frac{1}{z}\, dz = 0$$

$\Rightarrow \quad du - d(\log v) - d(\log z) = 0$

$\Rightarrow \quad d(u - \log v - \log z) = 0$

Integrating, we get $\quad u - \log v - \log z = \log k$

$\Rightarrow \quad u = \log(kvz) \quad \Rightarrow \quad kvz = e^u$

$\Rightarrow \quad ky = e^{x/z} \quad \Rightarrow \quad \mathbf{y = ce^{x/z}}. \qquad \left(\text{Putting } c = \dfrac{1}{k}\right)$

This is the required solution.

(*ii*) We have $\quad (yz + z^2)dx - xz\, dy + xy\, dz = 0 \qquad ...(1)$

Here $\quad P = yz + z^2, \quad Q = -xz, \quad R = xy$

$\therefore$ (1) is a homogeneous equation of degree 2.

$$\frac{\partial P}{\partial y} = z, \quad \frac{\partial P}{\partial z} = y + 2z, \quad \frac{\partial Q}{\partial x} = -z, \quad \frac{\partial Q}{\partial z} = -x, \quad \frac{\partial R}{\partial x} = y, \quad \frac{\partial R}{\partial y} = x$$

$$\therefore \quad P\left(\frac{\partial Q}{\partial z} - \frac{\partial R}{\partial y}\right) + Q\left(\frac{\partial R}{\partial x} - \frac{\partial P}{\partial z}\right) + R\left(\frac{\partial P}{\partial y} - \frac{\partial Q}{\partial x}\right)$$

$$= (yz + z^2)(-x - x) - xz(y - y - 2z) + xy(z + z)$$

$$= -2xyz - 2xz^2 + 2xz^2 + 2xyz = 0$$

$\therefore$ The equation (1) is integrable.

Let $x = uz$ and $y = vz$.

$\therefore$ $dx = u\,dz + z\,du$ and $dy = v\,dz + z\,dv$

$\therefore$ (1) reduces to $(vz^2 + z^2)\,[u\,dz + z\,du] - uz^2\,[v\,dz + z\,dv] + uv\,dz = 0$

Dividing by z^2, we get $(v + 1)\,[u\,dz + z\,du] - u\,[v\,dz + z\,dv] + uv\,dz = 0$

$$\Rightarrow \quad z(v + 1)\,du - uz\,dv + (uv + u - uv + uv)\,dz = 0$$

$$\Rightarrow \quad z[(v + 1)du - udv] + u(v + 1)dz = 0$$

$$\Rightarrow \quad \frac{(v + 1)du - u\,dv}{u(v + 1)} + \frac{1}{z}\,dz = 0$$

$$\Rightarrow \quad \frac{1}{u}\,du - \frac{1}{v + 1}\,dv + \frac{1}{z}\,dz = 0$$

$$\Rightarrow \quad d\,(\log u - \log\,(v + 1) + \log z) = 0$$

Integrating, we get $\log u - \log\,(v + 1) + \log z = \log k$

$$\Rightarrow \quad \frac{uz}{v + 1} = k \Rightarrow \frac{x}{\frac{y}{z} + 1} = k$$

$\Rightarrow$ $\mathbf{xz = k(y + z).}$

This is the required solution.

(*iii*) We have $(2xz - yz)dx + (2yz - xz)dy - (x^2 - xy + y^2)dz = 0$ ...(1)

There $P = 2xz - yz$, $Q = 2yz - xz$, $R = -(x^2 - xy + y^2)$

$\therefore$ (1) is a homogeneous equation of degree 2.

$$\frac{\partial P}{\partial y} = -z, \quad \frac{\partial P}{\partial z} = 2x - y, \quad \frac{\partial Q}{\partial x} = -z, \quad \frac{\partial Q}{\partial z} = 2y - x, \quad \frac{\partial R}{\partial x} = -2x + y, \quad \frac{\partial R}{\partial y} = x - 2y$$

$$\therefore \quad P\left(\frac{\partial Q}{\partial z} - \frac{\partial R}{\partial y}\right) + Q\left(\frac{\partial R}{\partial x} - \frac{\partial P}{\partial z}\right) + R\left(\frac{\partial P}{\partial y} - \frac{\partial Q}{\partial x}\right)$$

$$= (2xz - yz)(2y - x - x + 2y) + (2yz - xz)(-2x + y - 2x + y) - (x^2 - xy + y^2)(-z + z)$$

$$= 2z(2x - y)(2y - x) + 2z(2y - x)(y - 2x) = 0$$

$\therefore$ The equation (1) is integrable.

Let $x = uz$ and $y = vz$.

$\therefore$ $dx = u\,dz + z\,du$ and $dy = v\,dz + z\,dv$

$\therefore$ (1) reduces to

$$(2uz^2 - vz^2)\,[u\,dz + z\,du] + (2vz^2 - uz^2)[v\,dz + z\,dv] - (u^2z^2 - uvz^2 + v^2z^2)dz = 0$$

Dividing by z^2, we get

$$(2u - v)[u\,dz + z\,du] + (2v - u)[v\,dz + z\,dv] - (u^2 - uv + v^2)dz = 0$$

$$\Rightarrow \quad z(2u - v)du + z(2v - u)dv + (2u^2 - uv + 2v^2 - uv - u^2 + uv - v^2)dz = 0$$

$$\Rightarrow \quad z[(2u - v)du + (2v - u)dv] + (u^2 - uv + v^2)dz = 0$$

$$\Rightarrow \quad \frac{2u\,du - (v\,du + u\,dv) + 2v dv}{u^2 - uv + v^2} + \frac{1}{z}\,dz = 0 \qquad \textbf{(Note this step)}$$

$$\Rightarrow \quad \frac{d(u^2) - d(uv) + d(v^2)}{u^2 - uv + v^2} + \frac{1}{z}\,dz = 0$$

$$\Rightarrow \quad \frac{d(u^2 - uv + v^2)}{u^2 - uv + v^2} + \frac{1}{z}\,dz = 0$$

Integrating, we get $\quad \log (u^2 - uv + v^2) + \log z = \log c$

$$\Rightarrow \quad (u^2 - uv + v^2)\,z = c$$

$$\Rightarrow \quad \left(\frac{x^2}{z^2} - \frac{xy}{z^2} + \frac{y^2}{z^2}\right) z = c$$

$$\Rightarrow \quad \mathbf{x^2 - xy + y^2 = cz.}$$

This is the required solution.

(*iv*) We have $\quad yz = (y + z)dx + xz(x + z)dy + xy(x + y)dz = 0 \qquad \text{...(1)}$

Here $\quad P = yz(y + z),\ Q = xz(x + z),\ R = xy(x + y)$

$\therefore \quad$ (1) is a homogeneous equation of degree 3.

$$\frac{\partial P}{\partial y} = 2yz + z^2, \quad \frac{\partial P}{\partial z} = y^2 + 2yz, \quad \frac{\partial Q}{\partial x} = 2xz + z^2,$$

$$\frac{\partial Q}{\partial z} = x^2 + 2xz, \quad \frac{\partial R}{\partial x} = 2xy + y^2, \quad \frac{\partial R}{\partial y} = x^2 + 2xy$$

$$\therefore \quad P\left(\frac{\partial Q}{\partial z} - \frac{\partial R}{\partial y}\right) + Q\left(\frac{\partial R}{\partial x} - \frac{\partial P}{\partial z}\right) + R\left(\frac{\partial P}{\partial y} - \frac{\partial Q}{\partial x}\right)$$

$$= yz(y + z)(x^2 + 2xz - x^2 - 2xy) + xz(x + z)(2xy + y^2 - y^2 - 2yz) + xy(x + y)(2yz + z^2 - 2xz - z^2)$$

$$= 2xyz(y + z)(z - y) + 2xyz(x + z)(x - z) + 2xyz(x + y)(y - x)$$

$$= 2xyz[z^2 - y^2 + x^2 - z^2 + y^2 - x^2] = 0$$

$\therefore$ The equation (1) is integrable.

Let $\quad x = uz \quad$ and $\quad y = vz.$

$\therefore \quad dx = u\,dz + z\,du \quad$ and $\quad dy = v\,dz + z\,dv$

$\therefore \quad$ (1) reduces to

$$vz^2(vz + z)\,[u\,dz + z\,du] + uz^2(vz + z)[v\,dz + z\,dv] + u\,vz^2(uz + uz)\,dz = 0$$

Dividing by z^3, we get $\quad v(v + 1)[u\,dz + z\,du] + u(u + 1)\,[v\,dz + z\,dv] + uv(u + v)dz = 0$

$$\Rightarrow \quad zv(v + 1)du + zu(u + 1)dv + [uv(v + 1) + uv(u + 1) + uv(u + v)]dz = 0$$

$$\Rightarrow \quad z[v(v + 1)du + u(u + 1)dv] + 2uv(u + v + 1)dz = 0$$

$$\Rightarrow \quad \frac{v(v + 1)}{uv(u + v + 1)}\,du + \frac{u(u + 1)}{uv\,(u + v + 1)}\,dv + \frac{2}{z}\,dz = 0$$

$$\Rightarrow \quad \frac{v+1}{u(u+v+1)}\,du + \frac{u+1}{v(u+v+1)}\,dv + \frac{2}{z}\,dz = 0$$

$$\Rightarrow \quad \frac{(u+v+1)-u}{u(u+v+1)}\,du + \frac{(u+v+1)-v}{v(u+v+1)}\,dv + \frac{2}{z}\,dz = 0 \qquad \textbf{(Note this step)}$$

$$\Rightarrow \quad \left(\frac{1}{u} - \frac{1}{u+v+1}\right)du + \left(\frac{1}{v} - \frac{1}{u+v+1}\right)dv + \frac{2}{z}\,dz = 0$$

$$\Rightarrow \quad \frac{1}{u}\,du + \frac{1}{v}\,dv - \frac{1}{u+v+1}(du+dv) + \frac{2}{z}\,dz = 0$$

$$\Rightarrow \quad \frac{1}{u}\,du + \frac{1}{v}\,dv - \frac{1}{(u+v)+1}\,d(u+v) + \frac{2}{z}\,dz = 0$$

Integrating, we get $\log u + \log v - \log(u+v+1) + 2\log z = \log c$

$$\Rightarrow \quad \frac{uvz^2}{u+v+1} = c \quad \Rightarrow \quad uvz^2 = c(u+v+1)$$

$$\Rightarrow \quad \frac{x}{z}\cdot\frac{y}{z}\cdot z^2 = c\left(\frac{x}{z} + \frac{y}{z} + 1\right) \quad \Rightarrow \quad \mathbf{xyz = c(x+y+z).}$$

This is the required solution.

EXERCISE 4

Solve the following total differential equations :

1. $(2y + x - z)dx - (x + z)dy + 2(x + y)dz = 0$
2. $2(y + z)dx - (x + z)dy + (2y - x + z)dz = 0$
3. $(y^2 + yz)dx + (z^2 + 2x)dy + (y^2 - xy)dz = 0$
4. $z(z - y)dx + z(z + x)dy + x(x + y)dz = 0$
5. $z^2\,dx + (z^2 - 2yz)dy + (2y^2 - yz - zx)dz = 0$
6. $(x^2y - y^3 - y^2z)dx + (xy^2 - x^2z - x^3)dy + (xy^2 + x^2y)dz = 0$
7. $(2xyz + y^2z + yz^2)dx + (x^2z + 2xyz + xz^2)dy + (x^2y + xy^2 + 2xyz)dz = 0$
8. $(y^2 + yz + z^2)dx + (z^2 + zx + x^2)dy + (x^2 + xy + y^2)dz = 0.$

Answers

1. $(x + z)^2 = c(x + y)$
2. $(x + z)^2 = c(y + z)$
3. $y(x + z) = c(y + z)$
4. $z(x + y) = c(x + z)$
5. $xz + yz - y^2 = cz^2$
6. $x^2 + y^2 + yz + xz = cxy$
7. $xyz(x + y + z) = c$
8. $xy + yz + zx = c(x + y + z).$

---

12. (Method V) Solution of Integrable Total Differential Equation

Let $\quad P\,dx + Q\,dy + R\,dz = 0 \qquad \ldots(1)$

be an integrable total differential equation.

In this method, we consider any one of the variables, say z, as a constant for the moment.

$\therefore$ (1) reduces to $P\,dx + Q\,dy + R.0 = 0$

i.e., $\qquad P\,dx + Q\,dy = 0 \qquad \ldots(2)$

Let the solution of (2) be $u(x, y, z) = \phi(z)$, where $\phi(z)$ is any arbitrary function of z. We have taken $\phi(z)$ in the solution of (2) in place of an arbitrary constant because $\phi(z)$ is a constant with respect to x and y.

Let $\mu(x, y, z)$ be an integrating factor of (2).

$$\therefore \quad \frac{\partial u}{\partial x}dx + \frac{\partial u}{\partial y}dy = \mu P\,dx + \mu Q\,dy$$

$$\Rightarrow \quad \frac{\partial u}{\partial x} = \mu P \text{ and } \frac{\partial u}{\partial y} = \mu Q \qquad ...(3)$$

Differentiating $u(x, y, z) = \phi(z)$ w.r.t. x, y and z, we get

$$\frac{\partial u}{\partial x}dx + \frac{\partial u}{\partial y}dy + \frac{\partial u}{\partial z}dz = d\phi$$

$$\Rightarrow \quad \mu P dx + \mu Q\,dy + \frac{\partial u}{\partial z}dz = d\phi \qquad ...(4)$$

$$(1) \Rightarrow \quad \mu P dx + \mu Q\,dy + \mu R\,dz = 0 \qquad ...(5)$$

Subtracting (4) and (5), we get

$$d\phi = \frac{\partial u}{\partial z}dz - \mu R dz$$

i.e.,
$$d\phi = \left(\frac{\partial u}{\partial z} - \mu R\right)dz$$

This equation is solved to get the value of the function $\phi(z)$. The value of $\phi(z)$ is substituted in $u(x, y, z) = \phi(z)$ to get the solution of the given equation.

Example 1. *Solve the following total differential equations :*

(*i*) $2(y + z)dx - (x + z)dy + (2y - x + z)dz = 0$

(*ii*) $(2x^2 + 2xy + 2xz^2 + 1)dx + dy + 2z\,dz = 0$

(*iii*) $y^2z(x \cos x - \sin x)dx + x^2z(y \cos y - \sin y)dy + xy(y \sin x + x \sin y + xy \cos z)\,dz = 0$

(*iv*) $2yz\,dx + zx\,dy - xy(1 + z)dz = 0$.

Sol. (*i*) We have $2(y + z)dx - (x + z)dy + (2y - x + z)dz = 0 \qquad ...(1)$

Here $P = 2(y + z)$, $Q = -(x + z)$, $R = 2y - x + z$

$$\therefore \quad \frac{\partial P}{\partial y} = 2,\quad \frac{\partial P}{\partial z} = 2,\quad \frac{\partial Q}{\partial x} = -1,\quad \frac{\partial Q}{\partial z} = -1,\quad \frac{\partial R}{\partial x} = -1,\quad \frac{\partial R}{\partial y} = 2$$

$$\therefore \quad P\left(\frac{\partial Q}{\partial z} - \frac{\partial R}{\partial y}\right) + Q\left(\frac{\partial R}{\partial x} - \frac{\partial P}{\partial z}\right) + R\left(\frac{\partial P}{\partial y} - \frac{\partial Q}{\partial x}\right)$$

$$= 2(y + z)(-1 - 2) - (x + z)(-1 - 2) + (2y - x + z)(2 + 1)$$

$$= -6y - 6z + 3x + 3z + 6y - 3x + 3z = 0$$

$\therefore$ The equation (1) is integrable.

Considering z as a constant, we have $dz = 0$.

$\therefore \quad (1) \Rightarrow \quad 2(y + z)dx - (x + z)dy = 0$

$$\Rightarrow \quad \frac{dy}{dx} = \frac{2(y + z)}{x + z} \quad \Rightarrow \quad \frac{dy}{dx} - \frac{2}{x + z}y = \frac{2z}{x + z} \qquad ...(2)$$

This is a linear differential equation of first order.

$$\text{I.F.} = e^{\int -\frac{2}{x + z}dx} = e^{-2\log(x + z)} = \frac{1}{(x + z)^2}$$

$\therefore$ The solution of (2) is

$$y \cdot \frac{1}{(x+z)^2} = \int \frac{2z}{x+z} \cdot \frac{1}{(x+z)^2}\, dx + \phi(z),$$

where $\phi(z)$ is any arbitrary function of z.

$$\Rightarrow \quad \frac{y}{(x+z)^2} = 2z \cdot \frac{(x+z)^{-2}}{-2} + \phi(z)$$

$$\Rightarrow \quad \frac{y}{(x+z)^2} = -\frac{z}{(x+z)^2} + \phi(z) \quad \text{...(3)}$$

Differentiating (3) w.r.t. x, y and z, we get

$$\frac{1}{(x+z)^2}\, dy - \frac{2y}{(x+z)^3}\, dx - \frac{2y}{(x+z)^3}\, dz$$

$$= \frac{2z}{(x+z)^3}\, dx - \left(1 \cdot \frac{1}{(x+z)^2} - \frac{2z}{(x+z)^3}\right) dz + d\phi$$

$$\Rightarrow \quad \frac{dy}{(x+z)^2} - \frac{2y}{(x+z)^3}(dx+dz) = -\frac{dz}{(x+z)^2} - \frac{2z}{(x+z)^3}(dx+dz) + d\phi$$

$$\Rightarrow \quad \frac{2(y+z)}{(x+z)^3}(dx+dz) - \frac{dy}{(x+z)^2} - \frac{dz}{(x+z)^2} + d\phi = 0$$

$$\Rightarrow \quad 2(y+z)dx - (x+z)dy + (2y-x+z)dz + (x+z)^3 d\phi = 0$$

$$\Rightarrow \quad (x+z)^3\, d\phi = 0 \qquad \text{[By using (1)]}$$

$$\Rightarrow \quad d\phi = 0 \;\Rightarrow\; \phi = c$$

$$\therefore \quad (3) \Rightarrow \quad \frac{y}{(x+z)^2} = -\frac{z}{(x+z)^2} + c \;\Rightarrow\; \mathbf{y + z = c(x+z)^2}.$$

This is the required solution.

(*ii*) We have $(2x^2 + 2xy + 2xz^2 + 1)dx + dy + 2z\, dz = 0$ ...(1)

Here $P = 2x^2 + 2xy + 2xz^2 + 1$, $Q = 1$, $R = 2z$

$$\therefore \quad \frac{\partial P}{\partial y} = 2x, \quad \frac{\partial P}{\partial z} = 4xz, \quad \frac{\partial Q}{\partial x} = 0, \quad \frac{\partial Q}{\partial z} = 0, \quad \frac{\partial R}{\partial x} = 0, \quad \frac{\partial R}{\partial y} = 0$$

$$\therefore \quad P\left(\frac{\partial Q}{\partial z} - \frac{\partial R}{\partial y}\right) + Q\left(\frac{\partial R}{\partial x} - \frac{\partial P}{\partial z}\right) + R\left(\frac{\partial P}{\partial y} - \frac{\partial Q}{\partial x}\right)$$

$$= (2x^2 + 2xy + 2xz^2 + 1)(0-0) + 1.\,(0 - 4xz) + 2z(2x - 0)$$

$$= -4xz + 4xz = 0$$

$\therefore$ The equation (1) is integrable.

Considering x* as a constant, we have $dx = 0$

$\therefore$ (1) $\Rightarrow$ $dy + 2z\, dz = 0$

Integrating, we get $y + z^2 = \phi(x)$, ...(2)

where $\phi(x)$ is any arbitrary function of x.

---

*We have chosen x to be constant because P is more complex then Q and R.

Differentiating (2) w.r.t., x, y and z, we get

$$1.dy + 2z\,dz = d\phi \qquad ...(3)$$

$$(1) - (3) \Rightarrow \quad (2x^2 + 2xy + 2xz^2 + 1)\,dx = -\,d\phi$$

$$\Rightarrow \quad (2x^2 + 2xy + 2xz^2 + 1)dx = -\,\phi'(x)dx$$

$$\Rightarrow \quad \phi'(x) = -\,(2x^2 + 2xy + 2xz^2 + 1)$$

$$\Rightarrow \quad \phi'(x) = -\,(2x^2 + 2x(y + z^2) + 1)$$

$$\Rightarrow \quad \phi'(x) = -\,2x^2 - 2x\,\phi(x) - 1$$

$$\Rightarrow \quad \frac{d\phi}{dx} + 2x\phi = -\,(2x^2 + 1) \qquad ...(4)$$

(4) is a linear differential equation of first order.

$$\text{I.F.} = e^{\int 2x\,dx} = e^{x^2}$$

$\therefore$ Solution of (4) is

$$\phi(x)\,.\,e^{x^2} = \int -\,(2x^2 + 1)\,e^{x^2}\,dx + c$$

$$\Rightarrow \quad \phi(x)\,e^{x} = -\int x\,.\,2x\,e^{x^2}\,dx - \int e^{x^2}\,dx + c$$

$$= -\left[x\,.\,e^{x^2} - \int 1\,.\,e^{x^2}\,dx\right] - \int e^{x^2}\,dx + c = -\,xe^{x^2} + c$$

$$\therefore \quad \phi(x)e^{x^2} = -\,xe^{x^2} + c \Rightarrow (y + z^2)e^{x^2} = -\,xe^{x^2} + c$$

$$\Rightarrow \quad \mathbf{(x + y + z^2)e^{x^2} = c.}$$

This is the required solution.

(*iii*) We have $y^2z(x \cos x - \sin x)dx + x^2z\,(y \cos y - \sin y)dy + xy(y \sin x + x \sin y + xy \cos z)dz = 0$ ...(1)

It can be verified that the equation (1) is integrable.

Considering z as a constant, we have $dz = 0$.

$$\therefore \quad (1) \Rightarrow \quad y^2z\,(x \cos x - \sin x)dx + x^2z(y \cos y - \sin y)dy = 0 \qquad ...(2)$$

$$\Rightarrow \quad \frac{x \cos x - \sin x}{x^2}\,dx + \frac{y \cos y - \sin y}{y^2}\,dy = 0$$

$$\Rightarrow \quad d\left(\frac{\sin x}{x}\right) + d\left(\frac{\sin y}{y}\right) = 0$$

Integrating, we get
$$\frac{\sin x}{x} + \frac{\sin y}{y} = \phi(z), \qquad ...(3)$$

where $\phi(z)$ is any arbitrary function of z.

Differentiating (3) w.r.t. x, y and z, we get

$$\frac{x \cos x - \sin x}{x^2}\,dx + \frac{y \cos y - \sin y}{y^2}\,dy = \phi'\,(z)\,dz$$

$$\Rightarrow \quad y^2z(x \cos x - \sin x)dx + x^2z(y \cos y - \sin y)dy - x^2y^2z\,\phi'(z)dz = 0 \qquad ...(4)$$

$$(1) - (4) \Rightarrow \quad xy(y \sin x + x \sin y + xy \cos z)\,dz + x^2y^2z\,\phi'(z)dz = 0$$

$$\Rightarrow \quad y \sin x + x \sin y + xy \cos z = -xyz\, \phi'(z)$$

$$\Rightarrow \quad \frac{\sin x}{x} + \frac{\sin y}{y} + \cos z = -z\, \phi'(z)$$

$$\Rightarrow \quad \phi(z) + \cos z = -z\, \phi'(z)$$

$$\Rightarrow \quad \frac{d\phi}{dz} + \frac{1}{z}\phi = -\frac{\cos z}{z} \qquad ...(5)$$

This is a linear equation of first order.

$$\text{I.F.} = e^{\int \frac{1}{z} dz} = e^{\log z} = z$$

$\therefore$ The solution of (5) is $\quad \phi(z)\,.\,z = \int \left(-\frac{\cos z}{z}\right) z\, dz + c$

$$\Rightarrow \quad z\,\phi(z) = -\sin z + c$$

$$\Rightarrow \quad z\left(\frac{\sin x}{x} + \frac{\sin y}{y}\right) = -\sin z + c$$

$$\Rightarrow \quad \mathbf{\frac{\sin x}{x} + \frac{\sin y}{y} + \frac{\sin z}{z} = \frac{c}{z}}.$$

This is the required solution.

(*iv*) We have $\quad 2yz\, dx + zx\, dy - xy(1+z)dz = 0 \qquad ...(1)$

Here $\quad P = 2yz,\ Q = zx,\ R = -xy(1+z)$

$$\therefore \quad \frac{\partial P}{\partial y} = 2z, \quad \frac{\partial P}{\partial z} = \partial y, \quad \frac{\partial Q}{\partial x} = z, \quad \frac{\partial Q}{\partial z} = x, \quad \frac{\partial R}{\partial x} = -y - yz, \quad \frac{\partial R}{\partial y} = -x - xz$$

$$\therefore \quad P\left(\frac{\partial Q}{\partial z} - \frac{\partial R}{\partial y}\right) + Q\left(\frac{\partial R}{\partial x} - \frac{\partial P}{\partial z}\right) + R\left(\frac{\partial P}{\partial y} - \frac{\partial Q}{\partial x}\right)$$

$$= 2yz(x + x + xz) + zx(-y - yz - 2y) - xy(1+z)(2z - z)$$

$$= 4xyz + 2xyz^2 - 3xyz - xyz^2 - xyz - xyz^2 = 0$$

$\therefore$ The equation (1) is integrable. Considering z as a constant, we have $dz = 0$.

(1) $\Rightarrow \quad 2yz\, dx + zx\, dy = 0$

$$\Rightarrow \quad \frac{2}{x} dx + \frac{1}{y} dy = 0$$

Integrating, we get $\quad 2 \log x + \log y = \log \phi(z)$,

where $\phi(z)$ is any arbitrary function of z.

$$\therefore \quad x^2 y = \phi(z) \qquad ...(2)$$

Differentiating (2) w.r.t. x, y and z, we get

$$2xy\, dx + x^2\, dy = \phi'(z)dz$$

$$\Rightarrow \quad 2xy\, dx + x^2\, dy - \phi'(z)\, dz = 0$$

Multiplying by $\frac{z}{x}$, we get

$$2yz\, dx + xz\, dy - \frac{z}{x}\phi'(z)dz = 0 \qquad ...(3)$$

$$(1)-(3) \Rightarrow \quad -xy(1+z)dz + \frac{z}{x}\phi'(z)dz = 0$$

$$\Rightarrow \quad x^2y(1+z) = z\,\phi'(z) \quad \Rightarrow \quad \phi(z)(1+z) = z\,\phi'(z)$$

$$\Rightarrow \quad \frac{\phi'(z)}{\phi(z)} = \frac{1+z}{z} \quad \Rightarrow \quad \frac{\phi'(z)}{\phi(z)} = \frac{1}{z} + 1$$

Integrating, we get

$$\log \phi(z) = \log z + z + c$$

$$\Rightarrow \quad \log x^2y = \log z + z + c$$

$$\Rightarrow \quad \log \frac{x^2y}{z} = z + c.$$

This is the required solution.

EXERCISE 5

Solve the following total differential equations :

1. $(y + z)dx + (z + x)dy + (x + y)dz = 0$
2. $z(z - y)dx + z(z + x)dy + x(x + y)dz = 0$
3. $xz^3\,dx - zdy + 2y\,dz = 0$
4. $3x^2\,dx + 3y^2\,dy - (x^3 + y^3 + e^{2z})dz = 0$
5. $(x^2 + y^2 + z^2)dx - 2xy\,dy - 2xz\,dz = 0$
6. $(e^xy + e^z)dx + (e^yz + e^x)dy + (e^y - e^xy - e^yz)dz = 0$
7. $(z + z^3)\cos x\,dx - (z + z^3)dy + (1 - z^2)(y - \sin x)dz = 0$
8. $(2xz - yz)dx + (2yz - zx)dy - (x^2 - xy + y^2)dz = 0.$

Answers

1. $xy + yz + zx = c$
2. $(x + y)z = c(x + z)$
3. $2y - x^2z^2 = 2cz^2$
4. $x^3 + y^3 = e^{2z} + ce^z$
5. $y^2 + z^2 = x^2 + cx$
6. $e^xy + e^zx + e^yz = ce^z$
7. $(\sin x - y)(z^2 + 1) = cz$
8. $x^2 - xy + y^2 = cz.$

---

13. (Method VI) Solution of Integrable Non-Exact Total Differential Equation by the Method of Auxiliary Equations

Let $\quad P\,dx + Q\,dy + R\,dz = 0$ ...(1)

be an integrable non-exact total differential equation.

$$\therefore \quad P\left(\frac{\partial Q}{\partial z} - \frac{\partial R}{\partial y}\right) + Q\left(\frac{\partial R}{\partial x} - \frac{\partial P}{\partial z}\right) + R\left(\frac{\partial P}{\partial y} - \frac{\partial Q}{\partial x}\right) = 0 \quad ...(2)$$

Comparing (1) and (2), we have

$$\frac{dx}{\frac{\partial Q}{\partial z} - \frac{\partial R}{\partial y}} = \frac{dy}{\frac{\partial R}{\partial x} - \frac{\partial P}{\partial z}} = \frac{dz}{\frac{\partial P}{\partial y} - \frac{\partial Q}{\partial x}} \quad ...(3)$$

These are called the **auxiliary equations** of the given equation (1).

Since (1) is non-exact, $\frac{\partial Q}{\partial z} - \frac{\partial R}{\partial y}$, $\frac{\partial R}{\partial x} - \frac{\partial P}{\partial z}$ and $\frac{\partial P}{\partial y} - \frac{\partial Q}{\partial x}$ are not all zero.

Let $u(x, y, z) = a$ and $v(x, y, z) = b$ be two solutions* of (3).

In the next step, we form the equation Adu + Bdv = 0 and find the values of A and B by comparing this equation with the given equation (1). Using $u(x, y, z) = a$ and $v(x, y, z) = b$, we find the values of A and B in terms of u and v. These values of A and B are substituted in the equation Adu + Bdv = 0. The solution of this equation gives the solution of the given equation (1).

Example 1. *Solve the following total differential equations :*

(*i*) $(y^2 + yz)dx + (z^2 + zx)dy + (y^2 - xy)\,dz = 0$

(*ii*) $(x^2y - y^3 - y^2z)dx + (xy^2 - x^2z - x^3)\,dy + (xy^2 + x^2y)dz = 0.$

Sol. (*i*) We have $(y^2 + yz)dx + (z^2 + zx)dy + (y^2 - xy)dz = 0$ ...(1)

Here $\mathrm{P} = y^2 + yz, \quad \mathrm{Q} = z^2 + zx, \mathrm{R} = y^2 - xy$

$$\therefore \quad \frac{\partial \mathrm{P}}{\partial y} = 2y + z, \frac{\partial \mathrm{P}}{\partial z} = y, \frac{\partial \mathrm{Q}}{\partial x} = z, \frac{\partial \mathrm{Q}}{\partial z} = 2z + x, \frac{\partial \mathrm{R}}{\partial x} = -y, \frac{\partial \mathrm{R}}{\partial y} = 2y - x$$

Since $\dfrac{\partial \mathrm{P}}{\partial y} \neq \dfrac{\partial \mathrm{P}}{\partial x}$, the equation (1) is not exact.

$$\text{Now} \quad \mathrm{P}\left(\frac{\partial \mathrm{Q}}{\partial z} - \frac{\partial \mathrm{R}}{\partial y}\right) + \mathrm{Q}\left(\frac{\partial \mathrm{R}}{\partial x} - \frac{\partial \mathrm{P}}{\partial z}\right) + \mathrm{R}\left(\frac{\partial \mathrm{P}}{\partial y} - \frac{\partial \mathrm{Q}}{\partial x}\right)$$

$$= (y^2 + yz)(2z + x - 2y + x) + (z^2 + zx)(-y - y) + (y^2 - xy)(2y + z - z)$$

$$= 2y(y + z)(x - y + z) - 2yz(z + x) + 2y^2(y - x)$$

$$= 2y[(y + z)(z - y + x) - z(z + x) + y(y - x)]$$

$$= 2y[z^2 - y^2 + xy + xz - z^2 - xz + y^2 - xy]$$

$$= 2y \times 0 = 0$$

$\therefore$ The equation (1) is integrable.

The auxiliary equations of (1) are

$$\frac{dx}{\dfrac{\partial \mathrm{Q}}{\partial z} - \dfrac{\partial \mathrm{R}}{\partial y}} = \frac{dy}{\dfrac{\partial \mathrm{R}}{\partial x} - \dfrac{\partial \mathrm{P}}{\partial z}} = \frac{dz}{\dfrac{\partial \mathrm{P}}{\partial y} - \dfrac{\partial \mathrm{Q}}{\partial x}}$$

$$\Rightarrow \quad \frac{dx}{2z + x - 2y + x} = \frac{dy}{-y - y} = \frac{dz}{2y + z - z}$$

$$\Rightarrow \quad \frac{dx}{2(x - y + z)} = \frac{dy}{-2y} = \frac{dz}{2y}$$

$$\Rightarrow \quad \frac{dx}{x - y + z} = \frac{dy}{-y} = \frac{dz}{y} \qquad \text{...(2)}$$

Taking last two fractions of (2), we get $-dy = dz$.

$\Rightarrow \quad d(y + z) = 0$

Integrating, we get $y + z = a$

*Details regarding the methods of finding such solutions is given in chapter 23.

Taking 1, 0, 1 as multipliers, each fraction of (2)

$$= \frac{1.dx + 0.dy + 1.dz}{1.(x - y + z) + 0.(-y) + 1.(y)} = \frac{dx + dz}{x + z} \quad ...(3)$$

$\therefore$ We have $\dfrac{dx + dz}{x + z} = \dfrac{dy}{-y} \Rightarrow \dfrac{d(x + z)}{x + z} + \dfrac{dy}{y} = 0$

Integrating, we get

$$\log (x + z) + \log y = \log b \quad \text{or} \quad (x + z)y = b$$

Let $u(x, y, z) = y + z$ and $v(x, y, z) = (x + z)y$

$\therefore$ $du = dy + dz$ and $dv = y\,dx + (x + z)dy + y\,dz$

Let $\text{A}du + \text{B}dv = 0$ ...(4)

$\Rightarrow$ $\text{A}(dy + dz) + \text{B}(y\,dx + (x + z)dy + y\,dz) = 0$

$\Rightarrow$ $\text{B}y\,dx + (\text{A} + \text{B}(x + z))\,dy + (\text{A} + \text{B}y)dz = 0$ ...(5)

Comparing (1) and (5), we have

$$\text{B}y = y^2 + yz \quad ...(6)$$

$$\text{A} + \text{B}(x + z) = z^2 + zx \quad ...(7)$$

and

$$\text{A} + \text{B}y = y^2 - xy, \quad ...(8)$$

(6) $\Rightarrow$ $\text{B} = y + z = u$

(7) $\Rightarrow$ $\text{A} = z^2 + zx - (y + z)(x + z) = -xy - yz = -y(x + z) = -v$

Putting the values of A and B in (4), we get

$$-v\,du + u\,dv = 0$$

$$\Rightarrow \quad \frac{du}{u} - \frac{dv}{v} = 0$$

Integrating, we get

$$\log u - \log v = \log c$$

$$\Rightarrow \quad \frac{u}{v} = c \quad \text{or} \quad \frac{y + z}{(x + z)y} = c \quad \text{or} \quad \mathbf{y + z = c(x + z)y.}$$

This is the required solution.

(*ii*) We have $(x^2y - y^3 - y^2z)dx + (xy^2 - x^2z - x^3)dy + (xy^2 + x^2y)dz = 0$ ...(1)

Here $\text{P} = x^2y - y^3 - y^2z$, $\text{Q} = xy^2 - x^2z - x^3$, $\text{R} = xy^2 + x^2y$

$$\therefore \quad \frac{\partial \text{P}}{\partial y} = x^2 - 3y^2 - 2yz, \quad \frac{\partial \text{P}}{\partial z} = -y^2, \quad \frac{\partial \text{Q}}{\partial x} = y^2 - 2xz - 3x^2, \quad \frac{\partial \text{Q}}{\partial z} = -x^2, \quad \frac{\partial \text{R}}{\partial x} = y^2 + 2xy,$$

$$\frac{\partial \text{R}}{\partial y} = 2xy + x^2$$

Since $\dfrac{\partial \text{P}}{\partial y} \neq \dfrac{\partial \text{Q}}{\partial x}$, the equation (1) is not exact.

It can be verified that $\text{P}\left(\dfrac{\partial \text{Q}}{\partial z} - \dfrac{\partial \text{R}}{\partial y}\right) + \text{Q}\left(\dfrac{\partial \text{R}}{\partial x} - \dfrac{\partial \text{P}}{\partial z}\right) + \text{R}\left(\dfrac{\partial \text{P}}{\partial y} - \dfrac{\partial \text{Q}}{\partial x}\right) = 0$

$\therefore$ The equation (1) is integrable.

The auxiliary equations of (1) are

$$\frac{dx}{\frac{\partial Q}{\partial z}-\frac{\partial R}{\partial y}}=\frac{dy}{\frac{\partial R}{\partial x}-\frac{\partial P}{\partial z}}=\frac{dz}{\frac{\partial P}{\partial y}-\frac{\partial Q}{\partial x}}$$

$$\Rightarrow \quad \frac{dx}{-x^2-2xy-x^2}=\frac{dy}{y^2+2xy+y^2}=\frac{dz}{x^2-3y^2-2yz-y^2+2xz+3x^2}$$

$$\Rightarrow \quad \frac{dx}{-2x(x+y)}=\frac{dy}{2y(x+y)}=\frac{dz}{4x^2-4y^2-2yz+2xz}$$

$$\Rightarrow \quad \frac{dx}{-x(x+y)}=\frac{dy}{y(x+y)}=\frac{dz}{2x^2-2y^2-yz+xz} \qquad ...(2)$$

Taking first two fractions of (2), we get $\frac{dx}{x}+\frac{dy}{y}=0.$

Integrating, we get

$$\log x+\log y=\log a \quad \Rightarrow \quad xy=a$$

Taking 1, 1, 0 as multipliers, each fraction of (2)

$$=\frac{1.dx+1.dy+0.dz}{-x(x+y)+y(x+y)+0}=\frac{dx+dy}{(y-x)(x+y)}$$

Taking 1, 1, 1 as multipliers, each fraction of (2)

$$=\frac{1.dx+1.dy+1.dz}{-x(x+y)+y(x+y)+2x^2-2y^2-yz+xz}$$

$$=\frac{dx+dy+dz}{x^2-y^2+xz-yz}=\frac{dx+dy+dz}{(x-y)(x+y)+z(x-y)}$$

$$=\frac{dx+dy+dz}{(x-y)(x+y+z)}$$

$$\therefore \quad \frac{dx+dy}{(y-x)(x+y)}=\frac{dx+dy+dz}{(x-y)(x+y+z)} \quad \Rightarrow \quad \frac{d(x+y)}{x+y}+\frac{d(x+y+z)}{x+y+z}=0$$

Integrating, we get

$$\log(x+y)+\log(x+y+z)=\log b$$

$$\Rightarrow \quad (x+y)(x+y+z)=b$$

Let $u(x, y, z)=xy$ and $v(x, y, z)=(x+y)(x+y+z)$

$$\therefore \quad du=y\,dx+x\,dy$$

and

$$dv=(x+y)(dx+dy+dz)+(x+y+z)(dx+dy)$$

$$=(2x+2y+z)dx+(2x+2y+z)dy+(x+y)dz$$

Let $\quad A du+B dv=0 \qquad ...(3)$

$$\Rightarrow \quad A(y\,dx+x\,dy)+B((2x+2y+z)dx+(2x+2y+z)dy+(x+y)dz)=0$$

$$\Rightarrow \quad (Ay+B(2x+2y+z))dx+(Ax+B(2x+2y+z))dy+B(x+y)dz=0 \qquad ...(4)$$

Comparing (1) and (4), we have

$$Ay+B(2x+2y+z)=x^2y-y^3-y^2z \qquad ...(5)$$

$$Ax+B(2x+2y+z)=xy^2-x^2z-x^3 \qquad ...(6)$$

$$B(x + y) = xy^2 + x^2y \quad ...(7)$$

(7) $\Rightarrow$ $B = xy = u$

(5) – (6) $\Rightarrow$ $A(y - x) = x^2y - y^3 - y^2z - xy^2 + x^2z + x^3$

$$= (x^3 - y^3) + (x^2y - xy^2) + (x^2z - y^2z)$$

$$= (x - y)[x^2 + xy + y^2 + xy + z(x + y)]$$

$$= (x - y)[(x + y)^2 + z(x + y)] = (x - y)(x + y)(x + y + z)$$

$\Rightarrow$ $A = -(x + y)(x + y + z) = -v$

Putting the values of A and B in (3), we get

$$-v\,du + u\,dv = 0$$

$$\Rightarrow \quad \frac{du}{u} - \frac{dv}{v} = 0$$

Integrating, we get

$$\log u - \log v = \log c$$

$$\Rightarrow \quad \frac{u}{v} = c \quad \Rightarrow \quad u = cv \quad \Rightarrow \quad \mathbf{xy = c(x + y)(x + y + z).}$$

This is the required solution.

EXERCISE 6

Solve the following total differential equations :

1. $(y + z)dx + dy + dz = 0$
2. $(2xz - yz)dx + (2yz - zx)dy - (x^2 - xy + y^2)dz = 0$
3. $z(z - y)dx + z(z + x)dy + x(x + y)dz = 0$
4. $(y^2 + yz + z^2)dx + (z^2 + zx + x^2)dy + (x^2 + xy + y^2)dz = 0.$

Answers

1. $e^x (y + z) = c$
2. $x^2 + y^2 - xy = cz$
3. $(x + y)z = c(x + z)$
4. $xy + yz + zx = c(x + y + z).$

SECTION IV

22

Linear Differential Equations of the Second Order

1. Introduction

In this chapter, we shall learn the methods of solving linear differential equation of the second order.

If P, Q and R are functions of x then

$$\frac{d^2y}{dx^2} + P\frac{dy}{dx} + Q\,y = R$$

is a linear differential equation of the second order. There is no general method of solving a linear differential equation of the second order. We shall study certain procedures which at times will yield a solution.

2. Methods of Solving Linear Differential Equation of the Second Order

In the present text, we shall discuss the following methods of solving a linear differential equation of the second order :

(*i*) Method of changing the dependent variable

(*ii*) Method of changing the independent variable

(*iii*) Method of variation of parameters.

3. First Method of Solving a Linear Differential Equation of the Second Order by Changing the Dependent Variable (By Using a Particular Integral of a Corresponding Equation of the Given Equation)

Let $$\frac{d^2y}{dx^2} + P\frac{dy}{dx} + Q\,y = R \qquad ...(1)$$

be a linear differential equation of the second order where P, Q, and R are functions of x.

Let $y = uv$ where u and v are functions of x.

$$\therefore \quad \frac{dy}{dx} = u\frac{dv}{dx} + v\frac{du}{dx}$$

and $$\frac{d^2y}{dx^2} = \left(u\frac{d^2v}{dx^2} + \frac{du}{dx}\frac{dv}{dx}\right) + \left(v\frac{d^2u}{dx^2} + \frac{dv}{dx}\frac{du}{dx}\right) = u\frac{d^2v}{dx^2} + 2\frac{du}{dx}\frac{dv}{dx} + v\frac{d^2u}{dx^2}.$$

$$\therefore \quad (1) \Rightarrow \quad \left(u\frac{d^2v}{dx^2} + 2\frac{du}{dx}\frac{dv}{dx} + v\frac{d^2u}{dx^2}\right) + P\left(u\frac{dv}{dx} + v\frac{du}{dx}\right) + Quv = R$$

$$\Rightarrow \qquad u\frac{d^2v}{dx^2}+\left(2\frac{du}{dx}+Pu\right)\frac{dv}{dx}+\left(\frac{d^2u}{dx^2}+P\frac{du}{dx}+Qu\right)v=R$$

$$\Rightarrow \qquad \frac{d^2v}{dx^2}+\left(\frac{2}{u}\frac{du}{dx}+P\right)\frac{dv}{dx}+\frac{1}{u}\left(\frac{d^2u}{dx^2}+P\frac{du}{dx}+Qu\right)v=\frac{R}{u}$$

$$\Rightarrow \qquad \frac{d^2v}{dx^2}+P_1\frac{dv}{dx}+Q_1v=R_1 \qquad \text{...(2)}$$

where $\quad P_1=\frac{2}{u}\frac{du}{dx}+P,\ Q_1=\frac{1}{u}\left(\frac{d^2u}{dx^2}+P\frac{du}{dx}+Qu\right)\quad$ and $\quad R_1=\frac{R}{u}$.

Let u be any particular integeral of the equation $\frac{d^2y}{dx^2}+P\frac{dy}{dx}+Qy=0$.

$$\therefore \qquad \frac{d^2u}{dx^2}+P\frac{du}{dx}+Qu=0 \qquad \text{Also} \qquad Q_1=\frac{1}{u}\times 0=0$$

$\therefore$ (2) becomes $\quad \frac{d^2v}{dx^2}+P_1\frac{dv}{dx}=R_1$

$$\Rightarrow \qquad \frac{dp}{dx}+P_1p=R_1, \quad \text{where } p=\frac{dv}{dx} \qquad \text{...(3)}$$

(3) is a linear differential equation of first order.

$$\therefore \qquad \text{I.F.}=e^{\int P_1\,dx}=e^{\int\left(\frac{2}{u}\frac{du}{dx}+P\right)dx}=e^{2\log u+\int P\,dx}=u^2e^{\int P\,dx}$$

$\therefore$ The solution of (3) is

$$p\,.\,u^2e^{\int P\,dx}=\int\left(\frac{R}{u}\,.\,u^2e^{\int P\,dx}\right)dx+c_1$$

$$\Rightarrow \qquad \frac{dv}{dx}=u^{-2}e^{-\int P\,dx}\int\left(R\,u\,e^{\int P\,dx}\right)dx+c_1u^{-2}e^{-\int P\,dx}$$

Integrating this equation, we get the value of v.

Hence the solution $y = uv$ is known. This solution will involve two arbitrary constants.

4. Method of Finding a Particular Integral of $\frac{d^2y}{dx^2}+P\frac{dy}{dx}+Qy=0$

I. Let $\quad y = e^{mx}$ be a solution of $\quad \frac{d^2y}{dx^2}+P\frac{dy}{dx}+Qy=0$.

$\therefore \quad m^2e^{mx}+Pme^{mx}+Qe^{mx}=0 \quad$ or $\quad m^2+Pm+Q=0$.

$\therefore$ **e^{mx} is a solution if** $\quad$ **$m^2+Pm+Q=0$.**

Particular cases :

(*i*) e^x is a solution if $1 + P + Q = 0$

(*ii*) e^{-x} is a solution if $1 - P + Q = 0$

II. Let x^m be a solution of $\quad \dfrac{d^2y}{dx^2} + P\dfrac{dy}{dx} + Qy = 0.$

$\therefore \quad m(m-1)x^{m-2} + Pmx^{m-1} + Qx^m = 0$

or $\quad (m(m-1) + Pmx + Qx^2)\, x^{m-2} = 0$

or $\quad m(m-1) + Pmx + Qx^2 = 0.$

$\therefore$ **x^m is a solution if $m(m-1) + Pmx + Qx^2 = 0$.**

Particular cases :

(*i*) x is a solution if $P + Qx = 0$

(*ii*) x^2 is a solution if $2 + 2Px + Qx^2 = 0$.

$\therefore$ 1. If $m^2 + Pm + Q = 0$, then e^{mx} is a solution.

2. If $1 + P + Q = 0$, then e^x is a solution.

3. If $1 - P + Q = 0$, then e^{-x} is a solution.

4. If $m(m-1) + Pmx + Q\,x^2 = 0$, then x^m is a solution.

5. If $P + Qx = 0$, then x is a solution.

6. If $2 + 2Px + Qx^2 = 0$, then x^2 is a solution.

5. Working Rules for Solving $\dfrac{d^2y}{dx^2} + P\dfrac{dy}{dx} + Qy = R$

Step I. Find a particular integral of the equation $\dfrac{d^2y}{dx^2} + P\dfrac{dy}{dx} + Qy = 0$. Call this function as u. Sometimes it is given with the problem itself.

Step II. Take $y = uv$ and change the dependent variable y to v. The resultant equation will be a linear equation of first order with dependent variable $\dfrac{dv}{dx}$.

Step III. Put $p = \dfrac{dv}{dx}$. Solve the equation and find p. Put $p = \dfrac{dv}{dx}$ and integrate to find the value of v.

Step IV. Find $y = uv$. This gives the general solution of the given equation.

Example 1. *Solve*

$$x\frac{d^2y}{dx^2} - (2x-1)\frac{dy}{dx} + (x-1)\,y = 0.$$

Sol. We have

$$x\frac{d^2y}{dx^2} - (2x-1)\frac{dy}{dx} + (x-1)\,y = 0.$$

$$\Rightarrow \quad \frac{d^2y}{dx^2} - \frac{2x-1}{x}\frac{dy}{dx} + \frac{x-1}{x}\,y = 0 \qquad ...(1)$$

Comparing (1) with

$$\frac{d^2y}{dx^2} + P\frac{dy}{dx} + Q\,y = R, \text{ we get}$$

$$P = -\frac{2x-1}{x}, \quad Q = \frac{x-1}{x}, \quad R = 0$$

Here $\quad 1 + P + Q = 1 - \frac{2x-1}{x} + \frac{x-1}{x} = 0$

$\therefore$ e^x **is a particular integral of (1), whose right member is already zero.**
Let $y = e^x v$.

$$\therefore \quad \frac{dy}{dx} = e^x \frac{dv}{dx} + v e^x = e^x \left(\frac{dv}{dx} + v\right)$$

and
$$\frac{d^2y}{dx^2} = e^x\left(\frac{d^2v}{dx^2} + \frac{dv}{dx}\right) + e^x\left(\frac{dv}{dx} + v\right) = e^x\left(\frac{d^2v}{dx^2} + 2\frac{dv}{dx} + v\right).$$

$\therefore$ (1) becomes

$$e^x\left(\frac{d^2v}{dx^2} + 2\frac{dv}{dx} + v\right) - \frac{2x-1}{x} e^x \left(\frac{dv}{dx} + v\right) + \frac{x-1}{x} e^x v = 0$$

$$\Rightarrow \quad e^x \frac{d^2v}{dx^2} + \left(2e^x - \frac{2x-1}{x} e^x\right)\frac{dv}{dx} + \left(e^x - \frac{2x-1}{x} e^x + \frac{x-1}{x} e^x\right) v = 0$$

$$\Rightarrow \quad e^x \frac{d^2v}{dx^2} + \frac{e^x}{x}\frac{dv}{dx} = 0 \quad \Rightarrow \quad \frac{d^2v}{dx^2} + \frac{1}{x}\frac{dv}{dx} = 0$$

$$\Rightarrow \quad \frac{dp}{dx} + \frac{1}{x} p = 0 \text{ where } p = \frac{dv}{dx}$$

$$\Rightarrow \quad \frac{dp}{p} + \frac{dx}{x} = 0 \qquad \Rightarrow \quad \log p + \log x = \log c_1$$

$$\Rightarrow \quad px = c_1 \qquad \Rightarrow \quad \frac{dv}{dx} x = c_1$$

$$\Rightarrow \quad dv = c_1 \frac{dx}{x} \qquad \Rightarrow \quad v = c_1 \log x + c_2.$$

$$\therefore \quad y = e^x v = e^x (c_1 \log x + c_2)$$

$\therefore$ The general solution of (1) is $\mathbf{y = c_1 e^x \log x + c_2 e^x}$.

Example 2. *Solve* $(x+2)\frac{d^2y}{dx^2} - (2x+5)\frac{dy}{dx} + 2y = (x+1)e^x$.

(D.L.U. 2004 ; M.D.U. 2007)

Sol. Dividing by $x + 2$, the given equation in standard form is

$$\frac{d^2y}{dx^2} - \frac{2x+5}{x+2}\frac{dy}{dx} + \frac{2}{x+2} y = \frac{x+1}{x+2} e^x. \qquad ...(1)$$

Comparing (1) with $\quad \frac{d^2y}{dx^2} + P\frac{dy}{dx} + Qy = R$, we get

$$P = -\frac{2x+5}{x+2}, \quad Q = \frac{2}{x+2}, \quad R = \frac{x+1}{x+2} e^x$$

Here $\quad 4 + 2P + Q = 4 + 2\left(-\frac{2x+5}{x+2}\right) + \frac{2}{x+2} = 0$

$\therefore$ e^{2x} is a particular integral of (1) with its right member replaced by zero.

Let $\qquad y = e^{2x}\, v.$

$$\therefore \quad \frac{dy}{dx} = e^{2x}\left(\frac{dv}{dx} + 2v\right) \quad \text{and} \quad \frac{d^2y}{dx^2} = e^{2x}\left(\frac{d^2v}{dx^2} + 4\frac{dv}{dx} + 4v\right).$$

Putting the values of y, $\frac{dy}{dx}$ and $\frac{d^2y}{dx^2}$ in (1), we get

$$e^{2x}\left(\frac{d^2v}{dx^2} + 4\frac{dv}{dx} + 4v\right) - \frac{2x+5}{x+2}e^{2x}\left(\frac{dv}{dx} + 2v\right) + \frac{2}{x+2}e^{2x}\, v = \frac{x+1}{x+2}e^{x}$$

$$\Rightarrow \quad \frac{d^2v}{dx^2} + 4\frac{dv}{dx} + 4v - \frac{2x+5}{x+2}\left(\frac{dv}{dx} + 2v\right) + \frac{2}{x+2}v = \frac{x+1}{x+2}e^{-x}$$

$$\Rightarrow \quad \frac{d^2v}{dx^2} + \frac{2x+3}{x+2}\frac{dv}{dx} = \frac{x+1}{x+2}e^{-x}$$

$$\Rightarrow \quad \frac{dp}{dx} + \frac{2x+3}{x+2}p = \frac{x+1}{x+2}e^{-x} \qquad \text{...(2)}$$

where $\quad p = \frac{dv}{dx}$.

(2) is a linear differential equation of the first order.

$$\text{I.F.} = e^{\int \frac{2x+3}{x+2}dx} = e^{\int\left(2 - \frac{1}{x+2}\right)dx} = e^{2x - \log(x+2)} = e^{2x} \div e^{\log(x+2)} = \frac{e^{2x}}{x+2}.$$

$\therefore$ The solution of (2) is

$$p\,\frac{e^{2x}}{x+2} = \int \frac{x+1}{x+2}e^{-x} \cdot \frac{e^{2x}}{x+2}\,dx + c_1$$

$$\Rightarrow \quad p\,\frac{e^{2x}}{x+2} = \int \frac{x+1}{(x+2)^2}e^{x}\,dx + c_1 = \int\left(\frac{1}{x+2} - \frac{1}{(x+2)^2}\right)e^{x}\,dx + c_1$$

$$= \frac{e^{x}}{x+2} + c_1$$

$$\therefore \quad p = e^{-x} + c_1 e^{-2x}(x+2) \quad \text{or} \quad \frac{dv}{dx} = e^{-x} + c_1 e^{-2x}(x+2)$$

$$\Rightarrow \quad \int dv = \int (e^{-x} + c_1 e^{-2x}(x+2))\,dx + c_2$$

$$\therefore \quad v = -e^{-x} + c_1\int (x+2)e^{-2x}\,dx + c_2$$

$$= -e^{-x} + c_1\left[(x+2)\cdot\frac{e^{-2x}}{-2} - \int 1 \cdot \frac{e^{-2x}}{-2}\,dx\right] + c_2$$

$$= -e^{-x} - \frac{c_1}{2}(x+2)e^{-2x} - \frac{c_1}{4}e^{-2x} + c_2$$

$\therefore$ $$v = -e^{-x} - \frac{c_1}{4}(2x+5)e^{-2x} + c_2$$

$\therefore$ $$y = e^{2x} v = -e^{x} - \frac{c_1}{4}(2x+5) + c_2 e^{2x}$$

$\therefore$ The general solution of (1) is

$$\mathbf{y = -e^{x} - \frac{c_1}{4}(2x+5) + c_2 e^{2x}.}$$

Example 3. *Solve* $x^2 \frac{d^2y}{dx^2} - 2x(1+x)\frac{dy}{dx} + 2(1+x)y = x^3.$ (MDU 2006)

Sol. Dividing by x^2, the given equation in standard form is

$$\frac{d^2y}{dx^2} - \frac{2(1+x)}{x}\frac{dy}{dx} + \frac{2(1+x)}{x^2}y = x. \quad ...(1)$$

Comparing (1) with $\frac{d^2y}{dx^2} + P\frac{dy}{dx} + Qy = R$, we get

$$P = -\frac{2(1+x)}{x}, \quad Q = \frac{2(1+x)}{x^2}, \quad R = x.$$

Here $P + xQ = -\frac{2(1+x)}{x} + \frac{2(1+x)}{x} = 0.$

$\therefore$ x is a particular integral of (1) with its right member replaced by zero.

Let $y = xv$.

$\therefore$ $$\frac{dy}{dx} = x\frac{dv}{dx} + v \quad \text{and} \quad \frac{d^2y}{dx^2} = \left(x\frac{d^2v}{dx^2} + 1.\frac{dv}{dx}\right) + \frac{dv}{dx} = x\frac{d^2v}{dx^2} + 2\frac{dv}{dx}$$

Putting the values of y, $\frac{dy}{dx}$ and $\frac{d^2y}{dx^2}$ in (1), we get

$$x\frac{d^2v}{dx^2} + 2\frac{dv}{dx} - \frac{2(1+x)}{x}\left(x\frac{dv}{dx} + v\right) + \frac{2(1+x)}{x^2}xv = x$$

$\Rightarrow$ $$x\frac{d^2y}{dx^2} - 2x\frac{dv}{dx} = x \Rightarrow \frac{d^2y}{dx} - 2\frac{dv}{dx} = 1 \Rightarrow \frac{dp}{dx} - 2p = 1 \quad ...(2)$$

where $p = \frac{dv}{dx}$

(2) is a linear differential equation of the first order.

$$\text{I.F.} = e^{\int -2\,dx} = e^{-2x}$$

$\therefore$ The solution of (2) is

$$p \,.\, e^{-2x} = \int e^{-2x}\,dx + c_1 \quad \Rightarrow \quad pe^{-2x} = \frac{e^{-2x}}{-2} + c_1$$

$\Rightarrow$ $$\frac{dv}{dx} = -\frac{1}{2} + c_1 e^{2x} \quad \Rightarrow \quad \int dv = \int\left(-\frac{1}{2} + c_1 e^{2x}\right)dx + c_2$$

$$\therefore \quad v = -\frac{x}{2} + c_1 \frac{e^{2x}}{2} + c_2$$

$$\therefore \quad y = xv = -\frac{x^2}{2} + c_1 \frac{xe^{2x}}{2} + xc_2$$

$\therefore$ The general solution of (1) is

$$\mathbf{y = -\frac{x^2}{2} + \frac{c_1\, x\, e^{2x}}{2} + xc_2.}$$

Example 4. *Solve* $\frac{d^2y}{dx^2} - \cot x \frac{dy}{dx} - (1 - \cot x)\, y = e^x \sin x.$ (M.D.U. 2006)

Sol. We have

$$\frac{d^2y}{dx^2} - \cot x \frac{dy}{dx} - (1 - \cot x)\, y = e^x \sin x. \qquad \text{...(1)}$$

Comparing (1) with $\frac{d^2y}{dx^2} + P\frac{dy}{dx} + Q = R$, we get

$$P = -\cot x,\ Q = -1 + \cot x,\ R = e^x \sin x$$

Here $1 + P + Q = 1 - \cot x - 1 + \cot x = 0.$

$\therefore$ e^x is a particular integral of (1) with its right member replaced by zero.

Let $y = e^x v.$

$$\therefore \quad \frac{dy}{dx} = e^x\left(\frac{dv}{dx} + v\right) \quad \text{and} \quad \frac{d^2y}{dx^2} = e^x\left(\frac{d^2v}{dx^2} + 2\frac{dv}{dx} + v\right).$$

$\therefore$ (1) becomes

$$e^x\left(\frac{d^2v}{dx^2} + 2\frac{dv}{dx} + v\right) - (\cot x)\, e^x\left(\frac{dv}{dx} + v\right) - (1 - \cot x)\, e^x v = e^x \sin x$$

$$\Rightarrow \quad \frac{d^2v}{dx^2} + (2 - \cot v)\frac{dv}{dx} = \sin x \quad \Rightarrow \quad \frac{dp}{dx} + (2 - \cot x)p = \sin x \qquad \text{...(2)}$$

where $p = \frac{dv}{dx}.$

(2) is a linear differential equation of the first order.

$$\text{I.F.} = e^{\int (2 - \cot x)\, dx} = e^{2x - \log \sin x} = e^{2x} \div e^{\log \sin x} = \frac{e^{2x}}{\sin x}.$$

$\therefore$ The solution of (2) is

$$p\,\frac{e^{2x}}{\sin x} = \int \sin x \,.\, \frac{e^{2x}}{\sin x}\, dx + c_1 \quad \Rightarrow \quad p\,\frac{e^{2x}}{\sin x} = \frac{e^{2x}}{2} + c_1$$

$$\Rightarrow \quad \frac{dv}{dx} = \frac{\sin x}{2} + \frac{c_1 \sin x}{e^{2x}} \quad \Rightarrow \quad \int dv = \int\left(\frac{\sin x}{2} + \frac{c_1 \sin x}{e^{2x}}\right) dx + c_2$$

$$\Rightarrow \quad v = -\frac{\cos x}{2} + c_1 \int e^{-2x} \sin x\, dx + c_2.$$

Let $\quad I = \int e^{-2x} \sin x \, dx.$

$\therefore \quad I = e^{-2x} . - \cos x - \int -2e^{-2x} . - \cos x \, dx = -e^{-2x} . \cos x - 2 \int e^{-2x} \cos x \, dx$

$= e^{-2x} . \cos x - 2 \left[e^{-2x} . \sin x - \int -2e^{-2x} \sin x \, dx \right]$

$= e^{-2x} \cos x - 2 e^{-2x} \sin x - 2 + 4 \int e^{-2x} \sin x \, dx$

$\therefore \quad I = e^{-2x} (\cos x - 2 \sin x) - 4I$

$\therefore \quad I = \dfrac{e^{-2x}}{5} (\cos x - 2 \sin x).$

$\therefore \quad v = -\dfrac{\cos x}{2} + \dfrac{c_1 e^{-2x}}{5} (\cos x - 2 \sin x) + c_2$

$\therefore \quad \mathbf{y = e^x v = -\dfrac{e^x \cos x}{2} + \dfrac{c_1 e^{-x}}{5} (\cos x - 2 \sin x) + c_2 e^x.}$

This gives the general solution of the given equation.

Example 5. *Solve* $x^2 \dfrac{d^2y}{dx^2} + x \dfrac{dy}{dx} - y = 0$ *given that* $x + \dfrac{1}{x}$ *is one integral*.

Sol. We have $\quad x^2 \dfrac{d^2y}{dx^2} + x \dfrac{dy}{dx} - y = 0. \quad$ ...(1)

Let $\quad y = \left(x + \dfrac{1}{x} \right) v.$

$\therefore \quad \dfrac{dy}{dx} = \left(x + \dfrac{1}{x} \right) \dfrac{dv}{dx} + \left(1 - \dfrac{1}{x^2} \right) v$

and $\quad \dfrac{d^2y}{dx^2} = \left(x + \dfrac{1}{x} \right) \dfrac{d^2v}{dx^2} + \left(1 - \dfrac{1}{x^2} \right) \dfrac{dv}{dx} + \left(1 - \dfrac{1}{x^2} \right) \dfrac{dv}{dx} + \dfrac{2}{x^3} v$

$= \left(x + \dfrac{1}{x} \right) \dfrac{d^2v}{dx^2} + 2 \left(1 - \dfrac{1}{x^2} \right) \dfrac{dv}{dx} + \dfrac{2}{x^3} v.$

Putting the values of y, $\dfrac{dy}{dx}$ and $\dfrac{d^2y}{dx^2}$ in (1), we get

$$(x^3 + x) \frac{d^2v}{dx^2} + (2x^2 - 2) \frac{dv}{dx} + \frac{2}{x} v + (x^2 + 1) \frac{dv}{dx} + \left(x - \frac{1}{x} \right) v - \left(x + \frac{1}{x} \right) v = 0$$

$\therefore \quad (x^3 + x) \dfrac{d^2v}{dx^2} + (3x^2 - 1) \dfrac{dv}{dx} = 0$

$\Rightarrow \quad \dfrac{d^2v}{dx^2} + \dfrac{3x^2 - 1}{x(x^2 + 1)} \dfrac{dv}{dx} = 0 \quad \Rightarrow \quad \dfrac{dp}{dx} + \dfrac{3x^2 - 1}{x(x^2 + 1)} p = 0 \quad$ ...(2)

where $p = \dfrac{dv}{dx}$.

$$\Rightarrow \quad \frac{dp}{p} + \left(\frac{4x}{x^2+1} - \frac{1}{x}\right)dx = 0 \qquad \text{(On resolving into partial fractions)}$$

Integrating, we get

$$\log p + 2\log(x^2+1) - \log x = \log c_1$$

$$\Rightarrow \quad \frac{p(x^2+1)^2}{x} = c_1 \quad \Rightarrow \quad p = \frac{c_1 x}{(x^2+1)^2}$$

$$\Rightarrow \quad \frac{dv}{dx} = \frac{c_1 x}{(x^2+1)^2} \quad \Rightarrow \quad dv = \frac{c_1}{2} \cdot \frac{2x}{(x^2+1)^2}\,dx$$

Integrating, we get

$$v = \frac{c_1}{2} \cdot \frac{(x^2+1)^{-1}}{-1} + c_2 \quad \Rightarrow \quad v = -\frac{c_1}{2(x^2+1)} + c_2$$

$$\therefore \quad y = \left(x + \frac{1}{x}\right)v = \frac{x^2+1}{x}v = -\frac{c_1}{2x} + \frac{c_2(x^2+1)}{x}$$

$\therefore$ The general solution of (1) is $\mathbf{y = \frac{c_1'}{x} + c_2\left(x + \frac{1}{x}\right)}$.

Example 6. *Solve* $\sin^2 x \dfrac{d^2y}{dx^2} = 2y$ *given that* $y = \cot x$ *is a solution.* (MDU 2004)

Sol. We have

$$\sin^2 x \frac{d^2y}{dx^2} - 2y = 0 \quad \text{or} \quad \frac{d^2y}{dx^2} - \frac{2}{\sin^2 x}y = 0 \qquad \ldots(1)$$

Let $y = v \cot x$.

$$\therefore \quad \frac{dy}{dx} = -v\,\text{cosec}^2 x + \cot x \frac{dv}{dx}$$

and

$$\frac{d^2y}{dx^2} = v \cdot 2\,\text{cosec}^2 x \cot x - \text{cosec}^2 x \frac{dv}{dx} + \cot x \frac{d^2v}{dx^2} - \text{cosec}^2 x \frac{dv}{dx}$$

$$= 2v\,\text{cosec}^2 x \cot x - 2\,\text{cosec}^2 x \frac{dv}{dx} + \cot x \frac{d^2v}{dx^2}$$

Putting the values of y, $\dfrac{dy}{dx}$ and $\dfrac{d^2y}{dx^2}$ in (1), we get

$$2v\,\text{cosec}^2 x \cot x - 2\,\text{cosec}^2 x \frac{dv}{dx} + \cot x \frac{d^2v}{dx^2} - \frac{2}{\sin^2 x} v \cot x = 0$$

$$\Rightarrow \quad \cot x \frac{d^2v}{dx^2} - 2\,\text{cosec}^2 x \frac{dv}{dx} = 0$$

$$\Rightarrow \quad \cot x \frac{dp}{dx} - 2p\,\text{cosec}^2 x = 0 \qquad \ldots(2)$$

where $p = \dfrac{dv}{dx}$

$$\Rightarrow \quad \frac{dp}{dx} - \frac{2\,\text{cosec}^2 x}{\cot x}\,dx = 0.$$

Integrating, we get

$$\log p + 2\log\cot x = \log c_1$$

$$\Rightarrow \quad p\cot^2 x = c_1 \quad \Rightarrow \quad \frac{dv}{dx} = c_1\tan^2 x \quad \Rightarrow \quad dv = c_1(\sec^2 x - 1)\,dx$$

Integrating, we get

$$v = c_1(\tan x - x) + c_2$$

$$\therefore \quad y = v\cot x = c_1(1 - x\cot x) + c_2\cot x$$

$\therefore$ The general solution of (1) is $\mathbf{y = c_1\,(1 - x\cot x) + c_2\cot x.}$

EXERCISE 1

Solve the following differential equations by changing the dependent variable :

1. $\dfrac{d^2y}{dx^2} - \dfrac{3}{x}\dfrac{dy}{dx} + \dfrac{3y}{x^2} = 2x - 1$

2. $(x\sin x + \cos x)\dfrac{d^2y}{dx^2} - x\cos x\,\dfrac{dy}{dx} + y\cos x = 0$

3. $x^2\dfrac{d^2y}{dx^2} - (x^2 + 2x)\dfrac{dy}{dx} + (x+2)\,y = x^3 e^x$ (M.D.U. 2003)

4. $(x+1)\dfrac{d^2y}{dx^2} - 2(x+3)\dfrac{dy}{dx} + (x+5)y = e^x$

5. $x\dfrac{d^2y}{dx^2} - \dfrac{dy}{dx} + (1-x)y = x^2 e^{-x}$

6. $x\dfrac{d^2y}{dx^2} + (x-2)\dfrac{dy}{dx} - 2y = x^3$

7. $x\dfrac{dy}{dx} - y = (x-1)\left[\dfrac{d^2y}{dx^2} - x + 1\right]$

8. $\dfrac{d^2y}{dx^2} + (1 - \cot x)\dfrac{dy}{dx} - y\cot x = \sin^2 x$

9. $x\dfrac{d^2y}{dx^2} - 2(x+1)\dfrac{dy}{dx} + (x+2)y = (x-2)e^x$ (M.D.U. 2007)

10. $x^2\dfrac{d^2y}{dx^2} + x\dfrac{dy}{dx} - 9y = 0$ given that $y = x^3$ is a solution of the given equation.

Answers

1. $y = c_1 x^3 + c_2 x + x^3\log x + x^2$

2. $y = c_1\cos x + c_2 x$

3. $y = c_1 x e^x + c_2 x + (x-1)\,x e^x$

4. $y = c_1 e^x (x+1)^5 - \dfrac{1}{4} x e^x + c_2 e^x$

5. $y = c_1(2x+1)e^{-x} + c_2 e^x - \dfrac{1}{4}e^{-x}(2x^2 + 2x + 1)$

6. $y = c_1(x^2 - 2x + 2) + c_2 e^{-x} + x^3 - 3x^2 + 6x - 6$

7. $y = c_1 e^x + c_2 x - 1 - x^2$

8. $y = c_1(\sin x - \cos x) + c_2 e^{-x} - \dfrac{1}{10}(\sin 2x - 2\cos 2x)$

9. $y = c_1 x^3 e^x + c_2 e^x - \dfrac{1}{2}x^2 e^x + x e^x$

10. $y = c_1 x^3 + c_2 x^{-3}$.

6. Second Method of Solving a Linear Differential Equation of the Second Order by Changing the Dependent Variable (By the Removal of the First Derivative)

Let $$\frac{d^2y}{dx^2} + P\frac{dy}{dx} + Qy = R \qquad ...(1)$$

be a linear differential equation of the second order where P, Q and R are functions of x.

This method is used when a particular integral of the equation when R is replaced by zero is neither given nor is easily found.

Let $y = uv$ where u and v are functions of x.

$$\therefore \quad \frac{dy}{dx} = u\frac{dv}{dx} + v\frac{du}{dx} \text{ and } \frac{d^2y}{dx^2} = u\frac{d^2v}{dx^2} + 2\frac{du}{dx}\frac{dv}{dx} + v\frac{d^2u}{dx^2}.$$

$$\therefore \quad (1) \Rightarrow \left(u\frac{d^2v}{dx^2} + 2\frac{du}{dx}\frac{dv}{dx} + v\frac{d^2u}{dx^2}\right) + P\left(u\frac{dv}{dx} + v\frac{du}{dx}\right) + Quv = R$$

$$\Rightarrow \quad u\frac{d^2v}{dx^2} + \left(2\frac{du}{dx} + Pu\right)\frac{dv}{dx} + \left(\frac{d^2u}{dx^2} + P\frac{du}{dx} + Qu\right)v = R$$

$$\Rightarrow \quad \frac{d^2v}{dx^2} + \left(\frac{2}{u}\frac{du}{dx} + P\right)\frac{dv}{dx} + \frac{1}{u}\left(\frac{d^2u}{dx^2} + P\frac{du}{dx} + Qu\right)v = \frac{R}{u}$$

$$\Rightarrow \quad \frac{d^2v}{dx^2} + P_1\frac{dv}{dx} + Q_1 v = R_1 \qquad ...(2)$$

where $$P_1 = \frac{2}{u}\frac{du}{dx} + P,\ Q_1 = \frac{1}{u}\left(\frac{d^2u}{dx^2} + P\frac{du}{dx} + Qu\right) \text{ and } R_1 = \frac{R}{u}.$$

Let u be any function such that $$P_1 = \frac{2}{u}\frac{du}{dx} + P = 0. \qquad ...(3)$$

$$\therefore \quad \frac{du}{u} = -\frac{1}{2}P\,dx$$

Integrating, we get

$$\log u = -\frac{1}{2}\int P\,dx. \quad \therefore \quad u = e^{-\frac{1}{2}\int P\,dx}$$

$$(3) \Rightarrow \quad \frac{du}{dx} = -\frac{1}{2}uP \text{ and } \frac{d^2u}{dx^2} = -\frac{1}{2}\left(u\frac{dP}{dx} + P\frac{du}{dx}\right) = -\frac{1}{2}u\frac{dP}{dx} - \frac{1}{2}P\frac{du}{dx}.$$

$$\therefore \quad Q_1 = \frac{1}{u}\left(-\frac{1}{2}u\frac{dP}{dx} - \frac{1}{2}P\frac{du}{dx}\right) + \frac{P}{u}\frac{du}{dx} + Q$$

$$= -\frac{1}{2}\frac{dP}{dx} + \frac{P}{2u}\frac{du}{dx} + \frac{P}{u}\frac{du}{dx} + Q = -\frac{1}{2}\frac{dP}{dx} + \frac{P}{2u}\frac{du}{dx} + Q$$

$$= -\frac{1}{2}\frac{dP}{dx} + \frac{P}{2u}\left(-\frac{1}{2}uP\right) + Q = -\frac{1}{2}\frac{dP}{dx} - \frac{1}{4}P^2 + Q.$$

$$\therefore \quad Q_1 = Q - \frac{1}{4}P^2 - \frac{1}{2}\frac{dP}{dx}$$

Also, $$R_1 = \frac{R}{u} = \frac{R}{e^{-\frac{1}{2}\int P\,dx}} = R e^{\frac{1}{2}\int P\,dx}$$

$\therefore$ (2) becomes $$\frac{d^2v}{dx^2} + Q_1 v = R_1 \qquad ...(4)$$

where $Q_1 = Q - \frac{1}{4}P^2 - \frac{1}{2}\frac{dP}{dx}$ and $R_1 = R\, e^{\frac{1}{2}\int P\,dx}$.

Equation (4) is called the **normal form** of the equation (1). In the normal form (4), the first derivative term has been removed. This equation is solved to find the value of v.

Hence the solution $y = uv$ is known. This solution will involve two arbitrary constants.

Particular cases :

(*i*) If $Q_1 = \lambda$, a constant, then (4) becomes $\frac{d^2v}{dx^2} + \lambda v = R_1$.

This is a linear equation with constant coefficients.

(*ii*) If $Q_1 = \frac{\lambda}{x^2}$, then (4) becomes $x^2\frac{d^2v}{dx^2} + \lambda v = x^2 R_1$.

This is a homogeneous linear equation. The substitution $z = \log x$ will reduce it to one with constant coefficients.

Remark. In order to apply this method, it is advisable to calculate $Q - \frac{1}{4}P^2 - \frac{1}{2}\frac{dP}{dx}$. If this quantity is either a constant λ or $\frac{\lambda}{x^2}$, then this method should be used. In such a case the transformation $y = vu = ve^{-\frac{1}{2}\int P\,dx}$ will reduce the given equation to a linear equation with constant coefficients or to a homogenous linear equation.

7. Working Rules for Solving $\mathbf{\frac{d^2y}{dx^2} + P\frac{dy}{dx} + Q\,y = R}$

Step I. Find $Q_1 = Q - \frac{1}{4}P^2 - \frac{1}{2}\frac{dP}{dx}$. If this quantity is of the form λ or $\frac{\lambda}{x^2}$, then use this method.

Step II. Find $e^{-\frac{1}{2}\int P\,dx}$ and call this function as u.

Step III. Take $y = uv$. The equation with v as the dependent variable is $\frac{d^2v}{dx^2} + Q_1 v = R_1$, where $R_1 = \frac{R}{u}$. Solve this equation to find v.

Step IV. Find $y = uv$. This gives the general solution of the given equation.

Example 1. *Solve* $\frac{d^2y}{dx^2} - 2\tan x\frac{dy}{dx} + 5y = 0$. (M.D.U. 2003 ; K.U. 2006 S)

Sol. We have

$$\frac{d^2y}{dx^2} - 2\tan x\frac{dy}{dx} + 5y = 0. \qquad ...(1)$$

Here $\quad P = -2\tan x,\ Q = 5,\ R = 0.$

Now $\quad Q_1 = Q - \frac{1}{4}P^2 - \frac{1}{2}\frac{dP}{dx} = 5 - \frac{1}{4}(4\tan^4 x) - \frac{1}{2}(-2\sec^2 x)$

$= 5 - \tan^2 x + \sec^2 x = 6$, a constant.

Now $\quad \int P\,dx = \int -2\tan x\,dx = -2\log\sec x$

Let $\quad u = e^{-\frac{1}{2}\int P\,dx}$ $\quad \therefore \quad u = e^{-\frac{1}{2}(-2\log\sec x)} = \sec x.$

Let $\quad y = uv = (\sec x)\,v.$

$\therefore$ (1) reduces to $\dfrac{d^2v}{dx^2} + Q_1 v = R_1 \qquad ...(2)$

where $\quad R_1 = \dfrac{R}{u} = \dfrac{0}{\sec x} = 0.$

$\therefore$ (2) $\Rightarrow \quad \dfrac{d^2v}{dx^2} + 6v = 0 \quad \Rightarrow \quad (D^2 + 6)v = 0 \qquad ...(3)$

$\therefore$ The A.E. of (3) is $D^2 + 6 = 0. \quad \therefore \quad D = \pm\sqrt{6}\,i$

$\therefore \quad v = c_1 \cos\sqrt{6}\,x + c_2 \sin\sqrt{6}\,x$

$\therefore$ $y = uv$ implies $\quad \mathbf{y = \sec x(c_1 \cos\sqrt{6}\,x + c_2 \sin\sqrt{6}\,x).}$

This is the general solution of the given equation.

Example 2. *Solve* $\left(\dfrac{d^2y}{dx^2} + y\right)\cot x + 2\left(\dfrac{dy}{dx} + y\tan x\right) = \sec x.$ (K.U. 2007)

Sol. Given equation is same as

$$\cot x\,\frac{d^2y}{dx^2} + 2\frac{dy}{dx} + (\cot x + 2\tan x)\,y = \sec x.$$

Dividing by $\cot x$, we get

$$\frac{d^2y}{dx^2} + 2\tan x\frac{dy}{dx} + (1 + 2\tan^2 x)\,y = \sec x\tan x \qquad ...(1)$$

Here $\quad P = 2\tan x,\ Q = 1 + 2\tan^2 x,\ R = \sec x\tan x$

Now $\quad Q_1 = Q - \frac{1}{4}P^2 - \frac{1}{2}\frac{dP}{dx} = 1 + 2\tan^2 x - \frac{1}{4}\,.\,4\tan^2 x - \frac{1}{2}\,.\,2\sec^2 x$

$= 0$, a constant.

Now $\quad \int P\,dx = \int 2\tan x\,dx = 2\log\sec x$

Let $u = e^{-\frac{1}{2}\int P\,dx}$. $\therefore$ $u = e^{-\frac{1}{2}.2\log\sec x} = \cos x$.

Let $y = uv = (\cos x)\,v$.

$\therefore$ (1) reduces to $\dfrac{d^2v}{dx^2} + Q_1 v = R_1$ ...(2)

where $R_1 = \dfrac{R}{u} = \dfrac{\sec x \tan x}{\cos x} = \sec^2 x \tan x$.

$\therefore$ (2) $\Rightarrow$ $\dfrac{d^2v}{dx^2} + 0\,.\,v = \sec^2 x \tan x$ $\Rightarrow$ $\dfrac{d^2v}{dx^2} = \sec x\,(\sec x \tan x)$.

Integrating, we get $\dfrac{dv}{dx} = \dfrac{1}{2}\sec^2 x + c_1$

Integrating again, we get $v = \dfrac{1}{2}\tan x + c_1 x + c_2$.

$\therefore$ $y = uv$ implies $\mathbf{y = \cos x\left(\dfrac{1}{2}\tan x + c_1 x + c_2\right)}$.

This is general solution of the given equation.

Example 3. *Solve* $\dfrac{d}{dx}\left(\cos^2 x\,\dfrac{dy}{dx}\right) + y\cos^2 x = 0$. (K.U. 2004, 2006)

Sol. We have $\dfrac{d}{dx}\left(\cos^2 x\,\dfrac{dy}{dx}\right) + y\cos^2 x = 0$.

$\Rightarrow$ $\cos^2 x\,\dfrac{d^2y}{dx^2} + (-2\cos x \sin x)\,\dfrac{dy}{dx} + y\cos^2 x = 0$

Dividing by $\cos^2 x$, we get

$$\frac{d^2y}{dx^2} - 2\tan x\,\frac{dy}{dx} + y = 0 \qquad ...(1)$$

Here $P = -2\tan x$, $Q = 1$, $R = 0$.

Now $Q_1 = Q - \dfrac{1}{4}P^2 - \dfrac{1}{2}\dfrac{dP}{dx} = 1 - \dfrac{1}{4}(4\tan^2 x) - \dfrac{1}{2}(-2\sec^2 x)$

$= 2$, which is a constant.

Now $\int P\,dx = \int -2\tan x\,dx = -2\log\sec x$

Let $u = e^{-\frac{1}{2}\int P\,dx}$. $\therefore$ $u = e^{-\frac{1}{2}\int(-2\log\sec x)} = \sec x$.

Let $y = uv = (\sec x)\,v$

$\therefore$ (1) reduces to $\dfrac{d^2v}{dx^2} + Q_1 v = R_1$ ...(2)

where $R_1 = \dfrac{R}{u} = \dfrac{0}{\sec x} = 0$.

$\therefore$ (2) $\Rightarrow \dfrac{d^2v}{dx^2} + 2v = 0 \Rightarrow (D^2 + 2)v = 0$ ...(3)

$\therefore$ The A.E. of (3) is $D^2 + 2 = 0$. $\therefore$ $D = \pm\sqrt{2}\,i$

$\therefore$ $v = c_1 \cos\sqrt{2}\,x + c_2 \sin\sqrt{2}\,x$.

$\therefore$ $y = uv$ implies

$$\mathbf{y = \sec x\,[c_1 \cos\sqrt{2}\,x + c_2 \sin\sqrt{2}\,x\,]}$$

This is the general solution of the given equation.

Example 4. *Solve* $\dfrac{d^2y}{dx^2} + 2x\dfrac{dy}{dx} + (x^2 + 1)\,y = x^3 + 3x.$ (MDU 2005)

Sol. We have

$$\frac{d^2y}{dx^2} + 2x\frac{dy}{dx} + (x^2+1)\,y = x^3 + 3x. \quad ...(1)$$

Here $P = 2x,\ Q = x^2 + 1,\ R = x^3 + 3x$.

Now $Q_1 = Q - \dfrac{1}{4}P^2 - \dfrac{1}{2}\dfrac{dP}{dx} = x^2 + 1 - \dfrac{1}{4}\,.\,4x^2 - \dfrac{1}{2}\,.\,2 = 0$, a constant.

Now $\displaystyle\int P\,dx = \int 2x\,dx = x^2$

Let $u = e^{-\frac{1}{2}\int P\,dx}$. $\therefore$ $u = e^{-\frac{1}{2}x^2}$

Let $y = uv = e^{-\frac{1}{2}x^2}\,.\,v$

$\therefore$ (1) reduces to $\dfrac{d^2v}{dx^2} + Q_1 v = R_1$ ...(2)

where $R_1 = \dfrac{R}{u} = (x^3 + 3x)\,e^{x^2/2}$.

$\therefore$ (2) $\Rightarrow \dfrac{d^2v}{dx^2} + 0\,.\,v = (x^3 + 3x)\,e^{x^2/2} \Rightarrow \dfrac{d^2v}{dx^2} = (x^3 + 3x)\,e^{x^2/2}$.

Integrating, we get

$$\frac{dv}{dx} = \int (x^3 + 3x)\,e^{x^2/2}\,dx + c_1 = \int (x^2 + 3)\,x\,e^{x^2/2}\,dx + c_1$$

$$= \int (2t + 3)\,e^t\,dt + c_1 \quad \text{where } t = x^2/2$$

$$= (2t + 3)e^t - \int 2\,.\,e^t\,dt + c_1 = (2t + 3)e^t - 2\,e^t + c_1 = (2t + 1)\,e^t + c_1$$

$$= (x^2 + 1)\,e^{x^2/2} + c_1$$

Integrating again we get

$$v = \int (x^2 + 1)\,e^{x^2/2}\,dx + c_1 x + c_2\,.$$

Now, $\displaystyle\int x^2\,e^{x^2/2}\,dx = \int x\,(x\,e^{x^2/2})\,dx = x\,.\,e^{x^2/2} - \int 1\,.\,e^{x^2/2}\,dx$

$\therefore \quad \int (x^2+1) e^{x^2/2} dx = xe^{x^2/2}.$

$\therefore \qquad v = xe^{x^2/2} + c_1x + c_2$

$\therefore \quad y = uv$ implies $\quad y = e^{-x^2/2} [xe^{x^2/2} + c_1x + c_2]$

or $\quad \mathbf{y = x + (c_1x + c_2)\, e^{-x^2/2}}.$

This is the general solution of the given equation.

Example 5. *Solve*

$$x^2 (\log x)^2 \frac{d^2y}{dx^2} - 2x \log x \frac{dy}{dx} + (2 + \log x) - 2(\log x)^2\ y = x^2 (\log x)^3.$$

Sol. Dividing the given equation by $x^2 (\log x)^2$, we get

$$\frac{d^2y}{dx^2} - \frac{2}{x \log x}\frac{dy}{dx} + \frac{2 + \log x - 2(\log x)^2}{x^2 (\log x)^2} y = \log x \qquad ...(1)$$

Here $\quad P = -\frac{2}{x \log x}, \; Q = \frac{2 + \log x - 2(\log x)^2}{x^2 (\log x)^2}, \; R = \log x$

Now, $\quad Q_1 = Q - \frac{1}{4}P^2 - \frac{1}{2}\frac{dP}{dx}$

$$= \frac{2 + \log x - 2(\log x)^2}{x^2 (\log x)^2} - \frac{1}{4}\cdot\frac{4}{x^2 (\log x)^2}$$

$$-\frac{1}{2}.(-2)\left[\frac{1}{x}.(-1)(\log x)^{-2}.\frac{1}{x} + \frac{1}{\log x}\left(\frac{-1}{x^2}\right)\right]$$

$$= \frac{2 + \log x - 2(\log x)^2}{x^2 (\log x)^2} - \frac{1}{x^2 (\log x)^2} - \frac{1}{x^2 (\log x)^2} - \frac{1}{x^2 \log x}$$

$$= -\frac{2}{x^2}, \text{ which is of the form } \frac{\lambda}{x^2}.$$

Now $\quad \int P\, dx = \int -\frac{2}{x \log x} dx = -2 \log \log x$

Let $\quad u = e^{-\frac{1}{2}\int P\, dx}. \quad \therefore \quad u = e^{-\frac{1}{2}.-2 \log \log x} = e^{\log \log x} = \log x$

Let $\quad y = uv = (\log x)\, v.$

$\therefore$ (1) reduces to $\quad \frac{d^2v}{dx^2} + Q_1 v = R_1 \qquad ...(2)$

where $\quad R_1 = \frac{R}{u} = \frac{\log x}{\log x} = 1.$

$\therefore$ (2) $\Rightarrow \quad \frac{d^2v}{dx^2} - \frac{2}{x^2}v = 1 \quad \Rightarrow \quad x^2 \frac{d^2v}{dx^2} - 2v = x^2 \qquad ...(3)$

Let $\quad z = \log x. \quad \therefore \quad x = e^z$

$\therefore$ $$x^2 \frac{d^2}{dx^2} = D(D-1), \quad \text{where } D = \frac{d}{dz}.$$

$\therefore$ (3) $\Rightarrow$ $$(D(D-1)-2)v = e^{2z} \Rightarrow (D^2 - D - 2)\, v = e^{2z} \qquad ...(4)$$

$\therefore$ The A.E. of (4) is $D^2 - D - 2 = 0$. $\therefore$ $D = -1, 2$

$\therefore$ $$\text{C.F.} = c_1 e^{-z} + c_2 e^{2z} = c_1 (e^z)^{-1} + c_2 (e^z)^2 = c_1 x^{-1} + c_2 x^2$$

$$\text{P.I.} = \frac{1}{D^2 - D - 2} e^{2z} = z \cdot \frac{1}{2D-1} e^{2z} = z \cdot \frac{1}{2(2)-1} e^{2z} = \frac{z}{3} e^{2z} = \frac{\log x}{3} \cdot x^2$$

$\therefore$ $$v = \text{C.F.} + \text{P.I.} = c_1 x^{-1} + c_2 x^2 + \frac{x^2 \log x}{3}$$

$\therefore$ $y = uv$ implies $\mathbf{y = \log x \left(\frac{c_1}{x} + c_2 x^2 + \frac{x^2 \log x}{3} \right)}$.

This is the general solution of the given equation.

Example 6. *Solve*

$$\frac{d^2y}{dx^2} - 4x \frac{dy}{dx} + (4x^2 - 1)\, y = -3e^{x^2} (\sin 2x + 5e^{-2x} + 6).$$

Sol. We have

$$\frac{d^2y}{dx^2} - 4x \frac{dy}{dx} + (4x^2 - 1)\, y = -3e^{x^2} (\sin 2x + 5e^{-2x} + 6). \qquad ...(1)$$

Here $P = -4x$, $Q = 4x^2 - 1$, $R = -3\, e^{x^2} (\sin 2x + 5e^{-2x} + 6)$

Now $$Q_1 = Q - \frac{1}{4} P^2 - \frac{1}{2} \frac{dP}{dx} = 4x^2 - 1 - \frac{1}{4} \cdot 16x^2 - \frac{1}{2}(-4)$$
$$= 4x^2 - 1 - 4x^2 + 2 = 1, \text{ which is a constant.}$$

Now $$\int P\, dx = \int -4x\, dx = -2x^2$$

Let $u = e^{-\frac{1}{2}\int P\,dx}$. $\therefore$ $u = e^{-\frac{1}{2} \cdot -2x^2} = e^{x^2}$.

Let $y = uv = e^{x^2} v$.

$\therefore$ (1) reduces to $\dfrac{d^2v}{dx^2} + Q_1 v = R_1$ ...(2)

where $$R_1 = \frac{R}{u} = \frac{-3e^{x^2} (\sin 2x + 5e^{-2x} + 6)}{e^{x^2}} = -3(\sin 2x + 5e^{-2x} + 6)$$

$\therefore$ (2) $\Rightarrow$ $$\frac{d^2v}{dx^2} + 1 . v = -3(\sin 2x + 5e^{-2x} + 6)$$

$\Rightarrow$ $$(D^2 + 1)v = -3 \sin 2x - 15\, e^{-2x} - 18. \qquad ...(3)$$

$\therefore$ The A.E. of (3) is $D^2 + 1 = 0$. $\therefore$ $D = \pm i$

$\therefore$ $$\text{C.F.} = c_1 \cos x + c_2 \sin x$$

$$\text{P.I.} = \frac{1}{D^2+1}(-3\sin 2x - 15e^{-2x} - 18)$$

$$= -3\frac{1}{D^2+1}\sin 2x - 15\frac{1}{D^2+1}e^{-2x} - 18\frac{1}{D^2+1}e^{0x}$$

$$= -3\frac{1}{(-4)+1}\sin 2x - 15\frac{1}{(-2)^2+1}e^{-2x} - 18\frac{1}{0+1}e^{0x} = \sin 2x - 3e^{-2x} - 18.$$

$\therefore$ $v = \text{C.F.} + \text{P.I.} = c_1\cos x + c_2\sin x + \sin 2x - 3e^{-2x} - 18$

$\therefore$ $y = uv$ implies

$$\mathbf{y = e^{x^2}(c_1\cos x + c_2\sin x + \sin 2x - 3e^{-2x} - 18)}$$

This is the general solution of the given equation.

EXERCISE 2

Solve the following differential equations by removing the first derivative :

1. $\dfrac{d^2y}{dx^2} - 2\tan x\dfrac{dy}{dx} + 5y = e^x\sec x$

2. $\dfrac{d^2y}{dx^2} - \dfrac{2}{x}\dfrac{dy}{dx} + \left(n^2 + \dfrac{2}{x^2}\right)y = 0$ *(MDU 2005)*

3. $x\dfrac{d}{dx}\left(x\dfrac{dy}{dx} - y\right) - 2x\dfrac{dy}{dx} + 2y + x^2y = 0$

4. $\dfrac{d^2y}{dx^2} - \dfrac{2}{x}\dfrac{dy}{dx} + \left(1 + \dfrac{2}{x^2}\right)y = xe^x$

5. $\dfrac{d^2y}{dx^2} - \dfrac{1}{\sqrt{x}}\dfrac{dy}{dx} + \dfrac{x + \sqrt{x} - 8}{4x^2}y = 0$

6. $\dfrac{d^2y}{dx^2} - 2x\dfrac{dy}{dx} + (x^2 + 2)y = e^{\frac{1}{2}(x^2 + 2x)}$

7. $\dfrac{d^2y}{dx^2} - \dfrac{3}{x}\dfrac{dy}{dx} + \dfrac{3}{x^2}y = 2x - 1$

8. $\dfrac{d^2y}{dx^2} - 4x\dfrac{dy}{dx} + 4x^2y = xe^{x^2}$

9. $\dfrac{d^2y}{dx^2} - 4x\dfrac{dy}{dx} + (4x^2 - 1)y = -3e^{x^2}\sin 2x$

10. $x^2\dfrac{d^2y}{dx^2} - 2x(3x - 2)\dfrac{dy}{dx} + 3x(3x - 4)y = e^{3x}$.

Answers

1. $y = \sec x\left[c_1\cos\sqrt{6}\,x + c_2\sin\sqrt{6}\,x + \dfrac{e^x}{7}\right]$

2. $y = x[c_1\cos nx + c_2\sin nx]$

3. $y = x[c_1\cos x + c_2\sin x]$

4. $y = x\left[c_1\cos x + c_2\sin x + \dfrac{1}{2}e^x\right]$

5. $y = e^{\sqrt{x}}(c_1x^2 + c_2x^{-1})$

6. $y = e^{x^2/2}\left(c_1\cos\sqrt{3}x + c_2\sin\sqrt{3}x + \dfrac{1}{4}e^x\right)$

7. $y = c_1 x + c_2 x^3 + x^3 \log x + x^2$

8. $y = e^{x^2}\left(c_1 \cos\sqrt{2}x + c_2 \sin\sqrt{2}x + \frac{x}{2}\right)$

9. $y = e^{x^2}(c_1 \cos x + c_2 \sin x + \sin 2x)$

10. $y = \frac{e^{3x}}{x^2}\left(c_1 x^2 + c_2 x^{-1} + \frac{1}{3}x^2 \log x\right)$.

8. Method of Solving a Linear Differential Equation of the Second Order by Changing the Independent Variable

Let $$\frac{d^2y}{dx^2} + P\frac{dy}{dx} + Qy = R \quad \text{...(1)}$$

be a linear differential equation of the second order, where P, Q and R are functions of x.

Let $z = \phi(x)$ be a function of x.

$$\therefore \quad \frac{dy}{dx} = \frac{dy}{dz}\frac{dz}{dx} \quad \text{and} \quad \frac{d^2y}{dx^2} = \frac{dy}{dz}\cdot\frac{d^2z}{dx^2} + \frac{dz}{dx}\cdot\frac{d^2y}{dz^2}\frac{dz}{dx} = \frac{dy}{dz}\frac{d^2z}{dx^2} + \frac{d^2y}{dz^2}\left(\frac{dz}{dx}\right)^2$$

$$\therefore \quad (1) \Rightarrow \quad \frac{dy}{dz}\frac{d^2z}{dx^2} + \frac{d^2y}{dz^2}\left(\frac{dz}{dx}\right)^2 + P\frac{dy}{dz}\frac{dz}{dx} + Qy = R$$

$$\Rightarrow \quad \left(\frac{dz}{dx}\right)^2\frac{d^2y}{dz^2} + \left(\frac{d^2z}{dx^2} + P\frac{dz}{dx}\right)\frac{dy}{dz} + Qy = R$$

$$\Rightarrow \quad \frac{d^2y}{dz^2} + \frac{\frac{d^2z}{dx^2} + P\frac{dz}{dx}}{\left(\frac{dz}{dx}\right)^2}\frac{dy}{dz} + \frac{Q}{\left(\frac{dz}{dx}\right)^2}y = \frac{R}{\left(\frac{dz}{dx}\right)^2}$$

$$\Rightarrow \quad \frac{d^2y}{dz^2} + P_1\frac{dy}{dz} + Q_1 y = R_1 \quad \text{...(2)}$$

where $$P_1 = \frac{\frac{d^2z}{dx^2} + P\frac{dz}{dx}}{\left(\frac{dz}{dx}\right)^2}, \quad Q_1 = \frac{Q}{\left(\frac{dz}{dx}\right)^2} \quad \text{and} \quad R_1 = \frac{R}{\left(\frac{dz}{dx}\right)^2}.$$

We choose $z = \phi(x)$ so that $Q_1 = \pm a^2$ *i.e.*, $\frac{Q}{\left(\frac{dz}{dx}\right)^2} = \pm a^2$ or $\frac{dz}{dx} = \sqrt{\pm\frac{Q}{a^2}}$, the sign being that which makes square root real and a^2 is any positive constant. This positive constant may be taken as per our convenience. We may consistently take $a^2 = 1$.

Integrating $\frac{dz}{dx} = \sqrt{\pm \frac{Q}{a^2}}$, we get z in terms of x *i.e.* $z = \int \sqrt{\pm \frac{Q}{a^2}}\, dx$.

For this value of z, we calculate P_1. Let P_1 comes out to be a constant, say p_1.

$$\therefore \quad (2) \text{ becomes } \frac{d^2y}{dz^2} + p_1 \frac{dy}{dz} \pm a^2 y = \frac{R}{\left(\frac{dz}{dx}\right)^2}.$$

This is a linear differential equation with constant coefficients. This gives the value of y in terms of z. The value of z is put in terms of x so that we may get the value of y in terms of x.

9. Working Rules for Solving $\frac{d^2y}{dx^2} + P\frac{dy}{dx} + Qy = R$

Step I. Changing x to z, the given equation reduces to $\frac{d^2y}{dz^2} + P_1 \frac{dy}{dz} + Q_1 y = R_1$

where
$$P_1 = \frac{\frac{d^2z}{dx^2} + P\frac{dz}{dx}}{\left(\frac{dz}{dx}\right)^2}, \quad Q_1 = \frac{Q}{\left(\frac{dz}{dx}\right)^2} \quad \text{and} \quad R_1 = \frac{R}{\left(\frac{dz}{dx}\right)^2}.$$

Step II. Put $\frac{dz}{dx} = \sqrt{\pm \frac{Q}{a^2}}$ and choose sign out of '±' and the value of a^2. For this choice of $\frac{dz}{dx}$, the value of P_1 must come out to be a constant.

Step III. The resultant equation is a linear equation with constant coefficients. Solve this equation and get the value of y in terms of z.

Step IV. Put the value of z and get y in terms of x.

Example 1. *Solve* $\frac{d^2y}{dx^2} + \frac{2}{x}\frac{dy}{dx} + \frac{a^2}{x^4} y = 0$. (KU 2004 S)

Sol. We have
$$\frac{d^2y}{dx^2} + \frac{2}{x}\frac{dy}{dx} + \frac{a^2}{x^4} y = 0. \quad \text{...(1)}$$

Here $P = \frac{2}{x}$, $Q = \frac{a^2}{x^4}$, $R = 0$.

Let $z = \phi(x)$.

$$\therefore \quad (1) \Rightarrow \frac{d^2y}{dz^2} + P_1 \frac{dy}{dz} + Q_1 y = R_1 \quad \text{...(2)}$$

where
$$P_1 = \frac{\frac{d^2z}{dx^2} + P\frac{dz}{dx}}{\left(\frac{dz}{dx}\right)^2}, \quad Q_1 = \frac{Q}{\left(\frac{dz}{dx}\right)^2}, \quad R_1 = \frac{R}{\left(\frac{dz}{dx}\right)^2}$$

Let $\dfrac{dz}{dx} = \sqrt{\pm \dfrac{Q}{\lambda^2}}$. $\therefore$ $\dfrac{dz}{dx} = \sqrt{\pm \dfrac{a^2}{x^4 \lambda^2}}$. We choose '+' sign and $\lambda^2 = a^2$

$\therefore$ $\dfrac{dz}{dx} = \dfrac{1}{x^2}$. $\therefore$ $z = \dfrac{x^{-1}}{-1} = -\dfrac{1}{x}$ and $\dfrac{d^2z}{dx^2} = -\dfrac{2}{x^3}$

$$\therefore \quad P_1 = \frac{\dfrac{d^2z}{dx^2} + P\dfrac{dz}{dx}}{\left(\dfrac{dz}{dx}\right)^2} = \frac{-\dfrac{2}{x^3} + \dfrac{2}{x}.\dfrac{1}{x^2}}{\left(\dfrac{1}{x^2}\right)^2} = 0, \quad Q_1 = \frac{Q}{\left(\dfrac{dz}{dx}\right)^2} = \frac{\dfrac{a^2}{x^4}}{\left(\dfrac{1}{x^2}\right)^2} = a^2,$$

$$R_1 = \frac{R}{\left(\dfrac{dz}{dx}\right)^2} = \frac{0}{\left(\dfrac{1}{x^2}\right)^2} = 0.$$

$\therefore$ (2) becomes $\dfrac{d^2y}{dz^2} + 0.\dfrac{dy}{dz} + a^2y = 0$ or $(D^2 + a^2)\, y = 0.$ ...(3)

$\therefore$ The A.E. of (3) is $D^2 + a^2 = 0$. $\therefore$ $D = \pm ai$

$\therefore$ $y = c_1 \cos az + c_2 \sin az$

$\therefore$ $y = c_1 \cos\left(-\dfrac{a}{x}\right) + c_2 \sin\left(-\dfrac{a}{x}\right)$ $\left(\because\ z = -\dfrac{1}{x}\right)$

$\therefore$ $\mathbf{y = c_1 \cos\left(\dfrac{a}{x}\right) - c_2 \sin\left(\dfrac{a}{x}\right)}$.

This is the general solution of the given equation.

Example 2. *Solve*

$$(1+x^2)^2 \frac{d^2y}{dx^2} + 2x\,(1+x^2)\frac{dy}{dx} + 4y = 0 \cdot$$

Sol. Dividing by $(1 + x^2)^2$, the given equation becomes

$$\frac{d^2y}{dx^2} + \frac{2x}{1+x^2}\frac{dy}{dx} + \frac{4}{(1+x^2)^2}\,y = 0. \qquad ...(1)$$

Here $P = \dfrac{2x}{1+x^2}$, $Q = \dfrac{4}{(1+x^2)^2}$, $R = 0$.

Let $z = \phi(x)$.

$\therefore$ (1) $\Rightarrow$ $\dfrac{d^2y}{dz^2} + P_1\dfrac{dy}{dz} + Q_1\, y = R_1$ ...(2)

where $P_1 = \dfrac{\dfrac{d^2z}{dx^2} + P\dfrac{dz}{dx}}{\left(\dfrac{dz}{dx}\right)^2}$, $Q_1 = \dfrac{Q}{\left(\dfrac{dz}{dx}\right)^2}$, $R_1 = \dfrac{R}{\left(\dfrac{dz}{dx}\right)^2}$

Let $\dfrac{dz}{dx} = \sqrt{\pm \dfrac{Q}{a^2}}$. $\therefore$ $\dfrac{dz}{dx} = \sqrt{\pm \dfrac{4}{(1+x^2)^2\, a^2}}$. We choose '+' sign and $a^2 = 4$.

$$\therefore \quad \frac{dz}{dx} = \frac{1}{1+x^2}$$

$$\therefore \quad z = \int \frac{dx}{1+x^2} = \tan^{-1} x \quad \text{and} \quad \frac{d^2z}{dx^2} = -\frac{1}{(1+x^2)^2} \cdot 2x = -\frac{2x}{(1+x^2)^2}$$

$$\therefore \quad P_1 = \frac{\dfrac{d^2z}{dx^2} + P\dfrac{dz}{dx}}{\left(\dfrac{dz}{dx}\right)^2} = \frac{-\dfrac{2x}{(1+x)^2} + \dfrac{2x}{1+x^2} \cdot \dfrac{1}{1+x^2}}{\left(\dfrac{1}{1+x^2}\right)^2} = 0$$

$$Q_1 = \frac{Q}{\left(\dfrac{dz}{dx}\right)^2} = \frac{\dfrac{4}{(1+x^2)^2}}{\dfrac{1}{(1+x)^2}} = 4, \quad R_1 = \frac{R}{\left(\dfrac{dz}{dx}\right)^2} = \frac{0}{\dfrac{1}{(1+x^2)^2}} = 0.$$

$\therefore$ (2) becomes $\dfrac{d^2y}{dx^2} + 0 . \dfrac{dy}{dz} + 4y = 0$ or $(D^2 + 4)\, y = 0$ ...(3)

$\therefore$ The A.E. of (3) is $D^2 + y = 0$. $\therefore$ $D = \pm 2i$

$$\therefore \quad y = c_1 \cos 2z + c_2 \sin 2z$$

Putting $z = \tan x$, we get

$$\mathbf{y = c_1 \cos (2 \tan^{-1} x) + c_2 \sin (2 \tan^{-1} x).}$$

This is the general solution of the given equation.

Example 3. *Solve*

$$x \frac{d^2y}{dx^2} + (4x^2 - 1) \frac{dy}{dx} + 4x^3 y = 2x^2.$$

Sol. Dividing by x, the given equation becomes

$$\frac{d^2y}{dx^2} + \frac{4x^2 - 1}{x} \frac{dy}{dx} + 4x^2 y = 2x \qquad \text{...(1)}$$

Here $P = \dfrac{4x^2 - 1}{x}$, $Q = 4x^2$, $R = 2x^2$.

Let $z = \phi(x)$.

$$\therefore \ (1) \Rightarrow \quad \frac{d^2y}{dz^2} + P_1 \frac{dy}{dz} + Q_1 y = R_1 \qquad \text{...(2)}$$

where $$P_1 = \frac{\frac{d^2z}{dx^2} + P\frac{dz}{dx}}{\left(\frac{dz}{dx}\right)^2}, \quad Q_1 = \frac{Q}{\left(\frac{dz}{dx}\right)^2}, \quad R_1 = \frac{R}{\left(\frac{dz}{dx}\right)^2}$$

Let $\frac{dz}{dx} = \sqrt{\pm\frac{Q}{a^2}}$. $\therefore$ $\frac{dz}{dx} = \sqrt{\pm\frac{4x^2}{a^2}}$ We choose '+' sign and $a^2 = 1$.

$\therefore$ $\frac{dz}{dx} = 2x$. $\therefore$ $z = x^2$ and $\frac{d^2z}{dx^2} = 2$

$$\therefore \quad P_1 = \frac{\frac{d^2z}{dx^2} + P\frac{dz}{dx}}{\left(\frac{dz}{dx}\right)^2} = \frac{2 + \frac{4x^2 - 1}{x}.2x}{4x^2} = 2,$$

$$Q_1 = \frac{Q}{\left(\frac{dz}{dx}\right)^2} = \frac{4x^2}{4x^2} = 1, \quad R_1 = \frac{R}{\left(\frac{dz}{dx}\right)^2} = \frac{2x^2}{4x^2} = \frac{1}{2}.$$

$\therefore$ (2) becomes $\frac{d^2y}{dz^2} + 2\frac{dy}{dz} + 1.y = \frac{1}{2}$ or $(D^2 + 2D + 1)y = \frac{1}{2}$ ...(3)

$\therefore$ The A.E. of (3) is $D^2 + 2D + 1 = 0$. $\therefore$ $D = -1, -1$.

$\therefore$ C.F. $= (c_1 + c_2 z)e^{-z}$

$$\text{P.I.} = \frac{1}{(D+1)^2}\frac{1}{2} = \frac{1}{2}\cdot\frac{1}{(D+1)^2}e^{0z} = \frac{1}{2}\frac{1}{(0+1)^2}e^{0z} = \frac{1}{2}$$

$$\therefore \quad y = \text{C.F.} + \text{P.I.} = (c_1 + c_2 z)e^{-z} + \frac{1}{2}$$

Putting $z = x^2$, we get

$$\mathbf{y = (c_1 + c_2 x^2)\, e^{-x^2} + \frac{1}{2}}.$$

This is the general solution of the given equation.

Example 4. *Solve* $\frac{d^2y}{dx^2} - (8e^{2x} + 2)\frac{dy}{dx} + 4e^{4x}y = e^{6x}$. (MDU 2003)

Sol. We have

$$\frac{d^2y}{dx^2} - (8e^{2x} + 2)\frac{dy}{dx} + 4e^{4x}y = e^{6x}. \quad ...(1)$$

Here $P = -(8e^{2x} + 2)$, $Q = 4e^{4x}$, $R = e^{6x}$.

Let $z = \phi(x)$.

$$\therefore \ (1) \quad \frac{d^2y}{dz^2} + P_1\frac{dy}{dz} + Q_1 y = R_1 \quad ...(2)$$

where $$P_1 = \frac{\frac{d^2z}{dx^2} + P\frac{dz}{dx}}{\left(\frac{dz}{dx}\right)^2}, \quad Q_1 = \frac{Q}{\left(\frac{dz}{dx}\right)^2}, \quad R_1 = \frac{R}{\left(\frac{dz}{dx}\right)^2}.$$

Let $\dfrac{dz}{dx} = \sqrt{\pm \dfrac{Q}{a^2}}$. $\therefore$ $\dfrac{dz}{dx} = \sqrt{\pm \dfrac{4e^{4x}}{a^2}}$. We choose '+' sign and $a^2 = 1$.

$\therefore$ $\dfrac{dz}{dx} = 2e^{2x}$ $\therefore$ $z = \int 2e^{2x}\, dx = e^{2x}$ and $\dfrac{d^2z}{dx^2} = 4e^{2x}$.

$$\therefore \quad P_1 = \frac{\dfrac{d^2z}{dx^2} + P\dfrac{dz}{dx}}{\left(\dfrac{dz}{dx}\right)^2} = \frac{4e^{2x} - (8e^{2x} + 2)\,2e^{2x}}{4e^{4x}} = -4,$$

$$Q_1 = \frac{Q}{\left(\dfrac{dz}{dx}\right)^2} = \frac{4e^{4x}}{4e^{4x}} = 1,\ R_1 = \frac{R}{\left(\dfrac{dz}{dx}\right)^2} = \frac{e^{6x}}{4e^{4x}} = \frac{e^{2x}}{4}.$$

$\therefore$ (2) becomes $\dfrac{d^2y}{dz^2} - 4\dfrac{dy}{dz} + 1.y = \dfrac{e^{2x}}{4}$ or $(D^2 - 4D + 1)\,y = \dfrac{z}{4}$. ...(3)

$\therefore$ The A.E. of (3) is $D^2 - 4D + 1 = 0$. $\therefore$ $D = 2 \pm \sqrt{3}$

$$\therefore \quad \text{C.F.} = c_1 e^{(2-\sqrt{3})z} + c_2 e^{(2+\sqrt{3})z}$$

$$\text{P.I.} = \frac{1}{D^2 - 4D + 1}\frac{z}{4} = \frac{1}{4}\,(1 - (4D - D^2))^{-1} z$$

$$= \frac{1}{4}\,(1 + 4D - D^2 + \ldots\ldots)\,z = \frac{1}{4}\,(z + 4 - 0 + \ldots\ldots) = \frac{z}{4} + 1$$

$$\therefore \quad y = \text{C.F.} + \text{P.I.} = c_1 e^{(2-\sqrt{3})z} + c_2 e^{(2+\sqrt{3})z} + \frac{z}{4} + 1$$

Putting $z = e^{2x}$, we get

$$\mathbf{y = c_1 e^{(2-\sqrt{3})e^{2x}} + c_2 e^{(2+\sqrt{3})e^{2x}} + \frac{e^{2x}}{4} + 1.}$$

This is the general solution of the given equation.

Example 5. *Solve* $\dfrac{d^2y}{dx^2} + (3\sin x - \cot x)\dfrac{dy}{dx} + 2y\sin^2 x = e^{-\cos x}\sin^2 x.$

Sol. We have

$$\frac{d^2y}{dx^2} + (3\sin x - \cot x)\frac{dy}{dx} + 2y\sin^2 x = e^{-\cos x}\sin^2 x. \quad \text{...(1)}$$

Here $P = 3\sin x - \cot x$, $Q = 2\sin^2 x$, $R = e^{-\cos x}\sin^2 x$.

Let $z = \phi(x)$.

$$\therefore \quad (1) \Rightarrow \frac{d^2y}{dz^2} + P_1\frac{dy}{dz} + Q_1\, y = R_1 \quad \text{...(2)}$$

where $P_1 = \dfrac{\dfrac{d^2z}{dx^2} + P\dfrac{dz}{dx}}{\left(\dfrac{dz}{dx}\right)^2}$, $Q_1 = \dfrac{Q}{\left(\dfrac{dz}{dx}\right)^2}$, $R_1 = \dfrac{R}{\left(\dfrac{dz}{dx}\right)^2}$.

Let $\frac{dz}{dx} = \sqrt{\pm\frac{Q}{a^2}}$. $\therefore$ $\frac{dz}{dx} = \sqrt{\pm\frac{2\sin^2 x}{a^2}}$. We choose '+' sign and $a^2 = 2$.

$\therefore$ $\frac{dz}{dx} = \sin x$ $\therefore$ $z = -\cos x$ and $\frac{d^2z}{dx^2} = \cos x$.

$$\therefore \quad P_1 = \frac{\frac{d^2z}{dx^2} + P\frac{dz}{dx}}{\left(\frac{dz}{dx}\right)^2} = \frac{\cos x + (3\sin x - \cot x)\sin x}{\sin^2 x} = 3,$$

$$Q_1 = \frac{Q}{\left(\frac{dz}{dx}\right)^2} = \frac{2\sin^2 x}{\sin^2 x} = 2,\ R_1 = \frac{R}{\left(\frac{dz}{dx}\right)^2} = \frac{e^{-\cos x}\sin^2 x}{\sin^2 x} = e^{-\cos x}.$$

$\therefore$ (2) becomes $\frac{d^2y}{dz^2} + 3\frac{dy}{dz} + 2y = e^{-\cos x}$ or $(D^2 + 3D + 2)\,y = e^z$. ...(3)

$\therefore$ The A.E. of (3) is $D^2 + 3D + 2 = 0$. $\therefore$ $D = -1, -2$

$\therefore$ C.F. $= c_1 e^{-z} + c_2 e^{-2z}$

$$\text{P.I.} = \frac{1}{D^2 + 3D + 2}e^z = \frac{1}{1+3+2}e^z = \frac{e^z}{6}.$$

$$\therefore \quad y = \text{C.F.} + \text{P.I.} = c_1e^{-z} + c_2e^{-2z} + \frac{e^z}{6}$$

Putting $z = -\cos x$, we get

$$\mathbf{y = c_1e^{\cos x} + c_2e^{2\cos x} + \frac{e^{-\cos x}}{6}}.$$

This is the general solution of the given equation.

Example 6. *Solve* $(1+x)^2\frac{d^2y}{dx^2} + (1+x)\frac{dy}{dx} + y = 4\cos\log(1+x)$. (KU 2005)

Sol. Dividing by $(1 + x)^2$, the given equation becomes

$$\frac{d^2y}{dx^2} + \frac{1}{1+x}\frac{dy}{dx} + \frac{1}{(1+x)^2}y = \frac{4\cos\log(1+x)}{(1+x)^2} \quad ...(1)$$

Here $P = \frac{1}{1+x},\ Q = \frac{1}{(1+x)^2},\ R = \frac{4\cos\log(1+x)}{(1+x)^2}$.

Let $z = \phi(x)$.

$$\therefore \ (1) \Rightarrow \frac{d^2y}{dz^2} + P_1\frac{dy}{dz} + Q_1 y = R_1 \quad ...(2)$$

where $P_1 = \frac{\frac{d^2z}{dx^2} + P\frac{dz}{dx}}{\left(\frac{dz}{dx}\right)^2},\ Q_1 = \frac{Q}{\left(\frac{dz}{dx}\right)^2},\quad R_1 = \frac{R}{\left(\frac{dz}{dx}\right)^2}$.

Let $\dfrac{dz}{dx} = \sqrt{\pm\dfrac{Q}{a^2}}$. $\therefore$ $\dfrac{dz}{dx} = \sqrt{\pm\dfrac{1}{a^2(1+x)^2}}$. We choose '+' sign and $a^2 = 1$.

$\therefore$ $\dfrac{dz}{dx} = \dfrac{1}{1+x}$ $\therefore$ $z = \displaystyle\int \dfrac{dx}{1+x} = \log(1+x)$ and $\dfrac{d^2z}{dx^2} = -\dfrac{1}{(1+x)^2}$

$$\therefore \quad P_1 = \frac{\dfrac{d^2z}{dx^2} + P\dfrac{dz}{dx}}{\left(\dfrac{dz}{dx}\right)^2} = \frac{-\dfrac{1}{(1+x)^2} + \dfrac{1}{1+x}\cdot\dfrac{1}{1+x}}{\dfrac{1}{(1+x)^2}} = 0,$$

$$Q_1 = \frac{Q}{\left(\dfrac{dz}{dx}\right)^2} = \frac{\dfrac{1}{(1+x)^2}}{\dfrac{1}{(1+x)^2}} = 1,$$

$$R_1 = \frac{R}{\left(\dfrac{dz}{dx}\right)^2} = \frac{\dfrac{4\cos\log(1+x)}{(1+x)^2}}{\dfrac{1}{(1+x)^2}} = 4\cos\log(1+x).$$

$\therefore$ (2) becomes $\dfrac{d^2y}{dz^2} + 0.\dfrac{dy}{dz} + 1.y = 4\cos\log(1+x)$

$\therefore$ $(D^2 + 1)\,y = 4\cos z$ ...(3)

$\therefore$ The A.E. of (3) is $D^2 + 1 = 0$. $\therefore$ $D = \pm i$

$\therefore$ C.F. $= c_1\cos z + c_2\sin z$

$$\text{P.I.} = \frac{1}{D^2+1}4\cos z = 4\frac{1}{D^2+1}\cos z = 4\,.\,z\,\frac{1}{2D}\cos z = 2z\sin z$$

$\therefore$ $y = \text{C.F.} + \text{P.I.} = c_1\cos z + c_2\sin z + 2z\sin z$.

Putting $z = \log(1 + x)$, we get

$$\mathbf{y = c_1\cos\log(1+x) + c_2\sin\log(1+x) + 2\log(1+x)\sin\log(1+x).}$$

This is the general solution of the given equation.

EXERCISE 3

Solve the following differential equations by changing the independent variables :

1. $\dfrac{d^2y}{dx^2} + \tan x\dfrac{dy}{dx} + y\cos^2 x = 0$

2. $x^4\dfrac{d^2y}{dx^2} + 2x^3\dfrac{dy}{dx} + \lambda^2 y = 0$

3. $x^6\dfrac{d^2y}{dx^2} + 3x^5\dfrac{dy}{dx} + a^2y = \dfrac{1}{x^2}$ *(K.U. 2003 C)*

4. $\dfrac{d^2y}{dx^2} + \cot x\dfrac{dy}{dx} + 4y\,\text{cosec}^2 x = 0$

5. $\cos x\dfrac{d^2y}{dx^2} + \sin x\dfrac{dy}{dx} - 2y\cos^3 x = 2\cos^5 x$ *(K.U. 2005 S)*

6. $\dfrac{d^2y}{dx^2} - (1 + 4e^x)\dfrac{dy}{dx} + 3e^{2x}\,y = e^{2(x+e^x)}$ *(M.D.U. 2007 S)*

7. $\dfrac{d^2y}{dx^2} - \cot x \dfrac{dy}{dx} - y \sin^2 x = \cos x - \cos^3 x$

8. $(1+x)^2 \dfrac{d^2y}{dx^2} + (1+x)\dfrac{dy}{dx} + y = 4 \sin \log (1+x)$

9. $x \dfrac{d^2y}{dx^2} - \dfrac{dy}{dx} + 4x^3 y = x^5$

10. $x \dfrac{d^2y}{dx^2} - \dfrac{dy}{dx} - 4x^3 y = 8x^3 \sin x^2$.

Answers

1. $y = c_1 \cos \sin x + c_2 \sin \sin x$

2. $y = c_1 \cos (\lambda/x) + c_2 \sin (\lambda/x)$

3. $y = c_1 \cos \dfrac{a}{2x^2} + c_2 \sin \dfrac{a}{2x^2} + \dfrac{1}{a^2 x^2}$

4. $y = c_1 \cos \left(2 \log \tan \dfrac{x}{2}\right) + c_2 \sin \left(2 \log \tan \dfrac{x}{2}\right)$

5. $y = c_1 e^{\sqrt{2} \sin x} + c_2 e^{-\sqrt{2} \sin x} + \sin^2 x$

6. $y = c_1 e^{e^x} + c_2 e^{3e^x} - e^{2e^x}$

7. $y = c_1 e^{-\cos x} + c_2 e^{\cos x} - \cos x$

8. $y = c_1 \cos \log (1+x) + c_2 \sin \log (1+x) - 2 \log (1+x) \cos \log (1+x)$

9. $y = c_1 \cos x^2 + c_2 \sin x^2 + \dfrac{x^2}{4}$

10. $y = c_1 e^{x^2} + c_2 e^{-x^2} - \sin x^2$.

10. Method of Solving a Linear Differential Equation of the Second Order by Variation of Parameters

Let
$$\frac{d^2y}{dx^2} + P\frac{dy}{dx} + Qy = R \qquad ...(1)$$
be a linear differential equation of the second order where P, Q and R are functions of x.

Let y_1 and y_2 be two linearly independent solutions of the equation
$$\frac{d^2y}{dx^2} + P\frac{dy}{dx} + Qy = 0 \qquad ...(2)$$

$$\therefore \quad \begin{vmatrix} y_1 & y_2 \\ y_1' & y_2' \end{vmatrix} = y_1 y_2' - y_1' y_2 \neq 0 \qquad ...(3)$$

$\therefore$ $y = c_1 y_1 + c_2 y_2$ is the general solution of (2) and it is also the complementary function of the solution of (1).

Let $Y = Ay_1 + By_2$, where A and B are functions of x such that

$$A'y_1 + B'y_2 = 0 \quad ...(4) \qquad \text{and} \quad A'y_1' + B'y_2' = R \qquad ...(5)$$

$$\therefore \quad Y' = Ay_1' + By_2' + (A'y_1 + B'y_2) = Ay_1' + By_2' \qquad \text{(Using (4))}$$

and
$$Y'' = Ay_1'' + By_2'' + (A'y_1' + B'y_2') = Ay_1'' + By_2'' + R \qquad \text{(Using (5))}$$

Now $Y'' + PY' + QY$
$$= (Ay_1'' + By_2'' + R) + P(Ay_1' + By_2') + Q(Ay_1 + By_2)$$
$$= A(y_1'' + Py_1' + Qy_1) + B(y_2'' + Py_2' + Qy_2) + R = A.0 + B.0 + R = R$$

$\therefore$ $Y = Ay_1 + By_2$ is a particular integral of the equation (1).

$\therefore$ $y = c_1 y_1 + c_2 y_2 + (Ay_1 + By_2)$ is the general solution of the given differential equation (1).

The functions A and B satisfies the equations
$$A'y_1 + B'y_2 - 0 = 0$$
and
$$A'y_1' + B'y_2' - R = 0.$$

$$\therefore \quad \frac{A'}{-y_2R+0} = \frac{B'}{0+y_1R} = \frac{1}{y_1y_2' - y_1'y_2}$$

$$\therefore \quad A' = -\frac{y_2R}{y_1y_2' - y_1'y_2} \quad \text{and} \quad B' = \frac{y_1R}{y_1y_2' - y_1'y_2}.$$

A′ and B′ are meaningful because of (3).

Integrating above equations, we get the values of the functions A and B. Arbitrary constants are not used in the values of A and B because $Y = Ay_1 + By_2$ is taken as a particular integral of (1).

∴ The general solution of the equation (1) is known.

Remark. It is essential to have the coefficient of the second order term in the linear differential equation as unity, otherwise in the above discussion, 'Y' will not be a solution of the given equation.

11. Working Rules of Solving $\frac{d^2y}{dx^2} + P\frac{dy}{dx} + Qy = R$

Step I. Find two independent solutions of the equation $\frac{d^2y}{dx^2} + P\frac{dy}{dx} + Qy = 0$ and call these as y_1 and y_2.

Step II. Let $Y = Ay_1 + By_2$ be the required particular solution of the given equation.

Step III. Solve the equations $A' = -\frac{y_2R}{y_1y_2' - y_1'y_2}$ and $B' = \frac{y_1R}{y_1y_2' - y_1'y_2}$ to find the values of A and B.

Step IV. Write the general solution of the given equation as $y = c_1y_1 + c_2y_2 + Ay_1 + By_2$, where c_1 and c_2 are arbitrary constants.

Example 1. *Solve* $\frac{d^2y}{dx^2} + y = \text{cosec } x.$

Sol. We have $\frac{d^2y}{dx^2} + y = \text{cosec } x.$

$$\Rightarrow \quad (D^2 + 1)y = \text{cosec } x \qquad \ldots(1)$$

∴ The A.E. of (1) is $D^2 + 1 = 0$. ∴ $D = \pm i$

$$\therefore \quad \text{C.F.} = c_1 \cos x + c_2 \sin x$$

Let P.I. = $A \cos x + B \sin x$ ...(2)

$$\therefore \quad A' = -\frac{y_2R}{y_1y_2' - y_1'y_2}, \quad B' = \frac{y_1R}{y_1y_2' - y_1'y_2}$$

Here $y_1 = \cos x$, $y_2 = \sin x$, $R = \text{cosec } x$

$$\therefore \quad y_1y_2' - y_1'y_2 = \cos x \,.\, \cos x - (-\sin x) \sin x = 1$$

$$\therefore \quad A' = -\frac{\sin x \text{ cosec } x}{1} = -1 \quad \text{and} \quad B' = \frac{\cos x \text{ cosec } x}{1} = \cot x$$

$$\therefore \quad A = \int -1\, dx = -x \quad \text{and} \quad B = \int \cot x \, dx = -\log \sin x$$

$$\therefore \quad (2) \Rightarrow \quad \text{P.I.} = -1 \,.\, \cos x + (\log \sin x) \sin x$$

Using $y = \text{C.F.} + \text{P.I.}$, we have

$$\mathbf{y = c_1 \cos x + c_2 \sin x + (\log \sin x) \sin x - \cos x.}$$

This is the general solution of the given equation.

Example 2. *Solve* $\dfrac{d^2y}{dx^2} + n^2 y = \sec nx$.

Sol. We have $\dfrac{d^2y}{dx^2} + n^2 y = \sec nx$. ...(1)

$\Rightarrow \quad (D^2 + n^2)y = \sec nx$

$\therefore$ The A.E. is $D^2 + n^2 = 0. \quad \therefore \quad D = \pm ni$

$\therefore \quad \text{C.F.} = c_1 \cos nx + c_2 \sin nx.$

Let $\quad \text{P.I.} = A \cos nx + B \sin nx$ ...(2)

$$\therefore \quad A' = -\frac{y_2 R}{y_1 y_2' - y_1' y_2}, \quad B' = \frac{y_1 R}{y_1 y_2' - y_1' y_2}.$$

Here $\quad y_1 = \cos nx,\ y_2 = \sin nx,\ R = \sec nx$

$$\therefore \quad y_1 y_2' - y_1' y_2 = n \cos nx \cos nx - (-n \sin nx) \sin nx = n$$

$$\therefore \quad A' = -\frac{\sin nx \sec nx}{n} = -\frac{1}{n} \tan nx \quad \text{and} \quad B' = \frac{\cos nx \sec nx}{n} = \frac{1}{n}$$

$$\therefore \quad A = \int -\frac{1}{n} \tan nx\, dx = -\frac{1}{n} \cdot \frac{\log \sec nx}{n} = \frac{1}{n^2} \log \cos nx.$$

$$B = \int \frac{1}{n}\, dx = \frac{1}{n} x.$$

$$\therefore \quad (2) \Rightarrow \quad \text{P.I.} = \left(\frac{1}{n^2} \log \cos x\right) \cos nx + \left(\frac{1}{n} x\right) \sin nx.$$

Using $y = \text{C.F.} + \text{P.I.}$, we have

$$\mathbf{y = c_1 \cos nx + c_2 \sin nx + \frac{1}{n^2} (\log \cos nx) \cos nx + \frac{1}{n} x \sin nx.}$$

This is the general solution of the given equation.

Example 3. *Solve* $\dfrac{d^2y}{dx^2} - y = \dfrac{2}{1+e^x}$.

(K.U. 2003, 2006, 2006 S ; M.D.U. 2003, 2005, 2006)

Sol. We have $\dfrac{d^2y}{dx^2} - y = \dfrac{2}{1+e^x}$. ...(1)

$$\Rightarrow \quad (D^2 - 1)\, y = \frac{2}{1+e^x}$$

$\therefore$ The A.E. is $D^2 - 1 = 0. \quad \therefore \quad D = \pm 1$

$\therefore \quad \text{C.F.} = c_1 e^x + c_2 e^{-x}.$

Let $\quad \text{P.I.} = A e^x + B e^{-x}$ ...(2)

$$\therefore \quad A' = -\frac{y_2 R}{y_1 y_2' - y_1' y_2}, \quad B' = \frac{y_1 R}{y_1 y_2' - y_1' y_2}$$

Here $\quad y_1 = e^x, \quad y_2 = e^{-x}, \quad R = \dfrac{2}{1+e^x}$

$\therefore \quad y_1y_2' - y_1'y_2 = e^x(-e^{-x}) - e^x e^{-x} = -2$

$$\therefore \quad A' = -\frac{e^{-x}\left(\frac{2}{1+e^x}\right)}{-2} = \frac{e^{-x}}{1+e^x} \quad \text{and} \quad B' = \frac{e^x\left(\frac{2}{1+e^x}\right)}{-2} = -\frac{e^x}{1+e^x}$$

$$\therefore \quad A = \int \frac{e^{-x}}{1+e^x}\,dx = \int \frac{1}{e^x(1+e^x)}\,dx = \int\left[\frac{1}{e^x\,.\,1} + \frac{1}{(-1)(1+e^x)}\right]dx$$

$$= \int\left(\frac{1}{e^x} - \frac{1+e^x-e^x}{1+e^x}\right)dx = \int\left(e^{-x} - 1 + \frac{e^x}{1+e^x}\right)dx$$

$$= -e^{-x} - x + \log(1+e^x) = -e^{-x} - \log e^x + \log(1+e^x) = \log\frac{1+e^x}{e^x} - e^{-x}$$

$$B = \int -\frac{e^x}{1+e^x}\,dx = -\log(1+e^x)$$

$$\therefore \quad (2) \Rightarrow \text{P.I.} = \left(\log\frac{1+e^x}{e^x} - e^{-x}\right)e^x + (-\log(1+e^x))e^{-x}$$

$$= e^x \log\frac{1+e^x}{e^x} - 1 - e^{-x}\log(1+e^x).$$

Using y = C.F. + P.I., we have

$$\mathbf{y = c_1 e^x + c_2 e^{-x} + e^x \log\frac{1+e^x}{e^x} - 1 - e^{-x}\log(1+e^x).}$$

This is the general solution of the given equation.

Example 4. *Solve* $\frac{d^2y}{dx^2} + 4y = \tan 2x$. (K.U. 2004 ; M.D.U. 2004)

Sol. We have $\frac{d^2y}{dx^2} + 4y = \tan 2x$. ...(1)

$\Rightarrow \quad (D^2 + 4)\,y = \tan 2x$

$\therefore$ The A.E. is $D^2 + 4 = 0. \quad \therefore \quad D = \pm 2i$

$\therefore \quad$ C.F. $= c_1 \cos 2x + c_2 \sin 2x$

Let $\quad$ P.I. $= A\cos 2x + B\sin 2x$ ...(2)

$$\therefore \quad A' = -\frac{y_2R}{y_1y_2' - y_1'y_2},\quad B' = \frac{y_1R}{y_1y_2' - y_1'y_2}$$

Here $\quad y_1 = \cos 2x,\ y_2 = \sin 2x,\ R = \tan 2x$

$\therefore \quad y_1y_2' - y_1'y_2 = (\cos 2x)2\cos 2x - (-2\sin 2x)\sin 2x = 2$

$$\therefore \quad A' = -\frac{\sin 2x \tan 2x}{2} = -\left(\frac{1-\cos^2 2x}{2\cos 2x}\right) = \frac{\cos 2x - \sec 2x}{2}$$

and
$$B' = \frac{\cos 2x \tan 2x}{2} = \frac{\sin 2x}{2}$$

$$\therefore \quad A = \int \frac{\cos 2x - \sec 2x}{2}\, dx = \frac{1}{2}\left[\frac{\sin 2x}{2} - \frac{\log(\sec 2x + \tan 2x)}{2}\right]$$

$$= \frac{1}{4}(\sin 2x - \log(\sec 2x + \tan 2x))$$

$$B = \int \frac{\sin 2x}{2}\, dx = -\frac{\cos 2x}{4}$$

$$\therefore \quad (2) \Rightarrow \text{P.I.} = \frac{1}{4}(\sin 2x - \log(\sec 2x + \tan 2x))\cos 2x - \frac{\cos 2x}{4} . \sin 2x$$

$$= -\frac{1}{4}\cos 2x \log(\sec 2x + \tan 2x).$$

Using y = C.F. + P.I., we have

$$\mathbf{y = c_1 \cos 2x + c_2 \sin 2x - \frac{1}{4} \cos 2x \log (\sec 2x + \tan 2x)}$$

This is the general solution of the given equation.

Example 5. *Solve* $(1-x)\dfrac{d^2y}{dx^2} + x\dfrac{dy}{dx} - y = (1-x)^2,\ x \neq 1.$ (C.D.L.U. 2004 ; K.U. 2007)

Sol. Dividing by $1 - x$, the given equation becomes

$$\frac{d^2y}{dx^2} + \frac{x}{1-x}\frac{dy}{dx} - \frac{1}{1-x}y = 1 - x. \quad \text{...(1)}$$

Here $\quad P = \dfrac{x}{1-x},\ Q = -\dfrac{1}{1-x},\ R = 1 - x$

$$P + xQ = \frac{x}{1-x} + x\left(-\frac{1}{1-x}\right) = 0$$

$\therefore \quad y_1 = x$ is a solution of

$$\frac{d^2y}{dx^2} + \frac{x}{1-x}\frac{dy}{dx} - \frac{1}{1-x}y = 0 \quad \text{...(2)}$$

Also $\quad 1 + P + Q = 1 + \dfrac{x}{1-x} + \left(-\dfrac{1}{1-x}\right) = 0$

$\therefore \quad y_2 = e^x$ is a solution of (2)

Now, $\begin{vmatrix} y_1 & y_2 \\ y_1' & y_2' \end{vmatrix} = \begin{vmatrix} x & e^x \\ 1 & e^x \end{vmatrix} = (x-1)e^x \neq 0 \quad (\because \ x \neq 1)$

$\therefore \quad y_1$ and y_2 are linearly independent solutions of (2).

$\therefore \quad y = c_1 y_1 + c_2 y_2 = c_1 x + c_2 e^x$ is the general soultion of (2)

$\therefore \quad y = c_1 x + c_2 e^x$ is the C.F. of the solution of (1).

Let P.I. $= Ax + Be^x$ ...(3)

$$\therefore \quad A' = -\frac{y_2 R}{y_1 y_2' - y_1' y_2},\ B' = \frac{y_1 R}{y_1 y_2' - y_1' y_2}$$

$$\therefore \quad A' = -\frac{e^x(1-x)}{(x-1)e^x} = 1,\quad B' = \frac{x(1-x)}{(x-1)e^x} = -x e^{-x}$$

$$\therefore \qquad A = \int 1\, dx = x$$

$$B = \int -x\, e^{-x}\, dx = -\left[x.\frac{e^{-x}}{-1} - \int 1.\frac{e^{-x}}{-1} dx\right] = x\, e^{-x} + e^{-x} = e^{-x}(x+1)$$

$\therefore$ (3) $\Rightarrow$ P.I. $= x \,.\, x + e^{-x}(x+1) \,.\, e^{x} = x^2 + x + 1.$

Using y = C.F. + P.I., we have

$$\mathbf{y = c_1\, x + c_2\, e^x + x^2 + x + 1.}$$

This is the general solution of the given equation.

Example 6. *Verify that $y = x$ and $y = x^2 - 1$ are linearly independent solutions of the equation*

$$(x^2 + 1)\frac{d^2y}{dx^2} - 2x\frac{dy}{dx} + 2y = 0.$$

Using this information, find the general solution of the equation

$$(x^2 + 1)\frac{d^2y}{dx^2} - 2x\frac{dy}{dx} + 2y = 6(x^2 + 1)^2.$$

Sol. $y = x$ is a solution of

$$(x^2 + 1)\frac{d^2y}{dx^2} - 2x\frac{dy}{dx} + 2y = 0, \text{ if } (x^2 + 1) \,.\, 0 - 2x \,.\, 1 + 2x = 0, \text{ which is true.}$$

$y = x^2 - 1$ is a solution of $(x^2 + 1)\dfrac{d^2y}{dx^2} - 2x\dfrac{dy}{dx} + 2y = 0$

if $(x^2 + 1) \,.\, 2 - 2x \,.\, 2x + 2(x^2 - 1) = 0$

or if $2x^2 + 2 - 4x^2 + 2x^2 - 2 = 0$, which is true.

$\therefore$ $y = x$ and $y = x^2 - 1$ are both solution of

$$(x^2 + 1)\frac{d^2y}{dx^2} - 2x\frac{dy}{dx} + 2y = 0. \qquad \text{...(1)}$$

Dividing (1) by $x^2 + 1$, the functions $y = x$ and $y = x^2 - 1$ are also solutions of

$$\frac{d^2y}{dx^2} - \frac{2x}{x^2+1}\frac{dy}{dx} + \frac{2}{x^2+1}y = 0. \qquad \text{...(2)}$$

Let $y_1 = x$ and $y_2 = x - 1$.

Now $\begin{vmatrix} y_1 & y_2 \\ y_1' & y_2' \end{vmatrix} = \begin{vmatrix} x & x^2 - 1 \\ 1 & 2x \end{vmatrix} = 2x^2 - x^2 + 1 = x^2 + 1 \neq 0$

$\therefore$ y_1 and y_2 are linearly independent solutions of (1) and also of (2).

$\therefore$ $y = c_1 y_1 + c_2 y_2 = c_1 x + c_2 (x^2 - 1)$ is the general solution of (2).

The equation to be solved is

$$(x^2 + 1)\frac{d^2y}{dx^2} - 2x\frac{dy}{dx} + 2y = 6(x^2 + 1)^2. \qquad \text{...(3)}$$

$$\Rightarrow \quad \frac{d^2y}{dx^2} - \frac{2x}{x^2+1}\frac{dy}{dx} + \frac{2}{x^2+1}y = 6(x^2 + 1) \qquad \text{...(4)}$$

$\therefore$ C.F. of the solution of (4) $= c_1\, x + c_2\, (x^2 - 1)$

Let P.I. of (4) $= Ax + B(x^2 - 1)$ ...(5)

$\therefore \quad A' = -\dfrac{y_2 R}{y_1 y_2' - y_1' y_2}, \; B' = \dfrac{y_1 R}{y_1 y_2' - y_1' y_2}$

Here $\quad y_1 = x, y_2 = x^2 - 1, R = 6(x^2 + 1)$

$$A' = -\frac{(x^2-1)\,6\,(x^2+1)}{x^2+1} = 6(1 - x^2), \quad B' = \frac{x\,6\,(x^2+1)}{x^2+1} = 6x$$

$\therefore \quad A = \int 6(1-x^2)\,dx = 6x - 2x^3, \quad B = \int 6x\,dx = 3x^2$

$\therefore$ (3) $\Rightarrow$ P.I. of (4) $= (6x - 2x^3)x + 3x^2(x^2 - 1) = x^4 + 3x^2$

Using $\quad y =$ C.F. + P.I., we have

$$\mathbf{y = c_1x + c_2(x^2 - 1) + x^4 + 3x + 3x^2.}$$

This is the general solution of the equation (4) and hence of the given equation.

Remark. For applying the method of variation of parameters it is necessary to have the coefficient of $\dfrac{d^2y}{dx^2}$ as unity.

Example 7. *Solve* $x^2 \dfrac{d^2y}{dx^2} - 2x(1 + x) \dfrac{dy}{dx} + 2(1 + x)\,y = x^3$. (K.U. 2004 S ; M.D.U. 2006)

Sol. We have

$$x^2 \frac{d^2y}{dx^2} - 2x\,(1 + x) \frac{dy}{dx} + 2(1 + x)\,y = x^3.$$

$$\Rightarrow \quad \frac{d^2y}{dx^2} - \frac{2(1+x)}{x}\frac{dy}{dx} + \frac{2(1+x)}{x^2}y = x \quad \text{...(1)}$$

The equation (1) with right side '0' is

$$\frac{d^2y}{dx^2} - \frac{2(1+x)}{x}\frac{dy}{dx} + \frac{2(1+x)}{x^2}y = 0 \quad \text{...(2)}$$

We find the general solution of (2).

Let $\quad y = xV$. **(Note this step)**

$$\therefore \quad \frac{dy}{dx} = x\frac{dV}{dx} + V \text{ and } \frac{d^2y}{dx^2} = x\frac{d^2V}{dx^2} + 2\frac{dV}{dx}$$

Putting the values of $\dfrac{dy}{dx}$ and $\dfrac{d^2y}{dx^2}$ in (2), we get

$$x\frac{d^2V}{dx^2} + 2\frac{dV}{dx} - \frac{2(1+x)}{x}\left(x\frac{dV}{dx} + V\right) + \frac{2(1+x)}{x^2}\,xV = 0$$

$$\Rightarrow \quad x\frac{d^2V}{dx^2} - 2x\frac{dV}{dx} = 0 \quad \Rightarrow \quad (D^2 - 2D)V = 0 \quad \text{...(3)}$$

The A.E. is $\quad D^2 - 2D = 0. \quad \therefore \quad D = 0, 2$

$\therefore \quad V = c_1 e^{0x} + c_2 e^{2x} = c_1 + c_2 e^{2x}$

$\therefore \quad y = xV = x(c_1 + c_2 e^{2x}) = c_1 x + c_2 x e^{2x}$

Let $\quad y_1 = x \;$ and $\; y_2 = xe^{2x}$

$$\therefore \quad \begin{vmatrix} y_1 & y_2 \\ y_1' & y_2' \end{vmatrix} = \begin{vmatrix} x & xe^{2x} \\ 1 & (1+2x)e^{2x} \end{vmatrix}$$

$$= x(1 + 2x)e^{2x} - xe^{2x} = 2x^2 e^{2x} \neq 0$$

$\therefore$ y_1 and y_2 are independent solutions.

$\therefore$ $y = c_1y_1 + c_2y_2 = c_1x + c_2 xe^{2x}$ is the general solution of (2).

$\therefore$ C.F. of (1) $= c_1 x + c_2 x e^{2x}$

Let P.I. $= Ay_1 + By_2 = Ax + Bxe^{2x}$ ...(4)

$$\therefore \quad A' = -\frac{y_2 R}{y_1y_2' - y_1'y_2}, \quad B' = \frac{y_1 R}{y_1y_2' - y_1'y_2}$$

$$\therefore \quad A' = -\frac{xe^{2x} \cdot x}{2x^2 e^{2x}} = -\frac{1}{2} \text{ and } B' = \frac{x \cdot x}{2x^2 e^{2x}} = \frac{1}{2}e^{-2x} \quad (\text{Here } R = x)$$

$$\therefore \quad A = \int -\frac{1}{2}\, dx = -\frac{x}{2} \text{ and } B = \int \frac{1}{2} e^{-2x}\, dx = -\frac{e^{-2x}}{-4} = \frac{1}{4e^{2x}}$$

$$\therefore \quad (4) \Rightarrow \quad \text{P.I.} = -\frac{x}{2} \cdot x - \frac{1}{4e^{2x}} \cdot xe^{2x} = -\frac{x^2}{2} - \frac{x}{4}$$

$\therefore$ Using y = C.F. + P.I., we have

$$\mathbf{y = c_1x + c_2x\, e^{2x} - \frac{x^2}{4} - \frac{x}{4}.}$$

This is the general solution of the given equation.

EXERCISE 4

Solve the following differential equations by variation of parameters :

1. $\dfrac{d^2y}{dx^2} + y = \sec x$ (*KU 2003 C*)
2. $\dfrac{d^2y}{dx^2} + 9y = \sec 3x$
3. $\dfrac{d^2y}{dx^2} + n^2 y = \operatorname{cosec} nx$
4. $\dfrac{d^2y}{dx^2} + y = \tan x$
5. $\dfrac{d^2y}{dx^2} - 2\dfrac{dy}{dx} = e^x \sin x$
6. $\dfrac{d^2y}{dx^2} - 2\dfrac{dy}{dx} + y = 2x$
7. $x^2 \dfrac{d^2y}{dx^2} - 2x\dfrac{dy}{dx} + 2y = x \log x$
8. $x^2 \dfrac{d^2y}{dx^2} + x\dfrac{dy}{dx} - y = x^2e^x$
9. $\dfrac{d^2y}{dx^2} - 2\dfrac{dy}{dx} + 2y = e^x \tan x$
10. $\dfrac{d^2y}{dx^2} - 6\dfrac{dy}{dx} + 9y = x^{-2} e^{3x}$.

Answers

1. $y = c_1 \cos x + c_2 \sin x + x \sin x + \cos x \log \cos x$
2. $y = c_1 \cos 3x + c_2 \sin 3x + \dfrac{x}{3} \sin 3x + \dfrac{1}{9} \cos 3x \log \cos 3x$

3. $y = c_1 \cos nx + c_2 \sin nx - \dfrac{x}{n} \cos nx + \dfrac{1}{n^2} \sin nx \log \sin nx$

4. $y = c_1 \cos x + c_2 \sin x - \cos x \log(\sec x + \tan x) + 7$

5. $y = c_1 + c_2 e^{2x} - \dfrac{1}{2} e^x \sin x$

6. $y = (c_1 + c_2 x) e^x + 2x + 4$

7. $y = c_1 x + c_2 x^2 - \dfrac{1}{2} x (\log x)^2 - x \log x - x$

8. $y = c_1 x + \dfrac{c_2}{x} + e^x - \dfrac{e^x}{x}$

9. $y = e^x (c_1 \cos x + c_2 \sin x) - e^x \cos x \log(\sec x + \tan x)$

10. $y = (c_1 + c_2 x) e^{3x} - e^{3x} \log x.$

23

Ordinary Simultaneous Differential Equations

1. Introduction

Till now we have been discussing differential equations involving only two variables. In the present chapter, we shall consider differential equations involving more than two variables. If there is one independent variable then the equations are called ordinary differential equations and in case there are more than one independent variable then the equations are called partial differential equations. In the present text, we shall consider equations involving only one independent variable.

2. Ordinary Simultaneous Differential Equations

Since the number of dependent variables is more than one, only one such differential equation is not sufficient to find its solution. In the present case, we shall require systems of simultaneous differential equations containing at least two differential equations. We shall study the following types of simultaneous differential equations.

I. Simultaneous linear differential equations with constant coefficients.

II. Simultaneous total differential equations.

I. SIMULTANEOUS LINEAR DIFFERENTIAL EQUATIONS WITH CONSTANT COEFFICIENTS

3. Method of Solving Simultaneous Linear Differential Equations with Constant Coefficients

The number of simultaneous linear differential equations with constant coefficients in a given system must be equal to the number of dependent variables involved in the system.

For Example :

$$\left.\begin{aligned} \frac{dx}{dt}+\frac{dy}{dt}+2x+y=0 \\ \frac{dy}{dt}+5x+3y=0 \end{aligned}\right]$$

is a system of two equations and the number of dependent variables (x and y) is also two. Writing $\frac{d}{dt}$ as D, the above system can also be written as

$$\left.\begin{aligned} (D+2)x+(D+1)y=0 \\ 5x+(D+3)y=0 \end{aligned}\right].$$

By using differentiation, a set of equations is obtained from which all but one of the dependent variables, say x, can be eliminated. The equation resulting from the elimination of other variables is solved for this variable x. Then a relation between a second dependent variable and the independent variable can be deduced, either (1) by the method of eliminating and integration employed in the case of the first variable 'x' or (2) by substituting the value of 'x' in one of the equations involving 'x' and the second variable and the independent variable.

The complete solution consists of as many independent relations between the variables as there are dependent variables.

Remark. When the equations are written in the 'D' notation, there is a striking similarity between the procedure used here and the method of solving a system of n equations is n unknowns. This is due to the fact that the operator 'D' may at times be treated as a variable.

4. Simultaneous Linear Equations Involving two Dependent Variables

Let x, y be two dependent variables and 't' be the independent variable. A system of simultaneous linear equations with constant coefficients will contain two equations.

Let the simultaneous equations in symbolic form be

$$f_1(D)\,x + g_1(D)\,y = h_1(t) \qquad \text{...(1)}$$
$$f_2(D)\,x + g_2(D)\,y = h_2(t). \qquad \text{...(2)}$$

Let $\Delta = \begin{vmatrix} f_1(D) & g_1(D) \\ f_2(D) & g_2(D) \end{vmatrix}$

If $\Delta = 0$, then the given system is called a **dependent system.**

We shall consider only those systems for which $\Delta \neq 0$.

Operating (1) by $g_2(D)$, we get

$$g_2(D)\,f_1(D)x + g_2(D)\,g_1(D)\,y = g_2(D)\,h_1(t) \qquad \text{...(3)}$$

Operating (2) by $g_1(D)$, we get

$$g_1(D)\,f_2(D)x + g_1(D)\,g_2(D)\,y = g_1(D)\,h_2(t) \qquad \text{...(4)}$$

Subtracting (4) from (3), we get

$$(g_2(D)\,f_1(D) - g_1(D)\,f_2(D))\,x = g_2(D)\,h_1(t) - g_1(D)\,h_2(t).$$

In terms of determinants, we have

$$\begin{vmatrix} f_1(D) & g_1(D) \\ f_2(D) & g_2(D) \end{vmatrix} x = \begin{vmatrix} h_1(t) & g_1(D) \\ h_2(t) & g_2(D) \end{vmatrix}$$

Similarly, for y we have

$$\begin{vmatrix} f_1(D) & g_1(D) \\ f_2(D) & g_2(D) \end{vmatrix} y = \begin{vmatrix} f_1(D) & h_1(t) \\ f_2(D) & h_2(t) \end{vmatrix}$$

The number of arbitrary constants involved in the general solution should be equal to the degree of D in Δ.

Remark. We have $g_1(D)g_2(D) = g_2(D)g_1(D)$ because the linear differential equations are assumed to be with constant coefficients.

Example 1. *Solve the simultaneous equations :*

$$(D - 1)x + Dy = 2t + 1$$
$$(2D + 1)x + 2D\,y = t.$$

Sol. We have

$$(D - 1)\,x + Dy = 2t + 1 \qquad \text{...(1)}$$
$$(2D + 1)\,x + 2Dy = t \qquad \text{...(2)}$$

Multiplying (1) by 2 and subtracting (2) from it, we get

$$(2(D-1)-(2D+1))x = 4t+2-t$$

$$\Rightarrow \quad -3x = 3t+2 \quad \Rightarrow \quad x = -t-\frac{2}{3}$$

To find y, (1) $\Rightarrow$ $Dy = 2t+1-(D-1)x = 2t+1-(D-1)\left(-t-\frac{2}{3}\right) = t+\frac{4}{3}$.

$$\therefore \quad y = \int\left(t+\frac{4}{3}\right)dt + c_1 = \frac{t^2}{2}+\frac{4}{3}t+c_1$$

$\therefore$ The general solution of the given system is

$$\mathbf{x = -t-\frac{2}{3}, \; y = \frac{1}{2}t^2+\frac{4}{3}t+c_1.}$$

Remark. Here $\Delta = \begin{vmatrix} D-1 & D \\ 2D+1 & 2D \end{vmatrix} = (D-1)2D-(2D+1)D = -3D.$

Since, the degree of D in Δ is 1, the general solution will have one arbitrary constant.

Example 2. *Solve the simultaneous equations :*

$$\frac{dx}{dt} - 7x + y = 0$$

$$\frac{dy}{dt} - 2x - 5y = 0.$$

Sol. We have

$$Dx - 7x + y = 0 \quad \ldots(1)$$

$$Dy - 2x - 5y = 0 \quad \ldots(2)$$

(1) $\Rightarrow$ $(D-7)x + y = 0$ ...(3)

(2) $\Rightarrow$ $-2x + (D-5)y = 0$ ...(4)

Operating on (3) by D – 5 and subtracting (4) from it, we get

$$((D-5)(D-7)+2)x = 0 \quad \Rightarrow \quad (D^2-12D+37)x = 0$$

$\therefore$ The A.E. is $D^2 - 12D + 37 = 0$. $\therefore$ $D = 6 \pm i$

$$\therefore \quad x = e^{6t}(c_1 \cos t + c_2 \sin t)$$

To find y, (1) $\Rightarrow$ $y = 7x - Dx$

$$\therefore \quad y = 7e^{6t}(c_1 \cos t + c_2 \sin t) - 6e^{6t}(c_1 \cos t + c_2 \sin t) - e^{6t}(-c_1 \sin t + c_2 \cos t)$$

$$\therefore \quad y = e^{6t}[(c_1 - c_2)\cos t + (c_1 + c_2)\sin t]$$

$\therefore$ The general solution of the given system is

$$\mathbf{x = e^{6t}(c_1 \cos t + c_2 \sin t), \; y = e^{6t}((c_1 - c_2)\cos t + (c_1 + c_2)\sin t).}$$

Remark. For the above system :

$$\Delta = \begin{vmatrix} D-7 & 1 \\ -2 & D-5 \end{vmatrix} = (D-7)(D-5)+2 = D^2-12D+37.$$

Since, the degree of D in Δ is 2, the general solution will have 2 arbitrary constants.

Alternative method

Using determinants,

$$\Delta = \begin{vmatrix} D-7 & 1 \\ -2 & D-5 \end{vmatrix} = (D-7)(D-5)+2 = D^2-12D+37$$

For x $\Delta.x = \begin{vmatrix} 0 & 1 \\ 0 & D-5 \end{vmatrix} = 0 - 0 = 0$

$\Rightarrow \quad (D^2 - 12D + 37)x = 0.$

Example 3. *Solve simultaneously the differential equations :*

$$\frac{dx}{dt} + 4x + 3y = t$$

$$\frac{dy}{dt} + 2x + 5y = e^t. \qquad \text{(K.U. 2005, 2006 S, M.D.U. 2007)}$$

Sol. We have

$$Dx + 4x + 3y = t \qquad ...(1)$$

$$Dy + 2x + 5y = e^t \qquad ...(2)$$

(1) $\Rightarrow$ $(D + 4)x + 3y = t$ ...(3)

(2) $\Rightarrow$ $2x + (D + 5)y = e^t$ ...(4)

Operating on (3) by D + 5 and multiplying (4) by 3 and subtracting, we get

$$(D + 5)(D + 4)x - 6x = (D + 5)t - 3e^t$$

$\Rightarrow$ $(D^2 + 9D + 14)x = - 3e^t - 5t + 1$

$\therefore$ The A.E. is $D^2 + 9D + 14 = 0$. $\therefore$ $D = -2, -7$

$\therefore$ $\text{C.F.} = c_1e^{-2t} + c_2e^{-7t}$

$$\text{P.I.} = \frac{1}{D^2 + 9D + 14}(-3e^t + 5t + 1)$$

$$= -3\left(\frac{1}{D^2 + 9D + 14}e^t\right) + \frac{5}{14}\left(1 + \frac{9D + D^2}{14}\right)^{-1} t + \frac{1}{D^2 + 9D + 14}e^{0t}$$

$$= -3\left(\frac{1}{1 + 9 + 14}e^t\right) + \frac{5}{14}\left(1 - \frac{9}{14}D + \ldots\ldots\right)t + \frac{1}{0 + 0 + 14}e^{0t}$$

$$= -\frac{1}{8}e^t + \frac{5}{14}\left(t - \frac{9}{14}\right) + \frac{1}{14} = -\frac{1}{8}e^t + \frac{5}{14}t - \frac{31}{196}.$$

$\therefore$ $$x = \text{C.F.} + \text{P.I.} = c_1e^{-2t} + c_2e^{-7t} - \frac{1}{8}e^t + \frac{5}{14}t - \frac{31}{196} \qquad ...(5)$$

(1) $\Rightarrow$ $$3y = t - \frac{dx}{dt} - 4x = t - \left[-2c_1e^{-2t} - 7c_2e^{-7t} - \frac{1}{8}e^t + \frac{5}{14} - 0\right]$$

$$- 4\left[c_1e^{-2t} + c_2e^{-7t} - \frac{1}{8}e^t + \frac{5}{14}t - \frac{31}{196}\right]$$

$$= -2c_1e^{-2t} + 3c_2e^{-7t} - \frac{3}{7}t + \frac{5}{8}e^t + \frac{27}{98}$$

$\therefore$ $$y = -\frac{2}{3}c_1e^{-2t} + c_2e^{-7t} - \frac{1}{7}t + \frac{5}{24}e^t + \frac{9}{98}$$

$\therefore$ The general solution of the given system is

$$\mathbf{x = c_1e^{-2t} + c_2e^{-7t} - \frac{1}{8}e^t + \frac{5}{14}t - \frac{31}{196},}$$

$$\mathbf{y = -\frac{2}{3}\, c_1e^{-2t} + c_2e^{-7t} + \frac{5}{24}\, e^t - \frac{1}{7}\, t + \frac{9}{98}}.$$

Remark. For the above system :

$$\Delta = \begin{vmatrix} D+4 & 3 \\ 2 & D+5 \end{vmatrix} = (D + 4)(D + 5) - 6 = D^2 + 9D + 14.$$

Since, the degree of D in Δ is 2, the general solution will have two arbitrary constants.

Example 4. *Solve the simultaneous equations :*

$$\frac{dx}{dt} + y = \sin t\,, \qquad \frac{dy}{dt} + x = \cos t$$

given that x = 2 and y = 0 when t = 0.

Sol. Given equations are

$$Dx + y = \sin t \qquad ...(1)$$

$$x + Dy = \cos t \qquad ...(2)$$

Operating on (1) by D and subtracting (2) from it, we get

$$D^2x - x = D \sin t - \cos t \;\Rightarrow\; (D^2 - 1)\, x = 0$$

$\therefore$ The A.E. is $D^2 - 1 = 0$. $\therefore$ $D = \pm 1$

$$\therefore \qquad x = c_1e^t + c_2e^{-t} \qquad ...(3)$$

To find y, (1) $\Rightarrow$ $y = \sin t - D\,x$

$$\therefore \qquad y = \sin t - c_1e^t + c_2e^{-t} \qquad ...(4)$$

Now $x = 2, y = 0$ when $t = 0$.

$\therefore$ (3) $\Rightarrow$ $2 = c_1 + c_2$

(4) $\Rightarrow$ $0 = - c_1 + c_2$

Solving we get $c_1 = 1, c_2 = 1$.

$\therefore$ The required solution is $\mathbf{x = e^t + e^{-t}, y = \sin t - e^t + e^{-t}}$.

Example 5. *Solve the simultaneous equations :*

$$\frac{d^2x}{dt^2} + 4x + 5y = t^2\,, \qquad \frac{d^2y}{dt^2} + 5x + 4y = t + 1. \qquad \text{(D.L.U. 2004 ; K.U. 2006)}$$

Sol. We have

$$(D^2 + 4)x + 5y = t^2 \qquad ...(1)$$

$$5x + (D^2 + 4)y = t + 1 \qquad ...(2)$$

Operating (1) by $D^2 + 4$ and multiplying (2) by 5 and subtracting, we get

$$((D^2 + 4)^2 - 25)x = (D^2 + 4)t^2 - 5(t + 1) \;\Rightarrow\; (D^2 - 1)(D^2 + 9)x = 4t^2 - 5t - 3.$$

$\therefore$ The A.E. is $(D^2 - 1)(D^2 + 9) = 0$. $\therefore$ $D = \pm 1, \pm 3i$

$$\therefore \qquad \text{C.F.} = c_1e^t + c_2e^{-t} + c_3 \cos t + c_4 \sin 3t$$

$$\text{P.I.} = \frac{1}{D^4 + 8D^2 - 9}(4t^2 - 5t - 3) = \frac{1}{-9}\left(1 - \frac{8D^2 + D^4}{9}\right)^{-1}\left(4t^2 - 5t - 3\right)$$

$$= -\frac{1}{9}\left(1 + \frac{8}{9}D^2 + \frac{D^4}{9} + ...\right)(4t^2 - 5t - 3)$$

$$= -\frac{1}{9}\left((4t^2 - 5t - 3) + \frac{8}{9}(8)\right) = -\frac{1}{9}\left(4t^2 - 5t + \frac{37}{9}\right)$$

$$\therefore \quad x = c_1 e^t + c_2 e^{-t} + c_3 \cos 3t + c_4 \sin 3t - \frac{1}{9}\left(4t^2 - 5t - 3 + \frac{37}{9}\right) \quad ...(3)$$

$$(1) \Rightarrow \quad 5y = t^2 - \frac{d^2x}{dt^2} - 4x \quad ...(4)$$

$$(3) \Rightarrow \quad \frac{dx}{dt} = c_1 e^t - c_2 e^{-t} - 3c_3 \sin 3t + 3c_4 \cos 3t - \frac{1}{9}(8t - 5)$$

$$\therefore \quad \frac{d^2x}{dt^2} = c_1 e^t + c_2 e^{-t} - 9c_3 \cos 3t - 9c_4 \sin 3t - \frac{8}{9}$$

$$\therefore \quad (4) \Rightarrow \quad 5y = t^2 - c_1 e^t - c_2 e^{-t} + 9c_3 \cos 3t + 9c_4 \sin 3t + \frac{8}{9}$$

$$- 4\left(c_1 e^t + c_2 e^{-t} + c_3 \cos 3t + c_4 \sin 3t - \frac{4}{9}t^2 + \frac{5}{9}t - \frac{37}{81}\right)$$

$$= - 5c_1 e^t - 5c_2 e^{-t} + 5c_3 \cos 3t + 5c_4 \sin 3t + \frac{25}{9}t^2 - \frac{20}{9}t + \frac{220}{81}$$

$$\therefore \quad y = - c_1 e^t - c_2 e^{-t} + c_3 \cos 3t + c_4 \sin 3t + \frac{5}{9}t^2 - \frac{4}{9}t + \frac{44}{81}$$

$\therefore$ The general solution of the given system is

$$\mathbf{x = c_1 e^t + c_2 e^{-t} + c_3 \cos 3t + c_4 \sin 3t - \frac{1}{9}\left(4t^2 - 5t + \frac{37}{9}\right),}$$

$$\mathbf{y = c_1 e^t - c_2 e^{-t} + c_3 \cos 3t + c_4 \sin 3t + \frac{1}{9}\left(5t^2 - 4t + \frac{44}{9}\right).}$$

Remark. For the above system $\Delta = \begin{vmatrix} D^2 + 4 & 5 \\ 5 & D^2 + 4 \end{vmatrix} = (D^2 + 4)^2 - 25 = D^4 + 8D^2 - 9.$

Since, the degree of D in Δ is 4, the general solution will have four arbitrary constants.

Example 6. *Solve the simultaneous equations :* $\frac{dx}{dt} = 2y$, $\frac{dy}{dt} = 2z$, $\frac{dz}{dt} = 2x$.

Sol. We have

$$Dx - 2y + 0z = 0 \quad ...(1)$$

$$0x + Dy - 2z = 0 \quad ...(2)$$

$$2x + 0y - Dz = 0 \quad ...(3)$$

Using determinants, $\Delta = \begin{vmatrix} D & -2 & 0 \\ 0 & D & -2 \\ 2 & 0 & -D \end{vmatrix} = - D^3 + 8$

The degree of D in Δ is 3.

$\therefore$ The general solution will have three arbitrary constants.

For x $\quad \Delta . x = \begin{vmatrix} 0 & -2 & 0 \\ 0 & D & -2 \\ 0 & 0 & -D \end{vmatrix}$

$$\Rightarrow \quad (- D^3 + 8)x = 0$$

$\therefore$ The A.E. is $- D^3 + 8 = 0$. $\therefore$ $D = 2, - 1 \pm \sqrt{3}\, i$

$$\therefore \quad x = c_1 e^{2t} + e^{-t} (c_2 \cos \sqrt{3}\, t + c_3 \sin \sqrt{3}\, t)$$

$$(1) \Rightarrow \quad 2y = \frac{dx}{dt} = 2c_1 e^{2t} - e^{-t} (c_2 \cos \sqrt{3}\, t + c_3 \sin \sqrt{3}\, t)$$
$$+ e^{-t} (-\sqrt{3}\, c_2 \sin \sqrt{3}\, t + \sqrt{3}\, c_3 \cos \sqrt{3}\, t)$$
$$= 2c_1 e^{2t} - e^{-t} [(c_2 - \sqrt{3}\, c_3) \cos \sqrt{3}\, t + (c_3 + \sqrt{3}\, c_2) \sin \sqrt{3}\, t]$$

$$\therefore \quad y = c_1 e^{2t} - \frac{1}{2} e^{-t} [(c_2 - \sqrt{3}\, c_3) \cos \sqrt{3}\, t + (c_3 + \sqrt{3}\, c_2) \sin \sqrt{3}\, t]$$

$$(2) \Rightarrow \quad 2z = \frac{dy}{dt} = 2c_1 e^{2t} + \frac{1}{2} e^{-t} [(c_2 - \sqrt{3}\, c_3) \cos \sqrt{3}\, t + (c_3 + \sqrt{3}\, c_2) \sin \sqrt{3}\, t]$$
$$- \frac{1}{2} e^{-t} [-(\sqrt{3}\, c_2 - 3c_3) \sin \sqrt{3}\, t + (\sqrt{3}\, c_3 + 3c_2) \cos \sqrt{3}\, t]$$
$$= 2c_1 e^{2t} - e^{-t} [(c_2 + \sqrt{3}\, c_3) \cos \sqrt{3}\, t + (c_3 - \sqrt{3}\, c_2) \sin \sqrt{3}\, t]$$

$$\therefore \quad z = c_1 e^{2t} - \frac{1}{2} e^{-t} [(c_2 + \sqrt{3}\, c_3) \cos \sqrt{3}\, t + (c_3 - \sqrt{3}\, c_2) \sin \sqrt{3}\, t].$$

$\therefore$ The general solution of the given system is

$$\mathbf{x = c_1 e^{2t} + \frac{1}{2} e^{-t} (c_2 \cos \sqrt{3}\, t + c_3 \sin \sqrt{3}\, t),}$$

$$\mathbf{y = c_1 e^{2t} - \frac{1}{2} e^{-t} [(c_2 - \sqrt{3}\, c_3) \cos \sqrt{3}\, t + (c_3 + \sqrt{3}\, c_2) \sin \sqrt{3}\, t],}$$

$$\mathbf{z = c_1 e^{2t} - \frac{1}{2} e^{-t} [(c_2 + \sqrt{3}\, c_3) \cos \sqrt{3}\, t + (c_3 - \sqrt{3}\, c_2) \sin \sqrt{3}\, t].}$$

Example 7. *Solve the simultaneous equations :*

$$t \frac{dx}{dt} + y = 0, \quad t \frac{dy}{dt} + x = 0,$$

given that $x(1) = 1,\ y(-1) = 0.$ (M.D.U. 2006)

Sol. We have
$$t \frac{dx}{dt} + y = 0 \quad \text{...(1)}$$
$$t \frac{dy}{dt} + x = 0 \quad \text{...(2)}$$

Differentiating (1) w.r.t. t, we get

$$t \frac{d^2x}{dt^2} + \frac{dx}{dt} + \frac{dy}{dt} = 0 \quad \text{...(3)}$$

Multiplying (3) by t, we get

$$t^2 \frac{d^2x}{dt^2} + t \frac{dx}{dt} + t \frac{dy}{dt} = 0 \quad \text{...(4)}$$

$$(4)—(2) \Rightarrow \quad t^2 \frac{d^2x}{dt^2} + t \frac{dx}{dt} - x = 0 \quad \text{...(5)}$$

(5) is a homogeneous linear equation.

Let $z = \log t$. $\therefore$ $t = e^z$

$$\therefore \quad t\frac{d}{dt} = D, \quad t^2\frac{d^2}{dt^2} = D(D-1), \text{ where } D = \frac{d}{dz}$$

$\therefore$ (5) $\Rightarrow$ $D(D-1)x + Dx - x = 0$

$\Rightarrow$ $(D^2 - D + D - 1)x = 0$

$\Rightarrow$ $(D^2 - 1)x = 0$ ...(6)

The A.E., of (6) is $D^2 - 1 = 0$ $\therefore$ $D = \pm 1$

$\therefore$ The general solution of (6) is

$$x = C_1 e^z + c_2 e^{-z}$$

$\therefore$ The general solution of (5) is

$$x = c_1 t + \frac{c_2}{t}$$

(1) $\Rightarrow$ $$y = -t\frac{dx}{dt} = -t\left[c_1 - \frac{c_2}{t^2}\right] = -c_1 t + \frac{c_2}{t}$$

$\therefore$ The general solution of the given system is

$$x = c_1 t + \frac{c_2}{t} \quad ...(7)$$

$$y = -c_1 t + \frac{c_2}{t} \quad ...(8)$$

$$x(1) = 1 \Rightarrow c_1 + c_2 = 1$$

$$y(-1) = 0 \Rightarrow c_1 - c_2 = 0$$

$$\therefore \quad c_1 = \frac{1}{2} \text{ and } c_2 = \frac{1}{2}$$

$\therefore$ The required solution is

$$x = \frac{1}{2}t + \frac{1/2}{t}, \quad y = -\frac{1}{2}t + \frac{1/2}{t}$$

or $$\mathbf{x = \frac{1}{2}\left(t + \frac{1}{t}\right), y = \frac{1}{2}\left(-t + \frac{1}{t}\right).}$$

EXERCISE 1

Solve the following systems of simultaneous linear differential equations :

1. $\frac{dx}{dt} = \lambda y, \ \frac{dy}{dt} = \lambda x$ (M.D.U. 2004)

2. $\frac{dx}{dt} - 7x + y = 0, \ \frac{dy}{dt} - 2x - 5y = 0$ (M.D.U. 2005, 2006 ; K.U. 2007)

3. $\frac{dx}{dt} + 5x + y = e^t, \ \frac{dy}{dt} - x + 3y = e^{2t}$

4. $\dfrac{dx}{dt} + 2\dfrac{dy}{dt} - 2x + 2y = 3e^t,\ 3\dfrac{dx}{dt} + \dfrac{dy}{dt} + 2x + y = 4e^{2t}$ (K.U. 2005)

5. $\dfrac{dx}{dt} + \dfrac{dy}{dt} - 2y = 2\cos t - 7\sin t,\ \dfrac{dx}{dt} - \dfrac{dy}{dt} + 2x = 4\cos t - 3\sin t$

6. $\dfrac{dx}{dt} = ax + by,\ \dfrac{dy}{dt} = a'x + b'y$ (M.D.U. 2007) 7. $\dfrac{dx}{dt} + 4x + 3y = t^2,\ \dfrac{dy}{dt} + 2x + 5y = e^{2t}$

8. $\dfrac{d^2x}{dt^2} - \dfrac{dy}{dt} = 2x + 2t,\ \dfrac{dx}{dt} + 4\dfrac{dy}{dt} = 3y$ 9. $\dfrac{d^2x}{dt^2} + y = \sin t,\ \dfrac{d^2y}{dt^2} + x = \cos t$

10. $\dfrac{d^2x}{dt^2} - 3x - 4y = 0,\ \dfrac{d^2y}{dt^2} + x + y = 0.$

Answers

1. $x = c_1 e^{\lambda t} + c_2 e^{-\lambda t},\ y = c_1 e^{\lambda t} - c_2 e^{-\lambda t}$

2. $x = e^{6t}(c_1 \cos t + c_2 \sin t),\ y = e^{6t}((c_1 - c_2)\cos t + (c_1 + c_2)\sin t)$

3. $x = (c_1 + c_2 t)e^{-4t} + \dfrac{4}{25}e^t - \dfrac{1}{36}e^{2t},\ y = -(c_1 + c_2 + c_2 t)e^{-4t} + \dfrac{1}{25}e^t + \dfrac{7}{36}e^{2t}$

4. $x = c_1 e^{-6t/5} + \dfrac{1}{2}e^{2t} - \dfrac{3}{11}e^t,\ y = -8c_1 e^{-6t/5} + c_2 e^{-t} + \dfrac{15}{22}e^t$

5. $x = c_1 e^{\sqrt{2}t} + c_2 e^{-\sqrt{2}t} + 3\cos t,\ y = c_1(\sqrt{2}+1)e^{\sqrt{2}t} + c_2(1-\sqrt{2})e^{-\sqrt{2}t} + 2\sin t$

6. $x = c_1 e^{mt} + c_2 e^{nt},\ y = \dfrac{c_1}{b}e^{mt}(m-a) + \dfrac{c_2}{b}e^{nt}(n-a),$

where $m = \dfrac{a + b' + \sqrt{(a-b')^2 + 4a'b}}{2}$ and $n = \dfrac{a + b' - \sqrt{(a-b')^2 + 4a'b}}{2}$

7. $x = c_1 e^{-2t} + c_2 e^{-7t} + \dfrac{1}{14}\left(5t^2 - \dfrac{31}{7}t + \dfrac{209}{98}\right) - \dfrac{1}{12}e^{2t}$

$y = -\dfrac{2}{3}c_1 e^{-2t} + c_2 e^{-7t} - \dfrac{1}{7}\left(t^2 - \dfrac{9}{7}t + \dfrac{67}{98}\right) + \dfrac{1}{6}e^{2t}$

8. $x = (c_1 + c_2 t)e^t + c_3 e^{-3t/2} - t,\ y = (c_2(3-t) - c_1)e^t - \dfrac{1}{6}c_3 e^{-3t/2}$

9. $x = c_1 e^t + c_2 e^{-t} + c_3 \cos t + c_4 \sin t + \dfrac{t}{4}(\sin t - \cos t)$

$y = -c_1 e^t - c_2 e^{-t} + c_3 \cos t + c_4 \sin t + \dfrac{1}{4}(t+2)(\sin t - \cos t)$

10. $x = (c_1 + c_2 t)e^{-t} + (c_3 + c_4 t)e^t,\ y = -\dfrac{1}{2}(c_1 + c_2 + c_2 t)e^{-t} - \dfrac{1}{2}(c_3 - c_4 + c_4 t)e^t.$

II. SIMULTANEOUS TOTAL DIFFERENTIAL EQUATIONS

5. Total Differential Equation

Let P, Q, R be functions of three variables x, y and z. A differential equation of the form $P\,dx + Q\,dy + R\,dz = 0$ is called a **total differential equation.**

It can be proved that a total differential equation

$$P\,dx + Q\,dy + R\,dz = 0 \qquad \text{...(1)}$$

is exact if

$$\frac{\partial P}{\partial y} = \frac{\partial Q}{\partial x}, \qquad \frac{\partial Q}{\partial z} = \frac{\partial R}{\partial y}, \qquad \frac{\partial R}{\partial x} = \frac{\partial P}{\partial z}.$$

For example, consider the total differential equation

$$(y + 3z)\,dx + (x + 2z)\,dy + (3x + 2y)\,dz = 0 \qquad \text{...(2)}$$

Here $\qquad P = y + 3z,\ Q = x + 2z,\ R = 3x + 2y$

It can be checked that

$$\frac{\partial P}{\partial y} = \frac{\partial Q}{\partial x} = 1, \quad \frac{\partial Q}{\partial z} = \frac{\partial R}{\partial y} = 2, \quad \frac{\partial R}{\partial x} = \frac{\partial P}{\partial z} = 3.$$

$\therefore$ The equation (2) is exact and it can be written as

$$(y\,dx + x\,dy) + 2(y\,dz + z\,dy) + 3(x\,dz + z dx) = 0$$

$$\Rightarrow \quad d(xy) + d(2yz) + d(3zx) = 0 \quad \Rightarrow \quad d(xy + 2yz + 3zx) = 0$$

Integrating, we get $xy + 2yz + 3zx = c$.

This is the general solution of the total differential equation (2).

There are other methods also for solving a total differential equation of the form (1).

In the present chapter, we shall study the method of solving a system of two simultaneous total differential equations.

6. Method of Solving Simultaneous Total Differential Equations

We shall consider systems of two simultaneous total differential equations in three variables.

Let

$$P_1\,dx + Q_1\,dy + R_1\,dz = 0 \qquad \text{...(1)}$$

$$P_2\,dx + Q_2\,dy + R_2\,dz = 0 \qquad \text{...(2)}$$

be a system of two simultaneous total differential equations, where P_1, Q_1, R_1, P_2, Q_2, R_2 are all functions of x, y and z. Solving (1) and (2) for dx, dy, dz, we get

$$\frac{dx}{Q_1R_2 - Q_2R_1} = \frac{dy}{R_1P_2 - R_2P_1} = \frac{dz}{P_1Q_2 - P_2Q_1}$$

$$\Rightarrow \quad \frac{dx}{X} = \frac{dy}{Y} = \frac{dz}{Z} \qquad \text{...(3)}$$

where $\qquad X = Q_1R_2 - Q_2R_1,\ Y = R_1P_2 - R_2P_1,\ Z = P_1Q_2 - P_2Q_1.$

Thus, a system of two simultaneous total equations (1), (2) can always be put in the form (3).

The equations (3) are said to be completely solved when we get a solution of the form $u_1(x, y, z) = c_1$ and $u_2(x, y, z) = c_2$, where u_1 and u_2 are two independent solutions of the equation (3). The functions u_1 and u_2 are said to be independent if $\frac{u_1}{u_2}$ is not merely a constant.

There are number of rules of solving differential equation (3), depending upon the nature of the functions X, Y and Z.

Remark. In case any denominator of the equations $\frac{dx}{X} = \frac{dy}{Y} = \frac{dz}{Z}$ is zero, then the corresponding numerator is also taken equal to zero.

For example, if $X = 0$, then we take $dx = 0$ *i.e.,* $x =$ constant.

7. TYPE I

In this type we shall consider the solution of systems of equations $\frac{dx}{X} = \frac{dy}{Y} = \frac{dz}{Z}$ in which the equality of two fractions of the given system gives an equation involving only two variables. This equation is solved by usual methods. Same method is repeated by equating two other fractions of the given system.

Example 1. *Solve the following systems of simultaneous differential equations :*

(*i*) $\frac{dx}{y^2} = \frac{dy}{x^2} = \frac{dz}{x^2y^2z^2}$ (*ii*) $\frac{x\,dx}{y^2z} = \frac{dy}{xz} = \frac{dz}{y^2}$

(*iii*) $\frac{dx}{z} = \frac{dy}{0} = \frac{dz}{x}$ (*iv*) $\frac{dx}{x^2 + 2y^2} = -\frac{dy}{xy} = \frac{dz}{xz}$.

Sol. (*i*) We have

$$\frac{dx}{y^2} = \frac{dy}{x^2} = \frac{dz}{x^2y^2z^2}. \quad ...(1)$$

Taking first two fractions of (1), we get $x^2\,dx = y^2\,dy$.

Integrating, we get $\frac{x^3}{3} = \frac{y^3}{3} + c_1$ or $x^3 - y^3 = 3c_1$.

Taking last two fractions of (1), we get $\frac{dy}{x^2} = \frac{dz}{x^2y^2z^2}$

$$\Rightarrow \quad y^2\,dy = z^{-2}\,dz$$

Integrating, we get $\frac{y^3}{3} = \frac{z^{-1}}{-1} + c_2$ or $\frac{y^3}{3} + \frac{1}{z} = c_2$ or $y^3 z + 3 = 3c_2 z$

$\therefore$ The general solution of the given system is

$$\mathbf{x^3 - y^3 = 3c_1,\ y^3z + 3 = 3c_2\,z.}$$

(*ii*) We have $$\frac{x\,dx}{y^2z} = \frac{dy}{xz} = \frac{dz}{y^2} \quad ...(1)$$

Taking the first two fractions of (1), we get

$$x^2\,dx = y^2\,dy \quad \text{or} \quad 3x^2\,dx - 3y^2\,dy = 0 \quad ...(2)$$

Integrating (2), we have $x^3 - y^3 = c_1$

Taking the first and last fractions of (1), we get

$$xdx = zdz \quad \text{or} \quad 2xdx - 2zdz = 0 \quad ...(3)$$

Integrating (3), we have $x^2 - z^2 = c_2$.

$\therefore$ The general solution of the given system is

$$\mathbf{x^3 - y^3 = c_1, x^2 - z^2 = c_2.}$$

(*iii*) We have $\dfrac{dx}{z} = \dfrac{dy}{0} = \dfrac{dz}{x}$ ...(1)

Second fraction of (1) implies $dy = 0$

$\therefore \quad y = c_1$

Taking the first and third fractions of (1), we get

$$xdx = zdz \quad \text{or} \quad 2xdx - 2zdz = 0 \qquad ...(2)$$

Integrating (2), we have $x^2 - z^2 = c_2$

$\therefore$ The general solution of the given system is $\mathbf{y = c_1, x^2 - z^2 = c_2}$.

(*iv*) We have $\dfrac{dx}{x^2 + 2y^2} = -\dfrac{dy}{xy} = \dfrac{dz}{xz}$ ...(1)

Taking the first two fractions of (1), we get $\dfrac{dx}{x^2 + 2y^2} = -\dfrac{dy}{xy}$.

$$\Rightarrow \quad \frac{dx}{dy} = -\frac{x^2 + 2y^2}{xy} \quad \Rightarrow \quad 2x\frac{dx}{dy} = -\frac{2x^2}{y} - 4y$$

$$\Rightarrow \quad 2x\frac{dx}{dy} + x^2\left(\frac{2}{y}\right) = -4y \qquad ...(2)$$

Let $\quad z = x^2$.

$$\therefore \quad (2) \Rightarrow \quad \frac{dz}{dy} + z\left(\frac{2}{y}\right) = -4y$$

This is a linear differential equation of order one.

$$\text{I.F.} = e^{\int \frac{2}{y} dy} = e^{2 \log y} = y^2$$

$$\therefore \quad zy^2 = \int (-4y)y^2 \, dy + c_1 \quad \text{or} \quad x^2y^2 + y^4 = c_1$$

Taking the last two fractions of (1), we get $\dfrac{dy}{y} + \dfrac{dz}{z} = 0$ ...(3)

Integrating (3), we get

$$\log y + \log z = \log c_2 \quad \text{or} \quad yz = c_2$$

$\therefore$ The general solution of the given system is

$$\mathbf{x^2y^2 + y^4 = c_1, yz = c_2.}$$

8. TYPE II

In this type we shall consider the solution of system of equations $\dfrac{dx}{X} = \dfrac{dy}{Y} = \dfrac{dz}{Z}$ in which the equality of two fractions of the given system gives an equation involving only two variables. This equation is solved by usual methods. The solution of this equation is used to eliminate an undesired variable from the equation obtained by equating two other fractions of the given system. This equation is solved and the first arbitrary constant is also eliminated.

Example 2. *Solve the following systems of simultaneous differential equations :*

(*i*) $\dfrac{dx}{xy} = \dfrac{dy}{y^2} = \dfrac{dz}{z(xy - 2x^2)}$ (C.D.L.U. 2004; M.D.U. 2005, 2007)

(*ii*) $\dfrac{dx}{1} = \dfrac{dy}{2} = \dfrac{dz}{5z + \tan(y - 2x)}$

(*iii*) $\dfrac{dx}{y} = \dfrac{dy}{x} = \dfrac{dz}{xyz^2 (x^2 - y^2)}$

(*iv*) $\dfrac{dx}{xz(z^2 + xy)} = -\dfrac{dy}{yz(z^2 + xy)} = \dfrac{dz}{x^4}$.

Sol. (*i*) We have

$$\frac{dx}{xy} = \frac{dy}{y^2} = \frac{dz}{z(xy - 2x^2)}. \quad \text{...(1)}$$

Taking first two fractions of (1), we get $\dfrac{dx}{xy} = \dfrac{dy}{y^2}$

$$\Rightarrow \quad \frac{dx}{x} = \frac{dy}{y}$$

Integrating, we get $\log x = \log y + \log c_1$

$$\therefore \quad x = c_1 y$$

Taking last two fractions of (1), we get $\dfrac{dy}{y^2} = \dfrac{dz}{z(xy - 2x^2)}$ ...(2)

We eliminate x from (2).

$$(2) \Rightarrow \frac{dy}{y^2} = \frac{dz}{z(c_1 y \cdot y - 2c_1^2 y^2)} \qquad (\because \ x = c_1 y)$$

$$\Rightarrow \quad dy = \frac{dz}{z(c_1 - 2c_1^2)} \quad \Rightarrow \quad (c_1 - 2c_1^2)\, dy = \frac{dz}{z}$$

Integrating, we get $(c_1 - 2c_1^2)\, y = \log z + c_2$.

$$\Rightarrow \quad \left(\frac{x}{y} - \frac{2x^2}{y^2}\right) y = \log z + c_2 \quad \Rightarrow \quad x - \frac{2x^2}{y} = \log z + c_2.$$

$\therefore$ The general solution of the given system is

$$\mathbf{x = c_1 y, \ x - \frac{2x^2}{y} = \log z + c_2.}$$

(*ii*) We have $\dfrac{dx}{1} = \dfrac{dy}{2} = \dfrac{dz}{5z + \tan(y - 2x)}$ ...(1)

Taking the first two fractions of (1), we get

$$dy - 2dx = 0 \quad \text{...(2)}$$

Integrating (2), we have $y - 2x = c_1$ ...(3)

Taking the last two fractions of (1) and using (3), we have

$$\frac{dy}{2} - \frac{dz}{5z + \tan c_1} = 0 \qquad ...(4)$$

Integrating (4), we have $\frac{1}{2} y - \frac{1}{5} \log | 5z + \tan c_1 | = c_2$

or $$5y - 2 \log | 5z + \tan (y - 2x) | = 10c_2$$

$\therefore$ The general solution of the given system is

$$\mathbf{y - 2x = c_1,\ 5y - 2 \log | 5z + \tan (y - 2x) | = 10\ c_2}.$$

(*iii*) We have $$\frac{dx}{y} = \frac{dy}{x} = \frac{dz}{xyz^2(x^2 - y^2)} \qquad ...(1)$$

Taking the first two fractions of (1), we get

$$xdx = y\,dy \quad \text{or} \quad 2xdx - 2ydy = 0 \qquad ...(2)$$

Integrating (2), we have $x^2 - y^2 = c_1$ ...(3)

Taking the last two fractions of (1) and using (3), we have $ydy - \frac{dz}{c_1 z^2} = 0$ ...(4)

Integrating (4), we have $\frac{y^2}{2} - \frac{1}{c_1}\left(\frac{z^{-1}}{-1}\right) = c_2$

$$\Rightarrow \qquad \frac{y^2}{2} + \frac{1}{z(x^2 - y^2)} = c_2$$

$\therefore$ The general solution of the given system is

$$\mathbf{x^2 - y^2 = c_1,\ \frac{y^2}{2} + \frac{1}{z(x^2 - y^2)} = c_2}.$$

(*iv*) We have $$\frac{dx}{xz(z^2 + xy)} = -\frac{dy}{yz(z^2 + xy)} = \frac{dz}{x^4} \qquad ...(1)$$

Taking the first two fractions of (1), we get $\frac{dx}{x} + \frac{dy}{y} = 0$ ...(2)

Integrating (2), we have

$$\log x + \log y = \log c_1 \quad \text{or} \quad xy = c_1 \qquad ...(3)$$

Taking the first and third fractions of (1) and using (3), we get

$$\frac{dx}{z(z^2 + c_1)} = \frac{dz}{x^3} \quad \text{or} \quad x^3\,dx - (z^3 + c_1 z)\,dz = 0 \qquad ...(4)$$

Integrating (4), we have

$$\frac{x^4}{4} - \left(\frac{z^4}{4} + \frac{c_1 z^2}{2}\right) = c_2 \quad \text{or} \quad x^4 - z^4 - 2xyz^2 = 4c_2$$

$\therefore$ The general system of the given system is

$$\mathbf{xy = c_1,\ x^4 - z^4 - 2xyz^2 = 4c_2}.$$

9. TYPE III

In this type, we shall consider the solution of system of equations $\frac{dx}{X} = \frac{dy}{Y} = \frac{dz}{Z}$ by using the formula :

$$\frac{dx}{X} = \frac{dy}{Y} = \frac{dz}{Z} = \frac{X_1 dx + Y_1 dy + Z_1\, dz}{X_1 X + Y_1 Y + Z_1 Z},$$

where X_1, Y_1, Z_1, are some functions of x, y and z. If for some choice of X_1, Y_1, Z_1 the sum $X_1X + Y_1Y + Z_1Z$ is zero, then we have $X_1\, dx + Y_1\, dy + Z_1\, dz = 0$.

We integrate this equation to get one relation between x, y and z. The functions X, Y, Z, are called **multipliers.** By using different set of multipliers or by using two fractions of the given system, we find another independent solution of the given system.

Example 3. *Solve the following systems of simultaneous differential equations :*

(*i*) $\frac{dx}{mz - ny} = \frac{dy}{nx - lz} = \frac{dz}{ly - mx}$ (*ii*) $\frac{dx}{x(y^2 - z^2)} = \frac{dy}{y(z^2 - x^2)} = \frac{dz}{z(x^2 - y^2)}$ (K.U. 2006S)

(*iii*) $\frac{dx}{x(y^2 + z)} = \frac{dy}{-y(x^2 + z)} = \frac{dz}{z(x^2 - y^2)}$ (K.U. 2006, 2007) (*iv*) $\frac{dx}{y^2 + z^2} = \frac{dy}{-xy} = \frac{dz}{-xz}$

(*v*) $\frac{dx}{z^2 - 2yz - y^2} = \frac{dy}{xy + xz} = \frac{dz}{xy - xz}$. (M.D.U. 2006)

Sol. (*i*) We have

$$\frac{dx}{mz - ny} = \frac{dy}{nx - lz} = \frac{dz}{ly - mx}. \quad ...(1)$$

Choosing l, m, n as multiplier, each fraction of (1)

$$= \frac{l\,dx + m\,dy + n\,dz}{l(mz - ny) + m(nx - lz) + n(ly - mx)} = \frac{l\,dx + m\,dy + n\,dz}{0}.$$

$\therefore \quad l\,dx + m\,dy + n\,dz = 0$

Integrating, we get $lx + my + nz = c_1$

Choosing x, y, z as multiplier, each fraction of (1)

$$= \frac{x\,dx + y\,dy + z\,dz}{x(mz - ny) + y(nx - lz) + z(ly - mx)} = \frac{x\,dx + y\,dy + z\,dz}{0}.$$

$\therefore \quad x\,dx + y\,dy + z\,dz = 0$

Integrating, we get $\frac{1}{2}x^2 + \frac{1}{2}y^2 + \frac{1}{2}z^2 = c_2 \Rightarrow x^2 + y^2 + z^2 = 2c_2$.

$\therefore$ The general solution of the given system is

$$\mathbf{lx + my + nz = c_1,\ x^2 + y^2 + z^2 = 2c_2.}$$

(*ii*) We have

$$\frac{dx}{x(y^2 - z^2)} = \frac{dy}{y(z^2 - x^2)} = \frac{dz}{z(x^2 - y^2)} \quad ...(1)$$

Taking $\frac{1}{x}, \frac{1}{y}, \frac{1}{z}$ as multipliers, each fraction of (1)

$$= \frac{\frac{1}{x}\,dx + \frac{1}{y}\,dy + \frac{1}{z}\,dz}{(y^2 - z^2) + (z^2 - x^2) + (x^2 - y^2)} = \frac{\frac{1}{x}\,dx + \frac{1}{y}\,dy + \frac{1}{z}\,dz}{0}$$

$\therefore \quad \frac{1}{x}\,dx + \frac{1}{y}\,dy + \frac{1}{z}\,dz = 0$

Integrating, we get $\log|x| + \log|y| + \log|z| = \log c_1$

or $\quad |xyz| = c_1$ or $xyz = \pm c_1$ or $xyz = c_2$ (By putting $c_2 = \pm c_1$)

Taking x, y, z as multipliers, each fraction of (1)

$$= \frac{xdx + ydy + zdz}{x^2(y^2 - z^2) + y^2(z^2 - x^2) + z^2(x^2 - y^2)} = \frac{xdx + ydy + zdz}{0}$$

$\therefore \quad xdx + ydy + zdz = 0$ or $2xdx + 2ydy + 2zdz = 0$

Integrating, we get $x^2 + y^2 + z^2 = c_3$

$\therefore$ The general solution of the given system is $\mathbf{xyz = c_2, x^2 + y^2 + z^2 = c_3}$.

(*iii*) We have $\frac{dx}{x(y^2 + z)} = \frac{dy}{-y(x^2 + z)} = \frac{dz}{z(x^2 - y^2)}$ ...(1)

Taking $x, y, -1$ as multipliers, each fraction of (1)

$$= \frac{xdx + ydy + (-1)\,dz}{x^2(y^2 + z) - y^2(x^2 + z) - z(x^2 - y^2)} = \frac{xdx + ydy - dz}{0}$$

$\therefore \quad 2xdx + 2ydy - 2dz = 0$

Integrating, we get $x^2 + y^2 - 2z = c_1$ ...(2)

Taking $\frac{1}{x}, \frac{1}{y}, \frac{1}{z}$ as multiplier, each fraction of (1)

$$= \frac{\frac{1}{x}dx + \frac{1}{y}dy + \frac{1}{z}dz}{y^2 + z - x^2 - z + x^2 - y^2} = \frac{\frac{1}{x}dx + \frac{1}{y}dy + \frac{1}{z}dz}{0}$$

$$\therefore \quad \frac{1}{x}dx + \frac{1}{y}dy + \frac{1}{z}dz = 0$$

Integrating, we get $\log|x| + \log|y| + \log|z| = \log c_2$

or $\quad |xyz| = c_2$ or $xyz = \pm c_2$ or $xyz = c_3$ (By putting $c_3 = \pm c_2$)

$\therefore$ The general solution of the given equation is $\mathbf{x^2 + y^2 - 2z = c_1, xyz = c_3}$.

(*iv*) We have $\frac{dx}{y^2 + z^2} = \frac{dy}{-xy} = \frac{dz}{-xz}$ ...(1)

Taking the last two fractions of (1), we get $\frac{dx}{y} - \frac{dz}{z} = 0$

Integrating, we have $\log|y| - \log|z| = \log c_1$

or $\left|\dfrac{y}{z}\right| = c_1$ or $\dfrac{y}{z} = \pm c_1$ or $y = c_2 z$ ... (2) (By putting $c_2 = \pm c_1$)

Taking x, y, z as multipliers, each fraction of (1)

$$= \frac{xdx + ydy + zdz}{xy^2 + xz^2 - xy^2 - xz^2} = \frac{xdx + ydy + zdz}{0}$$

$\therefore$ $xdx + ydy + zdz = 0$ or $2xdx + 2ydy + 2zdz = 0$

Integrating, we have $x^2 + y^2 + z^2 = c_3$

$\therefore$ The general solution of the given system is

$$\mathbf{y = c_2 z, \quad x^2 + y^2 + z^2 = c_3.}$$

(v) We have $\dfrac{dx}{z^2 - 2yz - y^2} = \dfrac{dy}{xy + xz} = \dfrac{dz}{xy - xz}$ ...(1)

Taking the last two fractions of (1), we get

$$\frac{dy}{x(y+z)} = \frac{dz}{x(y-z)}$$

$\Rightarrow$ $\dfrac{dy}{y+z} = \dfrac{dz}{y-z}$

$\Rightarrow$ $(y - z)dy = (y + z)dz$

$\Rightarrow$ $y\,dy - z\,dz - (z\,dy + y\,dz) = 0$

Integrating, we get

$$\frac{y^2}{2} - \frac{z^2}{2} - yz = c_1$$

$\Rightarrow$ $y^2 - z^2 - 2yz = c_2$ (Putting $c_2 = 2c_1$)

Choosing x, y, z as multiplies, each fraction of (1)

$$= \frac{x\,dx + y\,dy + z\,dz}{x(z^2 - 2yz - y^2) + y(xy + xz) + z(xy - xz)}$$

$$= \frac{x\,dx + y\,dy + z\,dz}{0}$$

$\therefore$ $x\,dx + y\,dy + z\,dz = 0$

Integrating, we get

$$\frac{x^2}{2} + \frac{y^2}{2} + \frac{z^2}{2} = c_3$$

$\Rightarrow$ $x^2 + y^2 + z^2 = c_4$ (Putting $c_4 = 2c_3$)

$\therefore$ The general solution of the given system is

$$\mathbf{y^2 - z^2 - 2yz = c_2, \; x^2 + y^2 + z^2 = c_4.}$$

10. TYPE IV

In this type we shall consider the solution of system of equations $\frac{dx}{X}=\frac{dy}{Y}=\frac{dz}{Z}$ by using the formula :

$$\frac{dx}{X}=\frac{dy}{Y}=\frac{dz}{Z}=\frac{X_1dx+Y_1dy+Z_1\,dz}{X_1X+Y_1Y+Z_1Z},$$

where X_1, Y_1, Z_1 are some functions of x, y and z. If for some choice of X_1, Y_1, Z_1, the sum $X_1\,dx+Y_1\,dy+Z_1\,dz$ is the exact differential of a factor of $X_1X+Y_1Y+Z_1Z$, then the quotient $\frac{X_1dx+Y_1dy+Z_1\,dz}{X_1X+Y_1Y+Z_1Z}$ is equated with a suitable fraction of given equations to get one solution of the given equations. By using different set of multipliers or by using two fractions of the given equations, we find another independent solution of the given equations.

Example 4. *Solve the following systems of simultaneous equations :*

(*i*) $\frac{dx}{x^2-yz}=\frac{dy}{y^2-zx}=\frac{dz}{z^2-xy}$ (*ii*) $\frac{dx}{y+z}=\frac{dy}{z+x}=\frac{dz}{x+y}$ (M.D.U. 2006)

(*iii*) $\frac{dx}{\cos(x+y)}=\frac{dy}{\sin(x+y)}=\frac{dz}{z+\frac{1}{z}}$ (*iv*) $\frac{dx}{y^2+yz+z^2}=\frac{dy}{z^2+zx+x^2}=\frac{dz}{x^2+xy+y^2}$.

Sol. (*i*) We have

$$\frac{dx}{x^2-yz}=\frac{dy}{y^2-zx}=\frac{dz}{z^2-xy} \qquad ...(1)$$

Taking 1, 1, 1 and x, y, z as multipliers, each fraction of (1)

$$=\frac{1.dx+1.dy+1.dz}{x^2-yz+y^2-zx+z^2-xy}=\frac{x\,dy+y\,dy+z\,dz}{x^3-xyz+y^3-xyz+z^3-xyz}$$

$$\Rightarrow \quad \frac{dx+dy+dz}{x^2+y^2+z^2-yz-zx-xy}=\frac{x\,dx+y\,dy+z\,dz}{(x+y+z)(x^2+y^2+z^2-yz-zx-xy)}$$

$$\Rightarrow \quad (x+y+z)d(x+y+z)=\frac{1}{2}(2xdx+2ydy+2zdz)$$

$$\Rightarrow \quad (x+y+z)\,d(x+y+z)-\frac{1}{2}\,d(x^2+y^2+z^2)=0$$

Integrating, we get $\frac{1}{2}(x+y+z)^2-\frac{1}{2}(x^2+y^2+z^2)=c_1$

or $xy+yz+zx=c_1$

Taking 1, – 1, 0 and 0, 1, – 1 as multipliers, each fraction of (1)

$$=\frac{dx-dy+0}{(x^2-yz)-(y^2-zx)+0}=\frac{0+dy-dz}{0+(y^2-zx)-(z^2-xy)}$$

$$\Rightarrow \quad \frac{dx-dy}{x^2-y^2+z(x-y)}=\frac{dy-dz}{y^2-z^2+x(y-z)}$$

$$\Rightarrow \quad \frac{dx - dy}{(x-y)(x+y+z)} = \frac{dy - dz}{(y-z)(y+z+x)} \quad \Rightarrow \quad \frac{d(x-y)}{x-y} - \frac{d(y-z)}{y-z} = 0$$

Integrating, we get $\log|x-y| - \log|y-z| = \log c_2$

or $$\left|\frac{x-y}{y-z}\right| = c_2 \quad \text{or} \quad \frac{x-y}{y-z} = \pm c_2 \quad \text{or} \quad x - y = c_3(y-z). \quad \text{(By putting } c_3 = \pm c_2)$$

$\therefore$ The general solution of the given system is

$$\mathbf{xy + yz + zx = c_1, \; x - y = c_3(y - z).}$$

(*ii*) We have $$\frac{dx}{y+z} = \frac{dy}{z+x} = \frac{dz}{x+y} \quad ...(1)$$

Taking 1, – 1, 0 and 0, 1, – 1 as multipliers, each fraction of (1)

$$= \frac{dx - dy + 0}{(y+z)-(z+x)+0} = \frac{0 + dy - dz}{0 + (z+x)-(x+y)} \quad ...(2)$$

$$\Rightarrow \quad \frac{dx - dy}{-(x-y)} = \frac{dy - dz}{-(y-z)} \quad \Rightarrow \quad \frac{d(x-y)}{x-y} - \frac{d(y-z)}{y-z} = 0$$

Integrating, we get $\log|x-y| - \log|y-z| = \log c_1$

$$\Rightarrow \quad \left|\frac{x-y}{y-z}\right| = c_1 \quad \text{or} \quad \frac{x-y}{y-z} = \pm c_1 \quad \text{or} \quad x - y = c_2(y-z) \quad \text{(By putting } c_2 = \pm c_1)$$

Taking 1, – 1, 0 and 1, 1, 1 as multipliers, each fraction of (1)

$$\frac{dx - dy + 0}{(y+z)-(z+x)+0} = \frac{dx + dy + dz}{(y+z)+(z+x)+(x+y)}$$

$$\Rightarrow \quad \frac{d(x-y)}{-(x-y)} = \frac{d(x+y+z)}{2(x+y+z)} \quad \text{or} \quad 2\frac{d(x-y)}{x-y} + \frac{d(x+y+z)}{x+y+z} = 0$$

Integrating, we get $2\log|x-y| + \log|x+y+z| = \log c_3$

$$\Rightarrow \quad (x-y)^2\,|x+y+z| = c_3 \quad \Rightarrow \quad (x-y)^2(x+y+z) = \pm c_3$$

$$\Rightarrow \quad (x-y)^2(x+y+z) = c_4. \quad \text{(By putting } c_4 = \pm c_3)$$

$\therefore$ The general solution of the given system is

$$\mathbf{x - y = c_2(y - z), \; (x - y)^2(x + y + z) = c_4.}$$

(*iii*) We have $$\frac{dx}{\cos(x+y)} = \frac{dy}{\sin(x+y)} = \frac{dz}{z + \frac{1}{z}} \quad ...(1)$$

Taking 1, 1, 0 and 1, – 1, 0 as multipliers, each fraction of (1)

$$= \frac{dx + dy + 0}{\cos(x+y) + \sin(x+y) + 0} = \frac{dx - dy + 0}{\cos(x+y) - \sin(x+y) + 0} \quad ...(2)$$

(1) and (2) $\Rightarrow$ $$\frac{z\,dz}{z^2+1} = \frac{d(x+y)}{\cos(x+y) + \sin(x+y)}$$

$$\Rightarrow \quad \frac{1}{2}\cdot\frac{2z\,dz}{z^2+1} = \frac{dt}{\cos t + \sin t} = \frac{dt}{\sqrt{2}\sin(t + \pi/4)}, \quad \text{where} \quad t = x + y.$$

$$\Rightarrow \qquad \frac{1}{\sqrt{2}} \cdot \frac{2z}{z^2+1} dz - \operatorname{cosec}\left(t+\frac{\pi}{4}\right) dt = 0$$

Integrating, we get $\frac{1}{\sqrt{2}} \log | z^2 + 1 | - \log \tan \frac{1}{2}\left(t + \frac{\pi}{4}\right) = \log c_1$

$$\Rightarrow \qquad \frac{(z^2+1)^{1/\sqrt{2}}}{\tan\left(\frac{x+y}{2}+\frac{\pi}{8}\right)} = c_1$$

Also, (2) $\Rightarrow$ $\frac{\cos(x+y) - \sin(x+y)}{\cos(x+y) + \sin(x+y)} d(x+y) = d(x-y)$

$$\Rightarrow \qquad \frac{\cos t - \sin t}{\cos t + \sin t} dt - d(x-y) = 0, \text{ where } t = x + y.$$

Integrating, we get $\log | \cos t + \sin t | - (x - y) = \log c_2$

$$\Rightarrow \qquad | \cos t + \sin t | \, e^{y-x} = c_2$$

$$\Rightarrow \qquad (\cos(x+y) + \sin(x+y))e^{y-x} = \pm c_2$$

$$\Rightarrow \qquad (\cos(x+y) + \sin(x+y))e^{y-x} = c_3 \qquad \text{(By putting } c_3 = \pm c_2)$$

$\therefore$ The general solution of the given system is

$$\frac{(\mathbf{z}^2+1)^{1/\sqrt{2}}}{\tan\left(\frac{\mathbf{x}+\mathbf{y}}{2}+\frac{\pi}{8}\right)} = \mathbf{c}_1, \ (\cos(\mathbf{x}+\mathbf{y}) + \sin(\mathbf{x}+\mathbf{y}))\, \mathbf{e}^{\mathbf{y}-\mathbf{x}} = \mathbf{c}_3.$$

(*iv*) We have $\frac{dx}{y^2+yz+z^2} = \frac{dy}{z^2+zx+x^2} = \frac{dz}{x^2+xy+y^2}$ ...(1)

Taking 1, – 1, 0 and 0, 1, – 1 as multipliers, each fraction of (1)

$$= \frac{dx - dy + 0}{(y^2+yz+z^2) - (z^2+zx+x^2) + 0} = \frac{0 + dy - dz}{0 + (z^2+zx+x^2) - (x^2+xy+y^2)}$$

$$\Rightarrow \qquad \frac{dx - dy}{y^2 - x^2 + yz - zx} = \frac{dy - dz}{z^2 - y^2 + zx - xy}$$

$$\Rightarrow \qquad \frac{dx - dy}{(y-x)(x+y+z)} = \frac{dy - dz}{(z-y)(z+y+x)}$$

$$\Rightarrow \qquad \frac{d(x-y)}{-(x-y)} = \frac{d(y-z)}{-(y-z)} \quad \Rightarrow \quad \frac{d(y-z)}{y-z} - \frac{d(x-y)}{x-y} = 0$$

Integrating, we get $\log | y - z | - \log | x - y | = \log c_1$

$$\Rightarrow \qquad \left|\frac{y-z}{x-y}\right| = c_1 \quad \text{or} \quad \frac{y-z}{x-y} = \pm c_1$$

or $\qquad y - z = c_2 (x - y) \qquad$ (By putting $c_2 = \pm c_1$)

Taking 1, – 1, 0 and 1, 0, – 1 as multipliers, each fractions of (1)

$$= \frac{dx - dy}{(y^2 + yz + z^2) - (z^2 + zx + x^2)} = \frac{dx - dz}{(y^2 + yz + z^2) - (x^2 + xy + y^2)}$$

$$\Rightarrow \quad \frac{dx - dy}{y^2 - x^2 + yz - zx} = \frac{dx - dz}{z^2 - x^2 + yz - xy}$$

$$\Rightarrow \quad \frac{dx - dy}{(y - x)(y + x + z)} = \frac{dx - dz}{(z - x)(z + x + y)}$$

$$\Rightarrow \quad \frac{dx - dy}{-(x - y)} = \frac{dx - dz}{-(x - z)} \Rightarrow \frac{d(x - z)}{x - z} - \frac{d(x - y)}{x - y} = 0$$

Integrating, we get $\log | x - z | - \log | x - y | = \log c_3$

$$\Rightarrow \quad \left| \frac{x - z}{x - y} \right| = c_3 \quad \text{or} \quad \frac{x - z}{x - y} = \pm c_3$$

or $\quad x - z = c_4(x - y).$ (By putting $c_4 = \pm c_3$)

$\therefore$ The general solution of the given system is

$$y - z = c_2(x - y), \quad x - z = c_4 (x - y)$$

EXERCISE 2

Solve the following systems of simultaneous differential equations :

1. (i) $\frac{dx}{x} = \frac{dy}{y} = \frac{dz}{z}$ (ii) $\frac{dx}{\tan x} = \frac{dy}{\tan y} = \frac{dz}{\tan z}$

(iii) $\frac{dx}{1} = \frac{dy}{1} = \frac{dz}{\sin x}$ (iv) $\frac{dx}{a} = \frac{dy}{b} = \frac{dz}{c}$

(v) $\frac{dx}{yz} = \frac{dy}{zx} = \frac{dz}{xy}$ (vi) $\frac{dx}{x^2} = \frac{dy}{y^2} = \frac{dz}{z^2}$

(vii) $\frac{dx}{x - a} = \frac{dy}{y - b} = \frac{dz}{z - c}$ (viii) $\frac{dx}{y^2} = \frac{dy}{-xy} = \frac{dz}{x(z - 2y)}$

2. (i) $\frac{dx}{1} = \frac{dy}{3} = \frac{dz}{z + \cot (y - 3x)}$ (ii) $\frac{dx}{xz} = \frac{dy}{yz} = \frac{dz}{xy}$

(iii) $\frac{dx}{1} = \frac{dy}{-2} = \frac{dz}{3x^2 \sin (y + 2x)}$ (iv) $\frac{dx}{z} = \frac{dy}{-z} = \frac{dz}{x + y}$

(v) $\frac{dx}{-xy^2} = \frac{dy}{y^3} = \frac{dz}{axz}$ (vi) $\frac{dx}{xy} = \frac{dy}{y^2} = \frac{dz}{xyz - 2x^2}$

(vii) $\frac{dx}{x + y} = -\frac{dy}{x + y} = \frac{dz}{z}$ (*M.D.U. 2007*) (viii) $\frac{dx}{z} = -\frac{dy}{z} = \frac{dz}{z^2 + (x + y)^2}$ (*M.D.U. 2004*)

3. (i) $\dfrac{adx}{(b-c)yz} = \dfrac{bdy}{(c-a)zx} = \dfrac{cdz}{(a-b)xy}$ (ii) $\dfrac{yzdx}{y-z} = \dfrac{zxdy}{z-x} = \dfrac{xydz}{x-y}$

(iii) $\dfrac{dx}{z-y} = \dfrac{dy}{x-z} = \dfrac{dz}{y-x}$ (iv) $\dfrac{dx}{x(y^2-z^2)} = \dfrac{dy}{-y(z^2+x^2)} = \dfrac{dz}{z(x^2+y^2)}$

(v) $\dfrac{dx}{x(y-z)} = \dfrac{dy}{y(z-x)} = \dfrac{dz}{z(x-y)}$ (vi) $\dfrac{dx}{x^2(y-z)} = \dfrac{dy}{y^2(z-x)} = \dfrac{dz}{z^2(x-y)}$

(vii) $\dfrac{dx}{x-y} = \dfrac{dy}{x+y} = \dfrac{dz}{2xz}$ (viii) $\dfrac{dx}{z(x+y)} = \dfrac{dy}{z(x-y)} = \dfrac{dz}{x^2+y^2}$

4. (i) $\dfrac{dx}{x^2-y^2-z^2} = \dfrac{dy}{2xy} = \dfrac{dz}{2xz}$ (ii) $\dfrac{dx}{1+y} = \dfrac{dy}{1+x} = \dfrac{dz}{z}$

(iii) $\dfrac{dx}{x^2-y^2-yz} = \dfrac{dy}{x^2-y^2-zx} = \dfrac{dz}{z(x-y)}$ (iv) $\dfrac{dx}{xz} = \dfrac{dy}{yz} = \dfrac{dz}{xy}$

(v) $\dfrac{dx}{x(x+y)} = -\dfrac{dy}{y(x+y)} = -\dfrac{dz}{(x-y)(2x+2y+z)}$

(vi) $\dfrac{dx}{y(x+y)+az} = \dfrac{dy}{y(x+y)-az} = \dfrac{dz}{z(x+y)}$

(vii) $\dfrac{dx}{x} = \dfrac{dy}{z} = \dfrac{dz}{-y}$ (viii) $\dfrac{dx}{x^2+y^2} = \dfrac{dy}{2xy} = \dfrac{dz}{(x+y)z}$.

Answers

1. (i) $x = c_1 y,\ \ y = c_2 z$ (ii) $\sin x = c_1 \sin y,\ \sin y = c_2 \sin z$

(iii) $x - y = c_1,\ z + \cos x = c_2$ (iv) $bx - ay = c_1,\ cy - bz = c_2$

(v) $x^2 - y^2 = c_1,\ x^2 - z^2 = c_2$ (vi) $\dfrac{1}{y} - \dfrac{1}{x} = c_1,\ \ \dfrac{1}{z} - \dfrac{1}{y} = c_2$

(vii) $x - a = c_1(y-b),\ y - b = c_2(z-c)$ (viii) $x^2 + y^2 = c_1,\ yz - y^2 = c_3$

2. (i) $y - 3x = c_1,\ x - \log|z + \cot(y-3x)| = c_2$

(ii) $x = c_1 y,\ xy - z^2 = 2c_2$

(iii) $y + 2x = c_1,\ x^3 \sin(y+2x) - z = c_2$ (iv) $x + y = c_1,\ 2(x+y)x - z^2 = c_2$

(v) $xy = c_1,\ \log|z| + \dfrac{ax}{3y^2} + c_2 = 0$ (vi) $x = c_1 y,\ x - \log\left|z - \dfrac{2x}{y}\right| = c_2$

(vii) $x + y = c_1,\ x - (x+y)\log|z| = c_2$ (viii) $x + y = c_1,\ y + \dfrac{1}{2}\log(z^2 + (x+y)^2) = c_2$

3. (i) $ax^2 + by^2 + cz^2 = 2c_1,\ a^2x^2 + b^2y^2 + c^2z^2 = 2c_2$

(ii) $x + y + z = c_1,\ xyz = c_2$

(iii) $x + y + z = c_1,\ x^2 + y^2 + z^2 = 2c_2$ (iv) $x^2 + y^2 + z^2 = 2c_1,\ x = c_2 yz$

(v) $x + y + z = c_1,\ xyz = c_2$ (vi) $xyz = c_1,\ \dfrac{1}{x} + \dfrac{1}{y} + \dfrac{1}{z} + c_2 = 0$

(vii) $x + y - \log|z| = c_1,\ (x^2 + y^2)e^{-2\tan^{-1}(y/x)} = e^{-2c_2}$

(viii) $x^2 - y^2 - z^2 = 2c_1,\ 2xy - z^2 = 2c_2$

4. (*i*) $y = c_1 z,\ x^2 + y^2 + z^2 = c_2 z$ (*ii*) $(1 + x)^2 - (1 + y)^2 = 2c_1,\ x + y + 2 = c_2 z$

(*iii*) $z - x + y = c_1,\ x^2 - y^2 = c_2 z^2$ (*iv*) $xy - z^2 = c_1,\ x = c_2 y$

(*v*) $xy = c_1,\ (x + y)(x + y + z) = c_2$ (*vi*) $x + y = c_1 z,\ x^2 - y^2 - 2az = c_2$

(*vii*) $x = c_1 e^{\tan^{-1}(y/z)},\ y^2 + z^2 = 2c_2$ (*viii*) $x + y = c_1 z,\ y = c_2 (x^2 - y^2)$.

Hints

3. (*iv*) Try x, y, z and $\frac{1}{x}, -\frac{1}{y}, -\frac{1}{z}$ as multipliers.

(*viii*) Try $x, -y, -z$ and $y, x, -z$ as multipliers.

4. (*i*) Try x, y, z as multipliers. (*ii*) Try 1, 1, 0 as multipliers.

(*iii*) Try 1, – 1, 0 and $x, -y$, 0 as multipliers. (*iv*) Try $1/x,\ 1/y$, 0 as multipliers.

(*v*) Try 1, 1, 0 and 1, 1, 1 as multipliers. (*vi*) Try 1, 1, 0 and $x, -y$, 0 as multipliers.

(*vii*) We have $\dfrac{dx}{x} = \dfrac{0 + z\,dy - y\,dz}{z^2 + y^2} = \dfrac{\frac{1}{z}dy + \left(-\frac{y}{z^2}\right)dz}{1 + \left(\frac{y}{z}\right)^2} = \dfrac{d(y/z)}{1 + (y/z)^2}$

$\therefore \quad \log|x| - \tan^{-1}\dfrac{y}{z} = \log c.$

(*viii*) Try 1, 1, 0 and 1, – 1, 0 as multipliers.

Hints

EXAMINATION PAPERS

M.D.U., ROHTAK

B.A. Part I. Examination, April 2007

Paper II (B.M. 102)

Calculus and Ordinary Differential Equations

Time : Three Hours] *[Maximum Marks : 30*

Note : Attempt any **Five** questions in all, selecting at least **One** question from each section.

SECTION I

1. (*a*) If $y = \left[\log(x + \sqrt{1+x^2})\right]^2$, prove that

$$(1 + x^2)y_{n+2} + (2n + 1)xy_{n+1} + n^2 y_n = 0.$$ 3

(*b*) State and prove Taylor's theorem with Cauchy's form to Remainder after n terms. 3

2. (*a*) Show that the points of intersection of the curve $4x^3 - 2x^2y - 4xy^2 + 2y^3 + 6y^2 - 14xy + 4x^2 + 6y + 1 = 0$ and its asymptote lie on $8x + 2y + 1 = 0$. 3

(*b*) Find the radius of curvature for the parabola $\frac{2a}{r} = 1 + \cos\theta$.

Hence show that the square of the radius of curvature varies as the cube of the focal distance. 3

SECTION II

3. (*a*) Evaluate $\int_0^{\pi/2} \sin^m x \cos^n x \, dx$, where m and n are positive integers and hence find

$$\int_0^{\pi/2} \sin^5 x \cos^4 x \, dx.$$ 3

(*b*) Find the perimeter of the loop of the curve

$$9ay^2 = (x - 2a)(x - 5a)^2.$$ 3

4. (*a*) If A is the vertex, O the centre and P(x, y) any point on the hyperbola $\frac{x^2}{a^2} - \frac{y^2}{b^2} = 1$, prove that

$$x = a \cosh \frac{2S}{ab}, y = b \sinh \frac{2S}{ab},$$ where S is the sectorial area OPA. 3

(*b*) Find the area common to the circle $r = a$ and the cardioid
$$r = a(1 + \cos \theta).$$ 3

5. (*a*) Find the volume of the solid formed by the revolution about the x-axis of the curve
$$y^2(a + x) = x^2(3a - x).$$ 3

(*b*) Find the area of the surface formed by the revolution of $y^2 = 4ax$ about the x-axis by the arc from the vertex to one end to latus rectum. 3

SECTION III

6. (*a*) Solve the differential equation :
$$(x^2 + y^2 + 2x)dx + 2y\,dy = 0.$$ 3

(*b*) Solve the differential equation :
$$x^2\left(\frac{dy}{dx}\right)^2 - 2xy\frac{dy}{dx} + 2y^2 - x^2 = 0.$$ 3

7. (*a*) Find the orthogonal trajectories of the cardioid $r = a(1 - \cos \theta)$, where a is the parameter. 3

(*b*) Solve the differential equation :
$$\frac{d^2y}{dx^2} + 6\frac{dy}{dx} + 25\,y = 104e^{3x}.$$ 3

8. (*a*) Solve the following differential equation :
$$x^3\frac{d^3y}{dx^3} + 6x^2\frac{d^2y}{dx^2} + 4x\frac{dy}{dx} - 4y = 0.$$ 3

(*b*) Solve the differential equation :
$$(2x - 1)^3\frac{d^3y}{dx^3} + (2x - 1)\frac{dy}{dx} - 2y = 0.$$ 3

SECTION IV

9. (*a*) Solve the differential equation :
$$(x + 2)\frac{d^2y}{dx^2} - (2x + 5)\frac{dy}{dx} + 2y = (x + 1)e^x.$$ 3

(*b*) Apply the method of variation of parameters to solve :
$$\frac{d^2y}{dx^2} + 4y = \tan 2x.$$ 3

10. (*a*) Solve the following simultaneous equations :
$$\frac{dx}{dt} + 4x + 3y = t\ ;$$
$$\frac{dy}{dt} + 2x + 5y = e^t.$$ 3

(*b*) Solve the simultaneous equations :
$$\frac{dx}{xy} = \frac{dy}{y^2} = \frac{dz}{z(xy - 2x^2)}.$$ 3

M.D.U. ROHTAK

B.Sc. Part I. Examination, April 2007

Paper II (B.M. 102)

Calculus and Ordinary Differential Equations

Time : Three Hours] [*Maximum Marks : 45*

Note : Attempt **Five** questions, selecting at least **One** from each section.

SECTION I

1. (*a*) If $y = \dfrac{\sin^{-1} x}{\sqrt{1-x^2}}$, prove that

$$(1-x^2)y_{n+2} - (2n+3)\,xy_{n+1} - (n+1)^2\,y_n = 0.$$ 5

(*b*) Find all the asymptotes of the curve

$$x(y-x)^2 - x(y-x) = 2.$$ 4

2. (*a*) Find the radius of curvature of the curve $r^n = a^n \cos n\theta$. 5

(*b*) Find the points of inflexion of the curve

$$y = (x-2)^6\,(x-3)^5.$$ 4

SECTION II

3. (*a*) Connect

$$\int \sin^m x \cos^n x\,dx \quad \text{with} \quad \int \sin^{m-2} x \cos^n x\,dx \cdot$$

Hence evaluate

$$\int \sin^4 x \cos^2 x \,.\, dx\,.$$ 5

(*b*) Find the length of the arc of the parabola $y^2 = 4ax$ from the vertex to an extremity of the latus rectum. 4

4. (*a*) Find the intrinsic equation of the cardioid $r = a(1 + \cos\theta)$ and show that

$$s^2 + 9p^2 = 16a^2.$$ 5

(*b*) Show that the area of a loop of the curve

$$r = \sqrt{3}\cos 3\theta + \sin 3\theta \text{ is } \pi/3.$$ 4

5. (*a*) Show that the volume of the solid formed by the evolution of an arch of the cycloid : $x = a(\theta + \sin\theta)$, $y = a(1 - \cos\theta)$ about the tangent at the vertexin $\pi^2 a^3$. 5

(*b*) Find the surface of the solid generated by the revolution of the astrioid

$$x^{2/3} + y^{2/3} = a^{2/3}.$$ 4

SECTION III

6. (*a*) Solve the differential equation :

$$(xy^2 - x^2)dx + (3x^2y^2 + x^2y - 2x^3 + y^2)dy = 0.$$ 5

(*b*) Solve the differential equation :

$$x - yp = ap^2.$$ 4

7. (*a*) Find the orthogonal trajectories of the family of circles $x^2 + y^2 + 2gx + c = 0$, g being a parameter. 4

(*b*) Prove that :

$$\frac{1}{(D - \alpha)} Q = e^{\alpha x} \int Q . e^{-\alpha x} . dx.$$ 4

8. (*a*) Solve the differential equation :

$$\frac{d^2y}{dx^2} + 4x \frac{dy}{dx} + 2y = x - \sin x.$$ 5

(*b*) Solve the differential equation :

$$(D^2 - 3D + 2)y = \cos e^{-x}.$$ 4

SECTION IV

9. (*a*) Solve :

$$\frac{d^2y}{dx^2} - (1 + 4e^x) \frac{dy}{dx} + 3e^{2x} y = e^{2(x + e^x)}.$$ 5

(*b*) Solve the system of simultaneous linear differential equations :

$$\frac{dx}{dt} = ax + by, \frac{dy}{dt} = a'x + b'y.$$ 4

10. (*a*) Solve :

$$\frac{dx}{(x + y)} = \frac{dy}{-(x + y)} = \frac{dz}{z}.$$ 4

(*b*) Solve :

$$x \frac{d^2y}{dx^2} - 2(x + 1)\frac{dy}{dx} + (x + 2)y = (x - 2) . e^x.$$ 5

K.U., KURUKSHETRA

B.A./B.Sc. I. Examination, April 2007

Paper II (B.M. 102)

Calculus and Ordinary Differential Equations

Time : Three Hours] [*Maximum Marks : B.A–30*
B.Sc.–45

Note : Attempt **Five** questions, selecting at least **One** question from each section.

SECTION I

1. (*a*) Using the $\in - \delta$ definition of continuity, show that $\sin x$ is continuous for every value of x. 3(4½)

(*b*) If $y = (\sinh^{-1} x)^2$, prove that $(1 + x^2)y_{n+2} + (2n + 1)xy_{n+1} + n^2 y_n = 0$.
Hence, find $y_n (0)$ when $x = 0$. 3(4½)

2. (*a*) State and prove Taylor's theorem with Cauchy's form of remainder. 3(4½)

(*b*) Find the asymptotes of the curve $r \cos \theta = a \sin^2 \theta$. 3(4½)

3. (*a*) Show that the radius of curvature at the point $(a \cos^3 \theta, a \sin^3 \theta)$ on the curve $x^{2/3} + y^{2/3} = a^{2/3}$ is $3a \sin \theta \cos \theta$. 3(4½)

(*b*) Trace the curve $r = a (1 + \sin \theta)$. 3(4½)

SECTION II

4. (*a*) If $U_n = \int \cos n\theta \operatorname{cosec} \theta \, d\theta$, prove that

$$U_n - U_{n-2} = \frac{2\cos (n-1)\theta}{n-1}.$$ 3(4½)

(*b*) Find the length of complete cycloid given by $x = a(\theta + \sin \theta)$, $y = a(1 - \cos \theta)$. 3(4½)

5. (*a*) Find the intrinsic equation of the catenary $y = c \cosh x/c$, s being measured from the vertex $(0, c)$ of the catenary and show that $c\rho = c^2 + s^2$, ρ being the radius of carvature. 3(4½)

(*b*) Show that area of a loop of the curve $r = \sqrt{3} \cos 3\theta + \sin 3\theta$ is $\frac{\pi}{3}$. 3(4½)

6. (*a*) The cardioid $r = a(1 + \cos \theta)$ revolves about initial line. Find volume of the solid generated. 3(4½)

(*b*) Prove that the surface generated by the revolution of the tractrix $x = a \cos t + \frac{a}{2} \log \tan^2 \frac{t}{2}$, $y = a \sin t$ about its asymptote is equal to surface of sphere of radius a. 3(4½)

SECTION III

7. (*a*) Prove that the necessary and sufficient condition for equation Mdx + Ndy = 0 to be exact is $\frac{\partial M}{\partial y} = \frac{\partial N}{\partial x}$. 3(4½)

(*b*) Find orthogonal trajectory of the family of circles $x^2 + y^2 + 2gx + c = 0$, c being parameter. 3(4½)

8. (*a*) Solve the differential equation :

$$\frac{d^2y}{dx^2} + 2y = x^2e^{3x} + e^x \cos 2x. \qquad 3(4½)$$

(*b*) Solve :

$$x^2 \frac{d^2y}{dx^2} + 8x \frac{dy}{dx} + 13y = \log x. \qquad 3(4½)$$

SECTION IV

9. (*a*) Solve :

$$\left(\frac{d^2y}{dx^2} + y\right)\cot x + 2\left(\frac{dy}{dx} + y \tan x\right) = \sec x. \qquad 3(4½)$$

(*b*) Solve :

$$\frac{dx}{x(y^2 + z)} = \frac{dy}{-y(x^2 + z)} = \frac{dz}{z(x^2 - y^2)}. \qquad 3(4½)$$

10. (*a*) Solve the system of simultaneous linear differential equations :

$$\frac{dx}{dt} + 5x + y = e^t$$

$$\frac{dy}{dt} - x + 3y = e^{2t}. \qquad 3(4½)$$

(*b*) Apply the method of variation of parameters to solve the equation :

$$(1 - x)\frac{d^2y}{dx^2} + x\frac{dy}{dx} - y = (1 - x)^2. \qquad 3(4½)$$